Combustion Engineering

Combustion Engineering

Gary L. Borman
Kenneth W. Ragland

Department of Mechanical Engineering
University of Wisconsin-Madison

Boston Burr Ridge, IL Dubuque, IA Madison, WI New York San Francisco St. Louis
Bangkok Bogotá Caracas Lisbon London Madrid Mexico City Milan New Delhi
Seoul Singapore Sydney Taipei Toronto

WCB/McGraw-Hill

A Division of The **McGraw-Hill** *Companies*

COMBUSTION ENGINEERING

This book is printed on acid-free paper.

2 3 4 5 7 8 9 0 DOC/DOC 9 0 9 8

ISBN 0-07-006567-5

Editorial director: *Tom Casson*
Senior sponsoring editor: *Debra Riegert*
Marketing manager: *John Wannemacher*
Project manager: *Jim Labeots*
Production supervisor: *Heather D. Burbridge*
Senior designer: *Crispin Prebys*
Compositor: *Interactive Composition Corporation*
Typeface: *10/12 Times Roman*
Printer: *R. R. Donnelley & Sons Company*

Library of Congress Cataloging-in-Publication Data

Borman, Gary L.
Combustion engineering / Gary L. Borman, Kenneth W. Ragland.
p. cm.
Includes index.
ISBN 0-07-006567-5
1. Combustion engineering. I. Ragland, Kenneth W. II. Title.
TJ254.5.B67 1998
621.402′3—dc21 97-34198

http://www.mhhe.com

ABOUT THE AUTHORS

DR. GARY BORMAN retired from the faculty of the Mechanical Engineering Department at the University of Wisconsin-Madison in 1994 after 30 years of service. He was director of the Engine Research Center from 1986–1994 and currently is active in the ERC as an Emeritus Professor. His research concerns engine issues such as cycle analysis, droplet vaporization, emissions, and engine heat transfer. He taught courses in combustion, heat transfer, and engineering analysis. His honors include Phi Beta Kappa, Ralph R. Teetor Award, Arch T. Colwell Merit Award, Horning Memorial Award, Fulbright Scholar (1970 and 1990), Fellow of SAE, and Member of the National Academy of Engineering.

DR. KENNETH RAGLAND is a Professor of the Mechanical Engineering Department at the University of Wisconsin-Madison. He joined the department in 1968, and since 1995 he has served as Chair of the department. His research concerns solid fuel combustion rate processes, combustion in full scale power plants, and air pollution control. He teaches courses in combustion, fluid dynamics, and air pollution control. In 1967 he spent a year on the National Academy of Sciences exchange program at Novosibirsk University in the former Soviet Union. He is a member of ASME, ASEE, AWMA, and the Combustion Institute.

To our wives, Marlene and Nancy, for their patience and support during the long preparation of das buch.

CONTENTS

NOMENCLATURE

a	speed sound, m/s = 3.281 ft/s; radius, m = 3.281 ft
$\dot{ad}$	availability destruction rate, kW/kg · K = 0.2388 Btu/(lb_m · °R · s)
A	area, m^2 = 10.764 ft^2 = 1550 $in.^2$
$\overline{A}$	mean surface area per unit volume, m^{-1} = 0.3048 ft^{-1}
A_v	area per unit volume, m^{-1} = 0.3048 ft^{-1} = 0.0254 $in.^{-1}$
A_w	wrinkled flame area, m^2 = 10.764 ft^2
$[A]$	molar concentration of species A, kgmol/m^3 = 0.06243 lbmol/ft^3
b_t	probability distribution function of nondimensional temperature
B	fraction of air which flows through the dense phase in a fluidized bed
c	specific heat, kJ/kg · K = 0.2388 Btu/(lb_m · °R)
c_p	specific heat at constant pressure, kJ/kg · K = 0.2388 Btu/(lb_m · °R)
c_v	specific heat at constant volume, kJ/kg · K = 0.2388 Btu/(lb_m · °R)
C_d	discharge coefficient; drag coefficient
CN	cetane number
d	diameter, m = 3.28 ft = 39.37 in.
$\overline{d}$	surface mean diameter, m = 3.28 ft = 39.37 in.
D	fractal parameter
D_{AB}	binary diffusion coefficient, m^2/s = 10.764 ft^2/s= 1550 $in.^2$/s
$\mathbf{D}_{AB}$	multicomponent mixture diffusion coefficient, m^2/s = 10.764 ft^2/s = 1550 $in.^2$/s
E	activation energy, kJ/kg = 0.4303 Btu/lb_m
E'	ignition energy, mJ = 9.478×10^{-7} Btu
EA	excess air
f	fuel-to-air ratio by weight; fugacity
$\tilde{f}$	spray distribution function
f_s	stoichiometric fuel to air ratio by weight
F	equivalence ratio, f/f_s; force, N = 0.2248 lb_f
g	Gibbs free energy per unit mass, kJ/kg = 0.4303 Btu/lb_m
G	Gibbs free energy, kJ = 0.9488 Btu; API specific gravity
h	enthalpy per unit mass, kJ/kg = 0.4303 Btu/lb_m

h'	Planck's constant, kJ · s = 2.6356 × 10^{-4} Btu · h
$\tilde{h}$	convective heat transfer coefficient, W/m^2 · K = 0.17612 Btu/(h · ft^2 · °R)
$\tilde{h}^*$	convective heat transfer coefficient corrected for mass transfer effects, W/m^2 · K = 0.17612 Btu/(h · ft^2 · °R)
$\tilde{h}_D$	mass transfer coefficient, m/s = 1.1811 × 10^4 ft/h = 39.37 in./s
h_{fg}	latent heat of vaporization, kJ/kg = 0.4303 Btu/lb$_m$
H	heat of reaction, kJ/kg = 0.4303 Btu/lb$_m$
HV	heat release per unit mass of fuel, kJ/kg = 0.4303 Btu/lb$_m$
HHV	higher heating value per mass of fuel, kJ/kg = 0.4303 Btu/lb$_m$
HR	heat rate, Btu/kWh
HR''	heat input per unit flow area, W/m^2 = 0.3170 Btu/(ft^2 · h)
$\Delta h°$	enthalpy of formation, kJ/kg = 0.4303 Btu/lb$_m$
I	combustion intensity, kW/m^3 = 96.65 Btu/(h · ft^3)
$\tilde{k}$	thermal conductivity, W/m · K = 0.57782 Btu/(h · ft · °R)
k_a	attrition rate constant
k_i	kinetic rate constant, units vary
K	thermodynamic equilibrium constant; turbulent kinetic energy, kJ/m^3 = 0.02684 Btu/ft^3
L	length, m = 3.2808 ft = 39.370 in.
L_I	integral scale of turbulence, m = 3.2808 ft = 39.370 in.
L_K	Kolmogorov scale of turbulence, m = 3.2808 ft = 39.370 in.
LHV	lower heating value per mass of fuel, kJ/kg = 0.4303 Btu/lb$_m$
m	mass, kg = 2.2046 lb$_m$
$\dot{m}$	mass flow rate, kg/s = 2.2046 lb$_m$/s
$\dot{m}''$	mass flux, kg/m^2 · s = 0.20481 lb$_m$/(ft^2 · s)
M	molecular weight, kg/kgmol = lb$_m$/lbmol
MC	moisture content, dry basis
n	molar concentration, kgmol/m^3 = 10^{-3} gmol/cm^3 = 0.6242 lb mol/ft^3
n'	droplet or particle concentration, number/cm^3 = 16.387 number/in.^{3}
N	moles, mol; revolutions per minute
ΔN	fractional number of drops in size range

$\dot{N}$	molar flow rate, kgmol/s = 2.2046 lbmol/s
$\dot{N}''$	molar flux, kgmol/(m^2 · s) = 0.20481 lbmol/(s · ft^2)
p	pressure, kPa = 0.14504 lb$_f$/in.^{2}
P	probability distribution function
q	heat transfer rate, W = 3.4123 Btu/h
q''	heat transfer rate per unit area, W/m^2 = 0.3170 Btu/(ft^2 · h)
q'''	rate of heat release per unit volume, W/m^3 = 0.09663 Btu/(ft^3 · h)
Q_{12}	total heat input for process from state 1 to state 2, kJ = 0.94787 Btu
Q_p	heat transfer for reaction of constant pressure, kJ = 0.94787 Btu
Q_v	heat transfer for reaction at constant volume, kJ = 0.94787 Btu
r	reaction rate (rate of production or destruction of a chemical species per unit volume), kgmol/(m^3 · s) = 224.74 lbmol/(ft^3 · h)
r_v	compression ratio
r_B	fraction of heat extracted from fluidized bed
$\hat{R}$	universal gas constant, kJ/(kgmol · K)
R	specific gas constant ($\hat{R}/M$), kJ/(kg · K)
$R\cdot$	radical species
s	entropy per unit mass, kJ/(kg · K) = 0.2388 Btu/(lb$_m$ · °R)
sg	specific gravity, density at 20°C relative to water
$\dot{sp}$	entropy production rate, kW/(kg · K) = 0.2388 Btu/(lb$_m$ · °R · s)
t	time, s
T	temperature, K = 1.8°R
u	internal energy per unit mass, kJ/kg = 0.4303 Btu/lb$_m$
v	specific volume, m^3/kg = 0.062428 ft^3/lb$_m$
V	volume, m^3 = 35.314 ft^3 = 6.1023 × 10^{4}in.^{3}
$\dot{V}$	volume flow rate, m^3/s = 2118.84 ft^3/min
$\underline{V}$	velocity, m/s = 3.281 ft/s = 39.37 in./s
$\underline{V'}$	root mean square velocity of turbulent fluctuations, m/s = 3.281 ft/s
$\underline{\tilde{V}}$	velocity relative to a leading shock or detonation wave, m/s = 3.281 ft/s
$\underline{\breve{V}}$	velocity relative to detonation wave, m/s = 3.281 ft/s
$\underline{V_L}$	laminar burning velocity, m/s = 3.281 ft/s
$\underline{V_T}$	turbulent burning velocity, m/s = 3.281 ft/s

W	work, kJ = 0.94787 Btu; weight, N = 0.22481 lb_f
$\dot{W}$	power, W = 3.4123 Btu/h
WI	Wobble Index, kJ $m^{2/3}/kg^{3/2}$ = 1.7222 Btu $ft^{2/3}/lb_m^{2/3}$
x	distance, m = 3.28 ft = 39.37 in.
x_i	mole fraction of species i
X	fraction of fuel burned in bed
y_i	mass fraction of species i
z	axial distance, m = 3.28 ft = 39.37 in.; vaporization parameter
Z	correction factor for effect of mass transfer on heat transfer
α	thermal diffusivity = $\tilde{k}/\rho c_p$, m^2/s = 10.764 ft^2/s = 3.875 × 10^4 ft^2/h
β	droplet burning rate constant, m^2/s = 10.764 ft^2/s
γ	ratio of specific heats
δ	laminar flame thickness, m = 39.37 in.
ϵ	emissivity; porosity; void fraction; rate of dissipation of turbulent kinetic energy, kW/m^3 = 96.65 $Btu/(h \cdot ft^3)$
η	efficiency, effectiveness
θ	crank angle, degrees; degrees of angle, radians
μ	absolute viscosity, $N \cdot s/m^2$ = $5.8016 \cdot 10^{-6}(lb_f \cdot h)/ft^2$
ν	kinematic viscosity, m^2/s = 10.764 ft^2/s, frequency, s^{-1}
ρ	density, kg/m^3 = 0.062428 lb_m/ft^3
σ	surface tension, N/m = 0.06854 lb_f/ft; Stephan-Boltzmann constant, $W/(m^2 \cdot K^4)$
Σ	flame area per unit volume, m^{-1} = 0.3048 ft^{-1}
τ	characteristic time, s
ϕ	mass transfer correction factor used in solid fuel combustion
Φ	fraction of bed volume occupied by tubes
ω	rotation rate, radians/s

SUBSCRIPTS

a	air
$attr$	attrition
b	background (temperature)

B	bed; cylinder bore
c	char
C	carbon
d	diameter
df	dry fuel
D	dense phase
e	effective
f	fuel
g	gas
i	species; initial
j	jet
l	liquid
L	losses; laminar
m	mean
mf	minimum fluidization
o	outside; reference condition; orifice
p	products; particle; pintle
pm	porous media
pyr	pyrolysis
s	solid; surface; stoichiometric
T	turbulent
v	vapor; volatiles
w	water

OVERBARS

^	quantity per mole
-	average value

DIMENSIONLESS NUMBERS

Bi	Biot number $= \tilde{h}L/\tilde{k}_s$ or $\tilde{h}L/\tilde{l}e_2$
Da_I	First Damköhler number $= (L_I/\underline{V}')/(\delta/\underline{V}_L)$

Le	Lewis number = α/D_{AB}
Ma	Mach number = $\underline{V}/a$
$\breve{Ma}$	Mach number relative to a leading shock or detonation wave = $\breve{\underline{V}}/a$
Nu	Nusselt number = $\tilde{h}L/\tilde{k}_g$
Pe	Peclet number = $\underline{V}_L d/\alpha$
Pr	Prandtl number = $c_p\mu/\tilde{k}$
Re	Reynolds number = $\underline{V}L/\nu$
Sc	Schmidt number = ν/D_{AB}
Sh	Sherwood number = $\tilde{h}_D d/D_{AB}$
$\tilde{W}$	flow parameter = $(\rho_a \underline{V}_a)/(\rho_g \underline{V}_g)$
We	Weber number = $\rho \underline{V}^2 L/\sigma$
Z	Ohnesorge number = $\mu_l/\sqrt{\rho l \sigma d_j}$

VECTORS

Except for $\boldsymbol{D}_{\mathbf{AB}}$, symbols in bold are vectors.

ABBREVIATIONS

AFBC	atmospheric fluidized bed combustion
AFBR	apparent fuel burning rate
AHRR	apparent heat release rate
AKI	antiknock index
AMD	area mean diameter
API	American Petroleum Institute
AQIRP	Auto/Oil Air Quality Improvement Research Program
ASI	after start of injection (crankangles)
ASME	American Society of Mechanical Engineers
ASTM	American Society for Testing and Materials
ATDC	after top dead center (crankangles)
BDC	bottom dead center
BMEP	brake mean effective pressure

BSFC	brake specific fuel consumption
BTDC	before top dead center (crankangles)
CA°	crank angle degrees
CFD	computational fluid dynamics
CFR	Combustion Fuels Research Corp.
CI	cetane index, also compression ignition
CN	cetane number
CNF	cumulative number fraction
CR	compression ratio
CVF	cumulative volume fraction
CWF	coal-water fuel
DI	direct injected
DISC	direct injected stratified charge
DME	dimethyl ether
DNS	direct numerical simulation
EES	Engineering Equation Solver
EGR	exhaust gas recirculation
$EINO_x$	emission index for NO_x
EPA	(U.S.) Environmental Protection Agency
FSR	flame speed ratio
GC	gas chromatograph
HC	hydrocarbons (unburned)
HHV	higher heating value
HMMS	Hino Motors micro-mixing system
HRR	heat release rate
IDI	indirect injection
IMEP	indicated mean effective pressure
ISFC	indicated specific fuel consumption
IVC	intake value closing (crankangle)
LHV	lower heating value
LNG	liquified natural gas

LPG	liquified propane gas
MBT	maximum brake torque
MBTE	methyl-*t*-butyl ether
MIT	Massachusetts Institute of Technology
MMD	mass mean diameter
MMT	mass mean temperature
MS	mass spectrometer
NA	naturally aspirated
NASA	National Aeronautics and Space Administration (previously NACA)
NO_x	nitrogen oxide plus nitrogen dioxide
NTC	negative temperature coefficient
PAH	polycyclic aromatic hydrocarbons
PAN	peroxyacetyl nitrate
PDF	probability density function
PFBC	pressurized fluidized bed combustion
PMEP	pumping mean effective pressure
RCM	rapid compression machine
RDF	refuse derived fuel
RPM	revolutions per minute (rpm when used as unit)
SAE	Society of Automotive Engineers
SI	spark ignited
SMD	Sauter mean diameter
SOC	start of combustion
SOI	start of injection
SO_x	sulfur dioxide plus sulfur trioxide
TEL	tetraethyl lead
TML	tetramethyl lead
TDC	top dead center
VMD	volume mean diameter

PREFACE

Currently 90% of the energy used for transportation, power production, and heating is produced by combustion of liquid, solid, and gaseous fuels. Although significant increases in the price of oil and natural gas are certain to occur during the early part of the next century, it is unlikely that this percentage will change significantly for many years. Thus the study of combustion is of continuing importance, especially if we are to conserve our sources of energy and reduce air pollution in a world of increasing population and increasing energy needs.

The engineer intending to study combustion will find that current sources of information fall into two categories—literature on the scientific aspects of combustion, and literature on the design and performance of specific technologies such as engines, turbines, and furnaces. Although a considerable number of books on combustion are available, they tend to emphasize the scientific aspects with little detailed information on specific engineering applications. On the other hand, books on applications typically focus on one technology and contain only a few chapters on combustion as applied to that technology.

The authors feel that a need exists for a book which bridges the gap between the scientific monograph and the specific technology texts. This book attempts to meet this need by presenting a broad coverage of combustion technology with enough accompanying combustion theory to allow a rudimentary understanding of the phenomena. The goal is to give a broad engineering treatment of combustion technology with focus on fundamentals and gaseous, liquid, and solid fuel combustion systems. The level of the text is such that a senior in mechanical or chemical engineering should have little difficulty in comprehension. The book assumes a knowledge of thermodynamics, fluid mechanics, and heat transfer gained from at least nine semester credits of introductory courses. Students who have not been introduced to mass transfer will need some supplemental information, and an Appendix is provided for that purpose. Mathematical treatment has been kept to a minimum, although modeling concepts have been emphasized and state-of-the-art theoretical approaches have been explained in simple terms so as to indicate the current possibility of theoretical solutions to specific design questions and to motivate further study.

The book contains 17 chapters organized into 4 parts. The book begins with a historical overview of combustion and combustion technology. As indicated in the table below, there are eight chapters pertaining to fundamentals and eight chapters on combustion technology. The fundamentals are developed to the level needed for the technological applications chapters of the text. Each application chapter describes how the combustion system works, evaluates the thermodynamic and fluid dynamic states of the system, provides a physical description and simplified model of how the combustion proceeds in the system, and discusses the emissions from the system.

Part	Fundamentals	Applications
I–Basic concepts	Fuel properties (Ch. 2) Thermodynamics (Ch. 3) Chemical kinetics (Ch. 4)	
II–Gaseous fuels	Flames (Ch. 5) Detonations (Ch. 8)	Gas-fired furnaces (Ch. 6) Gasoline engines (Ch. 7)
III–Liquid fuels	Sprays (Ch. 9) Detonations (Ch. 13)	Oil-fired furnaces (Ch. 10) Gas turbines (Ch. 11) Diesel engines (Ch. 12)
IV–Solid fuels	Single particles (Ch. 14)	Fixed beds (Ch. 15) Suspension burning (Ch. 16) Fluidized beds (Ch. 17)

Ample text material and homework problems are provided for a three-credit course covering 45 periods of 50 minutes each over a 15-week semester. Our approach in teaching the course has been to give equal time to each of the four sections because we feel that at the senior/MS level breadth is important. However, if the students have had only one thermodynamics course, more time must be spent on Chapter 3, and it becomes very difficult to cover the entire book in a semester. Some instructors may prefer to emphasize fundamentals and transportation systems (Chapters 7, 11, and 12), or fundamentals and heat-power systems (Chapters 6, 10, 11, 15, 16, and 17). Although the material has been prepared for use as a classroom text, it is hoped that the practicing engineer will also find the material useful and the treatment amenable to self-study.

The text contains 60 example problems and 220 end-of-chapter problems. Appendixes contain additional fuel properties, thermodynamic properties, and transport properties for use with the problems. At this time in the United States practicing engineers work with both SI and English systems of units. Hence this text uses both sets of units, although SI units are used more extensively than English units. A separate problem solution manual is available for instructors using the text. This manual also contains additional remarks concerning course syllabi.

Some of the examples and problems require the use of a personal computer. For equation solving we recommend either EES or mathcad. EES, which stands for Engineering Equation Solver, provides extensive thermophysical properties in addition to a solver for coupled nonlinear equations and initial value problems. EES is available from McGraw-Hill. For details visit the McGraw-Hill web site at www.mhhe.com. For thermodynamic equilibrium calculations we recommend either STANJAN or PER. STANJAN is available to club members from Professor William Reynolds; instructors may send e-mail to wcr@thermo.stanford.edu to become a club member, and then distribute it to students. PER may be downloaded from the world wide web from www.engr.wisc.edu/centers/erc/per.

Because this book is concerned with combustion engineering, not reactive fluid mechanics, some of the traditional approaches are not used here. Examples of such approaches not given are: the Burke-Schumann analysis of a laminar jet diffusion flame, droplet vaporization and particle combustion using boundary layer equations, and the Rankine-Hugoniot relations for detonations. These classical approaches have been replaced by analyses more closely related to practical applications. For example, film theory is used to solve droplet vaporization. Although students taking this course typically have not had intermediate fluid dynamics, computational fluid dynamic modeling (CFD) is discussed in several chapters to indicate the rich possibilities of this approach for combustion systems. However, because the mathematical aspects of CFD are beyond the scope of this book, the theory is presented only in a heuristic fashion.

The authors wish to thank the many students and faculty who have provided valuable input along the way as this text has evolved. We especially thank David Foster, Glen Myers, Phillip Myers, Rolf Reitz, Christopher Rutland, Otto Uyehara, Kenneth Bryden and Danny Aerts at the University of Wisconsin-Madison, David Hofeldt at the University of Minnesota, Eric Van den Bulck at the University of Leuven in Belgium, Duane Abata at Michigan Technological University, and William L. Brown, Jr. at Caterpillar Corp. Sally Radeke faithfully did much of the early word processing during the extended period of manuscript preparation. In addition the authors wish to thank Ralph Aldredge, University of California, Davis; Kalyan Annamalai, Texas A&M University; Robert Brown, Iowa State University; Mario Colaluca, Texas A&M University; David Dowling, University of Michigan; John Johnson, Michigan Technological University; Thomas Litzinger, Pennsylvania State University; Kozo Saito, University of Kentucky; and Benjamin D. Shaw, University of California, Davis, for their helpful reviews.

Gary L. Borman
Kenneth W. Ragland

Combustion Engineering

PART I

Basic Concepts

Before beginning the study of combustion processes and combustion systems, the general nature of combustion is discussed, and a brief history of combustion technology is given. Then those aspects of fuel properties, thermodynamics, and chemical kinetics which will be used throughout the book are reviewed.

CHAPTER 1

Scope and History of Combustion

The primary purpose of this first chapter is to introduce the broad scope of phenomena which make up the subject of combustion and to give a brief overview of the historical development of combustion science and combustion technology. Combustion impacts many aspects of our lives, especially those dealing with the utilization of energy. For engineers the continuing challenge is to design safe, efficient, and nonpolluting combustion systems for many different types of fuels.

1.1 THE NATURE OF COMBUSTION

Combustion is such a commonly observed phenomenon that it hardly seems necessary to define the term. From a scientific viewpoint, combustion stems from chemical reaction kinetics. The term combustion is saved for those reactions which take place very rapidly with large conversion of chemical energy to sensible energy. Such a definition is not precise because the point at which a reaction is characterized as combustion is somewhat arbitrary. We would probably all agree that rusting of an automobile is not combustion even though the oxidation reaction may be much faster than we desire. However, if we were to powder the car metal into tiny particles we would find that the oxidation reaction could go very quickly.

Increasing surface area to increase reaction rate is a method frequently used in engineering practice. Examples are pulverized coal combustors and liquid spray combustors such as used in burners, diesel engines, and gas turbines. Of course, increases in the surface area can also produce undesirable results such as accidental explosions in flour mills.

Another common method of causing fast reactions is to increase the temperature. As the temperature is increased, the rate of a chemical exothermic reaction typically increases. The rate of such gaseous reactions is often proportional to $\exp(-C/T)$, where C is a constant and T is the reaction absolute temperature. Even a small increase in the temperature causes the exponential term in the rate equation to increase very rapidly and explains why heating the reactants to a sufficiently high temperature causes combustion. Since fuel reactions are exothermic, heating can cause a runaway condition. As the reactants are heated, thermal energy is released, and if this energy is released faster than it can be transported away by heat transfer, the temperature of the system will rise, causing the reaction to go faster and thus causing an even greater rate of energy release. Such positive feedback can cause the rate of reaction to rise very rapidly and cause an explosion.

If the temperature of a combustible mixture is raised uniformly, by adiabatic compression, for example, the reaction may take place homogeneously throughout the volume. However, this is not typical. The combustion commonly observed involves a *flame,* which is a thin region of rapid exothermic chemical reaction. For example, a Bunsen burner and a candle each exhibit a thin region in which the fuel and oxygen react, giving off heat and light.

In the case of the Bunsen burner, the reactants, consisting of a gaseous fuel such as methane and air, are well mixed before ignition. When ignited, the mixture flows into the thin cone-shaped reaction zone or *flame front* and exits as products. The conversion of chemical to sensible energy takes place in the flame and causes the flame temperature to be high. The products emerge from the flame at the flame temperature, but heat transfer and mixing with the surrounding room air cause the temperature to decrease as the products move away from the flame. The Bunsen burner is an example of premixed combustion and gives rise to a *premixed flame.* Burning in a carbureted, spark-ignition automobile engine is an another example where a premixed flame occurs. The Bunsen burner flame is a *stationary flame,* whereas the spark engine flame is an example of a *propagating flame.*

A candle flame is different from a Bunsen burner flame in that the fuel is not premixed with the oxidizer. The candle wax is heated by the flame, vaporizes, and mixes with air which is drawn into the flame by the buoyant motion of the upward-flowing products. This type of flame is called a *diffusion flame.* The fuel-air concentration in a diffusion flame ranges from a fuel-rich zone to pure air, and in the region between these zones some of the mixture strengths will be appropriate for reaction. For most fuels the high-temperature reaction zone transfers enough heat back to the fuel-rich zone that some solid carbon particles less than 1 micron in size are formed. These hot particles radiate and give rise to the yellow or red color luminosity. Combustion in spray combustors such as used in oil burners, gas turbines, and diesel engines primarily involves diffusion flames. Hence, flames may be of either the premixed or the diffusion type, and both types may occur at different times and locations in a practical combustion chamber.

Flames occur in both laminar and turbulent flows. Turbulence speeds up the rate at which reactants are prepared for combustion and increases the surface area of the reaction zone. Turbulence greatly increases the *burning velocity.* Here we refer to burning velocity as the rate of reaction zone travel relative to the unburned mixture. Understanding such turbulence effects is important because laminar flames are very slow, and it is only by the use of turbulence that most combustors can produce the heating rates required for a practical design.

The premixed laminar burning velocity for hydrocarbon-air flames is approximately 1/2 m/s, whereas the turbulent burning velocity is 2 to 10 times faster. In a quiescent combustible mixture a laminar flame will form from an ignition source such as a spark, and then accelerate to a turbulent flame. Given sufficient distance, the accelerating turbulent flame will abruptly switch over to a *detonation* which travels on the order of 2000 m/s. In contrast to a flame which exhibits a nearly constant pressure across the reaction zone, the pressure across a detonation rises by a factor of 15 or more. For example, if the reactants are at 1 atm, the products behind the detonation would be raised for a short period of time to 15 atm. Thus understanding the conditions which can lead to detonation is an important aspect of the safe design of combustion systems.

The nature of the combustion also depends on whether the fuel is gaseous, liquid, or solid. Gaseous fuels are easy to feed and mix, and are generally clean-burning. Liquid fuels are typically sprayed through a nozzle at high pressures, while solid fuels are usually crushed, pulverized, or chipped before feeding into the combustor. When heated, liquid fuels vaporize and then burn in the gaseous phase. Solid fuels, when heated, release moisture and gaseous volatiles, and the remainder is solid *char,* which is mostly porous carbon, and ash. Heavy liquid fuels also yield char and ash. The char burns out as a surface reaction, while the volatiles burn in the gas phase. Vaporization of liquid fuel sprays, and devolatilization of solid fuels, occur much more slowly than gas-phase chemical reactions. Char burn, in turn, occurs more slowly than devolatilization.

Since gaseous, liquid, and solid fuels require different preparation and react at different rates, each type of fuel presents different design challenges to build efficient, economical, and clean combustion systems. Also, the fuel properties influence the ash deposition, erosion, and corrosion of the system. Clearly, the fuel properties are interrelated with system design requirements.

Combustion emissions must satisfy governmentally imposed *emission standards* for selected compounds in the products, such as carbon monoxide, hydrocarbons, nitrogen oxides, sulfur dioxide, and particulate emissions. Emission standards are set at levels to try to keep the ambient air clean enough to protect human health and the natural environment. Low emissions can be achieved by a combination of fuel selection and preparation, combustion system design, and treatment of the products of combustion. There are challenging engineering trade-offs between low emissions, high efficiency, and low cost.

Recently, global warming has become a widespread concern. Carbon dioxide levels in the global atmosphere are increasing, and carbon dioxide emissions from combustion are a major contributor to the greenhouse effect, whereby long-wave radiation from the surface of the earth is trapped by the atmosphere. The relationship between CO_2 emissions and average global temperature rise is not clear at this time. However, it is well established that the CO_2 concentration in the atmosphere is increasing at an accelerating rate. Prior to the industrial revolution the CO_2 content of the atmosphere was fairly stable at 280 parts per million (ppm), based on measurements of air bubbles trapped in glacial ice corings. By 1900 the CO_2 level had reached 300 ppm. Accurate, direct measurements of atmospheric carbon dioxide concentrations were begun by Charles Keeling at the Mauna Loa Observatory in 1958. By 1958 the CO_2 concentration was 315 ppm; by 1980, 337 ppm; and by 1996, 362 ppm. Because the world population is expected to nearly double to around 10 billion people during the next several decades, the potential for future growth in CO_2 emissions cannot be ignored. The worldwide pressure for growth in fuel consumption and CO_2 emissions is tremendous, as evidenced, for example, by the fact that one-third of the people in the world still do not have any electricity.

A reduction in CO_2 emissions can be achieved by improvement in the overall efficiency of combustion systems, by using renewable fuels, and by replacing fossil fuels with other sources of energy such as solar photovoltaic, wind, geothermal, hydro, or nuclear power. In the business of providing heat and power to meet the needs of the global society, there are no easy choices but many difficult engineering challenges.

Improving the design of combustion systems requires an understanding of combustion from both a scientific and an engineering standpoint. Understanding the details of combustion requires utilizing chemistry, mathematics, thermodynamics, heat transfer, and fluid mechanics. For example, a detailed understanding of even the simplest turbulent flame requires a knowledge of turbulence and chemical kinetics which is at the frontiers of current science. However, the engineer cannot wait for such an understanding to evolve but must use a combination of science, experiment, and experience to find practical design solutions.

1.2 HISTORICAL PERSPECTIVE OF COMBUSTION SCIENCE

The student embarking on the study of a new field should have at least a rudimentary knowledge of the history of that subject. While it is not our purpose here to give a definitive history of combustion, some milestones are given which may help put the subject in perspective.

Although the scientific method was to await the 17th century, we may surmise that early humans found natural fire to be useful though frightening. Perhaps 30,000 years ago the first combustion engineer was born and found ways to produce fires artificially. The importance of this discovery is indicated by the central role of fire in early myths and religions. How could early humans have failed to be both exalted and mystified by fire when we in the twentieth century are still enthralled by the continuously changing flames of the campfire or fireplace?

With the advent of Greek science, fire took its place along with water, earth, and air as a fundamental element of the world. From Heraclitus to Aristotle to the Renaissance, fire played an important part in philosophy. While such philosophical thought had an enormous impact on history, it did little to bring about understanding of the nature of combustion. During the Renaissance, Bacon, Boyle, and Hooke among others began to observe the structure of flames. Because the release of combustion energy and the flow of energy by heat transfer was not understood, the phlogiston theory held forth until the 18th century. The mysterious phlogiston was thought to convect energy without the flow of matter. However, by the time of Lavoisier's untimely death (guillotined in the French Revolution), his *Réflexions sur le Phlogistique* had ended the idea of phlogiston. The scientific revolution was on its way.

In the early 1800s Benjamin Thompson (Count Rumford) and Humphry Davy invented the safety lamp for prevention of mine explosions, a major achievement in the application of flame quenching which saved countless lives. Thompson, an

American who went to England after studying at Harvard, published his observations on heat and friction in 1798. Forty-four years later Joule proved conclusively that heat was a form of energy and not a material substance. Thompson was interested in fuel economy and produced improved fireplaces and the Rumford boiler. After developing the first premixed gas burner in 1855, Robert Bunsen, a professor at the University of Heidelberg, measured flame temperature and flame speed and collected flame enthalpy data by use of a calorimeter. Meanwhile in France, Mallard conducted a pioneering study on flame propagation in 1868. The elementary model of flame propagation by Mallard and Le Châtelier, who were professors at École des Mines in Paris, set the foundation for flame theory in 1883. The distinction between flames (deflagrations) and detonations was made by Chapman and Jouguet in 1900, and they calculated the detonation velocity.

In the twentieth century rapid progress in the application of combustion science to technology took place, particularly in the period between World Wars I and II. Combustion was becoming a recognized area of study. The first Combustion Institute meeting (1928) and the text by Jost (1938) in Germany are hallmarks of this coming of age. The first theoretical treatment of diffusion flame height and shape was given by Burke and Schumann in 1928. Although the experimental works of this period are redoubtable and cover many subjects still pursued at present, they lacked accurate experimental techniques and instrumentation. The student wishing to sample such works will find browsing through the U.S. NACA Reports and Technical Notes of the 1930s to be quite enjoyable.

During World War II the investment in science and technology led to new instrumentation and to computers, which then expanded rapidly in the postwar era. The development of gas turbines and rockets gave a practical importance to the study of high-temperature kinetics and spray combustion. The use of shock tubes coupled with fast-response instrumentation led to rapid advancements in the chemical kinetics of combustion. The second combustion text by Lewis and von Elbe (1951) brought together the previously scattered knowledge that had been gained since the text of Jost. Combustion theory was also making advances, such as the pioneering work on laminar flame structure modeling carried out by Hirschfelder and von Kármán. Work on the structure of detonations, started during the war, gave birth to the Zeldovich–Doring–von Neumann theory.

The work in combustion modeling started in the 1940s and 1950s was hampered by the inability to compute rapidly and evaluate the results with sufficient accuracy. This changed in the 1960s and 1970s with the advent of high-speed computers and laser spectroscopy. In addition, the study of air pollution became a major research effort. The control of pollution required new knowledge of chemical kinetics and understanding the microscopic details of combustion. These new needs were further amplified by the sudden rise of fuel costs in the mid-1970s.

Today, computer hardware and software and combustion instrumentation techniques are evolving more rapidly than ever and are allowing a merger of chemical kinetics and fluid dynamics in order to investigate many aspects of com-bustion systems. The limited supply of fossil fuels and environmental concerns are providing continuing motivation to develop new understanding. But new scientific knowledge is not sufficient; as in the past, engineers must use this knowledge to produce practical devices. Engineers must learn to apply this knowledge, to utilize new computer models, and to bring forth a new generation of combustion technology.

1.3 HISTORICAL PERSPECTIVE OF FUELS

The history of combustion technology is closely related to fuel availability and utilization. Before 1850, wood was the primary worldwide fuel. The transition from an energy supply based on wood to one based predominantly on coal, then on petroleum and natural gas, is shown in Figure 1.1. As sources of petroleum and natural gas are gradually depleted, coal may once again become the major source of fuel. However, concern over global warming may limit the use of fossil fuels in the future. Renewable energy sources such as wind, biomass, photovoltaic, and geothermal energy may become more important in the future.

Commercial trade in coal started in England in the 13th century. The use of coal for iron smelting and for fueling steam engines driving locomotives and supplying shaft power to other machines was one of the important contributing factors in bringing about the Industrial Revolution. Coal surpassed wood as the main energy source in the United States by 1885. Coal gasification was used to generate a producer gas (sometimes called city gas) for lighting, since the nonaerated flame provided illumination that was far superior to candlelight. The advent of electric lighting in the 1920s expanded the use of coal for steam turbine power plants. The demand for electricity has continued to grow and has been met to a large extent with coal-fired power plants.

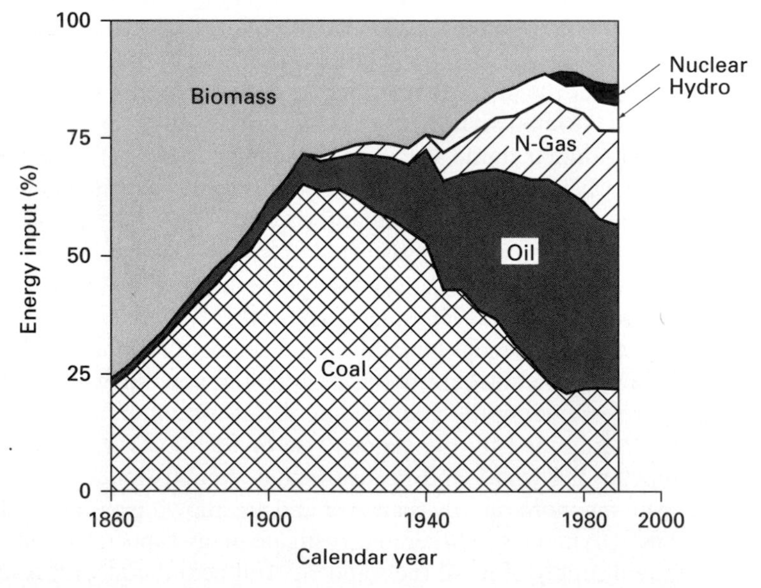

FIGURE 1.1
Contribution of each primary fuel to world energy input, 1860 to 1990 [Dupont-Roc et al., by permission of Shell International Limited].

The first oil strike at Titusville, Pennsylvania, in 1859 launched the oil industry. Refined crude oil gradually replaced whale oil as a lubricant and lamp fuel as whale kills could no longer keep up with the demand. The advent of the automobile in the early 1900s brought rapid expansion to the oil industry, and the race to find more and more oil for an ever-growing population of automobiles began. Also, firing fuel oil in utility, industrial, commercial, and residential furnaces and boilers became increasingly attractive. By 1950 petroleum replaced coal as the primary energy source in the United States.

Natural gas was at first regarded as a nuisance which was simply flared off at the oil well. Natural gas is a very clean-burning fuel but was difficult to transport. After World War II extensive pipelines were built to distribute the natural gas from the gas fields to individual users. Today there are over 900,000 miles of natural gas pipelines in the United States. Although the practice of flaring of natural gas continues, increasingly pipelines and shipment of liquefied natural gas are providing for better utilization of this valuable, nonrenewable resource.

Domestic oil production in the United States peaked in 1970. During 1973–74 the Organization of Petroleum Exporting Countries (OPEC) quadrupled the price of crude oil to \$12/barrel. The price then increased over time to more than \$30/barrel. Similarly, natural gas prices rose although less dramatically. Due to energy conservation and improved fuel efficiency, oil and gas use in the United States dropped an average of 1.2% per year from 1973 to 1986 (see Figure 1.2). During this period OPEC lost control of oil prices. Since 1986 oil and gas use have been increasing. Before 1973 energy use and gross national product (GNP) grew in lockstep. The effect of improved conservation and fuel efficiency was to save

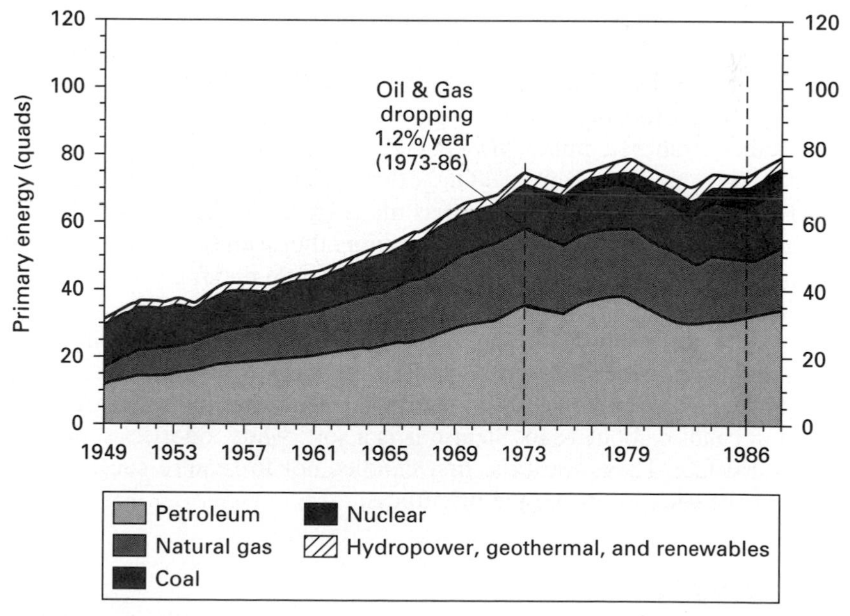

FIGURE 1.2
U.S. energy use from 1949 to 1988. One quad is 10^{15} Btu [Hollander].

28 quad, or 26% in 1990 compared with what would have been used in 1990 if existing trends prior to 1973 had continued.

1.4 HISTORICAL PERSPECTIVE OF COMBUSTION TECHNOLOGY

The first uses of combustion were for heating and lighting. Uses have progressed from open fires to fireplaces, stoves, burners and furnaces, boilers, and engines. A brief historical perspective of steam boilers, internal combustion engines, gas turbines, and rockets will be given, but first the history of lighting is briefly discussed.

Lighting

Prior to the first application of electricity to lighting in the late 19th century, all artificial light was produced by fire. Flames from wood, oil lamps, gas lamps, and candles were used. The prehistoric lamp consisted of a stone (later clay) reservoir for oil and a wick made from plant material to control the fuel flow by capillary action. The basic design changed little until a Swiss chemist, M. Argand, invented the tubular wick in 1784. Benjamin Thompson incorporated Argand's invention in his "astral" lamp design (1810), which also used a barometric fuel level supply and a rack-and-pinion wick adjustment. The flow of air through the center of the wick plus the flow around the wick exterior increased the flame volume and improved the brightness. The addition of a glass chimney helped to control the airflow. Thompson experimented with over 100 different lamp designs, and published a treatise called *The Management of Light in Illumination*. Figure 1.3 shows a diagram of one version of the Argand concept which incorporates astral lamp features. Note the holes around the bottom of the central stem which admit air to both the center of the wick and to the outer area of the wick. Fuel for lamps changed from animal fat to fish oil to whale oil, and by 1880 kerosene was used.

The Chinese probably first used gas for lighting by piping natural gas in bamboo tubes from salt mines. In 1784 Professor J. P. Minckelers at Louvain University, Belgium, pyrolyzed coal (and other solids) and lighted his lecture room with gas flames. The first gas mantle was made by C. A. Welsbach in 1890 using cotton fabric soaked with salt solution containing thoria and ceria, and burning out the organic matter. The flame stabilized on the mantle making it very bright, and greatly increasing the light output compared to a naked flame.

The tallow candle probably originated during the second century A.D. The whale oil industry introduced spermaceti (an oil obtained from the head of sperm whales) for candles in the mid-18th century. Stearin mixed with stearic acid was first used in candles about 1840. Stearin is is a soft, white, odorless solid found in many natural fats. These were the first candles not to require snuffing, that is, pinching of the wick to restore the brightness.

Steam Boilers

Early steam boilers consisted of little more than a kettle of water heated from the bottom. Boilers in the 1700s used the kettle principle but burned the fuel in an

FIGURE 1.3
This 1816 drawing shows an Argand lamp which incorporates an adjustable wick height and gives improved light by allowing air to flow inside the cylindrical wick and over its outer area giving two flames [Gregory].

enclosed furnace to direct more of the heat to the boiler kettle. To improve efficiency an integral furnace was developed by 1750 where the fuel was burned in a container enclosed within the water vessel. The flue gas tube wound through the water vessel to the atmosphere much like a coil in a still. A bellows was used to force air to the combustion zone.

As the demand for power increased, the single flue was replaced by many gas tubes, which increased the heating surface. Since the flue gas flows inside the tubes, this type of design is called a *fire-tube boiler*. The so-called Lancashire boiler designed by William Fairbain in 1845 is an example of a fire-tube boiler. However, many disastrous explosions resulted from direct heating of the pressure shell, which contained large amounts of saturated water. For example, in 1880 in the United States, 259 people were killed and 555 injured by boiler explosions [Stultz].

The idea that water, instead of flue gas, could flow through the inside of the tubes with heated gas outside, which would boost capacity and lead to safer operation, occurred to a number of early engineers, and the first *water-tube boiler* was patented by an American, James Ramsey, in 1788. These early water-tube designs were not successful because of construction problems, steam leaks, and internal deposits. It was not until 1856 that a truly successful water-tube boiler was designed by Stephen Wilcox. The Wilcox boiler had improved water circulation and increased surface area due to inclined water tubes, which connected water spaces at the front and rear and a steam chamber above. Wilcox's inherently safe design revolutionized the boiler industry. In 1866 Wilcox joined with George Babcock, and their company grew rapidly. Their first boiler is shown in Figure 1.4.

The first commercial steam turbine (5 hp) began operation in 1891 due to the pioneering work of Gustaf de Laval of Sweden. In the United States the first steam turbine electric generator (400 kW) went into operation due to the efforts of George Westinghouse. Steam turbines rapidly gained acceptance over steam engines, and in 1903 Chicago became the first city to have a central power station designed exclusively for steam turbines (5 MW each). By the early 1920s steam temperatures were about 600°F (300°C) and steam pressures were 13 to 20 atm. By 1929 an 80-atm boiler was built. Today 240-atm steam turbines are common, and engineers contemplate higher-pressure steam with temperatures of 1200°F (650°C) as a means of increasing the thermodynamic efficiency.

Fuel firing systems underwent a similar evolution. The hand-stoked furnace was replaced by the automatic spreader stoker in 1822, and the traveling-grate stoker was invented in 1833. Pulverized-coal firing advanced rapidly in the 1920s, as did stoker firing. Suspension firing became dominant in central power station boilers beginning in the 1930s. With the need to control sulfur dioxide emissions, the first fluidized-bed combustor for coal was introduced commercially in 1976. These events and others are summarized in Table 1.1.

Because of the ubiquitous nature of furnaces and boilers, smoke control was a major concern in cities by the 1920s if not before. Cyclone collectors began to be

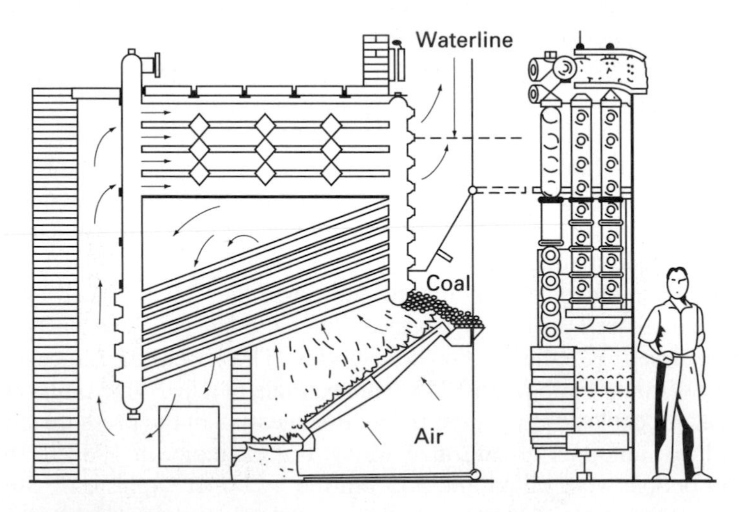

FIGURE 1.4
First water-tube boiler, by Babcock and Wilcox in 1867 [Stultz and Kitto, by permission of Babcock and Wilcox Co.].

TABLE 1.1
Developments in the utilization of coal [adapted from Elliott, by permission of National Academy Press]

Technological event	Year
Coal used to smelt iron	1621
Coke used in blast furnace	1718
Coal gas produced for lighting	1792
Spreader stoker invented	1822
Underfeed stoker invented	1833
Traveling-grate stoker invented	1833
Fischer-Tropsch synthetic gasoline conceived	1913
Pulverized-coal firing started	1914
Cyclone furnace invented	1947
Fluidized-bed combustion for central power stations began	1976

George Herman Babcock (1832–1893) and Stephen Wilcox (1830–1893), American engineers

George Babcock (left) spent one year at a technical institute in DeRuyter, New York, before going to work as a printer. In 1855 he patented (with his father) a polychromatic printing press, which won a prize in the Crystal Palace Exhibition in London. He also patented a bronzing machine and founded the journal *Literary Echo.* In 1859 Babcock sold his business and became a patent solicitor in Brooklyn. In the early 1860s he was chief draftsman of the Hope Iron Works in Providence, Rhode Island, and in the evenings he taught mechanical drawing at Cooper Union Academy. In 1866 Babcock became the sixth president of the American Society of Mechanical Engineers, and in 1881, the first President of the Babcock & Wilcox Co., a position he held until his death.

Stephen Wilcox (right) patented a letting-off motion for looms in 1853 and patented his first steam boiler in 1856 at the age of 26. In the 1860s he worked at the Hope Iron Works and became acquainted with Babcock. Wilcox was vice president of Babcock & Wilcox Co. from its incorporation until his death in 1893.

used for particulate control in the 1930s. The electrostatic precipitator was invented in 1910 by F. G. Cottrell and has been used for industrial and utility particulate control starting in the 1950s, as have fabric filters. In the 1970s control of large sources of particulates, carbon monoxide, nitrogen dioxide, and sulfur dioxide was mandated in the United States by the Clean Air Act.

Internal Combustion Engines

The concept of the internal combustion engine was conceived by Jean de Hautefeuille in 1676 and developed further by Huygens and Papin. In their engine the charge was exploded in the closed end of an open-topped cylinder fitted with a piston. Work was obtained as the atmosphere pushed the piston down during the cooling of the products. Papin later utilized this idea by replacing the charge with steam which condensed to give the vacuum, thus discovering the working principle of the Newcomen steam engine.

The idea of driving the piston by the burning products' expansion was first utilized in a production engine of 1/2 hp by Lenoir (1860), who ignited the mixture during the intake stroke by use of an electric spark. The idea of compression before ignition was not considered practical at that time. The Lenoir engine was powered with producer gas made from coal. In 1876 Nikolaus Otto built a 3-hp compression-ignition engine based on the four-stroke concept proposed by Beau de Rochas in 1862, and the basic cycle persists today. In 1878 Dugald Clark developed a two-stroke engine in order to obtain a higher power from the same size engine.

Combustion in the spark-ignited homogeneous-charge engine was at first dictated by mechanical considerations of strength, metal temperature, and lubrication constraints. During the era of cheap energy and prior to air pollution regulations, engine design evolved to higher and higher compression ratios using slightly richer than stoichiometric mixtures. Knocking combustion (explosion of the unburned gas giving pressure pulsations) was a serious problem, but was partially overcome by the use of tetraethyl lead (1923), which, when added to gasoline in small quantities, reduces the tendency to knock. Improvements in the quality of gasoline, lubricating oil, and metals allowed compression ratios to rise from the modest 3.6 : 1 of the 1915 Model T to 8.1 : 1 in engines of the early 1960s. The increase in compression ratio might have continued to approach the practical optimum of about 12, but new considerations of air pollution emissions, particularly nitrogen oxide (NO), were to halt this development. Catalysts developed to reduce the emissions of NO, CO, and unburned hydrocarbons were poisoned by lead in the gasoline. Thus considerations of knock again reduced the compression ratio because tetraethyl lead could no longer be used as a fuel additive.

With the growth of urban centers, industrial and automotive sources of air pollution became a serious concern. A most notable problem occurred first in southern California, where combustion emissions produced a heavy and health-threatening haze called smog. The term *smog* (smoke plus fog) originated in the early 1900s in Great Britain, where burning of high-sulfur coal plus natural fog produced deadly sulfuric acid aerosol. The smog in Los Angeles was quite different, however. In 1952 Professor Haagen-Smit from the California Institute of

Niklaus August Otto, German engineer (1832–1891)

In spite of a childhood interest in science and engineering, Otto left school at 16 to become a clerk. However, Otto spent all his spare time and money studying engineering, and was inspired earlier by news of an engine built by Lenoir. In 1861 at age 29 Otto built a small four-stroke internal combustion engine. The engine ran extremely roughly because of explosions. Nevertheless he obtained the backing of a wealthy industrialist and formed the Niklaus Otto Company. At the 1867 Paris Exhibition he won a gold medal in competition with 14 French engines. In 1877, fifteen years after his first attempt to produce a four-stroke engine, Otto produced and patented an 8-hp engine, which was a tremendous success. Others sought ways to exploit the new idea, and in 1886 the patent was invalidated. However, over 3000 engines were sold by this time, and the firm had plenty of work modifying the engines to run on liquid fuels. Otto died in 1891, a modest, retiring, and yet truly great engineer who developed the prototype for all modern internal combustion engines.

Technology showed that smog could be produced in the laboratory from automobile exhaust plus sunlight. The necessary ingredients for this type of smog are hydrocarbons, nitrogen oxides, air, and strong sunlight. Photochemical and chemical reactions in the atmosphere produce oxidants (ozone, nitrogen dioxide, and peroxyacetyl nitrate) and photochemical aerosols, which irritate the eyes and lungs. Due to public demand, federal emission controls for automobiles began with the 1972 models in the United States.

The history of the development of gasoline parallels the history of the Otto (spark-ignition) engine and includes some of the events mentioned above. Table 1.2 lists technological events for gasoline. Lead alkyl additive to prevent knock was introduced in 1923, and its use reached a maximum in 1970. Unleaded gasoline was required in 1974; phasedown of lead began in 1980; a total ban on lead was required by 1996. Although octane quality continues to be a major concern, starting in the 1970s environmental concerns also began to influence gasoline properties. The use of exhaust aftertreatment catalysts required the use of

TABLE 1.2
Major events affecting gasoline properties [Gibbs, modified and updated with permission from SAE paper 932828 © 1993, Society of Automotive Engineers]

Technological event	Year
Gasoline developed	1863
First four-stroke engine	1876
Thermal cracking*	1913
Lead alkyl antiknock used	1923
Octane scale developed	1927
Thermal reforming*	1931
Catalytic cracking*	1936
Polymerization*	1937
Akylation*	1938
Motor octane number method	1939
Catalytic reforming*	1940
Isomerization*	1943
Research octane number method	1951
Octane race begins	1953
Hydrocracking*	1959
Lead alkyl usage at maximum	1970
Antiknock index† at maximum	1972
Unleaded gasoline required	1974
Manganese banned	1977
Widespread use of oxygenates begins	1978
Lead phasedown begins	1980
Sulfur limited	1981
Vapor pressure limited—summer (phase I)	1989
Oxygenate content specified—winter	1992
Vapor pressure limited—summer (phase II)	1992
Gasoline reformulated‡	1995
Lead banned	1996

* Processing step introduced.

† See the section Gasoline Quality in Chapter 2.

‡ Benzene, aromatics, sulfur, olefins, vapor pressure limited; oxygen content specified; heavy metals removed.

nonleaded gasoline, the need to reduce unburned hydrocarbon emissions led to volatility controls, the desire to reduce CO in urban areas led to the use of oxygenates, and the current reformulation efforts are driven by continuing environmental concerns.

Compression-Ignition Engines

In 1893 a German patent was granted to Rudolf Diesel for the design of a "rational heat engine." Diesel planned a four-stroke engine which would incorporate the constant-temperature energy addition of the Carnot cycle. In his first engine, ammonia was injected into the cylinder at the end of the compression stroke to avoid premature ignition. Diesel's second engine (1896) had a water-jacketed cylinder and a pump to supply air to the cylinder to reduce exhaust smoke (Figure 1.5). In 1898 the first production engine ran on kerosene and gave an amazing 20 hp. By 1901 the external crosshead arrangement used in the original engine design was replaced by a trunk piston design innovated by the American Diesel Company. The M.A.N. Company built a 250-hp, two-cylinder engine in 1902, and the Sulzers Co. built a three-cylinder engine of 300 hp in 1906. Because such engines could not be directly reversed, marine applications used an electric drive for slow speed and maneuvering. A steam engine with oil-fired boilers and a diesel

FIGURE 1.5
Diesel's Series 250/400 engine used in tests at M.A.N., February 1897; its "twin" is on exhibit at the Deutsches Museum in Munich. The tests gave a brake horsepower of 17.8 at 154 rpm with a specific fuel consumption of 238 g/bhp · h, and a brake thermal efficiency of 26.2%. (M.A.N. Archives photo as reproduced in C. Lyle Cummins, *Diesel's Engine.*) [By permission of C. L. Cummins].

engine were shown side by side at the Turin Exhibition of 1911. The steam engine used several times more fuel per horsepower than the diesel engine. This greatly improved fuel economy foretold the demise of steam engines.

By 1910 German and British companies had developed diesel-powered submarines, and the first diesel-powered passenger ship appeared in 1921. Early four-stroke ship engines experienced considerable problems with fouling of the exhaust valves and ports with carbon. This problem led Sulzers to develop the two-stroke diesel. Engines with 2000 hp per cylinder were in operation before World War I, and by 1939 half of the world's shipping tonnage was diesel-powered.

High-speed diesel engines for commercial, farm, and industrial applications developed slowly because advances were needed in strength of materials and fuel-injection systems. Although the principle of airless injection of liquid fuel was pioneered by Herbert Akroyd Stuart as early as 1886, it was not until 1936 that

Rudolf Christian Karl Diesel, German engineer (1858–1913)

Rudolf Diesel was an outstanding student in mechanical engineering in Munich. Professor Linde, one of Diesel's instructors and a worldwide authority on heat engines and refrigeration, helped him get a job at the famous Sulzers factory in Switzerland, where he became a proficient machinist before becoming a factory manager. He began experimenting with a high-pressure ammonia engine. His 1893 prototype blew off the cylinder head, but four years later a reasonably reliable engine was produced. His new engine was soon accepted throughout the world, and many of his engines were made under license. His wife convinced him to name the engine after himself. However, Rudolf Diesel's enjoyment of fame and fortune was marred by ill health, probably brought on by exhausting legal battles over patent rights and unwise financial speculations. He lost a fortune, and while traveling on a ship to England, he disappeared overboard. In his notebook he had marked an X by that date. Diesel was a practical genius noted for his work on engines in the laboratory and for his study of heat engine cycles including the constant-pressure diesel cycle.

Robert Bosch introduced an ingenious method of metering which did not require a variable-stroke pump. Equally important, the Bosch Company had the ability to use the high-precision machining required to fabricate such systems.

Although diesels have displaced spark ignition engines in nearly all applications except automobiles and aircraft, a detailed understanding of the diesel combustion process is yet to be attained. Thus highly different combustion chamber designs have persisted, such as, for example, open-chamber designs where the fuel is directly injected into the cylinder, and prechamber designs where the fuel is injected into a small chamber which is attached to the cylinder by a small passage. The difficult requirements of fuel-air mixing are met either by high-pressure injection (1000 atm) or by lower-pressure injection (300 atm) combined with swirl imparted to the cylinder air by the induction process.

The modern diesel engine is a highly sophisticated device. It can produce high power with excellent fuel economy while still meeting strict gaseous emissions requirements. The recent addition of exhaust particulates regulation has, however, raised doubts concerning the future of the diesel engine for automobile and other light-duty transportation applications. Resolution of these many, often conflicting demands will require an improved understanding of the chemical kinetics and fluid mechanics which govern diesel combustion. The difficulty of meeting the dual demands of fuel economy and emissions has led to consideration of many different types of combustion engines. Included in these are various combinations of direct gasoline injection into the cylinder with spark ignition. Engines of this type have the potential to operate without problems of knock or compression ignition and thus could use a wide variety of fuels.

Gas Turbines

Although the use of flue gases or steam to drive a wheel dates to ancient times, the forerunner of the modern gas turbine can be traced to the patent of John Barber in 1791, which utilized a compressor, a combustor, and an impulse turbine. Early combustors typically employed an explosive, intermittent combustion in a closed space, which caused a flow through a nozzle to drive an impulse turbine. Although inefficient, this design persisted because development of continuous-flow machines was hampered by a lack of knowledge of aerodynamics, resulting in very inefficient compressors.

The first working gas turbine with a constant-pressure combustor was that of Aegidius Elling of Norway. He started working on gas turbines in 1882, and 21 years later Elling achieved a net power output of 11 hp with a six-stage centrifugal compressor and an axial impulse turbine with an inlet temperature of 400°C. In 1905 the Frenchmen Charles Lemale and René Armengaud used a 25-stage Brown Bovari centrifugal compressor (running at 4000 rpm, absorbing 325 hp, and giving a 3/1 pressure ratio), a high-temperature combustor, and a two-stage Curtis turbine. The thermal efficiency was 3.5%. By 1939 efficiency had improved dramatically, and a regenerative axial-flow compressor and turbine of Hungarian design gave an efficiency of 21%. The first gas turbine in the United States to generate electric utility power was installed in 1949.

The use of gas turbines in aircraft dates to the 1930 patent of Frank Whittle in England. The technical problems which had to be overcome were to make a combustor with about 20 times the combustion intensity of stationary gas turbines, to improve the compressor and turbine efficiency, and to overcome the mechanical failures that plagued turbines at that time. Meanwhile, Hans von Ohain, with the backing of aircraft manufacturer Ernst Heinkel, independently pioneered the aircraft gas turbine in prewar Germany. By 1939 both Whittle and von Ohain were well advanced toward a flying prototype. In August 1939 a Heinkel aircraft with von Ohain's engine flew aloft for 7 minutes. Two years later, in May 1941, a Gloster

Frank Whittle, British engineer (1907–1996)

In 1922, at age 15, Whittle became an apprentice in the British Royal Air Force (RAF). He graduated from the RAF College (Cranwell) in 1928, from the RAF Officers' School of Engineering in 1934, and from Cambridge University in 1936. From an early age he was fascinated by the idea of jet propulsion, and his 1928 thesis, "Future Developments in Aircraft Design," discussed the topic, which led to a patent in 1930. Unfortunately, Whittle's revolutionary invention failed to impress either the British Air Ministry or private companies, who remained skeptical because gas turbines had a long history of failure. During his senior year at Cambridge in 1936 he formed the Power Jets Co. with financial partners. For Whittle the period 1936–40 was a time of great technological highs and lows as the W1 engine was gradually made to operate at higher rpm with higher combustion intensity. Always short of cash because the RAF belittled the concept as impractical, recognition was slow at first. Although the design was shared with the United States, and progress was more rapid during the war, the British-American gas turbine–powered aircraft did not have an impact on the war. By the end of the war Whittle's small company and his patents were swept aside by the tides of history, but a new industry had been launched. After the war he became a technical adviser and consultant. In 1976 he emigrated to the United States. Whittle was awarded numerous honors.

Hans von Ohain, German physicist (1911–)

In 1935, at age 24, von Ohain received his doctoral degree from the University of Göttingen. Struck by the possibilities of jet propulsion while a student and encouraged by a professor who recognized his genius, one of his early engine designs earned him a job at the Heinkel Aircraft Co. Heinkel's backing allowed von Ohain to progress rapidly, and by 1937 (though entirely unaware of Whittle's work, as was Whittle of his) he successfully tested an engine in his workshop. In 1938 the German Air Ministry directed all private aviation companies to begin the development of jet engines, but a year later, as the country went to war, development efforts were shifted back to propeller-driven aircraft in a monumental lack of foresight that lost Germany its lead in jet engine development and possibly the war. After the war von Ohain came to the United States and continued his work at Wright-Patterson Air Force Base. By 1975 he was responsible for maintaining the quality of all Air Force research and development in turbojet propulsion, and upon his retirement in 1979 joined the University of Dayton Research Institute. He is the holder of 19 U.S. patents and many honors.

Sir Frank Whittle and Dr. Hans von Ohain were the 1991 recipients of the C. S. Draper Prize from the U.S. National Academy of Engineering "for engineering innovation and individual tenacity in the development and reduction to practice of the turbojet engine, thereby revolutionizing the world's transportation system, improving the world's economy, and transforming the relationship between nations and their peoples."

aircraft with an 850 lb_f thrust Whittle W1 engine flew aloft for 17 minutes, and during the next 12 days logged 10 hours of test flights. A General Electric version of the W1 was built and flown soon after that. The first gas turbine engine in Japan was developed independently by two naval officers, Tokiyasu Tanegashima and Osamu Nagano, starting in 1943. By late 1944 the Germans shared technical turbine information with the Japanese. The first Japanese test flight lasted 12 minutes, and eight days later the war ended. By the end of World II in 1945 both the English/American and the German jet aircraft could outperform propeller-piston planes on test flights; however, the engines were not durable and did not play a role in the war. The first commercial aircraft gas turbine engines entered service in 1953.

As improvements in overall engine design were made, requirements for the combustor increased from the 3/1 pressure ratio of early days to 30/1 in modern aircraft engines. Combustor temperatures have also increased so that cooling the turbine blades and combustor linings has become much more difficult. Fortunately, advances in film cooling of turbine blades, slot and port cooling of combustors, fuel atomization, and flow modeling have allowed the combustor design to keep pace with system demands.

Rocket Engines

The advent of rocket engines can be attributed to three men—Tsiolkovsky in Russia, Oberth in Germany, and Goddard in the United States—who began their work independently in about 1915. By 1945 Goddard had tested a multitude of rockets and obtained hundreds of patents. By 1933 in Russia Tsiolkovsky had tested a 1300-pound thrust kerosene and nitric acid rocket with regenerative cooling using the fuel. In 1928 in Germany, Opel drove a rocket-powered car which reached a speed of 125 mph, and the first rocket-powered plane was flown. By the end of World War II, Germany had developed the V-2 rocket engine, which used turbopump-fed liquid oxygen and alcohol. The V-2 was 46 ft long, weighed 27,000 lb_m, and produced 56,000 lb_f thrust. The range for a 2200-lb_m payload was 200 miles. Starting in the 1950s, the United States and the Soviet Union waged a missile race to deliver nuclear weapons, and for unmanned and manned space travel. For example, the first stage of the U.S. Saturn rocket weighed 6,000,000 lb_m and had 7,500,000 lb_f thrust.

Since rocket engines use specialized fuels rather than fossil fuels and use oxygen or other oxidizers rather than air, rocket engine combustion technology is not included in this text.

1.5 GENERAL REFERENCE SOURCES

Because of the widespread application of combustion, the amount of combustion science and combustion technology literature is huge. It is not our purpose in this text to try to supply the many references that are pertinent to each subject area.

Instead, a few key references to papers or texts of special interest are given at the end of each chapter. Below are a list of general American textbooks on combustion and a source list of monographs (single-topic books) and journals in the field of combustion.

American textbooks on combustion

Glassman, I., *Combustion,* Academic Press, New York, 3rd ed., 1997.
Kanury, A. M., *Introduction to Combustion Phenomena,* Gordon & Breach, New York, 1976.
Kuo, K. K., *Principles of Combustion,* Wiley-Interscience, New York, 1986.
Lewis, B., and von Elbe, G., *Combustion Flames and Explosions of Gases,* Academic Press, New York, 3rd ed., 1987.
Strehlow, R. A., *Combustion Fundamentals,* McGraw-Hill, New York, 1984.
Turns, S. R., *An Introduction to Combustion,* McGraw-Hill, New York, 1996.
Williams, F. A., *Combustion Theory,* Addison Wesley, Reading, MA, 2nd ed., 1985.

Journals and monographs

AGARD Combustion and Propulsion colloquiums
American Chemical Society monographs
American Institute of Aeronautics and Astronautics Journal
American Institute of Chemical Engineers journals
American Society of Mechanical Engineers journals
Biomass & Bioenergy (journal)
Combustion and Flame (journal)
Combustion Science and Technology (journal)
Fuel (British journal)
Institution of Mechanical Engineers (British journal)
Journal of Chemical Papers
Progress in Energy and Combustion Science (review journal)
Society of Automotive Engineers papers and transactions
Symposium (International) on Combustion (a volume has been published every other year since 1952)

1.6 SUMMARY

Combustion involves the rapid conversion of chemical energy to sensible energy. Combustion occurs as a flame with a thin reaction zone, as an extended exothermic reaction, or as a detonation. Flames are categorized as either premixed flames or diffusion flames. Designing today's combustion systems requires the use of thermodynamics, chemical kinetics, heat transfer, and fluid mechanics. The challenge is to achieve high efficiency, low emissions, and high reliability at a reasonable cost. A brief historical perspective on combustion science, fuels, and combustion technology was given. Brief biographies were given for Babcock, Wilcox, Otto, Diesel, Whittle, and von Ohain who were early combustion engineers who have left their mark on history. Last, various combustion reference sources were presented.

REFERENCES

Carvill, J., *Famous Names in Engineering,* Butterworths, London, 1981.

Charles Stark Draper Prize, National Academy of Engineering, Washington, D.C., 1991.

Cummins, C. L., *Internal Fire: The Internal Combustion Engine, 1673-1900,* Society of Automotive Engineers, Warrendale, PA, 1989.

Cummins, C. L., *Diesel's Engine, vol. 1: From Conception to 1918,* Carnot Press, Wilsonville, OR, 1993.

Dupont-Roc, G., Khor, A., Anastasi, C., "The Evolution of the World's Energy System," Shell International Ltd., SIL Shell Centre, London, 1996.

Elliott, M. A. (ed.), *Chemistry of Coal Utilization,* Wiley-Interscience, New York, 1981.

Gibbs, L. M., "How Gasoline Has Changed," SAE paper no. 932828, 1993.

Gibbs, L. M., "Gasoline Specifications, Regulations, and Properties," *Automotive Engineering,* SAE, October, 1996.

Gregory, G., *Dictionary of Arts and Sciences,* vol. II, W. T. Robinson, New York, 1821.

Hayward, A. H., *Colonial and Early American Lighting,* Dover, 1992. Originally published by Little, Brown and Co., New York, 1927.

Hollander, J. M., *The Energy-Environment Connection,* Island Press, Washington, DC, 1992.

Imanari, K., "First Jet Engine in Japan, NE20," *Global Gas Turbine News,* Am. Soc. Mech. Engrs., pp. 4-6, March/April 1995.

Jones, G., *The Jet Pioneers: The Birth of Jet Powered Flight,* Methuen, London, 1989.

Keeling, C. D., Bacastow, R. B., Whorf, T. P., and Mook, W. G., "Evidence of Accelerated Releases of Carbon Dioxide to the Atmosphere Inferred from Direct Measurements of Concentration and 13C/12C Ratio," *Proc., 82nd Ann. Mtg. Air and Waste Management Assoc.,* June 1989.

Rolt, L. T. C., *The Mechanicals,* Heinemann, London, 1967.

Stultz, S. C. and Kitto, J. B. (eds.), *Steam—Its Generation and Use,* Babcock and Wilcox Co., Barberton, OH, 40th ed., 1992.

Wilson, G. W., *The Design of High-Efficiency Turbomachinery and Gas Turbines,* Chap. 1, MIT Press, Cambridge, MA, 1984.

CHAPTER 2

Fuels

The purpose of this chapter is to introduce terminology associated with fuels commonly used in combustion systems and to summarize fuel properties for use in succeeding chapters. Fuels are those substances which, when heated, undergo chemical reaction with an oxidizer (typically oxygen in air) to liberate heat. Commercially important fuels contain carbon, hydrogen, and their compounds, which provide the heating value. Fuels may be classified as gaseous, liquid, or solid. The characteristics of each of these broad categories are discussed in this chapter.

To be practical sources of energy, fuels should be abundant and relatively inexpensive. Fuels for the combustion technology discussed in this book are either fossil fuels or biomass fuels. Fossil fuels are nonrenewable, whereas biomass fuels are renewable. Fossil fuels consist primarily of natural gas, petroleum-derived fuels, and coal. Biomass fuels consist primarily of wood, agricultural residues, and refuse. Worldwide production of fossil fuels for 1994 consisted of 170 quad of crude oil, 108 quad of coal, and 93 quad of natural gas (Figure 2.1). (Note that 1 quad $= 1 \times 10^{15}$ Btu $= 1.05 \times 10^{15}$ kJ.) In addition, biomass fuels currently provide approximately 17 quad/yr to world energy production. Fossil fuels provide 83% of world energy production, while the rest is provided by hydroelectric power, nuclear power, and biomass fuels.

The extent of global fossil fuel reserves is subject to debate. Natural gas and crude oil reserves are more limited than coal. The term *reserves* means estimated amounts in the ground that geological data demonstrate with reasonable certainty to be recoverable in future years under existing economic and operating conditions. Hence, this includes undiscovered reserves. The world production rates of natural gas, crude oil, and coal are shown in Table 2.1 along with the time to deplete the fuel reserves at the current production rate. Natural gas will be gone in 123 years, crude oil in 67 years, and coal in 230 years at the current rate of production. These estimates of the time to consume the fossil fuel reserves of the world are problematical because new exploration may expand or reduce the reserves figure, while

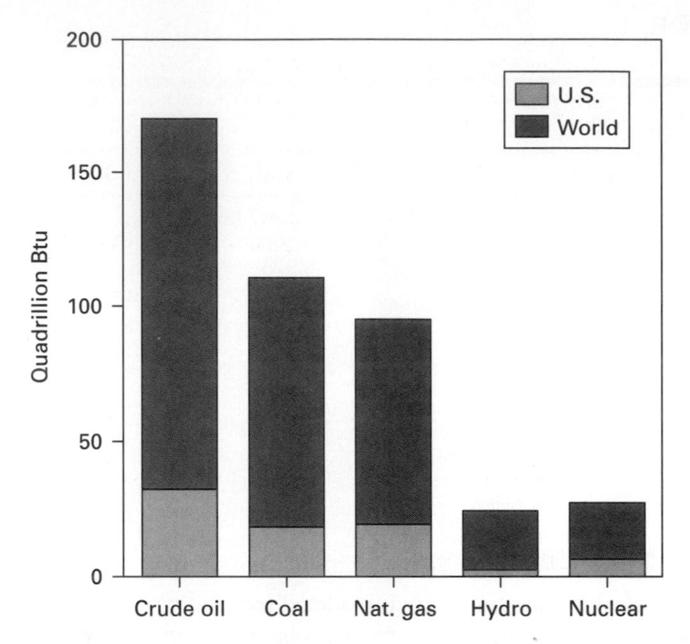

FIGURE 2.1
Primary energy production by source [Energy Information Administration, 1994].

TABLE 2.1
World fossil fuel production rate [Energy Info. Adm., 1994]

Fuel	1993 production	Supply at 1993 production rate
Natural gas	75 trillion ft^3	123 yr
Crude oil	23 billion barrels	67 yr
Coal	4.1 billion tons	230 yr

increasing consumption driven by rising population and unmet human needs will decrease the depletion time.

North America has 8% of the world's natural gas reserves, while the Middle East and Eastern Europe have 70% of the reserves. The Middle East has 67% of the crude oil reserves, and North America has 8%. In addition, Canada has an estimated 1 trillion barrels of oil enmeshed in tar sands (a tarry mixture of bitumen, sand, clay, and shale). North America, Eastern Europe, and China have the largest coal reserves.

In 1993 the United States consumed 25% of the world's production of natural gas, 26% of the world's crude oil production, and 21% of the coal production. In 1993 the United States imported 47% of its crude oil and 11% of its natural gas

TABLE 2.2
Average 1996 U.S. price (excluding taxes) of selected fuels and electricity [Energy Info. Adm., 1997]

Fuel	Unit price ($)	$/10⁵ Btu*
Natural gas (residential)	6.30/1000 ft³	0.63
Natural gas (industrial)	3.34/1000 ft³	0.33
Motor gasoline (unlead)	1.23/gal†	1.00
Kerosene	0.72/gal	0.52
No. 2 fuel oil	0.67/gal	0.47
No. 2 diesel	0.68 gal	0.49
Coal at power plants	29/ton	0.13
Wood (15% moisture)‡	40/cord	0.21
Electricity (residential)	0.084/kWh	2.46
Electricity (industrial)	0.047/kWh	1.38

* Does not consider efficiency of conversion.
† Includes taxes.
‡ Estimate for Wisconsin.

consumption, and exported 8% of its coal production. At the current rate of consumption the U.S. supply of natural gas is projected to last 26 to 66 years, crude oil 23 to 38 years, and coal 250+ years [EIA]. These estimates of the number of years that the fuel supply (recoverable resource) will last assume that the rate of consumption will not increase and that the undiscovered reserves have been properly estimated. The largest discoveries of natural gas in the United States were made during the 1950s. The largest discoveries of crude oil in the United States were made during the 1930s. It is unlikely that the U.S. resource base will increase significantly.

The price for fuels changes with supply and demand, taxes and international factors. The price of selected fuels is shown in Table 2.2. Price is a major criterion for selecting fuels, although convenience, emission control regulations, and system availability are also important considerations. A significant part of the price is due to transportation.

2.1 GASEOUS FUELS

The main gaseous fuels are natural gas and liquefied petroleum gas. Gaseous fuels are also produced from coal and wood, and from petroleum and natural gas.

Natural gas is found compressed in porous rock and shale formations sealed in rock strata below the ground. Natural gas frequently exists near or above oil deposits. Natural gas is a mixture of hydrocarbons and small quantities of various nonhydrocarbons existing in the gaseous phase or in solution with crude oil. Raw natural gas contains methane and lesser amounts of ethane, propane, butane, and pentane. Sulfur and organic nitrogen are typically negligible in natural gas. Carbon dioxide and nitrogen are sometimes present, although generally noncombustibles are very low. Natural gas is processed by compressing and cooling to condense the higher hydrocarbons. Dry pressurized natural gas is transmitted in pipelines. At some wells natural gas is liquefied (LNG) by cooling to −164°C and is transported from the wellhead to selected ports around the world in gigantic tankers.

Liquefied petroleum gas (LPG) consists of ethane, propane, and butane produced at natural gas processing plants, including plants that fractionate raw natural gas plant liquids. LPG also includes liquefied refinery gases such as ethylene, propylene, and butylene produced from crude oil at refineries. LPG is stored in tanks under pressure and is a gas at atmospheric pressure. At 38°C the maximum vapor pressure is 208 kPa for commercial LPG.

Producer gas is formed by passing a less than stoichiometric amount of air through a hot bed of coal, peat, wood, or agricultural residues. It may be used hot, or cleaned to remove the tars and soot which unavoidably occur. Carbon monoxide and hydrogen are the primary fuels, while up to 55% N_2 is present with lesser amounts of CO_2. Because of the nitrogen, air-blown producer gas is a low–heating content gas. A higher–heating content gas can be produced by blowing the hot fuel bed with substoichiometric oxygen rather than air, or by indirectly heating the solid fuel.

Water gas is made by blowing air through a bed of coal, coke, or wood char to raise the temperature of the bed, and then blasting the bed with steam. The steam reacts with the hot carbon endothermically as follows:

$$C + H_2O \rightarrow CO + H_2$$

In this way the CO and H_2 are increased and the N_2 decreased compared with producer gas, and the heating content is roughly doubled. There are many other types of manufactured gaseous fuel mixtures from various oils, oil shale, and coal and wood gasification processes.

Manufactured fuels such as hydrogen and acetylene are used when especially hot flames are desired. Hydrogen is manufactured by reforming natural gas, partial oxidation of liquid hydrocarbons, or coal gasification. For example, natural gas is converted to H_2, CO and CO_2 by reaction with steam over a catalyst at 800 to 900°C. Further shift of CO + H_2O to H_2 + CO_2 is carried out, and the gas is cooled and scrubbed to remove CO_2. Acetylene (C_2H_2) is made by adding water to calcium carbide.

Characterization of Gaseous Fuels

Important characteristics of gaseous fuels include the volumetric analysis, density, and heating value. The volumetric analyses of natural gas, LPG, and producer gas are shown in Table 2.3.

The *heating value* is the heat release per unit mass when the fuel initially at 25°C reacts completely with oxygen and the products are returned to 25°C. The heating value is reported as the *higher heating value* (HHV) when the water is

TABLE 2.3
Volumetric analysis of some gaseous fuels [Perry and Green, *Handbook of Chemical Engineering,* © 1984, by permission of The McGraw-Hill Companies]

Species	Natural gas	LPG	Coal producer gas	Wood producer gas
CO	—	—	20–30%	18–25%
H_2	—	—	8–20%	13–15%
CH_4	80–95%	—	0.5–3%	1–5%
C_2H_6	< 6	—	Trace	Trace
$> C_2H_6$*	< 4	100%	Trace	Trace
CO_2	< 5	—	3–9%	5–10%
N_2	< 5	—	50–56%	45–54%
H_2O	—	—	—	5–15%

* Contains hydrocarbons heavier than C_2H_6.

condensed or as the *lower heating value* (LHV) when the water is not condensed. The LHV is obtained from the HHV by subtracting the heat of vaporization of water in the products:

$$\text{LHV} = \text{HHV} - \frac{m_{H_2O}}{m_{\text{fuel}}} h_{fg} \tag{2.1}$$

where h_{fg} is the latent heat of vaporization of water at 25°C, which equals 2440 kJ/kg water (1050 Btu/lb$_m$). The water includes moisture in the fuel as well as water formed from hydrogen in the fuel. The heating value of a gaseous fuel may be obtained experimentally in a flow calorimeter and can be calculated from thermodynamics if the composition is known. The density and heating values of various gaseous fuels are given in Table 2.4.

TABLE 2.4
Heating value of some gaseous fuels [Perry and Green, *Handbook of Chemical Engineering,* © 1984, by permission of The McGraw-Hill Companies]

	HHV		LHV	
Fuel	**(Btu/ft^3)***	**(Btu/lb$_m$)**	**(Btu/ft^3)***	**(Btu/lb$_m$)**
Hydrogen (H_2)	319.4	61,030	270.0	51,593
Carbon monoxide (CO)	316.0	4,346	316.0	4,346
Methane (CH_4)	994.7	23,880	896.0	21,518
Ethane (C_2H_6)	1743	22,329	1594	20,431
Propane (C_3H_8)	2480	21,670	2283	19,944
Butane (C_4H_{10})	3216	21,316	2969	19,679
Ethylene (C_2H_4)	1576	21,646	1477	20,276
Acetylene (C_2H_2)	1451	21,477	1402	20,734
Natural gas (typical)	1030	23,300	935	21,150
Producer gas (typical)	170	2,500	155	2,280

*At 1 atm, 68°F.

EXAMPLE 2.1. Given the higher heating value of methane of 23,880 Btu/lb$_m$, calculate the lower heating value of methane, and check to see if it agrees with the value in Table 2.4.

Solution. One mole of methane (CH_4) yields two moles of water (H_2O). The mass of one lbmol of methane is 16 lb$_m$. The mass of 2 lbmol of water is 36 lb$_m$. From Eq. 2.1,

$$\text{LHV} = \text{HHV} - \frac{m_{H_2O}}{m_{\text{fuel}}} h_{fg}$$

$$\text{LHV} = 23{,}880\ \text{Btu/lb}_{CH_4} - (36\ \text{lb}_{H_2O}/16\ \text{lb}_{CH_4})(1050\ \text{Btu/lb}_{H_2O})$$

$$\text{LHV} = 21{,}518\ \text{Btu/lb}_{CH_4}$$

This agrees with the LHV of methane in Table 2.4.

2.2 LIQUID FUELS

Liquid fuels are derived primarily from crude oil. In the future, liquid fuels may increasingly be derived from oil shale, tar sands, coal, and biomass. Crude oil is a mixture of naturally occurring liquid hydrocarbons with small amounts of sulfur, nitrogen, oxygen, trace metals, and minerals. Crude oil is generally found trapped in certain rock formations that were originally part of the ocean floor. Organic marine matter on the ocean bottom was encased in rock layers at elevated pressure and temperature, and over millions of years gradually formed crude oil.

The ultimate analysis of crude oil does not vary greatly around the world, with roughly 84% carbon, up to 3% sulfur, and up to 0.5% nitrogen and 0.5% oxygen. Crude oil is sometimes burned directly; however, because of the wide range of densities, viscosities, and impurities, crude oil is generally refined. The refining processes of fractional distillation, cracking, reforming, and impurity removal are used to produce many products including gasoline, diesel fuels, gas turbine fuels, and fuel oils. Figure 2.2 shows typical end products from crude oil with the light, more volatile components at the top. Some adjustments of the end product amounts can be made at the refinery. For example, a particularly cold winter may require more heating fuel, typically resulting in a smaller production of gasoline. Figure 2.3 shows the trend of transportation fuel usage in the United States projected to the year 2000.

Liquid petroleum fuels can be broadly classified as true distillates or as ash-bearing fuels. True distillates are free from ash. Ash-bearing fuels contain significant amounts of minerals, which generally require removal in the refining process. Liquid petroleum fuels contain a variety of hydrocarbons with a wide range of properties. Before considering the properties and types of liquid fuels, the molecular structure of various fuel hydrocarbons is reviewed.

Molecular Structure

Chemically, crude oil consists primarily of alkanes (paraffins), cycloalkanes (naphthene), and aromatics. Petroleum fuels also contain alkenes (olefins), which are formed during the cracking part of the refining process.

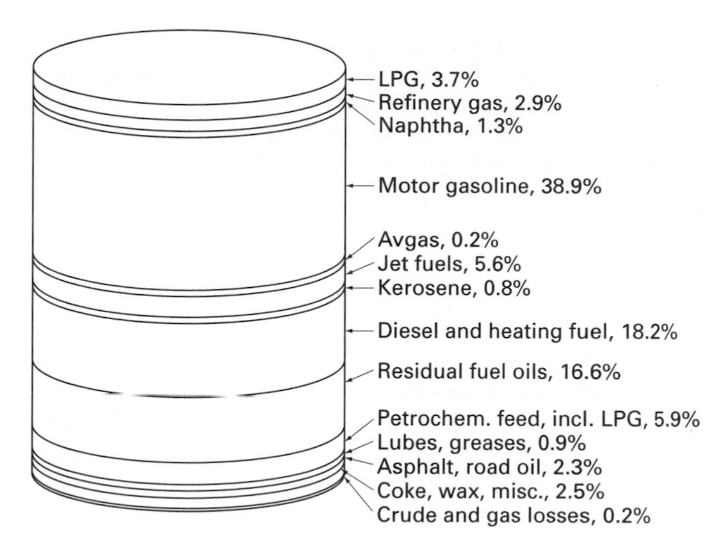

FIGURE 2.2
Typical end products from crude oil. A single refinery produces some, but not all, of the products shown. The percentages refer to overall production from total refinery output.

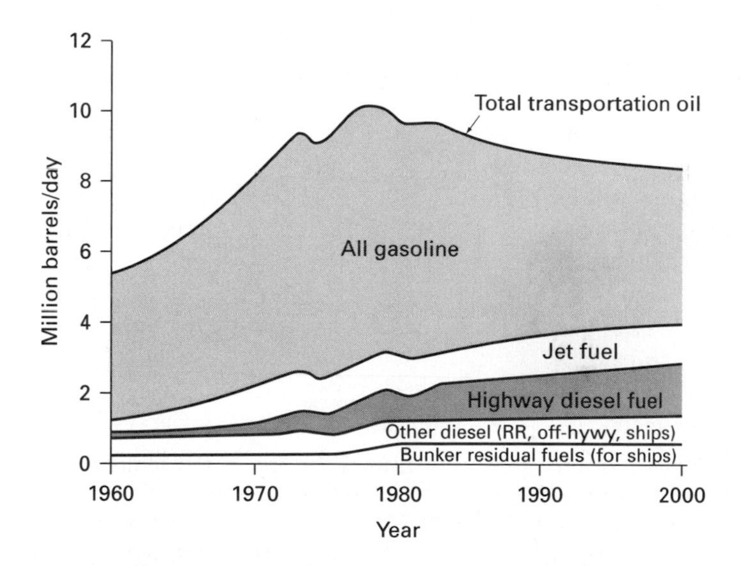

FIGURE 2.3
U.S. transportation oil consumption, data for 1990s from extrapolations [Davis et al.].

The general formula of alkanes is C_nH_{2n+2}. All the carbon bonds are shared with hydrogen atoms except for a minimum number of required carbon-carbon bonds. Alkanes (sometimes called paraffins) having a straight-chain structure are referred to as normal and designated by prefix *n*. The first four are gaseous at standard pressure and temperature:

Methane	CH_4
Ethane	$CH_3—CH_3$
n-Propane	$CH_3—CH_2—CH_3$
n-Butane	$CH_3—CH_2—CH_2—CH_3$

Alkanes containing more carbon atoms are designated by the following prefixes: 5, pent; 6, hex; 7, hept; 8, oct; 9, non; 10, dec; 11, undec; 12, dodec; etc. Alkanes are not necessarily straight-chained. If a side carbon chain or isomer exists, the name of the longest continuous chain of carbon atoms is taken as the base name. For example:

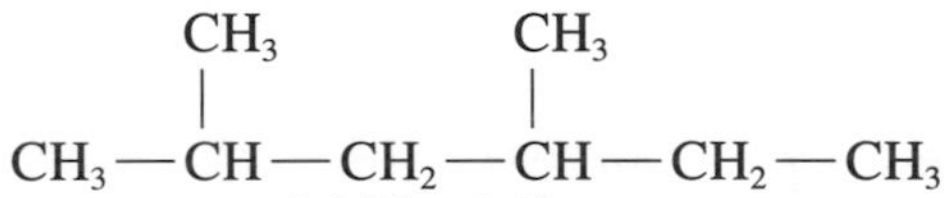

$$CH_3—CH(CH_3)—CH_2—CH(CH_3)—CH_2—CH_3$$

2,4-Dimethylhexane

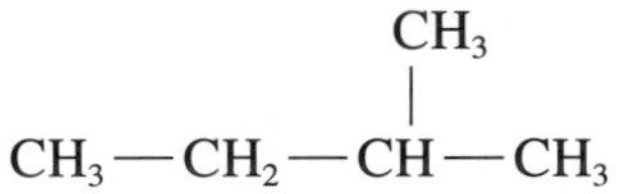

$$CH_3—CH_2—CH(CH_3)—CH_3$$

2-Methylbutane (isopentane)

Cycloalkanes (naphthenes) have the formula C_nH_{2n}, and the nomenclature follows the number of carbon atoms. For example,

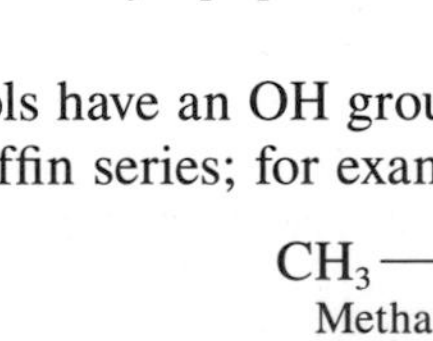

H_2C, CH_2, CH_2 (three-membered ring)

Cyclopropane

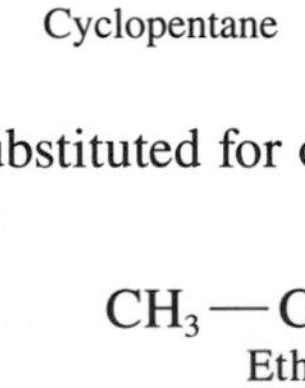

H_2C, CH_2, CH_2, CH_2, CH_2 (five-membered ring)

Cyclopentane

CH_2, H_2C, CH_2, H_2C, CH_2, CH_2 (six-membered ring)

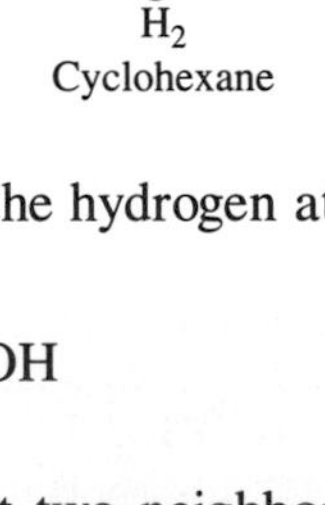

Cyclohexane

Alcohols have an OH group substituted for one of the hydrogen atoms in one of the paraffin series; for example:

$CH_3—OH$ Methanol $\qquad$ $CH_3—CH_2—OH$ Ethanol

Alkenes (olefins) also have the formula C_nH_{2n}, but two neighboring carbon atoms share a pair of electrons forming a double bond. The location of the double bond is indicated by a prefix. For example:

Ethylene	$CH_2{=}CH_2$
Propylene	$CH_3{-}CH{=}CH_2$
1-Butene	$CH_3{-}CH_2{-}CH{=}CH_2$
2-Butene	$CH_3{-}CH{=}CH{-}CH_3$

Diolefins have two double bonds and have the general formula C_nH_{2n-2}. Their names end with the letters "diene," as, for example, hexadiene:

$$CH_3{-}CH{=}CH{-}CH_2{-}CH{=}CH_2$$

1,4-Hexadiene

Double-bonded hydrocarbons may also be arranged in a ring structure. A basic building block is benzene, C_6H_6, which has three double bonds and is shown in two notations below. This class of compounds is called *aromatics*. Other aromatics may be formed by adding to the ring by displacing hydrogen, as, for example, toluene (methylbenzene):

Benzene Benzene Toluene

Two or more rings may be united to form many compounds, for example, naphthalene ($C_{10}H_8$) or anthracene ($C_{14}H_{10}$). These compounds, known as *polycyclic aromatic hydrocarbons* (*PAH*), can be formed in the refining processes and can also be formed during combustion. For example, benzo[a]pyrene is a known carcinogen which can be formed in flames.

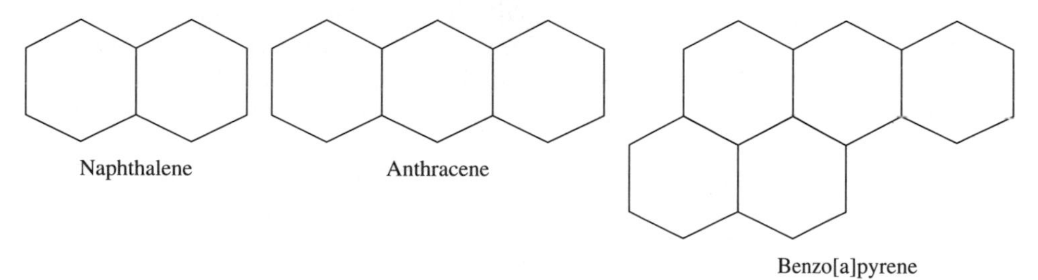

For detailed information and rules for hydrocarbon notation, see Chapter C of the *Handbook of Chemistry and Physics* [Lide].

Characterization of Liquid Fuels

Important properties of liquid fuels include the heating value, specific gravity, viscosity, flash point, autoignition temperature, distillation curve, sulfur content, vanadium and lead content, octane number (for gasolines), cetane number (for diesel fuels), and smoke point (for gas turbine fuels). Fuel property data are given in this section and in Appendix A. The methods for determination of these properties in the United States are specified by the ASTM standards (see references).

The higher heating value for liquid fuels is determined by combustion with pressurized oxygen in a bomb calorimeter. This device is a stainless steel container which is surrounded by a large water bath. The large bath ensures that the final product temperature will be only very slightly above the initial reactant temperature of 25°C. The combustion is carried out for a lean mixture to ensure complete combustion.

The specific gravity is the density of the fuel divided by the density of water at the same temperature. In some literature the American Petroleum Institute (API) specific gravity is used. The relation between API specific gravity (G) and conventional specific gravity (sg) at 16°C is

$$G = \frac{141.5}{\text{sg}} - 131.5 \tag{2.2}$$

Viscosity of a liquid is a measure of its resistance to flow. For a liquid fuel, viscosity indicates the ease at which it can be pumped and the ease of atomization. Viscosity of liquids decreases with increasing temperature. There are several standard tests for viscosity. Sometimes *pour point* is used as a simple indicator of viscosity. The pour point is an indication of the lowest temperature at which a fuel oil can be stored and still be capable of flowing under very low forces in a standard apparatus.

Flash point is an indication of the maximum temperature at which a liquid fuel can be stored and handled without serious fire hazard. The flash point is the minimum temperature at which fuel will rapidly catch fire when exposed to an open flame located above the liquid. An example of flash point concern is the ignitability of the mixture above liquid fuel in a partially full fuel tank. Gasoline, which has a flash point of −43°C, is typically so volatile that the mixture is too rich to burn. Diesel fuel (flash point of 52°C) is so nonvolatile that the mixture is too lean to burn. However, gasoline-diesel fuel mixtures or alternative fuels such as methanol may pose a danger.

Autoignition temperature is the lowest temperature required to initiate self-sustained combustion in a standard container in atmospheric air in the absence of a spark or flame. For example, the autoignition temperature of gasoline is 370°C. Flash point and autoignition temperatures for selected pure gases and liquids are given in Table 2.5. The autoignition temperature of the alkanes (paraffins) decreases with increasing molecular weight. Isooctane has a much higher autoignition temperature than octane. Ethane, ethylene, and ethanol have progressively decreasing autoignition temperatures. Carbon monoxide has a high autoignition temperature. In general, autoignition temperatures are an indication of the relative difficulty of combusting a fuel. Since the autoignition temperature varies with the geometry of the hot surface and other factors such as pressure, other tests such as octane number and cetane number are used for engine fuels.

TABLE 2.5
Flash point and autoignition temperature of pure fuels in air at 1 atm

Substance	Flash point (°C)	Autoignition (°C)
Methane	−188	537
Ethane	−135	472
Propane	−104	470
n-Butane	−60	365
n-Octane	10	206
Isooctane	−12	418
n-Cetane	135	205
Methanol	11	385
Ethanol	12	365
Acetylene	Gas	305
Carbon monoxide	Gas	609
Hydrogen	Gas	400

Source: Bartok and Sarofim, *Fossil Fuel Combustion: A Source Book,* ©1991, Wiley, by permission of John Wiley and Sons, Inc.

Octane number indicates the tendency of gasoline to knock (onset of autoignition) when the compression ratio in a spark-ignition engine is raised. The *octane number* of a fuel is measured by comparing the performance of the fuel with the performance of mixtures of isooctane and *n*-heptane in a standardized spark-ignition engine. Isooctane is arbitrarily set at 100 and *n*-heptane, which is more prone to knock, is arbitrarily set at zero. The octane number is the percentage of isooctane in the isooctane-heptane mixture which most nearly matches the performance of the test fuel. Two octane test methods are used for automobile gasoline: the research octane number, and motor octane number. The research method is run with 125°F inlet air at 600 rpm with 13° spark advance (before top dead center), and gives a higher octane rating than the 300°F inlet air at 900 rpm test with 19 to 26° spark advance of the motor method. The numerical difference between these two rating numbers is called the *sensitivity* and is zero by definition for the reference fuels. Sensitivity varies with the chemical composition of the fuel [Leppard].

In compression-ignition engines the time between start of injection and onset of combustion is known as the *ignition delay*. Cetane number (CN) ranks fuels according to their ignition delay when undergoing a standard test. Because cetane (*n*-hexadecane) is one of the fastest-igniting hydrocarbons in fuel, it is assigned a cetane number of 100. Isocetane (heptamethylnonane) ignites slowly and is arbitrarily assigned a cetane number of 15. A fuel is compared with mixtures of the reference fuels in a standardized prechamber diesel engine and rated by the mixture which most nearly matches the ignition delay of the test fuel. The *cetane number* of the reference mixture is defined by

$$\text{CN} = (\%\ n\text{-cetane}) + 0.15(\%\ \text{heptamethylnonane})$$

In the cetane number test the injection is fixed at 13° before top dead center and the compression ratio is changed until combustion of the test fuel starts at top dead center. The standard mixture is found which gives the same ignition delay at these fixed conditions of injection timing and compression ratio. The tests are run at

900 rpm with 212°F water temperature and 150°F inlet air. Because the test engine is a prechamber design, the cetane number is at best only a relative scale when applied to open-chamber engines. The test is particularly questionable for low-cetane fuels (CN < 35).

In many cases CN is computed from correlations based on the chemical structure and physical properties of the fuel rather than engine tests. For example, the aniline point (temperature at which aniline and the fuel are miscible), API gravity (G), and midboiling temperature (TM) are often used to correlate cetane number. The ASTM (D976) correlation is

$$\text{CN} = -420.34 + 0.016G^2 + 0.192G\log(\text{TM}) + 65.01(\log \text{TM})^2 - 0.0001809(\text{TM})^2 \tag{2.3}$$

where TM is °F and log is to base 10. Such calculated values of CN are more correctly designated as the cetane index (CI) to make clear that they are different from the measured value.

EXAMPLE 2.2. A truck fuel (D2) has an API gravity (ASTM method D287) of 35.7 and a 50% distillation temperature (ASTM method D86) of 529 K. Calculate its cetane number (CN).

Solution. From the data,

$$G = 35.7 \qquad \text{and} \qquad \text{TM} = 529\ \text{K} = 492°\text{F}$$

From Eq. 2.3,

$$\text{CN} = -420.34 + 20.39 + 18.45 + 471.11 - 43.79 = 45.8$$

The measured cetane number (ASTM method D613) gave 46.2. Given the accuracy of the cetane number as an indicator of actual engine ignition delay, the comparison should be considered excellent.

The *smoke point* measures the tendency of a liquid fuel to form soot. It is determined experimentally by burning the fuel in a special wick lamp and slowly increasing the height of the flame until smoke begins to appear. The maximum height of smokeless flame in millimeters is the smoke point. Hence, the higher the smoke point, the lower is the tendency of the fuel to form soot. Smoke point is used especially for gas turbine fuels.

Liquid Fuel Types

Automotive gasoline is a mixture of light distillate hydrocarbons made from blends of refined crude oil. Hence gasoline is a blend of paraffins, olefins, naphthenes, and aromatics which varies from company to company and by location and season of the year. Gasoline must be volatile enough to vaporize readily in the engine but not so volatile as to cause danger of detonation during handling. Gasolines have a boiling range of 25–225°C. *n*-Octane, which is sometimes used to represent gasoline, has a boiling point of 125.6°C.

Diesel fuel is a mixture of light distillate hydrocarbons with a higher boiling point range than gasoline. Larger, low-speed industrial and marine diesels often use

TABLE 2.6
Properties of automotive fuels

Fuel type	Automotive gasoline	No. 2 diesel fuel	Methanol	Ethanol
Specific gravity at 16°C	0.72–0.78	0.85	0.796	0.794
Kinematic viscosity at 20°C (m^2/s)	0.8×10^{-6}	2.5×10^{-6}	0.75×10^{-6}	151×10^{-6}
Boiling point range (°C)	30–225	210–235	65	78
Flash point (°C)	−43	52	11	13
Autoignition temperature (°C)	370	254	464	423
Octane no. (research)	91–100	—	109	109
Octane no. (motor)	82–92	—	89	90
Cetane no.	< 15	37–56	< 15	< 15
Stoichiometric air/fuel by weight	14.7	14.7	6.45	9.0
Heat of vaporization (kJ/kg)	380	375	1185	920
Lower heating value (MJ/kg)	43.5	45	20.1	27

a heavy fuel oil. Selected properties for diesel fuel and gasoline are given in Table 2.6. Comparison of gasoline and no. 2 diesel shows the higher density of diesel fuel which gives it more heating value on a volume basis. Note also the lower volatility and higher viscosity of diesel fuel. The autoignition temperature also partially reveals the reason for the large difference in cetane number when it is recalled that cetane number reflects the ease of compression ignition.

Gas turbine fuels are not limited by antiknock or ignition delay requirements and have a wide range of boiling points. Kerosene is very similar to aircraft jet A fuel. Jet B, which is also used as aircraft gas turbine fuel, has a lower boiling point range than jet A fuel. Turbine fuels limit the amounts of certain trace metals such as vanadium and lead, which tend to form deposits on the blades. The flash point of kerosene is 38°C, and the boiling point is 150–200°C. The tendency toward soot formation (smoke point) is controlled by limiting the aromatic content to 25% by volume. Similar considerations are used for industrial gas turbine fuel oils. Figure 2.4 shows temperature versus percent evaporated for a typical diesel fuel, for gasoline, and for jet A and jet B fuels.

Fuel oil covers a wide range of petroleum products, which have been divided into six grades, although grade 3 has been deleted (Table 2.7). Grade 1 is kerosene, and grade 2 is domestic fuel oil. Number 2 fuel oil boils between 218 and 310°C. Fuel oil properties are shown in Table 2.7. The heavier fuel oil grades are specified by viscosity, and are used in industrial and utility applications and some large engines. The heavier grades produce significant amounts of ash. Preheating of the oil is required for no. 6 and may be required for no. 5 and no. 4, depending on the climate. Number 6 fuel is a heavy residual fuel, which remains after all distillation processes are completed; it has high viscosity and tends to have relatively high amounts of asphaltenes, sulfur, vanadium, and sodium. Residual fuel oils are burned directly in some boilers, and after some treatment they can be used in heavy-duty industrial gas turbines. Sulfur is limited to 0.5% for no. 1 and no. 2 fuel oil, but can be as high as 4% for no. 6 fuel oil.

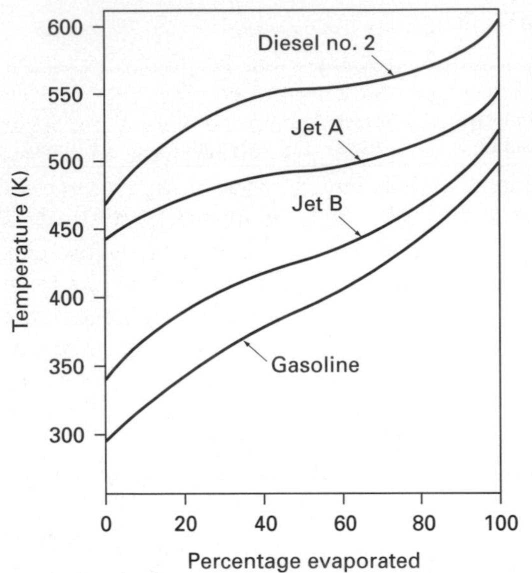

FIGURE 2.4
Typical distillation curves for various fuels using ASTM method D86.

TABLE 2.7
Typical properties of fuel oils

Grade no.	1	2	4	5	6
	Kerosene	**Distillate**	**Very light residual**	**Light residual**	**Residual**
Color	Clear	Amber	Black	Black	Black
Specific gravity, at 16°C	0.825	0.865	0.928	0.953	0.986
Kinematic viscosity at 38°C (m^2/s)	1.6×10^{-6}	2.6×10^{-6}	15×10^{-6}	50×10^{-6}	360×10^{-6}
Pour point (°C)	< -17	< -18	-23	-1	19
Flash point (°C)	38	38	55	55	66
Autoignition temperature (°C)	230	260	263	—	408
Carbon (%)	86.5	86.4	86.1	85.5	85.7
Carbon residue (%)	Trace	Trace	2.5	5.0	12.0
Hydrogen (%)	13.2	12.7	11.9	11.7	10.5
Oxygen (%)	0.01	0.04	0.27	0.3	0.38–0.64
Ash (%)	—	< 0.01	0.02	0.03	0.04
HHV (MJ/kg)	46.2	45.4	43.8	43.2	42.4

As petroleum resources decline in the future, alternative liquid hydrocarbon fuels will become increasingly important. Liquid fuels can be made from coal, biomass, oil shale, tar sands, and natural gas. Methanol (CH_3OH) can be made from natural gas, coal or wood, and ethanol (C_2H_5OH) can be made from sugar beets, corn, or wood, for example. Substitute diesel fuel can be made from vegetable oils such as rapeseed oil (rapeseed methyl diester) or soy beans, and natural gas. Whether or not this is done depends primarily on the economics of the process relative to petroleum. However, the demands of emissions regulation may also play a role. For example, dimethyl ether which can be made from natural gas is an excellent smokeless diesel fuel [Fleisch, et al.]. As indicated in Table 2.6, the heating value of methanol and ethanol is lower than other petroleum fuels because of the attached oxygen. When oil shale is heated, its resinous content decomposes into an oily liquid from which a crude oil (syncrude) may be obtained. Similarly, tar sands yield a syncrude which can be refined into various fuels.

Slurry fuels containing a mixture of pulverized coal and water are also being considered as a replacement for fuel oil. Coal-water slurries containing 25 to 30% water have been shown to be stable for pumping and burning in certain boilers. If the coal is beneficiated to remove much of the ash, it may be possible to burn coal-water slurries in industrial gas turbine engines. The properties of a coal-water slurry developed to replace no. 6 fuel oil are given in Table 2.8.

Gasoline Quality

Because gasoline for spark-ignition engines is the largest single product produced from crude oil (see Fig. 2.2), determination of its quality is of high importance. For most consumers, gasoline is judged in terms of octane number and price. However, determination of gasoline quality is quite complicated and involves many factors, and the variation of quality could be quite large if it were not for the fact that most refiners manufacture gasolines within the specifications established by the American Society for Testing Materials (ASTM D4814 specifications). Table 2.9 lists these test methods. Despite this formidable list of specifications, many other

TABLE 2.8
Typical properties of a coal-water slurry

Coal*	70% (wt)
Water	29%
Additives†	1%
Higher heating value	9500 Btu/lb_m
Specific gravity	1.2
Density	10 lb_m/gal
Flame temperature (20% EA)	3100°F
Viscosity at 70°F	1500 cP

* Desire volatiles $>$ 30% and ash $<$ 5% by weight.
† Stabilizer and dispersant.

TABLE 2.9
Gasoline specifications

Specification	ASTM* test method	Importance
Volatility distillation	D86	Cool weather drivability
		Carburetor icing
		Hot start and driveaway
		Vapor lock
		Carburetor evaporation losses
		Deposits in crankcase and combustion chambers, and on spark plugs
Vapor-to-liquid ratio	D2533, D5188	Vapor lock
Vapor pressure	D323, D5190	Low-temperature starting
		Fuel tank evaporative losses
		Vapor lock
Research octane no.	D2699	Low- to medium-speed knock
Motor octane no.	D2700	High-speed knock
		Part throttle knock
Antiknock Index	—	Average of research and motor octane no.
Corrosivity stability	D130	Fuel system durability
Existent gum	D381	Induction system deposits storage
Oxidation stability	D525	
Sulfur content	D1266	Exhaust emissions, engine deposits and wear
Antiknock additives	D2599	
Lead content	D3229	Catalyst deterioration
Manganese content	D3831	Engine hydrocarbon emissions
		Oxygen sensor deterioration

*American Society for Testing Materials.

properties are also important, such as density, surface tension, viscosity, compatibility with materials, and stoichiometric fuel-air ratio. The list is constantly increasing and changing. For example, the switch from carburetion to intake port fuel injection in automobiles brought about problems with fuel injector deposits and a rush to find additives to prevent such deposit buildup. At the same time specifications relating to carburetor performance have decreased in importance.

Fuel volatility affects the startup and transient performance of vehicles and the evaporative emissions during filling. The effects on performance can be severe if *vapor lock* is encountered (formation of vapor in the fueling system, which decreases the fuel flow); but even short of stall, the driver may encounter so-called *drivability* problems as exhibited by idle roughness, backfire, hesitation, stumble (a short, sharp reduction in acceleration after the vehicle is in motion), and *surge* (cyclic power fluctuations). Closed-loop fuel control, which is achieved by using feedback from an O_2 sensor in the exhaust to control the air and fuel flows, can alleviate these problems. However, during warmup the closed-loop control is not operative and the problems are similar to those of open-loop control systems. The total demerits assigned to such behavior can be correlated in terms of inlet air temperature and selected temperatures from the D86 curve (see Table 2.9). The

D86 curve is obtained by measuring the fraction of fuel which vaporizes at a given temperature. It should be noted that for real fuels the remaining liquid chemical composition changes as the more volatile fractions leave as vapor. Other methods of judging volatility are sometimes used, such as *Reid vapor pressure* and vapor-to-liquid volume ratio. In the Reid vapor pressure test the liquid is exposed to a volume of air which is four times the liquid volume. The rise in pressure at 100°F due to the fuel vapor is the Reid vapor pressure. None of these methods of volatility measurement duplicate engine conditions. In the D86 test, for example, the liquid is in contact with its own vapor rather than with air. From a combustion standpoint the volatility may cause variations of in-cylinder mixture homogeneity as well as fuel-to-air ratio cyclic variations.

To match seasonal and geographic considerations with volatility requirements, the system of six classes shown in Table 2.10 has been set up under the ASTM D4814 standard. Each geographical area of the U.S. is assigned a class on a monthly basis. The selection is based on altitude and expected ambient temperature; top classes (A) are for warm weather and the bottom (E) for cold weather. Matching of the volatility to climate considerations is complicated by variations in the fuel delivery system designs and underhood temperatures.

As previously stated, the most obvious gasoline quality parameter is octane number, because an octane index is posted [an average of research and motored number called the *antiknock index* (AKI)] and the driver can hear the knocking sound. The true evaluation of knock is the "road octane number" which depends heavily on the particular engine-vehicle combination. The *road octane* is obtained by determining the knock-limited spark advance for the test fuel under a given operating condition and then finding the octane number of the primary reference fuel that gives this same knock-limited spark advance at the same operating condition. At low speeds (1000 to 1200 rpm), the road octane correlates with research octane for a given vehicle. At medium engine speeds (1500 to 2500 rpm), the correlation is a combination of research and motored octane. At high engine speeds (3000 to 4000 rpm), road octane is fairly close to the motored octane number, but engines with heated intakes (for emissions reduction) and higher specific power output can have octane requirements above the motored value. Engines with in-cylinder deposits which insulate the chamber and cause the unburned mixture to undergo higher temperatures require higher octane fuels than the clean version of the same engine.

Vehicle octane requirements are lower at higher altitude, since the ambient pressure is lower. The range over U.S. conditions is about 87–92. Thus, fuels sold in high-altitude regions will probably cause knock at lower altitudes. The use of antiknock additives such as tetraethyl lead (TEL), tetramethyl lead (TML), and methyl-cyclo-pentadiene manganese tricarbonyl (MMT) has been outlawed due to the adverse effects of TEL and TML on catalysts used in aftertreatment devices and the adverse effects of MMT on unburned hydrocarbon emissions.

Alcohols and ethers have received attention as octane boosters. Ethanol and methanol tend to decrease miles per gallon because of their lower energy content on a volume basis. They decrease CO slightly but have negligible effects on HC and NO exhaust emissions. Methanol can cause vapor lock, corrosion of the fuel system,

TABLE 2.10
ASTM D4814 gasoline requirements [by permission of ASTM © 1996]

Gasoline volatility class	Vapor pressure psi: maximum	Distillation temperature, °F				Vapor lock protection class	Vapor-liquid ratio temperature °F	Vapor-liquid ratio maximum
		10% Evap maximum	50% Evap	90% Evap maximum	End point maximum			
AA	7.8	158	170–250	374	437	1	140	20
A	9.0	158	170–250	374	437	2	133	20
B	10.0	149	170–245	374	437	3	124	20
C	11.5	140	170–240	365	437	4	116	20
D	13.5	131	150–235	365	437	5	105	20
E	15.0	122	150–230	365	437	6	95	20

and swelling of some elastomers. Cars can be designed to run on these fuels, however, and pure methanol has received considerable attention as an alternative fuel. The higher alcohols and ethers have fewer undesirable effects. Addition of 7% methyl-*t*-butyl ether increases research octane by 2 to 3 and motor octane by 1 to 2 with negligible side effects. Although octane boosting additives are the most familiar to users, many other additives are used to improve fuel quality. Table 2.11 lists types of additives and their functions.

Gasoline Formulation and Emissions

Although 1990 motor vehicles emit 96% fewer hydrocarbons and carbon monoxide and 76% fewer nitrogen oxides than their uncontrolled counterparts of the late 1960s, motor vehicles continue to be a significant source of these emissions which plague most urban cities. It is thus clear that emissions must be further reduced by an aggressive program to better understand the fuel-vehicle system. In response to this need, three domestic automobile companies and 14 petroleum companies have agreed to form a joint program to bring about a more scientific basis for regulatory

TABLE 2.11
Gasoline additives

Additives	Type	Function
Oxidation inhibitors	Aromatic amines and phenols	Inhibit gum formation and oxidation
Corrosion inhibitors	Carboxylic acids and carboxylates	Inhibit corrosion of ferrous metals
Metal deactivators	Chelating agent	Inhibit gum formation Catalyzed by certain metals
Anti-icing additives	Surfactants and glycols	Prevent icing in carburetor and fuel system
Detergents	Amines and amine carboxylates	Prevent deposits in carburetor throttle body
Deposit control additives	Polybutene amines Polyether amines	Remove and prevent deposits throughout carburetor intake ports and valves
Blending agents	Ethanol, methanol, tertiary butyl alcohol, methyl tertiary ether	Extend gasoline supply, increase apparent octane quality with some loss in mileage
Antiknock compounds	Lead alkykl, organo-manganese compounds	Increase octane quality
Dyes		

Source: Courtney and Newhall, reprinted with permission from SAE paper 790809 © 1979, Society of Automotive Engineers, Inc.

decisions. The group is called the Auto/Oil Air Quality Improvement Research Program (AQIRP).

At this time the study is not complete; however, four general results are known [Morgan; Hochhauser]. First, reduction of gasoline sulfur from 466 ppm to 40 ppm does not affect engine out emissions significantly, but can improve catalytic converter efficiency, thus reducing emissions to the environment. For the 1989 model vehicles tested, the reductions were 16% less HC, 13% less CO and 9% less NO_x. Second, decreasing the 90% distillation temperature (T_{90}) of gasoline decreases HC emissions. A single-cylinder engine study by Quader et al. has shown this is caused by the combined effects of incomplete vaporization in fuels with heavy components and the slower oxidation of the heavy components. Although improved mixture preparation can reduce the effects, reduction of T_{90} should decrease hydrocarbon (HC) emissions for all engines regardless of fuel preparation. Third, the use of methyl-*t*-butyl ether (MBTE) at the 15% level shows reduction of CO and HC emissions for both catalyst and noncatalyst cars. Fourth, reduction of aromatics appears to give a significant reduction of CO and HC emissions, but changing aromatics content also typically changes mid-range volatility.

Although the intent of the AQIRP program is to reduce emissions, the effects on fuel economy should not be ignored. Reducing aromatics, reducing T_{90}, and adding oxygenates have each been shown to reduce volumetric fuel economy by a few percent for 1989 and older model year cars. Fuel economy is difficult to measure and is influenced by many factors. Fuel reformulation and automobile system design may be expected to evolve over a number of years to give a final best combination.

The specifications for gasoline used in U.S. emissions certification tests for automobiles are shown in Table 2.12. These specifications are necessary to ensure reliability of the emissions measurements used to certify cars.

Diesel Fuel Quality

Diesel fuel specifications are shown in Table 2.13. Fuel no. 1D is used for cold weather applications, and no. 2D is the most common fuel for diesel vehicles. Number 1D and no. 2D are very similar to heating oils no. 1 and no. 2, and are thus in competition when supplies are short. Number 4D is used for medium- to low-speed engines used for stationary applications. In terms of combustion considerations, the major factors are viscosity and cetane number. Emissions of particulates are adversely affected by higher sulfur and aromatic content. Although the primary effect of low cetane number is to cause cold starting problems, reduction of cetane number can also increase engine roughness, peak pressure, and NO emissions. Typically, highly turbocharged engines are more tolerant to low cetane number during steady-state operation.

As in the case of gasoline, diesel fuels can be improved by the addition of fuel additives. Table 2.14 lists some of the additive types and their functions. Barium smoke suppressants have been outlawed because of environmental concerns; thus, other direct methods are required to reduce exhaust particulates (smoke).

TABLE 2.12
Specifications of unleaded gasoline used in U.S. emissions certification [CFR]

Item	ASTM no.	Requirement
Octane, research (minimum)	D2699	93
Sensitivity (minimum)		7.5
Lead, organic (g/gal)		0.00–0.05
Distillation range		
Initial boiling pt.* (°F)	D86	75–90*
10% point (°F)	D86	120–135
50% point (°F)	D86	200–230
90% point (°F)	D86	300–325
End point (°F, max)	D86	415
Sulfur (wt %, max)	D1266	0.01
Phosphorus (g/gal, max)		0.005
Reid vapor pressure (lb/in.^{2})	D323	8.7–9.2[†,‡]
Hydrocarbon composition		
Olefins (%, max)	D1319	10
Aromatics (%, max)	D1319	35
Saturates	D1319	Remainder

* For testing at altitudes above 4000 ft the specified range is 75–105.

† For testing unrelated to evaporative emission control, the specified range is 8.0–9.2.

‡ For testing at altitudes above 4000 ft the specified range is 7.9–9.2.

TABLE 2.13
ASTM D975 diesel fuel specifications [by permission of ASTM © 1996]

	No. 1D	No. 2D	No. 4D
Flash point (°C, minimum)	38	52	55
Cloud point (°C)	Local requirement	Local requirement	Local requirement
Water and sediment (vol %, max)	0.05	0.05	0.05
Carbon residue 10% Btm (%, max)	0.15	0.35	—
Ash, wt (%, max)	0.01	0.01	0.10
Distillation 90% point (°C)	288 max	282–338	—
Viscosity at 40°C (cSt)	1.3–2.4	1.9–4.1	5.5–24
Sulfur (wt %, max)	0.05	0.05	2.0
Copper strip corrosion, max	No. 3	No. 3	—
Cetane no. (minimum)	40	40	30

TABLE 2.14
Automotive diesel fuel additives

Additive	Type	Function
Detergents	Polyglycols, basic nitrogen-containing surfactants	Prevent injector deposits, increase injector life
Dispersants	Nitrogen-containing surfactants	Peptize soot and products of fuel oxidant; increase filter life
Metal deactivators	Chelating agents	Inhibit gum formation
Rust and corrosion inhibitors	Amines, amine carboxylates, and carboxylic acids	Prevent rust and corrosion in pipelines and fuel systems
Cetane improvers	Nitrate esters	Increase cetane number
Flow improvers	Polymers, wax crystal	Reduce pour point modifiers
Antismoke additions or smoke suppressants	Organic barium compounds	Reduce exhaust smoke
Oxidation inhibitors	Low–molecular weight amines	Minimize deposits in filters and injectors
Biocides	Boron compounds	Inhibit growth of bacteria and microorganisms

Source: Courtney and Newhall, reprinted with permission from SAE paper 790809

Diesel Fuel Formation and Emissions

The major problem of U.S. heavy-duty truck engines for the 1990s is meeting the increasingly strict exhaust emissions standards. In 1998 emissions of NO_x and particulate will be lowered to two-thirds of 1990 levels for NO_x and one-sixth of present particulate levels. The sources of these emissions will be discussed in detail in Chapter 12; however, it is important here to recognize that the fuel formulation plays an important role in this reduction effort.

One well-recognized source of particulate emissions is sulfates (primarily sulfuric acid) produced from sulfur in the fuel. The U.S. EPA has specified that certification fuels contain 0.05% sulfur by weight or less after October 1, 1993. The effects of other reformulations are not so well documented, but typically proposed low-emissions fuels have higher cetane number, a lower boiling range, lower aromatics, and in some cases the addition of oxygenates. Federal regulation for 1994 highway diesel fuel mandates a minimum cetane number of 40 or a maximum aromatics content of 35%. Data show that significant emissions reductions are achieved by such reformulations. The problem is complex, however, because many fuel effects are engine-specific, implying that a better understanding of fuel effects on combustion is needed. The problem is also complicated by the fact that fuel changes, such as oxygenate additions, must comply with many other factors such as

TABLE 2.15
ASTM D1655 Aviation turbine fuel specifications, 1995 (abridged) [by permission of ASTM © 1995]

Property	Measurement units	Jet A or A1	Jet B
Acidity	mg KOH/g, max.	0.1	—
Mercaptan sulfur	wt %, max.	0.003	0.003
Total sulfur	wt %, max.	0.3	0.3
Aromatics	vol %, max.	22	22
Density	kg/m³ at 15°C	775 to 840	751 to 802
Distillation temperature			
10% recovered	°C	205	—
20% recovered	°C	—	145
50% recovered	°C	report	190
90% recovered	°C	report	245
Final boiling pt.	°C	300	—
Vapor pressure	kPa at 38°C, max.	—	21
Viscosity	cSt at −20°C, max.	8	—
Freezing point	°C, max.	−40 (Jet A) −47 (Jet A1)	−50
Flash point	°C, min.	38	—
Net heat of combustion	MJ/kg, min.	42.8	42.8
Smoke point	mm, min.	25	25
Existent gum	mg/100 ml, max.	7	7
Copper strip corrosion (cSt)	2 h at 100°C, max.	No. 1	No. 1

toxicity, odor, pumpability, and materials compatibility. In the final reformulation, cost-benefit considerations will, of course, play a major role.

Turbine Fuel Quality

The specifications for jet fuel are more exacting than those for gasoline or diesel fuel for reasons of safety and engine durability. Table 2.15 lists the ASTM specifications for nonmilitary turbine fuels. Important combustion considerations are soot formation and resulting thermal radiation, and nozzle spray characters resulting from fuel viscosity and surface tension. Considerable work has been expended on finding aircraft fuel additives which will reduce the spreading of fire when fuel is uncontained during an accident. Jet A is a kerosene type of fuel with a relatively high flash point. Jet B fuel is a wide boiling range volatile distillate. Jet A-1 fuel has a lower freezing point than Jet A.

2.3 SOLID FUELS

Naturally occurring solid fuels include wood and other forms of biomass, peat, lignite, bituminous coal, and anthracite coal. Municipal and certain industrial refuse also qualifies as a fuel.

In addition to carbon and hydrogen constituents, solid fuels contain significant amounts of oxygen, water, and ash, as well as nitrogen and sulfur. The oxygen is chemically bound in the fuel and varies from 45% by weight for wood to 2% for anthracite coal on a dry, ash-free basis (see Table 2.16). Hence, as the coalification process becomes more advanced, the amount of oxygen in the fuel decreases.

Moisture can exist in two forms in solid fuels—as free water, and as bound water. Free water is unbound and exists between the cell walls in wood or in the larger pores of low-grade coal and is drawn into the pores by capillary attraction. Bound water is held by physical adsorption and exhibits a small heat of sorption. Green wood typically consists of 50% water, and after logs are air-dried for one year, the moisture falls to 15–20%. Lignite coals contain 20 to 40% moisture, most of which is free water, whereas bituminous coals contain about 5% moisture as bound water. As we shall see, fuel moisture influences the rate of combustion and the overall efficiency of the combustion system.

Ash is the inorganic residue remaining after the fuel is completely burned. Wood usually has only a few tenths of a percent ash, while coal typically has 10% or more of ash. Typically the ash begins to soften at 1200°C and becomes fluid at 1300°C, although this varies significantly between fuels. Ash characteristics play an important role in system design in order to minimize slagging, fouling, erosion, and corrosion.

The composition of solid fuels is reported on an as-received basis, or on a dry basis, or on a dry, ash-free basis. The moisture content on an as-received basis is the mass of the moisture in the fuel divided by the mass of the moisture plus the mass of the dry fuel and ash. For example, if the as-received moisture content is 50%, on a dry basis it is 100%. If the moisture content is 20% on an as-received basis, it is 25% on a dry basis. The basis of the calculation must be specified. When computing the higher and lower heating values, the free and bound water and the water formed from hydrogen in the fuel must be considered.

Biomass

Biomass is cellulose material which can be broadly classified as woody and nonwoody biomass. Woody biomass may be further split into softwoods and

TABLE 2.16
Typical percent oxygen, water, and ash in solid fuels

Fuel	Oxygen (dry, ash-free)	Moisture (ash-free)	Ash (dry)
Wood	45%	15–50%	0.1–1.0%
Peat	35%	90%	0.1–10%
Lignite coal	25%	30%	> 5%
Bituminous coal	5%	5%	> 5%
Anthracite coal	2%	4%	> 5%
Refuse-derived fuel	40%	24%	10–15%

hardwoods. Nonwoody biomass that can be used as a fuel include agricultural residue such as bagasse, straws, stalks, husks, and pits. Also, manure can be used as a fuel. Tall grasses, such as switchgrass, can be grown as an energy crop. Switchgrass can be harvested several times per year and dried in the field. Wood fuel includes round wood (cord wood), limb wood, wood chips, bark, sawdust, forest residues, charcoal, pulp waste, and spent pulping liquor. Tree plantations growing new hybrid hardwoods, such hybrid poplar, yield 5 to 10 times the annual biomass of a natural forest.

Softwoods are evergreen trees with needles, sometimes called conifers because their seeds are formed in cones. Hardwoods refer to broad-leafed trees that shed their leaves at the end of each growing season. The hardwoods are generally denser than the softwoods. Softwoods are made up of vertically oriented, hollow, tubular fibers from 2 to 7 mm long. Large openings or holes scattered between groups of fibers are resin ducts. Hardwoods have shorter fibers and are more porous. Because of the fibers, wood is more difficult to pulverize than coal.

Bark differs from hardwood and softwood in both structure and composition. Structurally, bark appears more spongelike than as organized fiber. Bark pores twist and intermix in an irregular pattern. Bark contains more resin and more ash than wood. Wood and bark fuel can be utilized in the form of hog fuel (mill residue of bark, chips and fines of a wide size range), whole tree chips, saw fines, or sander dust, for example. Sometimes wood is pelletized to improve storage and shipping, or made into charcoal.

Charcoal is made by heating wood in the absence of air to produce char. Charcoal is a relatively clean-burning fuel. Charcoal can be pulverized easily and made into briquettes by the addition of a binder such as starch. Sometimes briquettes are manufactured with tubular holes for better air circulation.

Dry wood consists of cellulose, hemicellulose, lignin, resins (extractives), and ash-forming minerals. Cellulose ($C_6H_{10}O_5$) is a condensed polymer of glucose ($C_6H_{12}O_6$). The fiber wall consists mainly of cellulose and represents 40 to 45% of the dry weight of wood. Hemicellulose consists of various sugars other than glucose that encase the cellulose fibers and represent 20 to 35% of the dry weight of wood. Lignin ($C_{40}H_{44}O_6$) is a nonsugar polymer that gives strength to the wood fiber, accounting for 15 to 30% of the dry weight. Wood extractives include oils, resins, gums, fats, waxes, etc., that ordinarily do not exceed a few percent. Extractives in bark range from 20 to 40%, however. The constituents which make up the ash when wood burns amount to 0.2 to 1% by weight and are mainly calcium, potassium, magnesium, manganese, and sodium oxides and lesser amounts of other oxides such as iron and aluminum. The mineral matter is dispersed throughout the cells in molecular form. The ash from bark is greater than from wood and typically is 1 to 3%.

Peat

Peat is formed from decaying woody plants, reeds, sedges, and mosses in watery bogs, usually in northern climates. Peat forms in wet environments in which air is largely excluded. In the presence of bacterial action, chemical decomposition proceeds by a process called humification:

$$\underset{\text{Cellulose}}{2C_6H_{10}O_5} \xrightarrow[\text{Heating}]{\text{Bacterial action}} \underset{\text{Peat}}{C_8H_{10}O_5} + CH_4 + 2CO_2$$

Since the rate of formation of a peat bed is about 3 cm per 100 years, peat is not a renewable resource. Some of the hemicellulose and cellulose is decomposed into humic acid bitumens and other compounds. Peat is usually dark-brown in color and fibrous in character. Since freshly harvested peat typically contains 80 to 90% water, it must be dried before using as a fuel. Peat contains 1 to 10% mineral matter (ash).

Coal

Coal is a heterogeneous mineral consisting principally of carbon, hydrogen, and oxygen, with lesser amounts of sulfur and nitrogen. Other constituents are the ash-forming inorganic compounds distributed throughout the coal. Some coals melt and become plastic when heated and give off tars, liquors, and gases, leaving a residue called coke. Coke is a strong, porous residue, consisting of carbon and mineral ash that is formed when the volatile constituents of bituminous coal are driven off by heat in the absence of or in a limited supply of air. Coals that do not melt also give off tars, liquors, and gases when heated, and leave a residue of a friable char instead of coke.

Coal originated through the accumulation of wood and other biomass that was later covered, compacted, and transformed into rock over a period of hundreds of thousands of years. Most bituminous coal seams were deposited in wetlands that were regularly flooded with nutrient-containing water that supported abundant peat-forming vegetation. The lower levels of the wetlands were anaerobic and acidic, which promoted structural changes and biochemical decompositions of the plant remnants. This microbial and chemical alteration of the cellulose, lignin, and other plant substances, and later the increasing depth of burial, resulted in a decrease in the percentage of moisture and a gradual increase in the percentage of carbon. This change from peat through the stages of lignite, bituminous coal, and ultimately to anthracite (the process called coalification) is characterized physically by decreasing porosity and increasing gelification and vitrification. Chemically there is a decrease in volatile matter content, as well as an increase in the percentage of carbon, a gradual decrease in the percentage of oxygen, and, as the anthracite stage is approached, a decrease in the percentage of hydrogen.

The progressive changes involved in the coalification process are called an advance in rank of coal. Some investigators have held that high pressures exerted by massive overburdening strata have promoted the chemical changes of coalification as well as the physical compaction. Others, however, believe that the time-temperature factors have been predominant, and that it is probably the effect of increasing temperature at increasing depth in the earth's crust that accounts for the fact that coals of greater depth of burial have greater fixed-carbon content (i.e., more advanced rank).

The chemical composition of coal cannot be described in as straightforward manner as wood. Various physical-chemical methods have identified a vast array of organic compounds. Benzenoid ring units play an important role in the coal structure. Hydrogen, oxygen, nitrogen, and sulfur are attached to the carbon skeleton. Inorganic minerals form the ash which remains when coal is burned.

Nitrogen in coal is organic and varies up to a few percent by weight. Sulfur in coal consists of organic and inorganic forms. The organic sulfur which is bound into the coal varies widely from a small fraction of a percent to 8%. Inorganic sulfur is predominantly found as iron pyrite (FeS_2) and varies from zero up to a few percent. Pyritic sulfur may be removed by coal cleaning methods, while organic sulfur is distributed throughout and requires chemical degradation to release the sulfur.

The mineral matter in coal consists of minerals such as kaolinite, detrital clay, pyrite, and calcite, and hence includes oxides of silicon, aluminum, iron, and calcium. Lesser but significant amounts of magnesium, sodium, potassium, manganese, and phosphorus are found. Many trace elements are found at lower levels. Mineral matter in coal varies widely and is present in molecular form, as bands between layers of coal, and in some instances is added from the overburden during mining. A comparison between mineral matter in wood (pine) and a bituminous coal is given in Table 2.17.

Coal may be classified according to rank and grade. *Coal rank* expresses the progressive metamorphism of coal from lignite (low rank) to anthracite (high rank). Rank is based on heating value, which is calculated on a dry, ash-free basis for low-rank coals, and on percentage of fixed carbon, calculated on a dry ash-free basis, for higher-rank coals (see Table 2.18). As shown in Figure 2.5, the heating value and percentage of fixed carbon increase as the rank moves from lignite to low

TABLE 2.17
Representative mineral elements and chlorine in pine and bituminous coals

Element	Pine* (ppm)	Illinois coal (ppm)
Ca	760	> 5000
Na	28	200–5000
K	39	200–5000
Mg	110	200–5000
Mn	97	6–210
Fe	10	> 5000
P	40	10–340
Si	—	> 5000
Al	6	> 5000
Cl	48	200–1000

*Average values for whole wood, mixed pine species.

volatile bituminous coal and the volatile matter decreases. In Figure 2.5 the percentage of volatile matter plus the fixed carbon equals 100%. The percentage of oxygen, which is contained in the volatile matter, also decreases.

Lignite is a brownish-black coal of low rank. Lignite is also referred to as brown coal. Chemically, lignite is similar to peat, and it contains a large percentage

TABLE 2.18
Classification of coals by rank (dry, ash-free basis) [by permission of ASTM © 1990]

Rank	Fixed carbon (%)	Higher heating value (Btu/lb_m)
Meta-anthracite	> 98	
Anthracite	92–98	
Semianthracite	86–92	
Low volatile bituminous	78–86	
Medium volatile bituminous	69–78	
High volatile A bituminous		> 14,000
High volatile B bituminous		13,000–14,000
High volatile C bituminous		11,500–13,000
Subbutiminous A		10,500–11,500
Subbituminous B		9500–10,500
Subbituminous C		8300–9500
Lignite A		6300–8300
Lignite B		< 6300

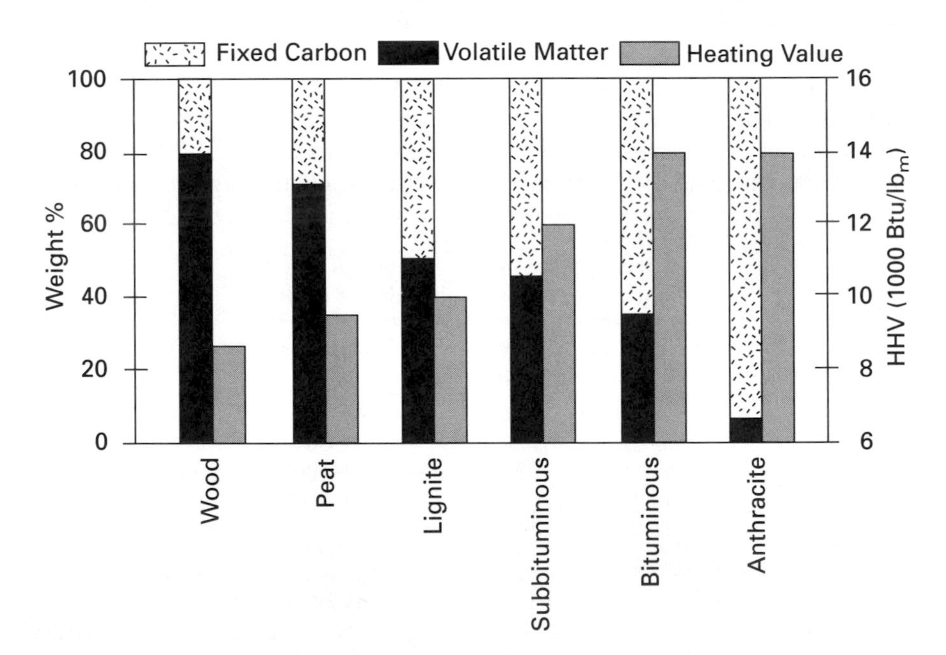

FIGURE 2.5
Typical volatile matter, fixed carbon, and higher heating value for wood, peat, and coal on a moisture and ash free basis.

of water and volatiles. Mechanically, lignite is easily fractured, but not spongy like peat. Subbituminous coal is dull-black, shows little woody material, and often appears banded. The coal usually fractures along the banded planes. Also, the moisture content of subbituminous coal is reduced. Bituminous coal is a dark black color and is often banded. The moisture content is low and the volatile content ranges from high to medium. Bituminous coal is more resistant to disintegration in air than are lignite and subbituminous coals. Anthracite coal is hard and brittle and has a bright luster. It has almost no volatiles or moisture. Anthracite is not banded.

The *grade of coal,* which is independent of rank, depends on the ash content, ash fusion temperature, sulfur content, and the presence of other deleterious constituents. Grade of coal is used more qualitatively than rank. The grade of coal can be improved by coal cleaning methods to remove some ash and pyrite sulfur. Mechanical cleaning processes include crushing, washing, dewatering, and drying. Coarse and medium sizes are cleaned by gravity separation, while the fine size is cleaned by froth floatation. Coal typically has a specific gravity of 1.1 to 1.3. When tiny particles are added to water to bring the specific gravity to about 1.5, then coal particles float while free mineral impurities tend to sink. In froth flotation the small coal particles are buoyed up to the top of a controlled surface froth, while the heavier impurities sink to the bottom. Mechanical cleaning can reduce the ash content by 10 to 70% and the sulfur content by up to 35%. Ultrafine grinding (micronization) of coal to an average particle size of 10 microns followed by washing has produced some coals with less than 1% ash. Chemical cleaning methods are being investigated to further clean or beneficiate coal.

Refuse Solid Fuel

Refuse solid fuel includes municipal solid waste and commercial, institutional, industrial, and agricultural waste. Usually these waste materials are disposed of by means of landfills and incineration without energy recovery. Recently, effort has been made to utilize these waste materials as fuels to produce steam. Since large quantities of these waste materials are generated, refuse represents a significant energy resource. In the United States over 100 million tons of combustible material is collected in municipal refuse annually.

Refuse can be burned directly in specially designed boilers, or the refuse can be processed to separate the combustibles from the noncombustibles. Processing includes shredding, magnetic separation, screening, and air classification. Hence, processing facilitates recovery of metals and glass as well as controlling the fuel size. The processed refuse is called *refuse-derived fuel* (RDF).

RDF can be processed into several different forms depending on how it is utilized. Coarse RDF is processed such that 95% passes through a 6-inch screen. Fluff RDF passes 95% through a 2-inch screen. Sometimes RDF is pulverized to pass a 10 mesh screen, and RDF can be densified into pellets or briquettes for better storage and shipping. Also RDF can be thermally converted to liquid and gaseous fuels.

Typical materials found in municipal solid waste in the United States are listed in Table 2.19. Approximately 30% of the unprocessed material is noncombustible, whereas in municipal RDF the ash varies from 10 to 15% typically. Moisture in municipal refuse varies from 15 to 25% depending on the season and weather. Municipal refuse is higher in chlorine, lead, and zinc but lower in sulfur than coal. Typical ultimate and proximate analysis of RDF are given in Table 2.20.

Analysis and Testing of Solid Fuels

Standard testing and analysis of coal is prescribed by ASTM standards. The proximate analysis, ultimate analysis, heating value, grindability, free swelling index, and ash fusion temperature are discussed here.

TABLE 2.19
Typical sorting analysis of raw municipal solid waste

Paper	43%
Plastics	3
Rubber and leather	2
Wood	3
Textiles	3
Yard waste	10
Food	10
Glass, ceramics	9
Ferrous metals	6
Nonferrous metals	1
Miscellaneous	10
	100%

TABLE 2.20
Representative proximate analysis, ultimate analysis, and heating value of solid fuels (dry, ash-free)

	Fuel type				
	Wood	Peat	Lignite	Bituminous coal	Refuse-derived fuel
Proximate analysis (wt %)					
Volatile matter	81	65	55	40	85
Fixed carbon	19	35	45	60	15
Ultimate analysis (wt %)					
Hydrogen	6	6	5	5	7
Carbon	50	55	68	78	52
Sulfur	0.1	0.4	1	2	0.3
Nitrogen	0.1	1	1	2	0.6
Oxygen	44	38	25	13	40
Higher heating value (Btu/lb)	8700	9500	10,000	14,000	9700

The *proximate* analysis (ASTM D3172) determines the moisture, volatile combustible matter, fixed carbon, and ash in a fuel sample. Although the determination is done quite accurately, the name *proximate* indicates the empirical nature of the method; a change in procedure can change the results. A sample of coal is crushed and dried in an oven at 105 to 110°C to constant weight to determine residual moisture. The sample is then heated in a covered crucible (to prevent oxidation) at 900°C to constant weight. This weight loss is referred to as *volatile matter*. The remaining sample is then placed in the oven at 750°C with the cover off so that the sample is combusted. The weight loss upon combustion is termed *fixed carbon* or *char*. The remaining residue is defined as *ash*. The components of a proximate analysis are rather arbitrary. There is no sharp distinction between free water and water chemically bound to the fuel. The split between volatile matter and fixed carbon depends on the rate of heating as well as the final temperature. Some of the ash can be volatilized during char determination. Nevertheless, the proximate analysis provides a useful comparison between fuels. The proximate analysis for biomass is limited to 600°C.

The *ultimate* analysis (ASTM D3176) provides the major elemental composition of the fuel, usually reported on a dry ash-free basis. Carbon and hydrogen are determined by burning the sample in oxygen in a closed system and quantitatively measuring the combustion products. The carbon includes organic carbon as well as carbon from the mineral carbonates. The hydrogen includes organic hydrogen as well as any hydrogen from the moisture of the dried sample and mineral hydrates. The extraneous carbon and hydrogen are usually negligible. Nitrogen and sulfur are determined chemically. Oxygen is determined by the difference between 100 and the sum of the percentages of C, H, N, and S. Sometimes chlorine is included in the ultimate analysis. The heating value is determined in a bomb calorimeter (ASTM D2105). Although the bomb calorimeter gives the heating value at constant volume, the difference between the constant-volume and constant-pressure heating values is essentially negligible.

EXAMPLE 2.3. A solid fuel contains 6% hydrogen, 30% moisture, and 10% ash and has a higher heating value of 5000 Btu/lb_m (all wt %, as-received). What is the dry, ash-free lower heating value?

Solution

$$\text{HHV (dry, ash-free)} = \frac{5000}{1 - 0.30 - 0.10} = 8333 \text{ Btu/lb}$$

Dry, ash-free fuel contains $6/(1 - 0.30 - 0.10) = 10\%$ hydrogen, and thus,

$$\frac{m_{H_2O}}{m_{fuel}} = \frac{0.1 \ lb_{H_2}/lb_{fuel}}{2 \ lb_{H_2}/mol_{H_2}}(1 \ mol_{H_2O}/mol_{H_2})(18 \ lb_{H_2O}/mol_{H_2O}) = 0.90 \ lb_{H_2O}/lb_{fuel}$$

Since the latent heat of vaporization of water is 1050 Btu/lb at 77°F, and using Eq. 2.1,

$$\text{LHV} = 8333 - (0.90)(1050) = 7388 \text{ Btu/lb fuel}$$

The *Hardgrove grindability* test (ASTM D409) is used to determine the relative ease of pulverization of coals. The coal is ground in a stationary grinding bowl which holds eight steel balls each 25 mm in diameter. The balls are driven by an upper grinding ring which is rotated at 20 rpm by a spindle which exerts a vertical force of 284.4 N. A 50-g sample is ground for 60 revolutions. The amount of coal passing a no. 200 sieve (75-micron opening) relative to a standard sample is the Hardgrove grindability index. The *free-swelling index* (ASTM D720) is an indication of the caking characteristics of coals when burned as a fuel. A1-g sample of ground coal is placed in a small crucible of a specific size. The crucible is covered and placed in an 820°C oven for 2.5 min. The coke button is then removed from the crucible and the increase in projected cross-sectional area is noted. A free-swelling index of 2 means that the projected area has doubled.

When ash is heated to a softened state, it has a tendency to foul boiler tubes and surfaces. The *ash fusion temperature* (ASTM D1857) is determined by heating a ground fuel sample at 850°C in air and then in oxygen to ensure complete oxidation of the fuel. The powdered ash is then mixed with a solution of dextrin to form a stiff paste, which is pressed into small cone-shaped molds. The ash is removed from the mold and placed in a furnace in which the temperature is slowly rising. The shape of the cone is observed and four temperatures are reported as indicated in Figure 2.6. The initial deformation temperature is when first rounding of the apex of the cone occurs; the softening temperature is when the cone has fused down into a spherical lump in which the height is equal to the width at the base. The hemispherical temperature is where the height is one-half the width of the base, and the fluid temperature is where the fused mass has spread out in a nearly flat layer with a maximum height of 1.6 mm. Ash fusion temperatures are reported for oxidizing atmospheres (air) and reducing atmospheres (60% CO and 40% CO_2). Ash fusion temperatures for a representative bituminous coal ash under oxidizing and reducing conditions are given in Table 2.21. The temperature differential between initial deformation and fluid temperature gives an indication of the type of deposit on furnace tube surfaces. A small temperature difference indicates that the surface

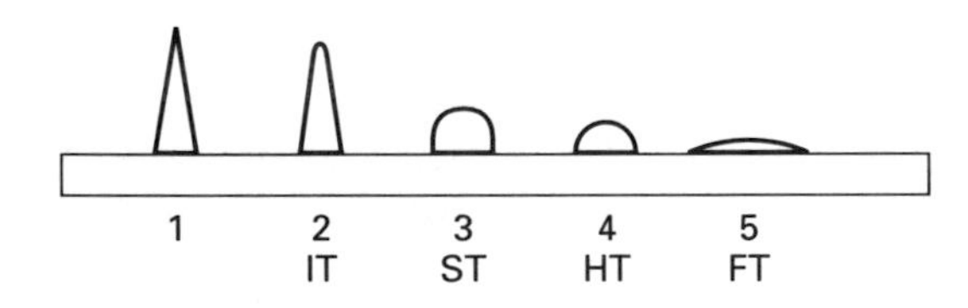

FIGURE 2.6
Determination of ash fusion temperature points: (1) before heating, (2) initial deformation temperature, (3) softening temperature (height equals width), (4) hemispherical temperature (height equals one-half width), (5) fluid temperature [ASTM D-1857, by permission of ASTM © 1990].

TABLE 2.21
Typical ash fusion temperatures (°C) for a representative bituminous coal

Deformation	Reducing	Oxidizing
Initial deformation temperature	1100	1125
Softening (height = width) temperature	1205	1230
Hemispherical (height = $\frac{1}{2}$ width) temperature	1230	1255
Fluid temperature	1330	1365

slag will be thin and tenacious. A larger temperature difference indicates the slag deposit may build up in thicker layers.

Ash fusion temperatures depend on the composition of the ash. Acidic oxide constituents such as SiO_2, Al_2O_3, and TiO_2 tend to produce higher melting temperatures, while basic oxides such as Fe_2O_3, CaO, MgO, Na_2O, and K_2O tend to produce lower melting temperatures. The combustion engineer should note that the ash fusion temperatures are typically below the flame temperature, and above the steam and wall temperatures. Herein lies one of the challenges of utilizing solid fuels.

2.4 SUMMARY

Properties of common gaseous, liquid, and solid fuels are presented in this chapter and in Appendix A for use in succeeding chapters. In addition to the chemical makeup of the fuel, the most important properties are higher and lower heating value (HHV and LHV), density, and autoignition temperature. Gasoline is characterized by octane number, diesel fuels by cetane number, gas turbine fuels by smoke number, and fuel oils by viscosity, as well as many other properties. Fuels for gasoline, diesel, and gas turbine engines have become highly specialized with many additives. Solid fuels such as coal, wood, and refuse are characterized by the ultimate analysis, proximate analysis (moisture, volatiles, fixed carbon, and ash), ash fusion temperature, and other specialized tests. With residual fuel oils and solid fuels, the sulfur, nitrogen, and ash content are particularly important. With biomass fuels moisture and mineral content are important considerations.

PROBLEMS

2.1. An industrial power plant has an average annual load of 100 MW (electrical). If the overall thermal efficiency is 33%, what is the annual cost of fuel for (*a*) natural gas, (*b*) no. 2 fuel oil, and (*c*) bituminous coal? Use the data of Table 2.2.

2.2. A house in Wisconsin uses 1200 therms of thermal energy during the heating season. Calculate the cost of fuel if the furnace is (*a*) natural gas with an efficiency of 70%, (*b*) fuel oil, efficiency 65%, (*c*) kerosene, efficiency 99.9% (unvented), and (*d*) wood (15% moisture), efficiency 50%. Use the data in Chapter 2.

2.3. Using Table 2.6 calculate the percent difference in lower heating value for methanol when introduced as a liquid at the standard temperature and when introduced as a vapor at the same temperature. Repeat for gasoline and compare with methanol.

2.4. Find the higher heating value on a volume and weight basis for a mixture of 50% methane and 50% hydrogen by volume. Use the data given in Chapter 2.

2.5. Natural gas can be simulated in the laboratory by a mixture of 83.4% (volume) methane, 8.6% propane, and 8.0% nitrogen. What is the higher heating value

(Btu/ft^3 std) for this mixture, and how does it compare with that given in Table 2.4 for natural gas?

2.6. Indicate the molecular structure of (*a*) *n*-hexane, (*b*) 3,4-diethylhexane, (*c*) 1-2,3,3-trimethylbutene, (*d*) methylnaphthalene, and (*e*) heptamethylnonane.

2.7. Kerosene has an API gravity of 42.5. What is the specific gravity?

2.8. If the price of gasoline, diesel fuel, and methanol were the same per gallon (after taxes), which would be the best buy for automobile use from a consumer standpoint? What factors other than density and LHV are involved here?

2.9. The selection of the distillation curve for automotive gasoline is a compromise between various operating requirements. For the distillation curve below, match the following operating characteristics to the position on the curve: (*a*) poor hot starting, vapor lock, and high evaporative losses; (*b*) combustion deposits, oil dillution; (*c*) poor long-trip economy; (*d*) poor warmup, rough acceleration, poor short-trip economy; (*e*) increased icing; (*f*) poor cold starting.

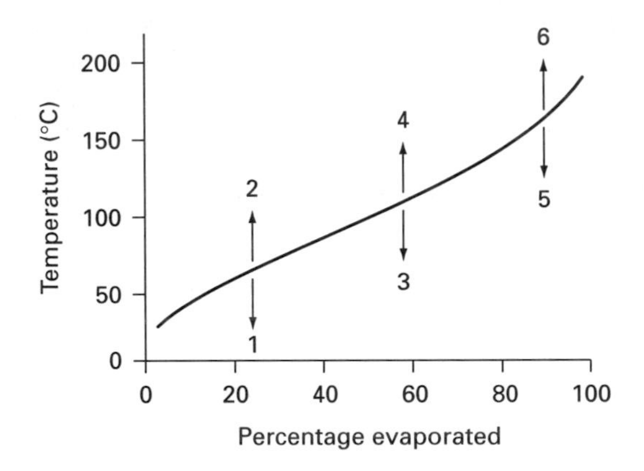

2.10. In view of the issues noted in Problem 2.9, discuss the problems which arise from a single-component fuel such as ethanol when used as an alternative to gasoline.

2.11. Wood has a dry, ash-free higher heating value of 8700 Btu/lb$_m$. Calculate and tabulate the ash-free HHV and LHV for wet wood as a function of fuel moisture content from 0 to 60% (as-received basis) using the data of Table 2.20.

2.12. If a bituminous coal contains 10% ash and 5% moisture and refuse-derived fuel contains 13% ash and 20% moisture all on an as-received basis, compare the as-received higher heating value and lower heating value of coal and refuse-derived fuel using the data of Table 2.20.

2.13. A coal-water slurry contains 70% dry powdered bituminous coal and 30% water. If the coal has a dry higher heating value of 13,000 Btu/lb, and contains 5% (wt) hydrogen, find the higher and lower heating value of the slurry. Neglect any ash in the coal.

2.14. For the coal-water slurry given in Table 2.8, the manufacturer claims that the heat loss due to the water is only 3%. Verify this assertion, if you can.

2.15. A proposed alternative diesel fuel contains 50% powdered bituminous coal and 50% methanol by weight. The coal contains 4% moisture, has been cleaned of ash, and has an as-received higher heating value of 25,000 kJ/kg. Assume the coal has 5% hydrogen as-received. Find the lower heating value of this fuel. Estimate the % volatiles and the % fixed carbon in this fuel.

REFERENCES

Annual Book of ASTM Standards, Part 26: Gaseous Fuels, Coal and Coke; Parts 23, 24, 25: Petroleum Products and Lubricants; Part 47: Test Methods for Rating Motor, Diesel and Aviation Fuels, Am. Soc. for Testing and Materials, Philadelphia, 1990.

API Research Project No. 44, National Bureau of Standards, Washington, D.C., December 1952.

ASTM Specifications for Petroleum Products, Am. Soc. for Testing and Materials, Philadelphia, 1996.

Bartok, W. and Sarofim, A. F., *Fossil Fuel Combustion: A Source Book,* Wiley, New York, 1991.

Bennethum, J. E. and Winsor, R. E., "Toward Improved Diesel Fuel," SAE paper no. 912325, 1991.

Benson, J. D., Burns, V. B., Gorse, R. A., Hochhauser, A. M., Koehl, W. J., Painter, L. J., and Reuter, R. M., "Effects of Gasoline Sulfur Level on Mass Exhaust Emissions—AQIRP," SAE paper no. 912323, 1991.

Burns, V. B., Benson, J. D., Hochhauser, A. M., Koehl, W. J., Kreucher, W. M., and Reuter, R. M., "Description of Auto/Oil Air Quality Improvement Research Program," SAE paper no. 912320, 1991.

CFR (Code of Federal Regulations) Part 86 Subpart A, para. 86.113-82, p. 482, "Protection of the Environment," July 1, 1985. Also in Springer (1992).

Courtney, R. L. and Newhall, H. K., "Automotive Fuels for the 1980's," SAE Paper no. 790809, 1979.

Davis S. C., Shonka, D. B., Anderson-Batiste, G. J., and Hu, P. S., Transportation Energy Data Book: Edition 10, Report ORNL-6565, Oak Ridge Nat. Lab., Oak Ridge, TN, 1989.

Energy Information Administration, "Energy INFOcard," U.S. Dept. of Energy, June 30, 1994.

Energy Information Administration, "Monthly Energy Review," U.S. Dept. Energy, DOE/EIA-0035(97/04), 1997.

Fleisch, T., McCarthy, C., Basu, A., Udovich, C., Charbonneau, P., Slodowske, W., Mikkelsen, S-E., and McCandless, J., "A New Clean Diesel Technology: Demonstration of ULEV Emissions on a Navistar Diesel Engine Fueled with Dimethyl Ether," SAE paper no. 950061, 1995.

"Forest Products World Outlook," Food and Agriculture Organisation of United Nations, Rome, paper 84, 1988.

Foster, A. D., "Gas Turbine Fuels," *Combustion,* vol. 44, no. 16, pp. 4–14 1973.

Francis, W., *Fuels and Fuel Technology,* Pergamon Press, Oxford, 1965.

Hancock, E. G., *Technology of Gasoline,* Critical Reports on Applied Chemistry, vol. 10, Blackwell Scientific Publications, London, 1985.

Hochhauser, A. M., Benson, J. D., Burns, V. B., Gorse, R. A., Koehl, W. J., Painter, L. J., Reuter, R. M., and Rutherford, J. A., "Fuel Composition Effects on Automotive Fuel Economy—Auto/Oil Air Quality Improvement Research Program," SAE paper no. 930138, 1993.

Kalghatgi, G. T., Golombok, M., and Snowdon, P., "Fuel Effects on Knock, Heat Release and "CARS" Temperatures in a Spark Engine," *Comb. Sci. and Tech.*, vols. 110–111, pp. 209–228, 1995.

Leppard, W. R., "The Chemical Origin of Fuel Octane Sensitivity," SAE paper no. 902137, 1990.

Lide, D. R. (ed.), *Handbook of Physics and Chemistry,* CRC Press, Boca Raton, FL, 1993.

Morgan, T. D. B., den Otter, G. J., Lange, W. W., Doyon, J., Barnes, T. R., and Yanashita, T., "An Integrated Study of the Effects of Gasoline Composition on Exhaust Emissions, Part 1: Program Outline and Results on Regulated Emissions," SAE paper no. 932678, 1993.

Owen, K. and Coley, T., *Automotive Fuels Handbook,* Society of Automotive Engineers, Warrendale, PA, 1990.

Perry, R. H. and Green, D. W., *Handbook of Chemical Engineering,* McGraw-Hill, New York, 1984.

Producer Gas: Another Fuel for Motor Transport, National Academy Press, Washington, D.C., 1983.

Quader, A. A., Sloane, T. M., Sinkevitch, R. M., and Olson, K. L., "Why Gasoline 90% Distillation Temperature Affects Emissions with Port Fuel Injection and Premixed Charge," SAE paper no. 912430, 1991.

Reed, R. J., *Combustion Handbook,* Vol. 1, *Combustion, Fuels, Stoichiometry, Heat Transfer, Fluid Flow,* North American Mfg. Co., 3rd ed., 1986.

Reed, T. B., and Das, A., *Handbook of Biomass Downdraft Gasifier Engine Systems,* US DOE SERI/SP-271-3022, 1988.

Singer, J. G. (ed.), *Combustion: Fossil Power Systems,* Chaps. 2 and 3, Combustion Engineering, Windsor, CT, 1981.

Springer, K. J., "Energy, Efficiency and the Environment, Three Big Es of Transportation," *Engineering for Gas Turbines and Power,* vol. 114, *ASME,* July 1992.

Stradling, R. J., Cowley, L. T., Lange, W. W., and Maillard, C., "The Influence of Fuel Properties and Test Cycle Procedures on the Exhaust Particulate Emissions from Light-Duty Diesel Vehicles," SAE paper no. 932732, 1993.

"Thermal Systems for Conversion of Municipal Solid Waste," Argonne National Laboratory, ANL/CNSV-TM-120 (6 vols.), 1983.

CHAPTER 3

Thermodynamics of Combustion

Thermodynamics deals with equilibrium states and how the chemical composition can be calculated for a system of known atomic or molecular composition if two independent thermodynamic properties are known. Although systems undergoing chemical reaction are generally far from chemical equilibrium, in many cases of interest a small enough control volume can be selected so that pressure and temperature are uniform within the control volume. The chemical composition at a given instant in time is dictated by the thermodynamic properties, chemical reaction rates, and fluid dynamics of the system. In this chapter it is assumed that chemical reactions are in equilibrium (i.e., not rate-controlled) and that there are no gradients and hence no fluid dynamics. The first law of thermodynamics, properties of mixtures, combustion stoichiometry, chemical energy, chemical equilibrium, and adiabatic flame temperature calculations are reviewed in this chapter.

3.1 REVIEW OF FIRST LAW CONCEPTS

For a thermodynamic system (i.e., a quantity of matter of fixed mass), the first law of thermodynamics in rate form states that the time rate of change of energy of the system is equal to the rate at which work is done on the system plus the rate at which heat is transferred to the system. If we neglect potential and kinetic energy within the system, the energy in the system consists of internal energy U due to (1) thermal energy due to translation, rotation, and vibration of the molecules, so-called sensible energy; and (2) chemical energy due to chemical bonds between atoms in the molecules. If we let $\dot{W}$ equal the rate work is done on the system (i.e., the power) and q denote the rate at which heat is transferred to the system (to conform to standard practice in heat transfer literature), then for a system of mass m, the first law of thermodynamics may be written as,

$$\frac{d(mu)}{dt} = \dot{W} + q \tag{3.1}$$

where in general $\dot{W}$ consists of shaft power, electrical power, and power due to moving boundaries. In Eq. 3.1 internal energy per unit mass is represented by the symbol u. If the work transfer for a system of volume V is restricted to mechanical work due to uniform system pressure against a moving boundary of area A moving outward with a velocity dx/dt, then the power is

$$\dot{W} = -pA\frac{dx}{dt} = -p\frac{dV}{dt} \tag{3.2}$$

Then the first law for a system becomes

$$\frac{d(mu)}{dt} = -p\frac{dV}{dt} + q \tag{3.3}$$

This form emphasizes that time is the independent variable and can be integrated numerically for complex systems, such as internal-combustion engines, to produce useful simulations.

The thermodynamic properties of the system can be evaluated if the whole system is assumed uniform. If the system is not uniform, it may be divided into subsystems of cells, each of which is assumed uniform. The properties of the whole system are then obtained by summing the cell values. For example, dividing the system into J cells each of volume V_j and assigning properties ρ_j, T_j, p_j to each cell, the system energy is,

$$mu = \sum_{j=1}^{J} \rho_j V_j u_j \tag{3.4}$$

Each cell may contain a mixture of gases. The method of determining thermodynamic properties of mixtures is discussed in Section 3.2, and the method of including chemical energy is presented in Section 3.4.

The average temperature $\overline{T}$ of a nonuniform mixture of ideal gases in J cells is obtained by applying the ideal gas law and assuming that the pressure is the same in each cell, so that

$$\sum_j p_j V_j = pV = \sum_j N_j \hat{R} T_j = N\hat{R}\overline{T} \tag{3.5}$$

Solving for $\overline{T}$,

$$\overline{T} = \frac{pV}{N\hat{R}} = \sum_j \frac{N_j T_j}{N} = \sum_j x_j T_j \tag{3.6}$$

where x_j is the mole fraction of species j. The average temperature defined in this way is the mole average temperature. If each cell has the same composition, then

$\overline{T} = pV/mR$ is the mass average temperature. However, even for a system of uniform composition,

$$mu(\overline{T}) \neq \sum_j m_j u_j(\overline{T})$$

if the specific heat c_v is a function of temperature so that u is a nonlinear function of T.

By integrating Eq. 3.3 with respect to time the closed-system energy balance becomes,

$$m(u_2 - u_1) = -W_{12} + Q_{12} \tag{3.7}$$

where $$W_{12} = \int_{t_1}^{t_2} p \frac{dV}{dt}\, dt \quad \text{and} \quad Q_{12} = \int_{t_1}^{t_2} q\, dt$$

For a uniform system with constant pressure the energy equation (3.3) simplifies to

$$\frac{d(mu + pV)}{dt} = q \tag{3.8}$$

or

$$\frac{d(mh)}{dt} = q \tag{3.9}$$

Integration of Eq. 3.9 gives

$$m(h_2 - h_1) = Q_{12} \tag{3.10}$$

If the chemical composition is constant, the chemical energy does not change, and for ideal gases,

$$u_2 - u_1 = \int_{T_1}^{T_2} c_v\, dT \tag{3.11}$$

$$h_2 - h_1 = \int_{T_1}^{T_2} c_p\, dT \tag{3.12}$$

where $c_p = c_v + R$.

For systems which have mass flow through the boundaries of the control volume (open systems), additional terms must be added to Eq. 3.3 to include the energy convected in and/or out due to the flow. The energy convected in and/or out consists of internal energy, kinetic energy, and potential energy evaluated over the upstream side of the flow surface. Also, the work done to move the fluid across the boundary must be included. The addition of the flow work and the internal energy per unit mass gives the enthalpy h. Neglecting the potential energy term for gas flows, the open-system version of Eq. 3.3 is

$$\frac{d(mu)}{dt} + \sum_i \dot{m}_i \left(h_i + \frac{V_i^2}{2} \right) = q + \dot{W} \tag{3.13}$$

The first term in this equation represents the time rate of change of energy in the system. In this equation $\dot{m}_i$ is the flow rate through flow area i at the system boundary, and h_i is the enthalpy of the fluid as it crosses boundary surface area i. The flow rates are given positive values for flow out of the system and negative values for flow into the system. The enthalpy includes sensible enthalpy and chemical energy of the mixture of gases at position i. The $V_i^2/2$ term is the kinetic energy of the flow crossing boundary surface area i, q is the rate of heat transfer to the system, and $\dot{W}$ is the power added to the system. In general, $\dot{W} = -p(dV/dt) + \dot{W}_s + \dot{W}_e$, where $\dot{W}_s$ = shaft power and $\dot{W}_e$ = electrical power.

For steady flow with one stream flowing in and one stream flowing out of the control volume, the open-system energy equation may be simplified to

$$\dot{m}\left(h_2 - h_1 + \frac{V_2^2}{2} - \frac{V_i^2}{2}\right) = q + \dot{W} \tag{3.14}$$

3.2 PROPERTIES OF MIXTURES

For mixtures of gases, the system mass is obtained from the sum of the masses of all the separate species:

$$m = \sum_i m_i \tag{3.15}$$

Since each species occupies the entire volume, the density of the mixture is the sum of the species densities:

$$\rho = \sum_i \rho_i \tag{3.16}$$

The mass fraction y_i is the mass of species i divided by total mass, or

$$y_i = \frac{m_i}{m} = \frac{\rho_i}{\rho} \tag{3.17}$$

and by definition,

$$\sum_i y_i = 1 \tag{3.18}$$

Similarly, the mole fraction is the moles of species i divided by the total moles, or the ratio of molar concentration of species i to total molar concentration:

$$x_i = \frac{N_i}{N} = \frac{n_i}{n} \tag{3.19}$$

and by definition,

$$\sum_i x_i = 1 \tag{3.20}$$

In chemical kinetics the molar concentration of a species is given by the chemical symbol of the species with square brackets around it; for example, [CO], moles/cm^3.

The molecular weight M is given by m/N, so that for a mixture,

$$M = \sum_i \frac{m_i}{N} = \sum_i x_i M_i \tag{3.21}$$

To find the relation between mole fraction and mass fraction, write

$$x_i = \frac{N_i}{N} = \frac{m_i/M_i}{m/M} \tag{3.22}$$

Thus,

$$x_i = \frac{My_i}{M_i} \tag{3.23}$$

The mixture internal energy u and enthalpy h per unit mass are given by

$$u = \sum_i y_i u_i \qquad \text{and} \qquad h = \sum_i y_i h_i \tag{3.24}$$

Similarly, the values per mole of mixture are

$$\hat{u} = \sum_i x_i \hat{u}_i \qquad \text{and} \qquad \hat{h} = \sum_i x_i \hat{h}_i \tag{3.25}$$

The pressure p of a mixture of ideal gases is equal to the sum of partial pressures which the component gases would exert if each existed alone in the mixture volume at the mixture temperature:

$$\sum_i p_i = \sum_i x_i p = p \tag{3.26}$$

The volume V of a mixture of ideal gases is equal to the sum of the partial volumes which the component gases would occupy if each existed alone at the pressure and temperature of the mixture:

$$\sum_i V_i = \sum_i x_i V = V \tag{3.27}$$

EXAMPLE 3.1. A gaseous mixture at 1000 K contains 25% CO, 10% CO_2, 15% H_2, 4% CH_4, and 46% N_2 by volume. Find the molecular weight M and the specific heat at constant volume, c_v, of the mixture in SI units.

Solution. From Eq. 3.21,

$$M = \sum_i x_i M_i = \sum_i \left(\frac{V_i}{V}\right) M_i$$

Similarly,

$$\hat{c}_v = \sum_i x_i \hat{c}_{vi}$$

Values of $\hat{c}_p$ are found in Appendix C, and $\hat{c}_v$ may be found by using $\hat{c}_v = \hat{c}_p - \hat{R}$. From the table of physical constants in the endpapers for this book, $\hat{R} =$ 8.314 kJ/(kgmol·K).

Species	x_i	M_i	x_iM_i	$\hat{c}_{pi}(1000)$	$\hat{c}_{vi}(1000)$	$x_i\hat{c}_{vi}$
CO	0.25	28	7.0	33.18	24.866	6.216
CO_2	0.10	44	4.4	54.31	45.996	4.600
H_2	0.15	2	3.0	30.20	21.886	3.283
CH_4	0.04	16	0.6	71.80	63.486	2.539
N_2	0.46	28	12.9	32.70	24.386	11.218
Sum	1.00		25.2			27.856

Thus, for the mixture, $M = 25.2$ kg/kgmol and $\hat{c}_v = 27.856$ kJ/(kgmol·K), and $c_v = \hat{c}_v/M = 1.105$ kJ/(kg·K).

3.3 COMBUSTION STOICHIOMETRY

When molecules undergo chemical reaction, the reactant atoms are rearranged to form new combinations. For example, hydrogen and oxygen react to form water:

$$H_2 + \tfrac{1}{2}O_2 \rightarrow H_2O$$

Two atoms of hydrogen and one atom of oxygen form one molecule of water, since the number of atoms of H and O must be the same on each side of the equation.

Such reaction equations represent initial and final results and do not indicate the actual path of the reaction, which may involve many intermediate steps and intermediate species. This overall or global approach is similar to thermodynamic system analysis, where only end states and not path mechanisms are used.

The relative masses of the molecules are obtained by multiplying the number of moles of each species by the respective molecular weights (which have units of kg/kgmol in the SI system of units). For the hydrogen-oxygen reaction above,

$$(1 \text{ kgmol } H_2)\left(\frac{2 \text{ kg}}{\text{kgmol } H_2}\right) + \left(\frac{1}{2} \text{ mol } O_2\right)\left(\frac{32 \text{ kg}}{\text{kgmol } O_2}\right)$$
$$= (1 \text{ kmol } H_2O)\left(\frac{18 \text{ kg}}{\text{kgmol } H_2O}\right)$$

and the mass of the reactants equals the mass of the products, although the moles of reactants do not equal the moles of products. For a fixed p and T and ideal gases,

$$1 \text{ volume } H_2 + \tfrac{1}{2} \text{ volume } O_2 = 1 \text{ volume } H_2O$$

Thus, when ideal gases react at constant T and p, the volume may change.

In most combustion calculations it is conventional to approximate dry air as a mixture of 79% (vol) N_2 and 21% (vol) O_2 or 3.764 moles of N_2 per mole O_2. The molecular weight of pure air is 28.96, because it actually contains small amounts of argon, carbon dioxide, and hydrogen. Thus, an apparent molecular weight of N_2 of 28.16 is used rather than 28.01. This value is obtained by substitution in Eq. 3.21, which defines M, to give

$$\frac{28.96 - (0.21)(32)}{0.79} = 28.16$$

In this text we shall assume dry, pure air for all calculations and use a molecular weight of 29.0. In practice for certain applications the effect of water vapor in the air may need to be considered. For example, at 80°F saturated water vapor in air occupies 6.47% by volume, and hence this air contains only 19.6% O_2.

Stoichiometric calculations are done by performing an atom balance for each of the elements in the mixture. The theoretical amount of air required to burn a fuel completely to products with no dissociation is defined as stoichiometric air. An example of a stoichiometric hydrogen-air reaction follows.

EXAMPLE 3.2. For a stoichiometric hydrogen-air reaction at 1 atm pressure, find (*a*) the fuel-to-air mass ratio f, (*b*) the mass of fuel per mass of reactants, and (*c*) the partial pressure of water vapor in the products.

Solution. React 1 mole H_2 with enough air to form the complete products H_2O and N_2:

$$H_2 + a(O_2 + 3.76N_2) \rightarrow bH_2O + 3.76aN_2$$

From an H balance: $2 = 2b$; $b = 1$. From an O balance: $2a = b$; $a = \frac{1}{2}$. From an N balance: $3.76/2 = 1.88$. Hence, the stoichiometric equation is

$$H_2 + \tfrac{1}{2}(O_2 + 3.76N_2) \rightarrow H_2O + 1.88N_2$$

(*a*) On a mass basis (1 kgmol H_2)(2 kg/kgmol) = 2 kg H_2 reacts with (4.76/2 kgmol air)(29.0 kg/kgmol) = 69.02 kg air, and

$$f = \frac{m_f}{m_a} = \frac{2}{69.02} = 0.0290$$

(*b*) The mass of H_2 per unit mass of reactant mixture is

$$\frac{m_f}{m_a + m_f} = \frac{f}{1 + f} = \frac{0.029}{1.029} = 0.0282$$

(*c*) The partial pressure of the water vapor in the products is obtained from the mole fraction:

$$x_{H_2O} = \frac{\text{moles water}}{\text{moles products}} = \frac{1}{2.88} = 0.347$$

and from Eq. 3.26,

$$\begin{aligned} p_{H_2O} &= x_{H_2O}p = 0.347p \\ &= 0.347 \text{ atm} \quad (\text{for } T > 163°F) \end{aligned}$$

For a fuel containing carbon, hydrogen, and oxygen which is burned to completion with a stoichiometric amount of air, atom balances on C, H, O, and N atoms yield the following general expression:

$$\begin{aligned} C_\alpha H_\beta O_\gamma + \left(\alpha + \frac{\beta}{4} - \frac{\gamma}{2}\right)(O_2 + 3.76N_2) \rightarrow \\ \alpha CO_2 + \left(\frac{\beta}{2}\right)H_2O + 3.76\left(\alpha + \frac{\beta}{4} - \frac{\gamma}{2}\right)N_2 \end{aligned} \tag{3.28}$$

where α, β, and γ are the number of carbon, hydrogen, and oxygen atoms in a molecule of fuel. Alternatively, α, β, and γ are the mole fractions of the carbon, hydrogen, and oxygen from the ultimate analysis of the fuel. The moles of stoichiometric air per mole of fuel are:

$$\frac{n_{as}}{n_f} = 4.76\left(\alpha + \frac{\beta}{4} - \frac{\gamma}{2}\right) \tag{3.29}$$

The stoichiometric fuel/air ratio by weight is

$$f_s = \frac{m_f}{m_{as}} = \frac{M_f n_f}{M_a n_{as}}$$

$$= \frac{M_f}{29.0(\alpha + \beta/4 - \gamma/2)(4.76)} \tag{3.30}$$

The percent *excess air* is the actual air used minus the stoichiometric air all divided by the stoichiometric air times 100:

$$\% \text{ excess air} = \frac{100(m_a - m_{as})}{m_{as}}$$

$$= \frac{100(n_a - n_{as})}{n_{as}}$$

$$= \frac{100(n_{O_2} - n_{O_2(s)})}{n_{O_2(s)}} \tag{3.31}$$

Percent *theoretical air* is the amount of air actually used divided by the stoichiometric air:

$$\% \text{ theoretical air} = \left(\frac{m_a}{m_{as}}\right)100 = \left(\frac{n_a}{n_{as}}\right)100 \tag{3.32}$$

Hence,

$$\% \text{ excess air} = \% \text{ theoretical air} - 100 \tag{3.33}$$

For example, 110% theoretical air is a lean mixture with 10% excess air; 85% theoretical air is a rich mixture which is 15% deficient in air.

Sometimes *equivalence ratio* is used instead of excess air to describe a combustible mixture. The equivalence ratio F is defined as the actual fuel/air mass ratio f divided by the stoichiometric fuel/air mass ratio f_s:

$$F = \frac{f}{f_s} \tag{3.34}$$

Excess air is directly related to equivalence ratio. Using Eq. 3.31, it follows that

$$\% \text{ excess air} = \frac{100(1 - F)}{F} \tag{3.35}$$

Equation 3.35 is plotted in Figure 3.1. For lean mixtures, excess air tends to infinity, and this is why the use of equivalence ratio is preferred for internal-

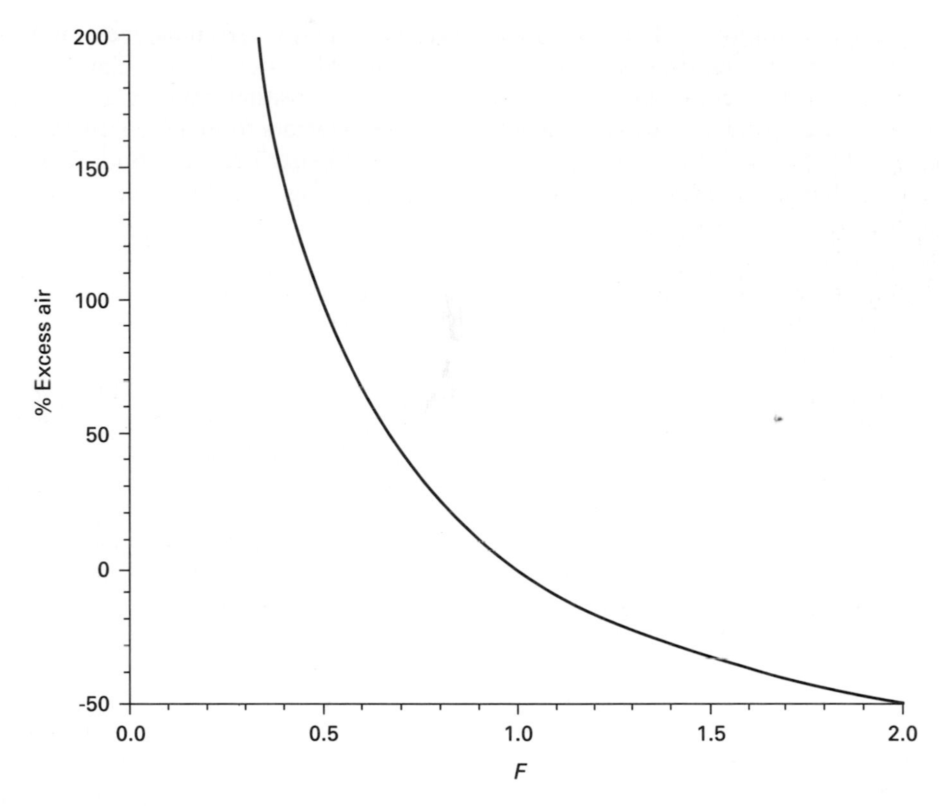

FIGURE 3.1
Excess air versus equivalence ratio.

combustion engines, which often run lean. Note that $1/(1 + f)$ is the mass of air to mass of mixture ratio, and $f/(1 + f)$ is the mass of fuel to mass of mixture ratio.

Excess air can be deduced by measuring the composition of the products. If the products are complete, then the numerator in Eq. 3.31 can be replaced by the moles of oxygen in the products, $(n_{O_2})_{prod}$. The stoichiometric moles of oxygen in the denominator of Eq. 3.31 can be obtained from the nitrogen in the products, $(n_{N_2})_{prod}$. Since gaseous measurements are usually done in terms of mole fraction, n's are replaced by x's by dividing the numerator and denominator by the total moles in the products. Referring to the products and noting that $(n_{N_2})_{prod} = (n_{N_2})_{reactants}$, the excess air becomes,

$$\begin{aligned}\%\text{ excess air} &= \left[\frac{(n_{O_2})_{prod}}{(n_{N_2})_{prod}/3.76 - (n_{O_2})_{prod}}\right]100 \\ &= \left[\frac{(x_{O_2})_{prod}}{(x_{N_2})_{prod}/3.76 - (x_{O_2})_{prod}}\right]100\end{aligned} \tag{3.36}$$

If the products contain species such as CO, H_2, and fuel fragments from incomplete combustion, it is still possible to obtain the excess air by measurements of product constituents. For example, exhaust gas emissions from engines can be used to compute the overall fuel/air ratio [Spindt]. Myers et al. have shown the errors in such calculations due to errors in fuel composition data. A graphical method of determining excess air in burners and combustors from measurement of oxygen (and CO) in the dry flue gases is presented in Figure 3.2. When both O_2 and CO_2 are measured and the fuel composition is known, Figure 3.2 provides a check on the measurements.

EXAMPLE 3.3. Measurements of the dry exhaust products from a burner which uses natural gas and air read 5% oxygen and 9% carbon dioxide. Find the excess air used for this burner.

Solution. From the measurements, $x_{O_2} = 0.05$, $x_{CO_2} = 0.09$, and $x_{N_2} = 0.86$ (by difference). From Eq. 3.36,

$$\%\text{ excess air} = \frac{0.05}{0.86/3.76 - 0.05} = 28\%$$

Note that the gas composition and excess air are consistent with Figure 3.2 for natural gas. To use Figure 3.2, connect the pivot point and the midrange of D by a straight line. Find 5% O_2 on the left-side scale, move horizontally to the intersection with the constructed straight line, and read off CO_2, N_2, and excess air values.

Typical values of the stoichiometric air-to-fuel ratio, the inverse ratio f_s, and the volume percent of CO_2 in the dry products are shown in Table 3.1 for various fuels. Concentrations in the dry products are used when water is condensed out of the products before gas analysis in order to protect the gas monitoring instruments.

TABLE 3.1
Stoichiometric complete combustion of several fuels in air

Fuel	m_{as}/m_f	f_s	CO_2 (% by volume in dry products)
Methane	17.2	0.0581	11.7
Gasoline	14.7	0.0680	14.9
Methanol	6.5	0.154	15.1
Ethanol	9.0	0.111	15.1
No. 1 fuel oil	14.8	0.0676	15.1
No. 6 fuel oil	13.8	0.0725	15.9
Bituminous coal*	10.0	0.100	18.2
Wood*	5.9	0.169	20.5

*Dry basis.

EXAMPLE 3.4. Bituminous coal is burned to completion with 50% excess air. Find the fuel-to-air ratio f and the volumetric analysis of the products. The as-received ultimate analysis of the coal is 70% (wt) carbon, 5% hydrogen, 15% oxygen, 5% moisture, and 5% ash.

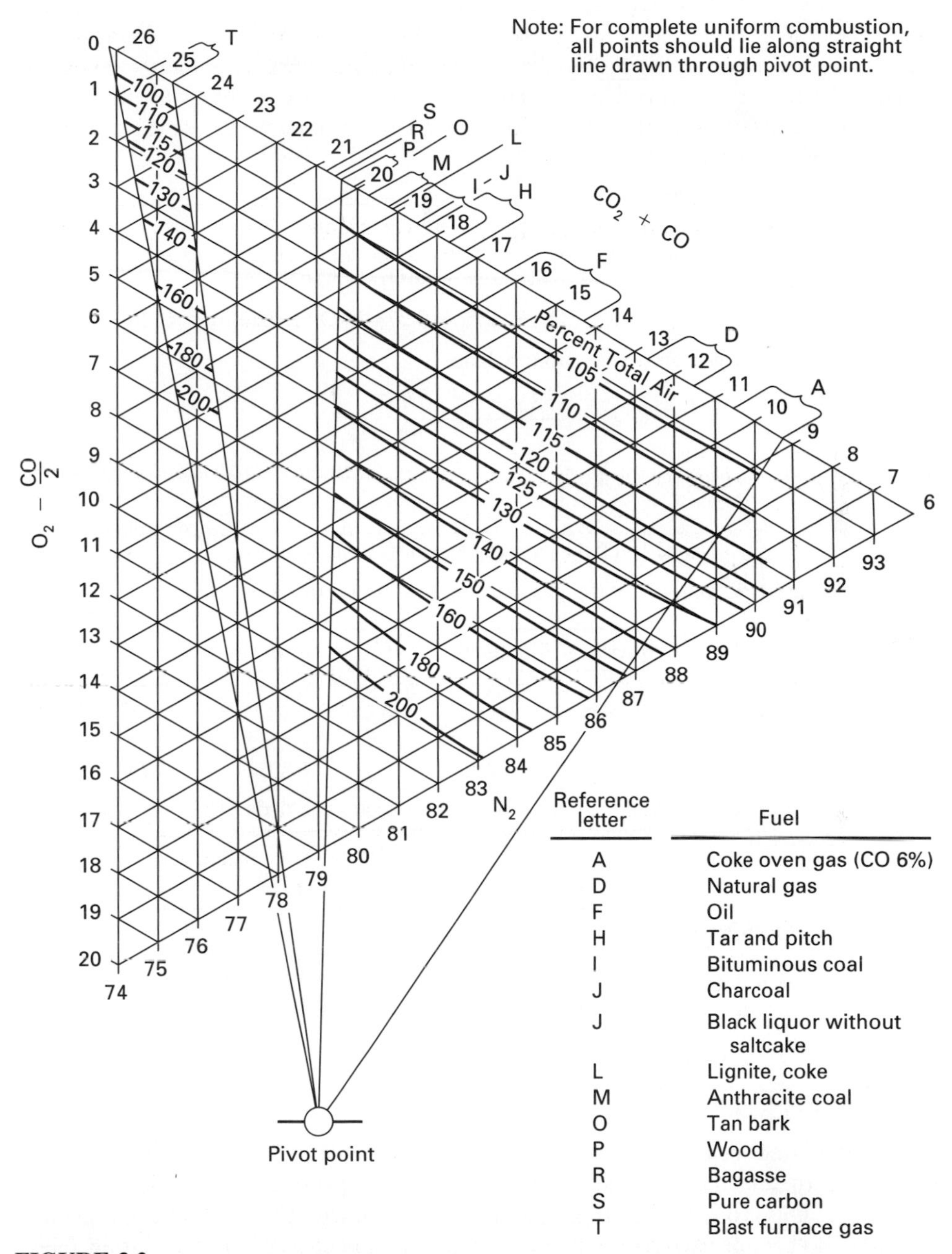

FIGURE 3.2
Nomograph for dry combustion products, percent by volume [Stultz and Kitto, by permission of Babcock and Wilcox].

Solution. For 100 kg of coal,

Species	m (kg)	M (kg/kgmol)	N (kgmol)
C	70	12	5.833
H	5	1	5.000
O	15	16	0.937
H_2O	5	18	0.278

The atom balance for 50% excess air is

$$5.833\text{C} + 5\text{H} + 0.937\text{O} + 0.278\text{H}_2\text{O} + 1.5\text{a}(\text{O}_2 + 3.76\text{N}_2) \rightarrow$$
$$5.833\text{CO}_2 + (2.5 + 0.278)\text{H}_2\text{O} + 0.5a\text{O}_2 + (1.5)(3.76)a\text{N}_2$$

Solving for a from an oxygen atom balance,

$$a = \frac{5.833 + 5}{4} - \frac{0.937}{2} = 6.614$$

Hence, the fuel-air mass ratio is

$$f = \frac{100 \text{ kg}}{[(6.614)(1.5)(1 + 3.76) \text{ kgmol}](29 \text{ kg/kgmol})} = 0.0730$$

The volumetric analysis of the gaseous combustion products is

Species	N (kgmol)	x_i	x_i (dry)
CO_2	5.833	0.1185	0.1256
H_2O	2.778	0.0564	—
O_2	3.307	0.0672	0.0712
N_2	37.303	0.7579	0.8032
Total	49.221	1.0000	1.0000

3.4 CHEMICAL ENERGY

The concepts of heat of reaction at constant pressure and constant volume, higher and lower heating value, heat of formation, and absolute enthalpy are reviewed in this section.

Heat of Reaction

In order to describe the chemical energy released when a fuel reacts with air to form products, the chemical species in the reactants and products and their states are specified. This is done by writing a balanced reaction with the phase of each species noted. The heat of vaporization of liquid fuels and heat of pyrolysis of solid fuels is typically small compared with the chemical energy released by combustion, but the effect of water condensation can be important, as illustrated, for example, by the substantial increase in furnace efficiency which is achieved when water is condensed in the heat exchanger of a home gas-fired furnace.

For very lean hydrocarbon-air mixtures where the temperature is low, the products may be assumed to be complete (usually CO_2, H_2O, O_2, and N_2). However, with high product temperatures and for rich mixtures, it is generally necessary to include other species and to assume chemical equilibrium to determine the species mole fractions. If the products are not in equilibrium, then a chemical kinetic analysis or direct measurement is required to determine the end state. For example, the amount of solid carbon produced by very rich-mixture combustion is typically not predicted correctly by equilibrium. The gaseous composition of rich-mixture products is also somewhat uncertain, even if equilibrium prevails, because the identity of the unburned hydrocarbon species is not easily determined. A solution to this problem is to complete the oxidation in a reactor and measure the additional energy release.

To understand the heat of reaction, consider the reaction of a fuel and air mixture of mass m. For a constant-volume combustion with heat transfer Q_v taken as negative if out of the system, the first law (Eq. 3.7) gives

$$m[(u_2 - u_1)_s + (u_2 - u_1)_c] = Q_v \tag{3.37}$$

where subscripts s and c refer to sensible and chemical energy. Note that in Eq. 3.37 the values of u at state 1 and state 2 are obtained by summing over all the n species, $i = 1, 2, \ldots, n$. Thus,

$$(u_1)_s = (u(T_1))_s = \sum_{i=1}^{n} y_i(u_i(T_1))_s = \sum_{i=1}^{n} y_i \int_{T_0}^{T_1} (c_{vi})_r \, dT = \int_{T_0}^{T_1} (c_v)_r \, dT$$

where T_0 is the reference temperature. The quantity $(c_v)_r$ is the specific heat of the reactant mixture. Similarly,

$$(u_2)_s = \int_{T_0}^{T_2} (c_v)_p \, dT$$

where T_0 is again the reference temperature and the quantity $(c_v)_p$ is the specific heat of the products mixture of m species.

If the heat transfer is just large enough to bring the products' temperature back to the reactant temperature, and if this temperature is taken as the reference temperature T_0 for the sensible energy, then $(u_2 - u_1)_s = 0$ by definition, and Q_v is the chemical energy released by the reaction. The quantity $[(1 + f)/f](-Q_v/m)$ is the lower heating value (LHV) of the fuel for constant-volume combustion. If the water in the products is condensed, the value of $[(1 + f)/f](-Q_v/m)$ becomes the higher heating value (HHV) of the fuel for constant-volume combustion.

If the reaction takes place at constant pressure and the total heat transfer is Q_p, then the energy equation (Eq. 3.10) becomes

$$m[(h_2 - h_1)_s + (h_2 - h_1)_c] = Q_p \tag{3.38}$$

Again, if $T_1 = T_2 = T_0$, then Q_p is the chemical energy released. For the constant-pressure case, if the moles of gaseous products N_p are larger than the moles of gaseous reactants N_r, then some of the chemical energy is expended to push aside the ambient pressure. Thus, for reactants and products which are ideal gases,

$$Q_p - Q_v = \Delta(pV) = (N_p - N_r)\hat{R}T_0 = \Delta N \hat{R} T_0 \tag{3.39}$$

For the reaction

$$C_\alpha H_\beta O_\gamma + \left(\alpha + \frac{\beta}{4} - \frac{\gamma}{2}\right)(O_2 + 3.76N_2) \rightarrow \alpha CO_2 + \left(\frac{\beta}{2}\right)H_2O + 3.76\left(\alpha + \frac{\beta}{4} - \frac{\gamma}{2}\right)N_2 \tag{3.28}$$

assuming the fuel is in the gas phase and assuming gaseous water,

$$\Delta N = \frac{\beta}{4} + \frac{\gamma}{2} - 1$$

so that $\Delta N = 0$ for methane (CH_4) and $\Delta N > 0$ if $\beta > 4$. Since both Q_p and Q_v are typically negative, $\Delta N > 0$ implies $|Q_v| > |Q_p|$, as anticipated. For most cases of interest, the difference $Q_p - Q_v$ is only a few kilocalories and is thus often neglected.

The heat of reaction may be calculated for reactions taking place at temperatures other than T_0, and for cases where the initial and final temperature are not equal, by use of the heat of reaction data taken at T_0. Consider the reaction at constant pressure with reactant temperature T_1 and product temperature T_2. Assume, for example, $T_2 > T_1 > T_0$. To use the $Q_p(T_0)$ value, imagine that first the reactants are cooled from T_1 to T_0, then the reaction takes place at T_0, and finally the products are heated from T_0 to T_2:

$$Q_p = m\int_{T_1}^{T_0} (c_p)_r\, dT + Q_p(T_0) + m\int_{T_0}^{T_2} (c_p)_p\, dT$$

or

$$Q_p - Q_p(T_0) = m(h_{s2} - h_{s1}) \tag{3.40}$$

where r refers to the reactants, p to the products, and h_s is the sensible enthalpy. Note that $Q_p(T_0)$ is negative for an exothermic reaction.

EXAMPLE 3.5. The higher heating value of gaseous methane and air at 25°C is 55.5 MJ/kg. Find the heat of reaction at constant pressure of a stoichiometric mixture of methane and air if the reactants and products are at 500 K.

Solution. The reaction is

$$CH_4 + 2(O_2 + 3.76N_2) = CO_2 + 2H_2O + 7.52N_2$$

Analysis of reactants and products using sensible enthalpies from Appendix C yields

Reactants	x_i	$\hat{h}_{si}$(MJ/kgmol)	M_i	y_i
CH_4	0.095	8.20	16	0.055
O_2	0.190	6.09	32	0.220
N_2	0.715	5.91	28	0.725

$N_r = 1 + 2(4.76) = 10.52$ $\qquad$ $M_r = \Sigma\, x_i M_i = 27.6$ kg/kgmol

$\hat{h}_{sr} = \Sigma\, x_i \hat{h}_{si} = 6.16$ MJ/kgmol $\qquad$ $h_{sr} = 6.16/27.6 = 223$ kJ/kg

The fuel-to-air mass ratio is $f = 0.055/(0.220 + 0.725) = 0.0582$.

Products	x_j	$\hat{h}_{sj}$	M_j
CO_2	0.095	8.31	44
H_2O	0.190	6.92	18
N_2	0.715	5.91	28

$$N_p = 10.52$$

$$\hat{h}_{sp} = 6.33 \text{ MJ/kgmol}$$

$$M_p = 27.6 \text{ kg/kgmol}$$

$$h_{sp} = 229 \text{ kJ/kg}$$

Since the water vapor does not condense, the lower heating value is used:

$$\text{LHV} = 55{,}000 \text{ kJ/kg fuel} - \frac{\left(\dfrac{2394 \text{ kJ}}{\text{kg water}}\right)\left(\dfrac{2 \text{ moles water}}{\text{mole fuel}}\right)\left(\dfrac{18 \text{ kg}}{\text{kgmol water}}\right)}{16 \text{ kg/kgmol fuel}}$$

$$= 50{,}113 \text{ kJ/kg fuel}$$

and $$\frac{Q_p(T_0)}{m} = -\frac{f}{1+\text{f}}(\text{LHV}) = -\frac{0.0582}{1 + 0.058}(50{,}113 \text{ kJ/kg fuel})$$

$$= -2747 \text{ kJ/kg reactants}$$

From Eq. 3.40, the heat of reaction per unit mass at 500 K is

$$Q_p = -2747 + (229 - 223) = -2741 \text{ kJ/kg reactants}$$

The negative sign indicates that the heat flows out of the system.

Heat of Formation and Absolute Enthalpy

The heat of reaction of fuels combusting in air (or oxygen) with the starting and ending points at 25°C and 1 atm gives the fuel heating value. These heating values are then tabulated for common fuels. Following this same practice for all possible reactions would, however, lead to an enormous amount of tabulated data. The solution to this problem comes by noting that we may simply add selected reactions and their heats of reaction to obtain any given reaction and its heat of reaction. The most basic set of reactions is thus for the formation of compounds from their elements. Once the data for this basic set are tabulated, any reaction can be constructed from them.

The *heat of formation* of a particular species is defined as the heat of reaction per mole of product formed isothermally from elements in their standard states. The standard state is chosen as the most stable form of the element at 1 atm and 25°C. For carbon the most stable form is solid graphite. For oxygen and nitrogen the standard state is gaseous O_2 and N_2. The heat of formation of elements in their standard states is assigned a value of zero. Heats of formation are given for pure fuels in Appendix Table A.1. Heats of formation for other species are given in the JANAF tables [Stull and Prophet]. The heat of formation is denoted as Δh° in this text. For fuels which are complex mixtures, such as gasoline or coal, for example, the heat of formation is generally not reported. However, the heat of formation can be calculated from the heating value and the hydrogen and carbon content of the fuel. This is typically necessary when using computer programs that use absolute enthalpies.

The absolute enthalpy of a substance is defined as the sensible enthalpy relative to the reference temperature T_0 plus the heat of formation at the reference temperature.

$$\hat{h} = \int_{T_0}^{T} \hat{c}_p \, dT + \Delta\hat{h}^\circ \tag{3.41}$$

Sensible enthalpies and heats of formation of selected species are tabulated in Appendix C for a reference temperature of 25°C (77°F). The absolute enthalpy of elements is equal to the sensible energy and therefore is always positive above the reference temperature. However, for compounds, since the heat of formation is typically a large negative number, the absolute enthalpy is usually negative up to a rather high temperature. For example, $\Delta\hat{h}^\circ$ of water is −241.83 MJ/kgmol, so that the absolute enthalpy is negative up to 5000 K.

EXAMPLE 3.6. In a flow calorimeter 24 mg/s of graphite particulate reacts completely with oxygen initially at 25°C to form carbon dioxide at 1 atm and 25°C. The rate of heat absorbed by the calorimeter water is 787.0 W. Find the heat of formation of CO_2.

Solution. The reaction is:

$$C(s) + O_2(g) \rightarrow CO_2(g)$$

Since for carbon $M = 12$ and since one mole of C yields one mole of CO_2,

$$\dot{N}_{CO_2} = \frac{(0.024 \text{ g C/s})(1 \text{ gmol } CO_2/\text{gmol C})}{12 \text{ g C/gmol C}} = 0.002 \text{ gmol } CO_2/\text{s}$$

is formed. The energy balance (Eq. 3.14 in molar form) for this problem is

$$(\dot{N}\hat{h})_r = (\dot{N}\hat{h})_p + q$$

And since the sensible energies are all zero at the reference temperature of 25°C and since the C and O_2 are in their standard states so that the heats of formation are zero,

$$0 = (\dot{N}\,\Delta\hat{h}^\circ)_{CO_2} + q$$

or

$$\Delta\hat{h}^\circ_{CO_2} = -787.0 \text{ W/(0.002 gmol/s)} = -393.5 \text{ MJ/kgmol}$$

This agrees with the value found in Appendix C for CO_2. This example indicates how a heat of formation could be determined experimentally.

EXAMPLE 3.7. The higher heating value of a dry ash-free bituminous coal is 12,500 Btu/lb$_m$ = 29,050 kJ/kg. The coal contains 70% (wt) carbon and 5% (wt) hydrogen on a dry, ash-free basis. Find the enthalpy of formation of this coal.

Solution. Consider coal plus air at T_0 reacting to produce products at T_0. An energy balance yields

$$(m\,\Delta h^\circ)_{coal} - (m\,\Delta h^\circ)_{CO_2} - (m\,\Delta h^\circ)_{H_2O(liquid)} = (m \text{ HHV})_{coal}$$

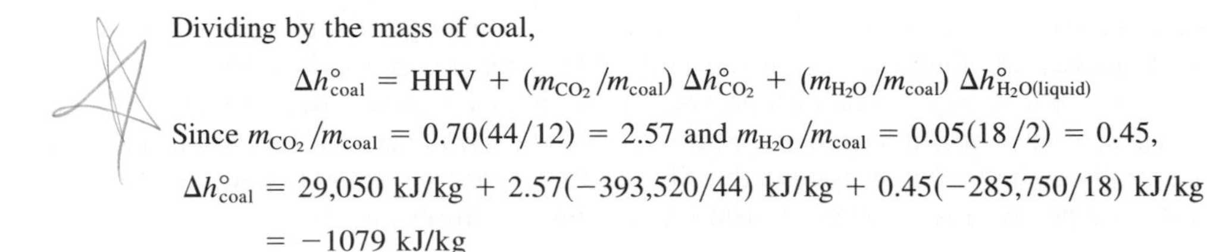

Dividing by the mass of coal,

$$\Delta h^\circ_{coal} = \mathrm{HHV} + (m_{CO_2}/m_{coal})\,\Delta h^\circ_{CO_2} + (m_{H_2O}/m_{coal})\,\Delta h^\circ_{H_2O(liquid)}$$

Since $m_{CO_2}/m_{coal} = 0.70(44/12) = 2.57$ and $m_{H_2O}/m_{coal} = 0.05(18/2) = 0.45$,

$$\Delta h^\circ_{coal} = 29{,}050\ \mathrm{kJ/kg} + 2.57(-393{,}520/44)\ \mathrm{kJ/kg} + 0.45(-285{,}750/18)\ \mathrm{kJ/kg}$$

$$= -1079\ \mathrm{kJ/kg}$$

3.5 CHEMICAL EQUILIBRIUM

To obtain thermodynamic equilibrium, it is necessary to have complete equilibrium between the molecular internal degrees of freedom, complete chemical equilibrium, and complete spatial equilibrium. Before discussing chemical equilibrium, internal and spatial equilibrium will be briefly discussed.

Internal molecular energies are the ways that molecules store energy. The major forms of energy for polyatomic molecules are translational, vibrational, and rotational energy, electronic level excitation, and nuclear spin. For most engineering combustion applications it is safe to assume equilibrium among the internal degrees of freedom. One case where it is not safe to assume equilibrium is in shock waves. The times for relaxation of the various internal degrees are typically: translation 10^{-13} s, rotation 10^{-8} s, and vibration 10^{-4} s. Thus, for a very sudden change in temperature the rotational energy will not reach equilibrium until about 10^{-8} s and the vibrational until about 10^{-4} s. Therefore, the translational energy will briefly overshoot the equilibrium value, and one cannot strictly assume internal equilibrium until 0.1 ms has elapsed.

Homogeneity is assumed for a purely thermodynamic system. This means that the system is described by a set of single-valued properties. If the system contains gradients, it can be divided into a number of subsystems such that each subsystem has negligible gradients. In a system with species gradients there will be mass transfer, and in the general case the equations of reacting fluid mechanics are required to completely describe the system.

Chemical equilibrium is achieved for constant temperature and pressure systems when the rate of change of concentration goes to zero for all species. In a complex reaction some species may come to equilibrium rapidly due to fast reaction rates or a very small change in concentration, while others approach equilibrium more slowly. For example, consider the flow of high-temperature hydrogen in a nozzle. Suppose that at the stagnation temperature the hydrogen is all dissociated to H atoms. If the expansion is slow, the reaction $2H \rightarrow H_2$ will follow the dropping temperature rapidly enough to give a series of equilibrium values (called *shifting equilibrium*). If the expansion is very rapid, hardly any reaction can occur and the result will be an essentially constant concentration of H atoms (*frozen equilibrium*). Between these two extremes the concentration will be determined by the rate of reaction and is said to be *kinetically limited.* This case is the subject of Chapter 4 and is important because chemical equilibrium does not exist in a flame zone. Temperature gradients are very steep in the flame zone, and many short-lived species are found there. In the postflame zone, many of the combustion products are in chemical equilibrium or possibly shifting equilibrium. An example of practical interest is nitrogen oxide production. The major species may follow an essentially equilibrium path while the NO reacts too slowly to stay in equilibrium as the temperature is lowered by heat transfer or expansion.

Chemical Equilibrium Criterion

When the products have reached chemical equilibrium, the problem is to determine the composition of the products at a known pressure and temperature, and a given reactant composition. Thermodynamics alone cannot determine what species may be in the product mixture. However, given an assumed set of constituents, thermodynamics can determine the proportions of each species which exist in the equilibrium mixture. Once the composition is determined, the thermodynamic properties of the mixture, such as u, h, etc., may be calculated.

For a system of J species in chemical equilibrium, the pressure and temperature do not change, which may be specified by stating that the Gibbs free energy of the system ($G = H - TS$) does not change:

$$(dG)_{T,p} = 0 \tag{3.42}$$

where

$$G = \sum_{j=1}^{J} N_j \hat{g}_j \tag{3.43}$$

and where $\hat{g}_j = \hat{h}_j - T\hat{s}_j$. Since

$$\hat{s}_j = \hat{s}_j^\circ - \hat{R} \ln\left(\frac{p_j}{p_0}\right) \tag{3.44}$$

and

$$\hat{s}_j^\circ = \int_{T_0}^{T} \frac{c_p}{T}\, dt \tag{3.45}$$

the Gibbs free energy for species j can be written as

$$\hat{g}_j = \hat{h}_j - T\hat{s}_j^\circ + \hat{R}T \ln\left(\frac{p_j}{p_0}\right) \tag{3.46}$$

Introducing $\hat{g}_j^\circ = \hat{h}_j - T\hat{s}_j^\circ$ and noting that $p_j/p = x_j$, then

$$\hat{g}_j = \hat{g}_j^\circ + \hat{R}T \ln(x_j) + \hat{R}T \ln\left(\frac{p}{p_0}\right) \tag{3.47}$$

Substituting Eq. 3.43 into Eq. 3.42, the equilibrium criterion becomes

$$d\left(\sum_{j=1}^{J} N_j \hat{g}_j\right) = 0 \tag{3.48}$$

Equation 3.48 is subject to atom balance constraints which hold the number of C, H, O, etc., atoms constant:

$$\sum_{j=1}^{\tilde{J}} \tilde{n}_{ij} N_j = A_i \tag{3.49}$$

where j refers to the species
i refers to the atoms
$\tilde{J}$ is the total number of elements in the system
$\tilde{n}_{ij}$ is the number of i atoms in species j
A_i is the moles of i atoms in the system

One approach to solving chemical equilibrium problems is to minimize G with constraints (Eq. 3.43). However, for reactions involving many species, there are too many degrees of freedom to make this a practical approach. A better approach is to expand Eq. 3.48 and note that $d\hat{g}_j = 0$ at constant p and T. Then the equilibrium criterion becomes

$$\sum_{j=1}^{J} \hat{g}_j \, dN_j = 0 \tag{3.50}$$

For a mixture containing I atom types and J species, it can be shown [Powell, 1959; Reynolds] that minimizing G using I Lagrange multipliers, λ_i (which in this context are called *element potentials*) results in the following equations that must be satisfied:

$$x_j = \frac{\exp\left[-\frac{\hat{g}_j^\circ}{\hat{R}T} + \sum_i^I \lambda_i \tilde{n}_{ij}\right]}{p/p_0} \tag{3.51}$$

and

$$\sum_{j=1}^{J} x_j = 1$$

where x_j is the mole fraction of species j ($j = 1, \ldots, J$), i = atom type, and the constraints of Eq. 3.49 must also hold. The STANJAN program [Reynolds] uses this element potential approach to solve chemical equilibrium problems.

A third approach, which is the more traditional approach, is to note that for a given reaction, $dN_j = a_j \, d\epsilon$, where the a_j are stoichiometric coefficients and ϵ represents the progress of the reaction. For example, for the reaction

$$a\mathrm{A} + b\mathrm{B} = c\mathrm{C} + d\mathrm{D}$$

it follows that

$$d(N_\mathrm{A}) = -a \, d\epsilon$$

$$d(N_\mathrm{B}) = -b \, d\epsilon$$

$$d(N_\mathrm{C}) = c \, d\epsilon$$

$$d(N_\mathrm{D}) = d \, d\epsilon$$

Then Eq. 3.50 becomes $(a\hat{g}_\mathrm{A} + b\hat{g}_\mathrm{B} - c\hat{g}_\mathrm{C} - d\hat{g}_\mathrm{D})\, d\epsilon = 0$. Using the definition of g and dividing by $\hat{R}T$,

$$\frac{a\hat{g}_\mathrm{A}^\circ + b\hat{g}_\mathrm{B}^\circ - c\hat{g}_\mathrm{C}^\circ - d\hat{g}_\mathrm{D}^\circ}{\hat{R}T} = \ln\left(\frac{p_\mathrm{C}^c \, p_\mathrm{D}^d}{p_\mathrm{A}^a \, p_\mathrm{B}^b}\right) + \ln(\mathrm{p}^\circ)^{a+b-c-d} \tag{3.52}$$

If the left-hand side of Eq. 3.52 is defined as $\ln K_p$ of the reaction and if p_0 is taken as 1 atm, then Eq. 3.52 for chemical equilibrium becomes

$$K_p = \frac{p_C^c\, p_D^d}{p_A^a\, p_B^b} = \frac{x_C^c\, x_D^d}{x_A^a\, x_B^b} \cdot p^{c+d-a-b} \tag{3.53}$$

The pressure p must be in atmospheres, and K_p is evaluated from thermodynamic data (Appendix C) as follows:

$$\ln K_p = \frac{a\hat{g}_A^\circ}{\hat{R}T} + \frac{b\hat{g}_B^\circ}{\hat{R}T} - \frac{c\hat{g}_C^\circ}{\hat{R}T} - \frac{d\hat{g}_D^\circ}{\hat{R}T} \tag{3.54}$$

When solving for equilibrium products using Eq. 3.54, the reactions to be considered are identified, and the equilibrium constants are evaluated at the specified temperature. Then the atom balance constraints are specified for the system and an equilibrium equation is written for each of the specified reactions using the form of Eq. 3.53. This set of equations is solved simultaneously to obtain the species mole fractions and other thermodynamic properties of the system. For combustion products, important gas phase equilibrium reactions include:

$$H_2O = H_2 + \tfrac{1}{2}O_2 \qquad \text{(i)}$$

$$CO_2 = CO + \tfrac{1}{2}O_2 \qquad \text{(ii)}$$

$$CO + H_2O = CO_2 + H_2 \qquad \text{(iii)}$$

$$H_2 + O_2 = 2OH \qquad \text{(iv)}$$

$$O_2 = 2O \qquad \text{(v)}$$

$$N_2 = 2N \qquad \text{(vi)}$$

$$H_2 = 2H \qquad \text{(vii)}$$

$$O_2 + N_2 = 2NO \qquad \text{(viii)}$$

Reactions (i), (ii), (v), (vi), and (vii) are dissociation reactions. Reaction (iii) is the so-called water-gas shift reaction. Reaction (iv) accounts for equilibrium OH formation, which is an important species in chemical kinetics reactions. Reaction (viii) accounts for equilibrium NO, an important air pollutant.

For a solid-gas equilibrium reaction, such as carbon-oxygen, carbon–water vapor, or carbon–carbon dioxide, for example, the equilibrium constant is determined using the Gibbs free energies of each constituent, and Eq. 3.53 is used in the same way as with gas-gas reactions. However, it should be noted that a solid has zero partial pressure.

The solution of chemical equilibrium problems is a challenging numerical computation when many species are involved, and when some important species may be many orders of magnitude smaller than other species. As noted above, if one uses the concept of equilibrium constants, then it is necessary to first determine the set of reactions that take place. This approach is used by the PER program [Olikara and Borman]. The PER program is faster than the NASA program [Gordon and McBride], because it is limited to products of combustion of hydrocarbon fuels in air. PER also gives the partials of h, u, and R with respect to T, p, and F. This is helpful when solving the unsteady energy equation using the shifting equilibrium approximation. For very rich combustion solid carbon will appear as a product if the carbon to oxygen ratio, C/O, is greater than 0.5. This is not correctly predicted by equilibrium thermodynamics and thus PER elects not to include free carbon. This limits the accuracy of PER for very rich mixtures.

With the STANJAN program [Reynolds] which uses the element potential method, the user selects the species to be included in each phase of the system, sets the atomic populations and two thermodynamic state parameters (such as p and T, p and s, h and p, or V and s), and then executes the program. The reactions are not specified, which is the beauty of this approach. STANJAN runs rapidly on a PC.

An example of thermodynamic equilibrium products from kerosene-air combustion is shown in Figures 3.3 and 3.4. Major products of lean combustion are H_2O, CO_2, O_2, and N_2; and for rich combustion the major products are H_2O, CO_2, CO, H_2, and N_2. At stoichiometric conditions at the flame temperature O_2, CO, and H_2 are present, whereas for the assumption of complete combustion, i.e., no dissociation, these three species are zero. Minor species of equilibrium combustion products at the flame temperature include O, H, OH, and NO. Carbon monoxide is a minor species in lean products, while O_2 is a minor species in rich products.

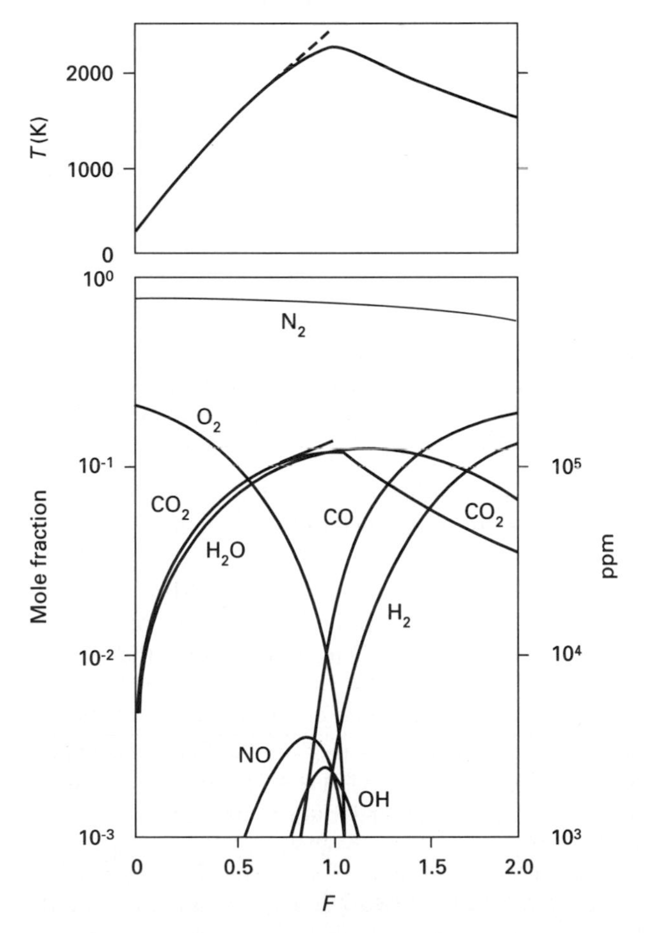

FIGURE 3.3
Equilibrium composition and temperature for adiabatic combustion of kerosene, $CH_{1.8}$, as a function of equivalence ratio [*Fundamentals of Air Pollution Engineering* by Flagan and Seinfeld, © 1988. Reprinted by permission of Prentice-Hall, Inc., Upper Saddle River, NJ.].

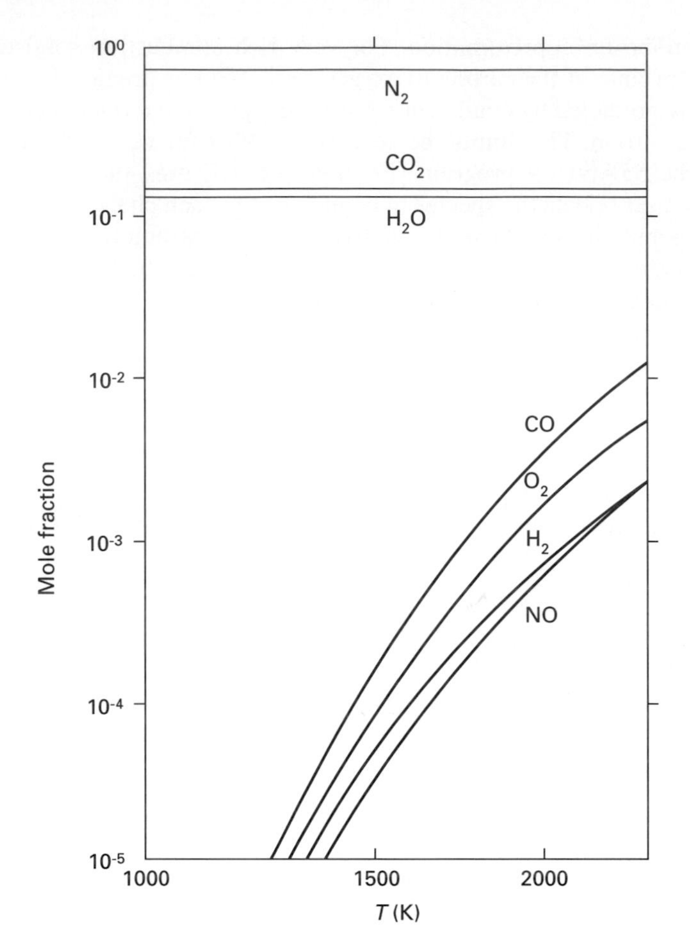

FIGURE 3.4
Variation of equilibrium composition with temperature for stoichiometric combustion of kerosene, $CH_{1.8}$ [*Fundamentals of Air Pollution Engineering* by Flagan and Seinfeld, © 1988. Reprinted by permission of Prentice-Hall, Inc., Upper Saddle River, NJ.].

Several examples of thermodynamic equilibrium calculations which can be done with a calculator rather than with a computer are given below. The equilibrium constant method is used because it is conceptually and computationally straightforward when only one equilibrium reaction is involved. Also, the equilibrium constant is useful because in Chapter 4 it will be related to the ratio of the forward to backward kinetic reaction rate constants. An example using the method of minimizing G is also given.

EXAMPLE 3.8. Equal moles of H_2 and O_2 react to produce H_2O, H_2, and O_2 at 2500 K and 1 atm. Find the percent by volume of H_2, O_2, and H_2O due to the equilibrium reaction $H_2O = H_2 + \frac{1}{2}O_2$ using the method of equilibrium constants.

Solution. Let z moles of H_2 react to form products. A hydrogen atom balance gives

$$2z = 2x_{H_2O} + 2x_{H_2} \qquad (a)$$

Since the moles of hydrogen atoms were given as equal to the moles of oxygen atoms, an oxygen atom balance gives

$$2z = x_{H_2O} + 2x_{O_2} \tag{b}$$

Dividing (*a*) by (*b*), the hydrogen to oxygen atom ratio is

$$1 = \frac{2x_{H_2O} + 2x_{H_2}}{x_{H_2O} + 2x_{O_2}} \tag{c}$$

The sum of the mole fractions of the products is 1:

$$x_{H_2} + x_{O_2} + x_{H_2O} = 1 \tag{d}$$

From Eq. 3.53,

$$K_p = \frac{x_{H_2}\sqrt{x_{O_2}}}{x_{H_2O}}\sqrt{p} \tag{e}$$

where from Eq. 3.54

$$\ln K_p = \frac{\Delta\hat{g}^\circ_{H_2O}}{RT} - \Delta\hat{g}^\circ_{H_2}/RT - \frac{1}{2}\frac{\Delta\hat{g}^\circ_{O_2}}{\hat{R}T}$$

Using data from Appendix C,

$$\ln K_p = -40.103 - (-20.198) - (-29.570)/2 = -5.120$$

and $\qquad K_p = 0.005976$

Substituting for K_p and using $p = 1$ atm in Eq. (*e*),

$$0.00598 = \frac{x_{H_2}\sqrt{x_{O_2}}}{x_{H_2O}} \tag{f}$$

Solving Eqs. (*c*), (*d*), and (*f*) simultaneously yields: $x_{H_2O} = 0.658$, $x_{O_2} = 0.335$, $x_{H_2} = 0.007$. Note that there is no OH, H, or O in this example because only equilibrium reaction (i) above was specified. To account for these additional species, equilibrium reactions (iv), (v), and (vii) above with associated equilibrium constants and equilibrium equations must be included in the analysis. The solution to the expanded problem (using STANJAN) is $x_{H_2O} = 0.6264$, $x_{O_2} = 0.3168$, $x_{H_2} = 0.00664$, $x_H = 0.00204$, $x_{OH} = 0.03994$, $x_O = 0.00809$.

EXAMPLE 3.9. Repeat Example 3.8 using the method of minimizing the Gibbs free energy. The mixture contains H_2, O_2, and H_2O, and

$$G = N_{H_2}\hat{g}_{H_2} + N_{O_2}\hat{g}_{O_2} + N_{H_2O}\hat{g}_{H_2O}$$

Solution. At 2500 K and 1 atm, this becomes

$$\frac{G}{\hat{R}T} = N_{H_2}\left[-20.198 + \ln\left(\frac{N_{H_2}}{N}\right)\right] + N_{O_2}\left[-29.570 + \ln\left(\frac{N_{O_2}}{N}\right)\right] + N_{H_2O}\left[-40.103 + \ln\left(\frac{N_{H_2O}}{N}\right)\right] \tag{a}$$

The atom balance constraints (Eq. 3.49) are,

$$\text{moles hydrogen: } 2N_{H_2} + 2N_{H_2O} = 1$$

$$\text{moles oxygen: } 2N_{O_2} + N_{H_2O} = 1$$

$$\text{moles: } N = N_{H_2} + N_{O_2} + N_{H_2O} \qquad (b)$$

Now minimize (*a*) subject to constraints (*b*) by varying one of the molar concentrations such as N_{O_2}. Using an equation solver such as EES yields,

$$(\hat{G}/\hat{R}T)_{\text{minimum}} = -55.853; \quad N = 1.505 \text{ gmol}$$

$$N_{H_2} = 0.010 \text{ gmol}; \; N_{O_2} = 0.505 \text{ gmol}; \; N_{H_2O} = 0.990 \text{ gmol}$$

Note that the selection of 1 mole of atomic hydrogen is arbitrary here, provided that the H/O atom ratio meets the specified ratio of 1, and in terms of mole fractions the result is the same as in Example 3.8.

EXAMPLE 3.10. Consider the reaction of carbon with stoichiometric air to produce CO_2, CO, and O_2 at 2200 K and 2 atm pressure. How much CO exists when the products are in equilibrium at 2200 K due to the dissociation of CO_2?

Solution. Let the reactants produce 1 mole of products. For this example assume that no solid carbon remains and the nitrogen does not react. The combustion reaction is

$$a(C + O_2 + 3.76N_2) = x_1CO_2 + x_2CO + x_3O_2 + x_4N_2 \qquad (a)$$

A carbon atom balance gives

$$a = x_1 + x_2 \qquad (b)$$

An oxygen atom balance gives

$$2a = 2x_1 + x_2 + 2x_3 \qquad (c)$$

A nitrogen balance gives

$$3.76a = x_4 \qquad (d)$$

The sum of the product mole fractions is 1:

$$x_1 + x_2 + x_3 + x_4 = 1 \qquad (e)$$

The dissociation reaction in the products is

$$CO_2 = CO + \tfrac{1}{2}O_2$$

From Eq. 3.53,

$$K_p = \frac{x_2\sqrt{x_3}\,\sqrt{p}}{x_1} \qquad (f)$$

where p must be in atmospheres and where

$$\ln K_p = \frac{\hat{g}_1^{\circ}}{\hat{R}T} - \frac{\hat{g}_2^{\circ}}{\hat{R}T} - \frac{\hat{g}_3^{\circ}}{2\hat{R}T}$$

$$= -53.737 + 34.062 + 29.096/2 \quad \text{(from Appendix C)}$$

$$= -5.127$$

and $\qquad K_p = 0.005934$

Substituting the value of K_p and $p = 2$ into Eq. (f),

$$x_1 = 238.3 x_2 \sqrt{x_3} \tag{g}$$

Solving Eqs. (b), (c), (d), (e), and (g) simultaneously,

$$x_1 = 0.1978 \qquad x_2 = 0.0111 \qquad x_3 = 0.0056 \qquad x_4 = 0.7855 \qquad a = 0.2089$$

Hence the CO is 1.11% by volume.

To verify this problem solution using STANJAN, select the products CO_2, CO, O_2, and N_2, and enter moles C = 1, O = 2, and N = 7.52 and $p = 2$, $T = 2200$ to obtain the equilibrium mole fractions.

EXAMPLE 3.11. Solid carbon reacts with steam at 1000 K and 1 atm to produce carbon monoxide and hydrogen. Find the equilibrium composition if the initial C/O atom mole ratio is 1/1 and the initial C/H atom mole ratio is 1/1. The reaction is

$$C(s) + H_2O = CO + H_2$$

Solution. The equilibrium constant for this reaction is obtained from

$$\ln K_p = -\frac{\hat{g}^{\circ}_{CO}}{\hat{R}T} - \frac{\hat{g}^{\circ}_{H_2}}{\hat{R}T} + \frac{\hat{g}^{\circ}_{C}}{\hat{R}T} + \frac{\hat{g}^{\circ}_{H_2O}}{\hat{R}T}$$

Using data from Appendix C,

$$\ln K_p = -[(-38.881) + (-17.491)] + (-1.520) + (-53.937) = 0.915$$

or

$$K_p = 2.50 \qquad \text{at} \qquad T = 1000 \text{ K}$$

Note that solid carbon is included in the calculation of K_p, but the right-hand side of Eq. 3.53 does not include solid carbon since it involves only the gas phase:

$$K_p = 2.50 = \frac{x_{CO} x_{H_2}}{x_{H_2O}} \tag{a}$$

The hydrogen-to-oxygen mole ratio is given as 1:

$$\frac{H}{O} = 1 = \frac{2x_{H_2} + 2x_{H_2O}}{x_{H_2O} + x_{CO}} \tag{b}$$

The sum of the gaseous mole fractions is 1:

$$x_{H_2O} + x_{CO} + x_{H_2} = 1 \tag{c}$$

Solving Eqs. (a) through (c) simultaneously gives the mole fractions in the gas phase:

$$x_{H_2O} = 0.073 \qquad x_{CO} = 0.642 \qquad x_{H_2} = 0.285$$

For the mixture of gas plus solid phases,

$$\frac{C}{O} = 1 = \frac{x'_{CO} + x'_{C}}{x'_{H_2O} + x'_{CO}} \tag{d}$$

$$\frac{C}{H} = 1 = \frac{x'_{CO} + x'_{C}}{2x'_{H_2O} + 2x'_{H_2}} \tag{e}$$

$$\frac{x'_{H_2O}}{x'_{H_2}} = \frac{x_{H_2O}}{x_{H_2}} = \frac{0.073}{0.285} \tag{f}$$

$$x'_{H_2} + x'_{H_2O} + x'_{CO} + x'_{C} = 1 \tag{g}$$

Solving equations (d) through (g) simultaneously, the mole fractions for the mixture are

$$x'_{H_2O} = 0.068 \qquad x'_{CO} = 0.599 \qquad x'_{H_2} = 0.265 \qquad x'_{C} = 0.068$$

To verify this problem using STANJAN, select the products C(s), H_2O, CO, and H_2, and enter moles H = 2, O = 1, and C(s) = 1 and $p = 1$, $T = 1000$ to obtain the equilibrium mole fractions.

Properties of Combustion Products

Once the mole fraction of each species has been determined, the internal energy, enthalpy, and average molecular weight of the products mixture may be obtained. For a given fuel each mole fraction of the products, x_i, is a function of T, p and f. Thus u and h become functions of T, p, and f even though each species internal energy, u_i, is only a function of T:

$$u = \frac{\Sigma\, x_i \hat{u}_i}{M} \tag{3.55}$$

When considering an unsteady energy balance, it may be necessary to evaluate the change of u with respect to T and p. Expanding Eq 3.55,

$$\frac{\partial u}{\partial T} = \frac{\Sigma\, [(\partial x_i/\partial T)\hat{u}_i + x_i\, \partial \hat{u}_i/\partial T]}{M} - \frac{(\partial M/\partial T)u}{M} \tag{3.56}$$

where $M = \Sigma\, x_i M_i$. Also,

$$\frac{\partial u}{\partial p} = \Sigma \left[\frac{(\partial x_i/\partial p)\hat{u}_i}{M}\right] - \frac{(\partial M/\partial p)u}{M} \tag{3.57}$$

Similar expressions may be written for the enthalpy and free energy. These relations are typically programmed into chemical equilibrium computer codes. Thus, when one calls a code such as PER, for example, for a given fuel and T, p, and F (recall that F is the fuel-air equivalence ratio as defined by Eq. 3.34), the code returns the equilibrium values of h, u, R, and $\partial h/\partial T$, $\partial h/\partial P$, $\partial h/\partial F$, $\partial u/\partial T$, $\partial u/\partial P$, $\partial u/\partial F$, $\partial R/\partial T$, $\partial R/\partial P$, $\partial R/\partial F$.

The effect of dissociation is not large except at high temperatures. This is illustrated in Figure. 3.5, where enthalpy is plotted versus temperature for the products of a stoichiometric methane and air reaction. The lines of constant pressure coincide at lower temperature where dissociation is small. At higher temperatures, the changing composition due to dissociation causes the lines to separate. For a fixed temperature, dissociation is largest at low pressures and becomes quite small at very high pressures. In general, Le Châtelier's rule states that if the moles of products exceed the moles of reactants, then an increase in pressure decreases the dissociation as shown by Eq. 3.53.

Figure 3.5 can be used to illustrate the effects of dissociation on a simple adiabatic throttling process. Neglecting changes in kinetic energy, $h_1 = h_2$. If

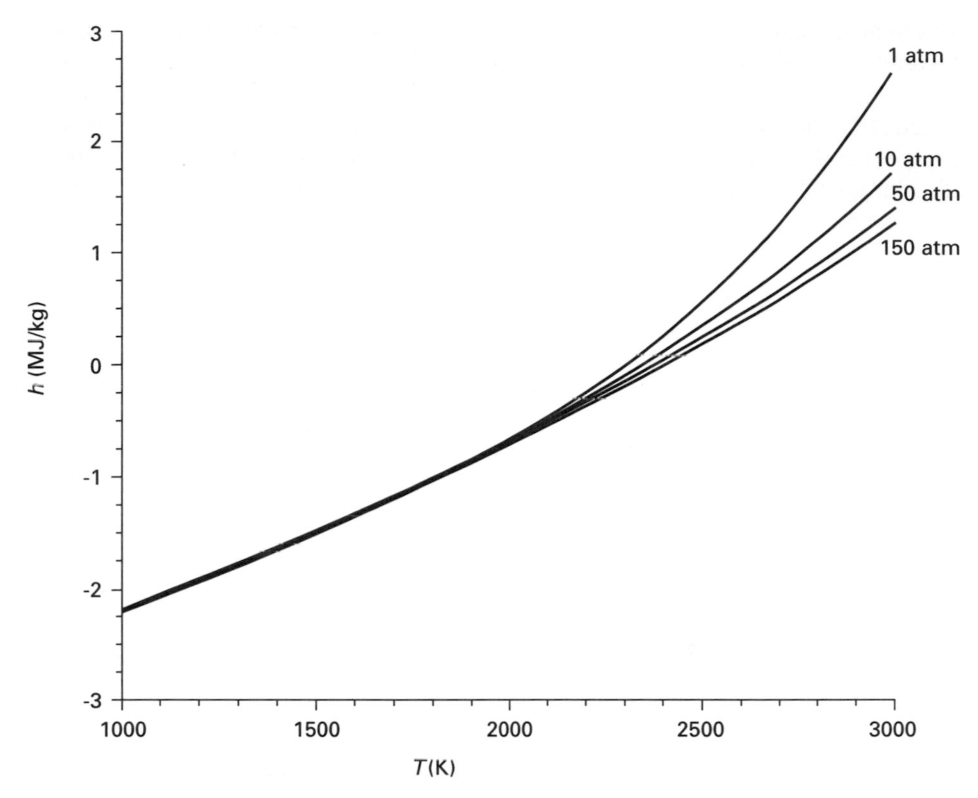

FIGURE 3.5
Absolute enthalpy of products from stoichiometric methane-air reaction.

enthalpy is a function of T only, then $T_1 = T_2$. This is true for the ideal gas mixture of Figure 3.5 in the nondissociated region. However, at the higher temperatures the mole fractions of the ideal gas constituents vary with both T and p, so that h is a function of both T and p. Throttling causes a drop in pressure resulting in some energy being used to break molecular bonds and lowering the downstream temperature. For example, throttling the products in Figure 3.5 from 3000 K and 150 atm to 1 atm gives a downstream temperature of 2700 K. Note that this energy is recovered by a constant pressure cooling process to a nondissociated state provided that we assume shifting equilibrium. The processes do however cause a loss in available energy. Because the throttled gas may go from state 1 to state 2 in a very short time the assumption of shifting equilibrium may not hold for all species; this is discussed further in Chapter 4.

A second example of dissociation effects is given by the expansion of products of combustion in a piston-and-cylinder device. Using the first law for the product system,

$$\frac{du}{dt} = -\frac{p}{m}\frac{dV}{dt} - \frac{q}{m} \tag{3.58}$$

where positive q is the heat transfer rate out of the system to the containing walls. Given the initial conditions and dV/dt, the first law may be solved to find T and p as a function of time if shifting equilibrium can be assumed. If the expansion is kinetically controlled, then reaction rate data, such as described in the next chapter, are required. Fortunately, the energy-containing species are typically in equilibrium for all but the most rapid expansions.

The equations to be solved are then

$$\frac{\partial u}{\partial T}\frac{dT}{dt} + \frac{\partial u}{\partial p}\frac{dp}{dt} = -\frac{RT}{V}\frac{dV}{dt} - \frac{q}{m} \tag{3.59}$$

$$p = \frac{mRT}{V}$$

$$\frac{dp}{dt} = \left(\frac{mR}{V}\right)\frac{dT}{dt} + mRT\frac{d(1/V)}{dt} + \frac{mT}{V}\frac{dR}{dt} \tag{3.60}$$

$$\frac{dR}{dt} = \frac{\partial R}{\partial T}\frac{dT}{dt} + \frac{\partial R}{\partial P}\frac{dP}{dt} \tag{3.61}$$

Algebraic manipulation of these equations produces a single differential equation of the form

$$\frac{dT}{dt} = \tau(T, t) \tag{3.62}$$

This equation can be solved numerically using commercially available ordinary differential equation solvers, provided the heat transfer q can be modeled accurately and an equilibrium subroutine is available to calculate u, R, and their partials in terms of T and p. This approach will be explored further in Chapter 7.

The above examples illustrate how equilibrium calculations can be used to solve problems involving dissociated products. In the next sections the same principles are applied to combustion problems.

3.6 FIRST LAW COMBUSTION CALCULATIONS

The use of the first law allows calculation of processes which go from one equilibrium state to another. If this concept is extended to a step-by-step process in time, the assumption is that the adjustments to the equilibrium state take place very rapidly within each time step (shifting equilibrium). Several simple cases are considered in this section. First consider an adiabatic, constant-pressure process during which reactants are converted to products. If the kinetic energy is small compared with the enthalpy and there is no shaft work, the energy equation (Eq. 3.14) is simply

$$h_r = h_p \tag{3.63}$$

and the problem is to determine the temperature of the products which will make this equality hold. This is the adiabatic flame temperature problem. Similarly, for constant volume,

$$u_r = u_p \tag{3.64}$$

and the problem is to find T and p of the products which make the equality hold. Because both p and T are now unknown, the extra equation $p = mRT/V$ is needed. Another type of problem is the expansion of the products in a nozzle or in a piston-cylinder system. The expansion is typically not adiabatic for either of these cases. The heat transfer can be included by using empirical formulations.

Adiabatic Flame Temperature

Consider a constant-pressure system with a deflagration. The adiabatic flame temperature is obtained from Eq. 3.63. Calculations of methane-oxygen-nitrogen flame temperatures using STANJAN are shown in Figure 3.6. As the pressure is increased, dissociation of the products is reduced and the temperature increases. The maximum temperature is at stoichiometric for no dissociation, but shifts slightly to the rich side when dissociation is included. Adiabatic flame temperatures for various representative fuels are shown in Table 3.2.

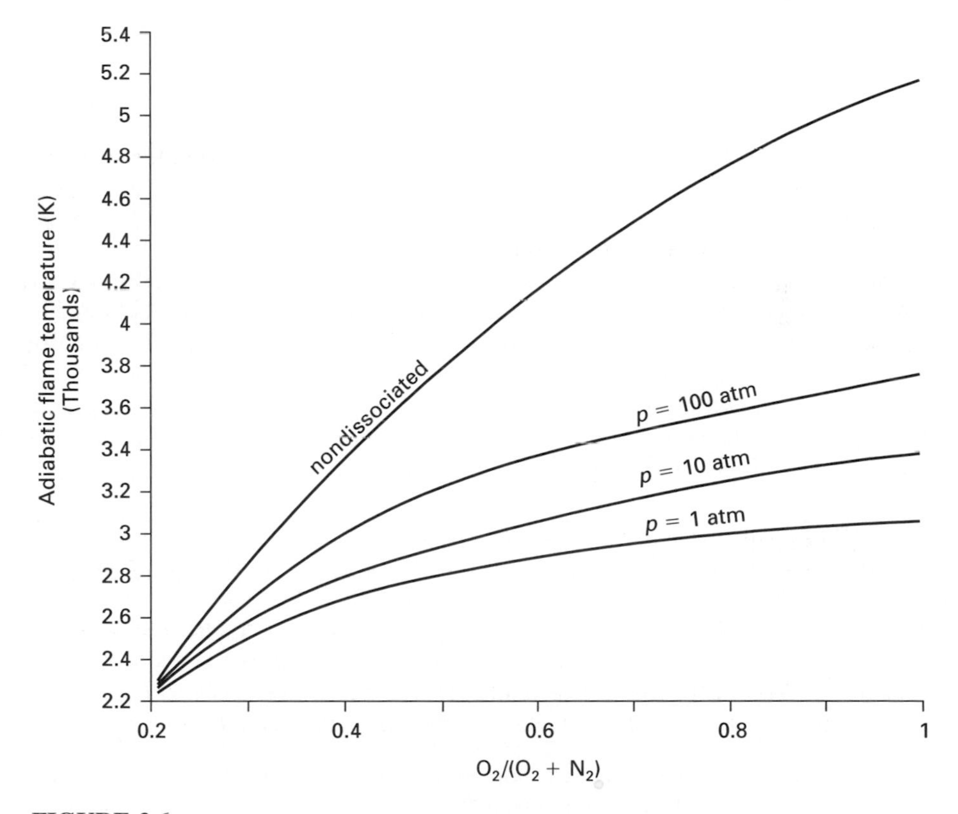

FIGURE 3.6
Adiabatic, constant-pressure flame temperature of stoichiometric methane-oxygen with nitrogen mixtures initially at 298 K.

TABLE 3.2
Adiabatic flame temperatures (K) of selected fuels with air at 1 atm; reactants initially at 298 K

	Equivalence ratio F		
Fuel	**0.8**	**1.0**	**1.2**
Gaseous fuels			
Methane	2020	2250	2175
Ethane	2040	2265	2200
Propane	2045	2270	2210
Octane	2150	2355	2345
Liquid fuels			
Octane	2050	2275	2215
Cetane	2040	2265	2195
No. 2 fuel oil	2085	2305	2260
Methanol	1755	1975	1810
Ethanol	1935	2155	2045
Solid fuel (dry)			
Bituminous coal	1990	2215	2120
Lignite	1960	2185	2075
Wood	1930	2145	2040
RDF*	1960	2175	2085
Solid fuels (25% moisture)			
Lignite	1760	1990	1800
Wood	1480	1700	1480
RDF*	1660	1885	1695

*Refuse-derived fuel.

For stoichiometric and lean mixtures at temperatures below 3500°F (1900 K), dissociation is low enough that the flame temperature may be calculated by assuming that combustion products go to completion. The mole fraction of each species is known from the complete reaction. The adiabatic flame temperature is obtained by equating the enthalpy of the reactants to the enthalpy of the products. The enthalpy of the reactants is known, and by trial and error, the temperature is determined which makes the enthalpy of the products equal to the enthalpy of the reactants. The procedure is shown in the following examples. For complex fuels the enthalpy of formation is first calculated from the heating value as shown in Example 3.7.

EXAMPLE 3.12. Find the adiabatic flame temperature of carbon monoxide burning with 50% excess air at 25°C and 1 atm. Neglect dissociation. The complete reaction is

$$CO + \frac{1.5}{2}(O_2 + 3.76N_2) \rightarrow CO_2 + 0.25O_2 + 2.82N_2$$

Solution. The energy equation (Eq. 3.63) is

$$[N(\hat{h}_s + \Delta\hat{h}^\circ)]_{CO} = [N(\hat{h}_s + \Delta\hat{h}^\circ)]_{CO_2} + [N(\hat{h}_s + \Delta\hat{h}^\circ)]_{O_2} + [N(\hat{h}_s + \Delta\hat{h}^\circ)]_{N_2}$$

Using Appendix C, the enthalpy of formation of CO is −110.50 MJ/kgmol, that of CO_2 is −393.51 MJ/kgmol, and O_2 and N_2 are zero.

$$(1)(0 - 110.53) = (1)(\hat{h}_{s,CO_2} - 393.52) + 0.25\hat{h}_{s,O_2} + 2.82\hat{h}_{s,N_2}$$

Using Appendix C and guessing $T = 2000$ K,

$$-110.53 - [91.45 - 393.52 + (0.25)(59.20) + (2.82)(56.14)] = 18.42$$

Guess $T = 2100$ K; then

$$-110.53 - [97.50 - 393.52 + (0.25)(62.99) + (2.82)(59.75)] = 1.25$$

Thus, by interpolation the adiabatic flame temperature is 2107 K.

When using STANJAN, first select the reactants and get the absolute enthalpy of the reactants. Enter $CO = 1$, $O_2 = 0.75$, $N_2 = 2.82$, $p = 1$, and $T = 298$. Get the enthalpy of the reactants $= -0.8439 \times 10^6$ J/kg. Then select the products case 5 with the enthalpy of the products $= -0.8439 \times 10^6$ J/kg and $p = 1$. To find the flame temperature of the complete reaction, select CO_2, O_2, and N_2. To include dissociation, select all of the products. When dissociation is included, STANJAN gives an adiabatic flame temperature of 2079 K. Dissociation always reduces the flame temperature. The effect of dissociation is greater with less excess air and thus higher flame temperature.

EXAMPLE 3.13. Find the adiabatic flame temperature of the bituminous coal of Example 3.4 burned with 50% excess air at 25°C and 1 atm. Neglect dissociation and neglect the ash. Repeat the problem using STANJAN and include dissociation.

Solution. From Example 3.4, 100 kg coal produced 5.833 moles CO_2, 2.778 moles H_2O, 3.307 moles O_2, and 37.303 moles N_2. From Example 3.7 the heat of formation of the coal was -1079 kJ/kg. The energy equation becomes

$$m\,\Delta h^\circ_{\text{coal}} = [N(\hat{h}_s + \Delta\hat{h}^\circ)]_{CO_2} + [N(\hat{h}_s + \Delta\hat{h}^\circ)]_{H_2O} + [N(\hat{h}_s + \Delta\hat{h}^\circ)]_{O_2} + [N(\hat{h}_s + \Delta\hat{h}^\circ)]_{N_2}$$

Substituting values,

$$(100)(-1081) = 5.83(\hat{h}_{CO_2} - 393.52) + 2.78(\hat{h}_{H_2O} - 241.83) + 3.31\hat{h}_{O_2} + 37.30\hat{h}_{N_2}$$

Solving by trial and error using Appendix C,

T **(K)**	**Right-hand side**
1900	-137×10^3 kJ
2000	58.7×10^3

Thus, by interpolation the adiabatic flame temperature is 1915 K.

To use STANJAN, first determine the absolute enthalpy of the reactants. From Example 3.7 the absolute enthalpy of the coal is -1079 kJ/kg. From Example 3.4 the fuel to air ratio is 0.073. Thus, the absolute enthalpy of the reactants is $-1079(0.073/1.073) = -73.41$ kJ/kg. Then select the all the products and enter moles of C = 5.833, H = 5.556, O = 21.093, and N = 74.606; and set $p = 1$, $h = -73{,}410$ J/kg. The adiabatic flame temperature including dissociation is 1902 K.

As a further simplification when only an approximate answer is desired, the enthalpy of the combustion products may be obtained from an average specific heat. Then the energy equation becomes

$$m_f[c_{p_f}(T_f - T_0) + \text{LHV}] + m_a c_{p_a}(T_a - T_0) = (m_f + m_a)c_p(T_{\text{flame}} - T_0) \quad (3.65)$$

Solving for the adiabatic flame temperature and assuming that the air and fuel enter at the reference temperature of 25°C = 77°F,

$$T_{\text{flame}} = T_0 + \frac{f}{1+f}\frac{\text{LHV}}{c_p} \tag{3.66}$$

where c_p is the average specific heat of the products. Hence, when comparing one fuel with another, a larger heating value does not necessarily imply a higher stoichiometric flame temperature, because the stoichiometric fuel-to-air ratio must also be considered. For example, carbon monoxide has one-third the heating value of natural gas and yet the stoichiometric flame temperature is several hundred degrees higher.

EXAMPLE 3.14. Calculate the adiabatic flame temperature using the approximate Eq. 3.66 for the data of Example 3.12.

Solution. From Chapter 2 the lower heating value of CO is 4346 Btu/lb$_\text{m}$ = 10,100 kJ/kg and

$$T_{\text{flame}} = 298\text{ K} + \frac{0.280}{1.280}\frac{10{,}100\text{ kJ/kg}}{c_p\text{ kJ/kg}\cdot\text{K}}$$

In order to obtain c_p let us use a value for nitrogen at 1000 K. From Appendix C, $c_p = 32.7/28 = 1.17$ kJ/kg · K. Thus $T_{\text{flame}} = 2186$ K. This compares with 2107 K, which was obtained more accurately in Example 3.12. Obviously, this method is only approximate, but it may be a sufficient estimate of the adiabatic flame temperature for some purposes. The true adiabatic flame temperature including dissociation is 2084 K for this problem.

Graphical Method for Determination of Flame Temperatures

Although data such as shown in Figure 3.6 and Table 3.2 can be easily generated using a computer and an equilibrium subroutine, it is instructive to examine the behavior graphically, as first suggested by Powell [1957]. In this graphical procedure the enthalpy of the products per unit mass of original air is plotted versus f for lines of constant temperature and a given constant pressure as shown in Figure 3.7. To understand the motivation for this plot, let us look again at the reaction equation.

Consider the lean combustion of hydrocarbon fuel of mass f with a unit mass of air. The complete products enthalpy will be a linear function of f for a fixed temperature of the products. To see this, we write

$$\begin{aligned}\left(\frac{f}{M_f}\right)C_\alpha H_\beta + \left(\frac{1}{4.76M_a}\right)(O_2 + 3.76N_2) &= \left(\frac{1}{4.76M_a}\right)O_2 + \left(\frac{3.76}{4.76M_a}\right)N_2 \\ &+ f\left[\left(\frac{\alpha}{M_f}\right)CO_2 + \left(\frac{\beta}{2M_f}\right)H_2O - \left(\frac{\alpha}{M_f} + \frac{\beta}{4M_f}\right)O_2\right]\end{aligned} \tag{3.67}$$

where M_f = molecular weight of the fuel and M_a = molecular weight of air. Note that f/M_f represents the moles of fuel to react with one mole of air.

Writing the enthalpy for the $1 + f$ mass units of the products in terms of species enthalpies and grouping the terms shows that $h = A(T) + fB(T)$ where

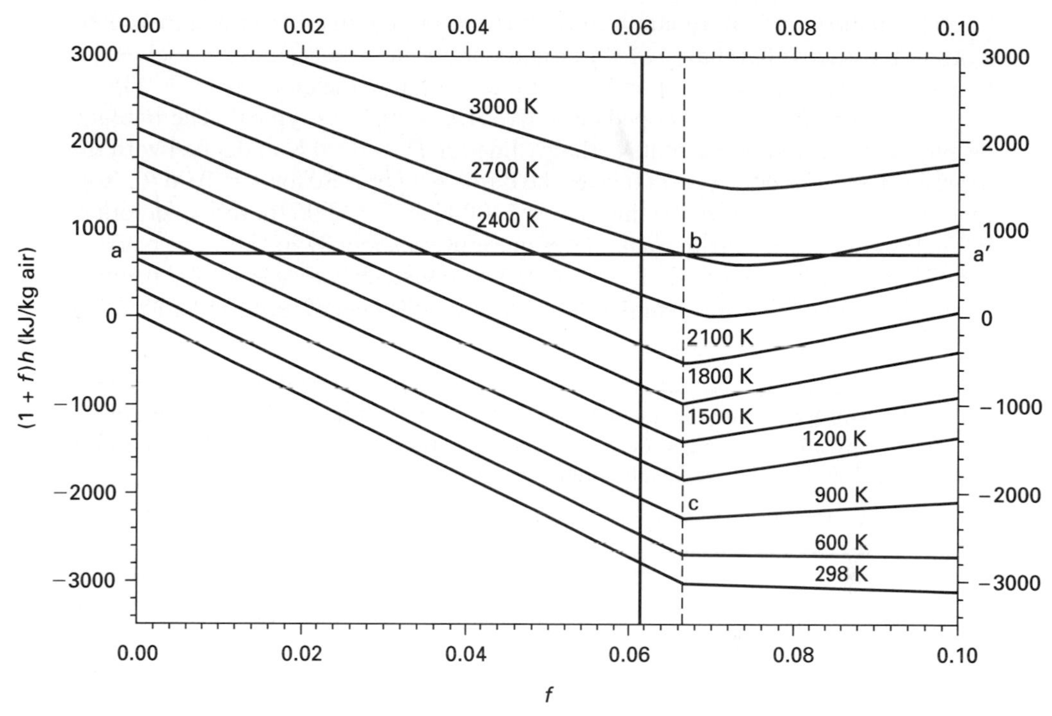

FIGURE 3.7
Reaction of $(CH_2)_x$ in air at a constant pressure of 20 atm showing product enthalpy per unit mass of air for lines of constant-temperature products versus fuel-to-air mass ratio.

functions A and B are calculated from the species enthalpy values. If we plot $(1 + f)h$ for the products versus f, lines of constant T will be straight lines (see Figure 3.7). The slope on the lean side is negative, and in fact the h values are negative for all but very high temperatures. The negative values are caused by the large negative heats of formation of CO_2 and H_2O. As the temperature is increased, the positive sensible values grow and eventually overcome the negative portion. The slope is negative on the lean side because increasing f increases CO_2 and H_2O while decreasing O_2. On the rich side the slopes are positive because H_2 and CO replace H_2O and CO_2 as f increases beyond stoichiometric. Note that $f = 0$ corresponds to h for pure air.

Thus far we have taken the products to be CO_2, H_2O, O_2, and N_2 for the lean combustion. At high temperatures the products will dissociate to form many other species, and the product species mole fractions at equilibrium will be a function of T, p, and f. Thus the lines of constant T are no longer straight and the plot can be drawn for only one pressure. For a constant-pressure combustor, such as a gas turbine combustor, the plot is still useful and we can follow the same procedures as outlined above. The effect of dissociation is to reduce the adiabatic flame temperature because some of the energy goes into breaking the molecular bonds. This energy is returned upon cooling if the products remain in equilibrium.

The flame temperature and heat of reaction are obtained from Figure 3.7 by plotting the reactant enthalpy of a unit mass of air plus f mass units of fuel versus f on top of the products lines. For a constant temperature the quantity $(1 + f)h_r = h_a + fh_f$ is linear in f. The slope is negative if $h_f < 0$, as is typical. The line a-a′ on Figure 3.7 is such a reactants enthalpy line for $T = 1000$ K and a fuel with the formula $(CH_2)_x$. Crossing point b gives the flame temperature for $f = 0.0676$. Note that the flame temperature for this case is 2700 K at $f = 0.0676$, the stoichiometric fuel/air ratio. The maximum flame temperature is about 2750 K for this reactant condition. If the products at b are cooled at constant pressure to the reactant temperature of 1000 K, the length of line b-c gives the heat of reaction for $(1 + f)$ mass units of reactant mixture.

EXAMPLE 3.15. Find the adiabatic flame temperature for a gaseous mixture of cyclooctane and 10% excess air which is compressed to 20 atm and 1000 K before combustion. Use the graphical method.

Solution. The enthalpy plot for products of cyclooctane-air combustion at 20 atm is given in Figure 3.7. From Eq. 3.35, 10% excess air represents $F = 0.9091$. From Appendix A, $f_s = 1/14.8 = 0.0676$. Thus $f = 0.0615$. To use the graph to determine the adiabatic flame temperature, note that $(1 + f)h_r = (1 + f)h_p$, and construct the line $(1 + f)h_r = h_a - h_{a0} + f[(\Delta h^\circ + c_p(T - T_0)]_{\text{fuel}}$ on the graph (see Figure 3.7). For 1000 K using Appendix A and Appendix B, the line is

$$1046 - 298 + f\left[\frac{-124{,}400}{112.2} + 1.173(1000 - 298)\right] = 748 - 285f.$$

For $f = 0.0615$, the crossing point yields a temperature of approximately 2650 K on Figure 3.7. (The STANJAN code gives a flame temperature of 2652 K.)

Combustion in a Closed Volume

Consider a constant-volume combustion process where the reactants are mixed uniformly and ignition takes place at the center so that the flame spreads spherically outward. The pressure rises and a small amount of heat is transferred from the unburned reactants to the bomb walls. As the reactants are converted to products, heat transfer increases from the products to the walls.

The classic thermodynamic calculation for a constant volume is to neglect heat transfer and compute the products mass average temperature just at the end of burning using $u_r = u_p$. The products pressure p_2 is calculated from the ideal gas relation, $p_2 = mR_2T_2/V$. Note that the dissociated products specific gas constant, R_2, is a function of T_2 and p_2 and is obtained by iteration. That is, for a mixed m, T_2, and V, p_2 is calculated using a guessed R_2, and then the equilibrium products are obtained. Once convergence to a consistent set of values occurs, the products u_2 is compared with the reactants u_1 to determine the correct T_2.

Calculation of the final products temperature in the above way is not adequate when investigating piston engines. The products of the first reactants to burn are compressed, while the last products are formed from a compressed mixture but are not themselves compressed. Hence, at the end of the burning, the products in the center, which burned first, are hotter than the last burned products in the outer layer. The measured pressure versus time history during combustion may be used to calculate the rate of mass burning.

To derive the governing equations for the thermodynamic history within the cylinder, divide the volume into two zones separated by a thin discontinuity, which is the flame. Neglecting blowby, the trapped mass, m, is a constant known from experiment for the given engine conditions. Designate the reactants zone as u for unburned and the products zone as b for burned. As the flame moves through the volume there is a mass flow of reactants, across the flame (i.e., relative to the flame):

$$\dot{m}_u = \rho_f \underline{V}_f A_f \tag{3.68}$$

Conservation of mass for the unburned and burned control volumes gives:

$$\frac{dm_u}{dt} = -\dot{m}_u = -\frac{dm_b}{dt} \tag{3.69a}$$

$$m_u + m_b = m \tag{3.69b}$$

$$V_u + V_b = V \tag{3.69c}$$

The energy equations for the unburned and burned control volumes follow from Eq. 3.13:

$$\frac{d(m_u u_u)}{dt} = -\dot{m}_u h_u - p\frac{dV_u}{dt} - q_u \tag{3.70a}$$

$$\frac{d(m_b u_b)}{dt} = \dot{m}_u h_u - p\frac{dV_b}{dt} - q_b \tag{3.70b}$$

where q_u and q_b are the rates of heat transfer out of each zone through the walls. Since the flame is a thin discontinuity, there is no heat transfer from the flame to the walls. Substituting Eqs. 3.69 into Eqs. 3.70, the energy equations become

$$\frac{d(m_u u_u)}{dt} = h_u\frac{dm_u}{dt} - p\frac{dV_u}{dt} - q_u \tag{3.71a}$$

$$\frac{d(m_b u_b)}{dt} = h_u\frac{dm_b}{dt} - p\frac{dV_b}{dt} - q_b \tag{3.71b}$$

The pressure is uniform throughout the volume since there is little pressure change across a flame, and

$$p = \frac{m_u R_u T_u}{V_u} = \frac{m_b R_b T_b}{V_b} \tag{3.72}$$

Equations 3.68 to 3.72 will be used in Chapter 7 to obtain the burning rate of the flame in the cylinder and investigate engine performance. At this point let us show that, neglecting heat loss, the reactant system undergoes an isentropic compression as the unburned volume decreases. Simplifying Eq. 3.71*a* for the reactants, assuming constant specific heats and using $c_p - c_v = R$,

$$\frac{m_u c_{v,u} dT_u}{dt} = \frac{R_u T_u dm_u}{dt} - m_u R_u T_u \left(\frac{d \ln V_u}{dt}\right) \tag{3.73}$$

Dividing Eq. 3.73 by $m_u R_u T_u$ and integrating term by term,

$$\int_1^2 \frac{c_{v,u}}{R_u} \frac{dt}{T} = \int_1^2 \frac{dm_u}{m_u} - \int_1^2 \frac{1}{V_u} \frac{dV_a}{dt} \tag{3.74}$$

and for constant $c_{v,u}/R_u = 1/(\gamma - 1)$,

$$\left(\frac{1}{\gamma - 1}\right) \ln\left(\frac{T_2}{T_1}\right) = \ln\left(\frac{m_{u_2}}{m_{u_1}}\right) - \ln\left(\frac{V_{u_2}}{V_{u_1}}\right) \tag{3.75}$$

or

$$\frac{T_2}{T_1} = \left(\frac{\rho_{u_2}}{\rho_{u_1}}\right)^{\gamma-1} \tag{3.76}$$

These ideas on combustion within a cylinder are explored further in Problems 3.31 and 3.32.

3.7 SECOND LAW ANALYSIS

In design practice, very little second law analysis has been carried out for combustion systems. This does not mean that an availability analysis is not useful, but perhaps reflects the fact that the irreversibility of combustion itself and irreversibilities of the heat transfer and fluid mechanisms found in practical systems are well known, but necessary, expedients. Fuel cells provide a way to reduce combustion irreversibilities, but fuel cells that use hydrocarbon fuels are not yet developed for commercial use. It is thus desirable, given the need for fuel economy, to understand how we may utilize combustion with the least destruction of available energy. Such analysis to evaluate losses has been carried out for power plants, diesel engines, and gas turbines.

Examples of attempts to improve second law efficiencies by reducing the available energy loss due to heat transfer are readily at hand. Regenerative cooling of rocket nozzles by using one of the propellants as a coolant is common. Cooling of turbine combustors by slot cooling with excess air is also common. The advantage of these techniques is obvious. Recently, reduction of losses due to heat transfer in reciprocating engines, especially diesels, has become a topic of considerable interest. The idea is to reduce the heat transfer from the combustion gas to the engine parts by introducing surfaces insulated on the hot side with ceramic material. Detailed first law analysis shows that most (80% or more) of the extra energy that does not go to heat transfer ends up in the exhaust gas. Thus the reduced heat transfer engine does not significantly improve the overall efficiency unless the exhaust energy is utilized in some manner. A turbine or a Rankine cycle which uses the exhaust energy are two ways of recovering some of the available work. Note, however, that the availability of the energy has been greatly reduced by the time it reaches the exhaust. A second practical problem with the low–heat transfer engine is that the volumetric efficiency will suffer because the chamber walls are hot and thus the charge density is decreased unless additional compression work is done by the turbocharger compressor.

Our brief discussion of the low–heat transfer engine clearly shows the need for both first law and second law analysis of the system. We shall not delve into such a second law system analysis, since it involves much more than combustion, but we shall look at the second law as applied to some simple cases.

Steady-Flow Burner

Consider a steady-flow, constant-pressure combustor with a reactant mixture entering at 1 and products leaving at 2 (Figure 3.8). Assume that heat transfer from the combustor to the atmosphere (where T_0 is the ambient temperature here) represents a direct loss of available energy—that is, we will not try to recover any useful work.

The reaction causes a rate of entropy production, $\dot{s}^*$, given by the steady-state, steady-flow entropy balance:

$$\frac{dS}{dt} = \dot{m}_1 s_1 - \dot{m}_2 s_2 - \frac{q}{T_0} + \dot{s}^* = 0 \tag{3.77}$$

Substituting $q/\dot{m}_1 = (h_1 - h_2)$ from the first law balance and solving for the entropy production,

$$\frac{\dot{s}^*}{\dot{m}_1} = s_2 - s_1 + \frac{h_1 - h_2}{T_0} \tag{3.78}$$

The availability destruction rate (irreversibility) $\dot{a}d$ is given by $T_0\dot{s}^*$ and thus by

$$\frac{\dot{a}d}{\dot{m}} = h_1 - h_2 + T_0(s_2 - s_1) \tag{3.79}$$

Recall that for ideal gases the absolute entropy is given by

$$s = \int_0^T \frac{c_p\, dT}{T} - R \ln\left(\frac{p}{p_0}\right) \tag{3.80}$$

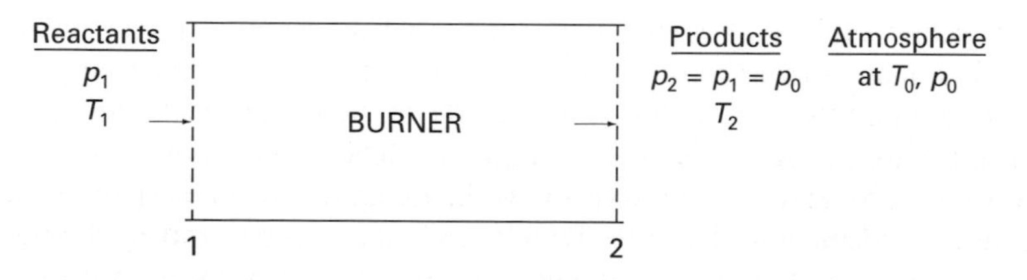

FIGURE 3.8
Control volume for availability balance.

As an example, consider the reactants and nondissociated products for the lean combustion of methane-air given by the chemical equation

$$CH_4 + 3(O_2 + 3.76N_2) = CO_2 + 2H_2O + O_2 + 11.322N_2$$

The reactants are at $T_1 = 537°R$ and 1 atm pressure. Let $q = 0$, and calculate the availability destruction for this reaction. From the first law balance, $h_1 - h_2 = 0$; and solving for T_2 we obtain $T_2 = 3214°R$. For the burner, Eq. 3.79 becomes

$$\dot{a}d = \dot{m}_1(s_2 - s_2)T_0 \tag{3.81}$$

The entropy per mole for each reactant and product species can be found in Appendix C, and the result is

$$\dot{m}_1 s_1 = 731 \text{ Btu/(h}\cdot°\text{R)}$$

$$\dot{m}_2 s_2 = 955 \text{ Btu/(h}\cdot°\text{R)}$$

Using $T_0 = 530°R$,

$$\dot{a}d = 119 \text{ Btu/h}$$

or
$$\left(\frac{1+f}{f}\right)\left(\frac{\dot{a}d}{\dot{m}}\right) = 7399 \text{ Btu/lb methane} = 17{,}195 \text{ kJ/kg methane}$$

This is about one-third of the availability of the reactant mixture per unit mass of methane. That is, if we could react the same mixture reversibly, we would not destroy available energy, but with the combustion process we have destroyed about one-third, leaving two-thirds available to do useful work.

Hydrocarbon Fuel Availability

The typical steady-flow combustor will have significant heat transfer losses and will have incomplete products containing unburned hydrocarbons, carbon monoxide, and other pollutants. Thus the previous example does not account for all of the availability destruction found in practical combustors. Also note that no work is produced during the combustion process. What happens in a simple combustor is a conversion of fuel chemical energy to sensible energy of the hot products. The maximum work that can be obtained by bringing these hot products to the dead state (in equilibrium with the environment) is determined by the availability of the products as they flow from the control volume. We now ask a more general question: What is the availability of the fuel? That is, what is the maximum work that can be obtained from a unit mass of fuel if we let it interact with the environment? To obtain the answer, one first needs to recognize that the fuel has a thermal-mechanical availability if it is treated as a chemically inert substance. This availability will be zero if the fuel is at the dead state. But even at the dead state the fuel has a "chemical availability" which can be obtained by letting the fuel react with atmospheric oxygen. We wish to define this availability for use in second law analysis just as the heating value was defined for use in first law analysis. To define the fuel chemical availability, take the fuel at a fixed reference state (25°C and 1 atm, ideal gas) and combine it with atmospheric oxygen at T_0 and its partial pressure in the atmospheric air. The complete combustion products leave the system at T_0 and with the partial pressure they have as they exist in the reference atmosphere.

For hydrocarbon fuels going to completion at $F \leq 1$, the reaction is

$$C_\alpha H_\beta + \left(\alpha + \frac{\beta}{4}\right)O_2 = \alpha CO_2 + \left(\frac{\beta}{2}\right)H_2O(vapor)$$

where the mole fractions of species in the reference atmosphere are taken as: $x(O_2) = 0.20428$, $x(H_2O) = 0.02471$, and $x(CO_2) = 0.00032$. The nitrogen and the trace species are not shown in the reaction because they cancel out. The products are diluted to the reference atmosphere.

The chemical availability for one mole of hydrocarbon fuel is

$$\hat{a}_f(\text{chemical at } T_0, p_0) = -\left[\alpha\hat{g}(CO_2) + \left(\frac{\beta}{2}\right)\hat{g}(H_2O\ vapor) - \hat{g}_f - \left(\alpha + \frac{\beta}{4}\right)\hat{g}(O_2)\right]_{T_0, p_0} + \hat{R}T_0 \ln\left[\frac{(0.20428)^{\alpha+\beta/4}}{(0.00032)^{\alpha}(0.02471)^{\beta/2}}\right] \tag{3.82}$$

where $\hat{g}$ is the Gibbs function (see Section 3.5) evaluated at $T_0 = 293$ K and $p_0 = 1$ atm. Typically (although not always), values of $\hat{a}_f$ are between the LHV and HHV values and are often approximated by the HHV of the fuel.

EXAMPLE 3.16. Calculate the chemical availability per kgmol of methane gas.

Solution. For methane, $\alpha = 1$ and $\beta = 4$. The reaction is

$$CH_4 + 2O_2 = CO_2 + 2H_2O(vapor)$$

By definition, the temperature is 293 K and the pressure is 1 atm. For the reactants, using data from Appendix C,

$$H_r = [(-74.87 - 0.18) + 2(0 - 0.15)](10^3) = -75{,}350 \text{ kJ}$$

$$S_r = \left[185.58 - 8.314 \ln\left(\frac{1}{1}\right)\right] + 2\left[204.56 - 8.3143 \ln\left(\frac{0.20463}{1}\right)\right]$$

$$= 621.08 \text{ kJ/K}$$

$$G_r = H_r - T_0 S_r = -75{,}350 - (293)(621.08) = -257{,}330 \text{ kJ}$$

For the products:

$$H_p = [(393.52 - 0.19) + 2(-241.83 - 0.17)](10^3) = -887{,}710 \text{ kJ}$$

$$S_p = \left[213.09 - 8.3143 \ln\left(\frac{0.00032}{1}\right)\right] + [-188.19 - 8.3143 \ln(0.02308)]$$

$$= 719 \text{ kJ/K}$$

$$G_p = H_p - T_0 S_p = -1{,}088{,}389 \text{ kJ}$$

Thus the availability of methane is,

$\hat{a}_f = -257{,}330 - (-1{,}088{,}398) = 831.1$ MJ/kgmol and $a_f = 51.8$ MJ/kg

This compares with LHV = 49.5 MJ/kg and HHV = 55.0 MJ/kg.

Calculation of the total availability of a fuel requires that we add the sensible availability to the chemical availability for a fuel at a temperature T and pressure p different from T_0 and p_0.

The concept of fuel availability is helpful when sorting out the relative values of the availability destruction terms. The availability destructions from the different causes (heat transfer, combustion, friction, etc.) are often computed as percentages of the fuel availability. An example of such an analysis, taken from the book *Availability Analysis* by M. J. Moran, is given in Table 3.3 for a turbojet engine. The major loss is the exhaust stream, which has a velocity of 772 m/s and a temperature of 985 K. The second greatest loss is due to combustion irreversibility.

TABLE 3.3
Availability analysis for a turbojet engine [by permission of M. J. Moran]

Location		Percent of fuel availability
Availability out nozzle exhaust		53
Availability destruction total		31
Diffuser	0.03	
Compressor	1.49	
Combustor	28.83	
Turbine	0.42	
Nozzle, jet pipe	0.32	
Thrust power		16

3.8 SUMMARY

Thermodynamics predicts the temperature, pressure, and composition (and other state variables) of a mixture due to a change in state, given a sufficiently long time to come to equilibrium. The time required to reach equilibrium is not predicted by thermodynamics. However, for unsteady systems at high temperature it is often appropriate to assume that shifting equilibrium holds. The first law of thermodynamics (conservation of energy) is formulated for mixtures of gases using Eq. 3.3 for closed systems and Eq. 3.13 for open systems with the mixture relationships of Section 3.2. The internal energy per unit mass, u, and enthalpy per unit mass, h, of a mixture contain sensible plus chemical energy due to each species in the mixture.

A mixture of substances undergoing chemical reaction follows conservation of mass. Stoichiometric calculations are done by performing atom balances for each element in the mixture. If the reaction is incomplete or if the products are dissociated, thermodynamic equilibrium calculations are required. For multiple reactions a computer program such as STANJAN or PER is used.

Chemical reactions for fuel combustion release heat and the heating value for fuels is determined in a calorimeter at standard temperature and pressure (25°C,

1 atm). Complete reaction of elements in their standard state with oxygen is defined as the heat of formation. The heat of reaction at conditions other than the standard state is calculated using equilibrium thermodynamics including absolute enthalpies which include the heats of formation. For a given heat of reaction the flame temperature is calculated from the equilibrium energy equation.

Thermodynamic first law analyses, which include the chemical energy and shifting equilibrium can be used to model the energy release rates in practical combustion systems. Second law analyses show where available energy is lost. Such analyses show that the combustion process is a major contributor to the destruction of available energy in practical combustors.

PROBLEMS

3.1. Calculate the stoichiometric fuel/air ratio by weight of octane (C_8H_{18}).
(a) How many kg of octane are in one cubic meter of stoichiometric mixture for octane and air at 300 K and 1 atm?
(b) How many pounds are in a cubic foot of this mixture at 80°F and 1 atm?

3.2. A four-cylinder, four-stroke automobile engine is running at 4000 rpm, and each cylinder has a displacement of 0.65 L. A stoichiometric mixture of octane and air at 20°C and 1 atm is drawn into the engine. The volumetric efficiency at wide-open throttle is 85%. Calculate the pounds mass of octane and the volume of air drawn in per minute. Estimate the gallons per hour of octane used by the engine. The volumetric efficiency is defined as the mass of charge ingested per intake event divided by the mass of charge with volume equal to the displacement and density equal to that entering the port.

3.3. To form a stoichiometric mixture of 80°F n-octane vapor and air, the air is brought in at 1 atmosphere and temperature T_0 and mixed with octane liquid. Assuming the process is adiabatic and isobaric, calculate T_0. The octane liquid is brought in at 77°F (537°R). Use the data in Appendix A.

3.4. An engineer, Pat, calculated the stoichiometric fuel/air ratio for methane and dry air. Pat then did an experiment on a burner with room air and methane gas. Pat measured the mass flow rate of methane and regulated the mass flow rate of air to give a stoichiometric ratio according to the calculations. Later another engineer, Fran, asked Pat if the humidity of the air had been measured. Pat said the air was 80°F and saturated, but didn't think that it mattered. Fran said, "I think the fuel to air ratio was not stoichiometric." Pat said, "I think the error is small." Who is correct? Justify your answer by calculation.

3.5. Rate methane, octane, and bituminous coal according to the mass of CO_2 produced per unit of energy released. Use the data in Chapter 2 and Appendix A.

3.6. A burner uses methane and air. A measurement of 3% oxygen (by volume) in the dry exhaust products is made. Find the excess air used for this burner by direct stoichiometry calculations, and verify your answer approximately using Figure 3.2.

3.7. Calculate the work required to isentropically compress 3.0 liters of stoichiometric octane-air to one-tenth cubic volume starting at 27°C and 1 atm. Use variable specific heat data provided in the appendixes. Calculate the resulting pressure and temperature. Repeat the calculation for air only. What can you conclude about the compression work of a lean versus a stoichiometric engine? Use c_v(octane) $= 0.367 + 0.00203T$ (kJ/kg · K), and c_v(air) $= 0.634 + 0.000128T$ (kJ/kg · K), where T is in degrees Kelvin.

3.8. A gaseous mixture from a wood gasifier contains 15% hydrogen, 4% methane, 25% carbon monoxide, 10% carbon dioxide, and 46% nitrogen by volume. What is the higher heating value in kJ/kg mixture and kJ/m^3 mixture?

3.9. For the mixture of Problem 3.8, find the specific heat on a mass basis and on a molar basis. The mixture is at 25°C. Does the specific heat of this mixture change with pressure?

3.10. Calculate the stoichiometric fuel-to-air ratio on a volume basis and weight basis for natural gas, bituminous coal, and refuse-derived fuel, given the following ultimate analysis (dry, ash-free):

Element	Natural gas	Bituminous coal	RDF
Carbon	75	80	52
Hydrogen	25	9	8
Oxygen	0	11	40

3.11. Natural gas can be simulated in the laboratory by a mixture of 83.4% by volume methane, 8.6% propane, and 8.0% nitrogen. For this mixture find the stoichiometric air/fuel weight ratio, and the molecular weight of the combustion products at 25% excess air.

3.12. Using the heat of formation of water vapor, calculate the lower heating value of H_2 in units of kJ/kg and Btu/lb$_m$.

3.13. Using the heat of formation of methane, carbon dioxide, and water vapor at 298 K, calculate the lower heating value of methane.

3.14. A fuel with a composition of $CH_{1.8}$ has a higher heating value of 44,000 kJ/kg. Find the lower heating value and the heat of formation of this fuel.

3.15. A fuel oil is 87% carbon and 13% hydrogen by weight. Find the difference between the higher and lower heating values.

3.16. A stoichiometric mixture of H_2 and O_2 reacts at 25 atm and 2500 K. Assume that the products are in equilibrium according to the reaction $H_2O = H_2 + \frac{1}{2}O_2$. (*a*) Calculate the mole fractions and mass fractions of the products using the equilibrium constant. (*b*) Use STANJAN with and without dissociation and compare the results.

3.17. Find the mole fraction and molar concentration (gmol/cm^3) of atomic oxygen due to dissociation of O_2 at 1500 K and 2500 K for pressures of 1 atm and 25 atm. The reaction is $O_2 = O + O$.

3.18. Repeat Problem 3.17 for H_2. The reaction is $H_2 = H + H$.

3.19. Repeat Problem 3.17 for N_2. The reaction is $N_2 = N + N$.

3.20. Carbon monoxide can react with water to form carbon dioxide and hydrogen if the temperature is high enough. Find the equilibrium composition at 1000 K and 2000 K, and at 1 atm, for a carbon to hydrogen atom ratio of 1 and a carbon to oxygen atom ratio of 0.5. The reaction is $CO + H_2O = CO_2 + H_2$. The use of STANJAN is suggested.

3.21. One mole of carbon dioxide reacts with one mole of solid carbon to form carbon monoxide. Find the equilibrium composition at 1000 K and 2000 K and 1 atm. The reaction is $CO_2 + C = 2CO$. The use of STANJAN is suggested.

3.22. Solid carbon reacts with steam to produce carbon monoxide and hydrogen. For a carbon to hydrogen atom ratio of 1/1 and a carbon to oxygen ratio of 1/1, find the equilibrium composition at 1500 K and at 1 atm. The reaction is $C(s) + H_2O = CO + H_2$. The use of STANJAN is suggested.

3.23. Calculate the adiabatic flame temperature for a stoichiometric hydrogen-air flame at 1 atm. The reactants are initially at 25°C. (*a*) Neglect dissociation and calculate using Appendix C, (*b*) Include dissociation and use STANJAN. (*c*) Repeat the calculation for a constant-volume reaction using STANJAN.

3.24. Calculate the adiabatic flame temperature of methane and air burning with 10% excess air at 1 atm pressure. The reactants enter at 500 K. (*a*) Neglect dissociation. (*b*) Repeat using STANJAN with and without dissociation.

3.25. Repeat Example 3.12 but use 25% excess air and assume that the air enters at 800°R. The pressure is 1 atm. Find the adiabatic flame temperature (*a*) neglecting dissociation and (*b*) including dissociation. Use STANJAN.

3.26. Use an equilibrium program such as STANJAN to verify Figure 3.3. Repeat several runs at 10 atm pressure, and comment on the effect of pressure on the concentrations of CO and NO and on flame temperature.

3.27. Assume you use a computer program such as PER that calculates the equilibrium composition of a mixture and also gives the quantities R, $\partial R/\partial T$, $\partial R/\partial p$, u, $\partial u/\partial p$, $\partial u/\partial T$. Consider the first law for a reacting mixture in a closed adiabatic system with only mechanical work. The energy equation is

$$m\frac{du}{dt} = -p\frac{dV}{dt}$$

and

$$pV = mRT$$

where

$$u = u(p, T)$$

$$R = R(p, T)$$

Obtain an expression for dT/dt in terms of V, dV/dt, and the output quantities of the computer program. *Hint:* Differentiate the ideal gas equation with respect to time t. Then substitute into the energy (first law) equation to eliminate dp/dt.

3.28. Using a computer program for the products property calculation, calculate the constant-pressure flame temperature for a methane-air mixture at 1 atm and 25°C, for $F = 0.8, 0.98, 1.02, 1.04, 1.06, 1.1$, and 1.2. Make an accurate plot of T versus F. Also plot $\partial(h_r - h_p)/\partial F$ and note its root corresponds to the maximum flame temperature.

3.29. In some burners the mixture may be stratified so that products at different fuel-air ratios are combined to form the final uniform equilibrium products. Consider two methane-air reactant systems each of 0.5 lb_m and each at 537°R and 1 atm. One system has an F of 0.8 and the other an F of 1.2. They are each burned adiabatically at constant pressure and then the products are mixed adiabatically at constant pressure. Compute the temperature of the mixed products. *Hint:* What equivalent homogeneous 1.0 lb_m mixture at 537°R and 1 atm would give the same outlet temperature?

3.30. To obtain a qualitative idea of the increase in adiabatic flame temperature caused by a rise in initial reactant temperature, neglect dissociation and assume that the reactants and products each have constant (but unequal) specific heats. Show that

$$\frac{T_{af} - T_0}{T_{af0} - T_0} = C\frac{T_r - T_0}{T_{af0} - T_0} + 1$$

where T_{af} = adiabatic flame temperature for a reactant temperature T_r
T_{af0} = adiabatic flame temperature for a reactant temperature T_0
C = ratio of reactants to products specific heats

Note that dissociation causes $\partial h/\partial T$ at a given T, p, and F to increase over its nondissociated value at the same T, p, F. Both the increase in product species specific heats with temperature and the increase in dissociation with temperature cause the rate of change $\partial T_{af}/\partial T_r$ to decrease as T_r is increased. This effect is partially offset by the increase in reactants specific heat with temperature.

3.31. Assume constant specific heats for the reactants and products, and further assume that the specific heats are equal. Neglect the change in number of moles caused by the reaction. Using these assumptions, take a case where the reactants are at 300 K and 1 atm and react adiabatically to produce products at 2300 K and 1 atm.

If the same reactants at 300 K and 1 atm react adiabatically at constant volume, what will be the products temperature? Assume $c_p/c_v = 1.35$. What will be the products pressure? Call this pressure p_2. Now calculate the isentropic compression temperature for the reactants compressed from 1 atm and 300 K to the pressure, p_2, and for the constant-pressure products at 1 atm and 2300 K compressed to p_2.

If the compressed reactants at p_2 are burned adiabatically at constant pressure p_2, what will be the products temperature? Now compare the two final temperatures, one for the case of constant-pressure combustion followed by the compression of the products, and the other for the compression of the reactants followed by constant-pressure combustion. What factors would tend to reduce the temperature gradient in the products in an actual constant-volume combustion?

3.32. Taking $c_{vu} = c_{v0} + b(T - T_0)$, find an expression relating m_u to T_u by integrating Eq. 3.74. Recall that $\gamma - 1 = R/c_v$ and express your result in terms of the ratio of specific heats, $c_p/c_v = \gamma$.

3.33. Write Eq. 3.67 for the specific case of ethylene (C_2H_4) gas. Then group the product terms and obtain specific expressions for $A(T)$ and $B(T)$. Find the values of $\partial H/\partial F$ at 500 K and 1000 K using the tables. Would the above expression also hold for 1-butene (C_4H_8)? Explain.

3.34. Find the equation of the reactant enthalpy at 500 K for f kg methane and 1 kg of air. The equation is linear.

3.35. Calculate the chemical availability of liquid methanol at 298 K and 1 atm, and compare the HHV and LHV for liquid methanol. The entropy of liquid methanol at the reference temperature is 112.5 kJ/kgmol.

3.36. Methanol vapor can be reformed by use of a heated catalyst to form H_2 and CO. Calculate the LHV and the chemical availability of the reformed methanol gas mixture. Compare these results and those of problem 3.35.

3.37. Rework Example 3.14 using Eq. 3.82 and values of the Gibbs free energy directly from App. C.

REFERENCES

Edo, T., and Foster, D., "A Computer Simulation of a Dissociated Methanol Engine," *VIth Int. Symp. Alcohol Fuels Technol.,* vol. 1, pp. 193–199, SAE, 1984.

Flagan, R. C., and Seinfeld, J. H., *Fundamentals of Air Pollution Engineering,* Prentice-Hall, Englewood Cliffs, NJ, 1988.

Flynn, P. F., Hoag, K. L., Kamel, M. M., and Primus, R. J., "A New Perspective on Diesel Engine Evaluation Based on Second Law Analysis," *SAE Trans.,* vol. 93, paper no. 840032, 1984.

Gordon, S., and McBride, B. J., "Computer Program for Calculation of Complex Chemical Equilibrium Compositions and Applications," NASA RP-1311, Part I, Analysis, 1994; Part II, Users Manual and Program Description, 1996.

Lide, D. R. (ed.), *CRC Handbook of Thermochemical Data,* CRC Press, Boca Raton FL, 1994. (Includes a 5.25-in. diskette.)

Moran, M. J., *Availability Analysis,* Prentice-Hall, Englewood Cliffs, NJ, 1982.

Myers, G. E., *Engineering Thermodynamics,* Prentice-Hall, Englewood Cliffs, NJ, 1989.

Myers, G. E., "Equilibrium via Element Potentials and Computers," ASME paper, WAM, 1992.

Myers, P., Myers, J., and Myers, M., "On the Computation of Emissions from Exhaust Gas Composition Measurements," *J. Engr. Gas Turbines and Power,* vol. 3, pp. 410–423, 1989.

Olikara, C., and Borman, G., "A Computer Program for Calculating Properties of Equilibrium Combustion Products with Some Applications to I.C. Engines," SAE paper no. 750468, 1975. The PER program can be downloaded from www.engr.wisc.edu/centers/erc/per.

Powell, H. N., "Applications of an Enthalpy-Fuel/Air Ratio Diagram to 'First Law' Combustion Problems," *Trans. ASME,* pp. 1129–1138, July 1957.

Powell, H. N., and Sarner, S. F., "The Use of Element Potentials in Analysis of Chemical Equilibrium," vol. 1, General Electric Co. report R59/FPD, 1959.

Reynolds, W. C., "The Element Potential Method for Chemical Equilibrium Analysis: Implementation of the Interactive Program STANJAN," Mechanical Engineering Dept., Stanford University, 1986. Instructors may contact Prof. Reynolds (wcr@thermo.stanford.edu) for a copy of the program.

Spindt. R. S., "Air-Fuel Ratios from exhaust Gas Analysis," SAE paper no. 650507, 1965.

Stull, D. R., and Prophet, H., JANAF Thermochemical Tables, NSRDS-NB537, National Bureau of Standards, Washington, DC, 1971.

Stultz, S. C., and Kitto, J. B. (eds.), *Steam: Its Generation and Use,* Chap. 9, Babcock and Wilcox Co., Barberton, OH, 40th ed., 1992.

Svehla, R. A., and McBride, B. J., "Fortran IV Computer Program for Calculation of Thermodynamic and Transport Properties of Complex Chemical Systems," NASA TN D-7056, 1973.

Wark, K., *Thermodynamics,* McGraw-Hill, New York, 3rd ed., 1977.

Zeleznik, F. J., and Gordon, S., "Calculation of Complex Equilibria," *Ind. Eng. Chem.,* vol. 60, pp. 27–57, 1968.

CHAPTER 4

Chemical Kinetics of Combustion

As explained in Chapter 3, the equilibrium composition of a reactive mixture can be predicted by the application of thermodynamics. Thermodynamics describes the potential for reaction, but it does not give the rate at which the reaction proceeds or tell how quickly equilibrium is approached when the temperature is changed. Chemical kinetics along with thermodynamics is needed to predict the reaction rates of well-mixed systems. If the system is not well mixed, mass transfer effects also come into play, as for example in nonpremixed flames and solid fuel combustion.

In this chapter the rudimentary concepts of elementary reactions, chain reactions, global reactions, surface reactions, and two important applications of chemical kinetics, the formation of nitrogen oxide and soot, are considered.

4.1 ELEMENTARY REACTIONS

Chemical reaction may occur when two or more molecules or atoms approach one another. The type of reaction which may take place depends on the intermolecular potential forces existing during a collisional encounter, the quantum states of the molecules, and the transfer of energy. A reaction which takes place by such a collisional process is called an *elementary reaction* to distinguish it from overall or *global reactions,* which are the end result of many elementary reactions. For example, we might write the global reaction

$$CO + \tfrac{1}{2}O_2 \rightarrow CO_2$$

However, the actual reaction mechanism takes place by virtue of concurrent elementary reactions. In the absence of water vapor in the mixture, the elementary reactions are

$$CO + O_2 \rightarrow CO_2 + O$$

and

$$CO + O + M \rightarrow CO_2 + M$$

Furthermore, additional oxygen atoms can be formed from dissociation of O_2:

$$O_2 + M \rightarrow O + O + M$$

where M is a third body such as N_2 or O_2.

When hydrogen is present in the mixture, as is always the case with combustion of hydrocarbon fuels, then the most important elementary reactions are

$$CO + OH \rightarrow CO_2 + H$$

$$CO + HO_2 \rightarrow CO_2 + OH$$

These reactions greatly accelerate the oxidation of CO, and detailed modeling requires another 20 elementary reactions involving H, OH, HO_2, H_2O_2, O, H_2O, H_2, and O_2 molecules, as shown in Table 4.1. When hydrocarbon radicals are present, additional reactions may involve CO.

The rate of reaction for each elementary reaction stems from individual encounters between molecules and does not depend on the mixture environment. This fact allows application of elementary reaction rates determined under idealized laboratory conditions, such as low pressure and presence of only reactant species, to complex situations where the pressure may be high and many other species may be present. In contrast, rate data obtained for a global reaction normally cannot be applied outside the range of the experimental conditions. From the viewpoint of the physical chemist an elementary reaction has the advantage of being amenable to evaluation by theoretical modeling; however, for engineering purposes experimental data for elementary reaction rates are almost always preferred.

TABLE 4.1
Elementary reactions involving hydrogen and oxygen

1. $H + O_2 \rightarrow O + OH$	11. $HO_2 + H \rightarrow OH + OH$
2. $O + H_2 \rightarrow H + OH$	12. $HO_2 + H \rightarrow H_2O + O$
3. $H_2 + OH \rightarrow H_2O + H$	13. $HO_2 + OH \rightarrow H_2O + O_2$
4. $O + H_2O \rightarrow OH + OH$	14. $HO_2 + O \rightarrow O_2 + OH$
5. $H + H + M \rightarrow H_2 + M$	15. $HO_2 + HO_2 \rightarrow H_2O_2 + O_2$
6. $O + O + M \rightarrow O_2 + M$	16. $H_2O_2 + OH \rightarrow H_2O + HO_2$
7. $O + H + M \rightarrow OH + M$	17. $H_2O_2 + H \rightarrow H_2O + OH$
8. $H + OH + M \rightarrow H_2O + M$	18. $H_2O_2 + H \rightarrow HO_2 + H_2$
9. $H + O_2 + M \rightarrow HO_2 + M$	19. $H_2O_2 + M \rightarrow OH + OH + M$
10. $HO_2 + H \rightarrow H_2 + O_2$	20. $O + OH + M \rightarrow HO_2 + M$

Source: Reprinted from *Prog. Energy Comb. Sci.*, vol. 10, Westbrook and Dryer, "Chemical Kinetic Modeling of Hydrocarbon Combustion," pp. 1–57, © 1984 with kind permission from Elsevier Science Ltd., The Boulevard, Lanford Lane, Kidlington OX516B, UK.

Through consideration of molecular processes it is possible to identify three major types of elementary reactions:

1. Bimolecular atom exchange reactions:

$$AB + C \xrightarrow{1} BC + A$$

2. Termolecular recombination reactions:

$$A + B + M \xrightarrow{2} AB + M$$

3. Bimolecular decomposition reactions:

$$AB + M \xrightarrow{3} A + B + M$$

The third body (M) is necessary to conserve both momentum and energy during the collisional process in the reactions of types 2 and 3. In decomposition the third body provides the energy needed to split the molecule, while in recombination the third body takes away the surplus energy. Three-body collisions are much less likely than binary collisions, and thus such reactions are often relatively slow.

In order to further understand the nature of these reactions, it would be necessary to examine the quantum mechanical processes which give rise to them. From a simple viewpoint, however, we may say that in each case the rate of reaction is proportional to the collision frequency, which in turn is proportional to the product of the reactant concentrations. Only a small fraction of the molecular collisions result in reaction. Those collisions that are energetic enough and in which the molecular orientation is favorable break a chemical bond. Molecular collision theory is contained in the kinetic theory of gases [Benson].

For an elementary reaction, the reaction rate depends on the reaction rate constant times the concentration of each of the reactants. For elementary reactions 1, 2, and 3 above, the rates of reaction are expressed respectively as

$$-\frac{d[AB]}{dt} = -\frac{d[C]}{dt} = \frac{d[BC]}{dt} = \frac{d[A]}{dt} = k_1[AB][C] \tag{4.1}$$

$$-\frac{d[A]}{dt} = -\frac{d[B]}{dt} = \frac{d[AB]}{dt} = k_2[A][B][M] \tag{4.2}$$

$$-\frac{d[AB]}{dt} = \frac{d[A]}{dt} = \frac{d[B]}{dt} = k_3[AB][M] \tag{4.3}$$

where the factors k_1, k_2, and k_3 are the reaction rate constants for the respective reactions. Note that rates are negative for species that are being consumed in the forward reaction. In this chapter we shall write the molar concentration n as [ ]. For example, $n_{AB} = [AB]$. Equations 4.1 to 4.3 give the concentration change due to chemical reactions only. The concentration can also change by system volume change and by addition of species due to mass flow into or out of the system.

It is observed from both kinetic theory of gases and experiments that the rate constant for an elementary reaction is an exponential function of temperature and is of the so-called Arrhenius form,

$$k = k_0 e^{-E/\hat{R}T} \tag{4.4}$$

where k_0 is the preexponential factor and E is the activation energy for the reaction. The activation energy is energy required to bring the reactants to a reactive state referred to as an *activated complex,* such that the chemical bonds can be rearranged to form products. Both E and k_0 have been determined from experiment for most elementary reactions which occur in combustion [Baulch].

For a reaction of the generic form

$$a\text{A} + b\text{B} \rightarrow c\text{C} + d\text{D}$$

where a, b, c, and d are stoichiometric coefficients, the rate of destruction of A and B and the rate of formation of C and D are given by:

$$\frac{d[\text{A}]}{dt} = -ak_f[\text{A}]^a[\text{B}]^b \tag{4.5a}$$

$$\frac{d[\text{B}]}{dt} = -bk_f[\text{A}]^a[\text{B}]^b \tag{4.5b}$$

$$\frac{d[\text{C}]}{dt} = ck_f[\text{A}]^a[\text{B}]^b \tag{4.5c}$$

$$\frac{d[\text{D}]}{dt} = dk_f[\text{A}]^a[\text{B}]^b \tag{4.5d}$$

Elementary reactions are reversible, so that the back reaction is

$$c\text{C} + d\text{D} \rightarrow a\text{A} + b\text{B}$$

For example, the back reaction rate of A is

$$\frac{d[\text{A}]}{dt} = ak_b[\text{C}]^c[\text{D}]^d \tag{4.6}$$

Combining the forward and backward reactions, the net reaction rate of A is

$$\frac{d[\text{A}]}{dt} = a(k_b[\text{C}]^c[\text{D}]^d - k_f[\text{A}]^a[\text{B}]^b) \tag{4.7}$$

At equilibrium, $d[\text{A}]/dt = 0$, and

$$\frac{k_f}{k_b} = \frac{[\text{C}]^c[\text{D}]^d}{[\text{A}]^a[\text{B}]^b} \tag{4.8}$$

Comparing Eq. 4.8 and Eq. 3.53, it follows that the ratio of the forward to backward kinetic rate constants equals the thermodynamic equilibrium constant based on concentrations, K_c:

$$\frac{k_f}{k_b} = K_c \tag{4.9}$$

where
$$K_c = K_p(\hat{R}T)^{a+b-c-d} \tag{4.10}$$

Thus, if only one of the rate constants is known, Eq. 4.9 can be used to obtain the other rate constant.

EXAMPLE 4.1. A closed chamber initially contains 1000 ppm of CO, 3% O_2, and the remainder N_2 at 1500 K and 1 atm pressure. Determine the time for 90% of the CO to react assuming only the elementary reaction

$$CO + O_2 \rightarrow CO_2 + O$$

with $k = 2.5 \times 10^6 \exp(-24{,}060/T)$ $(\text{gmol}^{-1} \cdot \text{m}^{-3} \cdot \text{s}^{-1})$.

Solution. Initially,

$$[CO] = n_{CO} = \frac{p_{CO}}{\hat{R}T}$$

Evaluating,

$$n_{CO} = \frac{(1000 \times 10^{-6})(101.3 \text{ kPa})}{\left(8.314 \dfrac{\text{kJ}}{\text{kgmol} \cdot \text{K}}\right)(1500 \text{ K})}$$

$$= 8.12 \times 10^{-6} \text{ kgmol/m}^3 = 0.00812 \text{ gmol/m}^3$$

and
$$[O_2] = n_{O_2} = \left(\frac{0.03}{1000 \times 10^{-6}}\right) n_{CO} = 0.244 \text{ gmol/m}^3$$

Thus,

$$k = 2.5 \times 10^6 \exp\left(\frac{-24{,}060}{1500}\right) = 0.270 \text{ gmol}^{-1} \cdot \text{m}^3 \cdot \text{s}^{-1}$$

From Eq. 4.5 the rate of destruction of CO is

$$\frac{d[CO]}{dt} = -k[CO][O_2]$$

In this case the O_2 is essentially constant, and integration yields,

$$\ln \frac{[CO]}{[CO]_i} = -k[O_2]t$$

or
$$t = \frac{-\ln(0.1)}{k[O_2]} = \frac{2.3}{(0.270 \text{ gmol}^{-1} \cdot \text{m}^3 \cdot \text{s}^{-1})(0.245 \text{ gmol} \cdot \text{m}^{-3})} = 35 \text{ s}$$

This is a very long reaction time; however, as noted above, water vapor and various radicals greatly speed up the oxidation rate of CO in combustion reactions so that the time to oxidize CO due to multiple elementary reactions is much shorter.

Most practical combustion applications involve fluid flow and heat transfer in addition to combustion. If the elementary reactions which apply to a given situation can be identified, and if valid rate constants can be obtained, then the problem can be analyzed by solving the kinetic rate equations simultaneously with the equations of mass, momentum, and energy conservation. The number of practical problems that have been solved following this route has been limited to date by both lack of kinetic data and the numerical difficulties of solving the equations. Work is continuing on both fronts, and significant progress has been made in solving large sets of chemical kinetic equations.

4.2 CHAIN REACTIONS

Many gas-phase reactions are initiated by the formation, at very low concentration, of an extremely reactive species which sets off a series of reactions leading to the formation of products. Such a process is referred to as a chain reaction, and typically it occurs after a short induction period to allow the formation of the reactive species. The chains are the way the radicals and atoms shuffle through the set of reactions. Although difficult to apply in detail to complex systems, the concepts strictly apply to all systems and are helpful in thinking about the mechanisms of reaction sets.

In *initiating reactions,* radicals are formed from stable species. Radicals are molecules with an unpaired electron such as O, OH, N, CH_3, or, in general, R•. For example,

$$CH_4 + O_2 \rightarrow CH_3 + HO_2$$

or

$$NO + M \rightarrow N + O + M$$

or generally

$$S \rightarrow R\bullet$$

where S is a stable species such as a hydrocarbon fuel or nitrogen oxide, for example. Radicals are often formed by the rupture of covalent bonds in which each fragment retains its contributing electron.

In *chain propagating* reactions, the number of radicals does not change but different radicals are produced. For example,

$$CH_4 + OH \rightarrow CH_3 + H_2O$$

or

$$NO + O \rightarrow O_2 + N$$

or generally

$$R\bullet + S \rightarrow R\bullet + S^*$$

where S* is an excited state of S or some new stable species.

In *chain branching* reactions, more radicals are produced than destroyed; for example,

$$CH_4 + O \rightarrow CH_3 + OH$$

or

$$O + H_2 \rightarrow OH + H$$

or generally

$$R\bullet + S \rightarrow \alpha R\bullet + S^* \qquad \alpha > 1$$

In *terminating reactions,* radicals are destroyed either by gas-phase reactions or by collisions with surfaces. For example,

$$H + OH + M \rightarrow H_2O + M$$

$$H + O_2 + M \rightarrow HO_2 \xrightarrow{\text{wall}} \tfrac{1}{2}H_2 + O_2$$

Reactions at the wall by which unstable species can disappear are important in determining explosion limits at low pressure.

A simple chain scheme applicable to combustion can be illustrated by the following generic set:

1. Initiation	$S \rightarrow R\bullet$
2. Branching, $\alpha > 1$	$R\bullet + S \rightarrow \alpha R\bullet + S^*$
3. Propagating, $\alpha = 1$	$R\bullet + S \rightarrow R\bullet + S^*$
4. Terminating	$R\bullet + S \rightarrow$ (a stable product)
5. Terminating	$R\bullet \xrightarrow{\text{wall}}$ destruction

Simple schemes can be found only for systems with a few species. However, the concepts of the chain formalism are often useful in sorting out the reactions of larger sets of rate equations. Combustion of only a few fuels, such as hydrogen, methane, methanol, and ethane, can be written in terms of elementary reactions at the present time. Kinetics specialists are continuing to refine these mechanisms and to build on them to obtain schemes for more complicated hydrocarbons [Frenklach et al., 1995]. As the schemes become properly evaluated, engineers can use them to help understand combustion processes, but caution is required to be sure that all of the important elementary reactions have been included. The importance of certain reactions depends on the temperature, and thus a reaction set that works well for flame kinetics may not properly predict ignition kinetics, which takes place at much lower temperatures. Similarly, three-body collision reactions become more important at higher pressures, and thus the kinetic mechanism can shift with pressure level.

Oxidation of a hydrocarbon fuel (RH) begins when an oxygen molecule of sufficient energy breaks a carbon-hydrogen bond to form radicals (hydrogen abstraction):

$$RH + O_2 \rightarrow R\bullet + HO_2 \qquad (a)$$

For example,

$$CH_3CH_3 + O_2 \rightarrow CH_3CH_2 + HO_2$$

An alternative initiation reaction is thermally induced dissociation:

$$RH + M \rightarrow R'\bullet + R''\bullet + M \qquad (b)$$

where R′• and R″• are two different hydrocarbon radicals. For example,

$$CH_3CH_2CH_3 + M \rightarrow CH_3CH_2 + CH_3 + M$$

The hydrocarbon radicals react rapidly with oxygen molecules to produce peroxy radicals:

$$R\bullet + O_2 + M \rightarrow RO_2 + M \qquad (c)$$

Peroxy radicals undergo dissociation at high temperature to form aldehydes and radicals:

$$RO_2 + M \rightarrow R'CHO + R''O \qquad (d)$$

The aldehydes may react with O_2:

$$RCHO + O_2 \rightarrow RCO + HO_2 \qquad (e)$$

which is a branching reaction since it increases the number of free radicals. In addition to O_2 and HO_2, O and OH react with RH (the parent hydrocarbon), R• (hydrocarbon fragment), or RCHO (aldehydes) and with each other.

The formation of carbon monoxide initially occurs by thermal decomposition of RCO radicals:

$$RCO + M \rightarrow R\bullet + CO + M \qquad (f)$$

Oxidation of CO to CO_2 is the last step in the combustion of organic fuels. The most important CO oxidation step is the reaction with the OH radical:

$$CO + OH \rightarrow CO_2 + H \qquad (g)$$

where the OH concentration involves all of the reactions in Table 4.1.

The initiation reaction (*a*) involves breaking a carbon-carbon bond or a carbon-hydrogen bond. The energy required for bond breakage can be estimated using the bond strengths summarized in Table 4.2. Hydrogen abstraction reactions involve breaking a carbon-hydrogen bond with a strength ranging from 364 to 465 kJ/gmol and forming HO_2, leading to a net energy of reaction of 168 to 269 kJ/gmol. Dissociation involves breaking a carbon-carbon bond, which requires 376 kJ/gmol for a single bond, 733 kJ/gmol for a double bond, and 965 kJ/gmol for a triple bond. Both reactions are endothermic, and the dissociation requires a higher enthalpy of reaction than hydrogen abstraction.

Methane oxidation kinetics have been studied extensively both experimentally and theoretically because of the widespread use of natural gas. Hunter et al. recently obtained data from a flow reactor for lean methane-air mixtures at temperatures from 930 to 1000 K and pressures of 6 to 10 atm. Concentration profiles for CH_4, CO_2, and six intermediate species were obtained. A kinetic model was developed which included 207 reactions with 40 species, and the major trends are

TABLE 4.2
Typical bond strengths at 298 K

Bond	kJ/gmol	Bond	kJ/gmol
Diatomic molecules		H—CHCH$_2$	465
H—H	436	H—C$_2$H$_5$	420
H—O	428	H—CHO	364
H—N	339	H—NH$_2$	449
C—N	754	H—OH	498
C—O	1076	H—O$_2$	196
N≡N	945	H—O$_2$H	369
N=O	631	HC≡CH	965
O=O	498	H$_2$C=CH$_2$	733
Polyatomic molecules		H$_3$C—CH$_3$	376
H—CH	422	O=CO	532
H—CH$_2$	465	O—N$_2$	167
H—CH$_3$	438	O—NO	305

Source: Reprinted with permission from Handbook of Chemistry and Physics, Lide (ed.), 74th ed., © 1993 by CRC Press, Boca Raton, FL.

shown in Figure 4.1. The primary attack on CH_4 is from OH, which produces methyl radicals (CH_3) and water. CH_3 oxidation takes place primarily by reaction with hydroperoxide (HO_2) via the reaction $CH_3 + HO_2 \rightarrow CH_3O + OH$. At atmospheric pressure other investigators have shown that methyl radical recombination and attack by oxygen atoms also play an important role. The C_2 hydrocarbon reactions appear to play a minor role for lean combustion, but they play an important role at rich conditions. Comparisons of various complex reaction schemes are

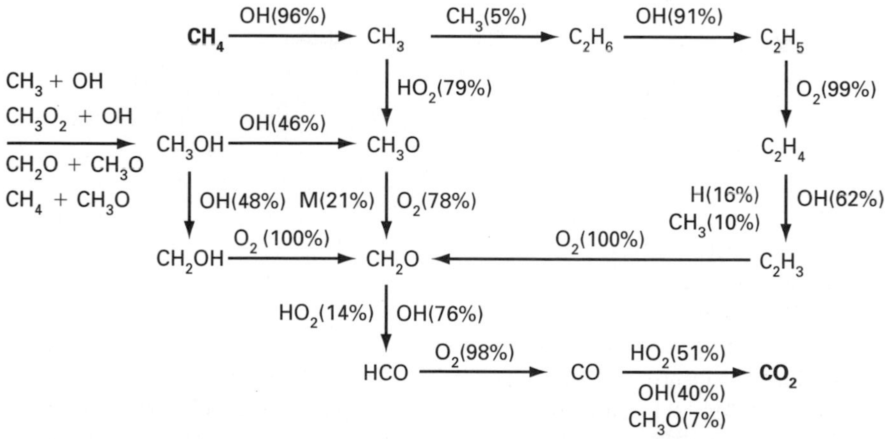

FIGURE 4.1
Reaction mechanism for methane oxidation; numbers in parentheses represent % of total species flux of the reactant through a given path [Reprinted by permission of Elsevier Science Inc. from "The Oxidation of Methane at Elevated Pressures: Experiments and Modeling," by Hunter et al., *Combustion and Flame,* vol. 97, pp. 201–224, ©1994 by The Combustion Institute].

difficult because each investigator tends to modify the reaction rates to find an optimum fit to the data. The database used typically includes data from shock-tube ignitions, flow-tube reactors, well-stirred reactors, and laminar flames.

A flow-tube reactor for obtaining chemical kinetic data is illustrated in Figure 4.2. The air is heated electrically, the fuel is mixed rapidly with the air, and the mixture flows rapidly to the lower-velocity test section. A movable sampling probe pulls and quenches a gas sample for analysis by gas chromatograph/mass spectrometer (GC/MS). Optical access is also provided. An example of data which were obtained from another flow tube reactor is shown in Figure 4.3. Note the intermediate breakdown products of ethanol which are formed and then consumed. The time is determined from the probe position, the flow rate, the major species, and the temperature.

The kinetics of propane and higher hydrocarbons is different from that of methane because ethyl radicals (C_2H_5) are produced, which oxidize more rapidly than methyl radicals. Ethyl radicals decompose rapidly to produce C_2H_4 and H atoms. The hydrogen atoms produce chain branching from $H + O_2 \rightarrow O + OH$. Then H, O, and OH accelerate the abstraction of hydrogen from propane and the

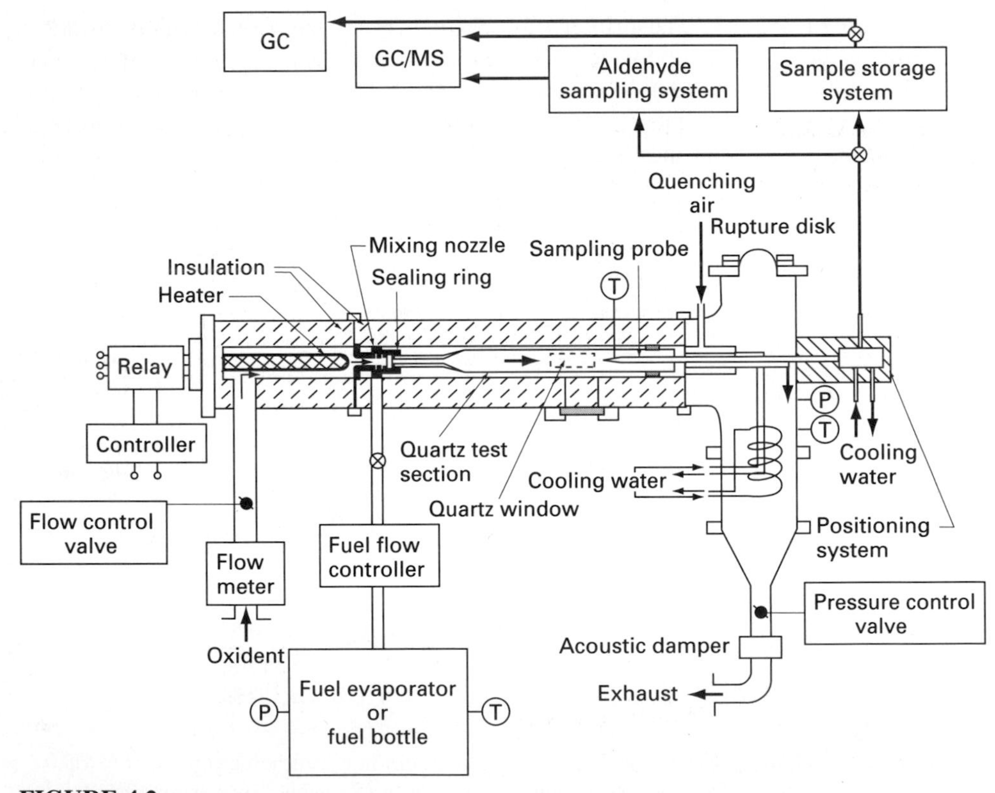

FIGURE 4.2
Flow reactor for chemical kinetic studies [Reprinted by permission of Elsevier Science Inc. from "The Oxidation of Methane at Elevated Pressures: Experiments and Modeling," by Hunter et al., *Combustion and Flame,* vol. 97, pp. 201–224, © 1994 by The Combustion Institute].

other hydrocarbons, thus producing a rapid chain reaction mechanism. A treatment of propane oxidation by Hoffman et al. includes low-temperature oxidation (500–1000 K) kinetics and extends the mechanism to include 493 reactions.

As mentioned, detailed oxidation mechanisms for specific hydrocarbons can involve several hundred reactions. As a broad simplification, it may be helpful to think of lean combustion reactions as proceeding along a primary path such as $RH \rightarrow R\bullet \rightarrow HCHO \rightarrow HCO \rightarrow CO \rightarrow CO_2$. For engineering purposes the oxidation of hydrocarbons is often treated by a few global reactions in order to expedite the analysis of practical combustion systems.

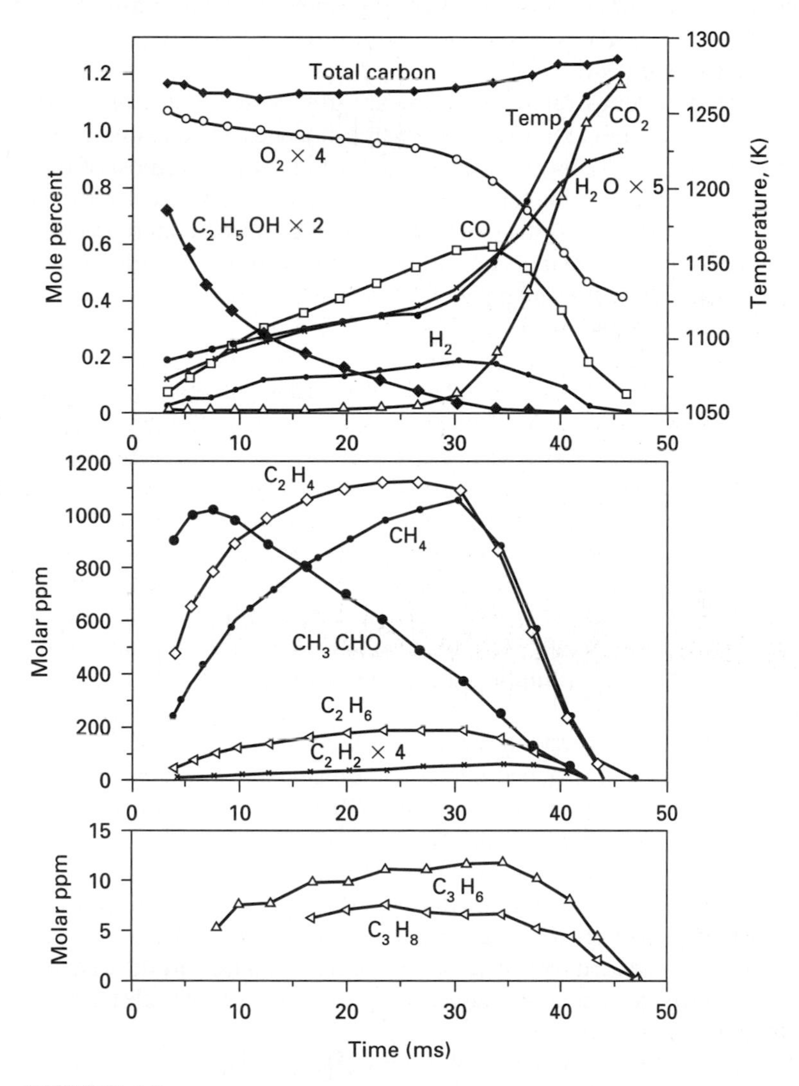

FIGURE 4.3
Flow reactor results of ethanol oxidation in air at 1 atm and $F = 0.61$ [Norton and Dryer, by permission of The Combustion Institute].

4.3 PREIGNITION KINETICS

When fuel-air mixtures are compressed either by flame propagation in a closed vessel or by a mechanical compression process, they may undergo reactions at high pressure (15–50 atm) and relatively low temperatures (< 1300 K). A similar regime can also be created by injection of fuel into air which has been rapidly compressed. In this case the mixing process, if rapid, causes the mixture to be formed at high pressure and relatively low temperature, after which it may undergo reaction. A discussion of the observed ignition and flame regimes is given in Chapter 5. This chapter is limited to an introductory discussion of the kinetics relevant to such processes.

The effects of temperature and pressure are to shift the prominent reaction pathways. If the elementary reaction scheme includes all possible reactions, the shift will be correctly followed by the computation. However, most of the schemes developed, especially for higher hydrocarbons, have been based on data at high temperatures and low pressures. Thus the schemes may be incomplete, i.e., lack reactions which become important at lower temperatures or higher pressures.

From observations under simple constant low pressure and temperature conditions, as discussed later in Chapter 5, temperature regimes of hydrocarbon oxidation have been established. At very low temperatures, below 520 K, the reaction rates are too slow to be of importance for time scales of practical interest. In the regime from 520 to 670 K, complex phenomena are observed in which an initially slow reaction regime sets the stage for a more rapid reaction. Relatively low rates of heat release are typically associated with these initially slow reactions. In the regime between 650 K and 700 K, increases in mixture temperature may actually inhibit the onset of rapid reaction, giving the so-called *negative temperature coefficient* (NTC). Parafins exhibit such NTC behavior while olefins and aromatics do not. This explains the observation that parafins reduce the octane parameter called sensitivity (see Chapter 2) while aromatics and olefins do not. Above 700 K the initially slow reactions are less important and the rapid reaction mechanism dominates.

In an attempt to understand this complex behavior from a kinetic standpoint, the reactions have been grouped into three temperature regimes: low (< 650 K), intermediate (700–1000 K), and high (> 1000 K). At high temperatures the most important chain branching is $H + O_2 \rightarrow O + OH$, and the branching agent is a hydrocarbon radical, R•. At intermediate temperatures, however, there are a number of other reactions that contribute to the radical pool; for example, radicals such as OH and HO_2 abstract hydrogen atoms from the fuel. In the case of the OH radical, water is produced by hydrogen abstraction, and this produces a heat release which can contribute to a temperature rise. The effect of HO_2 abstraction is illustrated by

$$C_nH_m + HO_2 \rightarrow C_nH_{m-1} + H_2O_2$$

and

$$H_2O_2 + M \rightarrow OH + OH + M$$

which shows how radical growth is produced by the branching agent hydrogen peroxide (H_2O_2). The alkyl radicals produced by abstraction decompose at high temperature, but in the intermediate regime they can also add O_2 as an important step, to produce a radical RO_2. The RO_2 radicals play an important role in the subsequent chain branching mechanisms. Because the O_2 addition reaction is reversible, an increase in temperature can reduce the concentration of RO_2 radicals. In addition, an alternative path by which an olefin plus HO_2 is formed comes into play. These reductions in RO_2 can, under some pressure conditions, lead to an overall reduction of reaction rate with increasing temperature as mentioned previously.

The kinetic mechanisms for the intermediate temperature regime are quite complex. For example, a scheme for 2-methylpentane involves 350 species and 2000 elementary reactions. Because of this complexity, recent work has concentrated on global mechanisms involving fewer than 20 reactions. These reduced mechanisms still require "calibrating" with experimental data, but at least provide some hope for application to computational fluid dynamics codes used to model practical combustion systems.

A reduced (four-step) mechanism for *n*-heptane ignition has been formulated by Muller and Peters, based on experimental data obtained in shock tubes. The mechanism reproduces the data reasonably well, as shown in Figure 4.4, but does not capture all of the negative temperature dependence details where an increase

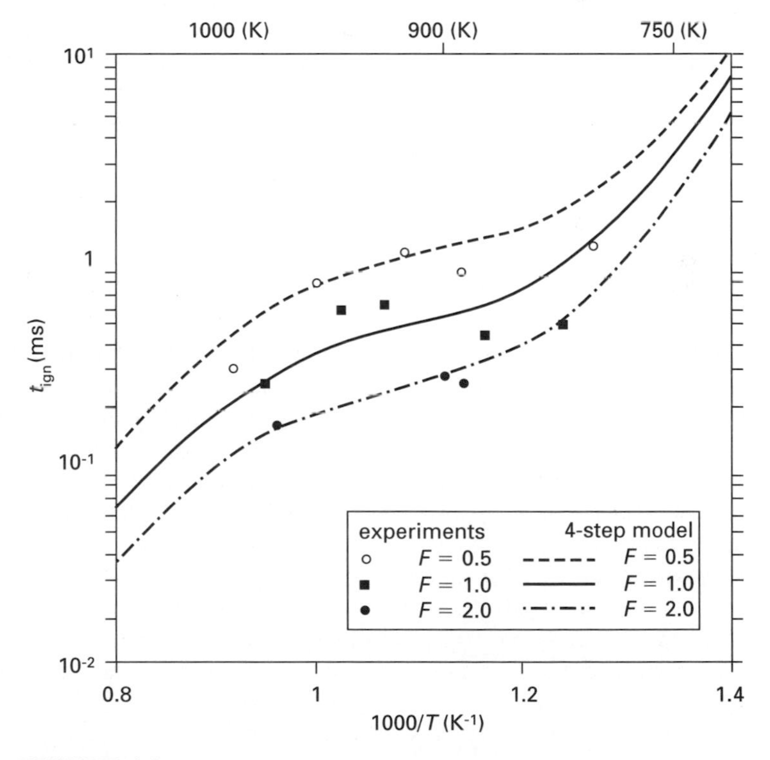

FIGURE 4.4
Comparison of calculated ignition delay times of the four-step model with experiments at 40 atm for three *n*-heptane/air equivalence ratios [Muller and Peters, by permission of The Combustion Institute].

in temperature increases the time for ignition to occur. For propane at 40 atm this region is in the temperature range of 900–1180 K, as opposed to the 650–700 K regime mentioned previously for lower pressure. The four-step scheme is given by the generic equations

$$C_7H_{16} \rightarrow 3C_2H_4 + CH_3 + H = X \tag{a}$$

$$X + 11O_2 = 7CO_2 + 8H_2O = P \tag{b}$$

$$C_7H_{16} + O_2 \rightarrow HO_2RO + H_2O = I \tag{c}$$

$$I + 9O_2 \rightarrow P \tag{d}$$

where X and I are intermediates, and P refers to the complete products. Only the third reaction is allowed to have a back reaction rate (which is not equal to the forward reaction rate). The back reaction comes into play above 830 K and decreases the importance of the third and fourth steps, which represent a degenerate branching mechanism which takes place at low temperature. Reaction steps (a) and (b) represent the high-temperature reactions where fuel decomposes into small hydrocarbons, which are then oxidized to carbon dioxide and water. Note that the oxidation of the low-temperature intermediates also goes to complete products even though we might expect that they would go to CO and then slowly go to CO_2 at low temperatures. The authors present simple Arrhenius rate equations for each of the four steps. In the negative temperature dependence range the rate of the n-heptane reaction is given by asymptotic analysis as

$$\frac{d[C_7H_{16}]}{dt} = -\left(\frac{p}{T}\right)^2 (3.75 \times 10^9)(e^{2370/T})[O_2]^2[C_7H_{16}] \tag{4.11}$$

4.4 GLOBAL REACTIONS

For hydrocarbon fuels the elementary reaction schemes are so complex that it is usually not feasible to consider all the chemically reacting species and their reaction rates when analyzing practical combustion systems. To simplify the chemistry it is useful to use an overall or global reaction scheme. The basic mechanism for lean combustion is removal of hydrogen from the fuel by OH and O to form water, intermediate hydrocarbons, hydrogen, and carbon monoxide. The hydrogen and carbon monoxide then oxidize to water and carbon dioxide. A global scheme for propane and other C_nH_{2n+2} fuels may be written as

$$C_nH_{2n+2} \rightarrow \left(\frac{n}{2}\right) C_2H_4 + H_2 \tag{a}$$

$$C_2H_4 + O_2 \rightarrow 2CO + 2H_2 \tag{b}$$

$$CO + \tfrac{1}{2} O_2 \rightarrow CO_2 \tag{c}$$

$$H_2 + \tfrac{1}{2} O_2 \rightarrow H_2O \tag{d}$$

Global rate equations in units of cm^3, gmol, s, cal, and K have been developed for each reaction species [Westbrook and Dryer]:

$$\frac{d[C_nH_{2n+2}]}{dt} = -10^{17.32}\exp\left(\frac{-49{,}600}{\hat{R}T}\right)[C_nH_{2n+2}]^{0.50}[O_2]^{1.07}[C_2H_4]^{0.40} \quad (4.12)$$

$$\frac{d[C_2H_4]}{dt} = -10^{14.70}\exp\left(\frac{-50{,}000}{\hat{R}T}\right)[C_2H_4]^{0.90}[O_2]^{1.18}[C_nH_{2n+2}]^{-0.37} \quad (4.13)$$

$$\frac{d[CO]}{dt} = -10^{14.6}\exp\left(\frac{-40{,}000}{\hat{R}T}\right)[CO]^{1.0}[O_2]^{0.25}[H_2O]^{0.50} \quad (4.14)$$

$$+ 5.0 \times 10^{8}\exp\left(\frac{-40{,}000}{\hat{R}T}\right)[CO_2]$$

$$\frac{d[H_2]}{dt} = -10^{13.52}\exp\left(\frac{-41{,}000}{\hat{R}T}\right)[H_2]^{0.85}[O_2]^{1.42}[C_2H_4]^{-0.56} \quad (4.15)$$

Use of global sets of this sort should be limited to the regimes for which they have been tested, which in this case is equivalence ratio F, 0.12 to 2; pressure, 1 to 9 atm; and temperature, 960 to 1540 K.

Even simplified schemes such as the four-reaction set given above are usually too complicated for practical use in present numerical fluid mechanical codes. Thus, single-step global rate equations have received renewed interest in recent years, such as

$$\text{Fuel} + \alpha O_2 \rightarrow \beta CO_2 + \gamma H_2O$$

where α, β, γ are stoichiometric coefficients. The associated global rate expression is then written as

$$r_{\text{fuel}} = \frac{d[\text{fuel}]}{dt} = -AT^n p^m \exp\left(-\frac{E}{\hat{R}T}\right) \cdot [\text{fuel}]^a[O_2]^b \quad (4.16)$$

where for many cases one may take $n = m = 0$ for a specified range of T and p. The global constants for Eq. 4.16 are given in Table 4.3.

The one-step reaction overestimates the heat release rate because the products are complete, while in the actual reaction CO oxidation continues after all the fuel has been oxidized. To improve this fault a two-step reaction may be used:

$$\text{Fuel} + \alpha O_2 \rightarrow \beta CO + \gamma H_2O \quad (a)$$

$$CO + \tfrac{1}{2} O_2 \rightarrow CO_2 \quad (b)$$

The reaction rate of the fuel is given by Eq. 4.16, and Table 4.3 gives the global constants. The rate expression for the CO reaction is given by Eq. 4.14. The alternative sets of data for methane and octane in Table 4.3 result from using different assumptions to fit the data.

TABLE 4.3
Global reaction constants for Eq. 4.16*

Fuel	*A* (one-step)	*A* (two-step)	E (kcal/gmol)	*a*	*b*
CH_4	1.3×10^9	2.8×10^9	48.4	−0.3	1.3
CH_4	8.3×10^6	1.5×10^7	30.0	−0.3	1.3
C_2H_6	1.1×10^{12}	1.3×10^{12}	30.0	0.1	1.65
C_3H_8	8.6×10^{11}	1.0×10^{12}	30.0	0.1	1.65
C_4H_{10}	7.4×10^{11}	8.8×10^{11}	30.0	0.15	1.6
C_5H_{12}	6.4×10^{11}	7.8×10^{11}	30.0	0.25	1.5
C_6H_{14}	5.7×10^{11}	7.0×10^{11}	30.0	0.25	1.5
C_7H_{16}	5.1×10^{11}	6.3×10^{11}	30.0	0.25	1.5
C_8H_{18}	4.6×10^{11}	5.7×10^{11}	30.0	0.25	1.5
C_8H_{18}	7.2×10^{11}	9.6×10^{12}	40.0	0.25	1.5
C_9H_{20}	4.2×10^{11}	5.2×10^{11}	30.0	0.25	1.5
$C_{10}H_{22}$	3.8×10^{11}	4.7×10^{11}	30.0	0.25	1.5
CH_3OH	3.2×10^{11}	3.7×10^{12}	30.0	0.25	1.5
C_2H_5OH	1.5×10^{12}	1.8×10^{12}	30.0	0.15	1.6
C_6H_6	2.0×10^{11}	2.4×10^{11}	30.0	−0.1	1.85
C_7H_8	1.6×10^{11}	1.9×10^{11}	30.0	−0.1	1.85

Source: Reprinted from *Prog. Energy Comb. Sci.*, vol. 10, Westbrook and Dryer, "Chemical Kinetic Modeling of Hydrocarbon Combustion," pp. 1–57, © 1984 with kind permission from Elsevier Science Ltd., The Boulevard, Lanford Lane, Kidlington OX516B, UK.

* Units are in cm^3, gmol, s, kcal, K.

EXAMPLE 4.2. For a stoichiometric propane-air reaction at 1 atm pressure and 1500 K, determine the fuel and CO_2 concentrations versus time using (*a*) a one-step global reaction and (*b*) a two-step global reaction.

Solution. The problem will be solved using the Engineering Equation Solver (EES) and the data from Table 4.3. The EES program is shown below, and the solution is plotted in Figures 4.5 and 4.6.

(a) One-step global reaction: C3H8+5(O2+3.76N2) = 3CO2+4H2O

```
          {constants and initial conditions}
p=1.0 {atm}
R= 82.05 {cm3 atm/K-gmol}
T=1500 {K}
xfi=1/24.8   {initial fuel mole fraction}
        {species and reaction rates}
n=p/(R*T)      {molar concentration, gmol/cm3}
nfi=xfi*n      {initial fuel mole concentration}
nCO2f=3*nfi    {final carbon dioxide concentration}
nO2=5*nf
nCO2=3*(nfi-nf)
rf=-8.6e+11*exp(-30,000/(1.987*T))*nf^0.1*nO2^1.65
{Integrate the reaction rate expression.}
nf=nfi+integral(rf,Time)
       {plotting parameters}
Z1=nf/nfi
Z2=nCO2/nCO2f
```

As shown in Figure 4.5 for this global one-step reaction, 90% of the propane reacts to form products in 0.8 ms at 1500 K and 1 atm.

(b) Two-step global reaction scheme:

$$C3H8+3.5(O2+3.76N2) = 3CO+4H2O$$

$$CO+0.5(O2+3.76N2) = CO2$$

```
    {constants and initial conditions}
p=1.0 {atm}
R= 82.05 {cm3 atm/K/gmol}
T=1500  {K}
n=p/(R*T)     {molar concentration, gmol/cm3}
xfi=1/24.8  {initial mole fraction of fuel}
nfi=xfi*n     {initial molar concentration of fuel}
nCO2i=0.0
nCO2f=3*nfi  {final CO2 conc.}
    {species and reaction rates; note rCO2= -rCO}
nO2=(3.5*nf)+(1.5*nfi-0.5*nCO2)
nCO=3*(nfi-nf) - nCO2
nH2O=4*(nfi-nf)
rf=-1.0e+12*exp(-30,000/(1.987*T))*nf^0.1*nO2^1.65
       {Using rCO =-rCO2 and Eq 4.14,}
```

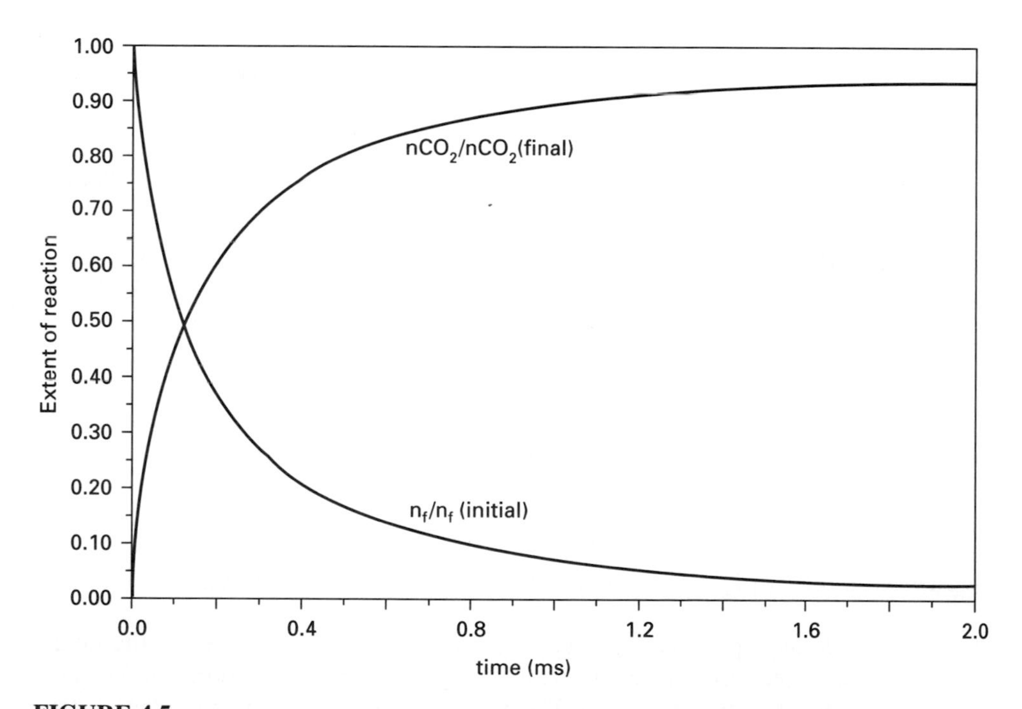

FIGURE 4.5
Calculated history of reaction using a one-step global reaction of stoichiometric propane-air reaction at 1 atm and 1500 K.

```
rCO2=10^14.6*exp(-40,000/(1.987*T))*nCO*nO2^0.25*nH2O^0.5-
   (5.0e+8)*exp(-40,000/(1.987*T))*nCO2
nf=nfi+integral(rf,time)
nCO2=nCO2i+integral(rCO2,time)
   {plotting parameters}
Z1=nf/nfi
Z2=nCO/nCO2f
Z3=nCO2/nCO2f
```

As shown in Figure 4.6 for this global two-step reaction, 90% of the propane reacts in 0.7 ms and 90% of the CO_2 is formed in 0.8 ms

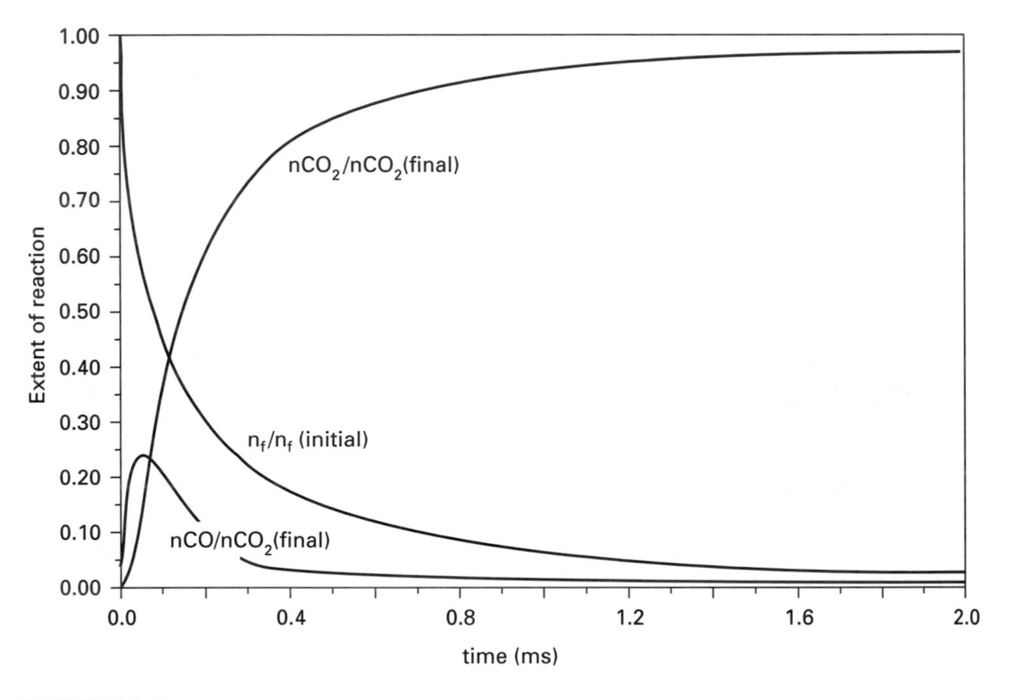

FIGURE 4.6
Calculated history of reaction using a two-step global reaction of stoichiometric propane-air reaction at 1 atm and 1500 K.

Comments on Applications of Chemical Kinetics to Modeling

Combustion in most practical systems is turbulent and highly coupled to the fluid velocity. Turbulence introduces questions concerning how the nonlinear kinetic equations should be time-smoothed. The precise model for reactive turbulent flows is still speculative. Current numerical fluid calculations generally use a one-step global kinetics model. In some computations kinetics is ignored altogether and the rate of turbulent mixing is used to determine the rate of reaction.

Use of reactive fluid models as a practical design tool is not possible until further research and evaluation is carried out. However, applications of kinetics have played an important role in understanding mechanisms for the formation of emissions. Particularly, NO_x emission control by combustion modifications has greatly benefited from kinetic modeling even though the ability to predict NO levels is limited to about a factor of 2 in gasoline engines, for example. In the case of unburned hydrocarbons, experimental evidence that wall quench layers do not contribute significantly to emissions has been backed up by numerical modeling. Kinetics have also been used to determine oxidation of hydrocarbons released from crevice volumes, as discussed further in Chapter 7. Carbon monoxide levels are often quite close to equilibrium, but the lower temperatures and slow combustion of very lean mixtures can produce higher than equilibrium levels of carbon monoxide. Kinetic models of CO oxidation can help to understand such phenomena. The kinetics of soot particles produced by rich burning is still not well understood and is complicated by heterogeneous fluid mixing effects. Understanding the mechanisms of knocking combustion in spark-ignition engines has been helped in recent years by improved kinetic models, although again the thermal phenomena such as wall heat transfer tend to obscure the understanding.

Rapid progress is being made in the ability to model hydrocarbon kinetics and also in the ability to numerically integrate the rate equations. During the last decade numerical methods have solved the problem of numerically integrating the kinetic rate equations. The kinetic equations can cause numerical difficulties because they are "stiff." That is, the radical species have much, much shorter time constants than the major species. Standard methods such as Runge-Kutta can be unstable for such stiff equations even when very small time steps are used. The use of implicit methods has solved this difficulty, and good general codes are available for solving sets of stiff ordinary differential equations [Gear; Brown]. Although the kinetic equations are first-order ordinary differential equations, the fluid mechanical equations are second-order partial differential equations. Hence, solving the full set of equations for a reacting fluid remains a challenge.

4.5 NITROGEN OXIDE KINETICS

Interest in nitrogen oxides is primarily due to their role in atmospheric pollution. Typically, combustion products contain nitrogen oxide (NO) at levels of several hundred to several thousand parts per million (ppm) and nitrogen dioxide (NO_2) levels in tens of ppm. In the atmosphere in the presence of ultraviolet sunlight, an equilibrium is established between nitrogen oxide, nitrogen dioxide, and ozone:

$$NO_2 + O_2 \overset{h\nu}{\rightleftarrows} NO + O_3$$

Addition of certain hydrocarbons slowly unbalances the above reaction; the hydrocarbons are oxidized, and reaction products such as nitrates, aldehydes, and PAN (peroxyacetyl nitrate) are formed. The NO is converted to NO_2, and as the NO is consumed, ozone begins to appear. Nitrogen dioxide forms a brownish haze. Nitrogen dioxide, ozone, and PAN are thought to have adverse health effects. Nitric acid, which forms in the atmosphere from nitrogen dioxide, contributes to acid rain. Hence, considerable effort has been aimed at understanding and controlling the formation of nitrogen oxides from mobile and stationary combustion systems, which are major sources of nitrogen oxide emissions.

It is interesting to note that the back reaction of NO + O_3 produces a photon emission. Reacting a mixture containing NO with ozone and measuring the photon production is the standard method of measuring NO concentration—the basis of the "chemiluminescent" analyzer.

The main sources of nitrogen oxide emissions in combustion are oxidation of molecular nitrogen in the postflame zone (termed *thermal* NO), formation of NO in the flame zone (*prompt* NO), and oxidation of nitrogen-containing compounds in the fuel (*fuel-bound* NO). The relative importance of these three sources of nitrogen oxide depends on the operating conditions and the type of fuel. For adiabatic combustion with excess oxygen in the postflame zone, thermal NO formation is the main source of NO emissions.

The basic mechanism for thermal NO production is given by the six reactions of the extended Zeldovich mechanism:

$$O + N_2 \underset{-1}{\overset{+1}{\rightleftarrows}} NO + N \tag{a}$$

$$N + O_2 \underset{-2}{\overset{+2}{\rightleftarrows}} NO + O \tag{b}$$

$$N + OH \underset{-3}{\overset{+3}{\rightleftarrows}} NO + H \tag{c}$$

The contribution of the third reaction pair is small for lean mixtures, but the reaction is normally included for rich combustion, where the O_2 concentration is low. The first forward reaction controls the system, but this reaction has a very high activation energy, and thus is slow at low temperature. As a result, the thermal NO is formed in the postflame products. Concentrations of 1000 to 4000 ppm are typically observed in uncontrolled combustion systems.

From reactions a, b, c with rate constants ± 1, ± 2, and ± 3, the rate of formation of NO is given by

$$\frac{d[\mathrm{NO}]}{dt} = k_{+1}[\mathrm{O}][\mathrm{N_2}] - k_{-1}[\mathrm{NO}][\mathrm{N}] + k_{+2}[\mathrm{N}][\mathrm{O_2}] - k_{-2}[\mathrm{NO}][\mathrm{O}] + k_{+3}[\mathrm{N}][\mathrm{OH}] - k_{-3}[\mathrm{NO}][\mathrm{H}] \tag{4.17}$$

In order to evaluate the rate of formation of NO from Eq. 4.17, the O, N, OH, and H concentrations must be determined. For very high-temperature applications, such as internal combustion engines, it may be safely assumed that the O, N, OH, and H remain in thermodynamic equilibrium in the postflame zone, and these values may be obtained from thermodynamic equilibrium. For moderately high temperatures, such as in furnaces, the N does not remain in thermodynamic equilibrium. However, it can be assumed that the N remains at a steady-state concentration. That is, the net rate of change of N is very small and may be set equal to zero in the following way. From reactions a, b, and c

$$\frac{d[\mathrm{N}]}{dt} = k_{+1}[\mathrm{O}][\mathrm{N}_2] - k_{-1}[\mathrm{NO}][\mathrm{N}] - k_{+2}[\mathrm{N}][\mathrm{O}_2] + k_{-2}[\mathrm{NO}][\mathrm{O}] - k_{+3}[\mathrm{N}][\mathrm{OH}] + k_{-3}[\mathrm{NO}][\mathrm{H}] = 0 \tag{4.18}$$

Solving for N at steady state,

$$[\mathrm{N}] = \frac{k_{+1}[\mathrm{O}][\mathrm{N}_2] + k_{-2}[\mathrm{NO}][\mathrm{O}] + k_{-3}[\mathrm{NO}][\mathrm{H}]}{k_{-1}[\mathrm{NO}] + k_{+2}[\mathrm{O}_2] + k_{+3}[\mathrm{OH}]} \tag{4.19}$$

Thus, the NO concentration in the postflame combustion products may be determined as a function of time by integrating Eq. 4.17. The equilibrium values of O_2, N_2, H_2O, OH, and H are obtained from a thermodynamic equilibrium program such as STANJAN. The concentration of N is obtained either from equilibrium or from Eq. 4.19. The rate constants in units of $\mathrm{cm}^3 \cdot \mathrm{mol}^{-1} \cdot \mathrm{s}^{-1}$ with the temperature in K are as follows [see Flagan and Seinfeld]:

$$k_{+1} = 1.8 \times 10^{14} \exp(-38{,}370/T)$$

$$k_{-1} = 3.8 \times 10^{13} \exp(-425/T)$$

$$k_{+2} = 1.8 \times 10^{10} T \exp(-4680/T)$$

$$k_{-2} = 3.8 \times 10^{9} T \exp(-20{,}820/T)$$

$$k_{+3} = 7.1 \times 10^{13} \exp(-450/T)$$

$$k_{-3} = 1.7 \times 10^{14} \exp(-24{,}560/T)$$

Calculations show that the rate of formation of NO is highly dependent on temperature, time, and stoichiometry. Let us consider two examples of NO formation which demonstrate this dependency: methane-air combustion products in a furnace, and octane-air combustion products related to spark-ignition engine conditions.

EXAMPLE 4.3. Methane and air initially at 298 K are burned adiabatically at 1 atm pressure and at theoretical air levels of 110, 100, 90, and 80%. Calculate and plot the NO concentration as a function of time.

Solution. First run an equilibrium program to get the adiabatic flame temperature and mole fractions of O_2, N_2, O, H_2O, OH, and H. The equilibrium results using STANJAN are:

Theoretical air		Mole fractions × 10^3					
(%)	T (K)	O_2	N_2	O	H_2O	OH	H
110	2146	16.8	717	0.245	171	2.75	0.134
100	2227	4.46	708	0.215	183	2.87	0.393
90	2204	0.257	692	0.0442	189	1.30	0.679
80	2098	0.00943	667	0.00418	186	0.358	0.576

The equilibrium species are then used to solve for NO using the extended Zeldovich mechanism. Since the O_2 and N_2 concentrations are much higher than the NO concentration, assume that the O_2 and N_2 remain constant. The problem is set up and solved using the Engineering Equation Solver (EES) program shown below, and the results are plotted in Figure 4.7.

```
{Example 4.3 THERMAL NO FORMATION}

{Constants and rate constants}
T=2098 {K}
p=1 {atm}
R=82.05 {cm3 atm/(gmol K)}
KA=1.8E14*exp(-38370/T)
KB=1.8E10*T*exp(-4680/T)
KD=3.8E13*exp(-425/T)
KE=3.8E9*T*exp(-20820/T)
KF=7.1E13*exp(-450/T)
KG=1.7E14*exp(-24560/T)
{mole fractions and moles from STANJAN}
XO2=.00943E-3
XN2=667E-3
```

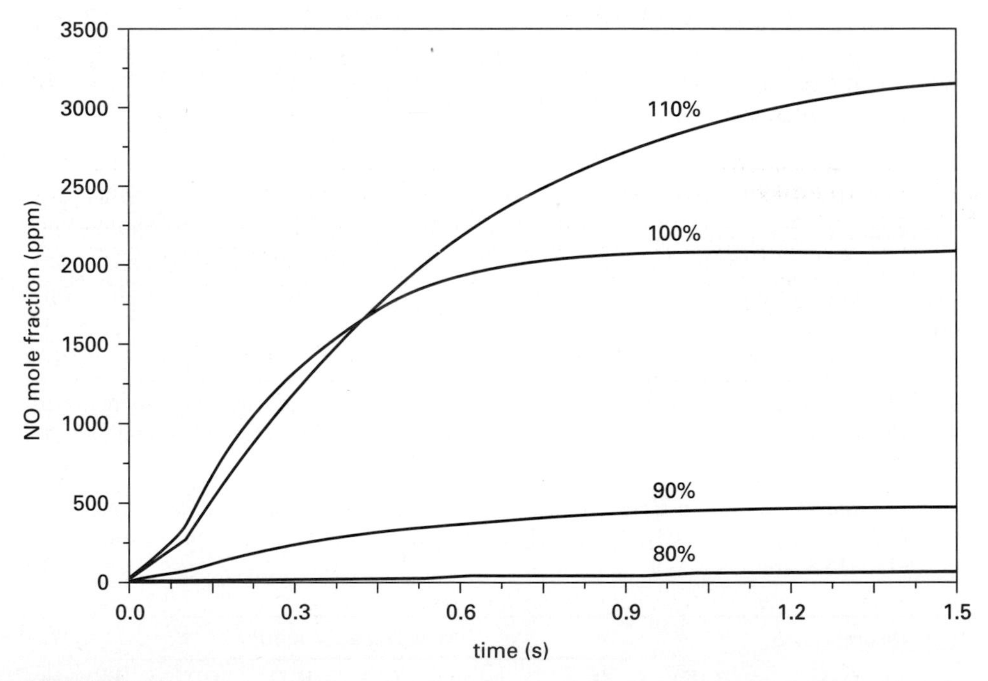

FIGURE 4.7
Formation of nitrogen oxide in adiabatic products of methane-air mixtures at 1 atm pressure and initial temperature of 300 K for 110%, 100%, 90%, and 80% theoretical air.

```
XO=0.00418E-3
XH2O=186E-3
XOH=0.358E-3
XH=0.576E-3
MOL=p/(R*T)
O2=XO2*MOL
N2=XN2*MOL
O=XO*MOL
H2O=XH2O*MOL
OH=XOH*MOL
H=XH*MOL
{Eq 4.21 for steady-state N}
N=(KA*O*N2+KE*NO*O+KG*NO*H)/(KB*O2+KD*NO+KF*OH)
{Integrate Eq 4.18 for NO vs time}
rNO=KA*O*N2+KB*N*O2-KE*NO*O-KD*NO*N+KF*N*OH-KG*NO*H
NO=integral(rNO,time)
CNO=NO/MOL*1.0E6
```

The lean mixture produces more NO than the stoichiometric case, although initially the rate of formation of NO is higher for the stoichiometric case. The fuel-rich cases have lower NO concentrations due to lack of oxygen and lower temperature. The NO levels reach equilibrium values in 0.5 to 1.5 s. At lower temperatures the approach to equilibrium is slower because the back reactions become more important. Thus NO emissions in actual systems are expected to be highly dependent on excess air.

EXAMPLE 4.4. A stoichiometric mixture of *n*-octane vapor and air is compressed polytropically starting at 1 atm and 298 K and then burned adiabatically at constant pressure. For compression ratios of 6, 10, and 14, calculate and plot the nitrogen oxide concentration in the postflame products as a function of time. Use a polytropic exponent of 1.3 to simulate heat loss during compression.

Solution. During compression, $pV^{1.3}$ = constant; and for an ideal gas of constant mass, pV/T = constant. Hence

$$\frac{p_2}{p_1} = \left(\frac{V_1}{V_2}\right)^{1.3}$$

and

$$\frac{T_2}{T_1} = \frac{p_2/p_1}{V_1/V_2}$$

Thus the initial conditions for the reaction are:

V_1/V_2	p_2 (atm)	T_2 (K)
6	10.3	510
10	20.0	594
14	30.9	658

Results from the STANJAN equilibrium program calculation of the adiabatic flame products are:

		Mole fractions × 10^3					
V_1/V_2	T (K)	O_2	N_2	O	H_2O	OH	H
6	2440	5.19	726	0.239	135	3.06	0.305
10	2506	5.13	726	0.238	124	3.19	0.293
14	2552	5.10	726	0.237	135	3.28	0.288
V_1/V_2	T (K)	0.28	726	0.288	H_2O	3.28	0.288

The results of running the NO formation program are plotted in Figure 4.8. Although the final equilibrium values of NO are only 33% higher for a compression ratio of 14 compared with 6, the initial rate of formation is much higher. For example, after 2.5 ms the NO levels are 1000, 2300, and 3450 ppm for compression ratios of 6, 10, and 14, respectively. Thus the NO emissions from a spark engine are related to the residence time of the products in the cylinder as well as the compression ratio.

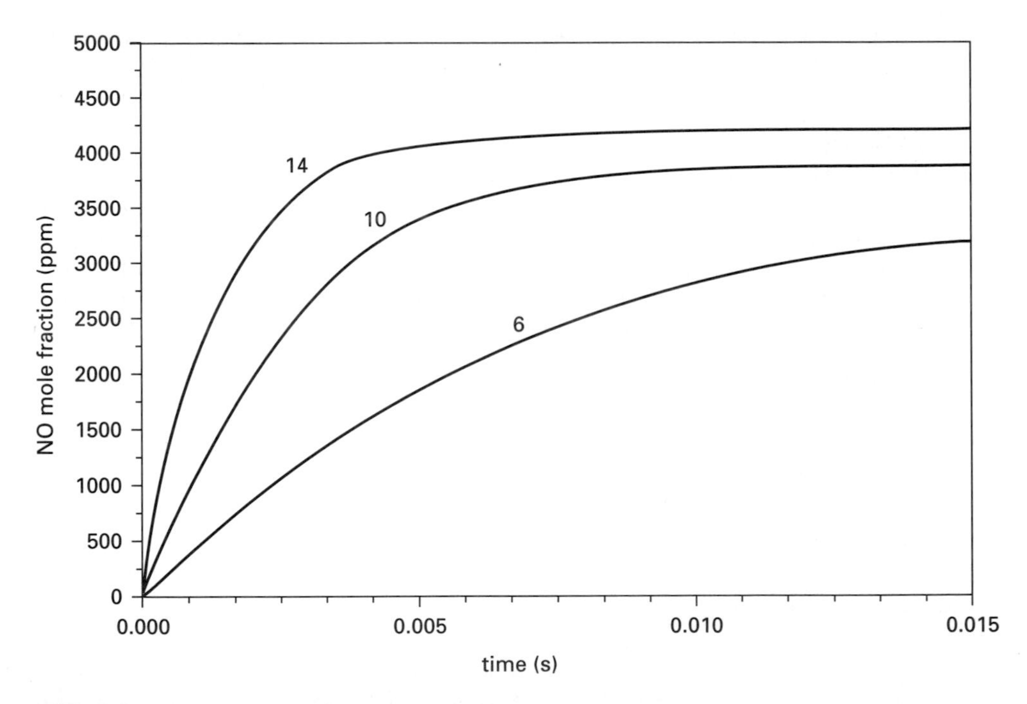

FIGURE 4.8
Formation of nitrogen oxide in products of stoichiometric octane-air initially at 1 atm and 298 K, compressed polytropically (n = 1.3) at compression ratios of 6, 10, and 14, and reacted adiabatically at constant pressure.

From Examples 4.3 and 4.4 it is evident that the way to reduce NO formation is to avoid excessively high temperatures and to reduce excess oxygen in the combustion products. Another important concept to be noted is that when combustion products are cooled rapidly by expansion or heat transfer, the NO concentration tends to freeze at a certain value because the kinetic rates slow down rapidly as the temperature drops. The concept of freezing is illustrated in the following example.

EXAMPLE 4.5. The combustion products of a stoichiometric *n*-octane–air mixture are contained in a cylinder fitted with a piston. At time $t = 0$ there is zero amount of NO and the products are at $T_0 = 2600$ K and $p_0 = 35$ atm. Then the products expand following the polytropic law,

$$\left(\frac{p}{p_0}\right)\left(\frac{V}{V_0}\right)^{1.35} = 1.0 \qquad 0 \leq t \leq t_0$$

as the volume increases with time, as

$$\frac{V}{V_0} = 1 + 4\left[1 + \sin\left(\frac{-\pi}{2} + \frac{\pi t}{t_0}\right)\right]$$

Calculate and plot the NO concentration versus time for $t_0 = 0.025$ s. Also plot the equilibrium NO for each temperature and show it on the same plot.

Solution. The solution procedure is similar to Example 4.4, but in this case the pressure and temperature are changing with time so that the equilibrium values of O and OH must be calculated for each time step and the new values used to solve the Zeldovich equations. The temperature is obtained from the ideal gas relation: $pV = mRT$, which implies that $T/T_0 = (V/V_0)(p/p_0)$. The details of the calculations are not shown here, but the results are plotted in Figures 4.9 and 4.10.

As the volume expands, the pressure and temperature drop. At first the NO is formed rapidly, but as the temperature drops, the reaction rates slow down, and the NO mole fraction freezes at a constant value of 2500 ppm. The initial theoretical equilibrium value of 4500 ppm is not attained, nor is the final equilibrium value of zero NO attained, because both the forward and backward reaction rates go to very low values and the NO reaction becomes frozen. When the expansion is faster ($t_0 = 0.01$ s), there is less time to form NO, and the mole fraction is frozen at a lower value.

Prompt NO and Fuel-Bound NO

Small amounts of NO (40–60 ppm) can be produced directly in a flame front. This rapidly formed or prompt NO is from reactions of hydrocarbon fragments with molecular nitrogen in the flame, such as

$$CH + N_2 \rightarrow HCN + N$$

$$CH_2 + N_2 \rightarrow HCN + NH$$

The species N, NH, and HCN then rapidly react to form NO in high-temperature flame regions where O and N can be present in excess of equilibrium concentrations. Prompt NO is usually small compared with thermally formed NO.

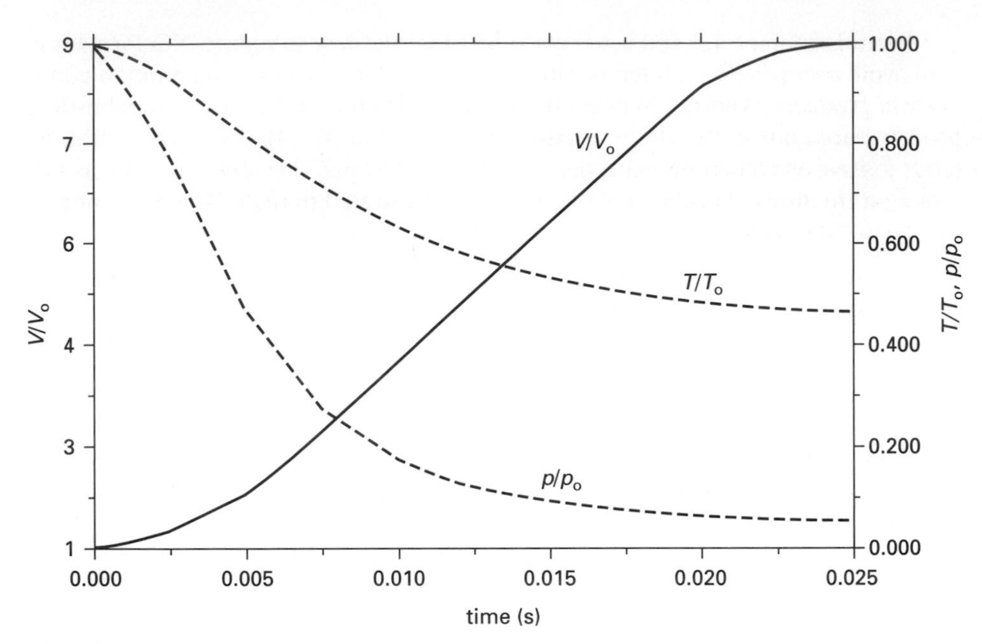

FIGURE 4.9
Decrease of pressure and temperature during the polytropic expansion used in Example 4.5.

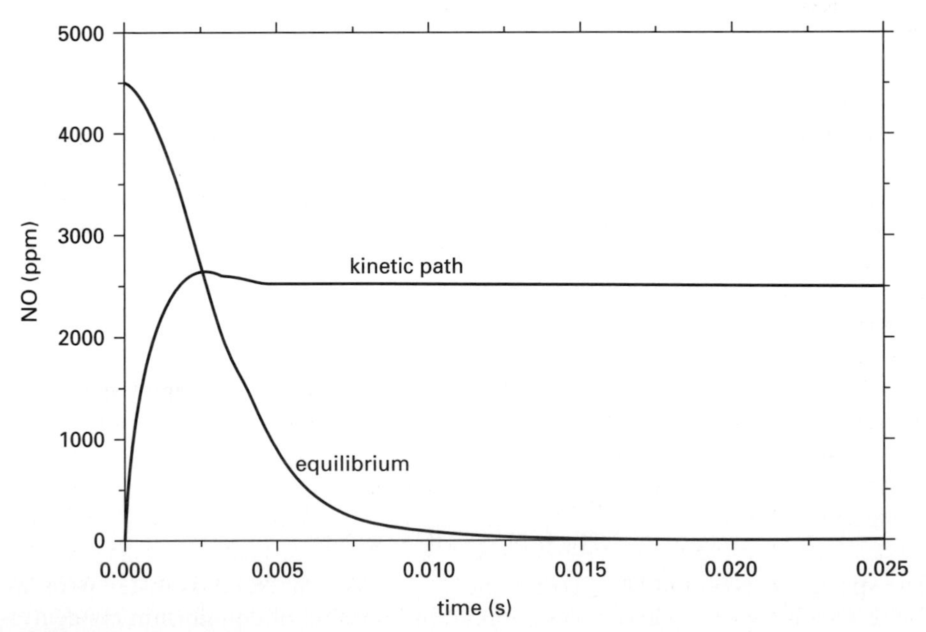

FIGURE 4.10
Nitrogen oxide formation for Example 4.5. The kinetic path is the correct path.

Fuel-bound NO is produced from organic nitrogen compounds in the fuel. Crude oil contains 0.1 to 0.2% organic nitrogen, which is concentrated in the residual fractions after refining. Coal typically contains 1.2 to 1.6% organic nitrogen. As these nitrogen-containing fuels are burned, hydrogen cyanide (HCN) is formed as an intermediate, which can further react with O, OH, and H to form NO in the postflame zone. In practical combustion systems, typically 20 to 50% of the fuel nitrogen is converted to NO.

4.6 REACTIONS AT SOLID SURFACE

A homogeneous reaction is one involving a single phase, while a heterogeneous reaction involves more than one phase. The oxidation of carbon monoxide to carbon dioxide on the surface of a platinum catalyst is a heterogeneous reaction. The oxidation of coal char is an example of a noncatalytic heterogeneous reaction. In diesel engines the oxidation of particulate carbon (soot) in the cylinder is an important step in emissions reduction.

In each case above a gas reacts on a solid surface. The reaction rate depends on the concentration of the gas phase, the temperature, diffusion rates, and the surface area accessible to the gas phase. For surface reaction rates to be significant, the accessible surface area must be large, such as the solid surface of a porous platinum-impregnated ceramic monolith or coal char. Typically the specific surface area in these examples is on the order of 100 m^2/g. Combustion of fuel sprays was not given as an example of a surface reaction because a fuel droplet does not burn at the surface, but rather the droplet vaporizes and then burns as a gas-gas reaction.

Heterogeneous reactions at a solid surface involve a series of mass transfer and chemical reaction steps: (1) the reactant gas diffuses through the external gaseous boundary layer associated with the surface, (2) the reactant gas molecules diffuse into the pores of the solid to reactive surface sites, (3) gas molecules are adsorbed on the surface and chemical reaction between gaseous and surface molecules occurs, and (4) the reaction product is desorbed from the surface. Steps 1 and 2 are external and internal mass transfer processes, while steps 3 and 4 are chemical kinetic processes.

The rate of diffusion of a gaseous species such as oxygen through the external boundary layer can be obtained from computational fluid dynamics computation or by means of mass transfer coefficients. The concentration of a gaseous species, say, oxygen, due to diffusion within the pores may also be obtained from basic computations, provided a suitable description of the pore structure is available. Once the gas reaches the interior surface, it attacks the surface if the site is active, such as oxygen reacting at a carbon edge to produce carbon monoxide. The reaction is described in terms of adsorption and desorption rate constants, which are expressed in the Arrhenius form (Eq. 4.4). If the site is inactive, the reaction does not proceed at that site.

In combustion engineering, the surface reaction rates of interest are char from coal, biomass, or heavy oil reacting with oxygen, water vapor, carbon monoxide, carbon dioxide, and hydrogen. Recall from Chapter 2 that char is what remains as a solid after pyrolysis and it contains carbon, dispersed inorganic minerals, and a small amount of hydrogen. For large fuel particles the fuel may outgas water vapor and carbon dioxide from the interior of the particle through the char surface during combustion of the char. The surface temperature may vary due to changes in convective and radiative heat transfer as well as outgassing, and exothermic and endothermic chemical reaction at the surface. Furthermore, the fraction of active sites is difficult to determine.

Char is a very porous surface, and it is useful to think of the pores as consisting of trunks and branches of trees (Figure 4.11). The porosity is established by the pyrolysis. The diameters of the branches range from angstroms to several microns. The total surface area of each tree is several orders of magnitude greater than that of the trunk. The trunks have a distribution of sizes, and the trees overlap so that there is an interconnected porosity. Oxygen and other gases diffuse into the branches and react with the surface. During combustion the porosity and pore surface area of the char may change with time and temperature. Thus, considering all the chemical and physical aspects of char reactivity, modeling the detailed processes occurring at the porous char surfaces from first principles is very difficult.

For engineering purposes, global heterogeneous reaction rates based on a global rate constant and the external solid surface area are typically used. Global heterogeneous reaction rates are obtained from laboratory experiments of single particles of specific fuels [Smoot and Smith]. As with global gas-phase reactions, the rate constants should not be extrapolated beyond the range of the experiments. Use of the external surface area rather than the internal pore surface is justified at

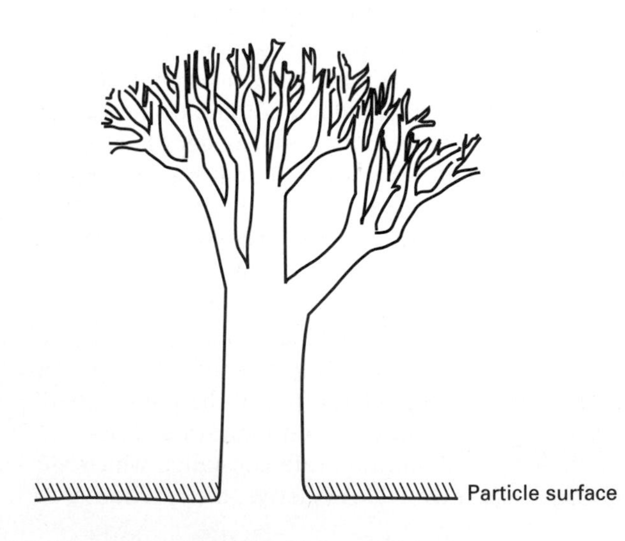

FIGURE 4.11
Schematic of char porosity based on the pore tree model showing continuous branching

high temperature because the bulk mass transfer through the external boundary layer is often the rate-limiting step. Global char reactions are discussed in Chapter 14.

4.7 SOOT KINETICS

The production of carbon particulates by rich premixed flames and by diffusion flames is typically observed as smoke. Luminosity of flames is another way to observe soot within the flame even when it does not escape the flame envelope. In such cases the soot is oxidized within the flame and especially at the flame tip.

In furnaces and boilers a certain amount of soot can be a desirable aspect of combustion as a means of increasing radiant heat transfer. However, in engines, which derive their power from expansion, soot causes additional heat loss, which is undesirable. The emission of soot particulates to the atmosphere is undesirable. In some cases the soot may contain condensed polycyclic aromatic hydrocarbons (PAHs), which have the potential to cause cancer. Thus, soot emissions from stacks and tailpipes need to be controlled.

Interestingly, soot particles formed by a variety of combustion devices and by various fuels are very similar. This leads to the assumption that the soot origins are similar in these devices. Mature soot particles contain primarily carbon and hydrogen. During the expansion process the particles may absorb additional hydrocarbons. The combustion-formed particles initially have approximately eight carbon atoms per hydrogen (99% carbon by weight, with a density of about 1.8 g/cm^3) and typically range in size from 20 to 50 nm diameter. For premixed hydrocarbon-air mixtures, equilibrium thermodynamics suggests that soot will appear for $C/O > 1$; however, experimentally soot appears at carbon-to-oxygen ratios of 0.4 to 0.6, indicating that kinetically rate-limited processes are important in the sooting process.

The soot formation process consists of four steps: inception, surface growth, oxidation, and coagulation. The coagulated particles may further form clusters of particles by agglomeration which may become as large as 1 μm. Inception is not well understood, but it is known that the fuel molecules break into smaller, lighter molecules as reactions begin. A precursor species, probably acetylene (C_2H_2), then causes the formation of larger ring structures, i.e., PAHs. Frenklach and Wang, have suggested three primary reaction classes:

1. Abstraction of a hydrogen atom from an aromatic molecule (AR) by a free hydrogen atom to form H_2 and an aromatic radical ($AR^{\bullet}$):

$$C_6H_6 + H \rightarrow C_6H_5 + H_2$$

2. Addition of an acetylene molecule to $AR^{\bullet}$:

$$AR^{\bullet} + C_2H_2 \rightarrow AR{-}C{\equiv}C{-}H + H$$

3. Cyclization to form aromatic rings:

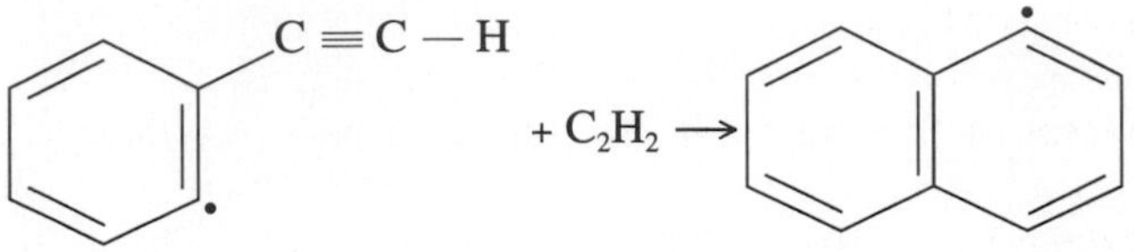

$$C \equiv C - H \quad + C_2H_2 \longrightarrow$$

This process repeats to form large PAHs. Frenklach contends that at low temperatures the growth is determined by the rate of the acetylene-addition step while at very high temperatures the radicals formed decompose faster than they grow. At normal flame temperatures the PAH growth is controlled by the super-equilibrium of hydrogen atoms. The various PAHs formed may coagulate; that is, they collide and stick together to form dimers, which in turn can form trimers and tetramers by collisions with PAHs or other dimers. The first clusters formed in this way may be called solid phase. Surface growth of these initial particles of 1 to 2 nm size then results in the soot particles. The growth rate by gaseous hydrocarbon reactions is of the order of 5×10^{-4} g/cm$^2 \cdot$s and seems to be governed by an Arrhenius-type reaction rate.

At the same time that growth occurs, the forming particles undergo oxidation from reactions with O_2, O, and OH. It is likely that in many cases of soot in flames, the OH oxidation dominates. However, the only rate equation available is that of Nagle and Strickland-Constable, which was obtained from oxidation of pyrolytic graphite in oxygen at temperatures of 1273–2273 K. Park and Appleton later found good correlations between the NS-C formula and their data obtained from carbon black in oxygen at 1700–4000 K [Neoh et al.]. None of these data were obtained at elevated pressures where the partial pressures of oxygen could be very high. The NS-C formula is given by

$$\frac{r(\text{carbon})}{12} = \frac{k_A\, p_{O_2} c}{1 + k_B\, p_{O_2}} + k_C\, p_{O_2}(1 - c) \qquad \text{g carbon/}(\text{cm}^2 \cdot \text{s}) \tag{4.20}$$

where $1/c = 1 + k_D/k_C p_{O_2}$

k_A	$= 20 \exp(-30{,}000/\hat{R}T)$	$\text{g} \cdot \text{atm}/(\text{cm}^2 \cdot \text{s} \cdot \text{atm})$
k_B	$= 21.3 \exp(4100/\hat{R}T)$	atm^{-1}
k_C	$= 4.46 \times 10^{-3} \exp(-15{,}200/\hat{R}T)$	$\text{g} \cdot \text{atm}/(\text{cm}^2 \cdot \text{s} \cdot \text{atm})$
k_D	$= 1.51 \times 10^{-5} \exp(-97{,}000/\hat{R}T)$	$\text{g} \cdot \text{atm}/(\text{cm}^2 \cdot \text{s})$
$\hat{R}$	$= 1.987$	cal/mol$\cdot$K
p_{O_2}	= partial pressure of oxygen	atm

The process of coagulation forms a larger sphere from two smaller ones, and its rate is given by

$$\frac{dn}{dt} = -kn^2 \tag{4.21}$$

where n is the particle number density and k is a rate constant which depends on pressure, temperature, and particle size. As particles grow to more than several tens of nm in diameter, they tend to agglomerate, forming chains rather than even larger spheres. How such chains oxidize is not understood. For example, they may behave differently for O_2 than for OH oxidation, because the O_2 oxidation is much slower and may allow diffusion of O_2 into the chain structure. However, agglomeration does not seem to change the oxidation rate despite the change in shape and surface area.

The kinetic models for soot are not yet complete, although modeling of simple cases shows promising results. The effects of temperature, pressure, and fuel type on the tendency to soot are thus not yet sorted out by the models, but empirical data indicate trends. Because oxygen competes with pyrolysis, premixed flames show decreasing soot with temperature, while diffusion flames show an increase. When sorting out fuels for the tendency to soot, it is important to recognize that the results are confounded by the flame temperature and that the nature of the effect is different for premixed and diffusion flames. For premixed flames the tendency to soot is in the following order:

Aromatics > Alcohols > Paraffins > Olefins > Acetylene

If the temperature is held constant by addition of diluent, the sooting tendency is correlated by the number of carbon bonds (double bonds count twice). For diffusion flames the fuel breaks down with little oxygen present and the sooting tendency is given in the order

Aromatics > Acetylene > Olefins > Paraffins > Alcohols

At the present time, the modeling of soot in practical combustion systems is greatly hampered by the ability to model the physical processes and the combustion itself. However, rapid progress is being made on understanding soot kinetics for simple systems so that we may expect soot models to be applied within this decade. Meanwhile empirical data on soot for specific systems such as diesel engines are being gathered at a furious rate due to the importance of meeting particulate emission standards. From these data it appears that at the high temperatures in diesel engines (2800 K flame temperature) the structure of the fuel is so broken down that increased aromatics do not increase particulates. This again shows the importance of temperature level when comparing sooting tendency of fuels. Fuels with lower aromatics are known to produce lower NO_x emissions, but this is probably a result of other changes such as heating value, density, H/C ratio, etc., rather than the direct effects of the aromatic structure.

4.8 SUMMARY

Chemical kinetics formulates the rate at which reactions occur in perfectly mixed systems at uniform temperature. Combustion generally proceeds by multiple elementary kinetic reactions, including chain reactions; but because of the complexities introduced by the fluid dynamics and heat transfer, and because there are so many competing elementary reactions, it is usually necessary in practice to use one- or two-step empirical global reaction rates when analyzing practical combustion systems. The exception is the Zeldovich mechanism for nitrogen oxide formation, which is well predicted by four (or sometimes six) elementary reactions in the postflame region. Surface reactions consume the char from coal and biomass. For engineering purposes a global reaction rate constant is used with the external surface area.

Soot is formed in diffusion flames and the rich zones of heterogeneous combustion. The soot particles form in four steps: inception, surface growth, oxidation, and coagulation. In practical systems much of the soot oxidizes before leaving the system, but the remaining soot plus attached hydrocarbons become a source of air pollution called particulates. These submicron-sized particles are thought to be a health hazard and their emission is regulated. Although some kinetic mechanisms for soot formation are available, practical models are phenomenological and in the development stage.

PROBLEMS

4.1. Consider the stoichiometric overall reaction for octane in air:

$$C_8H_{18} + 12.5(O_2 + 3.76N_2) \rightarrow 8CO_2 + 9H_2O + 47N_2$$

with the global reaction rate

$$r_f = -5.7 \times 10^{11} \exp(-30{,}000/\hat{R}T)[C_8H_{18}]^{0.25}[O_2]^{1.5}$$

where the units are cm, s, gmol, cal, and K. If the reactant mixture is suddenly brought to a temperature of 2000 K and 1 atm pressure, what is the initial rate of reaction? If the temperature is held constant at 2000 K and the volume is constant, what is the rate of reaction when one-half of the original fuel has been converted to products? The reaction rate has units of $\text{gmol}/(\text{cm}^3 \cdot \text{s})$.

4.2. Consider the following reactions given in Table 4.1 for H_2-O_2 combustion. Which of the reactions are initiating, and which are terminating, chain branching, and propagating?

$$H + O_2 \rightarrow O + OH \qquad (1)$$

$$H + H + M \rightarrow H_2 + M \qquad (2)$$

$$H + OH + M \rightarrow H_2O + M \qquad (3)$$

$$H_2 + OH \rightarrow H_2O + H \qquad (4)$$

$$H + O_2 + M \rightarrow HO_2 + M \qquad (5)$$

$$HO_2 + H \rightarrow H_2O + O \qquad (6)$$

$$O + H_2O \rightarrow OH + OH \qquad (7)$$

4.3. Reactions 1 and 5 of Problem 4.2 are in competition for H atoms. What is the effect of pressure on this competition?

4.4. The addition of a small amount of methane to an H_2-O_2 system may inhibit the overall rate even though the fuel adds energy and increases the temperature. Explain this effect by comparing the reaction rates of the primary branching reaction

$$H + O_2 \xrightarrow{1} OH + O$$

$$k_1 = 5.13 \times 10^{16} T^{-0.816} \exp(-8307/T)$$

and the reaction

$$H + CH_4 \xrightarrow{2} CH_3 + H_2$$

$$k_2 = 2.24 \times 10^4 T^3 \exp(-4420/T)$$

where T is in K.

4.5. The rate of reaction (2) of Problem 4.4 at 2000 K is 1.97×10^{13} gmol/cm$^3 \cdot$ s. If the activation energy is changed from 8750 to 10,000 cal/gmol, to what value must the preexponential constant be changed to keep k_2 the same at 2000 K? Compare the values of k_2 at 1000 K using these two expressions. Comment on the effects of fitting rates to high-temperature data and then extrapolating to low temperatures.

4.6. Consider the prediction of NO in the products of combustion during expansion in a piston-cylinder. Denote the rate due to reaction as $(d[NO]/dt)_R$ and explain why although $V = V(t)$,

$$\frac{d([NO]V)}{dt} = \left(\frac{d[NO]}{dt}\right)_R V$$

Find an expression for $(d[NO]/dt)$ and explain the significance of each term.

4.7. Write the first law energy balance for an adiabatic constant-mass homogeneous system undergoing chemical reactions. Assume i reaction rate equations and j species. Start by writing

$$\frac{dU}{dt} = -p\frac{dV}{dt}$$

and then express U in terms of each species in the mixture.

4.8. A mixture of gases containing 3% O_2 and 60% N_2 by volume at room temperature is suddenly heated to 2000 K at 1 atm pressure. Assume that the O_2 and N_2 remain constant. Find the initial rate of formation of NO (ppm/s). Indicate whether the NO formation rate increases or decreases as time increases. Use the Zeldovich mechanism and the information given in Section 4.5. There is no hydrogen in the mixture.

4.9. Repeat Problem 4.8 but use 1200 K.

4.10. Repeat Problem 4.8 but use 20 atm pressure.

4.11. For the conditions of Problem 4.8 find the NO concentration (ppm) versus time (s). Use an equation solver to obtain the answer.

4.12. Following the lead of Example 4.3, examine the effect of air preheated to 500 K on the NO concentration in methane-air products at 1 atm and 10% excess air.

4.13. Following the lead of Example 4.4, examine the effect of lean combustion on NO formation in *n*-octane vapor–air mixtures initially at 300 K and 1 atm, then compressed polytropically ($n = 1.3$) with $V_1/V_2 = 10$. Assume 180% theoretical air.

4.14. A lean mixture of methane and O_2 reacts at 1300 K and 1 atm to form CO and H_2O according to the global reaction

$$CH_4 + 3O_2 \rightarrow CO + 2H_2O + 1.5O_2$$

The CO reacts to form CO_2 by Eq. 4.14. Using an equation solver, calculate and plot the CH_4 and CO_2 mole fractions versus time. Use data from Table 4.3.

4.15. A pure carbon particle has a diameter of 0.01 μm and a density of 2000 kg/m^3. How many carbon atoms does the particle contain?

REFERENCES

Baulch, D. L., Cobos, Cox, Frank, Just, Kerr, Pilling, Troe, Walker, and Warnatz, "Kinetic Data for Combustion Modelling," in *Handbook of Chemistry and Physics,* D. R. Lide (ed.), 74th ed., CRC Press, Boca Raton, Florida, 1993; see also *Combustion and Flame,* vol. 98, pp. 59–79, 1994.

Benson, S. W., *The Foundations of Chemical Kinetics,* McGraw-Hill, New York, 1960.

Brown, P. N., Byrne, G. D., and Hindmarsh, A. C., "VODE, a Variable-Coefficient ODE Solver," *SIAM J. Sci. Stat. Computing,* vol. 10, pp. 1038–1051, 1989. The code is available from the netlib collection.

Cowart, J. S., Keck, J. C., Heywood, J. B., Westbrook, C. K., and Pitz, W. J., "Engine Knock Predictions Using a Fully-Detailed and a Reduced Chemical Kinetic Mechanism," *Twenty-Third Symp. (Int.) on Combustion,* pp. 1055–1062, The Combustion Institute, Pittsburgh, 1990.

Dagaut, P., Reuillon, M., and Cathonnet, M., "High Pressure Oxidation of Liquid Fuels from Low to High Temperature: *n*-Heptane and iso-Octane," *Comb. Sci. Tech.,* vol. 90, pp. 233–260, 1994.

Edelman, R. B., and Fortune, O. F., "A Quasi-Global Chemical Kinetic Model for Finite Rate Combustion of Hydrocarbon Fuels with Application to Turbulent Burning and Mixing in Hypersonic Engines and Nozzles," AIAA Paper no. 69–86, 1969.

Flagan, R. C., and Seinfeld, J. H., *Fundamentals of Air Pollution Engineering,* Chap. 2, Prentice-Hall, Englewood Cliffs, NJ, 1988.

Frenklach, M., Clary, D. W., Gardiner, W. C., and Stein, S. E., "Detailed Kinetic Modeling of Soot Formation in Shock Tube Pyrolysis of Acetylene," *Twentieth Sym. (Int.) on Combustion,* pp. 887–901, The Combustion Institute, Pittsburgh, 1984.

Frenklach, M., and Wang, H., "Detailed Modeling of Soot Particle Nucleation and Growth," *Twenty-Third Symp. (Int.) on Combustion,* pp. 1559–1566, The Combustion Institute, Pittsburgh, 1990.

Frenklach, M., Wang, H., Goldenberg, M., Smith, G. P., Golden, D. M., Bowman, C. T., Hanson, R. K., Gardiner, W. C., and Lissianski, V., "GRI-Mech—An Optimized Detailed Chemical Reaction Mechanism for Methane Combustion," Gas Research Institute topical report GRI-95/0058, 1995. This mechanism is accessible on the World Wide Web (http//www.me.berkeley.edu/gri_mech/).

Gear, C. W., *Numerical Initial Value Problems in Ordinary Differential Equations,* Prentice-Hall, Englewood Cliffs, NJ, 1971.

Glassman, I., *Combustion,* Academic Press, New York, 3rd ed., 1997.

Griffiths, J. F., "Reduced Kinetic Models and Their Application to Practical Combustion Systems," *Prog. Energy Comb. Sci.,* vol. 21, pp. 25–107, 1995.

Haynes, B. S., in Bartok, W., and Sarofim, W. S. (eds.), *Fossil Fuel Combustion: A Source Book,* Chap. 5: "Particulate Carbon Formation during Combustion," Wiley, New York, 1991.

Hoffman, J. S., Lee, W., Litzinger, T. A., Santavicca, D. A., and Pitz, W. J., "Oxidation of Propane at Elevated Pressures: Experiments and Modeling," *Comb. Sci. and Tech.,* vol. 77, pp. 95–125, 1991.

Hunter, T. B., Wang, H., Litzinger, T. A., and Frenklach, M., "The Oxidation of Methane at Elevated Pressures: Experiments and Modeling," *Comb. and Flame,* vol. 97, pp. 201–224, 1994.

Kee, R. J., Miller, J. A., and Jefferson, T. H., "CHEMKIN," Sandia National Laboratories report, SAND 80-8003, 1980.

Kuo, K. K., *Principles of Combustion,* Wiley, New York, 1986, Chap. 2, "Review of Chemical Kinetics."

Lide, D. R. (ed.), *Handbook of Physics and Chemistry,* CRC Press, Boca Raton, Florida, 74th ed., pp. 9-123–145, 1993.

Miller, J. A., and Fisk, G. A., "Combustion Chemistry," *Chemical Engineering News,* vol. 65, no. 35, pp. 22–46, 1987.

Muller, V. C., and Peters, N., "Global Kinetics for *n*-Heptane Ignition at High Pressures," *Twenty-Fourth Symp. (Int.) on Comb.,* The Combustion Institute, Pittsburgh, pp. 777–789, 1992.

Neoh, K. G., Howard, J. B., and Sarofim, A. F., "Soot Oxidation in Flames," in Siegla, P., and Smith, G. (ed.), *Particulate Formation during Combustion,* Plenum Press, New York, pp. 261–282, 1981.

Norton, T. S., and Dryer, F. L., "The Flow Reactor Oxidation of C_1–C_4 Alcohols and MTBE," *Twenty-Third Symp. (Int.) on Comb.,* The Combustion Institute, Pittsburgh, pp. 179–185, 1990.

Siegla, D. C., and Smith, G. W. (eds.), *Particulate Formation during Combustion,* Plenum Press, New York, 1981.

Simons, G. A., "The Pore Tree Structure of Porous Char," *Nineteenth Symp. (Int.) on Combustion,* The Combustion Institute, Pittsburgh, pp. 1067–1076, 1982.

Smoot, L. D., and Smith, P. J., *Coal Combustion and Gasification,* Plenum Press, New York, 1985.

Westbrook, C. K., and Dryer, F. L., "Chemical Kinetic Modeling of Hydrocarbon Combustion," *Prog. Energy Comb. Sci.,* vol. 10, pp. 1–57, 1984. (This review gives 494 references and extensive tables of reaction rate data.)

Westbrook, C. K., and Pitz, W. J., "A Comprehensive Chemical Kinetic Reaction Mechanism for Oxidation and Pyrolysis of Propane and Propene," *Comb. Sci. and Tech.,* vol. 77, pp. 117–152, 1984.

Wilk, R. D., Pitz, W. J., Westbrook, C. K., and Cernansky, N. P. "Chemical Kinetic Modeling of Ethene Oxidation at Low and Intermediate Temperatures," *Twenty-Third Symp. (Int.) on Combustion,* pp. 203–210, The Combustion Institute, Pittsburgh, 1990.

PART II

Combustion of Gaseous and Vaporized Fuels

Combustion of gaseous fuels may occur as premixed flames, as diffusion flames, as radiation-dominated reactions on surfaces of porous media, or as detonation waves. After a discussion of flames, the combustion processes in gas-fired furnaces and spark-ignition engines are discussed. Part II ends with a discussion of gaseous detonation waves.

CHAPTER 5

Flames

There are two basic idealized types of flames: premixed flames, and diffusion flames. *Premixed flames* arise from the combustion of gaseous reactants which are perfectly mixed prior to combustion. A premixed flame is a rapid, essentially constant-pressure, exothermic reaction of gaseous fuel and oxidizer which radiates light and heat and propagates as a thin zone with speeds of less than a few meters per second. Laminar premixed flames have a unique burning velocity for a given fuel-oxidizer mixture. Turbulence increases the burning velocity. *Diffusion flames* arise from the combustion of separate gaseous fuel and oxidizer streams which combust as they mix. Diffusion flames are dominated by the mixing of the reactants, which can be either laminar or turbulent, and reaction takes place at the interface between the fuel and oxidizer.

Most applications of combustion require the use of turbulent combustion to produce the volumetric rates of energy production needed for efficiency and compactness. As we shall see, however, the concepts of laminar flame propagation are an integral part of understanding and modeling turbulent flames. Turbulent flames found in practical devices cover a wide spectrum of phenomena which depend on the intensity of the turbulence, the temperature and pressure levels, and the reactant fuel-air ratio.

The most important applications of premixed combustion are gas-fired furnaces and stoves, gas-fired turbines, and automobile engines. These devices span an amazingly wide range of combustion time scales. For example, engines can be found that operate at speeds ranging from a few hundred revolutions per minute (rpm) to 12,000 rpm. The very short time scales are achieved by the use of high reactant temperatures and very high levels of turbulence. This chapter lays the background for understanding how this can be achieved.

For many applications the vaporization and mixing of fuel and air is achieved as an integral part of the combustion process. Fuel and air mixing rather than flame propagation rates often limit the energy release rate. For example, industrial gas-fired burners, diesel engines, oil-fired burners, and some gas turbine engines utilize diffusion flame combustion. The volatiles from solid fuels also burn in the diffusion mode. In this chapter the discussion will be resricted to gaseous fuels. The mixing of gases to produce a gaseous mixture is difficult, typically more difficult than mixing liquid fuel and hot air. One reason for this is that compression of liquids to create a high-pressure spray (as discussed in Chapter 9) can be done with very little expenditure of work compared with compression of gas.

The laminar premixed flame is the simplest type of flame, and the chapter begins here. Then, turbulent premixed flames, explosion limits, quench distance, ignition, and gaseous diffusion flames are discussed.

5.1 LAMINAR PREMIXED FLAMES

A combustion reaction started at a local heat source in a quiescent fuel-air mixture at ambient conditions will propagate initially as a laminar flame. Chemical reaction takes place in a relatively thin zone, and the flame moves at a fairly low velocity. For stoichiometric hydrocarbon mixtures in ambient air the flame is approximately 1 mm thick and moves at about 0.5 m/s. The pressure drop through the flame is very small (about 1 Pa), and the temperature in the reaction zone is high (2200–2600 K). Within the flame reaction zone a multitude of active radicals are formed in the high temperature part of the flame and diffuse upstream to attack the fuel. The radicals are then converted to products via chemical reactions such as suggested in Chapter 4. In addition, heat conducted from the high temperature to the lower temperature portions of the reaction zone sustains the flame.

The familiar Bunsen burner, shown schematically in Figure 5.1*a*, provides an example of a stationary laminar premixed flame, provided that the reactants flow through the tube under laminar conditions. Fuel enters under a slight positive pressure at the base of the burner and entrains air, which mixes in the burner tube. The flame zone is cone-shaped. Figure 5.1*b* shows the streamlines relative to the flame zone, and Figure 5.1*c* shows isotherms and streamlines for a slot burner. A slot burner with a rectangular cross section will produce a tent-shaped flame similar to the cone of a Bunsen burner. The observed peak temperature in the flame is slightly reduced from the adiabatic temperature due to radiation losses. A fuel-air mixture with a slower burning velocity will have a more pointed tip.

Referring to Figure 5.1*b*, the burning velocity can be measured by the relation

$$\underline{V}_{\text{flame}} = \underline{V}_{\text{tube}} \sin \alpha \tag{5.1}$$

The cone is not perfectly straight, but rather is rounded at the tip and is curved at the lip due to heat transfer to the tube, which serves to stabilize the flame. Also, the velocity in the tube is not perfectly uniform, due to boundary layer effects. Hence, the local velocity and angle should be used when applying Eq. 5.1, although an approximate value can be obtained simply by multiplying the mass average velocity

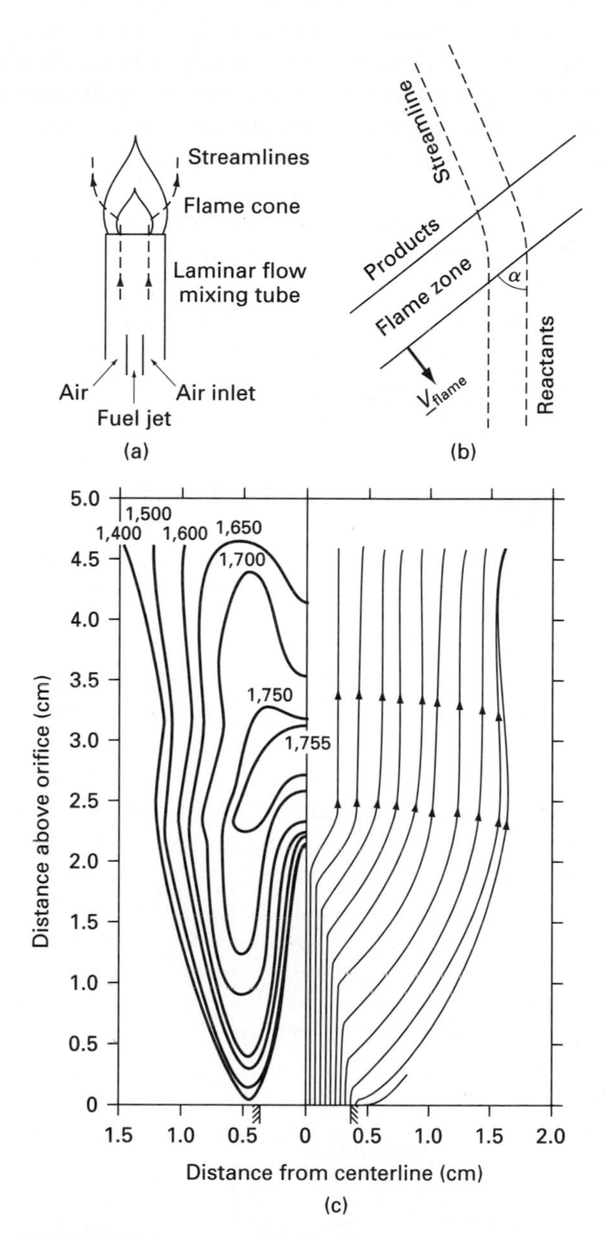

FIGURE 5.1
Bunsen burner flame: (*a*) schematic of burner, (*b*) flow diagram, and (*c*) streamlines and temperature (°C) for a laminar slot burner [Lewis and von Elbe, by permission of Academic Press, Ltd.].

in the tube by the ratio of the area of the cone to the cross-sectional area of the tube; this approximation has an error band of ±20%.

For each premixed fuel-air mixture there is a characteristic laminar burning velocity. The *burning velocity* is defined as the velocity of the flame relative to the unburned reactants. The laminar burning velocity depends on the fuel type, fuel-air mixture ratio, and initial temperature and pressure of the reactants. Before considering a mathematical model of a laminar flame, some basic data on laminar burning velocities are presented.

The reader should be forewarned that many factors may confound the experimental data, so that values reported in the literature may vary significantly. A discussion of such errors has been given by Andrews and Bradley for methane flames (1972b) and by Tseng et al. for propane, methane, ethane, and ethylene. A variation in reported propane burning velocity values of 0.35 to 0.45 m/s at 1 atm and 298 K reactant conditions is shown by Tseng et al.

Effect of Stoichiometry on Laminar Burning Velocity

The effect of fuel concentration on the laminar burning velocity is shown in Figure 5.2 for various fuels. It is seen that the laminar burning velocity for a

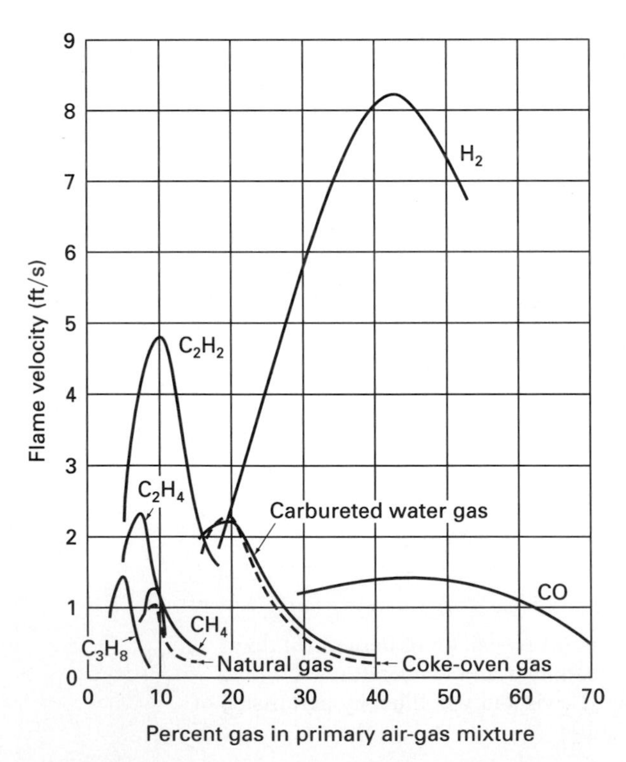

FIGURE 5.2
Variation of laminar burning velocity in air-gas mixtures with variation of gas content, at atmospheric pressure [Elliott and Denues].

particular fuel can vary by a factor of 3 depending on the fuel-air ratio. The rich and lean limits of flammability are also shown in this figure. Laminar flames will not occur above or below these limits. Hydrogen has the highest burning velocity and widest limits of flammability, while methane has the lowest burning velocity and the narrowest limits. The maximum burning velocities are found just to the rich side of stoichiometric. The flame temperature is highest near stoichiometric and lowest near the flammability limits (Figure 5.3). Typically, a higher laminar burning velocity is associated with a higher flame temperature.

Representative flammability limits for various fuels in air are given in Table 5.1. Most mixtures are flammable when the fuel-air volume ratio is 50 to 300% of the stoichiometric value, although hydrogen and acetylene are exceptions. *Flammability limits* are typically obtained for a given fuel by initiating a flame in a vertical glass tube of about 5 cm diameter and 1 m length, and observing if it propagates along the tube. Downward propagation should be used, since it gives a more conservative estimate. For an upward-propagating flame, the upward convection of the hot products tends to help the flame propagate and thus upward

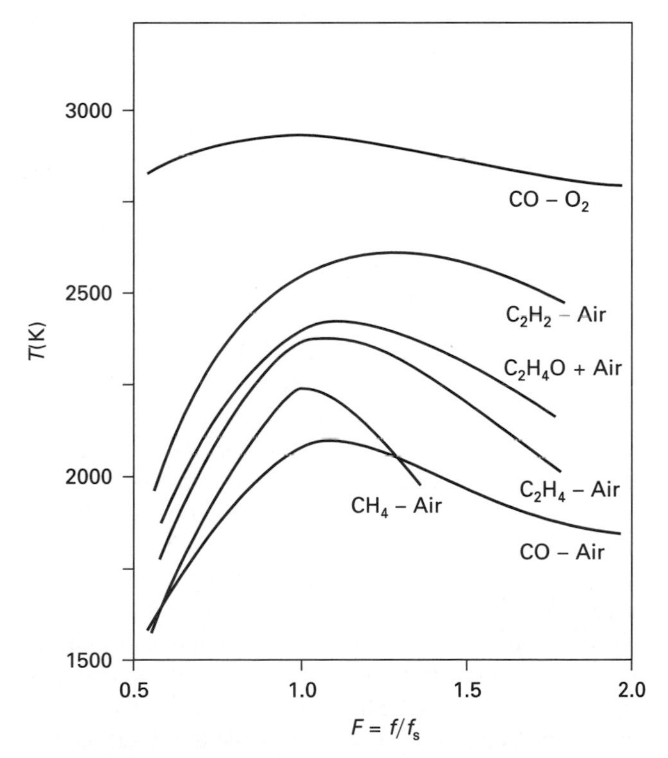

FIGURE 5.3
Flame temperature as a function of equivalence ratio for various fuels [Strehlow].

TABLE 5.1
Limits of flammability in standard air (% by volume)

Fuel vapor	Stoichiometric	Lean limit	Rich limit
Methane	9.47	5.0	15.0
Ethane	5.64	2.9	13.0
Propane	4.02	2.0	9.5
Isooctane	1.65	0.95	6.0
Carbon monoxide	29.50	12.5	74
Acetylene	7.72	2.5	80
Hydrogen	29.50	4.0	75
Methanol	12.24	6.7	36

Source: Bartok and Sarofim, *Fossil Fuel Conbustion: A Source Book,* © 1991, Wiley, by permission of John Wiley and Sons, Inc.

propagation gives slightly wider limits than either downward or horizontal propagation.

The effect of nonreactive additives such as nitrogen or argon is to reduce the flame temperature and the laminar burning velocity. The most common diluent addition is products of combustion. For example, in power plants a fraction of the combustion products are sometimes recirculated with the inlet air to reduce the amount of NO produced by decreasing the flame temperature. Similarly, in internal-combustion engines, a fraction of the residual products from the previous cycle mix with the new charge. Metghalchi and Keck [1982] offer the following simple correlation multiplier for the reduction of burning velocity due to addition of dry combustion products: $(1 - 2.1y_p)$, where y_p is the mass fraction of products mixed with the reactants. For example, 10% exhaust gas recirculation gives a 21% reduction in the laminar flame velocity.

Other additives may react directly or act as a catalyst agent. One striking example is the addition of small amounts (0.23%) of water vapor to a CO-O_2 mixture, which increases the burning velocity by a factor of 8. This is due to the formation of OH radicals, which enhance the flame propagation by increasing the reaction rate (see Section 4.1). Conversely, inhibitors such as heavy organic halides may greatly reduce flame velocity. Some effects are unexpected, such as the addition of 3.5% butane to a hydrogen-air mixture, which causes the burning velocity to decrease from 2.7 to 0.15 m/s.

Typically one would like to increase the burning velocity, especially of lean mixtures. Hydrogen has an exceptionally high burning velocity, almost 8 times that of methane. The addition of hydrogen can be accomplished in a practical way by adding rich products of combustion that contain hydrogen. The amount of hydrogen can be further increased by use of a water-gas shift catalyst, which reacts water vapor and carbon monoxide to form carbon dioxide and hydrogen. Such a scheme has recently been applied to a natural gas–fired engine by running one cylinder rich ($F = 1.4$) to supply the rich products, which are run through the water-gas shift catalyst and then mixed with the lean mixtures ($F = 0.6$) in the other cylinders.

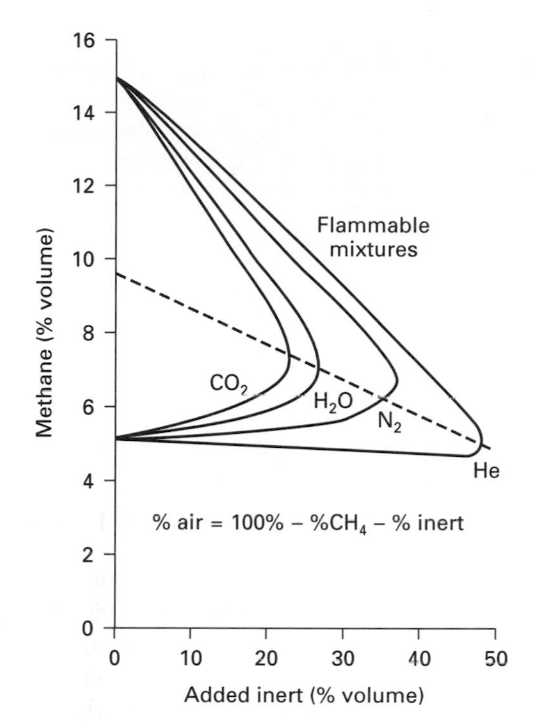

FIGURE 5.4
Limits of flammability of various methane–inert gas–air mixtures at 25°C and 1 atm [Zabetakis].

For a variety of fuels, the flame temperature at the lean limit is about 1200 to 1240°C (2190 to 2260°F). The lean flame temperature limit may be used to estimate the effect of adding inert gases to the reactant. Inert gases with high heat capacity have a greater effect than those with a low heat capacity for the same amount of addition. Some gases narrow the limits more than would be expected based on their heat capacity, as shown in Figure 5.4.

Effect of Reactant Pressure and Temperature on Laminar Burning Velocity

For slow-burning mixtures ($\underline{V}_L < 0.6$ m/s), the burning velocity decreases with increasing pressure. The observed pressure dependence can be expressed as a power law, $\underline{V}_L = ap^{\beta}$, where p is the pressure in atmospheres and β varies from 0 to -0.5. For example, the burning velocities for propane-air mixtures at various pressures are shown in Figure 5.5. For fast-burning mixtures ($\underline{V}_L > 0.6$ m/s), the value of β is either zero or slightly positive. Increased pressure increases the flame temperature because there is less dissociation, and hence the burning velocity should be increased. However, less dissociation means less active radicals are available to diffuse upstream to enhance flame propagation. Both effects are important.

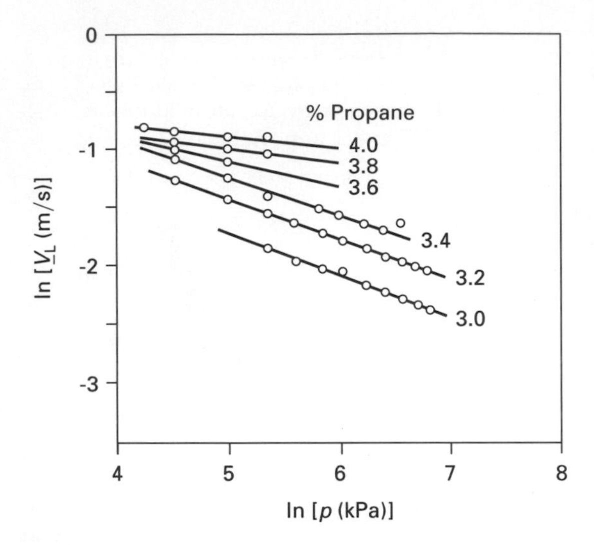

FIGURE 5.5
Influence of pressure on laminar burning velocity of propane-air mixtures at ambient temperatures [Egerton and Lefebvre, by permission of The Royal Society].

The burning velocity increases with the temperature of the reactants, provided the reactants do not partially react prior to the flame passage. The observed temperature dependence can also be expressed as a power law, and the burning velocity increases as the second or third power of the absolute temperature (see Table 5.2 and Figure 5.6). For example, the maximum burning velocity for propane-air goes from 40 cm/s to 140 cm/s as the reactant temperature is increased from 300 K to 617 K.

The reactant temperature must be less than the autoignition temperature. Autoignition temperatures for various fuels in ambient air are also shown in Table 5.3. The *autoignition temperature* is the temperature to which the gaseous mixture must be raised to achieve ignition, without regard to the ignition delay time. In a flame, the ignition temperature is not precisely the autoignition temperature, because of diffusion of active species from the reaction zone to the preheat zone.

TABLE 5.2
Effect of fuel-air preheat temperature on maximum laminar burning velocity in l atm air
$\underline{V}_L = a + b(T/T_0)^n$ cm/s; $T_0 = 300$ K

Fuel	a	b	n	T range
Methane	8	27.0	2.11	141–617 K
Propane	10	30.8	2.00	141–617 K
Ethylene	10	52.9	1.74	141–617 K
Benzene	30	13.5	2.92	300–700 K
n-Heptane	19.8	20.7	2.39	300–700 K
Isooctane	12.1	22.2	2.19	300–700 K

Source: Reprinted by permission of Elsevier Science Inc. from "Burning Velocity of Mixtures in Air with Methanol, Isooctane, and Indoline at High Pressures and Temperatures," by Metghalchi and Keck, *Combustion and Flame,* vol. 48, pp. 191–210, © 1982 by The Combustion Institute.

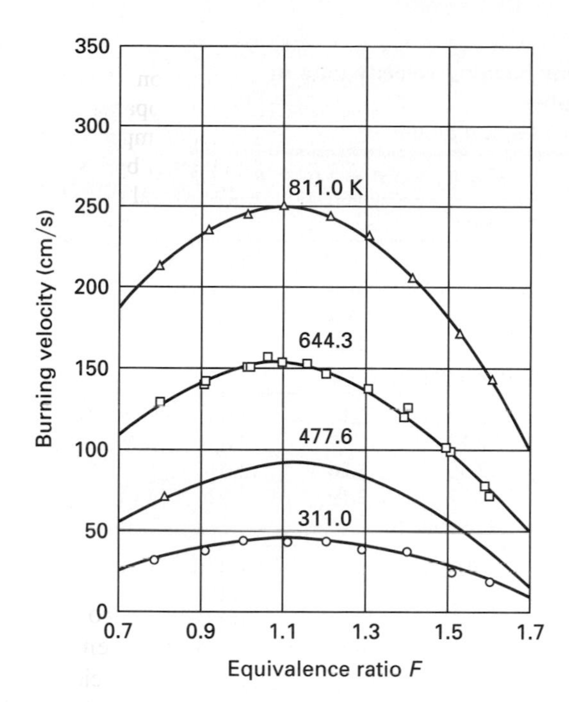

FIGURE 5.6
Laminar burning velocity versus equivalence ratio for propane-air at 1 atm and various reactant temperatures [Kuehl, by permission of The Combustion Institute].

Nevertheless, the autoignition temperature provides a general indication of the necessary temperature rise upstream of the flame front in order to sustain ignition and flame propagation.

The combined effect of increased pressure and reactant temperature on the laminar burning velocity is shown in Table 5.4. The data shown in this table were obtained using a closed, constant-volume chamber with a propagating spherical

TABLE 5.3
Ignition temperature and maximum laminar burning velocity at 20°C and 1 atm

Fuel	Autoignition temperature in air (°C)	Laminar burning velocity in air (cm/s)
Methane	537	34
Propane	470	39
n-Hexane	233	39
Isooctane	418	35
Carbon monoxide	609	39
Acetylene	305	141
Hydrogen	400	265
Methanol	385	48

Source: Bartok and Sarofim, *Fossil Fuel Combustion: A Source Book*, © 1991, Wiley, by permission of John Wiley and Sons, Inc.

TABLE 5.4
Empirical fit to laminar burning velocity data in constant-volume chamber
$\underline{V}_L = \underline{V}_0(T/298)^{\alpha}(p)^{\beta}$, for T(K) and p(atm)

Fuel	$F = 0.8$	$F = 1.0$	$F = 1.2$
	$\underline{V}_0$(cm/s)		
Methanol	25.6	32.7	38.1
Propane	23.2	31.9	33.8
Isooctane	19.2	27.0	27.6
RMFD-303*	19.1	25.2	28.1
	Temperature exponent, α		
Methanol	2.47	2.11	1.98
Propane	2.27	2.13	2.06
Isooctane	2.36	2.26	2.03
RMFD-303*	2.27	2.19	2.02
	Pressure exponent, β		
Methanol	−0.21	−0.13	−0.11
Propane	−0.23	−0.17	−0.17
Isooctane	−0.22	−0.18	−0.11
RMFD-303*	−0.17	−0.13	−0.087

Source: Reprinted by permission of Elsevier Science Inc. from "Burning Velocity of Mixtures in Air with Methanol, Osooctane, and Indoline at High Pressures and Temperatures," by Metghalchi and Keck, *Combustion and Flame,* vol. 48, pp. 191–210, © 1982 by The Combustion Institute.
*Synthetic gasoline (45% toluene, 14% undecene, and 41% isooctane).

flame. These spherical flames are unsteady and have changing flame area, and thus may give velocities different from a steady flat flame.

Reactant temperature and pressure influence the flammability limits. Increasing reactant temperature increases the lean limit. The lean limits are affected only slightly by pressure, whereas the rich limit is extended markedly when the pressure is increased, as seen in Table 5.5 for natural gas.

TABLE 5.5
Limits of flammability of natural gas in 20°C dry air versus pressure (% by volume) [Jones et al.]

Pressure (atm)	Lean limit	Rich limit
1	4.50	14.2
35	4.45	44.2
69	4.00	52.9
137	3.60	59.0
205	3.15	60 (est.)

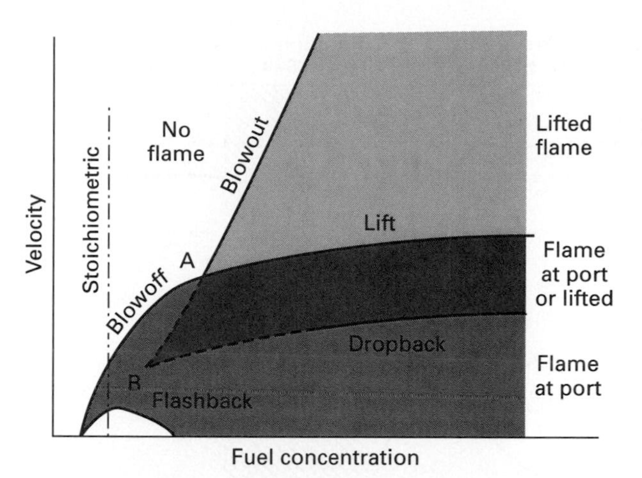

FIGURE 5.7
Characteristic stability diagram for a premixed, open burner flame [Wohl et al., by permission of The Combustion Institute].

Stabililization of a Premixed Flame

A premixed flame burning from the end of a cylindrical tube or nozzle exhibits a characteristic behavior depending on the fuel concentration and the gas velocity. When the approach velocity to a seated open flame is decreased until the flame velocity exceeds the approach velocity over some portion of the burner port, the flame flashes back into the burner (unshaded area of Figure 5.7). On the other hand, if the approach velocity is increased until it exceeds the flame velocity at every point, the flame will either be extinguished completely—i.e., blown off—or, for fuel-rich mixtures, it will be lifted until a new stable position in the gas stream above the burner is reached as a result of turbulent mixing and dilution with secondary air. The lift curve is a continuation of the blowoff curve beyond point A in Figure 5.7. The blowout curve corresponds to the velocity required to extinguish a lifted flame. Once the flame has lifted, the approach velocity must be decreased well below the lift velocity before the flame will drop back and be reseated on the burner rim. Between fuel concentrations A and B the blowout of a lifted flame occurs at a lower velocity than the flame blowoff from the port. Stabilization of burner flames is considered further in Chapter 6.

5.2 LAMINAR FLAME THEORY

Consider a laminar premixed flame free from the effects of any walls. The gas flow is uniform and the flame front is planar. Let the flame be stationary so that the gas flows into the flame front as noted in Figure 5.8. The burning velocity is always defined as relative to the unburned gas velocity, so in this case the burning velocity is the unburned gas speed. Let subscript r refer to the mixture of reactants and subscript p refer to the mixture of products. Conservation of mass, momentum, and

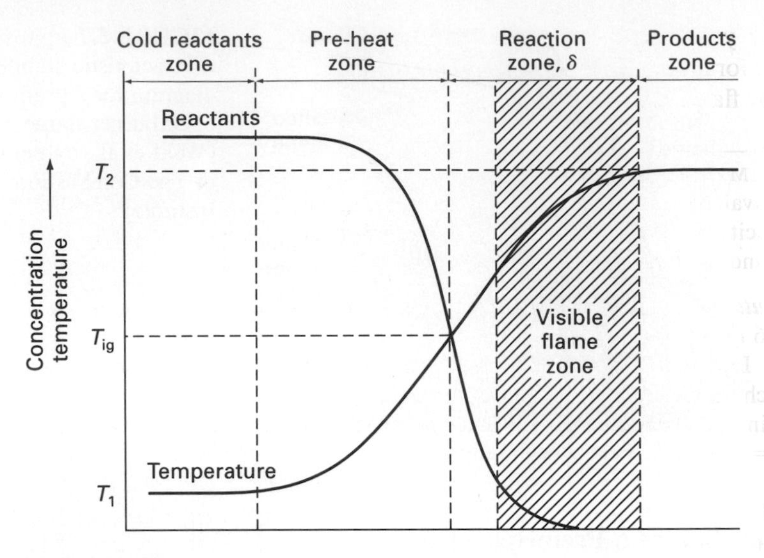

FIGURE 5.8
Typical laminar flame profile.

energy across the flame front are given by:

$$\rho_r \underline{V}_r = \rho_p \underline{V}_p \tag{5.2}$$

$$p_r + \rho_r \underline{V}_r^2 = p_p + \rho_p \underline{V}_p^2 \tag{5.3}$$

$$h_r + \frac{\underline{V}_r^2}{2} = h_p + \frac{\underline{V}_p^2}{2} \tag{5.4}$$

where h is the absolute enthalpy of the mixture.

The unknowns in Eqs. 5.2 to 5.4 are $\underline{V}_r$, $\underline{V}_p$, ρ_p, p_p, and h_p. The equation of state $p_p = \rho_p R_p T_p$ holds, and h_p is a function of T_p. However, we are still one equation short of solving for the state of the products and the burning velocity, $\underline{V}_r$. If the burning velocity is known from experiment or additional theory, then the state of the products can be determined from Eqs. 5.2 to 5.4. The burning velocities are low enough that the velocity terms in Eq. 5.4 are negligible compared with the enthalpy terms. Indeed, in Chapter 3 the adiabatic flame temperature was calculated by simply assuming that the enthalpy remains constant across the flame. Similarly, the pressure change across the flame is very small (see Example 5.1). Hence, the density decreases as the temperature increases, and by Eq. 5.2 the velocity increases in proportion to the temperature increase. Again, if the burning velocity $\underline{V}_r$ is known, then the state of the products can be directly determined, but Eqs. 5.2 to 5.4 alone do not yield the burning velocity.

Traditionally, Eqs. 5.2 to 5.4 have been combined into two equations and rearranged to give the so-called Rankine-Hugoniot equations [see, for example, Williams]. However, no new information is obtained by doing this, and the authors prefer to work directly with the basic conservation equations. Equations 5.2 to 5.4

also apply to detonations (see Chapter 8). For detonations $\underline{V}_r$ is supersonic, whereas for flames $\underline{V}_r$ is subsonic. This distinction is sometimes made by calling the subsonic flame a *deflagration*.

EXAMPLE 5.1. A laminar flame propagates through a propane-air mixture with an equivalence ratio of 0.9, a pressure of 5 atm, and a temperature of 300 K. The flame velocity is 22 cm/s. Find (*a*) the gas temperature, (*b*) the velocity, and (*c*) the pressure behind the flame.

Solution. The reactant mixture is $C_3H_8 + (5/0.9)(O_2 + 3.76N_2)$, or 1 mole of C_3H_8, 5.56 moles of O_2, and 20.89 moles of N_2.

Equations 5.2 to 5.4 are used to solve the problem. Since the velocity terms are much smaller than the enthalpy terms, Eq. 5.4 becomes $h_r = h_p$. Now use STANJAN to find T_p. Enter the moles of reactants, and $p_r = 5$ atm, $T_r = 300$ K. STANJAN gives $h_r = -0.1267 \times 10^6$ kJ/kg, $\rho_r = 5.97$ kg/m^3, $M_r = 28.318$ kg/kgmol.

(*a*) In Eq. 5.3 the momentum terms are much smaller than the pressure terms, and thus for STANJAN we pick the constant h, constant p process and select the equilibrium products (rather than sending the products to completion). STANJAN gives:

$$T_p = 2200 \text{ K} \qquad \text{and also } \rho_p = 0.784 \text{ kg/m}^3 \qquad M_p = 28.23 \text{ kg/kgmol}$$

(*b*) From Eq. 5.2, $\underline{V}_p = \underline{V}_r \rho_r/\rho_p = (0.22)(5.97/0.784) = 1.67$ m/s. This is the velocity relative to the flame front.

(*c*) From Eq. 5.3,

$$p_p - p_r = \rho_r(\underline{V}_r)^2 - \rho_p(\underline{V}_p)^2 = 5.97(0.22)^2 - 0.784\,(1.67)^2$$
$$= -1.90 \text{ Pa} = -0.0000187 \text{ atm}$$

and thus,

$$p_p = 4.999981 \text{ atm}$$

which justifies the assumption of essentially constant pressure.

Laminar Burning Velocity Theory

Since the integral form of the conservation equations for a laminar flame (Eqs. 5.2 to 5.4) are insufficient to determine the burning velocity, the differential form of the conservation equations through the flame zone must be used. The rate at which a laminar flame will propagate through a combustible mixture is determined by the appropriate chemical reaction rates as well as the heat and mass transfer rates throughout the flame zone. The process is a one-dimensional, steady, reacting, compressible-flow problem. The governing equations consist of the continuity equation, the species continuity equations, and the energy equation. The momentum equation is not needed because the pressure is essentially constant. The full form of the flame equations will be presented first, and then a simplified treatment will be given to gain physical insight into the analysis of flames. For a general derivation of the conservation equations for reacting flows, the reader may wish to consult Chapter 7 of Turns.

The steady-state, one-dimensional continuity equation for a differential fluid element in the flame states that the net change of mass flux through the flame is zero, and thus by integration the overall mass flux at any point in the reaction zone is equal to the reactant density times the burning velocity:

$$\frac{d\rho \underline{V}}{dx} = 0$$

and integrating,

$$\rho \underline{V} = \rho_r \underline{V}_r = \text{constant} \tag{5.5}$$

The continuity equation for each species i is complicated by two effects—mass diffusion and chemical reactions. The mass diffusion causes each species i to move relative to the mass average velocity of the mixture. The primary cause of this diffusional velocity $\tilde{\underline{V}}_i$ is the concentration gradient dy_i/dx. Such diffusion is called *ordinary diffusion.* Diffusion may also be caused by temperature and pressure gradients. Diffusion due to temperature gradients is called *thermal diffusion,* or thermophoresis or the Dufour effect. Even in flames, where the temperature gradients are very large, the thermal diffusion velocity is only 10 to 20% of the ordinary diffusion velocity and primarily affects low–molecular weight species such as hydrogen. Thus we shall neglect thermal diffusion here. Because the pressure gradients are very small in deflagrations, their effects on diffusion can also be neglected. The velocity of each species i is given by $\underline{V}_i = \underline{V} + \tilde{\underline{V}}_i$, and the species continuity equation is

$$\frac{d}{dx}\left[\rho y_i \left(\underline{V} + \tilde{\underline{V}}_i\right)\right] = \rho_r \underline{V}_r \frac{dy_i}{dx} + \frac{d}{dx}\left(\rho y_i \tilde{\underline{V}}_i\right) = M_i \hat{r}_i \tag{5.6}$$

The term on the right-hand-side of Eq. 5.6 is the net mass production rate of species i per unit volume within the differential volume due to chemical reactions involving species i. Recall from Chapter 4 that $\hat{r}_i$ is made up of as many terms as there are chemical reactions involving species i.

In flames there are many species, and thus the diffusional process is complicated. The species most affecting the reaction rates are typically radicals with very small concentrations. Thus, a good assumption for fuel-air flames is to approximate the diffusion in flames as a binary system consisting of the species and nitrogen. For a binary system consisting of species i and j, the ordinary diffusion velocity is given by Fick's first law of diffusion. For a one-dimensional system, this gives

$$\rho_i \tilde{\underline{V}}_i = -\rho D_{ij} \frac{dy_i}{dx}$$

or

$$\tilde{\underline{V}}_i = -\frac{D_{ij}}{y_i}\frac{dy_i}{dx} \tag{5.7}$$

The binary diffusion coefficient D_{ij} is a property of the mixture constituents and temperature. It and the basic concepts of ordinary diffusion are discussed further in Appendix E. A much more detailed discussion is given by Bird, Stewart, and Lightfoot.

The energy equation for the flame states that the net change in the convective flux of enthalpy of the mixture, plus the net flux of heat due to conduction, plus the net overall flux of enthalpy due to the diffusion of mass, equals the total heat release by chemical reaction within a differential fluid element. The radiation flux is neglected here; it is small for nonluminous flames, but can be significant for rich flames where carbon particles radiate, making the flame luminous. The dissipation terms in the energy equation due to work done by the viscous stresses are not included, since these terms are negligible because the velocity is very small. The energy equation is:

$$\rho_r \underline{V}_r c_p \frac{dT}{dx} - \frac{d}{dx}\left(\tilde{k}\frac{dT}{dx}\right) + \sum_{i=1}^{s} \rho y_i \underline{V}_i c_{pi} \frac{dT}{dx} = \sum_{i=1}^{s} M_i \hat{r}_i h_i \tag{5.8}$$

From Eq. 5.8 we see that upstream heat conduction and diffusion of active species promote chemical reactions and sustain the flame against the convective flow of heat.

There are two types of premixed flames: burner-stabilized flames, and freely propagating flames. The governing equations are the same for both types, since a flame-fixed coordinate system is used for the propagating flame. However, the boundary conditions are different. For burner-stabilized flames, the mass flux of reactants $\rho_r \underline{V}_r$, the inlet species mass flux fractions, and the inlet temperature are known. Vanishing gradients are imposed at the hot boundary. For freely propagating flames, $\rho_r \underline{V}_r$ must be determined as part of the solution. The additional boundary condition is provided by specifying the hot boundary temperature. Solution of the governing equations with appropriate boundary conditions is done with numerical computer codes such as the CHEMKIN package [Kee et al., 1980].

By way of example, results of a computer solution of a stoichiometric methanol-air laminar flame are presented in Figure 5.9. Eighty-four reactions involving 26 species were used. The temperature, velocity, density, and selected species profiles within the flame zone are shown. Notice the presence of active radicals such as HO_2 and CH_2O which have diffused ahead of the flame front. Another use of detailed laminar flame calculations is to test the validity of reaction schemes at high temperatures by comparing the model results with experimentally obtained profiles of temperature, species, and burning velocity.

The application of chemical kinetics has been greatly extended; for example Gottgens, Mauss, and Peters have performed laminar flame calculations for six different fuels using 82 elementary reactions to obtain expressions for the burning velocity and flame thickness as a function of initial temperature, pressure, and equivalence ratio. Their results are shown for a laminar propane-air flame in Figure 5.10. At high pressure, such as in an internal-combustion engine or gas turbine combustor, the flame thickness is considerably reduced.

Simplified Laminar Flame Model

In order to gain understanding of the laminar flame, let us consider a simpler, more approximate flame model following the work of Mallard and Le Châtelier in 1883.

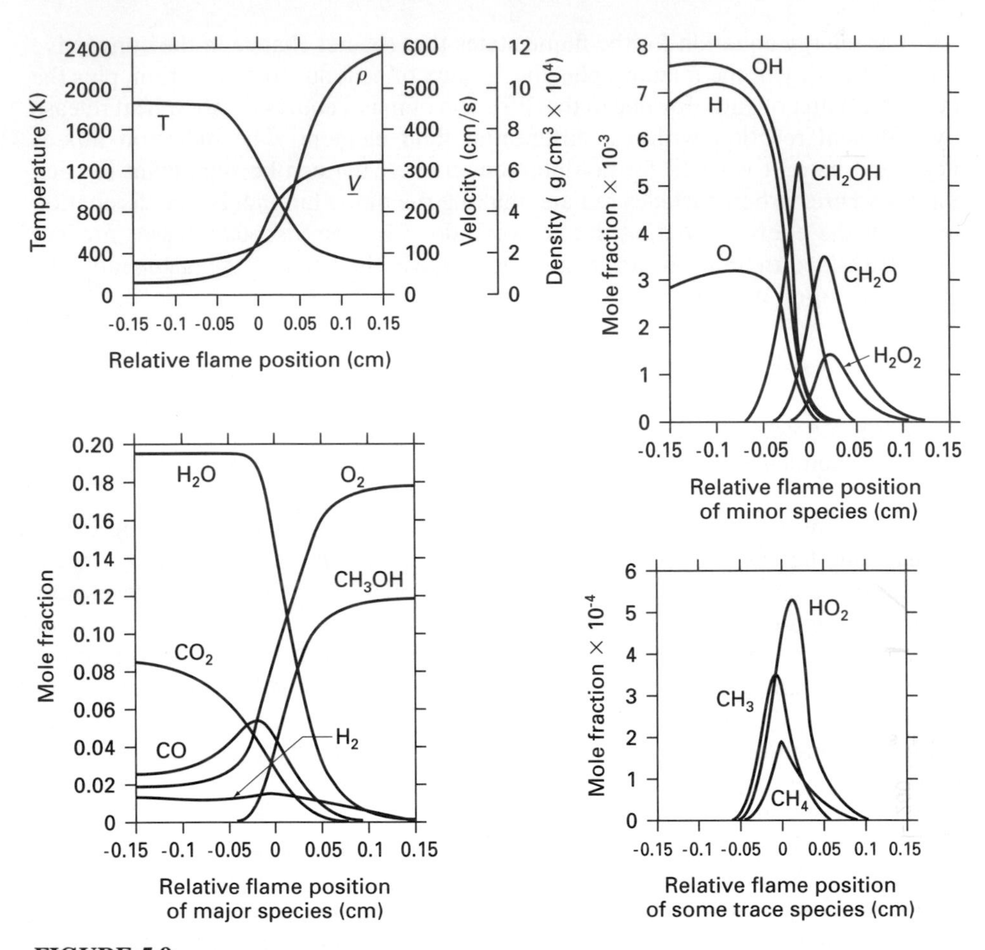

FIGURE 5.9
Calculated flame profiles for stoichiometric methanol-air, one-dimensional flame. The calculation used 26 species and 84 elementary reactions and gave a burning velocity of 44 cm/s [Westbrook and Dryer, © 1979, by permission of Gordon and Breach Publishers].

They reasoned that heat conduction from the flame to the reactants was the rate-limiting step. They assumed that the flame consisted of a preheat zone and a reaction zone, and they neglected any diffusion and chemical reactions in the preheat zone. The boundary between the two zones was assumed to be set by the ignition temperature, as shown in Figure 5.8. The reaction zone extends over a distance of δ. With these assumptions the energy equation (Eq. 5.8) in the preheat zone only becomes

$$\frac{d}{dx}\left(\tilde{k}\frac{dT}{dx}\right) = \rho_r \underline{V}_r c_p \frac{dT}{dx} \tag{5.9}$$

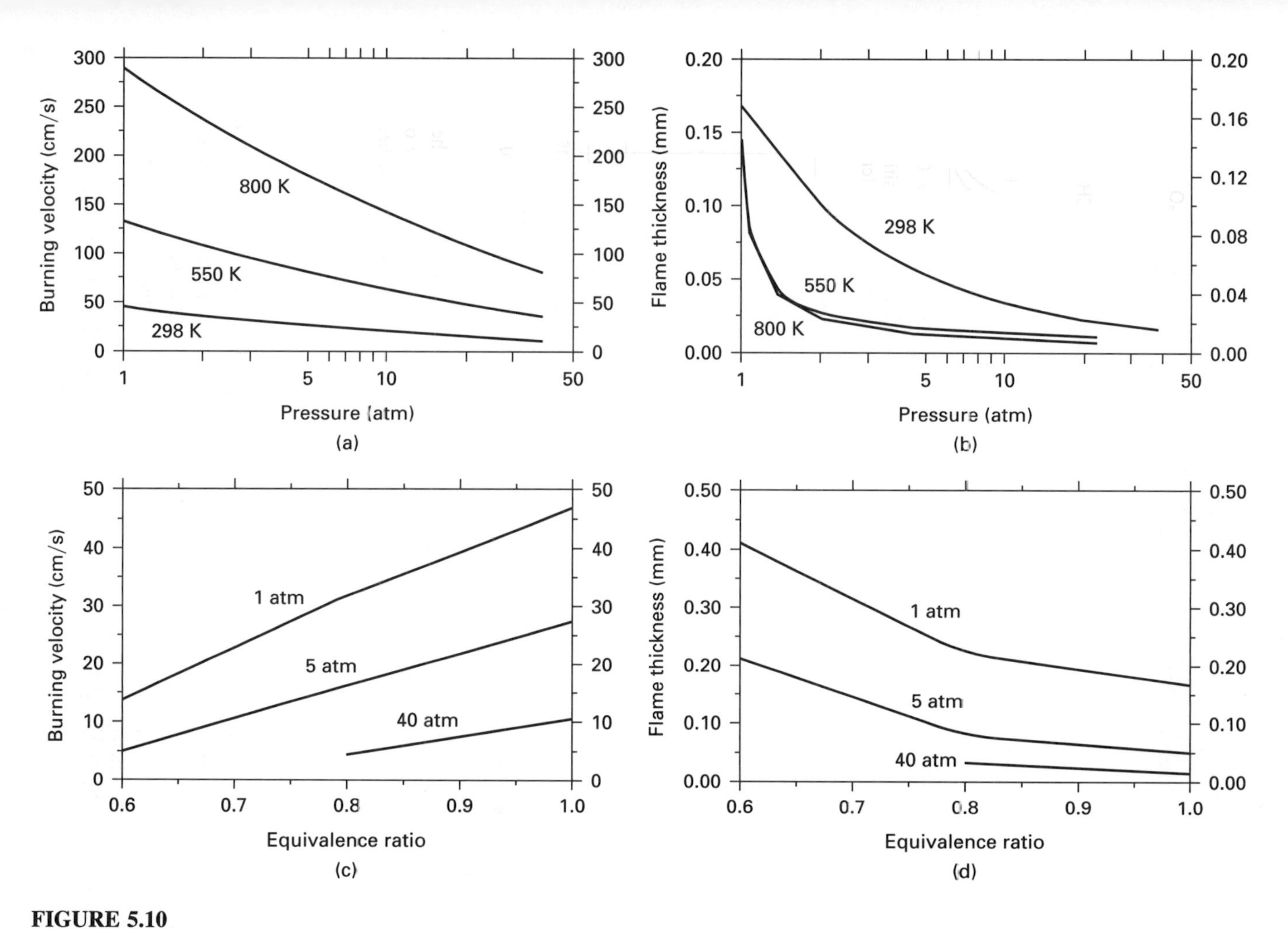

FIGURE 5.10
Modeled laminar burning velocity and flame thickness for propane-air: (*a, b*) stoichiometric; (*c, d*) initially at 298 K [Gottgens et al., by permission of The Combustion Institute].

Solution of Eq. 5.9, taking $\tilde{k}$ and c_p constant, is

$$T = C_1 + C_2 \exp\left(\frac{\underline{V}_r x}{\alpha}\right) \tag{5.10}$$

where α is the thermal diffusivity (m^2/s):

$$\alpha = \frac{\tilde{k}}{\rho c_p}$$

Assuming that chemical reaction starts at temperature T_{ig}, where $x = 0$, the boundary conditions are

$$T = T_r \qquad \text{at} \qquad x = -\infty \tag{5.11}$$

$$T = T_{\text{ig}} \qquad \text{at} \qquad x = 0 \tag{5.12}$$

and Eq. 5.10 becomes

$$T = T_r + (T_{\text{ig}} - T_r)e^{\underline{V}_r x/\alpha} \tag{5.13}$$

We desire a relation between the burning velocity $\underline{V}_r$ and the burned gas (products) temperature T_r. This is accomplished, following Mallard and Le Châtelier, by matching the slope of the temperature-distance curve at the ignition point on the unburned side with the slope on the reaction zone side and by making the assumption that

$$\left(\frac{dT}{dx}\right)_{x=0} = \frac{T_p - T_{\text{ig}}}{\delta} \tag{5.14}$$

Differentiating Eq. 5.13 yields

$$\left(\frac{dT}{dx}\right)_{x=0} = \frac{\underline{V}_r}{\alpha}(T_{\text{ig}} - T_r) \tag{5.15}$$

Equating Eq. 5.14 and Eq. 5.15 and using the laminar burning velocity notation for $\underline{V}_r$ gives the result of Mallard and Le Châtelier,

$$\underline{V}_L = \frac{(\alpha/\delta)(T_p - T_{\text{ig}})}{T_{\text{ig}} - T_r} \tag{5.16}$$

This result may be extended by noting that the reaction thickness depends on the reaction rate. Defining a characteristic reaction time as $\tau = [n_f]_i/\overline{r}_f$ where $\overline{r}_f$ is the average global reaction rate,

$$\delta = \underline{V}_L \tau = \frac{\underline{V}_L [n_f]_i}{\overline{r}_f} \tag{5.17}$$

Substituting Eq. 5.17 into 5.16,

$$\underline{V}_L = \left[\frac{(\alpha\overline{r}_f/[n_f]_i)(T_p - T_{\text{ig}})}{T_{\text{ig}} - T_r}\right]^{1/2} \tag{5.18}$$

Hence, the burning velocity depends on the thermal diffusivity, the average reaction rate, the initial fuel concentration, the flame temperature, ignition temperature, and initial temperature of the reactants. As noted previously, the actual ignition temperature in the flame is lower than the autoignition temperature because of diffusion of active species upstream from the reaction zone. Of course, the reaction rate is a strong function of temperature. Although Eq. 5.18 is incomplete because mass and heat transfer by diffusion have been neglected, it is nevertheless an interesting approximate model of the laminar flame. From Eqs. 5.17 and 5.18 the flame thickness is inversely proportional to the square root of the reaction rate, and when the equations are evaluated, the flame zone is indeed thin (i.e., less than 1 mm; see Problem 5.3).

EXAMPLE 5.2. Estimate the laminar burning velocity of a stoichiometric propane-air mixture initially at 298 K and 1 atm pressure using the thermal flame theory with a one-step global reaction rate.

Solution. From Table 3.2, the adiabatic flame temperature is $T_f = 2270$ K. From Table 5.3, $T_{ig} = (470 + 273) = 743$ K. In order to find the burning velocity from Eq. 5.18, the reaction rate is determined from Eq. 4.15 with the help of Table 4.3, which yields,

$$r_f = -8.6 \times 10^{11} \exp\left(\frac{-30{,}000}{1.987T}\right) \cdot (n_f)^{0.1}(n_{O_2})^{1.65}$$

The reaction is

$$C_3H_8 + 5(O_2 + 3.76N_2) = 3CO_2 + 4H_2O + 18.8N_2$$

Thus initially,

$$x_f = \frac{1}{(5)(4.76) + 1} = 0.0403 \qquad x_{O_2} = 5x_f$$

$$p_f = x_f p = 0.0403 \text{ atm}$$

and

$$n_f = \frac{p_f}{RT} = \frac{0.0403 \text{ atm}}{(82.05 \text{ cm}^3 \cdot \text{atm/gmol} \cdot \text{K})T}$$

$$n_{O_2} = 5n_f$$

For $T = 298$ K, $n_{fi} = 1.65 \times 10^{-6}$ gmol/cm^3 and $n_{O_2} = 8.25 \times 10^{-6}$ gmol/cm^3.

The question now is what temperature to choose to evaluate the average reaction rate and the properties. Let us try $T = (298 + 2270)/2 = 1284$ K and assume that the fuel and oxygen mole fractions are half the initial values. Then,

$$n_f = 1.91 \times 10^{-7} \text{ gmol/cm}^3 \qquad n_{O_2} = 9.55 \times 10^{-7} \text{ gmol/cm}^3$$

and $\qquad \bar{r}_f = -1.67 \times 10^{-4}$ gmol/cm$^3 \cdot$s

Using property values for air from Appendix B, $\tilde{k} = 0.0828$ W/m$\cdot$K, $\rho = 0.275$ kg/m^3, $c_p = 1.192$ kJ/kg$\cdot$K, and $\alpha = \tilde{k}/\rho c_p = 2.53$ cm^2/s.

From Eq. 5.18,

$$\underline{V}_L = \left[\frac{(\alpha\bar{r}_f/[n_f]_i)(T_p - T_{ig})}{T_{ig} - T_r}\right]^{1/2}$$

$$= \left[\frac{\dfrac{(2.53\ \text{cm}^2/\text{s})(1.67 \times 10^{-4}\ \text{gmol/cm}^3 \cdot \text{s})}{1.65 \times 10^{-6}\ \text{gmol/cm}^3}(2270 - 743)}{743 - 298}\right]^{1/2}$$

$$= 30\ cm/s$$

The correct burning velocity is 38 cm/s. Hence, the simplified thermal theory can be used to give an indication of the burning velocity, but a rigorous treatment requires numerical solution of the differential equations 5.5 through 5.8 with an appropriate set of reactions.

5.3 TURBULENT PREMIXED FLAMES

Combustion in furnace burners and engines utilizes turbulent flames, in which the heat release is much faster than with laminar flames. Turbulent flow increases the flame propagation, but there is no evidence that the turbulence substantially alters the chemistry. As the theory stands now, there is no practical universal method of predicting turbulent flame behavior from first principles.

Turbulent flames in burners are steady open flames, whereas flames in internal combustion engines are unsteady, enclosed flames. Hot-wire and laser Doppler anemometer measurements of turbulence upstream of the flame show no appreciable increase in turbulence ahead of the flame due to the flame for open flames, but indicate a small increase for enclosed flames such as a flame propagating in a tube. In order to determine the true flame velocity of enclosed flames, it is necessary to consider the flame area as well as the motion of the reactants. Two other factors must also be recognized in engines. First, the geometry of the cylinder coupled with the product expansion can cause increased turbulence. Second, the nature of the turbulence structure in engines is poorly known, in part because of the difficulty of defining turbulence in unsteady flows.

Turbulent flames may be categorized broadly as either weakly turbulent flames, wrinkled reaction sheets, or distributed reaction zones. The weakly turbulent flame is an extension of the laminar flame. The concept of the thin reacting flame sheet wrinkled by turbulence, although complicated, is less difficult to understand than the distributed flame front which can take place at higher turbulence levels in gas turbines, internal-combustion engines, and burners. In the distributed flame, the reaction zone is thought to contain small lumps of reactant which are entrained in the zone and burn up as they travel through the zone. The chemical reactions may not be completed in the reaction zone, and thus some additional reactions may take place in the "post-flame" products. Engulfment of reactants can take place in highly wrinkled flames, creating an occasional island or finger of burning reactant. Thus the distributed reaction is seen to be a limiting case where all of the fuel burns in entrained lumps.

Each of the three turbulent flame types will be discussed, but first the length scales and time scales appropriate to turbulent combustion are explained.

Turbulent Length and Time Scales

A semiquantitative diagram can be constructed to sort out the various flame regimes in terms of dimensionless parameters. To understand this it is first desirable to characterize the turbulent flow in terms of a few parameters. Two parameters which are appropriate are the turbulent intensity and the turbulent scale (the size of turbulent eddies). For steady flow, the *turbulent intensity,* $\underline{V}'$, is the root-mean-square of the velocity fluctuations, which are superimposed on the steady velocity. This concept is more difficult to apply to unsteady flows, where the low-frequency (below about 300 Hz) velocity changes represent unsteady bulk flows and are not traditionally treated as turbulence. Although this distinction for unsteady flow is not fundamental, it is a useful concept and a practical approach for modelers.

While measurements of $\underline{V}'$ can be carried out using laser-based instruments (laser Doppler velocimetry, for example) even in reacting flows, the experimental characterization of eddy size remains difficult. Basically, measurements of turbulent velocity made at two locations should correlate if they are inside an eddy. Thus such measurements give a statistical measure of the largest-size eddies, which is called the *integral scale,* L_I. The turbulent scales are distributed over a wide range, down to the smallest eddies, which exist only for a very short time due to viscous (molecular) dissipation. This smallest size is characterized by the *Kolmogorov scale,* L_K. Kolmogorov showed that

$$L_K \approx \left[\frac{\nu^3 L_I}{(\underline{V}')^3}\right]^{1/4} \tag{5.19}$$

For engineering applications concerned with the overall flow, the *Reynolds number* is the most useful measure of turbulence. However, the very local interaction of eddies with the reaction zone is of more concern here than is the global picture. Thus a *turbulent Reynolds number* based on the turbulent intensity and the integral scale is defined by

$$\text{Re}_I = \frac{\underline{V}'/L_I}{\nu} \tag{5.20}$$

Sometimes it is useful to use L_K in place of L_I, giving Re_K. Recall that the Reynolds number represents the ratio of inertia to viscous forces, which can be seen by writing,

$$\text{Re} = \frac{\rho L \underline{V}}{\mu} = \frac{\rho \underline{V}^2}{\mu \underline{V}/L} \tag{5.21}$$

Thus a large Re_I indicates that inertia will dominate over the dissipative effects of molecular viscosity, and Re_I is a measure of how much the larger eddies are damped by viscosity. For turbulence to occur at all requires $\text{Re}_I > 1$, and often in practice $\mathrm{Re}_I \gg 1$. For example, for hot gas flow at 1 atm pressure with $\nu \approx 1\ \mathrm{cm^2/s}$, $L_I \approx 0.1$ cm, and $\underline{V}' \approx 100$ cm/s, the turbulent Reynolds number is $\mathrm{Re}_I = (100)(0.1)/(1) = 10$. In some applications a pressure of 100 atm is common, and for this example Re_I becomes 1000. Empirical correlations of turbulent flame speeds have sometimes successfully used Re_I as a correlating parameter.

The characteristic *chemical reaction time* is an important time scale which can be estimated from the laminar flame thickness divided by the laminar flame speed, $\delta/\underline{V}_L$. This time scale is useful for characterizing the wrinkled reaction sheet regime. For turbulent eddies the "eddy turnover time," $L_I/\underline{V}'$, for the largest eddies gives a useful *turbulent time scale.* The ratio of the turnover time to the chemical reaction time is defined as a *Damköhler number,* Da_I. For high values of Da_I (10^3–10^4) the chemistry is very fast compared with the turbulence, and reaction sheets are observed. It is interesting to note that $\sqrt{\mathrm{Re}_I \mathrm{Da}_I} \propto L_I/\delta$ and thus the reaction sheet concept is associated with large values of L_I/δ. Also, if $\mathrm{Da}_K > 1$, then all of the eddies are larger than δ and the reaction sheet will wrinkle.

Using the concepts outlined above one may construct a diagram by plotting Da_I versus Re_I on a log-log scale—a diagram sometimes called a *Borghi diagram,* although the diagram has also been developed extensively by F. A. Williams. The diagram as shown in Figure 5.11 has been simplified by deleting the many

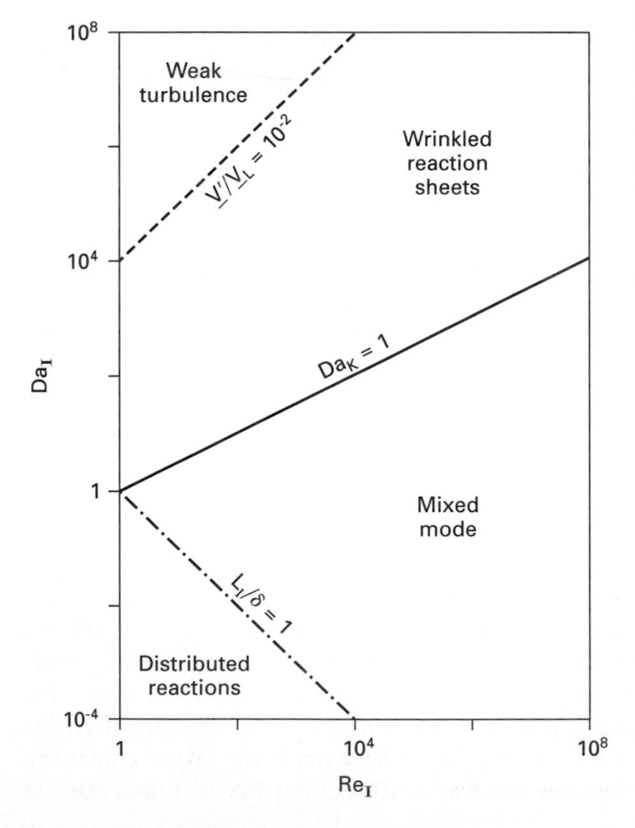

FIGURE 5.11
Diagram showing regions of premixed flame types on a plot of Damköhler number vs. Reynolds number [adapted from Williams].

constant-parameter lines which can be drawn to define various regions [see Williams]. We show only three distinct regions: weak turbulence, reaction sheets, and distributed reactions. The reaction sheet zone may be further characterized by the magnitude of $\underline{V}'/\underline{V}_L$. For values larger than unity we may expect the reaction sheets to be multiply connected, i.e., bending back on itself in some places. The region designated as "mixed mode" is currently not well understood. For this region the reaction sheet may contain holes as it responds to different scale eddies in the turbulent flow. An excellent phenomenological description of these various regimes, based on experimental observation, has been given by Ballal and Lefebvre.

Weak Turbulent Flames

The first published work on turbulent flames was by Damköhler, who studied propane-air flames using burners of small diameter (1.3 to 2.7 mm). The general trend found by Damköhler and most other later investigators was that the turbulent burning velocity depends on the laminar burning velocity and a turbulence factor which is a function of the intensity of turbulence. For cases where a Reynolds number may be identified (as in pipe flow), the turbulence factor may be correlated in terms of the approach Reynolds number. For very small-scale turbulence, where the scale of turbulence is of the same order as the flame thickness, the flame velocity increases but the flame stays relatively smooth. This suggests that the mechanism is one of increasing heat transfer and diffusion causing the burning velocity to increase. If that is the case, the ratio of turbulent to laminar burning velocity is given approximately by

$$\frac{\underline{V}_T}{\underline{V}_L} = \left(\frac{\alpha_T}{\alpha_L}\right)^{1/2} \tag{5.22}$$

The ratio of turbulent to molecular diffusivities, α_T/α_L, can be obtained from turbulence models. However, because of the high temperature in flames, the viscosity is very high and small eddies of size much less than δ are quickly dissipated. Thus there is reason to believe that eddies within the flame may not have the simple effect indicated by Eq. 5.22. The weak turbulence regime has not been studied as extensively as the reaction sheet regime, so a better understanding awaits further work.

Wrinkled Reaction Sheets

For larger-scale turbulence the flame becomes wrinkled, and thus the flame area becomes larger than that of a similar laminar flame. At these conditions ($Re_d > 4000$ for tube burners), the flame velocity is less dependent on the laminar burning velocity and thus less dependent on fuel-air ratio and fuel type. The velocity is 3 to 5 times the laminar burning velocity and the flame is brush-like.

The fact that turbulent flames are wrinkled with a relatively thick flame brush (5 to 6 mm) means that measuring the flame velocity is more difficult than for laminar flames. High-speed direct photographs show small "flamelets." These small tongues of flame fluctuate, thus making a simple area measurement difficult. Smoke particle track photography has been used to measure the flow angles and average flame area in a slot burner with some success, showing that the measured turbulent flame velocity divided by laminar flame velocity is essentially equal to the increase in flame area due to turbulence. An excellent two-dimensional technique uses a pulsed laser light sheet to cut through a flame in a mixture containing submicron particles of titanium dioxide (Figure 5.12). Mie scattering of the laser light allows imaging of the unburned gas, because the particles burn up in the flame. Use of multiple laser sheets can expand the technique to give rough three-dimensional visualizations [Mantzaras et al.].

The wrinkled turbulent flame has a surface which is continually changing with time; however, simple theory assumes a time-averaged surface area. The turbulent flame velocity is approximated to be locally equal to the laminar velocity, but because the flame surface is wrinkled, it has an increased area. For a one-dimensional turbulent flame of smoothed area A_s it follows that,

$$\underline{V}_T A_s = \underline{V}_L A_w \tag{5.23}$$

where A_w/A_s is the factor of increased area due to wrinkling. Various forms of geometric approximations can be used to show that

$$\frac{A_w}{A_s} = 1 + \frac{C\underline{V}'}{\underline{V}_L} \tag{5.24}$$

so that

$$\underline{V}_T = \underline{V}_L + C\underline{V}' \tag{5.25}$$

where $\underline{V}'$ is the root-mean-square turbulent fluctuation velocity and the proportionality constant C is between 1 and 2 in value.

For many years investigators have attempted to characterize the geometry of the wrinkled flame so as to be able to predict A_w and thus $\underline{V}_T$. A recent empirical method is to use the theory of fractals as a method of characterization in a statistical sense. A section of a wrinkled cross section of the flame such as one of those shown in Figure 5.12 is enclosed by a polygon of equal side length l. The wrinkled flame length L is measured as a function of l. For fractal behavior it is found that

$$L \propto l^{1-n_2} \tag{5.26}$$

where n_2 is the *fractal dimension* and lies between 1 and 2 in magnitude. A rough curve has a higher value of n_2 while a smooth curve has $n_2 = 1$. A similar concept may be extended to the area of a 3-D surface so that

$$A_w \propto l^{2-n_3} \tag{5.27}$$

where $2 \leq n_3 < 3$. In the absence of 3-D data for the wrinkled flame, the 2-D version is used to find n_2 and then the value of n_3 is approximated by $n_3 = n_2 + 1$. Values of $\underline{V}_T/\underline{V}_L$ obtained from fractal analysis in homogeneous-charge engines seem to agree with other estimated values, giving some hope that fractal analysis may be a useful tool in the correlation of wrinkled flames.

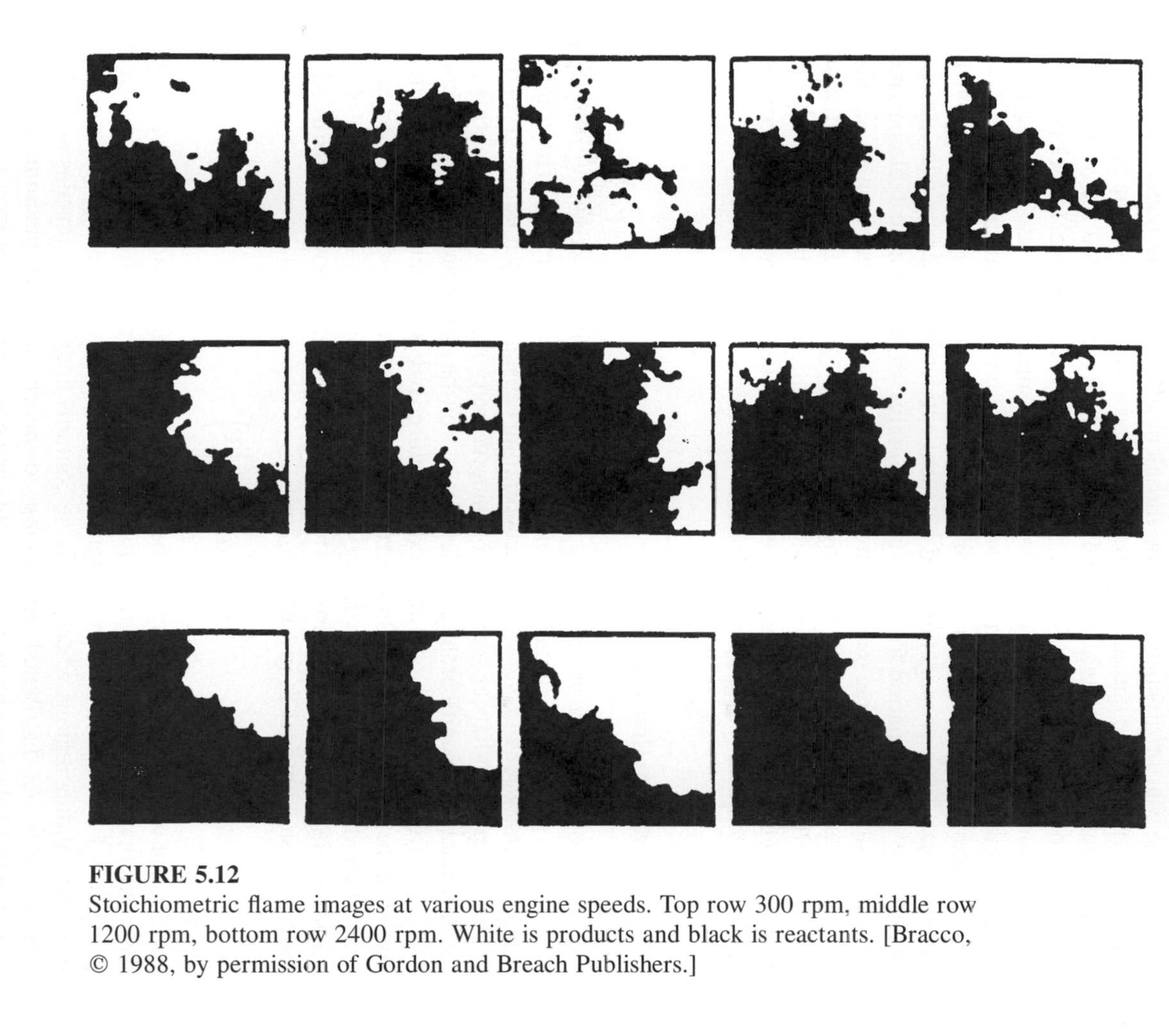

FIGURE 5.12
Stoichiometric flame images at various engine speeds. Top row 300 rpm, middle row 1200 rpm, bottom row 2400 rpm. White is products and black is reactants. [Bracco, © 1988, by permission of Gordon and Breach Publishers.]

For single wrinkled reaction sheets a number of recent works have used more sophisticated *flamelet* models to calculate the flame sheet area A_w per unit volume, called the *flame surface density* and often designated by the symbol Σ. In order to formulate such models it is necessary to also model the turbulence. The turbulence model often used is the so-called k-ϵ model in which a set of two differential equations is used to describe dk/dt and $d\epsilon/dt$, where k is the turbulent kinetic energy and ϵ is the rate of dissipation of the turbulent energy. A rate equation is then written for $d\Sigma/dt$. The ignition process is modeled separately up to the point where the flame can be treated as a wrinkled reaction sheet (see Section 5.6), i.e., where the flame ball is larger than L_I. If the flame is enclosed, the model requires modification as the flame approaches an enclosure surface. The rate equation for Σ contains terms for convection by the bulk flow and by turbulent diffusion, for growth due to the local strain caused by the turbulence, for consumption by the local laminar flame propagation, and for local extinction caused by overstraining at higher $\underline{V}'$. The equation for Σ typically contains about five adjustable parameters, and thus the resulting mass burning rate per unit volume based on $\rho \underline{V}_L \Sigma$ must be viewed as quasi-empirical. To bring these models to where they can be trusted for engineering use will require considerable validation by comparisons with experiment. The validations will also need to examine the turbulence model, because Σ is found to be very sensitive to the strain term.

Distributed Reaction Zones

The effect of wrinkling increases the flame area, but intense wrinkling may also cause small pockets of reactant to be entrained into the reaction zone. The entrainment of small turbulent eddies and the wrinkling caused by larger eddies gives rise to the observation that the turbulent burning velocity is not sensitive to either the laminar burning velocity or the turbulent scale. For the distributed regime the flame can no longer be viewed as a coherent structure, but is a thick zone of reactant eddies embedded in products. In this region, increases in turbulent scale can cause a decrease in burning velocity. Energy from combustion goes into motion of the eddies as well as directed motion of the flame. Figure 5.13 shows comparisons of data with an empirical fit given by

$$\underline{V}_T = 6.4\underline{V}' \left(\frac{\underline{V}_L}{\underline{V}'} \right)^{3/4} \tag{5.28}$$

Recent work on the dividing line between wrinkled flames and the mixed mode indicates that the wrinkled flame region may apply to even higher strain rates than previously believed and shown in Figure 5.11. This may increase the importance of understanding regions where the flame contains fingerlike peninsulas and some islands but is not distributed. An attempt to incorporate such phenomena into a fractal wrinkled flame model has recently been carried out [Matthews et al.]. Such models are particularly needed for highly turbulent combustion under conditions where products mixed with the reactants cause $\underline{V}_L$ to be lower so that $\underline{V}'/\underline{V}_L$ becomes high. An example is an automobile engine under conditions of rapidly

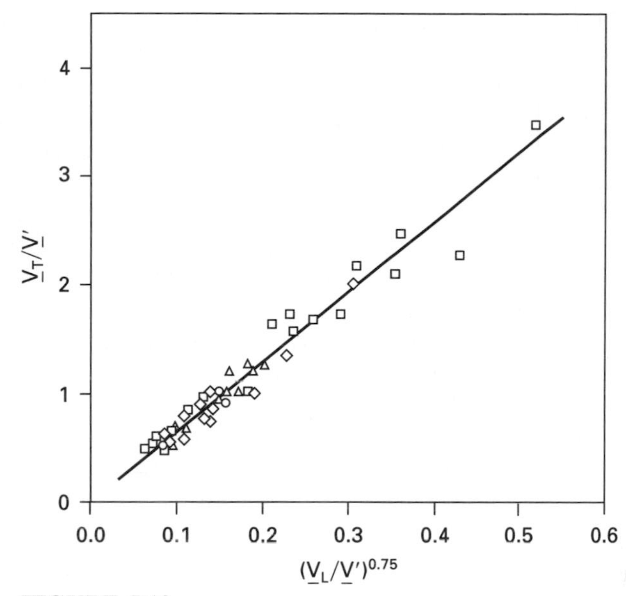

FIGURE 5.13
Comparison of distributed flame model with data [Gülder, 1991, by permission of Gordon and Breach Publishers].

closing throttle while the engine rpm is still high (i.e., driver takes foot off gas pedal while the car is traveling at high speed).

Much of the modeling of turbulent flames is still in the research stage, and better understanding of interaction between turbulence and flamelets is needed to advance the theory. One such source of understanding may come from *direct numerical simulation* (DNS) in which the fundamental conservation equations for a reacting fluid are solved directly without recourse to a turbulence model. Because of the complexity of such an approach, the application of DNS is currently very limited by computer speed and memory. However, the method has already produced helpful results for improved understanding [Rutland and Trouve].

5.4 EXPLOSION LIMITS

When a volume of premixed gas is suddenly heated by rapid compression, by rapid mixing, or by rapid heat transfer, an explosion may occur. It is often observed that there is a delay (a dwell period) before rapid reaction (i.e., explosion) takes place. For some pressures and temperatures the mixture will not explode at all. Thus a plot of the limit line separating explosions from no explosions on a p-T diagram can be created for a given mixture as shown in Figure 5.14. Explosion limits are different from flame propagation limits. In flame propagation, the high-temperature reaction zone propagates into the reactants and either is sustained or goes out. In the explosion limits, the mixture is homogeneous with no flame zone to provide a

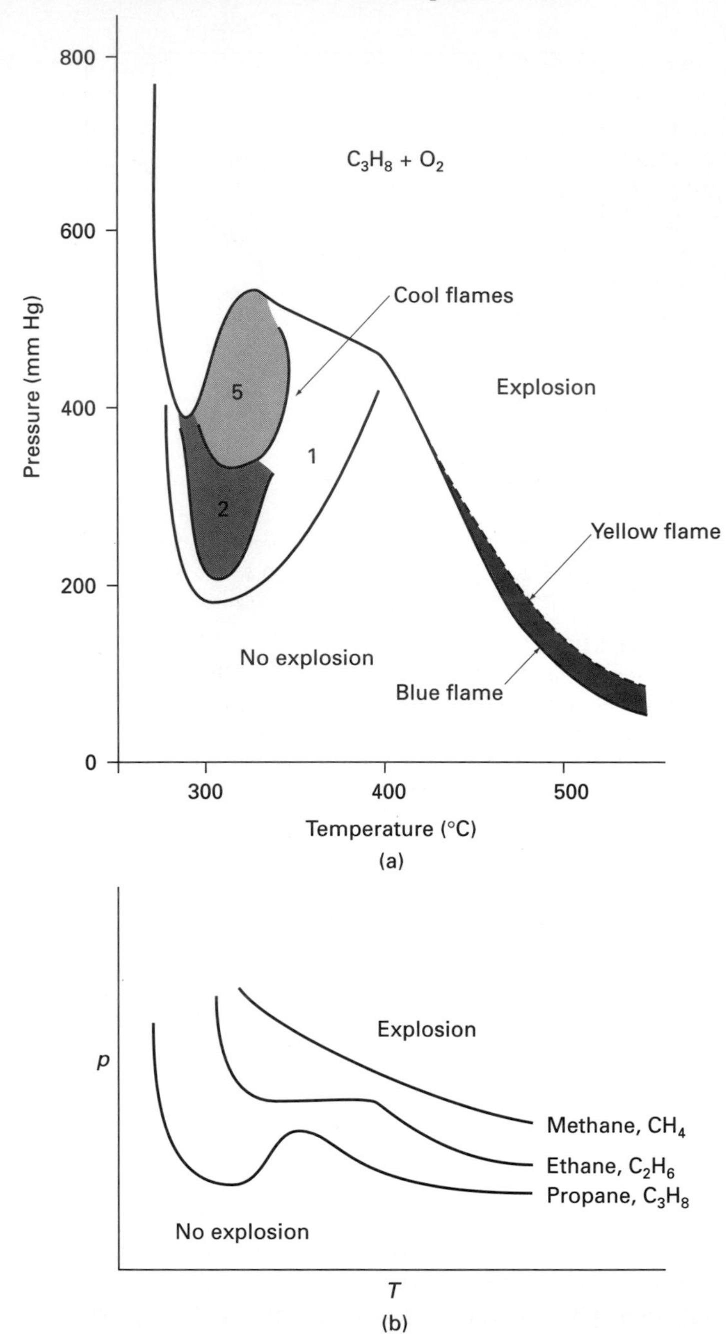

FIGURE 5.14
(*a*) The diagram shows the ignition behavior of an equimolar mixture of oxygen and propane. Cool flames of number indicated are observed in the regions labeled. For example, at 360 mm Hg pressure five cool flames are observed for 300–340°C, one cool flame for 345–385°C, and beyond 385°C a bright luminosity but no flame is observed until 425°C. At 425°C a bright blue flame is observed. At lower pressures the blue flame turns yellow as the temperature is increased slightly [Newitt and Thornes]. (*b*) Typical explosion diagram shapes for propane, ethane, and methane. Methane exhibits no cool flames.

source of radicals and high temperature. Thus, for homogeneous reactions the radicals, which are required to cause rapid reaction, must be built up from reactions within the mixture itself. If the destruction of the radicals predominates over their creation, the reaction will not provide an explosion. If the reactions release significant energy, then a thermal self-heating mechanism for causing explosion exists.

Explosion limit diagrams can be quite complex, exhibiting various zones or limit lines. During the dwell period most high–molecular weight hydrocarbon fuels have some p-T regions where cool flames are observed. The energy release due to these reactions is very low, and thus the term *cool* is used. The dwell or induction period may be quite long, with one or more chemiluminescent flames traveling through the mixture (Figure 5.14*a*). The cool flame period is followed by a thermal explosion. Such two-stage ignition is not observed for methane but is observed for propane and most other higher hydrocarbons, as indicated in Figure 5.14*b*.

The major application of explosion limits is to combustion in engines where the interest is in preventing *knock* in spark-ignition engines (see Chapter 7) and promoting ignition in compression-ignition engines (see Chapter 12). For the homogeneous spark-ignition engine, the last portion of reactants to burn (the "end-gas") can explode giving a knocking sound. The kinetic mechanisms for complex fuels are unknown, but some suggested synthetic two-stage processes have provided models for determining knock. Study of the end-gas temperature has indicated that for low-octane fuels, considerable energy is released in the end-gas prior to knock. Wall effects are also known to affect knock. The buildup of deposits on the engine surfaces causes an increase in octane requirement. This may indicate that reactions are initiated at the wall and propagate into the end-gas by reaction waves or turbulent diffusion.

The study of homogenous compression ignition is difficult to relate directly to ignition delay in a diesel engine, because temperature and mixture strength at the site of first ignition are not known accurately. Empirical expressions for diesel ignition delay are often correlated by equations containing an $\exp(C/T)$ term. This indicates that chemical kinetic rates dominate ignition delay in diesels.

5.5 FLAME QUENCHING

The basic problem of *flame quenching* is to determine the maximum size of a passage such as a tube diameter, orifice size, or distance between plates through which a flame will not propagate. If the flame is to propagate, the energy release due to chemical reaction must keep the reaction zone temperature high enough to sustain a rapid reaction. If heat transfer to the surrounding surfaces is high enough, the temperature will drop and the reaction will slow down. As the reaction slows down, the energy release rate is lowered, the temperature drops below the ignition temperature, and the flame will be quenched. The same phenomena can be observed in combustors where the boundary layer is cooled below the ignition temperature by metal surfaces.

For a flame propagating in a tube, the ratio of flame-exposed tube surface to flame volume is $4/d$. For a given mixture there will be some tube size, d_0, for which the flame will no longer propagate. The tube size is reduced until the flame no longer propagates. The tube may be replaced by parallel plates (a rectangular slot) or by a conical shape. In the case of the cone the flame is started at the large end and the location at which the flame goes out gives the quench diameter.

In the *flashback method* a mixture flows through an orifice (or tube) and is ignited downstream. If the flow velocity is decreased to below the burning velocity, the flame will try to burn back upstream through the orifice (the flame flashes back). For a small enough orifice the flame cannot pass through and is quenched. If the mixture velocity is increased, the flame height will increase until at some point the flame blows off. For blowoff the flame is no longer held or anchored by the heat transfer and fluid flow at the boundary layer which forms its base, and the base of the flame rises above the burner.

A method of extrapolation has been used in conjunction with the flashback method in several ways. Suppose that the orifice diameter is held constant and the flashback experiments performed with various fuel concentrations. At each concentration the mass flow rate at which flashback will take place is noted. A plot of fuel concentration versus mass flow at flashback is made and the curve extrapolated to zero mass flow. This method then gives the fuel concentration for flashback under zero-velocity conditions and the given orifice (tube) diameter. This method is superior to simply stopping the flow or adjusting the flow point, since it is known that a flame can be supported above a tube diameter slightly smaller than the quench distance. Another type of plot sometimes used is obtained by varying the ambient pressure for a fixed fuel concentration and orifice size. At each pressure the mass flow rate for flashback and blowoff is observed. If the data are plotted as flow velocity versus pressure, the flashback velocity and blowoff velocity coincide at a given low velocity and pressure. This gives the point at which the flame cannot be supported.

Experimental Results

Quench distance d_0 depends on the geometry of the walls, fuel type, stoichiometry, pressure, reactant temperature, and turbulence. Plots of quench distance versus equivalence ratio are essentially parabolic, with the minimum on the rich side of stoichiometric. On the lean side, the quench distance is inversely proportional to burning velocity. On the rich side, quench distance generally decreases slightly with increasing molecular weight of the fuel. Figures 5.15 and 5.16 show data for parallel-plate quenching of propane-air mixtures. The quench distance is roughly inversely proportional to pressure. For stoichiometric propane-air, d_0 is proportional to $p^{-0.88}$, and for hydrogen-air d_0 is proportional to $p^{-1.14}$. As a rough approximation, if the burning velocity is proportional to the s power of pressure, then d_0 is proportional to $p^{-(1+s)}$ for low pressures. The effect of reactant temperature T_0 on quench distance is very approximately given by $d_0 \underline{V}_L / T_0 = \text{constant}$.

The effect of turbulence is less well known, but increased turbulence should increase the heat transfer and thus increase the size of the quench distance. Other

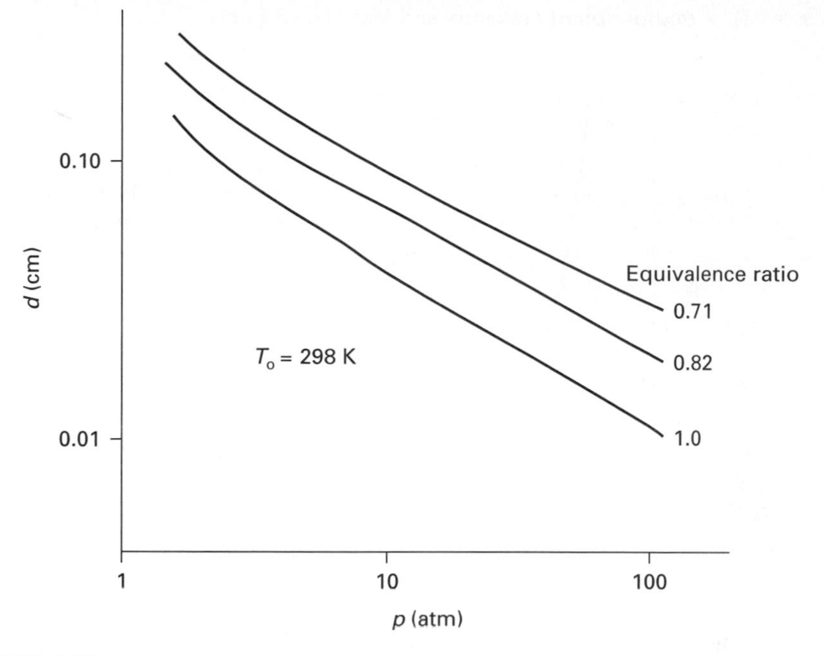

FIGURE 5.15
Effect of pressure on quench distance for propane-air mixtures [Reprinted by permission of Elsevier Science Inc. from "Quenching Distances of Propane-Air Flames," by Green and Agnew, *Combustion and Flame,* vol. 15, pp. 191–198, © 1970 by The Combustion Institute].

hydrodynamic effects such as buoyancy are known to affect the quench distance. For lean flames in tubes, the quench distance is 10% larger for downward than for upward propagation. For rich flames, instability can occur and the quench phenomena are complex.

Wall effects are mainly due to heat transfer and not diffusion of species. Coating of walls seems to have little effect on the quench. Heating of the wall can reduce the quench thickness by reducing heat transfer, but since the wall temperature is typically low compared with the flame temperature, the effect is small. Most data on quench distances are taken at subatmospheric pressures. The sample data in Table 5.6 were extrapolated to 1 atm for parallel-plate quench distances. These values increase with increased air-fuel ratios and decrease with increased pressure. The quench distance for parallel plates is 65% less than the circular quench diameter.

TABLE 5.6
Quench distance (mm) for various stoichiometric fuel-air mixtures at 1 atm and 20°C

Hydrogen	0.6	Isooctane	2.0
Methane	1.9	Methanol	1.8
Propane-air	2.1	Propane-He-O_2*	2.5

Source: Bartok and Sarofim, *Fossil Fuel Combustion: A Source Book,* © 1991, Wiley, by permission of John Wiley and Sons, Inc.

* He replaced N_2 in air proportions.

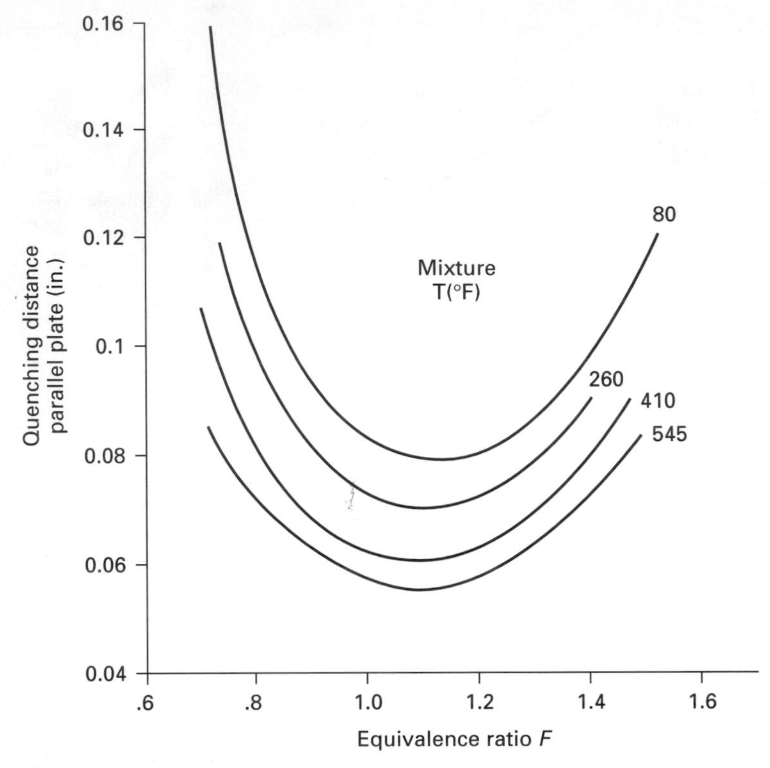

FIGURE 5.16
Quench distance for a slit of width d_{11} versus propane-air equivalence ratio F for various reactant temperatures. Pressure is 0.97 atm, and flow is laminar [Friedman and Johnston, by permission of Journal of Applied Physics].

Quench Theory

The theory of flame quenching is based primarily on the calculation of heat transfer from the flame reaction zone to the quench-producing surface. Many geometries may be considered; the most common are: (*a*) a flame approaching a wall head on, and (*b*) a flame traveling between two walls or through a hole. Although not associated with extinction, the case of a flame traveling parallel to a wall also produces a quench layer which typically disappears as it mixes with the hot post-flame products.

For the case of a flame approaching a wall head on, the process becomes unsteady after the time when the edge of the preheat zone encounters the wall boundary layer. The heat transfer then begins to increase, causing the reaction zone temperature to decrease and the flame to slow down. The rate of energy release, $\rho_1 \underline{V}_L c_p \Delta T_f$, where ΔT_f is the temperature rise through the flame, thus decreases. The reaction ceases at some point, leaving a layer of partially reacted mixture of thickness d at the wall surface. This process can be modeled for the laminar case using the one-dimensional energy equation for the flame in unsteady form, provided the chemical kinetics for the flame reactions are known for the whole temperature range from adiabatic flame temperature down to the temperature at quench.

A qualitative grouping of the parameters can be obtained from a simple phenomenological model. Consider the fraction formed by the rate of energy release by reaction divided by the heat conducted to the wall at the quench conditions:

$$\frac{\rho_1 \underline{V}_L c_p \, \Delta T_f}{\tilde{k}_1 \, \Delta T_c/d_0}$$

Here $\partial T/\partial x$ at the wall has been set equal to $\Delta T_c/d_0$, ΔT_f is the temperature rise due to combustion, and ΔT_c is the characteristic temperature difference for the heat transfer to the surface based on the quench distance d_0. Introducing the thermal diffusivity, the ratio becomes

$$\frac{(\underline{V}_L d_0/\alpha_1) \, \Delta T_f}{\Delta T_c}$$

The quantity $\underline{V}_L d_0/\alpha_1$ is the dimensionless Peclet number. Correlations of the form Pe = constant, although quite approximate, are often used to represent the experimental data. For example, for a head-on wall quench,

$$\frac{d_0 \underline{V}_L}{\alpha_1} = 8 \tag{5.29}$$

To obtain a more accurate theory which can be used over a wide range of conditions, it is necessary to start from the unsteady-flame equations coupled with a detailed chemical kinetic model as in the computations of Westbrook et al. Figure 5.17 shows the computed results for a stoichiometric methanol-air flame at 10 atm pressure. The figure shows species profiles for both the unquenched flame and the flame at the time of quenching. The region near the wall where $T < 800$ K is depleted of active radicals for the quenching condition. In this calculation catalytic effects of the wall were neglected, but would have only added to the rapid depletion of radicals. An important point to note is that the fuel and stable intermediates in the wall region can be rapidly consumed as they are transported to the higher (1500 K) temperature region by diffusion. In the quiescent case the fuel concentration at the wall was significantly decreased in less than 2 ms.

Similar computations of quench can be carried out for the turbulent flow case which is of more practical importance. Unfortunately, such models have not yet incorporated the wrinkled flame theory. We may expect, however, that the fuel in the postquench period will be very rapidly consumed by turbulent transport to the higher-temperature regions. For low turbulence where the turbulent intensity, $\underline{V}'$ is less than twice the laminar burning velocity, experiments have shown that the head-on wall quench distance d_0 is given by

$$\frac{d_0(\underline{V}_L - 0.16\underline{V}')}{\alpha_1} = 8 \tag{5.30}$$

As might be expected based on the increase of heat transfer, the effect of turbulence is to increase d_0.

The case of a flame traveling between walls is more difficult to model, as it is two-dimensional. A two-dimensional model using simple one-step kinetics has

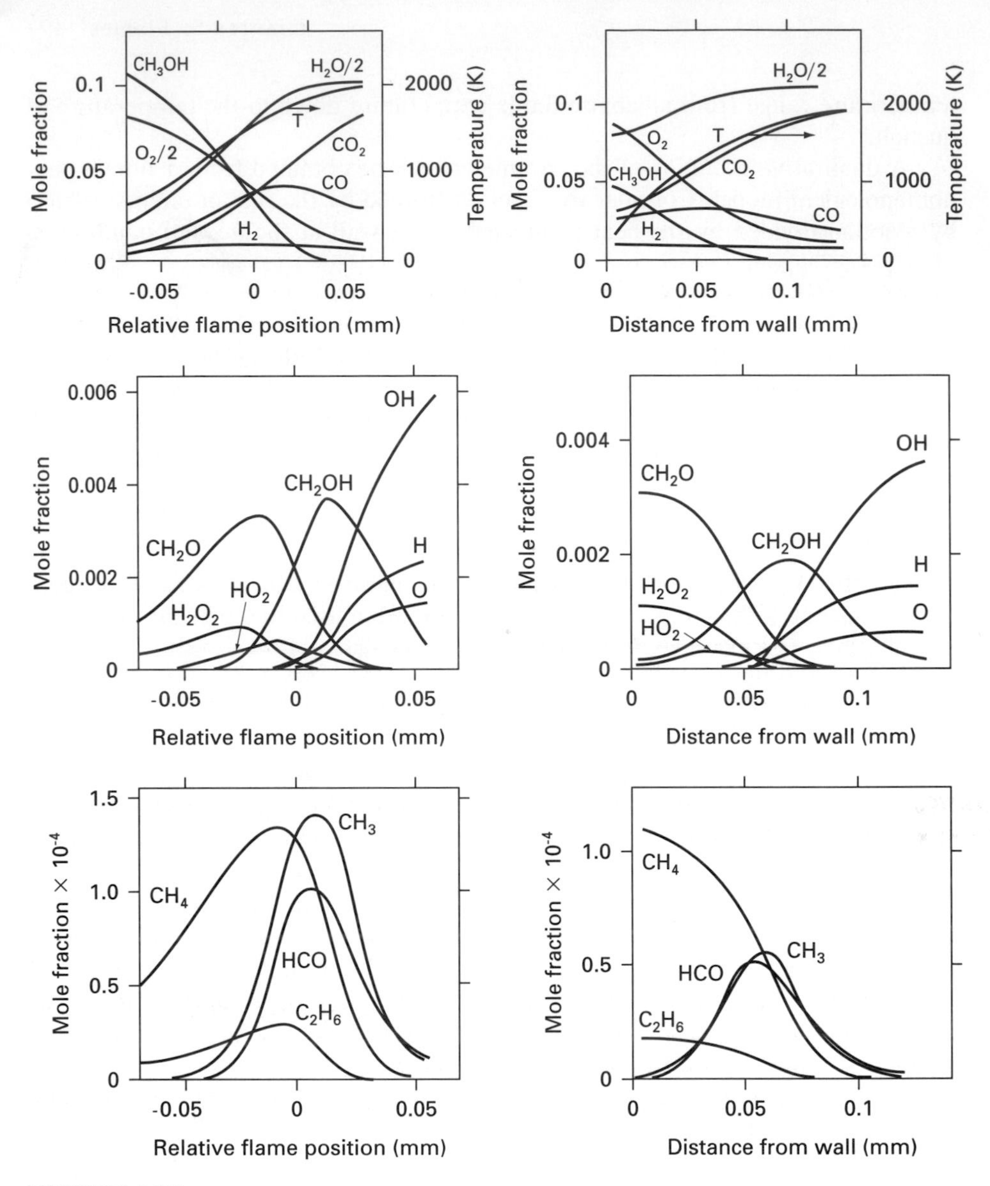

FIGURE 5.17
Premixed laminar flame of stoichiometric methanol-air at 10 atm pressure: (left column) temperature and species profiles for an unquenched flame; (right column) temperature and species profiles at time of flame quench. Wall temperature is 300 K [Reprinted by permission of Elsevier Science Inc. from "A Numerical Study of Laminar Wall Quenching," by Westbrook et al., *Combustion and Flame,* vol. 40, pp. 81–99, © 1981 by The Combustion Institute].

shown that the Peclet number at quench is a function of mixture ratio, but constant over a range of subatmospheric pressures. It is doubtful, however, that such an approach is adequate for a wide range of mixture temperatures and pressures.

Although the process of heat transfer from the flame to a solid surface is typically the cause of flame quenching, flames may undergo extinction by other means. For example, in the case of flames propagating in near-limit mixtures in engines, the flame velocity is low and the piston expansion can cause the products temperature to drop below the flame temperature. Thus, the flame reaction zone can lose heat to the product gases and the flame can be quenched. A similar phenomenon can be induced by heat transfer from the products to a solid surface.

Another cause of extinction may be flow parallel to the flame surface caused by the geometry of the flow or by the curvature of the flame front. In these cases the flame is strained. For the case of flame curvature, which leads to a change in flame area as the flame propagates, the strain may be expressed as a *flame stretch.* The amount of stretch is typically defined as the rate of change of the logarithm of the flame area, $d(\ln A)/dt = (dA/dt)/A$. Large enough values of stretch may cause extinction, especially if the flame is propagating in a very lean mixture.

Extinction by flame stretch can be illustrated by counterflowing streams of mixture in which a stagnation plane is formed in the products between two parallel flames. The outflow of products parallel to the flames causes stretching and extinction. A similar experiment by Law et al. used a single stream of lean propane-air mixture stagnating against a wall. In this case the wall could cause heat transfer from the products, but the result was insensitive to wall temperature, indicating that stretch caused extinction. The maximum flame temperature at extinction was about 1230 K over the range of reactant temperatures and mixture velocities tested. Increases in reactant temperature, wall temperature, or mixture velocity caused the flamc distance from the wall at extinction to decrease.

5.6 IGNITION

Ignition of a premixed gas is done with a spark, a glow plug, another flame, or possibly a laser. The basic idea is to raise a small amount of reactant to a temperature high enough to cause continued flame propagation after the ignition energy source is removed. The volume of gaseous reactant heated during ignition must be large enough so that when the ignition source is removed, the heat loss to the surroundings will not exceed the chemical energy release rate. In the case of spark plugs, the heat loss is to the electrodes and to the surrounding mixture. Looked at in the opposite way, the problem becomes to determine if the heat transfer will cause the flame to be quenched.

The most common type of igniter is the spark plug. Either an induction coil or a capacitor is used to store the electrical energy which is released during the rapid discharge. Figure 5.18 shows the relationship between minimum ignition energy and electrode spacing for two simple electrode configurations. The flanged electrode gives a well-defined minimum gap or quench distance. For gaps less than this minimum distance, the heat transfer to the plates causes quenching of the ignition. For the free electrodes, the minimum gap is not sharply defined, but the heat transfer to the electrodes per unit volume of spark nevertheless increases as the gap is decreased, causing the required energy to rise.

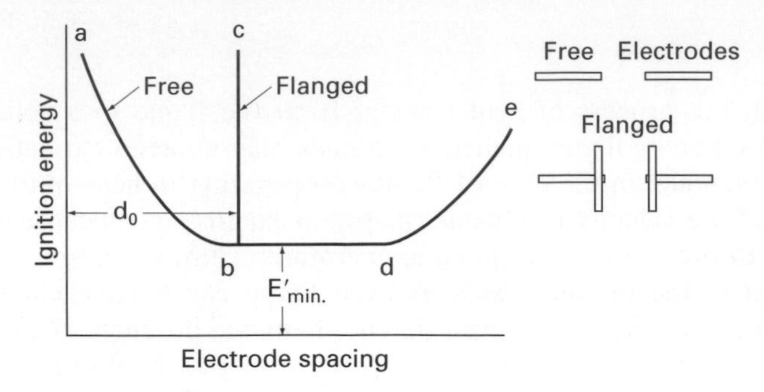

FIGURE 5.18
Minimum ignition energies for various electrode configurations.

The minimum spark energy is constant over the range of gap distances from b to d of Figure 5.18. The electrical resistance of the spark is primarily in the regions near the electrodes, and thus the increased gap from b to d will not greatly change the plasma characteristics. Similarly, the heat transfer in this region is primarily to the electrodes or flange and thus does not change very much. For the very wide spacing (d to e), the heat transfer to the surrounding mixture becomes important and the required energy rises.

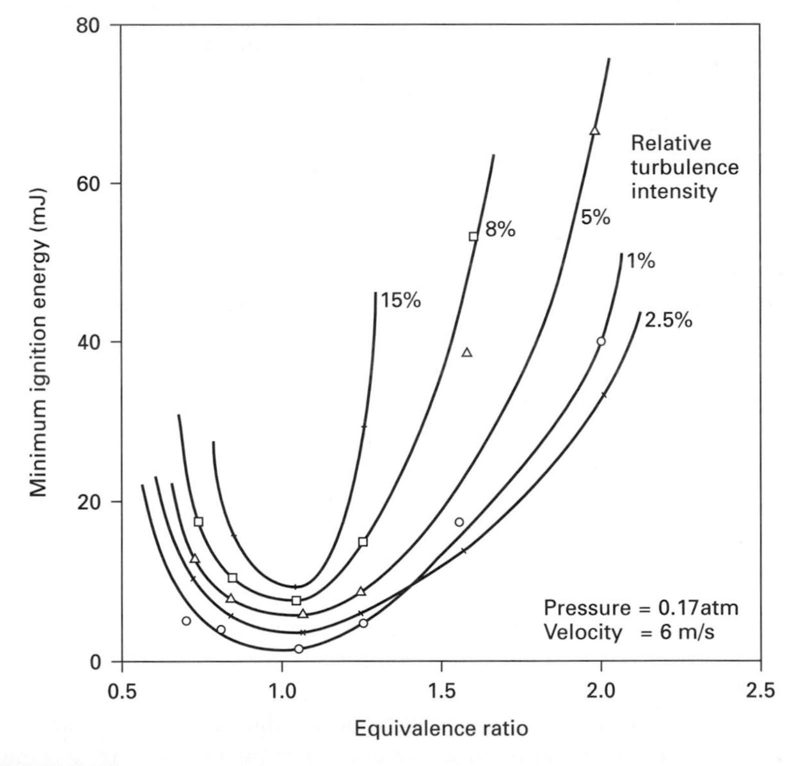

FIGURE 5.19
The effect of equivalence ratio and relative turbulence intensity for a propane-air mixture at a pressure of 0.17 atm. The relative intensity is the intensity ($\underline{V}'$) divided by the time-averaged flow velocity and expressed here as a percentage [Ballal and Lefebvre, 1975, by permission of The Royal Society].

Turbulence and mixture composition influence the minimum ignition energy (Figure 5.19). The minimum energy was found to increase with increasing flow velocity and increasing turbulence intensity. For the data of Figure 5.19 the turbulence intensity ranged from 0.15 to 0.90 m/s, and the reactant mixture temperature was 290 K. Two effects come into play as the turbulence intensity is increased: burning velocity increases, and this decreases the required energy; but heat transfer also increases, and this increases the required energy. The heat transfer increase dominates.

A simple criterion for successful ignition is that the rate of heat release in the small flame zone surrounding the spark should exceed the total rate of heat loss. This implies that the characteristic dimension of the flame kernel should equal or exceed the quenching distance. Taking a sphere of diameter d as the kernel geometry, the energy deposition to achieve ignition is

$$\mathrm{E}'(\text{minimum}) = \frac{c_p \rho \,\Delta T \pi d^3}{6} \tag{5.31}$$

where ΔT is the temperature rise due to combustion. Minimum ignition energy data are given in Table 5.7.

Using the experimental data for the gap d and minimum spark energy E', one can plot $\ln\,(E'/\rho c_p\,\Delta T)$ versus $\ln(d)$. Such a plot produces a straight line as predicted from the above relationship. Typically, the quench distance and the distance to obtain the optimum gap are taken equal. Ballal and Lefebvre, however, found that $d(\text{gap}) = 5d(\text{quench})$.

Low-pressure experiments, such as those just discussed, are not easily applied to an engine. From direct engine combustion experiments, we know that if the velocity past the plug is high, the kernel may detach from the plug and move downstream before it develops into a full-fledged flame front. In such cases the effective point of ignition is no longer the spark plug location. Such high flow velocities are not typical, but can be produced by high swirl. Under some conditions the plug may act as a flame holder and the flame may rotate, thus producing an "apostrophe" shaped burned-gas volume, the narrow end being at the spark plug. For more typical cases, the spark is located on axis and the flame kernel is only

TABLE 5.7
Minimum ignition energy in air at 1 atm and 20°C [Bartok and Sarofim, by permission of John Wiley & Sons]

Fuel	E' (10^{-5} J)
Methane	33
Ethane	42
Propane	40
n-Hexane	95
Isooctane	29
Acetylene	3
Hydrogen	2
Methanol	21

slightly distorted by the local turbulent velocity. However, this distortion changes the surface-to-volume ratio of the kernel and the wall surface area in contact with the kernel; both of these effects change the heat transfer. Such fluctuations in heat transfer affect the kernel growth rate, especially for lean mixtures, and lead to cyclic variations of the start of rapid combustion.

In a piston engine the cylinder gas turbulent intensity is closely related to the large-scale velocity patterns in the cylinder. These patterns are initiated by the flow as it enters the cylinder from the intake port. Swirling flow around the cylinder axis and vertical vortex motions called *tumble* are two common structures. These large-scale patterns are modified by piston velocity and chamber shape as they undergo compression. The decay of these large flows provides the major source of turbulence at the time of ignition. In the case of low swirl, tumble motions dominate. Tumble motions break down under compression more readily than swirl motions and thus give higher turbulence intensity at the time of ignition. Creation of a definite swirl pattern, however, reduces cyclic variations, while low swirl rates allow the less repeatable tumble motions to dominate, thus giving large cyclic variations. Thus the desire to create high burning velocities may lead to flows which cause cyclic variability because of the flow interaction with the flame kernel.

From basic experiments it can be concluded that the conditions which are most conducive to spark ignition and which lower the minimum spark energy are:

Low burning velocity
High initial temperature
High mean reaction rate
Low volumetric heat capacity ρc_p
Low thermal conductivity
High total pressure
Nearly stoichiometric mixture
Low intensity of turbulence
Electrode separation distance close to quench distance

In spark-ignition, homogeneous-charge engines, the lean limit of flame propagation is found to be extended by:

Increasing mixture homogeneity
Decreasing charge dilution
Increasing compression ratio
Decreasing engine speed
More central spark location
Use of multiple spark plugs

Although the minimum ignition energy is a fraction of a millijoule for a quiescent mixture, an order of magnitude more energy is typically required for a flowing mixture. Conventional spark-ignition engine ignition systems deliver 30 to 50 mJ of energy to the spark to be sure of positive ignition. Practical considerations concerning spark ignition are discussed in Chapter 7.

5.7 DIFFUSION FLAMES

Diffusion flames take place when the sources of fuel and oxidizer are physically separate so that the energy release rate is limited primarily by the mixing process. There is no fundamental flame speed as in the case of premixed flames, and the flames are not one-dimensional. Chemical kinetics plays a secondary role in the behavior of diffusion flames. Diffusion flames occur with flowing gases, with vaporization of liquid fuels, and with devolatilization of solid fuels. Flames from liquid and solid fuels will be considered in later chapters.

A candle flame, shown in Figure 5.20, is an example of a diffusion flame. Wax is melted, flows up the wick, and is vaporized. Air flows upward due to natural convection. The reaction zone is between the air and fuel zones. Air diffuses inward and fuel diffuses outward. In hydrocarbon flames, soot particles are produced, giving rise to intense luminosity.

The rest of the chapter will consider: (1) a jet of fuel flowing into quiescent air, creating a flare flame; (2) two concentric parallel, laminar-flowing streams of fuel and oxidizer which form a laminar diffusion flame; (3) a counterflow diffusion flame; and (4) a brief discussion of turbulent diffusion flame modeling.

Free Jet Flames

Consider a gaseous fuel which jets upward from a nozzle of diameter d_j into stagnant air as shown in Figure 5.21. As the velocity of the fuel jet is increased, the character of the flame changes. At low jet velocity the mixing rate is slow and the flame is long and smooth (laminar). The laminar flame height increases linearly

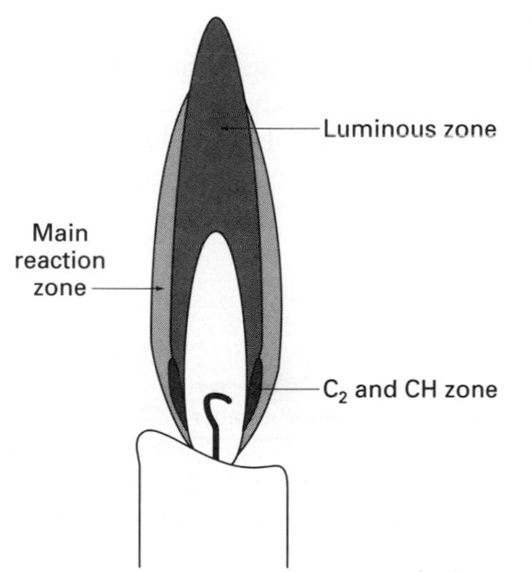

FIGURE 5.20
Candle flame in air showing reaction zones [after Gaydon and Wolfhard].

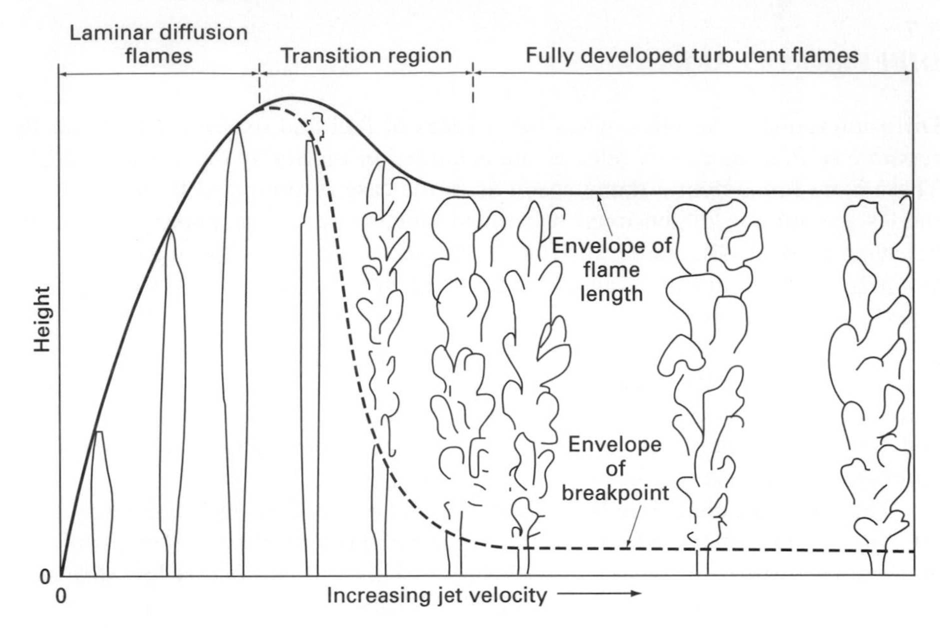

FIGURE 5.21
Free jet diffusion flame transition from laminar to turbulent flow [Hottel and Hawthorne, by permission of The Combustion Institute].

with jet velocity up to a point where the flame becomes brushlike (turbulent). The flame height decreases due to more rapid turbulent mixing. In the stable, fully developed turbulent region, the flame height is independent of jet velocity. The turbulent flame emits more noise than a laminar flame, and the yellow luminosity due to soot formation is decreased. The flame length is proportional to $\underline{V}_j d_j^2$ for laminar flow and proportional to d_j for turbulent flow.

The transition to a fully developed turbulent flame may be characterized by a transition Reynolds number. Interestingly, this transition Reynolds number is different for different fuels, indicating that the chemical kinetics as well as fluid mechanics plays a role in the combustion. Several transition Reynolds numbers are given in Table 5.8.

TABLE 5.8
Jet diffusion flame transition to turbulent flow [Hottel and Hawthorne, by permission of The Combustion Institute]

Fuel into air	Transition Reynolds number
Hydrogen	2000
City gas	3500
Carbon monoxide	4800
Propane	9000–10,000
Acetylene	9000–10,000

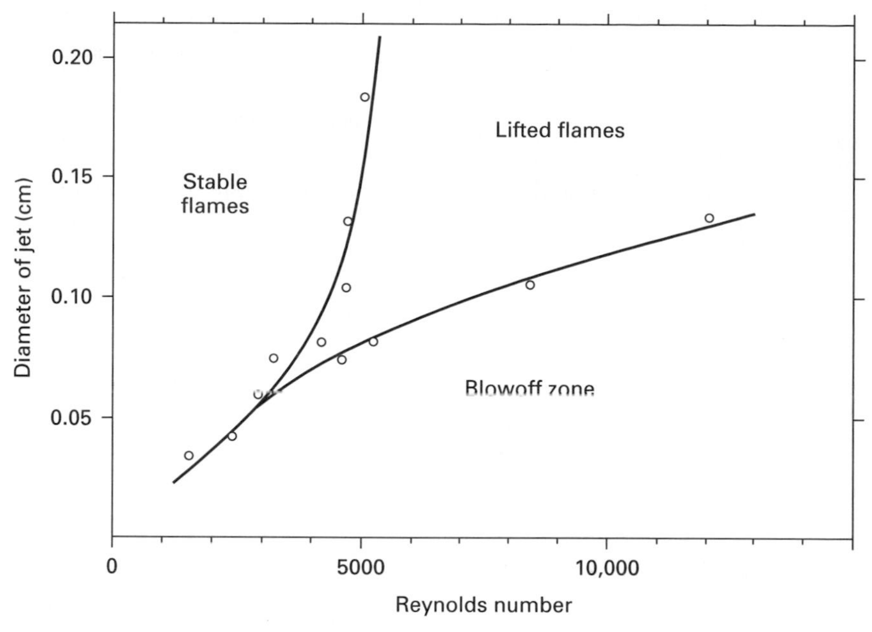

FIGURE 5.22
Stability regimes for free jet diffusion flame of ethylene-air mixture [Scholefield and Garside, by permission of The Combustion Institute].

As the jet velocity is further increased, a point is reached where the flame lifts off from the nozzle and exhibits a nonburning region at the bottom. A further increase in jet velocity causes the flame to blow off completely. A stability diagram showing the stable, lifted, and blowoff regions for an ethane-air flame is shown in Figure 5.22.

Concentric Jet Flame

Mixing of jets can be controlled by using two concentric tubes. Fuel flows in the inner tube and exits into a larger concentric tube which has flowing air. The velocity in each tube can be adjusted, and the overall fuel-to-air flow rate can be adjusted by selecting the tube diameters. Equal velocities produce a laminar diffusion flame, and different velocities produce a shear flow which can produce turbulence. The fuel moves outward into the air because of the concentration gradient, while the air moves inward. Fuel and oxygen are consumed in a flame zone when the fuel-air ratio is sufficient for combustion. The product species and inerts diffuse both inward and outward. An experimental setup to study such a flame is shown in Figure 5.23.

The laminar concentric jet burner of Figure 5.23 provides an opportunity to study the chemical kinetics of diffusion flames. These flames exhibit a range of fuel-to-oxygen ratios within the flame zone. Just to the fuel side of the flame zone, the fuel is at a high temperature and undergoes pyrolysis reactions which produce

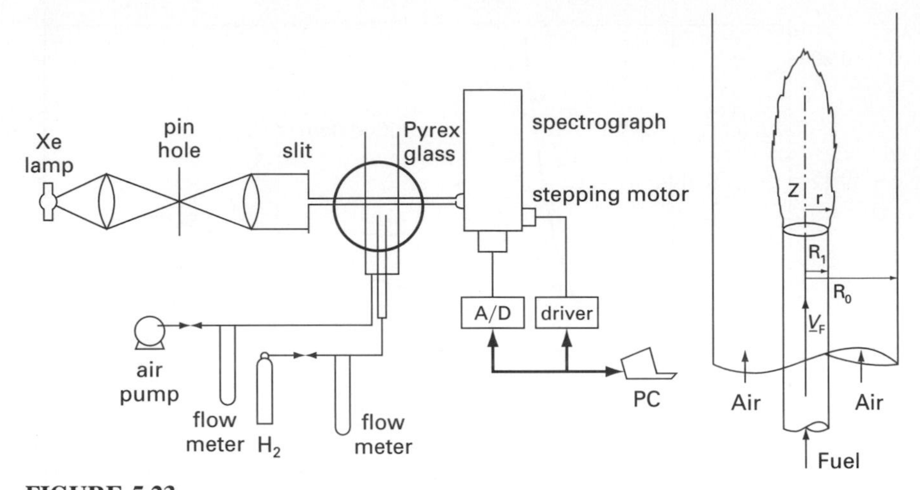

FIGURE 5.23
Concentric jet diffusion flame test setup [Fukutani et al., by permission of The Combustion Institute].

soot particles. Test results on a methane-air laminar diffusion flame are shown in Figure 5.24. The flame produces CO on the fuel side, and as the CO oxidizes to CO_2, the maximum flame temperature is reached. The NO peaks at the point of maximum flame temperature. The dotted lines indicate regions of soot where the sampling probe clogged. Considerable energy is radiated by this high-temperature soot, causing a reduction in peak flame temperature.

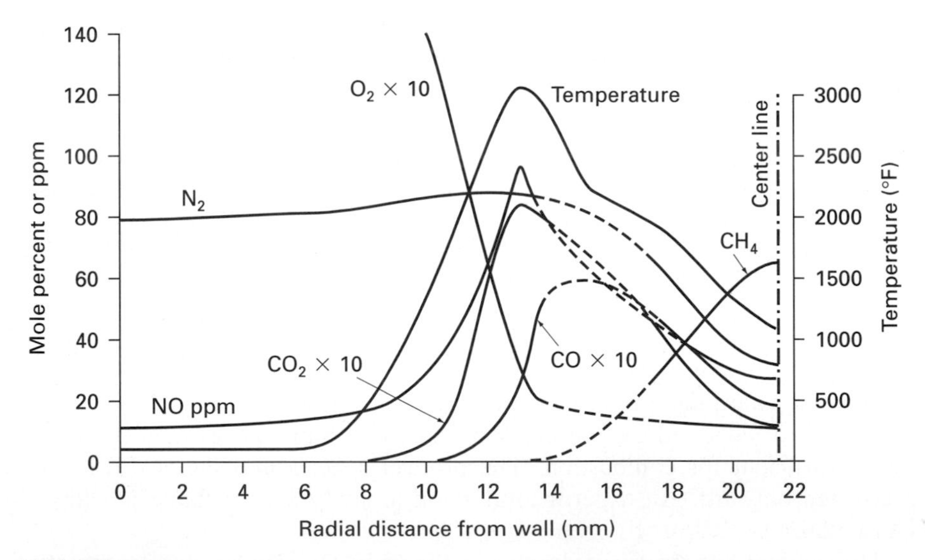

FIGURE 5.24
Methane-air diffusion flame temperature and concentration profiles, concentric jet burner (methane velocity 0.22 m/s, air velocity 0.41 m/s, overall fuel-air ratio 0.577, standard temperature and pressure initially, 17 mm above fuel inlet) [Tuteja].

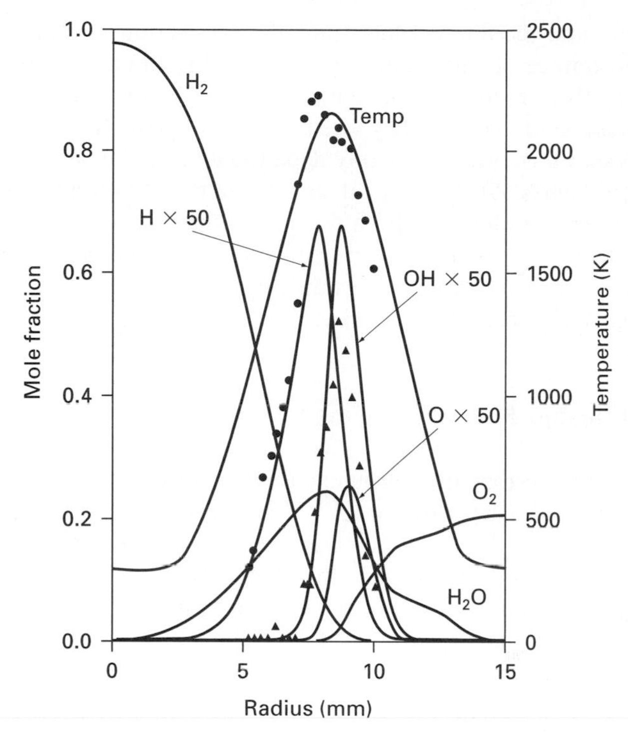

FIGURE 5.25
Concentration and temperature profiles in a hydrogen-air diffusion flame at 2 cm above fuel inlet; the hydrogen tube had a 5-cm radius. Circles are measured temperatures and triangles measured OH concentrations [Fukutani et al., by permission of The Combustion Institute].

Theoretical analysis of hydrocarbon diffusion flames is currently hampered by the ability to properly model the sooting reactions over a wide range of conditions. Thus, study of the simpler hydrogen-air diffusion flame offers a model which is free of soot chemistry and soot radiation. Fukutani et al. have used a 21-reaction scheme to compute the temperature and species concentrations in an axisymmetric hydrogen-air diffusion flame. Figure 5.25 shows some of the computed profiles and experimental data points. The hydrogen and oxygen each disappear at about the location of peak temperature, and the reactions are spread over a distance of several millimeters. Peak reaction rate values were found to occur at different locations as shown by the figure.

Counterflow Diffusion Flames

Diffusion flames in the forward stagnation region of porous cylinders have been studied for cylinders fed with either gaseous fuel or liquid fuel. Many studies have concentrated on the soot formation processes in the region between the forward stagnation point and the flame zone, since the pioneering work of Tsuji and Yamaoka. The advantage of this geometry is the ability to make line-of-sight measurements along the direction of the cylinder axis with only small effects from the end regions. It is found that fuel flow rate has a very small effect on flame temperature and that soot formation is weakly dependent on fuel flow rate. In a study of fuel dilution with inerts, Axelbaum et al. concluded that dilution influences the soot inception most and confounds the effects of temperature when used to study temperature effects. It should be noted that in such burners fuel flux and local fuel concentration are closely coupled. Thus the application of such results to practical cases should be approached with caution.

Turbulent Diffusion Flame Modeling

Most practical combustors use turbulent flow in order to obtain high rates of combustion energy release per unit volume. In many cases the fuel enters as a spray or jet so that much of the combustion is diffusional in nature and the rate of burning is limited by the mixing rate, because the chemical reactions are typically much faster than the mixing rate. Modeling of such combustion is quite complex and not yet well established in its approach even for simpler cases which do not involve sprays or the geometry of practical combustors. Since the present models are in the research stage of development, the discussion is limited to experimental observations and simple flow geometries.

In the past the ability to measure turbulent flames was typically limited to time-averaged measurements and imaging of the basic flame shape. Such data gave overall characterization of the flame zones, but did little to reveal the structure which is dominated by the turbulent fluctuations. Even point measurements which give the history of velocity, species, or temperature at a single location do not reveal the structure, although much insight and statistical correlation data can be obtained in this way. It is only recently that planar imaging of species profiles has been possible with a resolution of about 50 microns. The main optical methods utilize Rayleigh scattering, Raman scattering, and laser-induced fluorescence [Eckbreth]. Combination of the point and imaging techniques is giving a much improved understanding.

The simplest models for laminar diffusion flames employ the concept of rapid chemistry so that all the reactions take place at a reaction sheet where fuel and oxygen are in stoichiometric proportion. Fuel is on one side of the reaction sheet and oxygen on the other, and it is assumed that no fuel or oxygen passes through the reaction sheet. Burke and Schumann [1928] solved such a problem for the case of the axisymmetric fuel jet (Figure 5.23) and obtained analytical expressions for the flame height and shape. Extension of the laminar models to finite-rate reactions for hydrogen-air is readily done, as shown in Figure 5.25. Application to methane and other gaseous hydrocarbon fuels adds the significant problem of soot formation and radiation heat loss from the soot particles. The effect of the radiation on the flame structure can be neglected, except near extinction; however, the energy loss may change the flame temperature enough to affect the NO formation.

Because practical combustors are typically turbulent, a great deal of effort has been expended on the development of modeling methods for turbulent nonpremixed flames. Models have been developed for simple flows such as turbulent steady-flow fuel jets under conditions of fast chemistry (high Damköhler number). Optical measurements of these fuel jets show vortices and flames. Laser Doppler velocimetry, particle image velocimetry, and various particle light-scattering techniques have been used to obtain probability distribution functions (PDFs) of important variables such as velocity components. Similarly, other optical techniques such as Raman-Rayleigh scattering of laser light and coherent anti-Stokes Raman spectroscopy (CARS) have been used to obtain PDFs for temperature and major species concentrations [Eckbreth]. However, the steep gradients and particle radiation can cause significant experimental error. Nevertheless, such PDF data are useful for understanding the flame structure and modeling these flames using the PDF modeling methods, which have been developed extensively for simple flow geometries and fast chemistry.

Multipoint Raman scattering has been applied to combustion in engines at the Sandia Combustion Research Facility to produce a trace of mole fractions of major species. As shown in Figure 5.26, light scattered from a 532-nm pulsed laser beam is collected and focused onto the entrance slit of an imaging spectrograph, where it forms an image of a length of the original beam. The imaging spectrograph simply disperses the image at the slit spectrally, resulting in multiple images at the detector plane. Each detector plane image corresponds to light scattered from a particular molecular species. By examining the intensity at various positions along

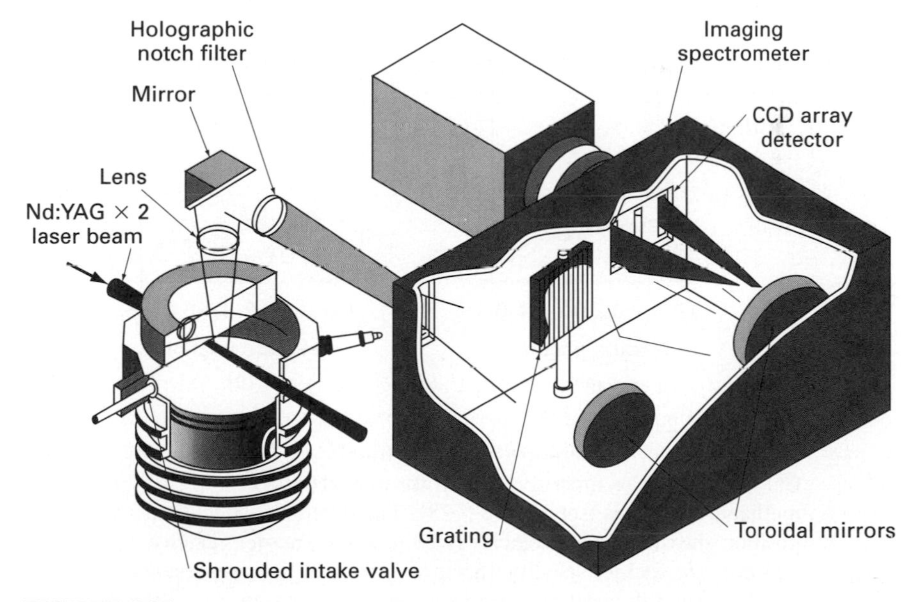

FIGURE 5.26
Setup for multipoint Raman scattering to measure major species mole fractions during combustion in an engine [Miles and Dilligan].

these images it is possible to determine the composition of the in-cylinder gases at the corresponding positions along the laser beam. With broadband collection over a 100-nm range, the major species of combustion—CO_2, O_2, N_2, CO, fuel, and H_2O—can all be monitored simultaneously.

As an example, direct scatter plot data obtained using laser diagnostics are shown in Figure 5.27. The data are from an intense zone of turbulent mixing of

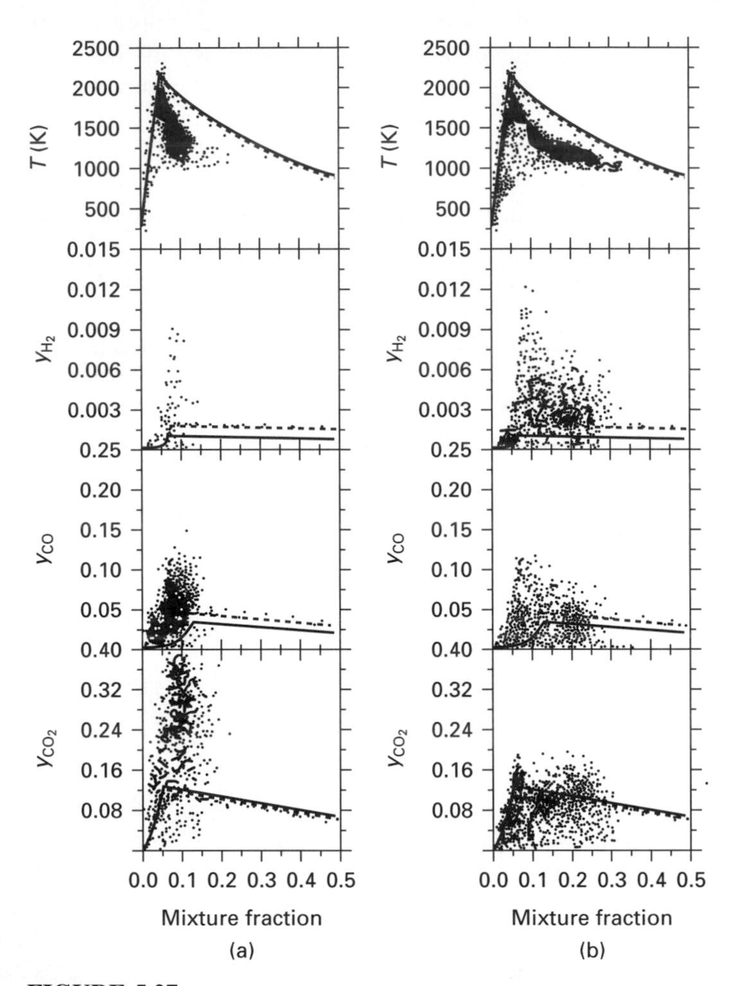

FIGURE 5.27
Scatter plots for temperature and mass fractions of H_2, CO, and CO_2 plotted versus mixture fraction for nonpremixed methane-air flames (see Figure 5.28). The data are collected at position shown. The air velocity is 20 m/s. (*a*) Fuel jet velocity 124 m/s; (*b*) fuel jet velocity 155 m/s. The solid lines represents a fully burned flamelet. The broken lines represent an "intermediate" flamelet, and the dotted lines denote a fully stretched flamelet [Masri et al., by permission of The Combustion Institute].

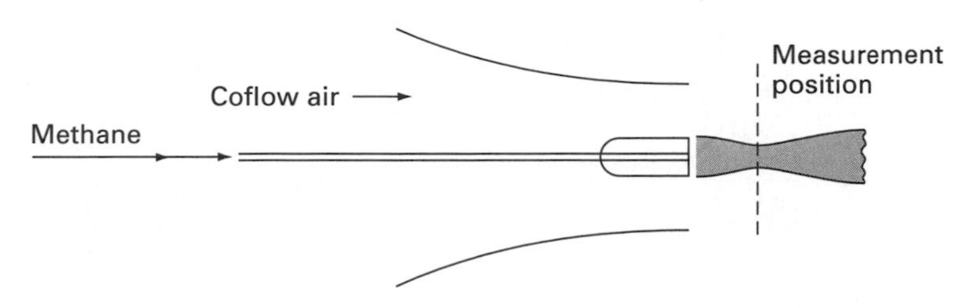

FIGURE 5.28
Schematic of the bluff-body stabilized burner setup to study diffusion flames [Masri et al., by permission of The Combustion Institute].

coflowing jets of air and methane which is stabilized behind a bluff body, as shown in Figure 5.28. The scatter plots of Figure 5.27 show how the temperature and product species are distributed as a function of fuel mixture fraction. For the condition shown, the flame was dominated by a luminous recirculation zone. The solid and broken lines represent calculated laminar flamelet predictions for various values of the flame stretch. The derived PDF of nondimensional temperature, b_T, which is a measure of the reactedness, is shown in Figure 5.29. For a fully reacted mixture, $b_T = 1$. The conditional PDF for three mixture fraction ranges is shown. In Figure 5.29*a* the conditional PDFs are centered around high values of reactedness, while at the higher fuel jet velocity of Figure 5.29*b*, the PDFs are bimodal.

If the Damköhler number based on the Kolmogorov scale is large, the chemistry is fast enough to assume thermodynamic equilibrium. Then the mixture faction y, defined as the fraction of total mass at a given location that came from the fuel, may be used to obtain all the state variables. The y PDF, $P(y)$, is defined such that $P(y)\,dy$ is the probability that y lies in the range dy about y. The problem becomes one of calculating the spatial distribution of P. Given $P(y)$, it is possible to calculate the average value $\overline{y}$ from

$$\overline{y} = \int_0^\infty yP(y)\,dy \tag{5.32}$$

Similarly, other functions of y such as $T(y)$ can be averaged.

If the functional form of P is selected to mathematically approximate available data, then the problem becomes one of turbulent mixing. The theory of this method and that of more advanced methods [Pope, 1985; Hulek and Lindstedt, 1996] of PDF modeling are beyond the scope of this text. At present the methods appear too complicated for application to complex geometries, although recent work has extended the methods to bluff-body flows.

The methods of turbulent flow analysis without combustion can be applied to combustion if the PDFs are disregarded. These methods time-average the conservation equations of fluid mechanics. The resulting partial differential equations are solved using supplemental equations provided by a turbulence model. The k-ϵ model is a popular model for turbulence which uses differential equations for the turbulent kinetic energy k and the average rate of dissipation of turbulent kinetic energy ϵ [see Warsi]. With the resulting equation set it is still necessary to provide

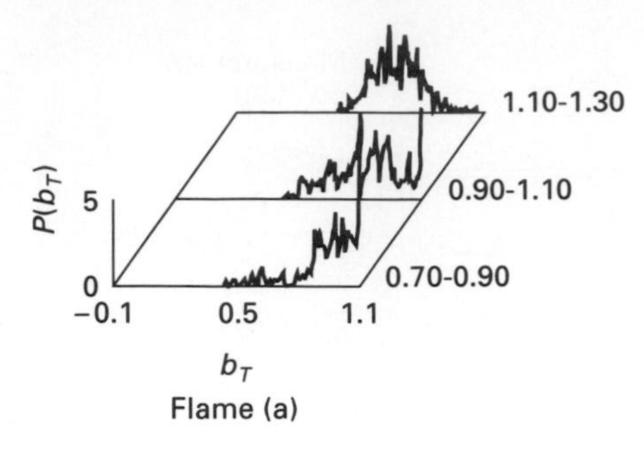

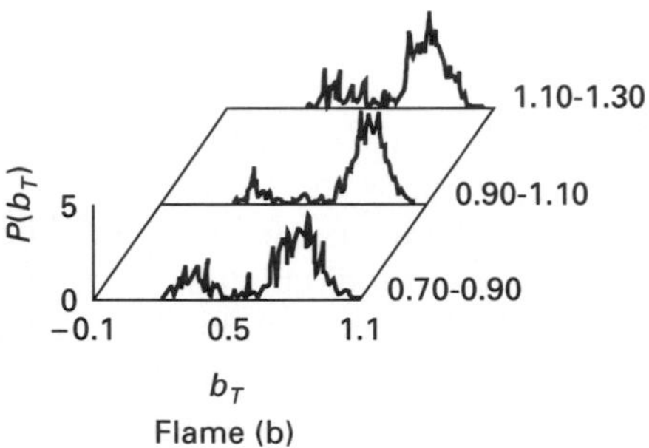

FIGURE 5.29
Probability density functions for nondimensional temperature for three mixture fraction ranges (0.7–0.9, 0.9–1.1, 1.1–1.3) as derived from Figure 5.27; $b_T = (T - T_m)/(T_a - T_m)$, where T is the measured temperature, $T_m = 300$ K, and T_a is adiabatic flame temperature. (a) Fuel jet velocity 124 m/s; (b) fuel jet velocity 155 m/s. [Masri et al., by permission of The Combustion Institute].

a combustion model. Flamelet models [see Turns and Chapter 12] are of possible utility in some regimes, but may not be adequate for practical combustors with complex flows.

Given this situation, we next describe a semiempirical characteristic time approach which can be applied to either premixed or nonpremixed combustion. The model originally was given by Abraham, Bracco, and Reitz in 1985, and has undergone revisions by Reitz and co-workers. In this model the rate of reaction of species i is given by

$$\frac{dy_i}{dt} = -\frac{y_i - y_i^*}{\tau_c} \tag{5.33}$$

where y_i^* is the local equilibrium value of the mass fraction of species i and τ_c is the characteristic time for achievement of thermodynamic equilibrium. The characteristic time is taken to have two components: the laminar time scale τ_L, and the turbulent time scale τ_T:

$$\tau_c = \tau_L + C_1 \tau_T \tag{5.34}$$

The laminar time scale is obtained by using a one-step global reaction approach to give τ_L:

$$\tau_L = A[n_f]^m[n_{O_2}]^n \exp\left(\frac{E}{\hat{R}T}\right) \tag{5.35}$$

The values of A, E, m, and n depend on the fuel used. The turbulent time scale is proportional to the eddy turnover time, and using k and ϵ from the turbulence model to define τ_T:

$$\tau_T = \frac{C_2 k}{\epsilon} \tag{5.36}$$

The quantity C_1 is called the delay coefficient and simulates the influence of turbulence on combustion:

$$C_1 = C_3(1 - e^{-a}) \tag{5.37}$$

where

$$a = \frac{y_{CO_2} + y_{H_2O} + y_{CO} + y_{H_2}}{1 - y_{N_2}} \tag{5.38}$$

The parameter a is a measure of the completeness of combustion at a given location. Given the empirical nature of the model it requires adjusted values of A, E, C_2, and C_3 for a given application. The model has been found to work reasonably well for CFD applications to engine combustion, but should be considered an expedient to be used until practical, fundamentally based models can be developed.

5.8 SUMMARY

In premixed flames the gaseous fuel and air are mixed prior to combustion. In the laminar premixed flame the temperature rises rapidly, forming a preheat region followed by ignition and combustion. The laminar premixed flame is very thin (less than 1 mm thick at 1 atm), travels at relatively low velocity relative to the unburned mixture (0.5 m/s), and has a negligibly small pressure change through it, and in its strictly one-dimensional form the burning velocity is a property of the mixture. The burning velocity increases approximately as the square of mixture temperature and decreases slightly with pressure. Empirical formulas are available for burning velocity, but it can also be calculated from theory if the chemical kinetics of the fuel are known.

The turbulent burning velocity is higher than the laminar burning velocity due to the turbulent intensity of the flow. There are three general types of turbulent premixed flames: weakly turbulent flames, wrinkled reaction sheets, and distributed reaction zones.

Diffusion flames arise at the interface between unmixed portions of gaseous fuel and air. In the case of nongaseous fuel, the fuel vaporizes, creating the source of gaseous fuel. The flame front fuel-to-air ratio is distributed over a narrow region; if taken as a flame sheet, the ratio is typically taken as stoichiometric. Pyrolysis of fuel takes place on the fuel side of the flame zone, leading to the formation of solid carbon, soot particles. For free fuel jet flames in abundant air the flame height is proportional to $\underline{V}_j d_j^2$ for laminar flow and to d_j for turbulent flow.

Flames can be quenched by heat transfer to surfaces, rapid expansion, and flame curvature. The quench distance for an orifice may be defined in terms of the smallest diameter which will allow flame propagation. For fuel mixtures at 1 atm pressure this diameter is of the order of 2 to 3 mm. Empirical formulations of quench distance in laminar flows are typically expressed as Peclet number equals a constant of order 10; thus the quench distance is inversely proportional to pressure. Data for turbulent flows are scarce, but show an increase in quench distance with increasing turbulence intensity.

Forced ignition is typically used to start combustion in premixed charges. A spark, a glow plug, another flame, a jet of high-cetane fuel, or a laser can be used, but spark plugs are by far the most commonly used. There is a minimum spark energy which will allow the flame kernel to grow; however, practical systems typically use a powerful spark of 30 to 50 mJ.

PROBLEMS

5.1. A stoichiometric methane-air flame has a laminar burning velocity of 0.33 m/s and a temperature of 2200 K. The reactants are initially at 300 K and 1 atm. Find the velocity of the combustion products relative to the flame front and the pressure change across the flame. Assume that the reaction goes to completion and there is no dissociation.

5.2. Repeat Problem 5.1 for a stoichiometric methanol-air flame. The burning velocity is 0.48 m/s and the flame temperature is 2000 K.

5.3. The laminar burning velocity of a stoichiometric mixture of propane-air at 1 atm and 298 K is 0.4 m/s. Estimate the average reaction rate and the reaction zone thickness. Assume the thermal conductivity of the mixture is 0.026 W/m·K at 298 K and increases as $T^{1/2}$. Assume an appropriate average mixture temperature and ignition temperature.

5.4. The laminar burning velocity of a stoichiometric mixture of propane-air at 1 atm and 298 K is 0.4 m/s. If the nitrogen is replaced with helium, estimate the burning velocity using the thermal flame theory assuming no change in temperature. The thermal conductivity of the helium mixture is 0.36 W/(m·K) at 298 K, and assume it increases as the square root of T. Use $c_p = (5/2)R$ for helium and $(7/2)R$ for nitrogen. Actually, the flame temperature will increase, because the specific heat of the helium is lower than that of nitrogen. Explain quantitatively how this would change the burning velocity estimated above.

5.5. For the conditions of Problem 5.3, what is the approximate thickness of the preheat zone? Use Eq. 5.13 and select a criterion that the temperature is 99% of the reactant temperature.

5.6. Review the list of six parameters in Section 5.6, which extend the lean flame limit in engines and then explain each one in qualitative terms.

5.7. For methane, the maximum laminar burning velocity in pure oxygen is 11 m/s, whereas in air the maximum burning velocity is 0.45 m/s. Explain what causes this increase in the burning velocity.

5.8. For the conditions of Problem 5.3, calculate the flame quench distance using Eq. 5.29 and compare with Table 5.6.

5.9. Explain why the quench distance for a hydrogen-air flame given in Table 5.6 is 3 times smaller than the quench distance for a propane-air flame. Be somewhat quantitative in your explanation.

5.10. Using Table 5.4 for a stoichiometric propane-air laminar flame in a constant-volume chamber, calculate the parallel-plate quench distance for the following three cases: (*a*) $p = 1$ atm, $T = 298$ K, (*b*) $p = 1$ atm, $T = 558$ K, (*c*) $p = 100$ atm, $T = 298$ K. Assume the mixture is air when calculating the transport properties. Compare your calculations for quench distance with the data given in Chapter 5.

5.11. Using the data given in Table 5.4, estimate the laminar burning velocity of a stoichiometric mixture of air and gasoline at 20 atm pressure and 700 K in a closed vessel. How much can turbulence increase this flame speed if $\underline{V}_T = \underline{V}_L + 2\underline{V}'$ and the turbulent intensity $\underline{V}' = 1.2$ m/s?

REFERENCES

Abraham, J., and Bracco, F. V., "Comparisons of Computed and Measured Premixed Charge Engine Combustion," *Comb. and Flame,* vol. 60, pp. 309–322, 1985.

Aly, S. L., and Hermance, C. E., "A Two-Dimensional Theory of Laminar Flame Quenching," *Comb. and Flame,* vol. 40, pp. 173–185, 1981.

Andrews, G. E., and Bradley, D., "The Burning Velocity of Methane/Air Mixtures," *Comb. and Flame,* vol. 19, p. 275, 1972a.

Andrews, G. E., and Bradley, D., "Determination of Burning Velocities: A Critical Review," *Comb. and Flame,* vol. 18, p. 133, 1972b.

Axelbaum R. L., Flower, W. L., and Law, C. K., "Dilution and Temperature Effects of Inert Addition on Soot Formation in Counter Flow Diffusion Flames," *Comb. Sci. Tech.,* vol. 61, pp. 51–73, 1988.

Ballal, D. R. and Lefebvre, A. H., "Turbulent Structure and Propagation of Turbulent Flames," *Proc. Royal Society (London),* vol. 344A, pp. 217–234, 1975a.

Ballal D. R., and Lefebvre, A. H., "The Influence of Flow Parameters on Minimum Ignition Energy and Quenching Distance," Fifteenth Symp. (Int.) on Combustion, The Combustion Institute, Pittsburgh, pp. 217–234, 1975b.

Barnard, J. A., and Bradley, J. N., *Flames and Combustion,* Chapman & Hall, London, 1985.

Barnett, H. C., and Hibbard, R. R. (eds.), "Basic Considerations in the Combustion of Hydrocarbon Fuels with Air," NASA report 1300, Lewis Propulsion Laboratory, Cleveland, 1959.

Bartok, W., and Sarofim, W. S. (eds.), *Fossil Fuel Combustion: A Source Book,* Wiley, New York, 1991.

Bird, R. B., Stewart, W. E., and Lightfoot, E. N., *Transport Phenomena,* Wiley, New York, 1960.

Bracco, F. V., "Structure of Flames in Premixed-Charge IC Engines," *Comb. Sci. Tech.,* vol. 58, pp. 209–230, 1988.

Burke, S. P., and Schumann, J. E. W., "Diffusion Flames," *Indust. Eng. Chem.,* vol. 20, p. 998, 1928.

Dugger, G. L., and Heimel, S., "Flame Speeds of Methane-Air, Propane-Air and Ethylene-Air Mixtures at Low Initial Temperatures," NACA TN 2624, 1952.

Eckbreth, A. C., *Laser Diagnostics for Combustion, Temperature, and Species,* Abacus Press, Cambridge, MA, 1988.

Egerton, A. C., and Lefebvre, A. H., "Flame Propagation: The Effect of Pressure on Burning Velocities," *Proc. R. Soc.* (London Ser. A.) vol. 222, pp. 206–223, 1954.

Elliott and Denues, U.S. Bureau of Standards, *Journal of Research,* vol. 17, pp. 7–43, 1936.

Friedman, R., and Johnston, W. C., "The Wall Quenching of Laminar Propane Flames as a function of Pressure, Temperature and Air-Fuel Ratio," *J. Appl. Phys.,* vol. 21, pp. 791–795, 1950.

Fukutani, S., Kunioshi, N., and Jinno, H., "Flame Structure of an Axisymmetric Hydrogen-Air Diffusion Flame," *Twenty-Third Symp. (Int.) on Combustion,* pp. 567–573, The Combustion Institute, Pittsburgh 1991.

Gaydon, A. G., and Wolfhard, H. G., *Flames, Their Structure, Radiation and Temperature,* Chapman & Hall, London, 4th ed., 1979.

Gottgens, J., Mauss, F., and Peters, N., "Analytical Approximations of the Burning Velocity and Flame Thickness of Lean Hydrogen, Methane, Ethylene, Ethane, Acetylene, and Propane Flames," *Twenty-Fourth Symp. (Int.) on Combustion,* The Combustion Institute, Pittsburgh, pp. 129–135, 1993.

Green, K. A., and Agnew, J. T., "Quenching Distances of Propane-Air Flames," *Comb. and Flame,* vol. 15, pp. 191–198, 1970.

Gülder, O. L., "Laminar Burning Velocities of Methanol, Isooctane and Isooctane/Methanol Blends," *Comb. Sci. Tech.,* vol. 33, pp. 179–192, no. 1–4, 1983.

Gülder, O. L., "Turbulent Premixed Flame Propagation Models for Different Combustion Regimes," *Twenty-Third Symp. (Int.) on Combustion,* pp. 743–750, The Combustion Institute, Pittsburgh, 1991.

Heimel, J. A., and Weast, R. C., "Effect of Initial Mixture Temperature on the Burning Velocity of Benzene-Air, *n*-Heptane-Air and Isooctane-Air Mixtures," *Sixth Symp. (Int.) on Combustion,* Reinhold, New York, 1957.

Hottel, H. C., and Hawthorne, W. R., "Diffusion in Laminar Flame Jets," *Third Symp. (Int.) on Combustion,* pp. 254–266, The Combustion Institute, Pittsburgh, 1949.

Hulek, T. and Lindstedt, R. P., "Computations of Steady-State and Transient Premixed Flames Using PDF Methods," *Combustion and Flame,* vol. 104, pp. 481–504, 1996.

Jones, G. W., "Inflammability of Natural Gas: Effect of High Pressure on the Limits," U.S. Bureau of Mines report 3798, 1945.

Jones, G. W., Kennedy, R. E., and Spolan, I., "The Effect of High Pressures on the Flammability of Natural Gas-Air Nitrogen Mixtures," U.S. Bureau of Mines Report of Investigation no. 4557, 1949.

Kee, R. J., Grcar, J. F., Smooke, M. D., and Miller, J. A., "A Fortran Program for Modeling Steady Laminar One-Dimensional Premixed Flames," Sandia National Lab. report SAND85-8240 UC-4, 1985.

Kee, R. J., Miller, J. A. and Jefferson, T. H., "CHEMKIN: A General Purpose Problem-Independent, Transportable, Fortran, Chemical Kinetic Program Package," Sandia National Lab. report SAND80-8003, 1980.

Kuehl, D. K., "Laminar Burning Velocities of Propane-Air Mixtures," *Eighth Symp. (Int.) on Combustion,* Williams & Wilkins, Baltimore, pp. 510–521, 1962.

Law, C. K., Ishizuka, S., and Mizomoto, M., "Lean-Limit Extinction of Propane/Air Mixtures in the Stagnation Point Flow," *Eighteenth Symp. (Int.) on Combustion,* pp. 1791–1798, The Combustion Institute, Pittsburgh, 1981.

Lefebvre, A. H., and Reid, R., "The Influence of Turbulence on the Structure and Propagation of Enclosed Flames," *Comb. and Flame,* vol. 10, pp. 355–366, 1966.

Lewis, B., and von Elbe, G., *Combustion, Flames, and Explosions of Gases,* Academic Press, Orlando 3rd ed., 1987.

Libby, P. A. and Williams, F. A. (eds.), *Turbulent Reacting Flows,* Academic Press, London, 1993.

Liñán, A., and Williams, F. A., *Fundamental Aspects of Combustion,* Oxford University Press, New York, 1993.

Mantzaras, J., Felton, P. G., and Bracco, F. V., "Three-Dimensional Visualization of Premixed-Charge Engine Flames," SAE paper no. 881635, 1988.

Masri, A. R., Dibble, R. W., and Barlow, R. S., "Raman-Rayleigh Measurements in Bluff Body Stabilized Flames of Hydrocarbon Fuels," *Twenty-Fourth Symp. (Int.) on Combustion,* pp. 317–324, The Combustion Institute, Pittsburgh, 1992.

Matthews, R. D., Hall, M. J., Dia, W., and Davis, G. C., "Combustion Modeling in SI Engines with a Peninsula-Fractal Combustion Model," SAE paper no. 960072, 1996.

Metghalchi, M., and Keck, J. C., "Laminar Burning Velocity of Propane-Air Mixtures at High Temperature and Pressure," *Comb. and Flame,* vol. 38, pp. 143–154, 1980.

Metghalchi, M., and Keck, J. C., "Burning Velocity of Mixtures of Air with Methanol, Isooctane, and Indolene at High Pressures and Temperatures," *Comb. and Flame,* vol. 48, pp. 191–210, 1982.

Miles, P., and Dilligan, M., "Raman Scattering Measures Fluid Composition in IC Engines," *Combustion Research Facility News,* Sandia National Labs., vol. 18, no. 3, 1996.

Newitt, P. M., and Thornes, L. S., "Oxidation of Propane. Part I, The Products of the Slow Oxidation at Atmospheric and Reduced Pressures," *J. Chem. Soc.,* Part II, London pp. 1656–1665, 1937.

Pope, S. B., "PDF Methods for Turbulent Reactive Flows," *Prog. Energy and Comb. Sci.,* vol. 11, pp. 119–192, 1985.

Rutland, C. J., and Trouve, A., "Direct Simulations of Turbulent Premixed Flames with Non-unity Lewis Numbers," *Comb. and Flame,* vol. 94, pp. 41–57, 1993.

Scholefield, D. A., and Garside, J. E., "The Structure and Stability of Diffusion Flames," *Third Symp. (Int.) on Combustion,* Williams & Wilkins, Baltimore, pp. 102–110, 1949.

Strehlow, R. A., *Fundamentals of Combustion,* International Textbook Co., Scranton, PA, 1968.

Tseng, L. K., Ismail, M. A., and Faith, G. M., "Laminar Burning Velocities and Markstein Numbers of Hydrocarbon/Air Flames," *Combustion and Flame,* vol. 95, pp. 410–426, 1993.

Tsuji, H., and Yamaoka, I., "The Counterflow Diffusion Flame in the Forward Stagnation Region of a Porous Cylinder," *Eleventh Symp. (Int.) on Combustion,* The Combustion Institute, Pittsburgh, pp. 979–984, 1967.

Turns, S. R., *An Introduction to Combustion,* McGraw-Hill, New York, 1996.

Tuteja, A. D., "The Formation of Nitric Oxide in Diffusion Flames," Ph.D. thesis, Mech. Engr. Dept., University of Wisconsin–Madison, 1972.

Warsi, Z. U. A., *Fluid Dynamics—Theoretical and Computational Applications,* Chap. 6, CRC Press, Boco Raton, FL, 1993.

Westbrook, C. K. Adamczyk, A. A., and Lavoie, G. A., "A Numerical Study of Laminar Wall Quenching," *Comb. and Flame,* vol. 40, pp. 81–99, 1981.

Westbrook, C. K., and Dryer, F. L., "A Comprehensive Mechanism for Methanol Oxidation," *Combustion Science and Technology,* vol. 20, pp. 125–140, 1979.

Williams, F. A., *Combustion Theory,* Benjamin/Cummings div. of Addison-Wesley, Menlo Park, CA, 2nd ed., 1985.

Wohl, K., Kapp, N. M., and Gazley, C., "The Structure of Open Flames," *Third Symp. (Int.) on Combustion,* Williams & Wilkins, Baltimore, pp. 1–16, 1949.

Zabetakis, M. G., "Flammability Characteristics of Combustible Gases and Vapors," U.S. Bureau of Mines, bulletin 627, 1965.

Zurloye, A. O., and Bracco, F. V., "Two-Dimensional Visualization of Premixed Charge Flame Structure in an I. C. Engine," SAE paper no. 870454, 1987.

CHAPTER 6

Gas-Fired Furnace Combustion

Gaseous fuels are the easiest fuels to utilize in furnaces and boilers. No fuel preparation is necessary, the gases easily mix with air, combustion proceeds rapidly, and emissions are low compared with liquid and solid fuels. The drawbacks are that gaseous fuels can be expensive and they may not be available, particularly if no natural gas pipeline hookup is available. Since gaseous fuels are convenient to use, they are best suited for small combustion systems such as residential, commercial, and small industrial furnaces. Large boilers are sometimes fired with gaseous fuels, but this is generally discouraged because of cost and to conserve the resource.

Essentially, all gas-fired furnaces use a burner to mix the fuel and air before combustion. In this chapter premixed burners, nozzle-mixed burners, pulse furnaces, and the use of low-grade gaseous fuels are discussed. First, however, let us examine the furnace efficiency using energy balance calculations.

6.1 ENERGY BALANCE AND FURNACE EFFICIENCY

The fuel and air requirements for a given heat output, the volume of exhaust products, the air blower power required, and the overall efficiency of the furnace may be obtained from an analysis of the generic system shown in Figure 6.1. This analysis will point out that the highest furnace efficiency is obtained by operating the burner as close to stoichiometric as possible.

The fuel and air flow rates required for a given heat output are calculated from an energy and mass balance for a control volume placed around the combustion chamber–heat exchanger–mixer:

$$\dot{N}_a \hat{h}_a + \dot{N}_f \hat{h}_f = q + q_L + \sum_{j=1}^{J} \dot{N}_j \hat{h}_j \tag{6.1}$$

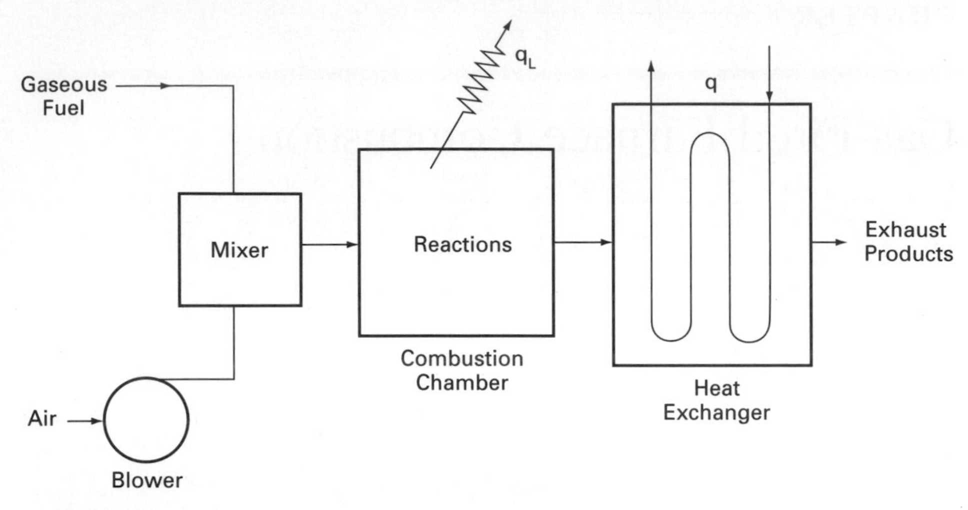

FIGURE 6.1
Schematic of gas-fired furnace, q is the heat flow from the products to the heat exchanger tubes.

where p stands for products. In terms of the sensible enthalpy, fuel heating value, and enthalpy of vaporization of water, the energy balance is

$$\dot{N}_a\hat{h}_{sa} + \dot{N}_f(\hat{h}_{sf} + \widehat{\text{HHV}}) = q + q_L + \sum_{j=1}^{J} \dot{N}_j\hat{h}_{sj} + (\dot{N}\hat{h}_{fg})_{\text{H}_2\text{O}} \qquad (6.2)$$

Equation 6.2 assumes that the combustion is complete, i.e., 100% combustion efficiency, which is reasonable for a gaseous fuel. Solving for the useful heat output,

$$q = \dot{N}_f\left[\hat{h}_{sf} + \widehat{\text{HHV}} + \left(\frac{\dot{N}_a}{\dot{N}_f}\right)\hat{h}_{sa} - \sum_{j=1}^{J}\left(\frac{\dot{N}_j}{\dot{N}_f}\right)h_{sj} - \left(\frac{\dot{N}_{\text{H}_2\text{O}}}{\dot{N}_f}\right)\hat{h}_{fg}\right] - q_L \qquad (6.3)$$

The moles of air and product species per mole of fuel are obtained from a chemical atom balance for the combustion process (Section 3.3). Hence, the heat output can be obtained if the fuel flow rate is specified; and conversely, if the heat output is specified, the required fuel flow rate may be calculated.

EXAMPLE 6.1. Propane is burned to completion in a furnace with 10% excess air. The fuel and air are at 77°F. If 5% of the heat is lost through the walls of the furnace and the combustion products exit the furnace to the stack at 340°F, what is the useful heat output of the furnace per pound of propane?

Solution. The reaction is

$$C_3H_8 + (1.05)(5)(O_2 + H_2O) = 3CO_2 + 4H_2O + (1.05)(3.76)N_2 + (0.05)(5)O_2$$

From Table 2.4, the HHV is 21,670 Btu/lb or 953,480 Btu/lbmol. The enthalpy of the products is derived from Appendix C:

Products	$\dot{N}_j/\dot{N}_f$	kBtu/lbmol	kBtu/lbmol fuel
CO_2	3	2.528	7.584
H_2O	4	2.142	8.568
O_2	0.25	1.882	0.470
N_2	19.74	1.830	36.282
Sum			52.904

From Eq. 6.3 the useful heat output per mole of fuel is

$$\frac{q}{\dot{N}_f} = 953{,}480 - 52{,}904 - (4)(1050)(18) - \frac{0.05q}{\dot{N}_f}$$

Thus,

$$\frac{q}{\dot{N}_f} = 785{,}690 \text{ Btu/lbmol fuel}$$

and

$$\frac{q}{\dot{m}_f} = 17{,}856 \text{ Btu/lb fuel}$$

The useful heat output is 82.4% of the higher heating value and 89.5% of the lower heating value.

The volumetric analysis of the products when burning natural gas is shown in Figure 6.2. The fuel for these calculations contains 83% methane and 16% ethane. Some external heat is required for mixtures more than 30% deficient in air (dotted lines). Dashed lines show trends with poor mixing or wall quenching. Volumetric analysis is on a dry product basis to allow comparison with gas analyzers that require the moisture to be condensed. Note that the assumption that the combustion is complete is valid only on the lean side of the stoichiometric.

The mass flow rate is obtained from the molar flow rate by using the molecular weight:

$$\dot{m} = \dot{N}M \tag{6.4}$$

The volume flow rate is obtained from the molar flow rate and the equation of state:

$$p\dot{V} = \dot{N}\hat{R}T \tag{6.5}$$

The power required by the air blower is obtained from an energy balance around the blower. The result is

$$\dot{W}_a = \frac{\dot{V}_a \, \Delta p_a}{\eta_b} \tag{6.6}$$

where Δp_a is the pressure rise across the blower, which essentially equals the air pressure supplied to the burner less any loss in the ducting and air preheater, and η_b is the efficiency of the blower.

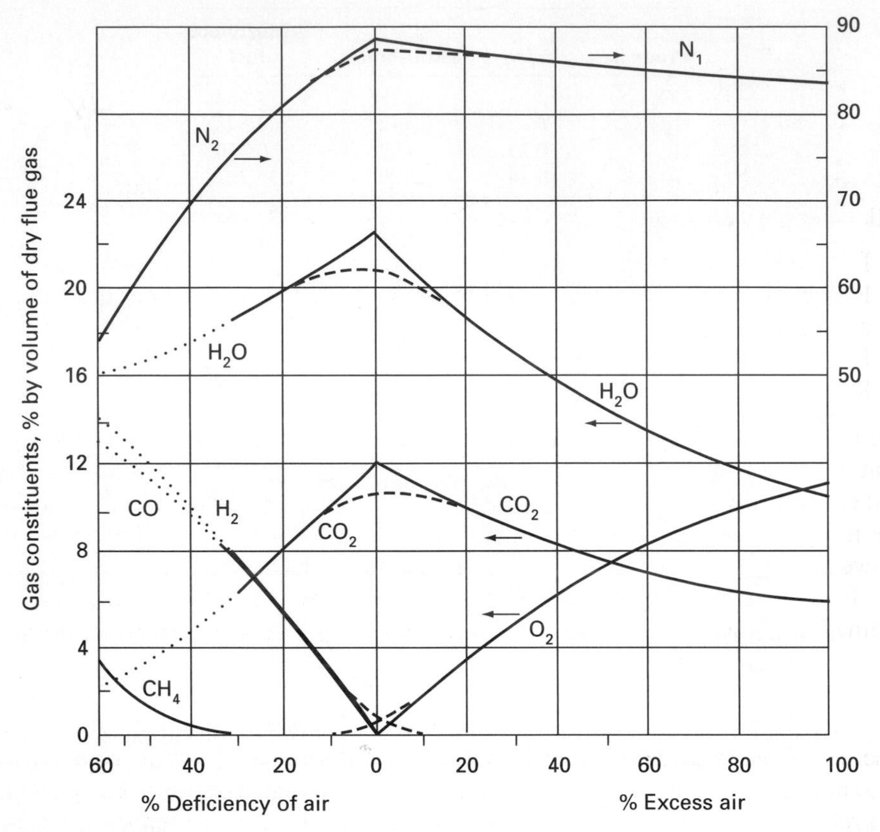

FIGURE 6.2
Volumetric analysis of natural gas combustion products [*North American Combustion Handbook,* by permission of North American Mfg. Co.].

Furnace Efficiency

The efficiency of the furnace is defined as the ratio of the useful heat output to the energy input, and by convention is always based on the higher heating value. Hence, in general for any type of fuel,

$$\eta = \frac{q}{\dot{m}_f \, \text{HHV} + \dot{W}_a} \tag{6.7}$$

The furnace efficiency may be determined directly by measuring the useful heat output and the fuel flow rate, or it can be determined indirectly from the products of combustion. Sometimes the fuel flow rate is not measured, and then the indirect method is used. Also, from the combustion system viewpoint, it is useful to use the indirect method of evaluation of efficiency to assess where the losses arise and thus indicate how the efficiency may be increased. Substituting Eq. 6.3 into

Eq. 6.7 and introducing the mole fraction $x_j = N_j/N_p$, where $N_p = \Sigma_p N_j$:

$$\eta = \frac{\hat{h}_{sf} + \hat{h}_{sa}\left(\frac{N_a}{N_f}\right) + \widehat{\text{HHV}} - \left(\frac{N_p}{N_f}\right)\sum_j x_j \hat{h}_{sj} - \left(\frac{N_{H_2O}}{N_f}\right)\hat{h}_{fg} - \frac{q_L}{\dot{N}_f}}{\widehat{\text{HHV}} + \dot{W}_a/\dot{N}_f} \tag{6.8}$$

Inspection of Eq. 6.8 shows that the furnace efficiency can be increased by the following measures:

1. Decrease the temperature of the exhaust products.
2. Reduce the excess air, which will reduce the moles of products per mole of fuel and also reduce the blower power.
3. Reduce the extraneous heat loss.
4. Reduce the blower power requirements.

Item 1 can be done by increased heat transfer to the load, perhaps by cleaning the heat transfer surfaces. Item 2 can be accomplished only if there is adequate mixing and can only be done to the point where the CO starts to increase. Item 4 indicates the motivation for aerated burners and the pulse furnace, which do not need a blower.

In order to evaluate Eq. 6.8 based on the composition of the products and the ultimate analysis (i.e., elemental analysis) of the fuel, it is convenient to write

$$\frac{N_p}{N_f} = \frac{N_p}{N_C} \cdot \frac{N_C}{N_f} = \frac{N_C/N_f}{x_{CO_2} + x_{CO}} \tag{6.9}$$

where the subscripts p, f, and C refer to products, fuel, and carbon, respectively. Also note that the enthalpy of water contains the latent heat of vaporization, which is a relatively large term. Hence, if the fuel analysis is known, then the efficiency can be determined from gas sampling of the products. The blower power per unit fuel flow is usually small compared with the heating value, and hence, the efficiency can be determined without knowing the fuel and air flow rates. Of course, if the fuel and air flow rates are known, then the indirect method for efficiency determination may be compared with the direct method.

EXAMPLE 6.2. Stack gas analysis of a natural gas–fired furnace gave the following volumetric analysis: 4% O_2, 10% CO_2, 17% H_2O, 86% N_2, all on a dry basis. The fuel was 84% CH_4 and 16% C_2H_6 by volume, and the higher heating value was 23,300 Btu/lb$_m$. The fuel and air entered the furnace at 77°F, and the stack gas temperature was 340°F. No blower was used, and heat losses were negligible. What is the operating efficiency of this furnace, and what is the excess air?

Solution. The excess air is (Eq. 3.36):

$$\text{EA} = \frac{(x_{O_2})100}{x_{N_2}/3.76 - x_{O_2}} = \frac{(4)(100)}{(86/3.76) - 4} = 21.2\%$$

The wet product analysis by volume is

$$4/1.17 = 3.42\%\ O_2; \qquad 8.55\%\ CO_2; \qquad 14.53\%\ H_2O; \qquad 73.50\%\ N_2$$

The moles of carbon in the fuel per mole of fuel are $N_C/N_f = 0.84 + (0.16)(2) = 1.16$.

From Eq. 6.9,

$$\frac{N_p}{N_f} = \frac{1.16}{0.0855} = 13.6$$

Using Appendix A at 800°R,

$$\hat{h}_{sp} = (0.0342)(1.882) + 0.0855(2.525) + 0.1453(2.142)$$
$$+ (0.7350)(1.838) = 1.94 \text{ kBtu/lbmol}$$

The molecular weight of the fuel is

$$M_f = (0.84)(16) + (0.16)(30) = 18.24 \text{ lbm/lbmol}$$

Hence the molar higher and lower heating values are

$$\widehat{\text{HHV}} = (23.3 \text{ kBtu/lb}_\text{m})(18.24 \text{ lbm/lbmol})$$
$$= 425.0 \text{ kBtu/lbmol}$$

$$\widehat{\text{LHV}} = 425 - (0.1453 \text{ mol}_{\text{H}_2\text{O}}/\text{mol}_\text{prod})(13.6 \text{ mol}_\text{prod}/\text{mol}_\text{fuel})$$
$$\cdot (1.03 \text{ kBtu/lb}_{\text{H}_2\text{O}})(18 \text{ lb}_\text{m}/\text{lbmol})$$
$$= 388.4 \text{ kBtu/lbmol}$$

Substituting into Eq. 6.8 using the lower heating value, and noting that the enthalpy of the fuel and air is zero at 77°F, the furnace efficiency is

$$\eta = \frac{388.4 - (13.6)(1.94)}{425.0} = 0.852$$

6.2 BURNER TYPES

The various types of burners can be categorized as premixed burners with entrained air, premixed burners with pressurized air, and nozzle-mixed burners. In addition, interesting new low-emissions gas burners are being introduced.

The burner design influences the combustion stability, efficiency, safety, reliability, and emissions. The shape and distribution of the flame should match the furnace combustion chamber. In general, it is desirable that the flame fill the combustion volume but not impinge on the furnace walls. For a given burner, increasing the mixture pressure will broaden the flame. Increasing primary air will shorten the flame. Rapid mixing produces a short, bushy flame, while delayed mixing and low velocities result in long, slender flames.

Premixed Burners with Entrained Air

Many domestic and commercial gas appliances and some industrial gas units employ atmospheric gas burners. An air blower is not used in this type of burner. Rather, the combustion air from the ambient surroundings is entrained by the fuel flow in a manner similar to a Bunsen burner. The fuel enters at low pressure (0.5

to 15 kPa; 2 to 60 in. H_2O) but at high velocity through an orifice as shown in Figure 6.3 for a single-port burner. Primary air is drawn in through shutter openings due to the momentum of the fuel jet and is controlled by adjusting the air shutters. A venturi throat improves the entrainment. Mixing occurs in the extended tube leg and the throat. The mixture flows through the burner head port and burns as an attached flame at the burner port. The momentum of the flame entrains secondary air, which completes the complete combustion. The book by Jones is a good reference for design information for premixed burners.

Entrained-air burners such as shown in Figure 6.3 typically operate at 40 to 60% primary aeration. The ratio of the area of the venturi throat to the area of the fuel orifice is typically 50 to 100. The burner port area to the venturi throat is typically 1.25 to 1.5. Burner port loadings for a single-port burner with 50% primary aeration are in the range of 9 to 14 W/mm^2 port area. The burner port loading, which is set by the fuel flow rate, must be matched to primary aeration (which is the percent of theoretical air) in order to achieve a clean, stably burning flame. Figure 6.4 shows a burner operating diagram where the area of satisfactory operation is shown as a function of primary air and burner port loading. Three areas of unsatisfactory operation are noted: (1) *flame lift* at high primary aeration due to the increased flow rate not being balanced by a similar increase in burning velocity, (2) *lightback or flashback* at low heat input where burning velocity is greater than the gas velocity, and (3) *yellow tipping* and incomplete combustion at low primary aeration due to fuel-rich combustion with insufficient secondary air entrainment. The exact size and location of the stable area are dependent on the particular burner configuration. Typically for small single burners without swirl, the flow velocity at the burner head should be between 2 and 5 times the laminar burning velocity.

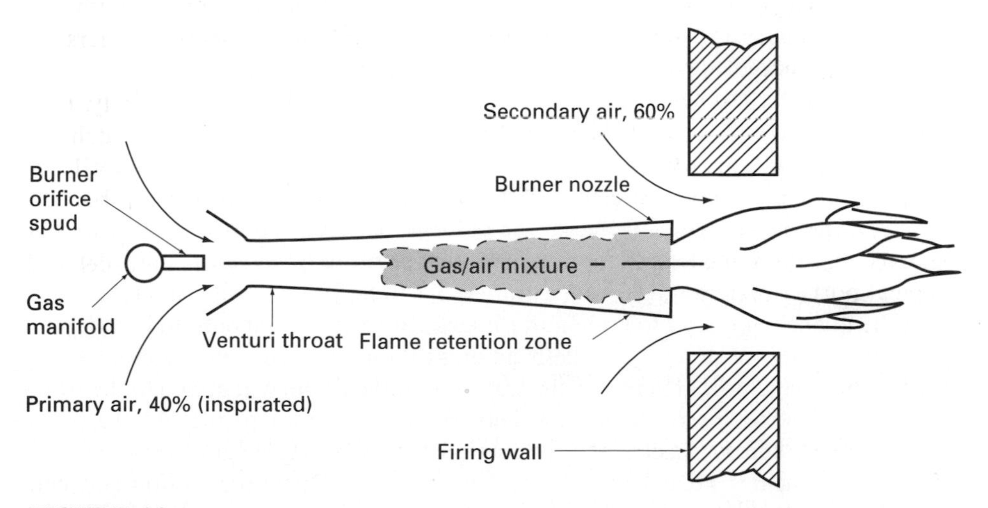

FIGURE 6.3
Single-port atmospheric-type gas burner.

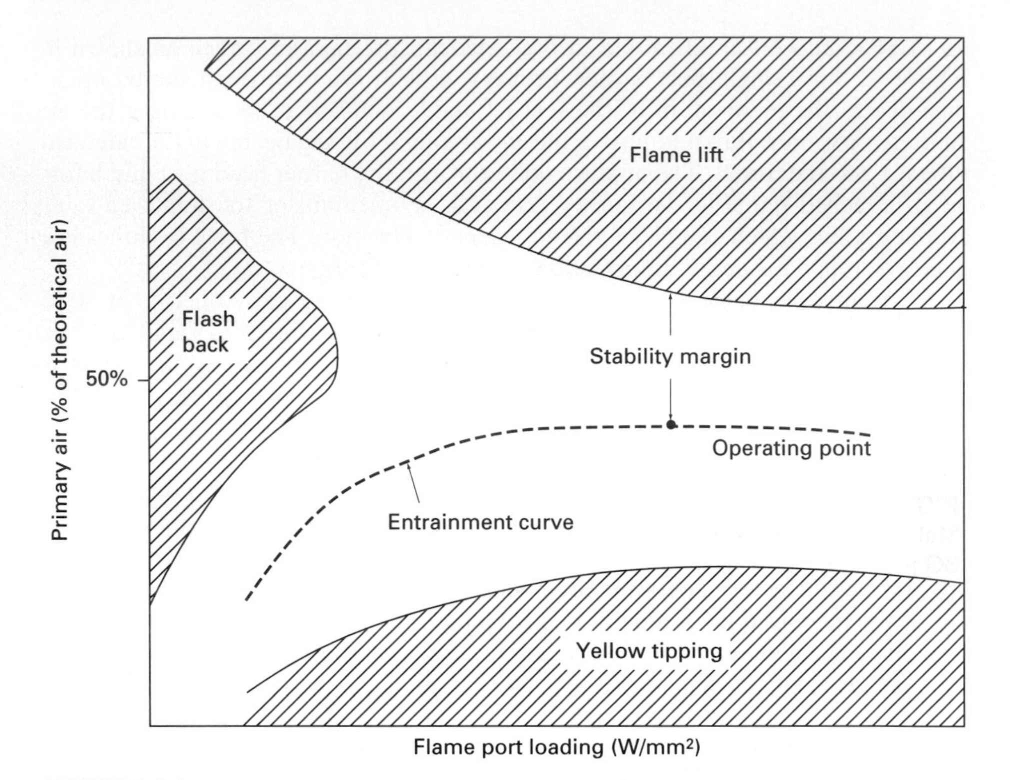

FIGURE 6.4
Stability diagram for single-port burner showing liftoff, flashback, and yellow tipping regions [Jones, by permission of BG plc, Research Technology].

Domestic gas appliances generally use a multiple-port burner head to spread out the flame (Figure 6.5). At low aerations (up to about 65% primary air) flame interaction induces a single inner cone over the array of flame ports. As the primary aeration is increased, separate inner cones appear over each port. The same general parameters as for the single-port aerated burner apply, although the stability diagram is slightly different, as shown by Figure 6.6.

The limit of flame lift for multiple-port burners depends on the heat input rate, port diameter and depth, and interport spacing. An increase in heat input rate increases the tendency to lift due to the balance between flame velocity and gas velocity. For the same heat input rate, flame lift is more likely with smaller ports. Typical port diameters range from 1 mm to 3.5 mm. The port size should be less than the quench diameter to prevent flashback. Burner ports should have a depth-to-diameter ratio of at least 2 to help suppress flame lift. This ensures that some streamlines in the port are close to the wall so that the flame will attach to the wall. An increase in the interport spacing increases the tendency to lift. As the ports become more widely spaced, the interaction between neighboring flames is reduced. For spacings greater than 6 mm between ports, little interaction between flames is observed. Reduction in spacing from 6 to 1.5 mm can double the primary aeration at which lift occurs.

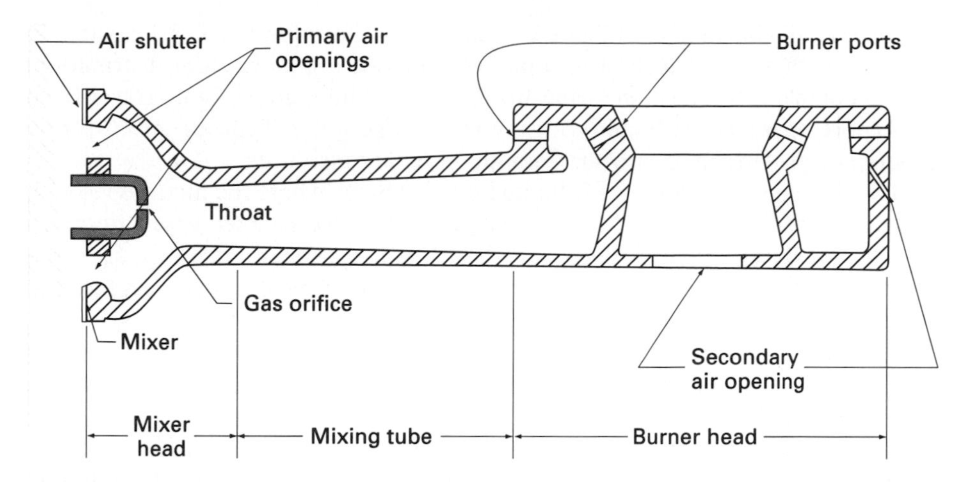

FIGURE 6.5
Multiple-port atmospheric-type gas burner [Weber and Vandaveer, by permission of BG plc, Research Technology].

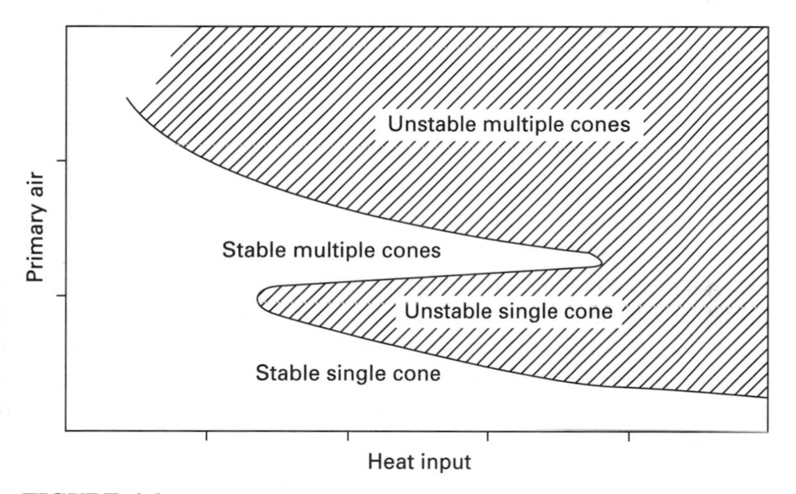

FIGURE 6.6
Stability diagram for multiple-port gas burner showing liftoff region [after Jones, by permission of BG plc, Research Technology].

Where high burner port output is required, a flame retention head is used as a simple means of delaying flame liftoff. Several small holes with approximately 10% of the area of the main port hole are drilled approximately 2 mm edge to edge from the main hole. The small retention flames are more stable because of their low efflux velocity. Burner port loading at flame lift can be increased up to a factor of 5 by this technique. For example, burner port loadings of 35 W/mm^2 can be obtained at 50% aeration, when the natural gas supply pressure is 20 kPa.

In general a burner should have a wide range of turndown (i.e., be capable of operating with a wide range of heat output), provide a uniform heat distribution, provide complete combustion, avoid lifting of the flame, allow flame travel from port to port, operate quietly, and withstand the temperature. Typically atmospheric burners are made from cast iron. Ceramic parts are also used.

The minimum gas flow rate is limited by flashback, where the mixture velocity is less than the flame speed. Maximum flame speeds for selected gases are given in Table 6.1. Gases with high flame speed, such as hydrogen or acetylene, are more prone to flashback at low firing rates. However, fast-burning gases are less likely to lift off compared with natural gas or LPG, for example. Hence, a burner operating on hydrogen or acetylene should operate at higher flow rates than when operating on natural gas or LPG. A heat output turndown ratio of about 4 to 1 is reasonable for atmospheric burners.

Secondary air comprises the remainder of the total air requirements. The design challenge is to obtain good mixing of the secondary air into the flame to complete combustion. Excess air should be held as low as possible to achieve maximum furnace efficiency without smoke or CO emissions.

Premixed Burners with Pressurized Air

Greater energy density—and thus higher temperature—and better control of flame shape can be achieved when both the fuel and the air are introduced into the burner under pressure. There are a variety of burner styles such as the furnace burner, the spear-flame burner, the ribbon burner, the flat-flame burner, and the hand torch. Since the mixture flow rate is higher in the pressurized burner, the flame must be stabilized against blowoff. This can be achieved by various design features such as swirl, hot ceramic surfaces, or a ring of lower-velocity jets around a higher-velocity central core, and by using a very large flame angle. On the other hand, the burner design must also guard against flashback because of the intense heat from the flame. Sufficient heat transfer must occur by conduction and radiation from the burner surfaces so that the gas mixture near the burner walls is cooled sufficiently to reduce or retard the flame speed. This precludes the use of preheated air in these types of burners.

TABLE 6.1
Maximum flame speed in small burners [Jones, by permission of BG plc, Research Technology]

Fuel	% fuel	% theoretical air	Flame speed (cm/s)
Hydrogen	42	58	225
Acetylene	9.5	80	145
Ethylene	7.2	90	70
Propane	4.7	85	45
Carbon monoxide	45	50	43
Methane	9.8	95	37
Pentane	2.9	88	35

Nozzle-Mix Burners

As the size of the burner is increased, it becomes feasible to mix the fuel and air externally instead of internally within the burner. Hence, the danger of flashback is eliminated. Flame stabilization is achieved with an external refractory nozzle, sometimes called a tunnel, as shown in Figure 6.7. By reradiating heat from the flame, the refractory nozzle maintains the flame temperature at the base or root of the flame. Mixing is achieved by impingement of the fuel and air jets and by introducing swirl to the air. Swirl tends to spread the flame radially and shorten the length of the flame. Flame stability is enhanced by swirl, which recirculates hot products of incomplete combustion back into the root of the flame. Good mixing is essential to reduce soot formation and to minimize excess air requirements. Preheating the air by means of a heat exchanger in the exhaust gas and operating the burner with low excess air improve the overall efficiency.

In large industrial and utility applications, the gas burner typically uses primary and secondary air, and the fuel is introduced into the nozzle through tubes called *spuds,* as shown in Figure 6.8. The primary air flows from a plenum called a *windbox* to the core of the burner. The secondary air, which is the majority of the air, enters the periphery of the burner and flows over adjustable swirl vanes. The swirl flow, which is a rotating flow about the central axis of the nozzle, spreads the flame and enhances mixing, as noted in Figure 6.9. The swirl vanes are adjusted until internal recirculation is achieved (typically a swirl ratio of about 0.5) which enhances mixing and flame stability. The *swirl ratio* is defined as the ratio of the flux of angular momentum to the flux of axial momentum times the radius of the burner outlet. The internal recirculation zone is formed by a positive radial pressure gradient which is caused by the swirl.

Because of the large volume flow rates, it is desirable to keep the pressure drop across the burner as low as possible. Typically, 0.3 psig and 600°F air is supplied at full load, while the gaseous fuel is at 15 psig and ambient temperature. The gas is injected at sonic velocity, while the inlet air velocity is about 200 ft/s at full load. Often it is desirable to operate at part load, and a burner turndown of up to 10:1 is generally possible.

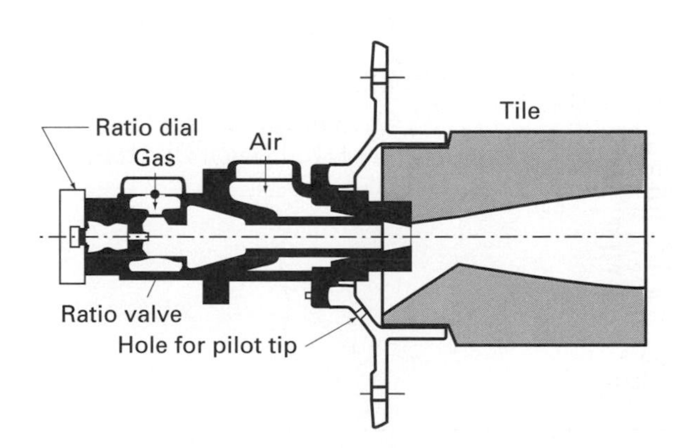

FIGURE 6.7
Small nozzle-mix type gas burner.

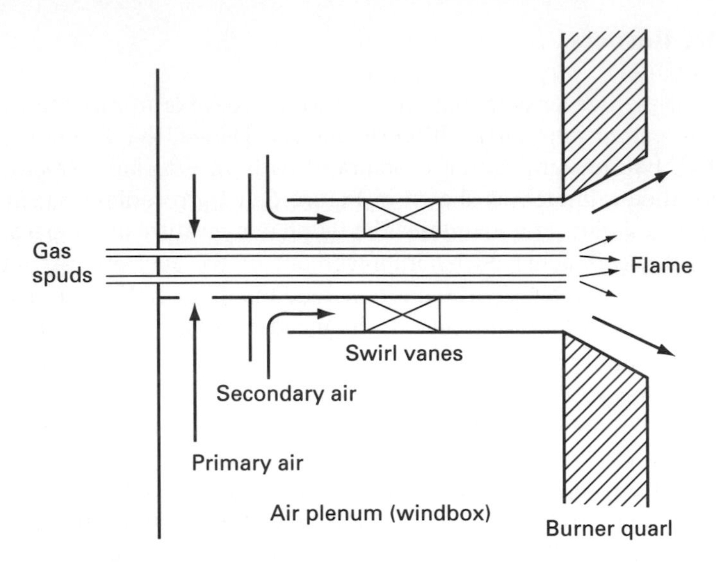

FIGURE 6.8
Large nozzle-mix type gas burner.

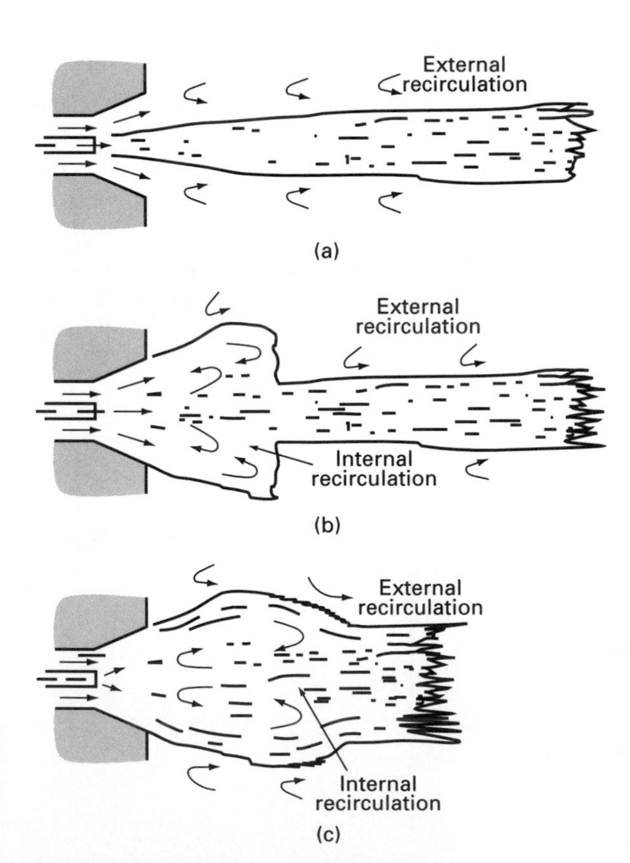

FIGURE 6.9
Flame types: (*a*) long jet flame, no swirl (type-0 flame); (*b*) combination jet flame and partial internal recirculation zone, intermediate swirl (type-1 flame); and (*c*) flame with closed internal recirculation zone, high swirl (type-2 flame) [Reprinted from *Prog. Energy Comb. Sci.*, vol. 18, Weber and Dugue, "Combustion Accelerated Swirling Flows in High Confinements," pp. 349–367, © 1992 with kind permission from Elsevier Science Ltd., The Boulevard, Langford Lane, Kidlington OX516B, UK].

6.3 PULSE COMBUSTION FURNACE

Pulse-type burners for residential and commercial furnaces, water heating, and drying are an outgrowth of the pulse-jet, which was made infamous by the German V-1 buzz bombs during World War II. The motivation is that the combustion products oscillate and hence have high heat transfer rates, which potentially can result in smaller furnace size. Overall efficiencies of 96% are achieved by pulse furnaces because the water in the combustion products is condensed, the excess air is low, and extraneous heat losses are minimized. Actually, such high efficiency can be achieved in a conventional furnace by these same measures. However, the size tends to be larger, and a loss of flue draft necessitates the use of a blower. The pulsing nature of the combustion produces a flow of combustion products without a continuous blower.

Although there are several different types of pulse combustors (for example, the Helmholtz resonator, the quarter-wave or Schmidt tube, the Rijke tube, and the Reynst pulse pot [Zinn]), the underlying principle controlling their operation is the same—namely, that the periodic addition of energy must be in phase with the periodic pressure oscillations. In spite of this seeming simplicity, the processes that occur in a pulse combustor are complicated; they involve a three-dimensional, transient flow field that is highly turbulent and has variable physical properties, a resonant acoustic pressure field, and a large transient energy release whose characteristic time is on the same order as the characteristic times for the chemical reactions and the acoustic resonance.

The pulse combustion burner is shown schematically in Figure 6.10. Gaseous fuel is supplied through a gas distributor head. Flapper valves for the air and fuel are used to check the backflow of combustion products. The combustion chamber has a larger diameter than the exhaust pipe. The combustion products flow in an oscillatory fashion into a heat exchanger, where the temperature of the products is lowered to 100°F. Water condenses and is dumped to a drain, while the exhaust products are vented directly outside through a small tube.

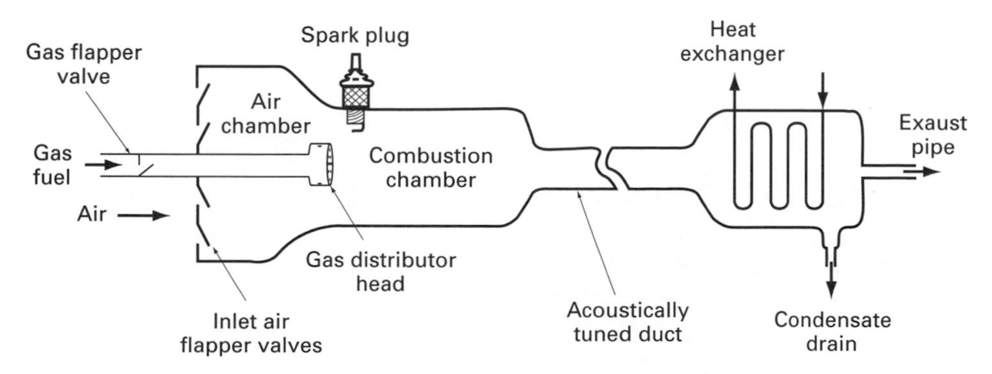

FIGURE 6.10
Schematic of pulse-jet furnace.

Figure 6.11 illustrates the operational sequence of a pulse combustor. Combustion is initiated by supplying air with a small blower and gas under line pressure through flapper valves. The fuel-air mixture is ignited by an electric spark. Ignition creates a positive pressure, closing the flapper valves. The positive pressure is created by the acceleration of the flame propagation away from the spark. Combustion products expand through the open exhaust pipe, and the expansion results in a negative pressure at the inlet end of the system. The flapper valves open as the pressure drops, and fresh air and fuel flow into the combustion chamber. The cycle repeats itself without the need of the air blower and spark once started. Pulse combustion burners using natural gas operate at frequencies of 25 to 100 Hz depending on the geometry selected.

When the pulse combustion process is observed in slow motion through a quartz wall, it looks like a cloud of hot gas which moves back and forth, gradually working its way from the front of the combustion chamber to the back. A thin flame front is not observed. The combustion temperature is lower because of the high heat transfer; and because of the turbulence, the required excess air is low. The relatively low-temperature combustion, low excess air, and good mixing through turbulence result in low nitrogen oxide emissions, while carbon monoxide levels remain at or below those of conventional furnaces.

Two design problems occur with pulse combustors. Pulse combustion is inherently noisy. Expansion chambers and mufflers at the inlet and outlet help to reduce

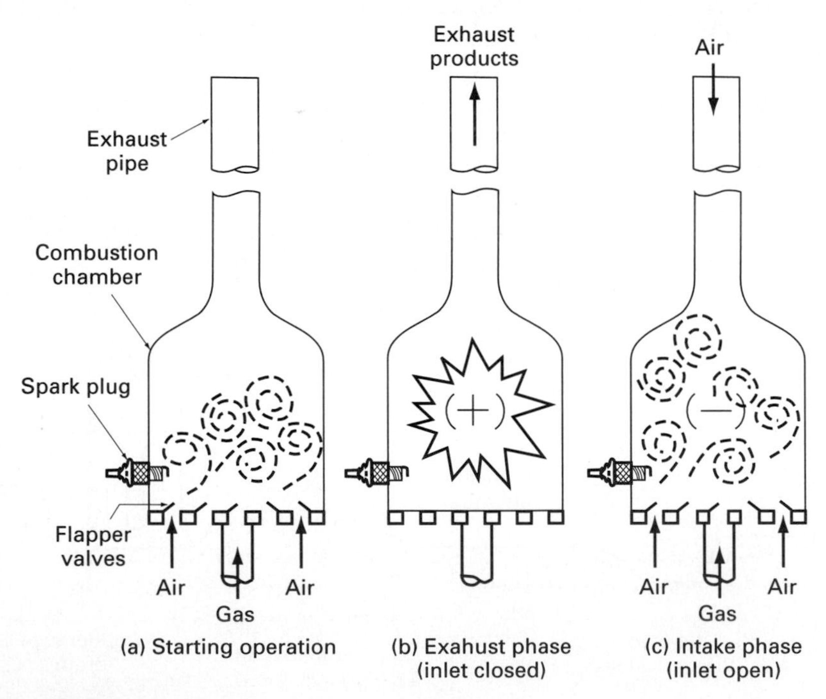

FIGURE 6.11
Operating cycle of pulse-jet combustor.

the noise. Second, high heat release rates of the oscillatory combustion result in hot spots on the tube walls, but careful use of fins and airflow can minimize local hot spots. Convective heat transfer rates from oscillating flows can be much higher than those of steady turbulent flows at the same Reynolds number. Convective heat transfer increases with increases in the amplitude and frequency of the velocity oscillations and decreases with increases in the mean flow rate for a fixed oscillation amplitude and frequency.

6.4 FUEL SUBSTITUTION

Substitution of one fuel for another more readily available fuel is sometimes necessary for a particular burner. Perhaps a manufactured gas from coal or wood gasification is to be substituted for natural gas. The burner should be adjusted to preserve the heat rate and flame stability and shape where possible. If the substitute fuel is similar, such as substituting propane for methane, it may be sufficient to adjust the airflow to the proper equivalence ratio.

If the substitute fuel is dissimilar, the fuel flow rates may have to be adjusted by changing the fuel pressure or orifice size. The heat rate equals the fuel volume flow rate times the heating value of the fuel per unit volume. The fuel volume flow rate is given by

$$\dot{V}_f = \sqrt{\frac{\Delta p}{\rho_f}} A_f \tag{6.10}$$

where Δp = pressure drop across the fuel orifice
A_f = effective area of fuel orifice
ρ_f = density of the fuel

Hence, for a fixed orifice size and fuel pressure, the heat rate q is given by

$$q = \frac{K_1 \text{ HHV}}{\sqrt{\rho_f}} \tag{6.11}$$

where K_1 is the system constant. The term $\text{HHV}/\sqrt{\rho_f}$ is a measure of the interchangeability of fuels and is termed the Wobbe index, WI:

$$\text{WI} = \frac{\text{HHV}}{\sqrt{\rho_f}} \tag{6.12}$$

If the Wobbe index of the substitute fuel is significantly different from that of the design fuel, the burner should be modified. In addition, the flame length, flashback, and blowoff characteristics should be considered. The density of a fuel cannot be changed appreciably by increasing the pressure, because this will increase the flow rate, moving the burner out of the stable design region. For low–heating value fuels such as from air-blown coal or wood gasification, the fuel flow rate must be considerably greater than for natural gas for the same heat output. The volume of products will also be greater, which may require larger ductwork.

EXAMPLE 6.3. Compare the Wobbe index for wood producer gas with the index for natural gas and comment on the interchangeability of these fuels.

Solution. From Table 2.4 the volumetric higher heating value of wood producer gas and natural gas are respectively 170 Btu/ft^3 and 1030 Btu/ft^3. The density is inversely proportional to the molecular weight, and from Table 2.3, the molecular weight is approximately 25, whereas for natural gas the molecular weight is about 16. Thus, using Eq. 6.12,

$$\frac{(\mathrm{WI})_{\text{producer gas}}}{(\mathrm{WI})_{\text{nat.gas}}} = \frac{170\sqrt{25}}{1030\sqrt{16}} = 0.20$$

Thus, not only will the flow rate of fuel need to be increased to maintain the same heat output, but the burner will need to be modified to ensure stability, since the Wobbe index of the producer gas is one-fifth that of natural gas.

6.5 EMISSIONS FROM GAS-FIRED FURNACES

In a conventional gas-fired furnace, emissions of particulate matter, carbon monoxide, sulfur dioxide, and hydrocarbons are typically low. Nitrogen oxide emissions, on the other hand, are high unless the traditional designs are modified. Emission factors, which are uncontrolled emissions per unit of input fuel for typical furnaces, as judged by the U.S. Environmental Protection Agency, are given in Table 6.2 for natural gas. Particulate emissions are due to soot particles formed in fuel-rich zones with inadequate mixing with air. Sulfur dioxide arises due to the trace amounts of sulfur in natural gas. Carbon monoxide and hydrocarbon emissions indicate pockets of gas with insufficient time at high temperature and insufficient oxygen to complete the reactions. Nitrogen oxide emissions indicate that the temperature was

TABLE 6.2
Uncontrolled emission factors for natural gas combustion systems [EPA]

	Type of installation		
Pollutant (kg/10^6 m^3 of natural gas*)	**Utility boiler**	**Industrial furnace**	**Domestic furnace**
Particulates	16–80	16–80	16–80
Sulfur dioxide	9.6	9.6	9.6
Carbon monoxide	640	560	320
Nitrogen oxides (as NO_2)†	8800	2240	1600
Nonmethane organics	23	44	84
Methane	4.8	48	43

*At standard temperature and pressure.

†Assumes no burner modification.

too high for too long a time with oxygen present. The requirements for low particulates, CO, and hydrocarbons conflict with the conditions for low NO_x.

The units in Table 6.2 are useful for environmental considerations, but are obscure with respect to thermodynamic and kinetic considerations. Let us compare the NO_x emission factor with the equilibrium NO_x levels in an adiabatic methane-air flame in order to see how close the factors are to the maximum possible formation of NO_x. By way of example, consider the utility boiler with 5% excess air preheated to 600 K. STANJAN gives an adiabatic flame temperature of 2343 K, NO concentration of 4200 ppm, and NO_2 concentration of 1.0 ppm. The practice, as noted in Table 6.2, is to weight NO_x as NO_2 (because much of the NO will be converted to NO_2 in the atmosphere). The density of NO_2 at this temperature is 0.249 kg/m³. For 5% excess air the volume of products per volume of methane is 11.0. Hence, the maximum emission factor is

$$(4201 \times 10^{-6}\ m^3_{NO_2}/m^3_{prod})(11.0\ m^3_{prod}/m^3_{CH_4})(0.249\ kg_{NO_2}/m^3_{NO_2})$$
$$= 11{,}507\ kg_{NO_2}/10^6\ m^3_{CH_4}$$

The uncontrolled utility boiler has about 76% of the maximum theoretical NO_x level. Uncontrolled industrial and residential furnaces have lower NO_x emissions because they do not use preheated air and have shorter residence times.

Emission standards for combustion systems are set at the national and state levels to protect health and the environment. Generally, new sources must meet more strict standards than old sources. Emission standards for new and modified gas-fired furnaces and boilers are summarized in Table 6.3. These are nitrogen oxide and particulate standards. It is presumed that there is no significant amount of sulfur and organic nitrogen in the fuel. Comparing the emission factors in Table 6.2 and the emission standards in Table 6.3 indicates the amount of emissions control required. This is illustrated in Example 6.4, where required degree of NO control is calculated. Nitrogen oxide emission control from gas-fired furnaces and boilers is typically achieved by combustion modification.

TABLE 6.3
U.S. emission standards for gas furnaces and boilers built after 1973 (kg/10^6 kJ heat input) [EPA]

Size	NO (as NO_2)	Particulates
Large*	0.09	0.014
Small†	—	0.068

* $> 250 \times 10^6$ kJ/h.
† $< 250 \times 10^6$ kJ/h.

EXAMPLE 6.4. A large utility burns natural gas using a conventional burner with no additional emission controls. Using Tables 6.2 and 6.3, what degree of emission control is required to meet the emission standards for nitrogen oxides and particulates? Use a natural gas heating value of 1050 Btu/ft³ = 39,100 kJ/m³.

Solution. Uncontrolled NOx emissions are:

$$\frac{8800 \text{ kg}/10^6 \text{ m}^3}{39{,}100 \times 10^6 \text{ kJ}/10^6 \text{ m}^3} = 0.225 \text{ kg}/10^6 \text{ kJ}$$

$$\%\text{controlled required} = \frac{0.225 - 0.09}{0.225} = 60\%$$

From Table 6.2 uncontrolled particulate emissions are at most 80/39,100 = 0.0020 kg/10^6 kJ. Thus, no particulate controls are required.

Combustion Modification to Control NO_x

In Chapter 4, it was shown that the key to preventing NO formation is to reduce the peak temperatures. Low NO emissions in domestic burners have been achieved by suspending ceramic rods in the flame zone just above the burner ports. The ceramic inserts lower the peak flame temperature by radiating the heat away. NO emissions are reduced about 25% when the inserts are added.

For small and medium-size industrial burners, a new concept is the so-called ceramic fiber burner, which is a porous matrix of ceramic and metallic fibers formed into a cylindrical tube, as shown in Figure 6.12. Premixed fuel and air flow uniformly through the porous material and, aided by catalytic reaction, burn within the fiber matrix. The low conductivity of the fibers and convective cooling by the outflowing reactants allow the burner to operate safely and with no flashback

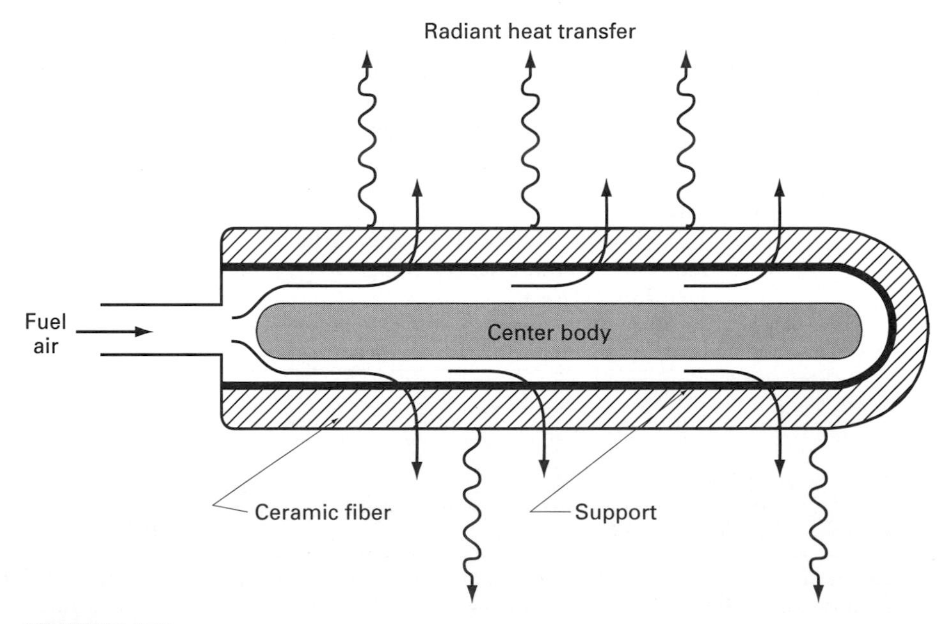

FIGURE 6.12
Cross section of a flameless fiber matrix burner.

tendencies at surface velocities below the mixture flame speed. The surface temperature is maintained below 1366 K (2000°F) for low NO_x formation and above 1256 K (1800°F) to react the CO. The surface temperature is set by the surface velocity of the reactants. Heat release rates of 790 kW/m^2 (250,000 Btu/h · ft^2) are possible with a turndown of 6:1. Typically 70% of the heat release is transferred by radiation. Nitrogen oxide emissions are reduced to a level as low as 15 ppm by the fiber matrix burner. Since there are no hot spots, furnace capacity can possibly be increased. However, since the heat release per unit burner area is reduced, the fiber matrix burner area must have a larger surface area than a flame. Excess air levels may be held to 10% without increasing CO or hydrocarbon emissions [GRI].

For large gas-fired industrial and utility boilers, two methods have been shown to significantly reduce NO emissions: two-stage combustion, and flue gas recirculation. The motivation for these methods is seen in Figure 6.13, which shows the adiabatic flame temperature and the equilibrium NO concentration as a function of excess air for a methane-air flame. Notice that for a rich flame and a very lean flame the NO concentration is relatively low, whereas at 10–20% excess air the NO is maximum. As noted previously for best efficiency the system should operate with 10% excess air or less overall.

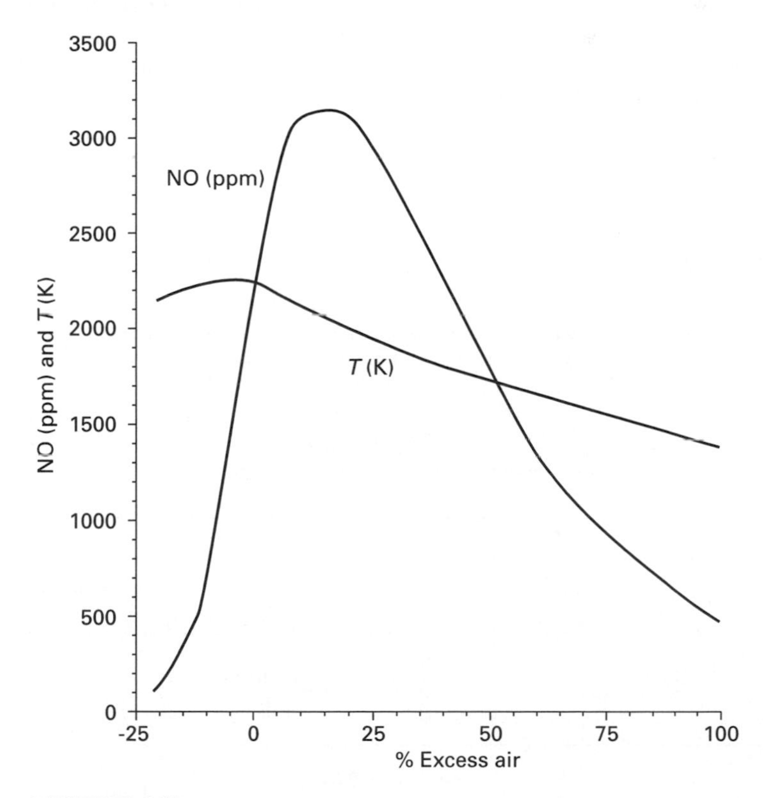

FIGURE 6.13
Equilibrium NO concentration and adiabatic flame temperature for methane-air at 20°C and 1 atm.

In one approach to two-stage combustion, the burner air is reduced to 85–90% of stoichiometric so that the near flame is fuel rich, and due to the sooty nature of the flame, heat is radiated away and the temperature drops. Secondary air is then directed into the fuel-rich zone from separate overfire air jets located above and below the burner to complete burnout and bring the overall stoichiometry back to 110% theoretical air. In this way, the flame zone is extended and peak temperatures are reduced. The burner must be designed to operate at lower airflows, and complete mixing is achieved with the overfire air jets.

In another approach to two-stage combustion in an industrial burner, the fuel rather than the air is staged, as shown in Figure 6.14. For a 300-kW burner, D_0 is 87 mm. The primary natural gas jets from 24 holes mix with swirling combustion air and form a lean flame starting in the burner nozzle. Secondary natural gas flows through eight pipes and feeds into the flame further downstream as it is pulled in by the recirculating flow (indicated in Figure 6.9). Burner flames of this geometry have been investigated in detail at the International Flame Research Foundation in the Netherlands by Weber et al (1992). Fuel gas staging resulted in reduction in NO_x emissions as large as 80%; the swirl ratio was 0.56, the peak flame temperature was about 1900 K, and the NO_x was as low as 33 ppm.

In flue gas recirculation, 10–20% of the cooled combustion products are ducted back to the burner and mixed with the primary air. Additional ducting, fan power, and controls are needed. The oxygen content of the vitiated air should be maintained to at least 17% to prevent burner instability (lifted flame and blowoff). The flow rate of the vitiated air is increased to maintain the full rated output of the

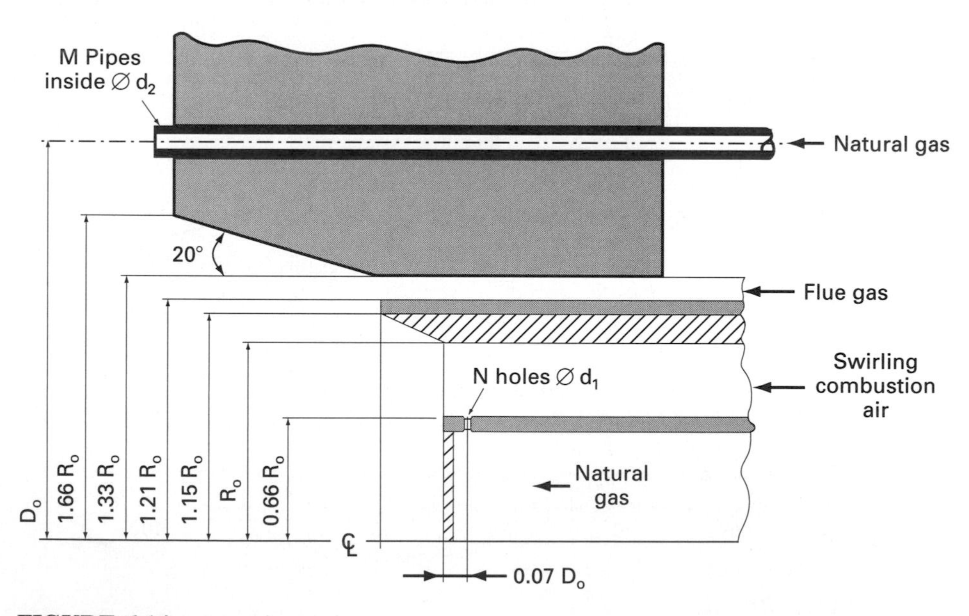

FIGURE 6.14
Industrial natural gas burner with staged fuel and flue gas recirculation [Sayre et al., by permission of The Combustion Institute].

furnace. The increased heat capacity of the recirculated flue gas helps to reduce peak flame temperatures. Instead of mixing the flue gas with the primary combustion air upstream of the burner, the flue gas can be brought in through an annulus in the burner, as indicated in Figure 6.14.

A low NO_x, radially stratified industrial burner has been further developed by Toquan and Beer et al. which utilizes a fuel jet surrounded by three concentric air nozzles. As shown in Figure 6.15, the fuel flows through a spud in the central core of the burner. The three concentric jets are arranged so that the innermost nozzle has the highest amount of swirl and the outermost jet has the lowest swirl. The flow settings which resulted in low NO_x emissions are shown in Table 6.4. This

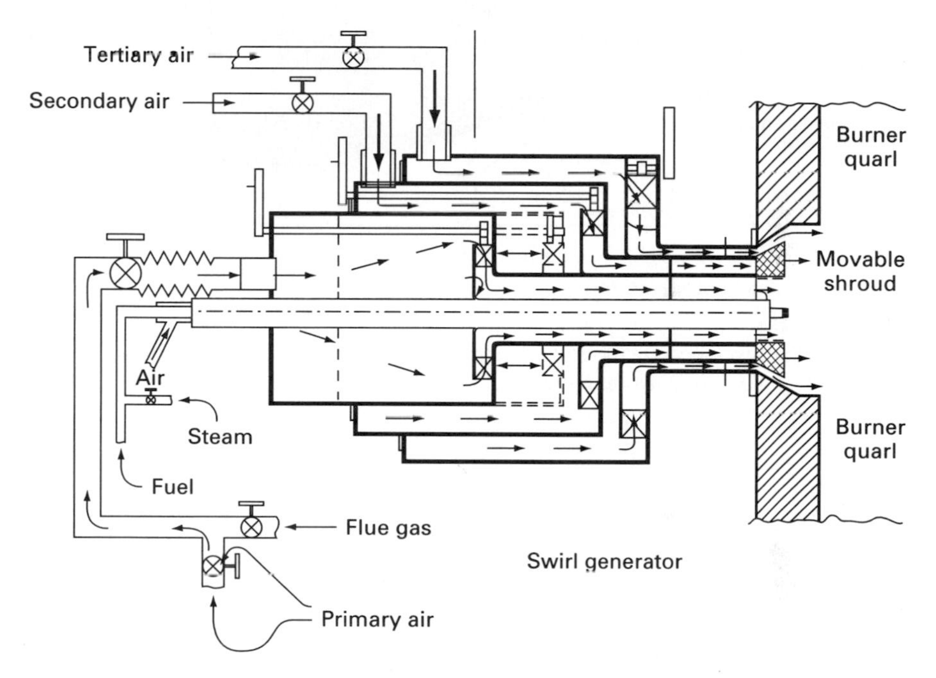

FIGURE 6.15
Schematic of radially stratified flame burner for low NO_x [Toquan et al., by permission of The Combustion Institute].

TABLE 6.4
Operating parameters of a radially stratified low NO_x burner [Toquan et al., by permission of The Combustion Institute]

Nozzle	Velocity (m/s)	Swirl no.*	% of total air/gas
Primary	22	2.8	15% of air
Secondary	53	1.8	30% flue gas recir.
Tertiary	34	1.2	85% of air
Fuel	78	0	100% nat. gas

*Ratio of axial flux of angular momentum to axial flux of linear momentum times radius of the inlet.

arrangement for swirl delays the formation of the internal recirculation zone (shown in Figure 6.16), which allows the flame temperature to decrease by radiation before intense mixing and burnout in the internal recirculation zone, where the combustion is completed. The use of recirculated flue gas in the primary (inner) and secondary air nozzles further reduces the NO_x emissions as shown in Figure 6.17. This radially stratified burner design allows for more flue gas

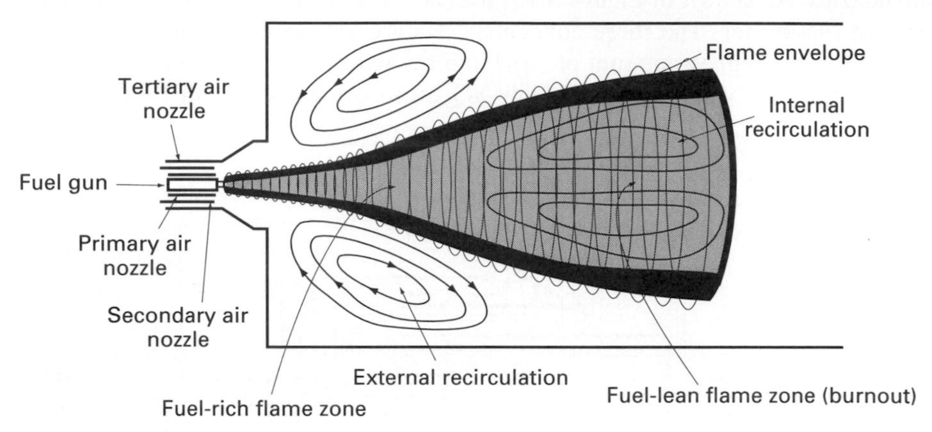

FIGURE 6.16
Flow patterns developed by the radially stratified burner [Toquan et al., by permission of The Combustion Institute].

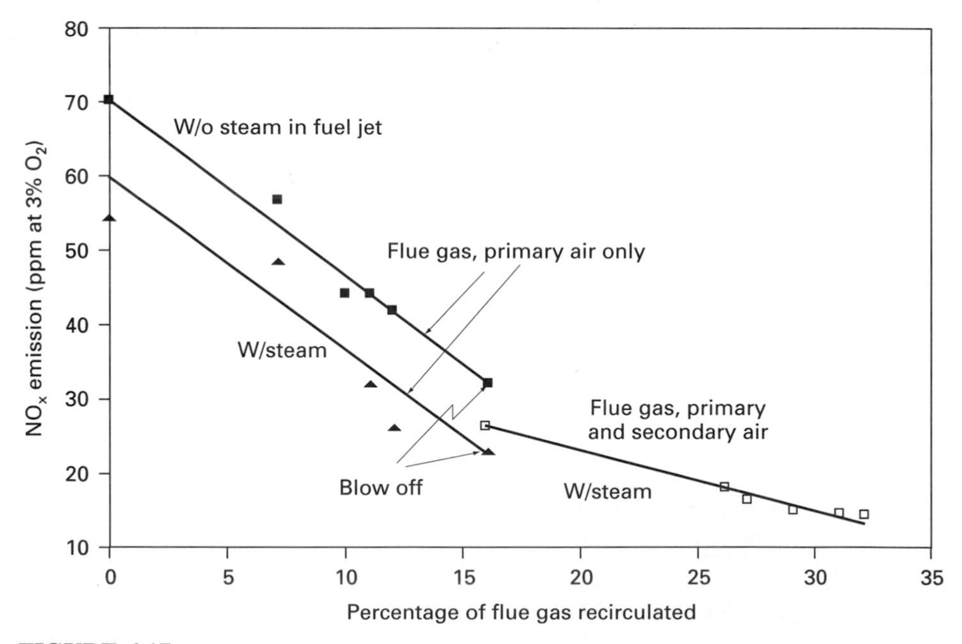

FIGURE 6.17
Effect of burner flue gas recirculation and steam injection on NO_x emissions for the radially stratified burner. The steam/fuel ratio was 0.12 kg/kg [Toquan et al., by permission of The Combustion Institute].

recirculation without flame instability than was possible with the more standard burner. As noted in Figure 6.17, further reduction in NO_x down to 15 ppm was achieved by means of steam injection in the fuel jet without flame instability or increase in CO.

6.6 SUMMARY

The burner is the heart of the gas-fired furnace. The small, single-port burner operates like a Bunsen burner, but with 40 to 60% primary (premixed) air, and the remaining 60 to 40% secondary air is entrained into the flame. A stable flame with an energy flux of 9 to 14 W/mm^2 of port area is possible. Domestic appliances and furnaces use multiple-port, naturally aerated, premixed burners. The port size should be less than the quench diameter for safety. Large industrial and utility burners use a blower and mix the fuel and air with swirl in a nozzle which stabilizes the turbulent flame.

As shown by an energy balance on the system, gas-fired furnaces must operate at as low excess air as possible (to about 3% excess air), without increasing CO and soot emissions, in order to achieve high furnace efficiency. This requires good mixing of fuel and air either before the burner for small systems or in the burner nozzle for large systems. Unfortunately, as the excess air is reduced, the flame temperature increases and NO_x emissions are increased.

Radiation is used to reduce the flame temperature, and hence NO_x, by introducing ceramic rods in the flame zone or by means of the new ceramic fiber burners. Large burners use two-stage combustion and flue gas recirculation to reduce NO_x emissions. Two-stage (rich-lean) combustion can be accomplished with external air jets or internally with radial stratification produced by swirl.

The pulse furnace has been developed for small systems as a small, compact furnace with high efficiency and low emissions. In general, when changing fuels and maintaining the same heat output, the burner orifices must be changed if the Wobbe index ($WI = HHV/\sqrt{\rho_f}$) of the fuel changes appreciably.

PROBLEMS

6.1. The useful heat output of a residential furnace is 100,000 Btu/h. The fuel is methane, and heat losses are 10% of the useful output. The stack temperature is 340°F and 40% excess air is used. Both air and fuel enter the furnace at 77°F and 1 atm. Combustion is assumed to be complete. Calculate:

(*a*) The flow rate of methane (scfm)
(*b*) The flow rate of inlet air (scfm)
(*c*) The flow rate of exhaust products (acfm)
(*d*) Furnace efficiency

Note: scfm = ft^3/min at standard temperature (77°F) and pressure (1 atm); acfm = ft^3/min at actual temperature and pressure.

6.2. Repeat Problem 6.1 but assume that combustion products leave at 100°F.

6.3. Repeat Problem 6.1 but assume that combustion products leave at 100°F, that heat losses are negligible, and excess air is 5%.

6.4. An entrained-air burner uses methane gas. The air and fuel enter at 298 K. The stack temperature is 400 K, and the CO_2 concentration is 9% by volume. The CO concentration is 10 ppm by volume. Assuming the extraneous heat losses are negligible, what is the efficiency of the furnace? The heating value of methane is 23,900 Btu/lb_m = 55.5 MJ/kg.

6.5. A furnace is rated at 100,000 Btu/h input using natural gas. Assuming that the burner head consists of $\frac{1}{8}$-in.-dia holes with $\frac{3}{8}$ in. between hole centers, what is the size of the burner head, and how many holes are required? What is the heat release rate per ft^2 of burner head? What is the range of velocities in the burner ports for the stable region of operation, and how does this compare with the laminar flame speed? Assume that the temperature of the burner head is 600°F.

6.6. A utility generates 500 MW of electricity while operating on natural gas. The overall thermal efficiency of the plant is 33%. Air is supplied to the burners at 0.3 psig and 80°F. Find the cost of natural gas per hour assuming a price of $\$0.55/10^6$ Btu. Also, find the power required to deliver the burner air across a 0.3 psig pressure rise at 80°F. Assume that 10% excess air is required for clean combustion and the fan efficiency is 90%. Assume that the stoichiometric air/fuel weight ratio is 17.2/1.

6.7. For Problem 6.6, what are the NO_x and particulate emission rates (ton/year), assuming that the plant is just meeting the federal emission standards.

6.8. Producer gas from a wood gasifier consists of 20% CO, 13% H_2, 3% CH_4, 10% CO_2, 10% H_2O, and 44% N_2 by volume. The wood gasifier is used to retrofit an industrial natural gas-fired furnace rated at 500,000 Btu/h on an input basis. Calculate the standard volume flow rate of air and fuel (scfm) for both methods of firing, assuming 10% excess air in both cases. How would the burner need to be modified for the retrofit? How does the standard volume flow rate of combustion products compare with natural gas firing for this case? Assume that the volumetric analysis for natural gas is that given in Example 6.1. Use Table 2.4 for heating value data.

6.9. A burner operating on methane produces flue gas with 250 ppm of NO and 57 ppm of NO_2, measured on a wet basis. The burner is supplied with 20% excess air. What is the NO_2 emission factor in ng/J? The higher heating value is 39.7 MJ/m^3. Assume complete combustion and a stack temperature of 400 K. *Hint:* Calculate $m_{NO_2}/V_{products}$ using the perfect gas law and find the volume of products per volume of methane.

6.10. Outside air is brought directly into a gas furnace through a well-insulated duct. The furnace burner was originally set to give a stoichiometric mixture for 70°F air. On a −10°F winter day, what will be the equivalence ratio of the burner?

6.11. Determine the adiabatic flame temperature and equilibrium NO concentration for a methane-air flame which has 15% excess air by volume and has 20% of the combustion products recirculated back into the inlet of the burner. The inlet air and recirculated products are at 250°C and 1 atm.

REFERENCES

Barnard, J. A., and Bradley, J. N., *Flame and Combustion,* Chapman & Hall, London 2nd ed., 1985.

Breen, B. P., *Thirteenth Symp. (Int.) on Combustion,* The Combustion Institute, Pittsburgh, p. 398, 1970.

Dec, J. E., and Keller, J. D., "Pulse Combustor Tail-Pipe Heat Transfer Dependency on Frequency, Amplitude and Mean Flow Rate," Report SAND88-8618, Sandia National Laboratory, Livermore, CA, 1988.

EPA, "Compilation of Air Pollution Emission Factors," U.S. Environmental Protection Agency, EPA-AP-42, 1985.

Faulkner, E. A., *Guide to Efficient Burner Operation: Gas, Oil and Dual Fuel,* The Fairmont Press, Atlanta, 1987.

GRI, "Pulse Combustion," *Gas Research Institute Digest,* vol. 3, pp. 1–11, 1984.

Jones, H. R. N., *The Application of Combustion Principles to Domestic Gas Burner Design,* British Gas monograph, E. & F. N. Spon Ltd., London, 1989.

North American Combustion Handbook, 2nd ed., North American Manufacturing Co., Cleveland, 1978; 3rd ed., vol. I, 1986.

Peters, A. A. F., and Weber, R., "Mathematical Modeling and Scaling of Fluid Dynamics and NO_x Characteristics of Natural Gas Burners," *Int. ASME/EPRI Power Generation Conference,* Phoenix, 1994.

Rhine, J. M., *Modelling Gas-Fired Furnaces and Boilers and Other Industrial Heating Processes,* British Gas monograph, McGraw-Hill, New York, 1991.

Sayre, A., Lallemant, N., Dugue, J., and Weber, R., "Effect of Radiation on Nitrogen Oxides Emissions from Non-Sooty Swirling Natural Gas Flames," *Twenty-Fifth Symp. (Int.) on Combustion,* The Combustion Institute, Pittsburgh, pp. 235–242, 1994.

Schreiber, R., Krill, W., Kesselring, J., Vogt, R, and Lukasiewicz, M., "Industrial Applications of the Radiant Ceramic Fiber Burner," D06-83, International Gas Research Conference, London, 1983.

Segeler, L. G. (ed.), *Gas Engineers Handbook,* The Industrial Press, New York, 1965.

Toquan, M. A., Beer, J. M., Jahnsohn, P., Sun, N., Testa, A., Shihadeh, A., and Teare, J. D., "Low NO_x Emission from Radially Stratified Natural Gas-Air Turbulent Diffusion Flames," *Twenty-Fourth Symp. (Int.) on Combustion,* The Combustion Institute, Pittsburgh, pp. 1391–1397, 1992.

Weber, E. J. and Vandaveer, F. E., "Gas Burner Design," Chap. 12 in C. G. Segeler (ed.), *Gas Engineers Handbook,* The Industrial Press, New York, 1965.

Weber R., and Dugue, J., "Combustion Accelerated Swirling Flows in High Confinements," *Prog. Energy Comb. Sci.,* vol. 18, pp. 349–367, 1992.

Weber, R., and Visser, B. M., "Assessment of Turbulent Modeling for Engineering Prediction of Swirling Vortices in the Near Burner Zone," *Int. J. Heat and Fluid Flow,* vol. 11, pp. 225–235, 1990.

Zinn, B. T., "Pulse Combustion: Recent Applications and Research Issues," *Twenty-Fourth Symp. (Int.) on Combustion,* The Combustion Institute, Pittsburgh, pp. 1297–1305, 1992.

CHAPTER 7

Premixed-Charge Engine Combustion

Most light-duty transportation and off-road vehicles, and small utility devices, use premixed-charge reciprocating engines with liquid fuels. Premixed-charge engines have been included in this section of the book because the primary nature of combustion in these devices is associated with an essentially homogeneous mixture of fuel vapor and air. Although large stationary engines sometimes use gaseous fuel directly, with either spark ignition or ignition by injection of a small amount of high-cetane (diesel) fuel, the discussion here is limited to automotive spark-ignition engines. The emphasis is on the combustion aspects of these engines and thus does not dwell on such topics as cycle analysis, engine performance, or mechanical design. To aid the reader who is not familiar with internal combustion engines, a compilation of terminology is given at the end of this chapter.

7.1 INTRODUCTION TO SPARK-IGNITION ENGINE COMBUSTION

For homogeneous-charge engines, the energy released by combustion is a function of the heating value of the fuel, the fuel-air ratio, and the trapped mass of the charge in the cylinder. To change load for a fixed fuel-air ratio, the charge flow is throttled so that the pressure in the cylinder drops below atmospheric pressure during the intake stroke. Throttling of the charge by varying the maximum lift of the intake valve has also been used, but thus far only for research purposes. In either case the throttling increases the work required to pull the charge into the cylinder and thus reduces engine efficiency.

Thermodynamic analysis shows that the thermal efficiency of the cycle increases with increasing compression ratio; however, the compression ratio of spark-ignition engines is limited by abnormal combustion. Increasing the compression ratio increases the temperature and pressure of the unburned charge. During combustion the unburned portion is further compressed by the expansion of the burned products. Figure 7.1 shows a typical configuration when the charge is compressed and combustion is nearly completed. The area marked b at the left is the combustion products zone. The unburned reactants zone, marked u, is at the right. As the last portion of unburned mixture is compressed, it may undergo rapid reaction causing rapid autoignition of the whole unburned volume. This last portion of gas to burn is called the *end-gas.* Rapid burning of the end-gas causes a blast wave to develop, and the pressure waves so produced cause the engine to vibrate. The sound caused by the vibration of the engine is known as *knocking,* and the sudden rapid burning of the end-gas is called *knocking combustion,* or simply *knock.*

Knock causes reduced efficiency, increased heat transfer, and, if severe, engine damage. The limiting compression ratio is directly related to the octane number of the fuel. With current designs limited by emission standards as well as fuel octane number, automobile engine compression ratios are in the 7.5 to 9.4 range. Engines with higher compression ratios typically require a higher-octane gasoline than that currently available as unleaded regular. Knock is promoted by increased compression ratio, advancing the ignition timing (i.e., moving the spark timing to allow a longer time before the piston reaches top dead center), opening the throttle to wide open, slightly rich mixtures, increased inlet air temperature, increased coolant temperature, and buildup of deposits on the cylinder walls. Knock is discouraged by part throttle, lean or over-rich mixtures, increased engine speed, and decreased inlet pressure. Reducing the time for end-gas reaction by chamber shape and spark

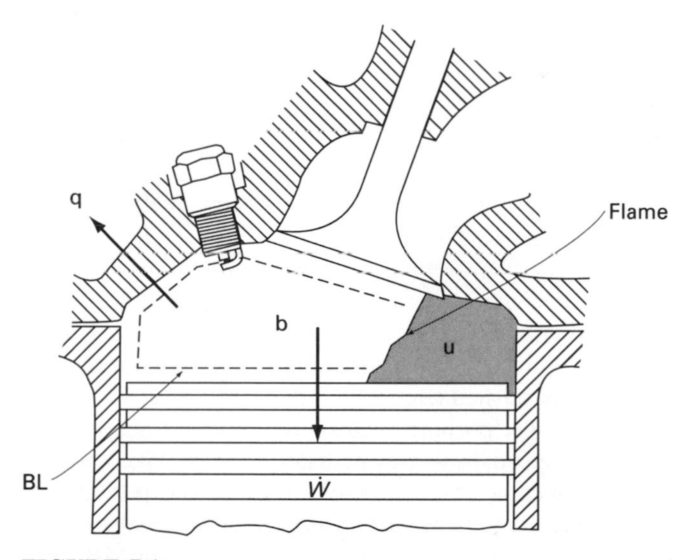

FIGURE 7.1
Three-zone combustion model; u denotes unburned gas, b the adiabatic burned core, and BL the burned gas boundary layer [Heywood et al., reprinted with permission from SAE paper 790291 © 1979, Society of Automotive Engineers, Inc.].

plug location, and locating the exhaust valve away from the end-gas, also discourage knock.

Engines that do not knock when new may knock as deposits form on the combustion chamber surfaces. Fuel additives that remove these deposits may prevent an octane requirement increase, but may also cause increased liner wear. One fuel provider has recently claimed a fuel formulation that causes the deposits to have increased thermal conductivity and thus reduce the required octane increase.

Knock effects can be reduced by use of a knock sensor and spark timing control, which retards the timing when engine vibration at knock frequency is detected by an accelerometer mounted on the block. Retarding the spark reduces fuel economy, but the loss is typically small because the knock is limited to brief periods during vehicle acceleration.

The simple Otto cycle assumes that the charge burns instantaneously at constant volume and at top dead center (TDC) when the volume is a minimum. This would give maximum work if no dissociation of products and no heat transfer were to take place. In the real engine the peak pressure is limited by the mechanical design, and heat transfer plays a role in determining the efficiency. Furthermore, the flame travel speed is finite, so that a 60-crank-angle-degree burning duration is typical. For best efficiency the peak pressure takes place 5–20° after TDC. Timing of the spark is thus determined by the burning rate curve and the engine speed. The burning rate for a given engine design and engine speed is primarily a function of the fuel-air ratio of the mixture. Thermodynamics shows that lean mixtures give improved thermal efficiency because of the advantageous ratio of specific heats; however, this gain is offset by the lower flame speed which accompanies lean mixtures. Decreased flame speed causes late burning, which means that more of the combustion energy goes into the exhaust rather than into work. Lean burn engines have lower power output per unit engine volume, which may lead to a greater fraction of the fuel energy going to overcome mechanical friction and added vehicle weight.

Figure 7.2 shows a typical diagram of cylinder pressure and temperature for an engine with an 8.1:1 compression ratio, stoichiometric fuel-air ratio, part throttle, and 1400 rpm engine speed. The quantity x is the mass fraction burned for 22° before TDC spark timing. This timing gave the maximum brake torque and is thus called *MBT timing*. The mass average product temperature, $\overline{T}_b$, peaks at about 2500 K. A computed adiabatic core temperature is also shown. This temperature was calculated by assuming that heat was removed only from the boundary layer. In this example the fuel burned over an interval of 60°, or 8.4 ms.

Flame speed is almost proportional to engine speed because the turbulence, which increases flame speed, increases with engine rpm. The combustion rate does not quite keep up with rpm, however, because the initial period of flame growth tends to be rather constant with engine speed. Thus it is necessary to advance the spark timing as the speed is increased in order to maintain MBT timing.

Although the engine mechanism repeats precisely each cycle, the combustion does not. In particular, the initial flame growth period for very dilute mixtures can vary considerably. Cyclic variability increases with decreasing load because of the increased residual gas fraction. Differences in flow pattern and turbulence are

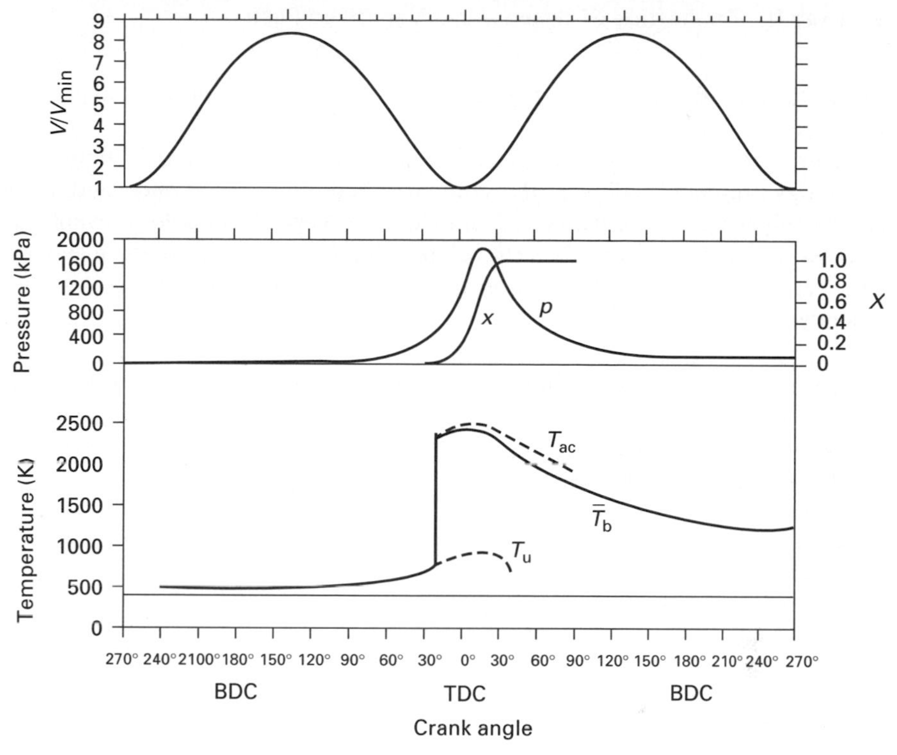

FIGURE 7.2
Engine cylinder volume relative to TDC volume, pressure, computed mass fraction burned, and computed temperatures all versus crank angle. See text for operating conditions [Heywood et al., reprinted with permission from SAE paper 790291 © 1979, Society of Automotive Engineers, Inc.].

thought to be a major contributing factor to the cycle-to-cycle variations of burning rate. Spark effects such as duration and shape also influence the early period of combustion. Residual mixing, inlet pressure fluctuations, and inhomogeneity of charge can also contribute, but seem to be less important. Although cyclic variations are most apparent during the first 1% of burning for lean mixtures, variations in burning rate throughout the burning period contribute to cyclic variability for normal mixtures. Despite the fact that peak pressure may vary by as much as 50% from cycle-to-cycle, the variation in work done by the gas is rather small, typically about ±10% variation around the average value. This is because almost all of the fuel energy is released, even on slow burning cycles, and the work output (integral of $p\,dV$) is not highly sensitive to the shape of the burning rate curve. Exceptions to this are those cycles which either misfire or give partial burns. In the misfire case, the small volume of flame formed by the spark (the flame kernel) does not grow to become a stable flame and almost all of the charge for that cycle is unburned. In the case of partial burns, the flame does stabilize, but then extinguishes, probably due to a decrease in flame temperature caused by piston expansion. It should be noted that dilution, which causes slow burning, can be from either the introduction of a lean fuel-air mixture and/or the presence of large amounts of residual products, such as occurs at idle.

Before proceeding further, the reader may wish to review the engine cycle terminology presented following the summary section of this chapter, and to consider the following three examples. Example 7.1 reviews terminology, events, and processes. Example 7.2 considers the relationship between compression ratio and octane requirement, while the relationships between compression ratio and engine efficiency are considered in Example 7.3.

EXAMPLE 7.1. Using the drawing of cylinder pressure versus volume (p-V diagram) for an S.I. engine, identify the indicated events and processes according to the numbers shown in the figure. If necessary, first read the review of terminology following the summary section of this chapter.

Events

(*a*) Maximum cylinder volume
(*b*) Intake valve closes
(*c*) Spark discharge starts
(*d*) Clearance volume
(*e*) Peak pressure of combustion
(*f*) End of combustion heat release
(*g*) Exhaust valve opens
(*h*) BDC of expansion stroke
(*i*) Intake valve opens
(*j*) Exhaust valve closes

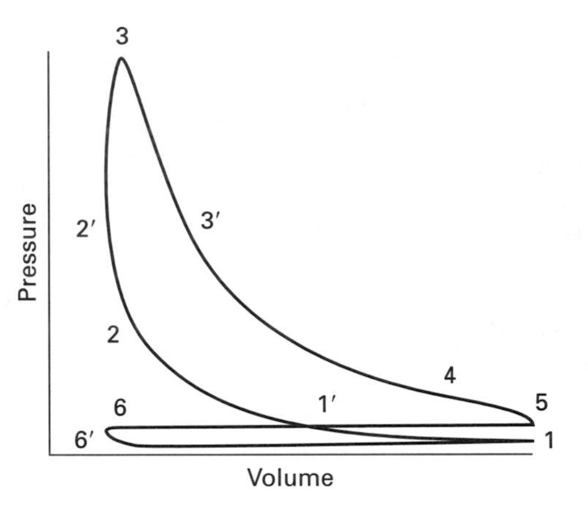

Process parameters

(*k*) Displacement volume
(*l*) Compression ratio
(*m*) Closed cycle portion of engine cycle
(*n*) Gas exchange portion of engine cycle
(*o*) Combustion duration
(*p*) Trapped mass in cylinder
(*q*) Theoretical mass in cylinder
(*r*) Volumetric efficiency (%)
(*s*) Blowdown portion of exhaust
(*t*) Valve overlap period
(*u*) Pumping work
(*v*) Indicated work
(*w*) Indicated mean effective pressure
(*x*) Net indicated work

Solution. (*a*) 1; (*b*) 1′; (*c*) 2; (*d*) 2′, minimum cylinder volume; (*e*) 3; (*f*) 3′; (*g*) 4; (*h*) 5; (*i*) 6; (*j*) 6′; (*k*) 1 − 2′; (*l*) $d/(d+k)$; (*m*) 1′ − 4; (*n*) 4 − 1′; (*o*) 2 − 3′; (*p*) $m(1')$; (*q*) $\rho_0(d+k)$; (*r*) $100p/q$; (*s*) 4 − 5; (*t*) 6 − 6′; (*u*) $\int_5^1 p\,dV$; (*v*) $\int_1^5 p\,dV$; (*w*) v/k; (*x*) $u + v$.

Notes. (*d*) Minimum cylinder volume; (*t*) some of the products (residuals) in the cylinder blow back into the intake port; (*u*) defined this way even though the valve events are not at TDC and BDC; the pumping work is typically large at part throttle (about 40% of indicated work at highway load conditions, but small, $< 10\%$, at wide-open throttle; (*x*) larger than the engine shaft work (brake work) because of friction and work used to drive accesories.

EXAMPLE 7.2. The accompanying diagram [Baker, et al., reprinted with permission from SAE paper 770191 © 1977, Society of Automotive Engineers, Inc.] shows indicated fuel consumption versus compression ratio for a 400 cubic inch displacement engine at highway driving conditions. Comment on the shape of the curve and the probable fuel octane implications.

Solution. The data for indicated specific fuel consumption (ISFC) show that increasing the compression ratio improves the fuel economy, but the curve flattens at high compression ratios. The octane requirement will increase with compression ratio, so that while regular gas at 87 octane will be satisfactory at 8:1 compression, at 9.5:1 an octane of 89 will probably be required, with an increased cost of about 10%. The decreased fuel consumption at high compression ratios will be even less than shown here because engine friction increases with increasing compression ratio.

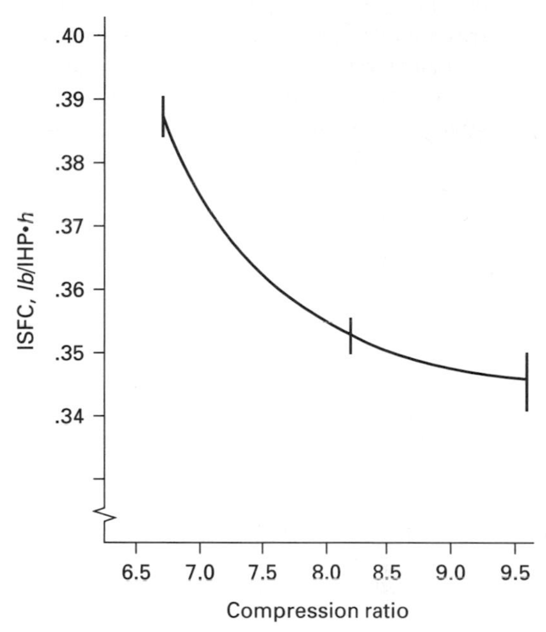

EXAMPLE 7.3. The ISFC levels off in the diagram of Example 7.2 so that increases in compression ratio (CR) beyond 11:1 are not advantageous even if high-octane fuel were available at no cost penalty. Data from various engines show that the best CR ranges from 10:1 to 17:1, and that the amount of improvement with increasing CR depends upon engine size, speed, load, and fuel-air ratio. Studies of combustion chamber shape indicate that surface area to volume ratio is also a factor, but that combustion duration is not a strong factor, although combustion duration does affect thermal efficiency for a fixed compression ratio, with best efficiency for durations in the range of 30 to 40 crankdegrees.

Discuss why these parameters have an effect on ISFC, and why there is a peaking of efficiency with CR.

Answer. Two basic factors limit the effect of CR on ISFC: heat loss and incomplete combustion. For a fixed engine size and geometry, heat loss is a function of charge density, combustion temperature, engine speed, and surface area per unit volume. The engine size influences the surface area to volume ratio. For a pancake geometry with a bore-to-stroke ratio of unity, the surface-to-volume ratio at TDC is $2(CR + 1)/L_B$, where L_B is the cylinder bore. Either increasing CR or decreasing the bore increases the potential for heat transfer loss at TDC. An increase in CR also increases charge density and temperature at TDC. Increased density increases the heat transfer coefficient, and increased charge temperature may increase the temperature difference between the gas and the wall. Note that one would also expect the heat loss per cycle to increase with decreasing rpm and with increasing fuel burned per cycle.

The effects of unburned fuel are complex because the crevices include the volume(s) between the rings. The charge in the crevices is likely to be close to the surface temperatures. Thus the trapped mass in the crevices is proportional to the cylinder pressure and is increased by increases in CR. Typically, some of this fuel is oxidized as it escapes from the crevices, but the timing is late in the expansion stroke and thus cannot contribute much to the piston work. The plot shows the variation of mass trapped in the crevice as a function of compression ratio for a typical engine. The mass is normalized by the mass trapped for CR = 9.5. Note the increase of 40% in trapped mass when going from CR = 7 to CR = 12. The curve was computed for the authors courtesy of K. Min and W. K. Cheng using the MIT model.

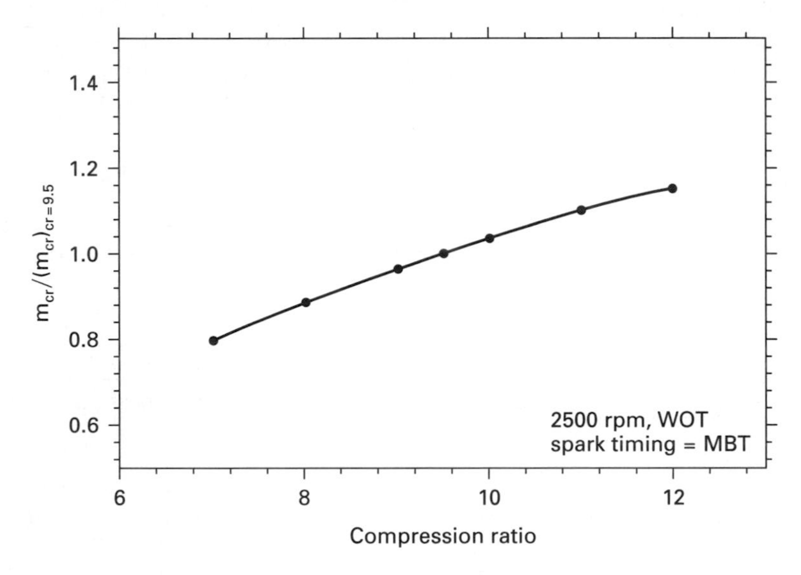

Calculated charge trapped in the crevice compared to the value at a compression ratio of 9.5 versus compression ratio for a 85.6 mm bore, 86 mm stroke and 496 cm^3 displacement engine.

7.2 CHARGE PREPARATION

The historic method of charge preparation for premixed-charge engines is the carburetor. The purpose of the carburetor is to provide a mixture of finely divided fuel and air which will quickly form a uniform fuel vapor and air mixture at the desired fuel-to-air ratio. For good fuel economy the mixture should be as lean as possible, while for power a slightly rich mixture is desired. For perfect operation the carburetor would provide such mixtures as required during the driving cycle. Before 1970, in days of inexpensive fuel, large engines, and no emissions standards, carburetors were the source of many tales; perhaps even now you have heard that patents for "perfect" carburetors had been suppressed. Such imaginative stories are of course now less prevalent as automakers strive to find more fuel-efficient cars which can also meet emissions standards. Currently, most cars require a "three-way" catalyst (a catalyst that converts NO_x, CO, and unburned hydrocarbons to N_2, CO_2, and H_2O) to meet the emissions standards, which in turn requires that the fuel-air ratio be kept within a narrow band around stoichiometric. In order to meet such fuel metering requirements, carburetors are rapidly being replaced by fuel injectors. It appears that with closed-loop fuel control, using an oxygen sensor in the exhaust to provide the control signal to an onboard computer, more and more engines will have port injection. With this method each cylinder has its own fuel injector, thus ensuring precise control and a uniformity of mixture between cylinders. If very lean engines can be built which meet emissions standards without aftertreatment, they will also undoubtedly require port injection and a closed-loop control system. For small utility engines (such as lawn mowers), cost has discouraged the use of fuel injection. Figure 7.3 shows a conceptual diagram of a simple carburetor used in many small utility engines. With the advent of nationwide

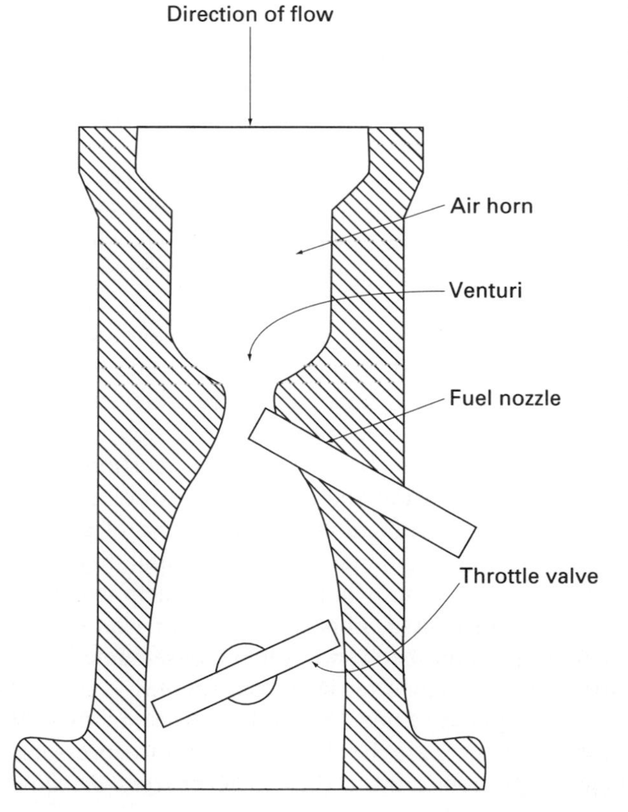

FIGURE 7.3
Idealized carburetor; real carburetors have additional complex features. The venturi reduces the pressure in the throat region, which induces the fuel flow. The throttle regulates the mixture flow rate.

small-engine emission standards in 1996, the use of fuel injection may also be used even for small engines.

Complete vaporization and mixing of the fuel and air charge is probably never achieved in any of the carburetor or manifold injection systems. This means that the charge entering the cylinder contains fuel droplets as well as inhomogeneous mixtures of fuel vapor and air. Tests in the laboratory, where complete premixing of the charge can be accomplished prior to its entering the engine, do not show substantial improvement in engine performance except at very lean conditions and during cold start. However, low-quality mixtures do tend to increase the rate of deposit buildup in the cylinder. Liquid fuel droplets that enter the cylinder vaporize rapidly during the compression stroke, and the highly turbulent flow within the cylinder promotes rapid mixing. However, during cold starting, liquid that reaches the chamber surfaces forms a layer which does not vaporize and mix, thus making the gas-phase mixture leaner at the time of ignition.

Swirl is often imparted to the charge as it enters the cylinder, causing the droplets to orbit the cylinder and affecting the pattern of turbulence. The term *swirl* is a somewhat generic term. Typically, swirl is essentially a solid-body rotation of gas around the cylinder bore axis. However, not all swirling flows are constant–angular velocity flows. In addition, the chamber geometry and combustion distort and change the original swirl velocity profiles. Recently, more complex inlet configurations which divert some of the airflow causing swirl with less pressure drop than conventional-shaped ports have shown promise for lean engine operation. The use of four valve heads has allowed one port to be straight, giving tumble flow, and the other inlet port to be shaped to give swirl. Use of controls then allows the amount of tumble and swirl flow to be changed as a function of operating conditions.

Heating the intake port and mixing the inlet air with hot exhaust gas both promote vaporization, but the higher-temperature charge also decreases the trapped mass and increases the combustion temperature level. The decrease in trapped mass can be overcome by turbocharging or a tuned intake. The increased temperature level causes increases in NO emissions which can be partly overcome by use of the three-way catalyst.

Before leaving the subject of charge preparation, it should be mentioned that the presence of liquid droplets in the premixed charge may have some benefits. First, if horizontal stratification is desired in the cylinder, it may be achieved by late port injection. The fuel then may enter the cylinder primarily as droplets, which circulate in the swirling flow as they vaporize and stay in the upper portion of the chamber. This gives a layer of air near the piston and a layer of mixture near the head. If an overall lean mixture is used, stratification allows burning of a more nearly correct (richer) mixture, thus increasing flame speed and ignitability. Second, the presence of droplets or tiny vapor islands in the flow may help to extend the lean limit of combustion. This is, of course, important only for lean engine operation. Third, the presence of droplets has been shown to increase flame speed slightly, which is again of particular importance for lean burn engines.

Thus far the discussion of charge preparation has been focused on fuel-air mixing. However, equally important features are the charge temperature, pressure, and turbulent state at the time of ignition. During the intake process the flow through the poppet valve(s) causes the flow to be turbulent. It has been shown that if a variable valve lift is used, the intake valve can replace the throttle plate as a means of reducing the trapped charge at part load operation. Such throttling further increases the turbulence at intake valve closing, but, as we shall see, this may not increase the burning rate. The flow into the cylinder is particularly complicated by the off center valve position, which directs a portion of the flow against the cylinder wall. The resulting flow consists of large vortex patterns, which may be tumbling as well as participating in the swirl pattern. Current knowledge of such complex flows is meager, but is being greatly increased by experimental studies using laser Doppler velocimetry and other flow-imaging methods such as particle image velocimetry, and by three-dimensional numerical modeling of the flows using computational fluid dynamics (CFD) codes.

As the trapped charge is compressed, it follows an almost net-adiabatic process. At first, heat flows from the chamber surfaces to the gas, and then as the gas temperature rises due to compression, heat flows out of the gas to the chamber surfaces. A significant portion of the charge is trapped in crevices (around the valves and spark plug and above the top piston ring) and in the thermal boundary layer. The volume of this gas is quite small, but its mass is significant because its density is much higher than that of the core gas. Recent estimates indicate as much as 7 to 9% of the fuel is unburned, although much of it oxidizes later in the cylinder and exhaust port during the exhaust stroke.

The details of the turbulence in the engine at the time of ignition are not very well known, due to the difficulty of obtaining optical access to the cylinder, the unsteady nature of the flow during a given cycle, and the large cycle-by-cycle variations which always take place. Since the flow is three-dimensional and unsteady, it is difficult to distinguish between bulk flow variations and those due to turbulence. One method is to use the frequency of the velocity fluctuations to define the turbulence. In this method velocity fluctuations above an arbitrary cutoff frequency are defined as turbulence, and those below the cutoff frequency are considered the bulk flow. This method has the advantage of not being affected by cycle-by-cycle changes in flow pattern. A less accurate method is to ensemble-average the data over many cycles so that random variations should average out. Thus, comparison between the ensemble-averaged cycle data and data from a single cycle should reveal the turbulent fluctuations. A third method is to conditionally average the data so that cycles with similar large flow patterns are grouped together. The group ensemble average is then compared with individual cycles in the group to define the turbulent fluctuations.

When using conventional CFD codes, the equations are first time-smoothed. The resulting extra terms due to the turbulent fluctuations must then be modeled. Current practice is to use a quasi-empirical rate of turbulent kinetic production equation (the turbulent kinetic energy is given the symbol k) and a rate of turbulent energy dissipation equation (ϵ) which is also quasi-empirical. The resulting solution of this so-called k-ϵ model gives the turbulent intensity for each computational cell. The computations do not give cycle-by-cycle variations, because the inlet flow is also computed from ensemble-averaged equations. Thus, precise comparisons between experiment and model are difficult.

Both models and experiments show that the small-scale turbulence generated during intake does not survive the compression process but that the larger-scale motions, which are affected by the compression process and the chamber geometry, generate the turbulence important at the start of combustion. Thus, a large tumbling vortex may be generated during intake, spin up to higher rotation speed as it is compressed (due to conservation of angular momentum), and thus provide a growth in turbulence generation by shear. Because the turbulence so generated is nonuniformly distributed, the flame moving through the compressed mixture may encounter large changes in turbulent intensity both over its surface at a given instant and as a function of time. The flame motion and product expansion also affect the turbulence during combustion, so that the picture is quite complex.

Given the complexity of the interaction between the in-cylinder flow, the geometry of the chamber, the location and timing of the spark plug, and the engine combustion, it is difficult to determine the flow structure at the time of spark that will maximize engine performance while meeting emissions requirements. Tumble motions give a high intensity of turbulence near TDC, but for this very reason they are quickly dissipated and thus do not provide a continuing source of turbulence. Swirling flows are more stable and thus provide a continuing source of turbulence, but typically do not provide the high early (TDC) turbulence of tumble flows. This makes an engine design which can control swirl and tumble, desirable.

It should be noted that while turbulence intensity in the bulk gas governs the flame travel, turbulence near the surfaces influences heat transfer losses and thermal efficiency.

7.3 IGNITION AND BURNING RATE ANALYSIS

The basic concepts of spark ignition have been covered in Chapter 5. To review, recall that the flame kernel growth is strongly affected by the flow field in the region of the spark. Both bulk gas motion and smaller-scale turbulence influence the growth during the first few crankdegrees after ignition. For strong flows, the kernel may detach from the plug and move downstream as it grows. The influence of cyclic changes in the growth rate is important for lean-limit combustion, where the timing is quite important. In such combustion a slow start may cause the flame to go out later as it propagates during the rapid-expansion phase of piston motion. For normal fuel-air ratios, the effect of the kernel growth period is less important than the effect that cyclic changes in the flow have on the fully developed flame propagation.

As observed in Chapter 5, two spark plugs are better than one, but typically the advantage has not been great enough to be cost-effective for engines with after-treatment devices. Very lean burn engines may need to incorporate multiple plugs or unconventional plug designs. Surface plugs, plasma plugs, and long-duration discharges are currently under investigation. These plugs thus far use much more energy than conventional plugs. The basic idea is to provide a larger initial kernel size, thus reducing the area-to-volume ratio. The larger kernel surface area may also be affected by a larger range of turbulent eddy scales, thus increasing the initial flame speed.

EXAMPLE 7.4. The early flame development period (say, 0–2% of the mass burned) is about 30% of the total burn duration, and this period is a significant contribution to cyclic variation in the burn duration. Does this also mean that the 90% burn duration correlates with the expansion velocity of the kernel during its growth to about 5 mm size?

Answer. No, there are many other factors which determine combustion duration such as chamber geometry and turbulent intensity. The figure shown is taken from Bianco, Cheng, and Heywood. The data support the conclusion that the 90% burn duration does not correlate well with kernel growth rate. [Bianco et al., reprinted with permission from SAE paper 912402 © 1991, Society of Automotive Engineers, Inc.]

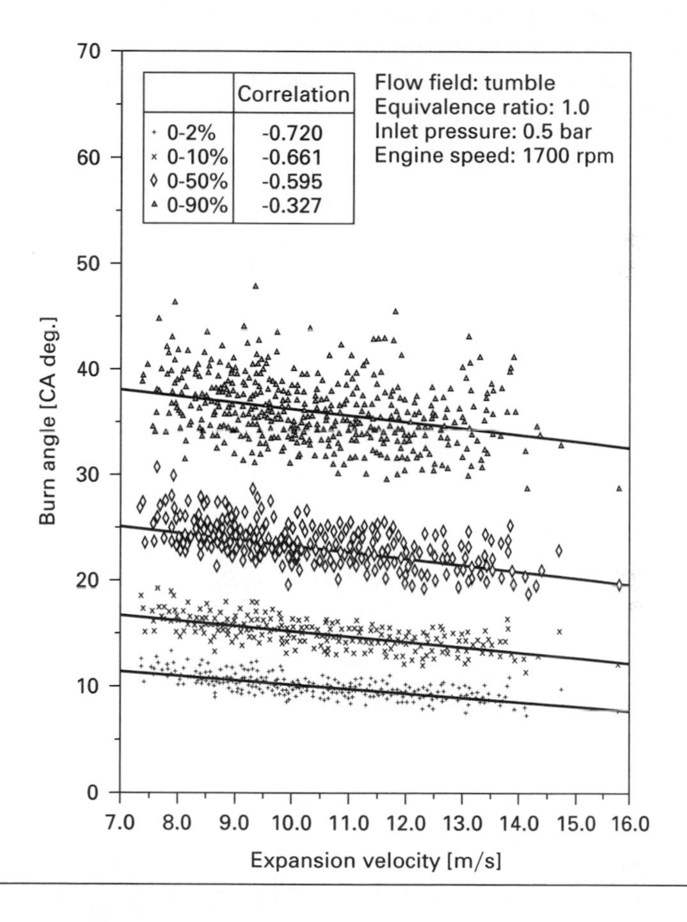

Burning Rate Analysis

Because of the complicated flow pattern in the cylinder, the difficulty in modeling turbulence and turbulent flames, and the inability to model the chemical kinetics of practical fuels, it is desirable to use experimental data to determine the burning rate. A simple burning rate analysis can be based on a two-zone thermodynamic model which uses pressure-time data. The two zones are the unburned charge and the combustion products. The flame is taken to be a very thin zone of zero volume. The products are assumed to follow shifting equilibrium. The first portion to burn is compressed by the following combustion, so that it is at a higher temperature than the last portion to burn. This effect can be more accurately included in the model by dividing the product gas into many subzones; however, the mass-averaged product temperature is generally adequate even for calculation of NO kinetics. A more important consideration in computing the products temperature is the thermal boundary layer. Although the boundary layer can be treated as a third zone, as shown in Figure 7.1, the mixing rate between the core gas and the very thin zone of boundary layer gas is unknown unless a computational fluids model is used. In the two-zone model the heat transfer is removed instantly from the entire mass of each zone rather than from the boundary layer.

The equations for the two-zone model consist of energy equations for the unburned zone (u) and burned zone (b), a mass conservation equation, a volume conservation equation, and two ideal gas equations. These equations were given in Chapter 3 and are reproduced here for convenience:

$$\frac{d(m_u u_u)}{dt} = h_u \frac{dm_u}{dt} - p\frac{dV_u}{dt} - q_u \tag{3.71a}$$

$$\frac{d(m_b u_b)}{dt} = h_u \frac{dm_b}{dt} - p\frac{dV_b}{dt} - q_b \tag{3.71b}$$

$$m_u + m_b = m \tag{3.69b}$$

$$V_u + V_b = V \tag{3.69c}$$

where q_u and q_b are the rates of heat transfer out of each system through the walls. The flame is assumed to be thin. Because all heat exchange between zones occurs within the flame, no term for such exchange appears in the equations. The pressure is uniform throughout, and

$$p = \frac{m_u R_u T_u}{V_u} = \frac{m_b R_b T_b}{V_b} \tag{3.72}$$

After much algebra and using the dot notation for the time derivative, the equations to be solved for the burning rate are

$$\frac{dT_u}{dt} = \frac{T_u\, d(\ln p)/dt + (q_u / m_u R_u)}{\dfrac{1}{R_u}\dfrac{\partial u_u}{\partial T_u} + 1} \tag{7.1}$$

$$\frac{dm_b}{dt} = \frac{(m_u R_u \dot{T}_u - p\dot{V})\left(\dfrac{1}{R_u}\dfrac{\partial u_b}{\partial T_b} + 1\right) - \dot{p}V\left(\dfrac{1}{R_b}\dfrac{\partial u_b}{\partial T_b} + \dfrac{m_b}{V}\dfrac{\partial u_b}{\partial p} + \dfrac{V_u}{V}\right) + q_b}{(u_b - u_u) + \left(\dfrac{R_u T_u}{R_b T_b} - 1\right)\left(T_b \dfrac{\partial u_b}{\partial T_b}\right)} \tag{7.2}$$

$$\frac{dV_b}{dt} = V_u\left(\frac{\dot{m}_b}{M_u} - \frac{\dot{T}_u}{T_u} + \frac{\dot{p}}{p}\right) + \dot{V} \tag{7.3}$$

The calculation requires p and $\dot{p}$ from measurement data, V and $\dot{V}$ from the geometry of the cylinder-piston mechanism, and the property relationships for u_b,

$\partial u_b/\partial T_b$, and $\partial u_b/\partial p$ from thermodynamic equilibrium calculations. In these equations the $\partial R_b/\partial T_b$ and $\partial R_b/\partial p$ were approximated as zero. The initial condition is taken at the time of spark. A small, but finite, volume of products at the adiabatic flame temperature is taken as the initial products zone. At the end of combustion some of the charge is in crevices and is thus unburned. Upon estimating this unburned mass, the final mass of product gas is known. A more accurate analysis would model the mass in the crevice volume and subtract it from m so that m becomes a function of time.

The heat transfer terms require an empirical coefficient, the surface temperatures, and the wetted areas for each zone. Various formulas have been given for the convective heat transfer coefficient, $\tilde{h}$, but they do not account for the rather large observed spatial variation of $\tilde{h}$. The formula of Annand is often cited, but does not account for the effects of swirl. The Annand formula is similar to a pipe flow formula and is given by

$$\frac{\tilde{h}L_B}{k} = a\left(\frac{\overline{V}_p L_B}{\nu}\right)^b \tag{7.4}$$

where L_B is the engine bore, $\overline{V}_p$ the average piston velocity, and the empirical constants have the values $a = 0.5$ to 0.8 and $b = 0.7$. An overall energy balance between the start of burning and the end of burning gives the total heat transfer for this period, which can be used to correct the average value of the heat transfer coefficient.

Woschni has suggested a similar equation for $\tilde{h}$ with a more complicated definition of $\overline{V}_p$. His formula, was originally developed for diesels, but may be suitable for S.I. engines. Calculation of the wetted areas can only be approximated unless the flame geometry is assumed.

The amount of instantaneous heat flux data available for comparison with Eq. 7.4 is quite meager. Figure 7.4 shows data taken at several locations on the cylinder head. The flux jumps up as the flame passes the measuring position. The arrows show the calculated time of flame arrival for each position. The lag in flux may be caused by deposits covering the surface transducer or may be the result of inaccuracy of the computation, which assumed a spherical flame front with center at the spark plug. Recent advanced CFD models for in-cylinder heat transfer have been used to show good agreement between the heat transfer calculated from boundary layer model within the CFD program and the data of Figure 7.4 [Reitz].

Analysis of pressure data using the two-zone model requires pressure derivative data ($\dot{p}$) obtained from digitized pressure–crank angle data. Pressure waves generated by the flame motion are not compatible with the assumption of uniform pressure and thus must be removed by either a digital filter or a smoothing technique. The typical method is to use a spline technique which combines smoothing with the fitting functions. Even small variations in $\dot{p}$ must be smoothed as the calculation method amplifies $\dot{p}$ oscillations into much larger $\dot{m}_b$ oscillations. The data should be taken at intervals of 0.1 to 0.5 crankdegrees with at least 12-bit accuracy to provide adequate resolution. In most analyses ensemble-averaged pressure records are used. Ensemble averaging over 100 or more cycles removes white noise and averages the pressure correctly for the computation of time-averaged power (work per unit time).

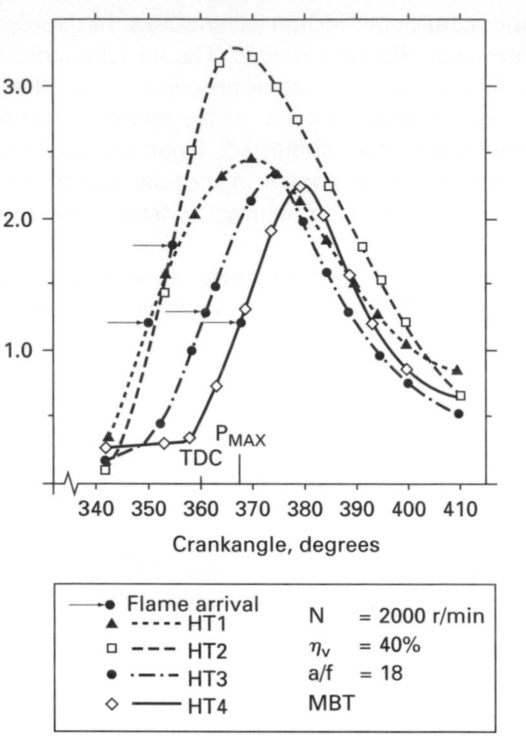

FIGURE 7.4
Heat flux at several locations on cylinder head [Alkidas and Myers, *J. Heat Transfer,* © 1982, by permission of ASME].

To examine cycle-by-cycle variations it is necessary to compute the burning rate for individual cycles. Considerable smoothing may be necessary for such analysis. In particular, the last portion of burning is strongly influenced by the heat transfer correlation so that the shape of the $\dot{m}_b$ curve may be quite inaccurate for the last 20% of burning. An alternative method is to use a single zone model which introduces approximate expressions for the internal energy and heat transfer in terms of the mass-averaged temperature. Such analysis is very fast and can be applied to pressure data from a single cycle.

The analysis method allows the experimentalist to better understand the consequences of changes in design. Diagrams of mass fraction burned, m_b/m, and product gas temperature, T_b, as previously shown in Fig. 7.2, can be compared for various chamber geometries, cylinder flow patterns, spark locations, and so on. The temperature of the products system can be used as input to the Zeldovich mechanism to predict NO_x trends. The analysis is thus useful and can be extended for prediction of the effects of design changes so as to avoid costly engine testing.

Burning Rate Prediction

The use of pressure data to determine the burning rate leaves many questions still unanswered and is not predictive. For a predictive method, the chamber geometry, turbulence parameters, bulk velocity, and chemical kinetics must be introduced. In lieu of theory, a strictly empirical method of fitting the burning rate to engine parameters can be used. The most successful approach used to date is to define the flame velocity $\underline{V}_f$, one-dimensionally from

$$\underline{V}_f = \frac{\dot{m}_b}{A_f \rho_u} \tag{7.5}$$

The flame velocity is relative to the unburned gas and is thus much slower than the laboratory coordinate flame velocity, which includes the motion due to product expansion. The one-dimensional surface that seems to best approximate the flame geometry is a sphere with its origin at the spark plug or at the point where the kernel first begins to grow rapidly. Figure 7.5 shows flame profile data obtained from high-speed motion pictures of the flame using a piston with a quartz window. In this example the cylinder bore was 92 mm and the stroke was 76 mm. The engine was at part throttle so that the volumetric efficiency was about 50%. The indicated mean effective pressure was 524 kPa, and the thermal efficiency was 29.2% for the condition of Figure 7.5a and 32.5% for that of Figure 7.5b. From the two conditions shown plus others it is concluded that a spherical flame model reasonably approximates a curve fit through the data. The center of the sphere is not coincident with the spark plug location, because of the swirl.

The two-zone model can be made predictive if the flame velocity relative to the unburned mixture, as defined by Eq. 7.5, is known in terms of engine parameters.

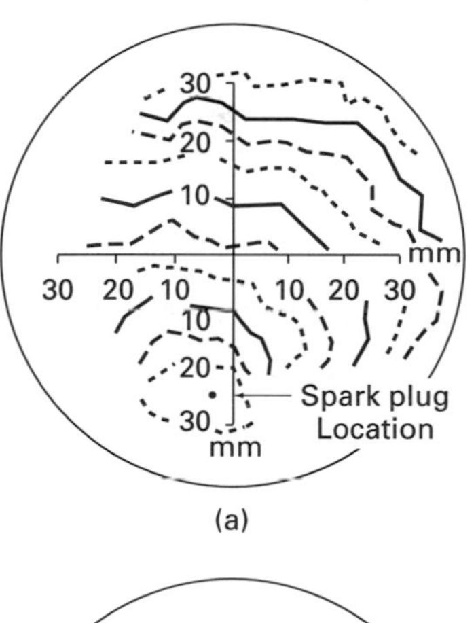

FIGURE 7.5
Flame shape in wedge-shaped chamber. (*a*) 1000 rpm, $F = 1.0$, 23° spark advance, no EGR, swirl = 1200 rpm clockwise. (*b*) 1400 rpm, F = 0.92, 32° spark advance, 5% EGR, swirl = 1680 rpm clockwise [Mattavi et al., "Engine Improvements Through Combustion Modeling," in Mattavi and Amann (eds.), *Combustion Modeling in Reciprocating Engines,* © 1980, with permission of Plenum Press].

Such correlations are discussed in Section 7.4. Given such a correlation and defining A_f as the surface area of a smooth spherical flame front which encloses the burned volume, V_b, one may calculate $\dot{m}_b$ from Eq. 7.5. Thus, the unknowns are reduced by 1 and the pressure can be calculated, rather than supplied from data. The appropriate equations to be solved for this predictive version are

$$\frac{1}{T_u} \cdot \frac{dT_u}{dt} = \frac{\left[\frac{h_u - u_u}{R_b T_b} + (\beta_1 - 1) - \frac{\beta_1 V_u m_b}{V_b m_u}\right]\frac{d \ln m_b}{dt} - \left(\frac{\beta_1}{V_b}\right)\frac{dV}{dt} - \frac{q_b}{pV_b} - \left(\frac{\beta_1 V_u}{V_b} + \beta_2\right)\frac{q_u}{pV_u}}{\frac{\beta_1 V_u}{R_u V_b}\frac{\partial u_u}{\partial T_u} + \frac{\beta_2}{R_u}\frac{\partial h_u}{\partial T_u}} \tag{7.6}$$

$$\frac{dp}{dt} = \frac{p}{R_u T_u}\frac{\partial h_u}{\partial T_u}\frac{dT_u}{dt} + \frac{q_u}{V_u} \tag{7.7}$$

$$\frac{1}{T_b}\frac{dT_b}{dt} = \frac{d \ln p}{dt}\left(1 + \frac{V_u}{V_b}\right) + \frac{d \ln m_b}{dt}\left(\frac{m_b V_u}{m_u V_b} - 1\right) + \frac{1}{V_b}\frac{dV}{dt} - \frac{V_u}{V_b T_u}\frac{dT_u}{dt} \tag{7.8}$$

$$\frac{1}{V_b}\frac{dV_b}{dt} = \frac{d \ln m_b}{dt} + \frac{d \ln T_b}{dt} - \frac{d \ln p}{dt} \tag{7.9}$$

where

$$\beta_1 = \frac{1}{R_b}\frac{\partial u_b}{\partial T_b} + 1$$

$$\beta_2 = \frac{1}{R_b}\frac{\partial u_b}{\partial T_b} + \frac{1}{R_b}\left(\frac{\partial u_b}{\partial p}\right)\frac{p}{T_b}$$

The above equations, when coupled to flame speed correlations such as discussed in the next section, allow prediction of the effects of engine geometry and have been used to improve designs as illustrated in Section 7.6. The weakness lies in the indirect coupling to the chamber fluid mechanics. Such coupling to the fluid mechanics requires the computational fluids approach discussed in Section 7.5.

7.4 FLAME STRUCTURE AND CORRELATIONS

The actual flame in the cylinder is not smooth. The flame zone is a 5- to 6-mm-thick zone containing a very thin reaction zone separating layers or even islands of unburned or partially burned mixture. Both theory and experiment indicate that the flame structure in engines is a highly wrinkled, thin reaction sheet. The turbulent Reynolds number is in the range of 100–1000, the Damköhler number is of the order of 100, and the smallest turbulent eddies are larger than the laminar flame thickness. The flame envelope surface area is much smaller than the reaction sheet area, a ratio of 0.1 being typical at 2000 rpm. The propagation rate of the reaction sheet is proportional to the laminar flame speed, and since the laminar speed is a property of the mixture, it is reasonable to define a flame speed ratio (FSR) as the actual flame speed divided by the laminar burning velocity relative to the unburned gas:

$$\mathrm{FSR} = \frac{\underline{V}_f}{\underline{V}_L} \tag{7.10}$$

Groff and Matekunas obtained the following FSR correlation based on analysis of engine data at the 50% mass burned point:

$$\mathrm{FSR}_0 = 2 + 1.21\left(\frac{\underline{V}'}{\underline{V}_L}\right)\left(\frac{p}{p_m}\right)^{0.82} S \tag{7.11}$$

where $\underline{V}'$ is the root-mean-square of the velocity fluctuations, p_m is the motoring pressure corresponding to the given throttle and engine speed setting, and S is an empirical factor related to spark timing advance:

$$S = 1 + 0.05\theta^{0.4} \tag{7.12}$$

where θ is the crank angle (in degrees) of spark advance before TDC. Motoring the engine means turning the engine by use of an electric motor attached to the shaft without the addition of fuel.

The FSR_0 values vary from 3 to 35 for the data used in the study. At the higher values the first term is negligible and the flame speed is independent of the laminar speed and directly proportional to the turbulent intensity. Physical explanations of the factors S and p/p_m should be considered speculative, but are nevertheless interesting. The factor S accounts for changes in turbulent intensity $\underline{V}'$ with θ, and the factor p/p_m accounts for increases in $\underline{V}'$ due to flame augmentation of turbulence. However, recent measurements of $\underline{V}'$ ahead of the flame indicate that the flame causes only a small increase in $\underline{V}'$ in the direction normal to the flame front. Components parallel to the front were found to be important only during the very last portion of burning.

Equation 7.12 was obtained for the middle portion of the burning period. Typically, the FSR rises to a maximum near 50% burned and then falls off again. Correction of the analysis for a thick flame brush causes the initial rise to be steeper because the flame area is increasing during that time. Conversely, the thick flame brush correction causes the later portion of the FSR to flatten because the area is decreasing during that time. Thus, if a 4-mm flame brush thickness is assumed, Groff and Matekunas found that Eq. 7.12 could be used for flame radius values $r_f > 30$ mm. For $r_f < 30$ mm, they suggest a corrected FSR given by

$$\mathrm{FSR} = \mathrm{FSR}_0\left[1.07\left(\frac{r_f}{30}\right)^{1.04}\right] \tag{7.13}$$

We note that the correction for flame thickness was somewhat arbitrary and that the behavior of the FSR near the end of burning may be influenced by heat transfer, chamber shape, spatial variations of $\underline{V}'$, and as yet not understood coupling effects between the flame motion and the turbulence. It must be recognized that by the time 80% of the mass is burned, the end-gas takes up only 10% of the chamber volume and consists of a thin volume bounded by the chamber surface on one side and the flame on the other.

The correlations given above, or other semi-empirical flame theory equations, allow prediction of the effects of chamber geometry and spark plug location on the combustion behavior. Using the flame speed correlation and the spherical flame geometry allows the two-zone model to be used to calculate pressure, temperature, heat transfer, etc. From such analysis one can find an optimum chamber geometry for a given set of design constraints. The problem with the correlation is that the turbulent intensity $\underline{V}'$ must be known. As a first approximation, the turbulent intensity is proportional to engine speed. However, a more detailed prediction which includes inlet geometry, valve geometry, compression ratio, and chamber geometry requires use of CFD modeling.

7.5 COMPUTATIONAL FLUID DYNAMICS MODELING

Although the zero-dimensional heat release analysis and the one-dimensional flame analysis discussed in the previous sections can be used to improve engine designs, the process lacks a direct means of coupling the engine fluid mechanics to the combustion process. The turbulent intensity which appears in Eq. 7.11 must be supplied from other measurements or theory. Even when $\underline{V}'$ is known, the effects of the cylinder flow on the flame shape are not taken into account. What is needed to overcome these shortcomings is a means of modeling the engine fluid flow three-dimensionally starting with the intake flow through the moving valves and including mixing and turbulence effects. Although this a complex calculation requiring hours of supercomputer CPU time, much progress has been made in the last decade. Thus far, however, the calculations have worked with ensemble-averaged equations so that cyclic variability of the in-cylinder flows is not predicted. Because a portion of the cyclic variability of combustion stems from the ignition kernel development, it seems feasible to capture some of these effects by a separate model of the ignition process, and using the results in the CFD code to generate various starting geometries for combustion.

In order to model the combustion process, the CFD models should predict the fluid velocity and turbulence at the time of spark, the mixture inhomogeneity due to the fuel preparation process, and residual mixing with the fresh charge. These codes should also contain models for the crevice and wall heat transfer. The most important crevice is the one between the cylinder and liner including the volume above the top ring and the volume(s) between the rings. A small amount of fuel is also trapped in the oil layer on the liner, and this too can be modeled. The heat transfer models are subgrid models, because grids adjacent to the surfaces are larger than the boundary layer thickness. Gridding the boundary layer would require much finer grids than are currently practical. The subgrid model most often used is the law-of-the-wall model developed for steady turbulent flows. However, for the high turbulence with unsteady bulk flows and rapidly varying pressure of engines, the law-of-the-wall models are inadequate. Special models have been developed recently which account for some of the engine effects and predict the heat transfer to within 25%.

Once the fluid mechanics model is in place, the combustion model simulates the rate of burning and the flame shape. The combustion model can be divided into three phases: the ignition or kernel growth phase, the fully developed flame phase, and the end-gas burnup phase.

Kernel Growth Phase

The initial spark kernel is a high-temperature ionized gas the size of the spark gap. When the electrical energy is no longer supplied, the small kernel cools rapidly and the energy supplied by chemical reaction of combustion becomes important. Until the kernel has grown to the size of the largest turbulent eddies (~2 mm), only the smaller eddies can influence its surface wrinkling. Because of the small size of the kernel, stretch effects and heat loss effects are very important. The heat loss to surfaces is highly influenced by the wetted area of the kernel and the local flow as it grows in size. If the wetted area is large due to the kernel being pushed against a surface, the kernel growth will be inhibited by the increased heat transfer and the loss in flame area. Because current CFD models do not account for cyclic variations, only an average kernel growth can be predicted. A flow geometry which exposes the spark plug to a consistent flow velocity will reduce cyclic variability. Detailed models which predict the ensemble-averaged behavior of the ignition now exist (see Herweg and Maly).

Fully Developed Flame Phase

Once the kernel has grown to about 5 mm in radius, which takes 5–10 crank-degrees at normal engine speeds, the combustion has moved into the fully developed flame phase. The mass fraction burned at the end of the kernel growth phase is still very small. All of the various models for turbulent flames discussed in Chapter 5 have been attempted for the fully developed flame phase. The early models tended to use a simple characteristic time model coupled with one-step kinetics, which works reasonably well provided the constants are properly adjusted. Since this model is empirical and gives a thick reaction zone, various wrinkled-flame models have been tried. One wrinkled-flame approach which retains an empirical formula is to find a correlation for the wrinkled flame area in terms of fractal dimensions obtained from experimental flame contours. The outer and inner cutoff distances are taken to be the integral and Komogorov scales. Then,

$$\mathrm{FSR} = \left[\frac{L_K}{L_I}\right]^{2-D} \tag{7.14}$$

where a correlation is needed for D in order to define the exponent. A correlation was obtained by Wu et al.,

$$D = \frac{2.0\underline{V}_L}{\underline{V}' + \underline{V}_L} + \frac{2.3\underline{V}'}{\underline{V}' + \underline{V}_L} \tag{7.15}$$

For typical engine values at 1500 rpm,

$$D = \frac{2.0(0.5)}{2.0 + 0.5} + \frac{2.3(2.0)}{2.0 + 0.5} = 2.28$$

and

$$\text{FSR} = \left[\frac{0.03}{2.0}\right]^{2-2.28} = 3.24$$

While the fractal model is relatively simple, it hinges on the empirical correlation for D and the estimated cutoff distances. A more fundamental approach is to use a flame area evolution model or a PDF model. These models, which were discussed in Chapter 5, have recently been applied to engines, but it is too early to determine if they will prove successful. Because they depend heavily on the turbulence models, it may be difficult to determine the causes of various faults as they appear. If these models can be made practical, they will help to tie changes in engine flows caused by port designs and combustion chamber geometry to the combustion rates. This may in turn produce innovations in design.

End-Gas Burnup Phase

Slower-burning engines typically show an FSR that increases with mass-burned fraction and then levels off, staying constant until near the end of burning. Modern fast-burn engines which utilize tumble, compact chamber geometry, and central spark plug location typically show a rapid decrease in FSR long before wall quenching reduces the flame velocity. For these engines, the end-gas is typically spread out over a large area of the chamber surface so that the wrinkled flame is close to the chamber surfaces while there is still a large amount of unburned charge. Many factors contribute to the reduction in FSR, such as flame stretch, reentrainment of products from quenched portions of the flame, and effects of geometry on turbulent scales.

7.6 CHAMBER DESIGN

Modern chamber designs have tended toward a more compact design which gives a larger flame area and faster burning. A centrally located spark plug gives the best flame area and keeps the hot products away from the surface for as long as possible so that heat transfer is reduced. This is important, since a 10% reduction in heat transfer gives about a 3% increase in mean effective pressure. A strong tumble flow is also often used, since it not only produces more turbulence, and thus higher flame speed, but also reduces cycle-to-cycle variations. The reduction of cyclic variability allows the spark timing to be optimum over a greater fraction of the cycles. Such fast-burn designs are also more tolerant to higher values of exhaust gas recirculation.

As an example of the analysis method and its impact on chamber design, the work of Mattavi et al. using the two-zone model is presented. Two designs—the wedge chamber (*a*) and open chamber (*b*)—are shown in Figure 7.6. Curve fitting of measurements produced the flame velocity and radius diagrams shown in Figure 7.7. The open chamber, with its more central spark plug position, had a shorter maximum flame travel distance at a given crank angle than the wedge chamber. The higher flame speed of the wedge chamber was attributed to higher turbulent intensity. Figure 7.8 shows the burning velocity as a function of flame radius, and Figure 7.9 shows the FSR versus flame radius. Note the strong dependence on engine speed, as expected. Figure 7.10 summarizes the comparisons. The open chamber, with larger flame area and shorter flame travel, gave a higher mass burning rate. The performance data given in Figure 7.11 show the superiority of the open chamber. These data also show that the open chamber can be run at the same specific NO_x level ($EINO_x$, emission index for NO_x in g NO_x per kg fuel) as the wedge by increasing the exhaust gas recirculation (EGR) for the open chamber from point B to point C.

The wedge chamber, with its higher flame speed, could be improved by providing a larger flame area, and Figure 7.12 shows this design modification. Predictive modeling using the FSR correlations previously obtained showed that the modified design gave a higher mass burning rate and a smaller area exposed to the product gas. The result was lower heat transfer, higher mean effective pressure, and higher

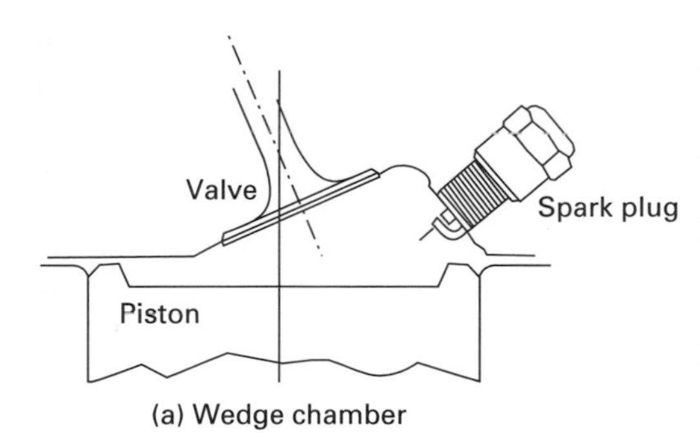

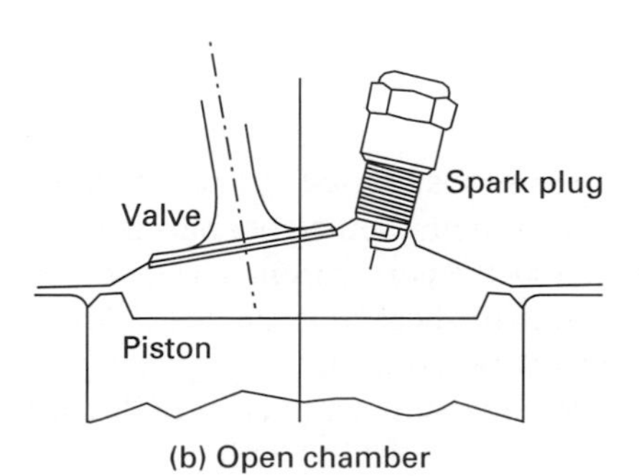

FIGURE 7.6
Two cylinder head-piston designs: (*a*) wedge chamber and (*b*) open chamber [Mattavi et al., "Engine Improvements Through Combustion Modeling," in Mattavi and Amann (eds.), *Combustion Modeling in Reciprocating Engines*, © 1980, with permission of Plenum Press].

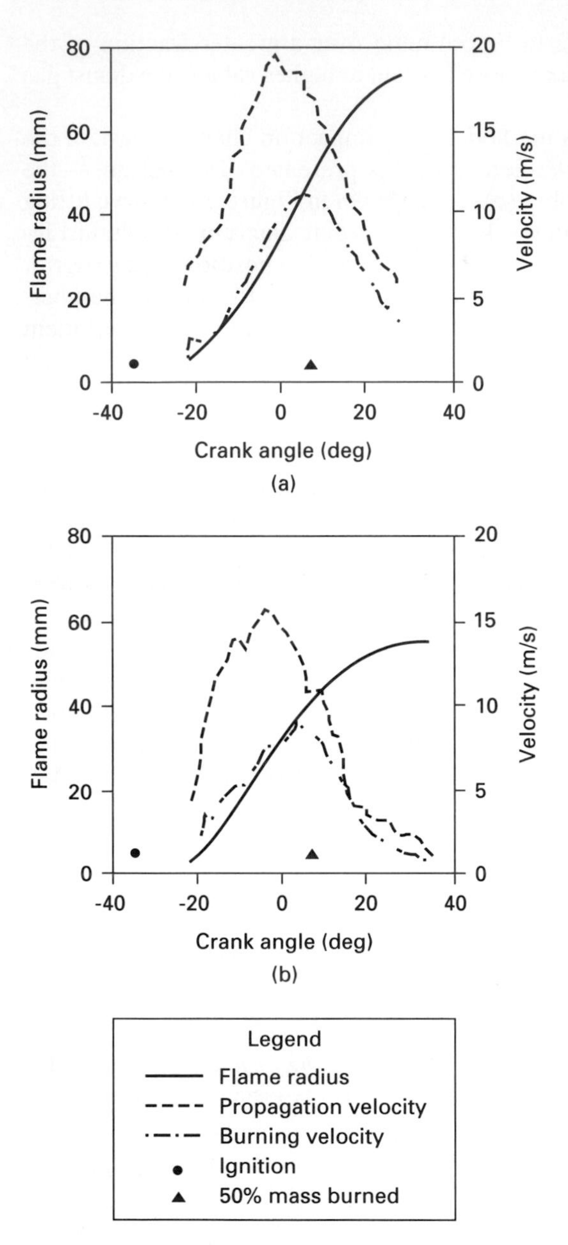

FIGURE 7.7
Flame radius vs. crank angle for (*a*) wedge chamber and (*b*) open chamber [Mattavi et al., "Engine Improvements Through Combustion Modeling," in Mattavi and Amann (eds.), *Combustion Modeling in Reciprocating Engines,* © 1980, with permission of Plenum Press].

thermal efficiency. Building and testing of the design was carried out based on the favorable predictions. The experimental results showed an even higher burning rate than predicted due to a higher FSR than predicted by the correlation. The modified design gave higher NO_x at a fixed EGR due to the lower heat transfer, but this was overcome by increasing the EGR by 1% and increasing the spark advance by 3°. The modified chamber proved to have higher efficiency and tolerance to EGR than either the old wedge or the open chamber.

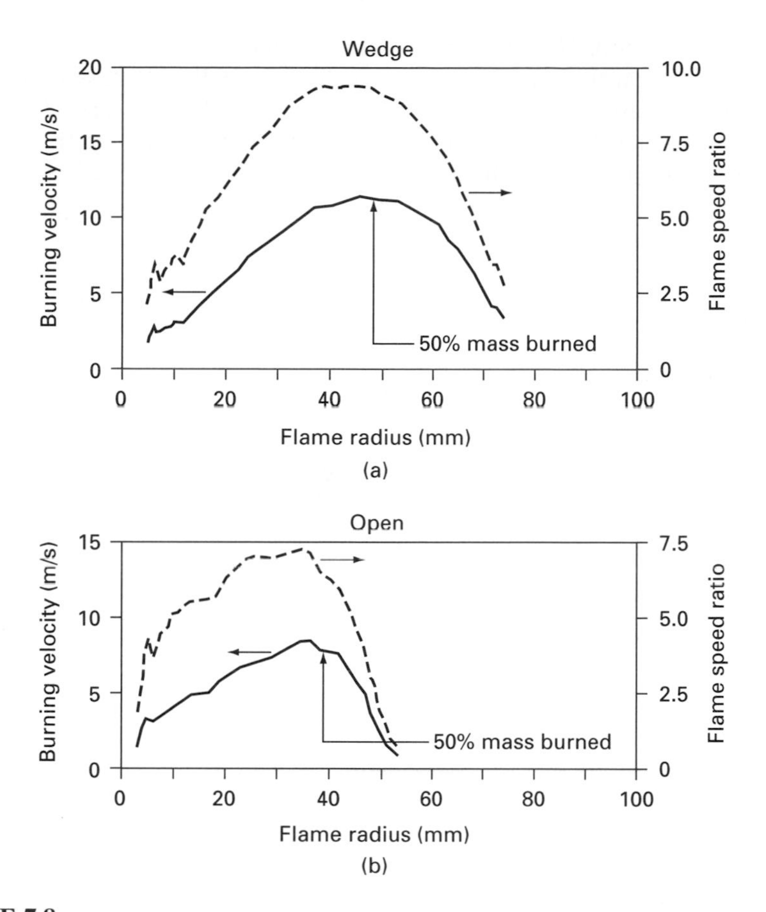

FIGURE 7.8
Flame speed ratio and flame speed vs. flame radius for (*a*) wedge chamber and (*b*) open chamber [Mattavi et al., "Engine Improvements Through Combustion Modeling," in Mattavi and Amann (eds.), *Combustion Modeling in Reciprocating Engines,* © 1980, with permission of Plenum Press].

The above example is an excellent indication of how combustion analysis can be applied to engine design. However, one must recognize that many other factors must be analyzed before such a modified design is proven worthy of production. For example, the analysis must include effects on hydrocarbon emissions, drivability, and manufacturing considerations.

Obtaining swirl without reduction in volumetric efficiency is important in reducing pumping loss at wide-open throttle and allowing for maximum power. Recent designs have tended toward four valves to allow better breathing and a central spark plug location. The use of ports which give tumble flows has also increased the flame speed.

Engine knock is another important consideration in chamber design. As mentioned previously, the end-gas should be in a cool part of the chamber (away from the exhaust valve, for example) and should undergo as little time as possible at the compressed condition. Unfortunately, many fast-burn geometries show a rapidly decreasing FSR near the end of burning as well as a rapidly decreasing flame area. In many designs fast burn is obtained by increased swirl; however, this increases heat transfer, which tends to offset gains in efficiency made by the increased compression ratio possible with less knock.

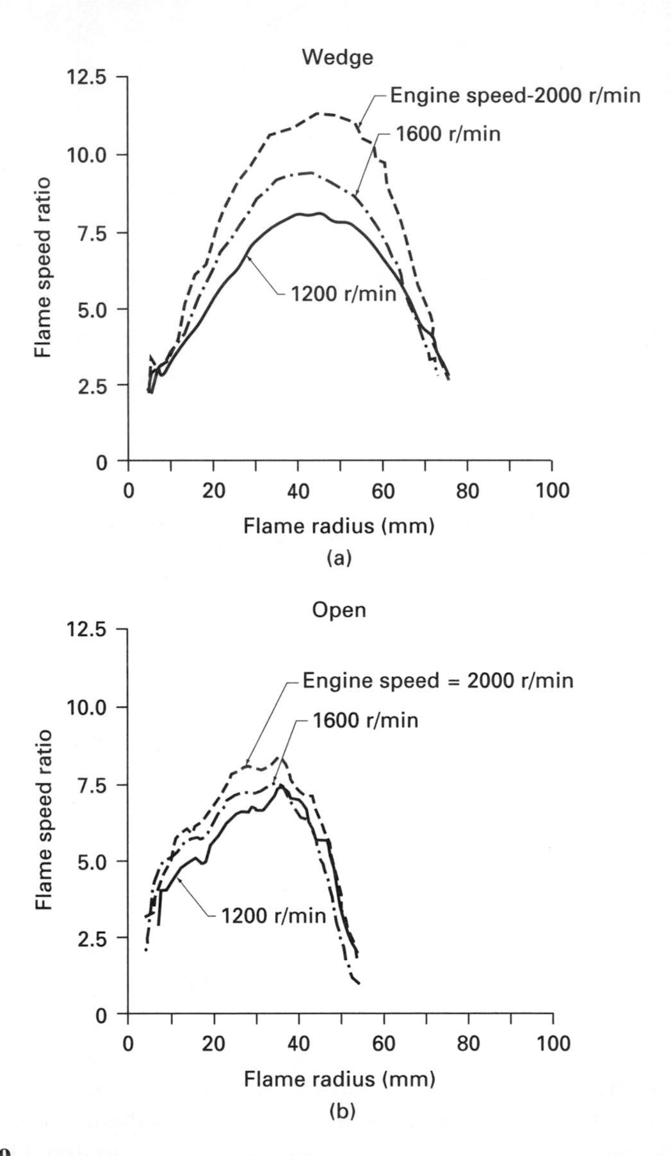

FIGURE 7.9
Effect of engine speed on flame speed ratio for (*a*) wedge chamber and (*b*) open chamber [Mattavi et al., "Engine Improvements Through Combustion Modeling," in Mattavi and Amann (eds.), *Combustion Modeling in Reciprocating Engines,* © 1980, with permission of Plenum Press].

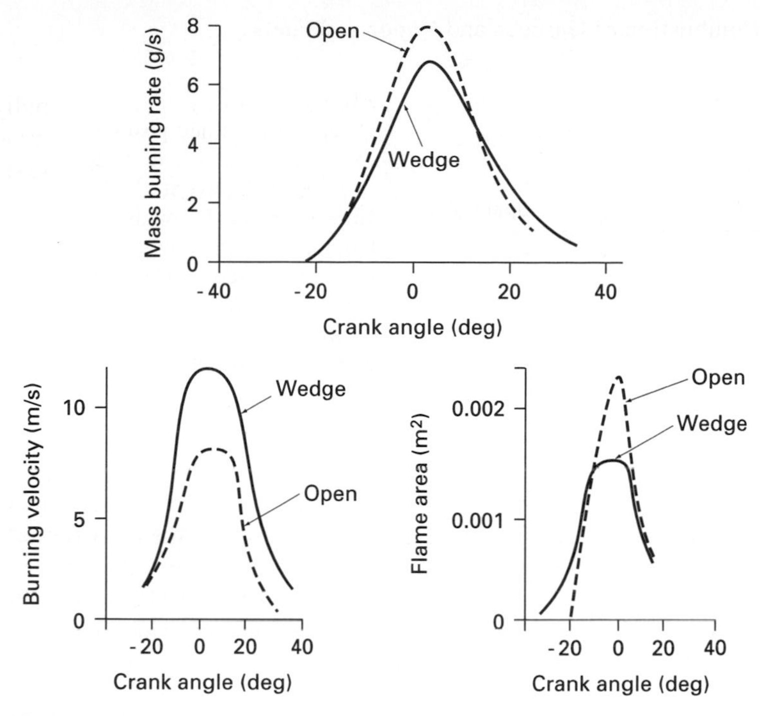

FIGURE 7.10
Summary of flame propagation results for the wedge and open chambers [Mattavi et al., "Engine Improvements Through Combustion Modeling," in Mattavi and Amann (eds.), *Combustion Modeling in Reciprocating Engines,* © 1980, with permission of Plenum Press].

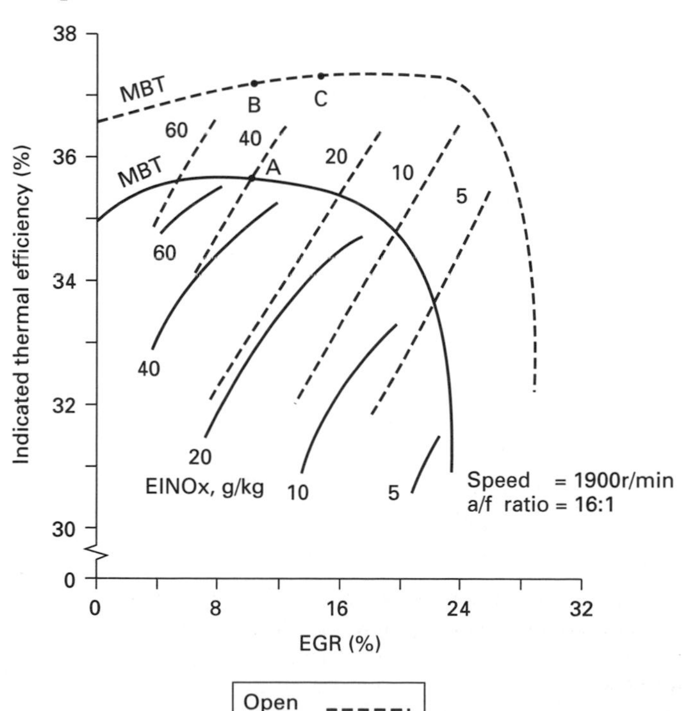

FIGURE 7.11
Thermal efficiency maps as functions of exhaust gas recirculation and spark timing for the wedge and open chambers [Mattavi et al., "Engine Improvements Through Combustion Modeling," in Mattavi and Amann (eds.), *Combustion Modeling in Reciprocating Engines,* © 1980, with permission of Plenum Press].

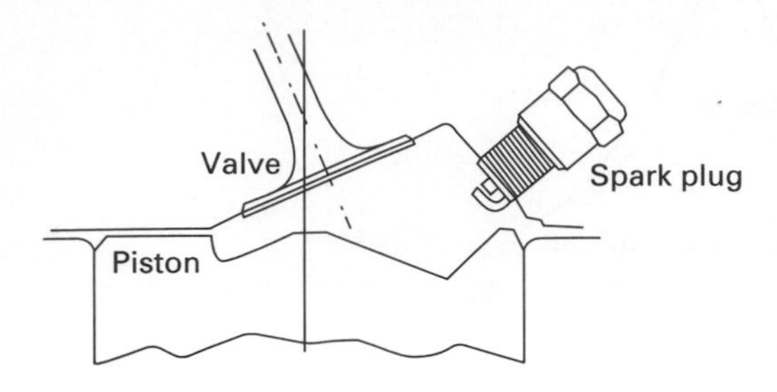

FIGURE 7.12
Design of modified piston to increase flame area in wedge chamber [Mattavi et al., "Engine Improvements Through Combustion Modeling," in Mattavi and Amann (eds.), *Combustion Modeling in Reciprocating Engines,* © 1980, with permission of Plenum Press].

7.7 EMISSIONS CONTROL

Starting in 1968, automobiles driven in the United States were required to meet exhaust emissions standards set by Congress and administered by the Environmental Protection Agency. The first step was installation of a positive crankcase ventilation valve to prevent venting of hydrocarbon fumes to the atmosphere. As more strict standards for unburned hydrocarbons (HC), carbon monoxide (CO), and oxides of nitrogen (NO_x) were set, modifications to the combustion system were required.

Without going into detail, the standards are set by regulating the grams of each pollutant produced during a specified driving cycle. The standards are expressed in terms of grams produced divided by the total miles driven during the cycle (g/mi). Table 7.1 gives the history of US emissions standards for automobiles. Separate, more stringent standards are required in California. Figure 7.13 shows the reduction in emissions produced by these standards over the period from precontrol to 1982. Note that currently CO and HC emissions have been reduced by 96% and NO_x has been reduced by 90% from the values before controls.

TABLE 7.1
History of U.S. automobile tailpipe emission standards (g/mile)

Year	CO	NO_x	HC
1960*	84.0	4.1	10.6
1968–1971**	34.0	—	4.1
1972–1974**	28.0	3.1	3.0
1975–1976	15.0	3.1	1.5
1977–1979	15.0	2.0	1.5
1980	7.0	2.0	0.41
1981–1982	3.4	1.0	0.41
1983–1993	3.4	1.0	0.41
1994–1996+	3.4	0.4	0.25
2004 or 2006	1.7	0.2	0.125

*Typical values before emission standards.

**Adjusted assuming present test procedures, which began in 1975.

+Tier 1; a phase-in of doubled useful life with less stringent standards was also allowed.

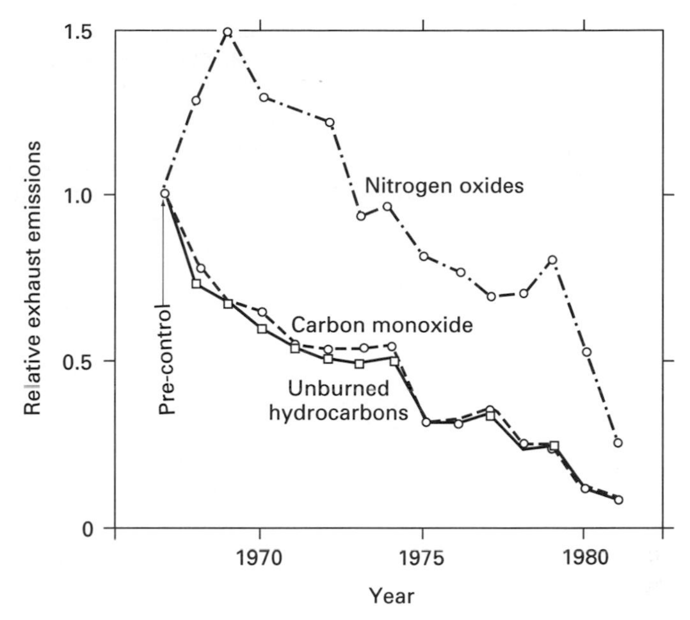

FIGURE 7.13
Relative tailpipe emissions of U.S. in-use passenger cars.

Before the advent of aftertreatment, staying on the lean side of stoichiometric reduced CO. Better mixture preparation and control, especially during starting and idling, helped reduce HC emissions. At the time, and until relatively recently, a major cause of HC emissions during normal engine operation was thought to be quench layers on the engine surfaces. It is now known that HC in the quench layers is oxidized by mixing and diffusion so that they are not an important source. However, the crevice volumes around the piston, spark plug, and valves are a major source. A secondary source may be hydrocarbons absorbed in the lubricating oil on the cylinder walls and in deposits on the head and piston. Because of the belief in quench layers as a source, combustion chambers were designed to give minimum surface-to-volume ratios. Such compact chambers have the desirable effect of reducing combustion duration if the spark plug can be centrally located.

Figure 7.14 shows the fate of gasoline in a typical spark-ignition automobile engine as estimated by Cheng and co-workers at MIT. The left-side path shows the mechanism by which 9% of the fuel escapes normal combustion. Fuel can follow this path either directly as fuel or as fuel-air mixture. With port injection, about one-third of the unburned fuel escapes directly as fuel and two-thirds as fuel-air mixture. Under warmed-up conditions, only 20% of this unburned fuel escapes from the engine to enter the catalytic converter. The catalyst reduces the vehicle-out HC emissions to 0.1–0.4% of the fuel used by the engine. The 7% of the fuel which does not leave the engine but is not consumed by normal combustion is either oxidized in the cylinder and exhaust port/manifold or is recycled in the residual gas. It is estimated that the net fuel economy loss is about 6% due to the 9% of fuel which does not burn in the normal combustion process. For cold starting and the 15 minutes or so of driving before warmed-up conditions are reached, the amount

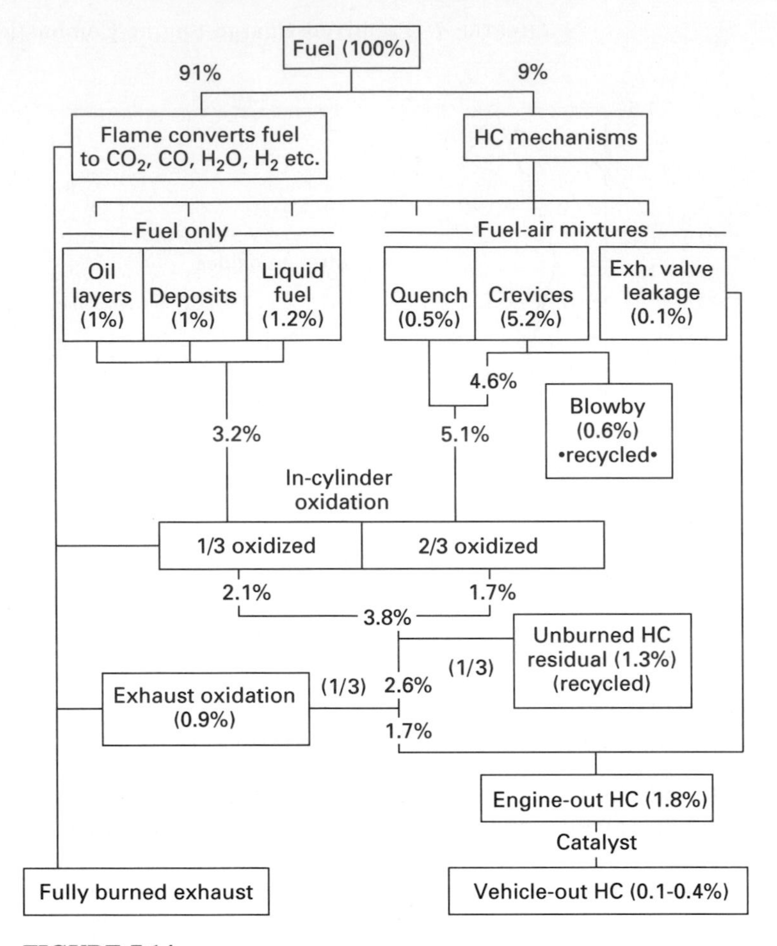

FIGURE 7.14
Complete flow-chart for gasoline fuel, which enters each cylinder, through both the normal combustion process (left side) and the unburned hydrocarbon mechanisms (right side). Sources or processes by which some of the fuel escapes normal combustion are shown as boxes at the top of the diagram. In-cylinder oxidation, retention in the exhaust port and manifold, and catalyst are shown in sequence. Numbers in parentheses denote the HC emission index (% of gasoline entering the cylinder each cycle) for each step in the total process [Cheng et al., reprinted with permission from SAE paper 932708 © 1993, Society of Automotive Engineers, Inc.].

of unburned fuel is undoubtedly higher, contributing to reduced fuel economy and high HC emissions for the first fraction of a minute before the catalyst reaches its warmed-up state.

The formation of NO follows the extended Zeldovich mechanism introduced in Chapter 4. Because the reactions are slow, the amount of NO produced is controlled by the rate of reaction. Thus, lower temperatures, which greatly reduce the reaction rate, lower the amount of NO produced before expansion lowers the temperature to about 1800 K, where the reactions freeze. Retarding the timing and addition of diluent both accomplish such a reduction in temperature and thus a reduction in NO. Of course, retarding the timing from the maximum brake torque (MBT) setting increases the specific fuel consumption. Figure 7.15 shows a computation of NO using the Zeldovich mechanism and a thermodynamic combustion model. The mass that burns during early combustion forms NO at a higher rate than the mass that burns near the end of combustion. In both cases the rate of reaction is too slow to follow shifting equilibrium.

Reduction of fuel-air ratio to leaner and leaner conditions reduces CO and HC emissions as long as flame speed and ignitability do not become a problem. Lean operation also reduces NO formation because of the lower flame temperatures. Not only are emissions reduced, but fuel economy is improved by lean operation. However, lower flame speed creates drivability problems and in the limit partial

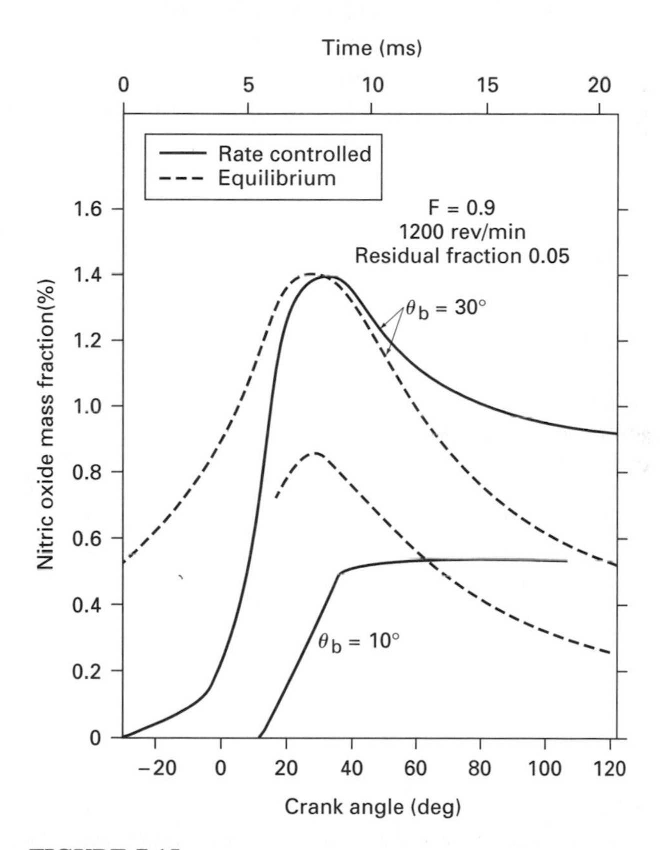

FIGURE 7.15
NO % mass as a function of time in the burned gas for two elements which burn at different times: $\theta_b = -30°$ is the first part of the charge to burn; at $\theta_b = 10°$ one-third of the charge is burned [Lavoie et al., ©1970, by permission of Gordon and Breach Publishers].

burning of the charge. The partial burning takes place because the slow-moving flame burns during rapid expansion of the piston, which causes a lowering of the temperature to a point where the flame will no longer propagate. A second effect of lean operation is to lower the exhaust gas temperature. If further oxidation of HC and CO is to take place in the exhaust by use of an oxidizing catalyst, the temperature of the exhaust must be kept up so that the catalyst stays above 400°C.

In 1973, when the federal standard for NO_x was reduced to 3.1 g/mi, and before catalytic converters were introduced, the use of exhaust gas recirculation (EGR) was begun as a means of NO_x, control. Figure 7.16 shows the effect of EGR at constant vehicle speed. The increase in best-economy spark advance with increasing EGR indicates that the flame speed is reduced by the dilution effect. For the conditions and engine of Figure 7.16, the thermal efficiency increases to a maximum at 10% EGR. One reason for the increase is reduced pumping loss caused by the increased intake manifold pressure. A second reason is that the ratio of specific heats increases due to the lower temperatures with dilution. For EGR levels beyond 10% the data show a decrease in efficiency, mainly caused by the effects of reduced flame speed. The addition of EGR reduces the flame temperature, and thus we see a continuous decrease in NO_x. However, a continued increase in EGR eventually causes the engine to surge up and down in speed (the so-called drivability problem) and to emit increasing amounts of unburned HC. These are the same problems encountered when leaning the mixture with excess air. Of course, more EGR tolerance or increased tolerance to leaning could be achieved if the engine design were to increase the intensity of turbulence, and thus the flame speed, as the

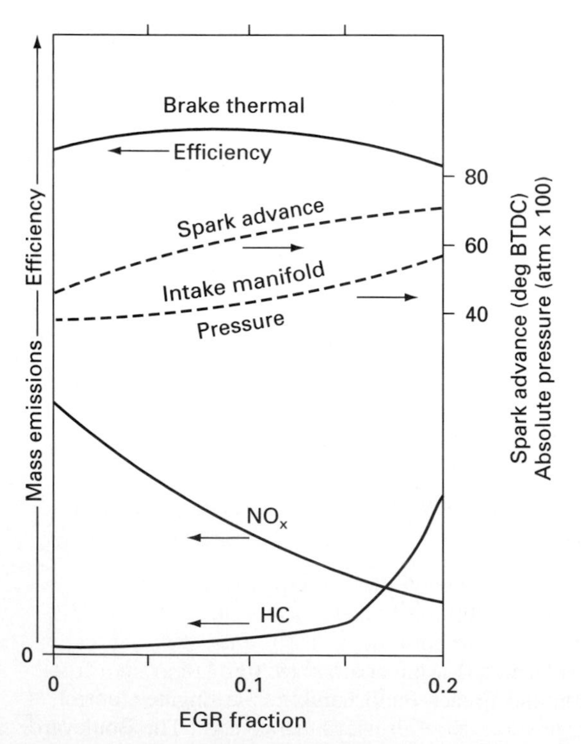

FIGURE 7.16
Effect of adding exhaust gas recirculation at a constant road speed.

mixture is leaned out. Hope for such an improvement is offered by a recent design which controls the amount of swirl in the chamber by using a lean-mixture oxygen sensor in the exhaust as the control function.

In 1975, when oxidizing catalytic converters were introduced, it was necessary to use unleaded gasoline, because the catalyst is poisoned by the lead and rendered ineffective. Because no equally cheap, effective, and environmentally acceptable substitute was found for lead, the octane number of regular unleaded fuel decreased. This decrease in octane number required the continuation of lower compression ratios which had been introduced earlier as an aid to NO_x control. Prior to 1970 the compression ratios were in the range of 9.5–9 and were limited by knock. In 1970, compression ratios were dropped to the 8–8.5 level. The sharp reduction in compression ratio plus timing adjustments caused an estimated 15% decrease in engine fuel efficiency prior to 1975. With the advent of the catalyst, and a simultaneous reduction in car weight and size, fuel economy began to increase again in 1975.

With a further decrease in the NO_x standard to 1.0 g/mi in 1981, U.S. cars began increasingly to use a three-way catalyst which oxidizes CO and HC and reduces NO_x. This catalytic converter requires an absence of oxygen to reduce NO_x. Figure 7.17 shows the catalyst efficiency as a function of fuel-air ratio, and

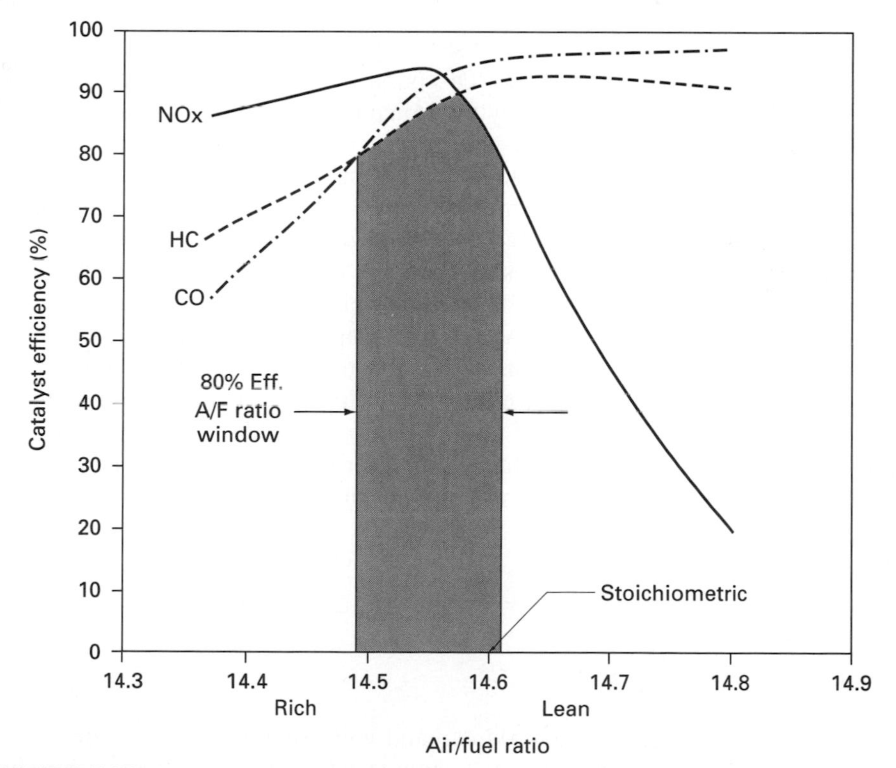

FIGURE 7.17
Emissions conversion efficiency characteristics of a three-way catalyst [Reprinted from *Prog. Energy Comb. Sci.,* vol. 6, Kummer, "Catalysts for Automobile Engine Control," pp. 177–199, © 1981, with kind permission from Elsevier Science Ltd., The Boulevard, Langford Lane, Kidlington OX51GB, UK.].

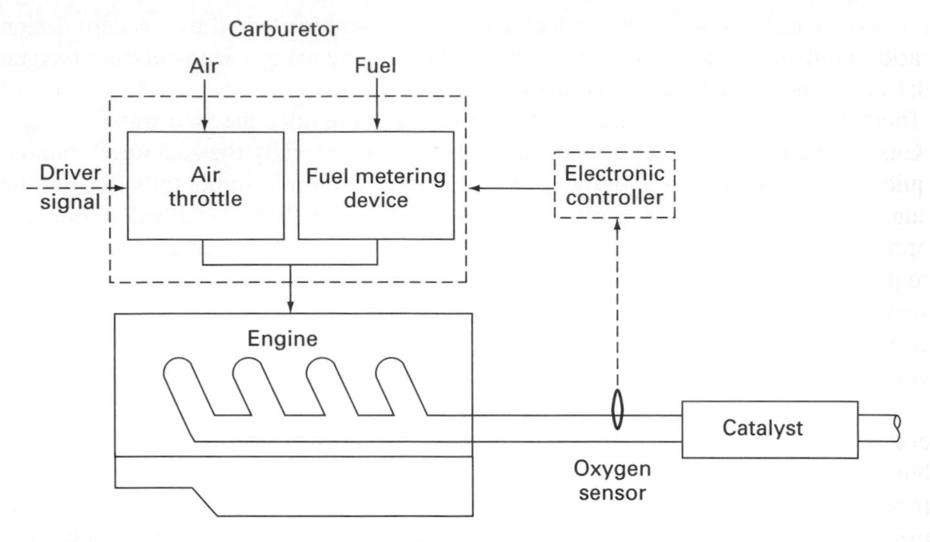

FIGURE 7.18
System for closed-loop control of air-fuel ratio.

Figure 7.18 shows a schematic of the closed-loop control system used to achieve the required stoichiometric fuel-air ratio. Of course, the fuel efficiency benefits of a lean-burn engine are not achieved by the stoichiometric engine with EGR.

At the present time, use of the three-way catalytic converter system with computer control and in-port fuel injection has provided an excellent, but costly, means of solving the dual goals of emissions control and high fuel economy. However, the aftertreatment system cost in terms of both consumer cost and use of precious noble metals is high. It is clear that a major goal should be to find a combustion engine which reduces these costs while reducing both emissions and fuel consumption. Some current trends in developments with alternative engine designs directed to meeting this objective are discussed in Section 7.9. Methods of improving today's gasoline engine are discussed in the next section. As discussed in Chapter 2, the reformulation of gasoline offers an additional approach. Currently, regions of air quality nonattainment require oxygenated gasoline in winter to reduce CO emissions, and lower-volatility gasoline in summer to reduce evaporative hydrocarbon emissions.

7.8 ENGINE EFFICIENCY

Although fuel is still plentiful, world demand will eventually cause shortages of high-quality fuel. Thus, it is necessary to conserve fuel even now. In addition, fuel conservation has a desirable effect on the balance of payments for fuel-importing countries and reducing emissions of greenhouse gases.

An obvious method of improving fuel economy is to increase compression ratio. If knock were not a problem, the indicated thermal efficiency of cars with an 8:1 compression ratio could be increased by about 12% (from, say, 40% thermal efficiency to 45%) by increasing the compression ratio, as previously mentioned. Knocking may be avoided by using electronic control of the spark timing, which quickly retards spark timing when knock is detected by a sensor mounted on the engine head. The system thus attempts to keep the timing as close as possible to the optimum advance. Of course, such gains are small if spark retard is continually required to prevent knock, and thus the possible increase in compression ratio is very modest. Although knock is the primary limit on compression ratio, increased compression ratio also has the effect of reducing exhaust temperature, which complicates catalytic oxidation of unburned hydrocarbons in the converter.

Charge dilution with either air or exhaust gas increases indicated thermal efficiency. Dilution air increases efficiency more than exhaust gas recirculation, but also gives higher amounts of exhaust NO_x than dilution of a stoichiometric mixture with exhaust gas. Figure 7.19 shows the NO_x efficiency trade-off for the two methods. The air addition gives a maximum NO_x at above 0.9 equivalence

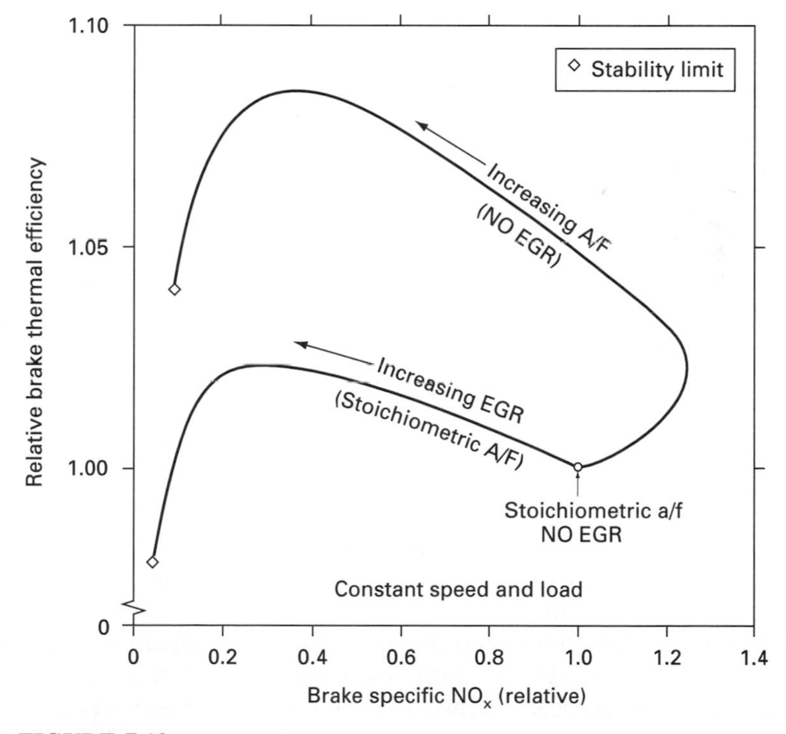

FIGURE 7.19
Typical efficiency vs. NO_x trade-off for dilution with excess air and with exhaust gas recirculation [Amann, by courtesy of General Motors Corp.].

ratio and then gives decreasing NO_x. The air addition method is superior to the stoichiometric-EGR approach, but cannot be used with the three-way-catalyst system. The maximum of either curve could be moved upward and to the left by increasing the flame speed of dilute mixtures and decreasing the cycle-by-cycle variations which increase with dilution.

7.9 ALTERNATIVE AUTOMOBILE ENGINES

The most competitive automobile engine to the four-stroke homogeneous-charge spark-ignited engine is the four-stroke indirect-injection compression-ignition (diesel) engine. Before the rapid rise in gasoline prices in the 1970s, diesel engines enjoyed only a small part of the U.S. market due to higher cost, weight, noise, and exhaust odor of the diesel and the difficulty in starting during cold weather. In fact, the dieselization of cars was rather modest at that time even in Europe despite higher fuel costs. Exceptions were European taxi fleets, where the diesel's excellent fuel economy at idle is an advantage, and countries where tax policy favored diesel fuel. In the United States, the diesel gained some popularity, reaching a peak of 6% of the market in 1981, due to the lower price of diesel fuel and the higher fuel economy of diesel engines. However, increased spark-ignition engine fuel economy, increased diesel fuel tax, numerous mechanical difficulties of some models, and problems with water in fuel and fuel waxing at low temperatures caused diesel popularity to rapidly decrease back to its original low level. An added problem of these engines was brought about by the introduction in the 1980s of diesel exhaust particulate standards. These problems greatly lowered U.S. public interest in diesels as a passenger car power plant. In Europe, however, the growth in diesel car sales has been significant due to higher fuel prices and tax incentives for diesel fuel. If a lean-burn catalyst aftertreatment can be developed for NO_x, the lean-burn engine, diesel or gasoline, offers both fuel economy advantages and the ability to meet emissions standards. If such an aftertreatment is not developed, the alternatives discussed below may yet prove practical.

In sorting out ideas for power plants, it is possible to classify them as having internal or external combustion, continuous or intermittent combustion, compression or spark ignition, and homogeneous or stratified charge. The external continuous combustion engines most developed are the steam engine and the Sterling engine. The Rankine cycle (steam) engine for automobiles has been abandoned because of low fuel economy. The Sterling cycle has outstanding steady-state thermal efficiency, low exhaust emissions, quiet operation, and multifuel capability. However, the cost, hydrogen sealing problem, required warmup time, and poor idle fuel economy make this engine an unlikely strong competitor to the reciprocating spark-ignition engine. Similarly, the continuous internal combustion engine, represented by the gas turbine, has serious fuel economy problems. The turbine engine also has a problem with low initial acceleration due to the rotating inertia of its parts. The use of ceramics in turbines may make them more competitive in fuel economy, but presents a very difficult problem in materials technology.

Of the engine types not yet discussed, the internal combustion intermittent-firing spark-ignited stratified-charge engine offers the most hope as a serious contender. Three versions of this type of engine are shown in Figure 7.20. The divided-chamber (three-valve) engine has been marketed commercially. The direct-injection engine has received considerable interest, and a complex design has been introduced in Japan as discussed later.

In the three-valve engine, the mixture in the prechamber is rich, while that in the main chamber is lean. Burning of the rich mixture causes a jet of burning gases to penetrate the lean mixture, giving a much higher rate of burning than could take place by spark-igniting the lean mixture. The overall lean burning brings about a dramatic reduction in NO_x for air-fuel ratios greater than 20. The problem with the engine is unburned hydrocarbon emissions, which are very high at light loads. Mixture forced into the crevice volumes at light load exits without burning and gets into the exhaust. Exhaust aftertreatment is difficult at light loads because of the low exhaust gas temperature.

In the axially (or horizontally) stratified-charge engine, timed port injection and in-cylinder swirl are used to stratify the mixture at part load. Because only air is in contact with the piston, the unburned hydrocarbon problem due to crevice quenching is reduced. However, the mixing of some fuel into the bottom layer of air can cause HC and CO problems at light load.

The direct-injection stratified-charge engine (DISC engine) has been developed in several versions. In the early-injection engine, the fuel evaporates and mixes before ignition by the spark. In the late-injection engine, the fuel is sprayed onto the spark plug at about the time of ignition. The spark plug acts as a flame holder and the swirling flow brings fresh air to the combustion zone. In a third method, fuel is sprayed onto the piston surface, where it evaporates and mixes to form a combustible cloud. These engines all offer reduced throttling losses and

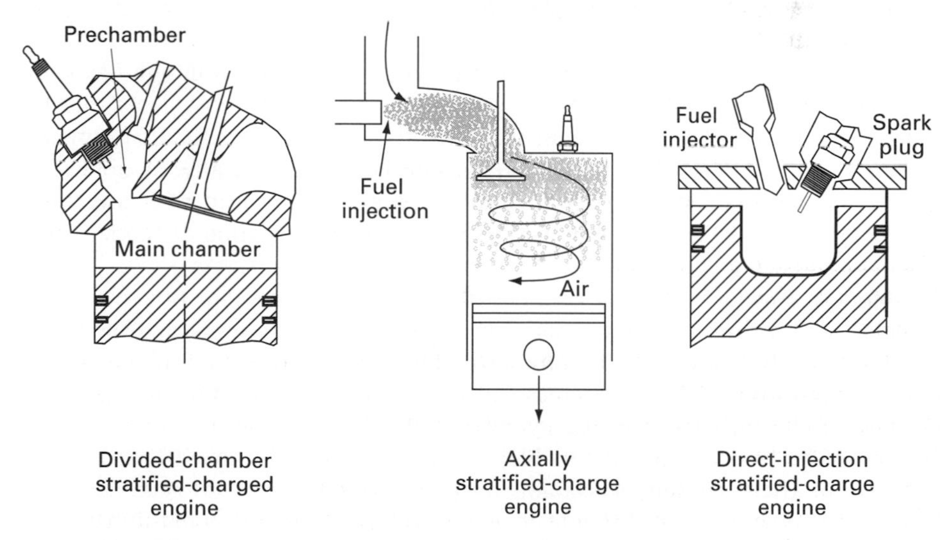

FIGURE 7.20
Types of spark-ignited-stratified charge engines [Amann, by courtesy of General Motors Corp.].

overall lean combustion at part load. The late-injection engine is also quite fuel-tolerant, since it suffers from neither octane nor cetane requirements.

Thus far, none of the stratified-charge designs discussed above has shown the ability to meet United States emissions standards. Because of the use of direct injection, they have the problems of reliability and cost associated with in-cylinder fuel injectors.

Although it uses an in-cylinder fuel injector, a two-stroke design direct-injection engine has received considerable notice. The main feature of this engine is an air-assisted fuel injector which uses very low pressure fuel injection with atomization aided by a flow of compressed air. By using direct injection, the problem of short-circuiting hydrocarbons into the exhaust during scavenging is avoided. However, the fuel must be evaporated and mixed in the very short time between exhaust closing and the start of combustion.

In 1997 several direct injection spark ignited engines which burn gasoline were introduced in Japan. These engines meet Japanese emission standards, but not United States standards. The engines combine a number of combustion designs depending on speed and load. At part load the fuel is injected before ignition and burned stratified so that the overall fuel-air ratio is very lean. At higher loads the injection is early so that the engine runs as a conventional stoichiometric homogeneous charge engine. At intermediate conditions fuel is introduced in two stages, early during intake and then later during compression. This provides a transition from the very lean ($F < 0.5$) to stoichiometric ($F = 1.0$) operation. When the engine is running on a stoichiometric mixture the three way catalyst is in operation. During lean operation a second special NO_x storage-reduction catalyst is also used in series. In order to recycle this catalyst the engine is switched from lean to stoichiometric operation at a frequency of about once per minute. To accomplish all of this, many controls are used including an electronically controlled throttle, controlled EGR, electronic high pressure (12 MPa) injection, swirl control port, and many sensors.

None of the engine designs discussed is without fault. Each design has some limitation in compression ratio or emissions. It would seem most likely that the design which offers the higher number of controllable parameters should ultimately be best suited to today's computer control technology. We have already noted that fast burn chambers and improved breathing can allow increased EGR levels and can give improved fuel economy. However, clearly the selection of the fast burn chamber geometry is complicated by many other constraints.

Modern engines have been greatly improved by port fuel injection, electronic controls, high-energy ignition systems, fast-burn chamber designs, and the use of three-way catalysts with closed-loop control. Further improvements are possible by reduction of throttling losses; decrease in heat transfer; improved breathing, which gives desirable turbulence in the cylinder without cyclic variations or large increases in heat transfer; and, if possible, efficient operation without aftertreatment. Removal of aftertreatment is desirable primarily to reduce cost. It is unlikely that large improvements will be brought about purely by changes in combustion characteristics; however, even small improvements will have a large overall effect, because of the large number of vehicles. Innovative design still offers hope for as yet unrealized gains.

7.10 SUMMARY

Four-stroke automotive spark-ignition engines use port fuel injection to obtain the fuel control required for the dual requirements of fuel economy and emissions standards. Although the mixture entering the cylinder is not homogeneous, the mixture at time of combustion is assumed to be homogeneous. Ignition is by a high-energy spark. Combustion efficiency is limited by the finite combustion duration, heat transfer, dissociation of the products, and mixture trapped in crevices and thus not burned. Engine efficiency is further limited by system issues which are not discussed here, such as throttling losses, friction, pumping losses, and incomplete expansion.

The combustion duration consists of three parts: the ignition and flame kernel development period, the fully developed flame propagation period, and the end-gas burnup period. The flame kernel development is strongly influenced by the turbulent velocity field in the region of the spark plug. Distortion of the kernel by the velocity causes cycle-by-cycle variations in the kernel flame area as well as heat transfer from the kernel to the chamber and plug surfaces. The time to reach the fully developed flame phase thus varies from cycle to cycle so that optimum spark times can only be achieved in an average sense. Because of the finite burning time, the optimum spark timing is before top dead center even though this means work is done to move the piston against the additional pressure created by combustion. The fully developed flame is in the wrinkled reaction sheet regime for wide-open throttle. At part throttle, especially at idle conditions, charge dilution by residual products from the previous cycle reduces the laminar flame speed and may cause the flame to fall into the distributed flame regime.

Quasi-one-dimensional modeling, based on thermodynamic analysis of cylinder pressure, shows that the ratio of turbulent to laminar burning velocity is proportional to turbulent intensity. Turbulent intensity is proportional to the average piston velocity and at a given engine speed depends strongly on the large-scale flow patterns set up by the inflow fluid mechanics and their subsequent breakdown into turbulence under the influence of the compression process and cylinder head geometry. Currently, CFD analysis using either fractal analysis of flame geometry or flamelet theory allows calculation of the flame travel for the ensemble-averaged case, but is limited by the ability to model engine turbulence. The final phase of combustion occurs when the flame approaches the chamber surfaces; it is not well understood.

The best engine designs use a compact combustion chamber with central, high-energy spark and a combination of tumble and swirl motion in the inlet charge to optimize turbulence. The crevice volumes are minimized, but still represent a major (7 to 9%) loss of charge to the initial combustion event. Compression ratio is limited by knocking of the end-gas and is thus limited by available fuel octane number.

Exhaust emissions of CO, NO_x, and unburned hydrocarbons are currently regulated to very low levels, and a catalyst is used to achieve these standards. The catalyst efficiency is high only in a small region around stoichiometric, and thus modern engines cannot run lean to achieve improved efficiency. Homogeneous lean-burn engines are also limited by the decreased flame speed inherent in lean mixtures. Various attempts are under way to produce stratified-charge engines which can operate overall lean, but burn a less lean mixture within the cylinder. Lean-burn engines are also under consideration because of the potential for development of a lean-burn catalyst.

Review of S.I. four stroke engine terminology

		Geometry
CR	Compression ratio	$\dfrac{\text{Maximum cylinder volume}}{\text{Minimum cylinder volume}}$
V_c	Clearance volume	Minimum cylinder volume, which takes place at TDC.
V_d	Displacement volume	The swept volume as the piston moves over its stroke, L_s from TDC to BDC.
$\overline{\underline{V}}_p$	Mean piston speed	For a given engine rotational speed ω_e, $\overline{\underline{V}}_p = 2L_s\,\omega_e$ (m/s).
rpm	Revolutions/min	Typically, engine speed is given as rpm $= 60\omega_e$.
L_B	Cylinder bore	Inside diameter of cylinder liner (or sleeve): $V_d = \pi L_B^2\, L_s/4$.
L_r	Connecting rod length	The connecting rod connects the crank of length $L_s/2$ to the piston. Typically, $2L_r/L_s$ is 3 to 4.
L_s	Stroke length	$0 \leq L_s(\theta) \leq L_s$.
V_r	Crevice volumes	The crevice volumes are those volumes which fill with mixture, but are inaccessible to normal combustion. These volumes include the volume between rings.
$\underline{V}_p$	Piston speed	$L_s(\theta) = (L_s/2)\cos\theta + [L_r^2 + (L_s/2)^2 \sin^2\theta]^{1/2}$ $\underline{V}_p(\theta) = (dL_s(\theta)/d\theta)(d\theta/dt)$ $= \overline{\underline{V}}_p(\pi/2)(\sin\theta)[1 + \cos\theta/(4(L_r/L_s)^2 - \sin^2\theta)^{1/2}]$. The maximum value of V_p ranges from 8 to 15 m/s and is limited by stress and friction.
		Timing
θ	Crank angle	For the four-stroke cycle, $0 \leq \theta \leq 720°$. Often θ is expressed in degrees before top dead center (BTDC) and after top dead center (ATDC) with $\theta = 0°$ at TDC of the compression stroke.
θ_s	Spark timing	Crank angle when spark discharge begins. MBT timing is that which gives maximum brake torque for a given operating condition. Timing is *retarded* from MBT if it is moved closer to TDC and *advanced* if it is moved further back in the compression stroke away from TDC.
	Valve timing	Intake and exhaust valve openings and closing crank angles are abbreviated as IVO, IVC, EVO, EVC. The interval during which both intake and exhaust are open is called the *valve overlap period.*

Gas exchange

m_t	Trapped mass	The mass of gases in the cylinder at IVC.
η_v	Volumetric efficiency	Ratio expressed as a percentage of trapped mass to mass of intake air at inlet density, ρ_0, that would fill the cylinder at BDC: $\eta_v = 100\ m_t/[\rho_0(V_c + V_d)]$.
X_r	Residual fraction	The mass fraction of products from the previous cycle retained in the trapped mass.
EGR	Exhaust gas recirculation	Some exhaust products are recirculated with the fresh charge to reduce emissions.
	Blowback	Products which flow from the cylinder to the intake port near the start or end of the intake valve open period.
	Blowby	Gas which escapes from the cylinder, mainly due to leakage past the piston rings.
	Blowdown	The period of rapid exhaust between EVO and BDC of the expansion stroke.

Performance parameters

(Engine performance is rated by indicated or brake values. Indicated values are based on the cylinder gas as a system, while brake values treat the engine as a system.)

W_i	Indicated work	By convention, the net work computed from the $p\ dV$ integral from BDC compression to BDC expansion.
W_p	Pumping work	By convention, the net work computed from the $p\ dV$ integral from BDC of exhaust stroke to BDC of intake stroke.
W_{in}	Net indicated work	$W_{in} = W_i + W_p$ and expresses the net work done by the cylinder gas against the moving piston during the two revolutions of the cycle.
IMEP	Indicated mean effective pressure	This parameter normalizes the work output by engine size. For each cylinder, IMEP $= W_i/V_d$.
PMEP	Pumping mean effective pressure	Similarly, PMEP $= W_p/V_d$.
W_b	Brake work	Defined based on engine shaft output and is less than the indicated work because of pumping, friction, and accessory power.
BMEP	Brake mean effective pressure	The brake equivalent to imep, BMEP $= W_b/nV_d$, where n = number of cylinders which contribute to W_b.
$\dot{W}_b$	Brake power	The rate at which the engine does work against a load. $\dot{W}_b = 2\pi\omega T$, where T is the shaft torque exerted by the engine.
W_i	Indicated power	n(IMEP)$\omega/2$, where $\omega/2$ is the number of power strokes per unit time. For a multicylinder engine, the IMEP must be the average value.

Performance parameters

BSFC	Brake-specific fuel consumption	$\dot{m}_f/\dot{W}_b$
ISFC	Indicated specific fuel consumption	$\dot{m}_f/\dot{W}_b$
η_c	Combustion efficiency	The engine is treated as a steady-flow device with $\dot{m}_a$, $\dot{m}_f$ entering at ambient T_0 and $\dot{m}_e$ exhaust gases leaving the engine. The enthalpy of the gases leaving is evaluated at T_0. $\eta_c = \dfrac{\dot{m}_f h_f(T_0) + \dot{m}_a h_a(T_0) - \dot{m}_e h_e(T_0)}{\dot{m}_f \text{LHV}}$ This definition accounts for incomplete conversion of products, and for lean to stoichiometric mixtures is 0.95 to 0.98.
η_t	Thermal efficiency	$\dfrac{\dot{W}_b}{\eta_c\, \dot{m}_f \text{LHV}}$
η_r	Second law efficiency	$\dfrac{\dot{W}_b}{\dot{m}_e\, G_{\text{prod}}(p_0, T_0) - (\dot{m}_a + \dot{m}_f)\, G_{\text{react}}(p_0, T_0)}$ For most hydrocarbon fuels the denominator can be approximated by $\dot{m}_f$ HV.

PROBLEMS

7.1. Calculate the dewpoint temperature for a stoichiometric mixture of octane and air at atmospheric pressure. Use the data in Appendix A.

7.2. If a gas disk which is 3 mm in diameter and 0.5 mm thick at atmosphere pressure and temperature is spinning around its axis at 1000 rpm, what is its rotational speed after being adiabatically compressed to 25 atm pressure? (The gas disk is compressed in both the radial and axial direction and is not confined to constant radius as it is compressed. The disk is supposed to represent an idealized turbulent eddy.) Discuss your result in terms of turbulence in engines.

7.3. For the following data, estimate the temperature of the core gas in a pancake-shaped combustion chamber with a 100-mm bore and 10-mm height, mass average gas temperature = 2500 K, pressure = 1800 kPa, average, surface temperature = 500K, average, boundary layer thickness = 0.5 mm. Assume $T_\delta = (T_w + T_c)/2$.

7.4. A fairly general plot of mass versus volume burned fraction is shown in the figure. Plot flame radius normalized by cylinder radius versus mass burned fraction for a pancake-shaped chamber. Approximate the flame shape as a cylinder with the same axis as the combustion chamber and assume that the flame starts in the center of the cylinder.

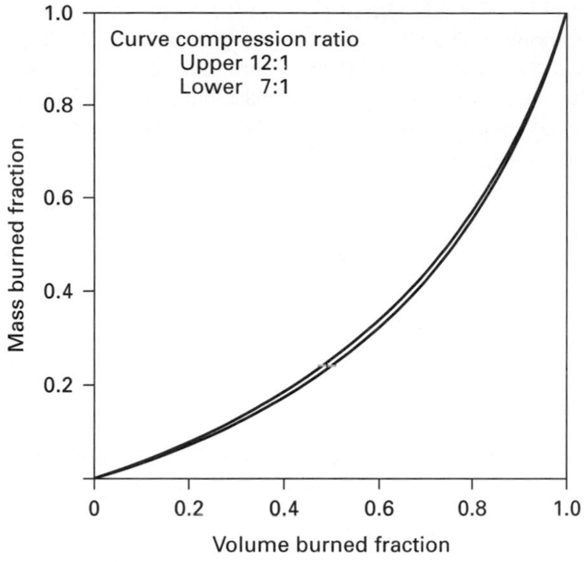

"Universal" mass versus volume burned curve

7.5. During combustion in S.I. engines, the ratio ρ_u/ρ_b is about 4. Using this, show that

$$\frac{V_b}{V} = \frac{4m_b/m}{1 + 3m_b/m}$$

Compare values from this formula with the data shown in the figure of problem 7.4.

7.6. It is suggested that turbulence could be increased in the end gas by a piston with a raised rim at the radius corresponding to 80% volume burned as indicated in the accompanying sketch. Discuss what effects this might have on flame propagation.

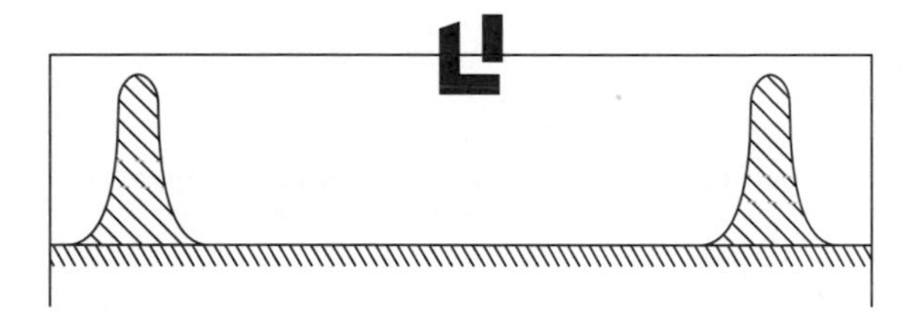

7.7. Consider a central spark location in a "double-hemi" chamber (see sketch). Would a long-reach spark plug location (B) decrease the heat transfer significantly over a conventional location (A)?

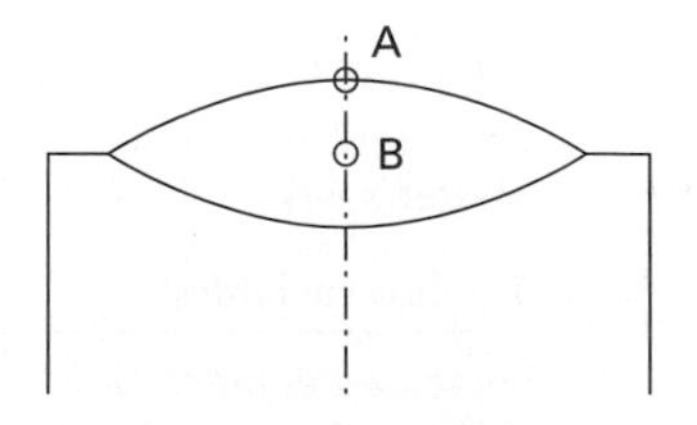

7.8. If engine flame speed is only slightly affected by laminar flame speed at high turbulence levels, why do lean mixtures burn more slowly in a highly turbulent engine?

7.9. Return to Problem 7.4 and sketch to scale the flame and cylinder wall at the 80% mass burned point. Use a view similar to that of Problem 7.6 but without the raised rims. Assume the engine has a 125-mm bore and 125-mm stroke with an 8:1 compression ratio. The crank angle at 80% mass burned is 20 CA° after TDC. At this crank angle the volume is at 1.25 of its TDC value. Remember, the wrinkled flame covers a thickness of about 5–6 mm.

7.10. The following typical S.I. engine values for 1500 rpm and a stoichiometric mixture are given by J. Heywood in his presentation "Combustion Modeling in S.I. Engines," *COMODIA , 94,* Yokohama, Japan, 1994:

$$\mathrm{Re}_I = 300, \quad L_I = 2\text{ mm}, \quad \delta = 0.02\text{ mm}, \quad \underline{V}' = 2\text{ m/s}, \quad \underline{V}_L = 0.5\text{ m/s}$$

Using these values, compute Da_I, L_I/δ, L_K, L_K/δ, and $\underline{V}'/\underline{V}_L$. Show the region of Figure 5.11 in which these values fall.

7.11. Sketch to scale a wrinkled flame as a 0.1-mm-thick sheet which is in the form of an isosceles triangles. The wrinkled flame thickness is 6 mm, and its flame speed ratio (FSR) is 20.

7.12. If the flame speed ratio is known from a correlation of burning rate data, discuss in detail how you would calculate the pressure history in the cylinder of a spark-ignition engine. Assume the compressed gas is a pancake shape with the spark plug at the center. Assume the flame is cylindrical so as to avoid the complexity of the more realistic spherical geometry. Give all necessary equations.

7.13. Given the results of a computer model such as discussed in Problem 7.12, indicate how you would calculate the NO produced by the engine.

7.14. Discuss the reasons for the trends of knock promotion that are given in Section 7.1, giving physical arguments.

7.15. Fill in the blanks in the table using the symbols provided, and discuss your answers in terms of governing phenomena.

Effect of operating variables on ignition and flame propagation

	Effect	
	Ignition	**Flame propagation**
Ignition system variables		
Increased spark energy		
Hotter spark		
Engine variables		
Increased compression		
Increased mixture swirl		
Increased residual gas		
Increased flame propagation distance		

Symbols: ↑(helps); ↓(hinders); —(no effects).

7.16. The survey of cyclic variability by Ozdor et al. lists the following factors which influence variability and the best design or condition for minimum variability. In each case discuss briefly the mechanism you think is operative and if you think the factor is likely to be important.

Factor	**Best condition**
1. Fuel	Fuel that provides highest burning velocity
2. Air-fuel ratio	Slightly rich of stoichiometric
3. Spark plug	High breakdown energy; long duration; wide spark gap; thin or sharp-pointed electrode; gap discharge direction perpendicular to the mean flow, which is 3 to 5 m/s
4. Turbulence	Small scale at time of ignition, with high turbulent intensity over the whole cylinder volume

7.17. Typically, when peak pressure of cycle n is plotted versus peak pressure of cycle $n + 1$, there is no correlation; that is, the variations are random. However, for very lean fuel-air ratios at part load or idle conditions, it is observed that a very low peak pressure on cycle n is followed by a high peak pressure on cycle $n + 1$. This is called a *prior-cycle effect*. Discuss why the typical case should be expected to be random and what might cause the prior-cycle effect.

REFERENCES

Abraham, J., Williams, F. A., Bracco, F. V., "A Discussion of Turbulent Flame Structure in Premixed Charges," SAE paper no. 850345, 1985.

Alkidas, A. C., and Myers, J. P., "Transient Heat-Flux Measurements in the Combustion Chamber of a Spark-Ignition Engine," *ASME J. Heat Transfer,* vol. 102, no. 4, pp. 62–67, 1982.

Amann, C. A., "The Powertrain, Fuel Economy and the Environment," GMR-4949, General Motors research publication, Warren, MI, 1985.

Annand, W. J. D., "Heat Transfer in the Cylinders of Reciprocating Internal Combustion Engines," *Inst. Mech. Engrs.,* vol. 177, no. 36, pp. 973–996, 1963.

Arcoumanis, C., Hu, J., Vafidis, C., and Whitelaw, J. H., "Tumbling Motions: A Mechanism of Turbulence Enhancement in S.I. Engines," SAE paper no. 900060, 1990.

Baker, R. E., Daby, E. E., and Pratt, W., "Selecting Compression Ratios for Optimum Fuel Economy with Emissions Constraints," SAE paper no. 770191, 1977, also in SP-414 and SAE Transactions, vol. 86, 1977.

Benson, R. S., and Whitehouse, N. D., *Internal Combustion Engines,* vols. I and II, Pergamon Press, Oxford, 1979.

Bianco, Y., Cheng, W. K., and Heywood, J. B., "The Effects of Initial Flame Kernel Conditions on Flame Development in SI Engines," SAE paper no. 912402, 1991.

Borman, G., and Nishiwaki, K., "I.C.E. Heat Transfer," *Prog. Energy Comb. Sci.,* vol. 13, pp. 1–46, 1987.

Cheng, W. K., and Diringer, J. H., "Numerical Modeling of SI Engine Combustion with a Flame Sheet Model," SAE paper no. 910268, 1991.

Cheng, W. K., Min, K., Hochgreb, S., Heywood, J. B., and Norris, M., "An Overview of Hydrocarbon Emissions Mechanisms in Spark-Ignited Engines," SAE paper no. 932708, 1993.

Crousae, W. H., and Anglin, D. L., *Automobile Engines,* McGraw-Hill, 5th ed., New York, 1975.

Groff, E. G., and Matekunas, F. A., "The Nature of Turbulent Flame Propagation in a Homogeneous Spark-Ignited Engine," SAE paper no. 800133, 1980.

Harada, J., Tomita, T., Mizuno, H., Mashiki, Z., and Ito, Y., "Development of Direct Injection Gasoline Engine," SAE paper no. 970540, 1997.

Herweg, R., and Maly, R. R., "A Fundamental Model for Flame Kernel Formation in S.I. Engines," SAE paper no. 922243, 1992.

Heywood, J. B., *I.C. Engine Fundamentals,* McGraw-Hill, New York, 1988.

Heywood, J. B., Higgins, J. M., Watts, P. A., and Tabaczynski, R. J., "Development and Use of a Cycle Simulation to Predict SI Engine Efficiency and NO_x Emissions," SAE paper no. 790291, 1979.

Hillard, J. C., and Springer, G. S. (eds.), *Fuel Economy in Road Vehicles Powered by Spark Ignition Engines,* Plenum Press, New York, 1984.

Kravchik, T., Sher, E., and Heywood, J. B., "From Spark Ignition to Flame Initiation," *Comb. Sci. and Tech.,* vol. 108, pp. 1–30, 1995.

Krieger, R. B., and Borman, G. L., "The Computation of Apparent Heat Release for Internal Combustion Engines," ASME 66-WA/DGP-4, 1966.

Kummer, J. T., "Catalysts for Automobile Emission Control," *Prog. Energy Comb. Sci.,* vol. 6, pp. 177–199, 1981.

Lavoie, G., Heywood, J., and Keck, J. "Experimental and Theoretical Study of Nitric Oxide Formation in Internal Combustion Engines," *Comb. Sci. Tech.,* vol. 1, p. 313, 1970.

Mattavi, J. A., Groff, E., Lienesch, J., Matekunas, F., and Noyes, R., "Engine Improvements Through Combustion Modeling," in Mattavi and Amann, (eds.), *Combustion Modeling in Reciprocating Engines,* Plenum Press, New York, 1980.

Muranake, S., Takagi, Y., and Ishida, T., "Factors Limiting the Improvement in Thermal Efficiency of S.I. Engines at Higher Compression Ratios," SAE paper no. 870548, 1987.

Obert, E. F., *Internal Combustion Engines and Air Pollution,* Intext Education Publishers, New York, 1973.

Ozdor, N., Dulger, M. and Sher, E., "Cyclic Variability in Spark Ignition Engines: A Literature Survey," SAE paper no. 940987, 1994.

Patterson, D. J., "Cylinder Pressure Variations: A Fundamental Combustion Problem," SAE paper no. 660129, 1966.

Poulos, S. G., and Heywood, J. B., "The Effect of Chamber Geometry on Spark-Ignition Engine Combustion," SAE paper no. 830334, 1983.

Reitz, R. D., "Assessment of Wall Heat Transfer Models for Premixed Charge Engines," SAE paper no. 910267, 1991.

Smith, J. R., "Turbulent Flame Structure in a Homogeneous Charge Engine," SAE paper no. 820043, 1982.

Woschni, G., "A Universally Applicable Equation for the Instantaneous Heat Transfer Coefficient in the Internal Combustion Engine," SAE paper no. 670931, 1968.

Wu, C. M., Roberts, C. E., Matthews, R. O., and Hall, J. M., "Effects of Engine Speed on Combustion in S.I. Engines: Comparisons of Predictions of a Fractal Burning Model with Experimental Data," SAE paper no. 932714, 1993.

CHAPTER 8

Detonation of Gaseous Mixtures

A *detonation* is a combustion wave which travels at supersonic speeds, producing high temperature and pressure for short periods of time. Each fuel and oxidizer mixture has a unique detonation speed. Detonations are important because, given sufficient volume and time, a propagating flame (deflagration) can change into a detonation. Examples are explosion due to leakage of natural gas in a building, a coal mine explosion due to natural gas, and a hydrogen-oxygen explosion in an overheated nuclear reactor. Study of detonations can lead to ways to prevent accidents and may be relevant to the phenomenon of knock in internal combustion engines. As will be discussed later in this chapter, attempts have been made to utilize the intense combustion of a detonation, but no practical applications have yet been developed. Let us consider first the transition from a flame to a detonation and then consider the features of steady-state detonations.

8.1 TRANSITION TO DETONATION

Consider a combustible mixture of gaseous fuel and air in a long tube. The mixture is ignited at the closed end of the tube. A flame forms and begins to propagate along the tube at the laminar flame speed. The propagating flame loses its smooth shape and becomes wrinkled. As a result of the increase in effective flame surface, the flame accelerates with respect to the unburned gas. The wrinkled, fluctuating flame front generates turbulence and weak pressure pulses which run ahead of the flame front and gradually preheat the gas ahead of the flame, causing the flame to speed up. High-speed schlieren photography of the transition from a flame to a detonation shows that as the flame accelerates, the pressure pulses become stronger, coalesce, and further preheat the gas ahead of the flame. Eventually, a pocket of gas ahead of the flame reaches the autoignition temperature and produces

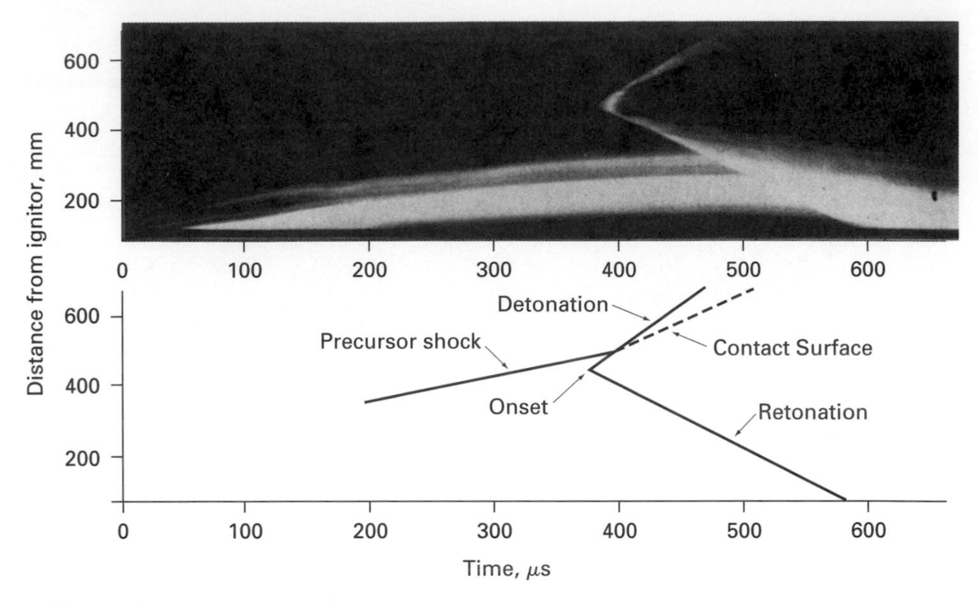

FIGURE 8.1
Streak self-light photograph and interpretation of the onset of detonation.

a local explosion. The rapidly expanding gases produce a shock wave, which interacts with the walls, sending a forward-propagating shock, which rapidly ignites the fuel ahead, and a backward-moving shock, which dies out. The forward-moving shock-combustion complex is a detonation, and the rearward moving shock is called a *retonation,* as shown in Figure 8.1. Note that the velocity is obtained from the slope of the various lines indicated in Figure 8.1.

Transition to detonation also occurs in large-diameter tubes and in spherical geometry, such as focused laser ignition at the center of a spherical chamber or a hot jet of gas. Transition from ignition to detonation depends on the strength of the ignition source and the fuel mixture. Typically, a distance of 1–4 m and times of 2–200 ms are required for spark initiation of detonation of gaseous fuels initially at standard temperature and pressure. Transition can be accelerated by induced turbulence ahead of the flame such as, for example, from rough or irregular walls or turbulence from jets.

8.2 STEADY-STATE DETONATIONS

A detonation may be viewed macroscopically as a shock wave followed by combustion. The heat release is coupled to the shock and drives the shock. The propagation velocity depends on the heat release due to combustion. Representative detonation velocities, which are given in Table 8.1, are about 1000 times greater than the corresponding laminar burning velocities. The detonation velocity is higher for

TABLE 8.1
Detonation velocities for various premixed gases initially at 1 atm and 25°C [Soloukhin]

Mixture	Detonation velocity (m/s)
Stoichiometric H_2-O_2	2840
Stoichiometric H_2-air	1970
1.12 stoichiometric H_2-O_2	3390
0.37 stoichiometric H_2-O_2	1760
Stoichiometric CH_4-O_2	2320
Stoichiometric CH_4-air	1800
1.5 stoichiometric CH_4-O_2	2530
1.2 stoichiometric CH_4-O_2	2470
Stoichiometric C_2H_2-air	1870
Stoichiometric C_2H_2-O_2	2430
Stoichiometric CO-O_2	1800
Stoichiometric C_3H_8-O_2	2350
Stoichiometric C_3H_8-air	1800

oxygen-fuel mixtures than air-fuel mixtures, as is the case with flames. Detonation causes a pressure jump of 10 to 30 times the initial pressure, which, although transient, can be very destructive. The detonation velocity also depends on the fuel-air ratio and exhibits rich and lean mixture limits which are somewhat narrower than the flammability limits.

Detailed measurements of the structure of gaseous detonation waves show that the picture is really more complicated than a normal shock wave followed by rapid combustion. Indeed, it was shown theoretically (after the initial experimental observations in the 1960s) that combustion behind a normal shock wave is unstable and warps the shock in a cellular manner. The shock front is actually composed of a three-dimensional grid of shock intersections which move transverse to the wave front, as the front moves forward. This view helps to explain the rapid reactions that occur in a detonation, because the local temperatures and pressures are higher than the one-dimensional model would suggest. Experimentally, the three-dimensional, cellular nature of the detonation front may be observed best at low pressure, where the cells are larger. Mixtures near the limits of detonability have larger cells than stoichiometric mixtures and at the limit take on a spinning motion normal to the direction of propagation.

High-speed, self-luminous photographs of the structure of an acetylene-oxygen detonation are shown in Figure 8.2. The luminous zones are caused by triple-shock intersections which sweep across the detonation front, consuming the fuel. Imprints on a soot-blackened wall from a reflected detonation, as in Figure 8.3, are used to investigate the cellular nature of the detonation wave. Specially designed pressure transducers have been used to measure the pressure spikes in a detonation wave front, and pressures more than twice that predicted by normal shock theory are observed.

The cellular nature of the detonation front is caused by the rapid heat release, which warps the front, causing curved shocks that interact by means of triple-shock interactions, as indicated in Figure 8.4. (It can be shown that double-shock

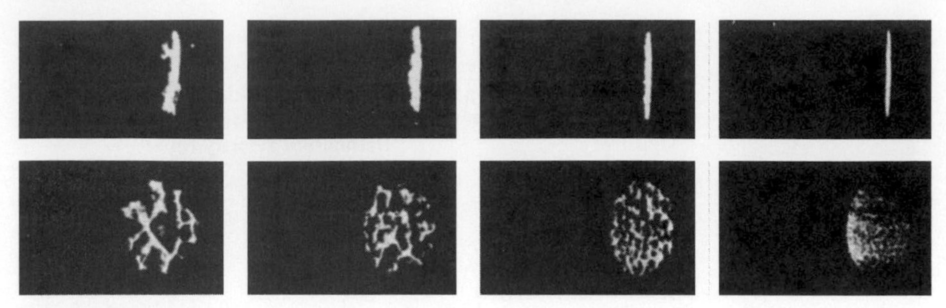

FIGURE 8.2
Self-luminous photographs of an acetylene-oxygen detonation wave. Top row taken at right angles to front, and bottom row at oblique angle [Soloukhin].

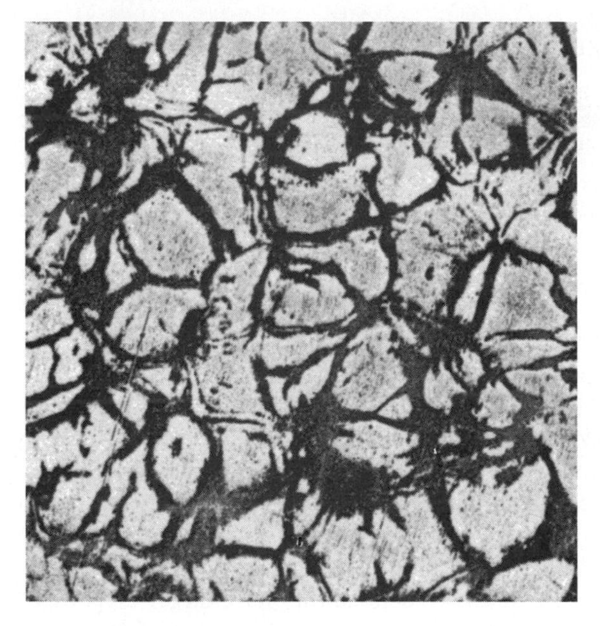

FIGURE 8.3
Imprints from reflection of a detonation off a soot-covered end wall [Shchelkin and Troshin].

interactions cannot exist.) The incident shock is oblique and directs the flow toward the point of intersection. The transmitted shock, called a Mach-stem shock, that it intersects is nearly normal to the flow and hence is stronger. The reflected shock is required to balance the pressure, and the slipstream signifies regions of different temperatures but constant pressure. The triple points propagate along curved paths and periodically collide, leaving a pattern on smoked foil as indicated in Figure 8.5. The incident and reflected shock waves extend above the trajectory of the triple point, and the transmitted shock and slipstream extend below. The incident shock is curved and extends to another triple point. The transmitted and reflected shocks are also curved and are intensified where the triple points collide and decay away from the intersections. Anderson and Dabora have obtained Rayleigh scattering photographs of the detonation structure and compared them with schlieren

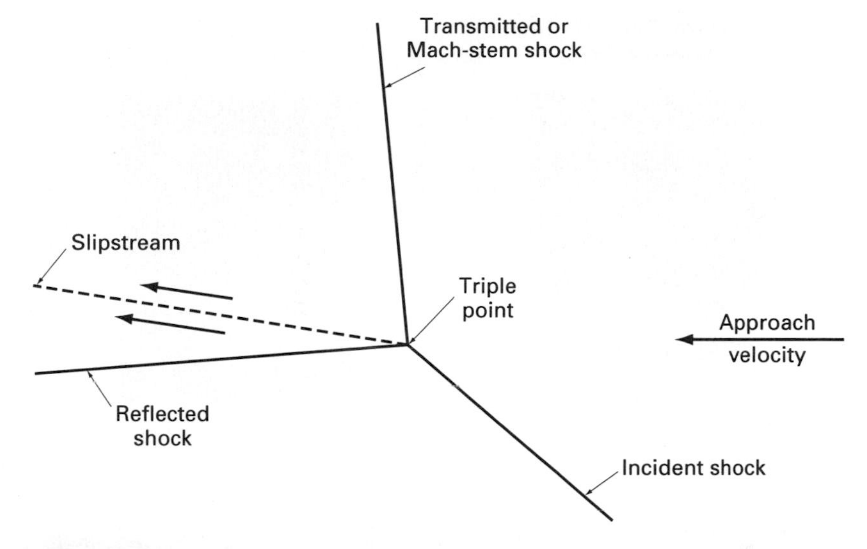

FIGURE 8.4
Schematic diagram of the triple-shock configuration in the detonation front [Strehlow, 1968].

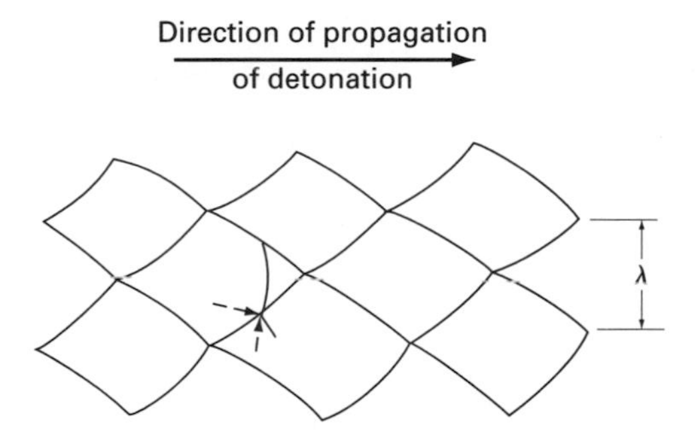

FIGURE 8.5
Soot track formed by triple point of a detonation sweeping along a sidewall. The characteristic cell size λ is noted [Strehlow et al., 1967, by permission of The Combustion Institute].

photographs and the numerical analysis of Kailasanath et al. (Figure 8.6). Rayleigh scattering uses a pulsed laser and measures the gas density, whereas schlieren images measure density gradients and are sensitive to shock waves and turbulence. The Rayleigh images clearly show the high-density triple-point regions behind the leading shock front.

Most of the heat release occurs near the triple points. The distance between triple points defines a detonation cell size λ. The detonation cell size is a fundamental characteristic length of a detonation wave which can be used to correlate mixture limits, as well as initiation energy and quenching behavior. Cell size data for various fuels and equivalence ratios at atmospheric pressure are shown in Figure 8.7. The solid lines are based on the theoretical correlation that the cell size λ equals a constant times the distance for the fuel to react according to a one-dimensional model calculation (ℓ, in Figure 8.7). The constant A is fitted at

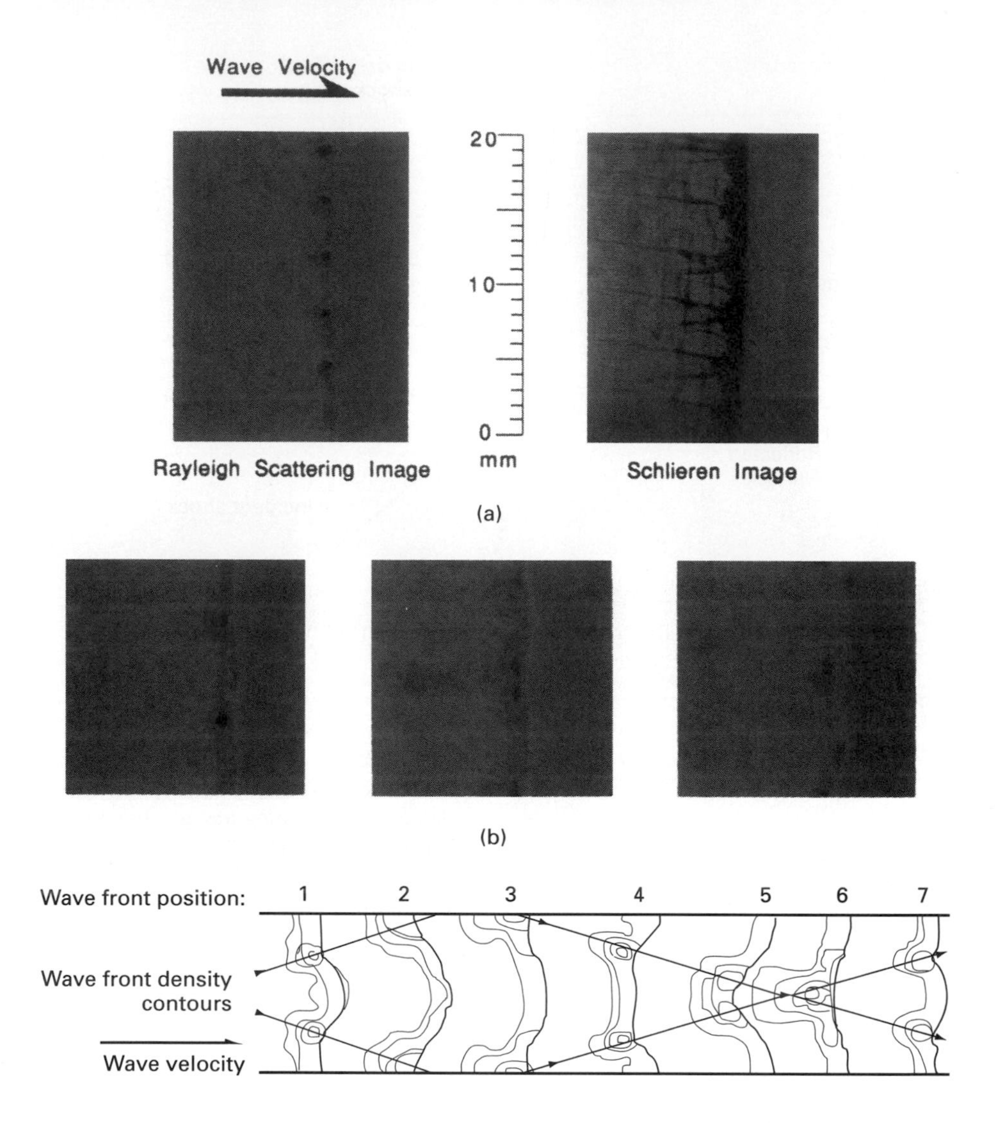

FIGURE 8.6
Structure of a detonation wave in hydrogen-oxygen-argon mixture at 0.374 atm: (*a*) Rayleigh and schlieren images acquired simultaneously from the same wave front; (*b*) comparison of cell structure from Rayleigh images with density contours from the numerical analysis of Kailasanath et al. The straight lines represent triple-point trajectories [Anderson and Dabora, by permission of The Combustion Institute].

stoichiometric and is different for each fuel (e.g., $A = 10.1$ for C_2H_2, $A = 52.2$ for H_2). Cell sizes for detonations in pure oxygen are smaller than in air, and diluents such as argon or carbon dioxide alter the cell size.

The fuel-air mixture limits to sustain a detonation can be estimated from Figure 8.7. Near the rich and lean limits, the detonation cell size becomes so large that the shock waves decay and eventually the detonation reverts to a flame. For

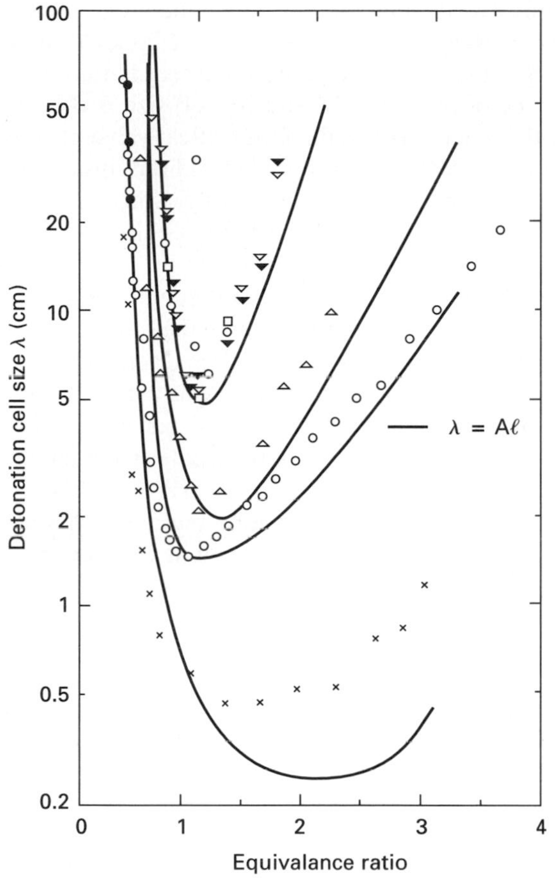

Fuel	Smoked foil	Pressure oscillation
CH_4	◇	
C_4H_{10}	○	
C_3H_8	▽	▼
C_2H_6	□	

Fuel	Smoked foil	Pressure oscillation
C_2H_4	△	
H_2	○	●
C_2H_2	×	

FIGURE 8.7
Detonation cell size of fuel-air mixtures at 1 atm [Lee, with permission from the *Annual Review of Fluid Mechanics,* vol. 16, © 1984, by Annual Reviews Inc.].

hydrogen-air mixtures, the rich and lean limits are approximately $F = 0.4$ to $F = 3.5$; for acetylene-air, $F = 0.4$ to $F = 3.0$; for ethane-air, propane-air, and butane-air, $F = 0.7$ to $F = 2.0$. Detonation limits in small-diameter tubes are narrower because of interaction with the walls. Observations show that the tube diameter must be approximately 13 times greater than the characteristic cell size to sustain detonation.

Direct initiation of detonation means that a strong blast wave such as from a solid explosive or a focused laser is used to start the process and that the energy decays asymptotically to a steady detonation wave of constant cell size. If the ignition energy is less than a certain critical value, the reaction zone progressively decouples from the blast wave as it decays, and a deflagration (flame) results. Experiments have shown that the ignition energy required for direct initiation of detonation is proportional to the cube of the cell size. Hence, cell size is the key indicator of detonability. If the ignition energy is less than the critical value, it does not mean that detonation cannot occur. Rather, it means that indirect instead of direct initiation occurs. Indirect initiation is the mechanism by which a flame accelerates to a detonation, as discussed in Section 8.1, and this requires a longer transition distance than direct initiation.

Quenching of a detonation occurs, at least temporarily, when a detonation wave in a tube suddenly emerges into an unconfined volume containing the same mixture when $d < 13\lambda$, where d is the tube diameter and λ is the cell size. Quenching here means reversion to a deflagration. This correlation holds for a wide range of fuels and mixture ratios. This correlation also holds for flow through an orifice. For noncircular orifices, $d_{eff} < 13\lambda$, where d_{eff} is the average of the smallest and largest openings. For example, for a square orifice of side W, $d_{eff} = 0.5(W + \sqrt{2}W) = 1.2W$. Since quenching here means quenching of the detonation but not the flame, it is always possible that given sufficient distance, the flame can again accelerate into a detonation.

8.3 ONE-DIMENSIONAL MODEL FOR PROPAGATION VELOCITY AND PRESSURE AND TEMPERATURE RISE ACROSS A DETONATION

Since detonations have a three-dimensional microstructure, a one-dimensional model of the structure of a detonation is not realistic for analysis of the chemical kinetics. Nevertheless, a one-dimensional gas dynamic model is reasonably accurate for predictions of the detonation velocity and the post-reaction pressure and temperature, and thus is useful for engineering purposes. By writing one-dimensional equations for conservation of mass, momentum, and energy across the detonation wave, expressions for the detonation velocity, average pressure rise, and average temperature rise across the detonation may be obtained.

Consider a plane detonation wave that travels into a premixed combustible gas which is at rest. The gas can be in a very large volume, or it can be in a tube. In either case, the reaction zone is thin, so that losses to the walls may be neglected (except for very small tubes). The detonation wave sweeps over the reactants, converts them to products at elevated pressure and temperature, and sets the products into motion. At a fixed position in space (or in a tube), we have an unsteady problem. Hence, it is convenient to transform the velocities to a wave-fixed coordinate system by superimposing a velocity $-\underline{V}_D$ onto the gas in the room (or tube) as indicated in Figure 8.8. The transformation between the wave-fixed coordinates and room-fixed coordinates is

$$\check{\underline{V}}_r = \underline{V}_D \tag{8.1}$$

$$\check{\underline{V}}_p = \underline{V}_D - \underline{V}_p \tag{8.2}$$

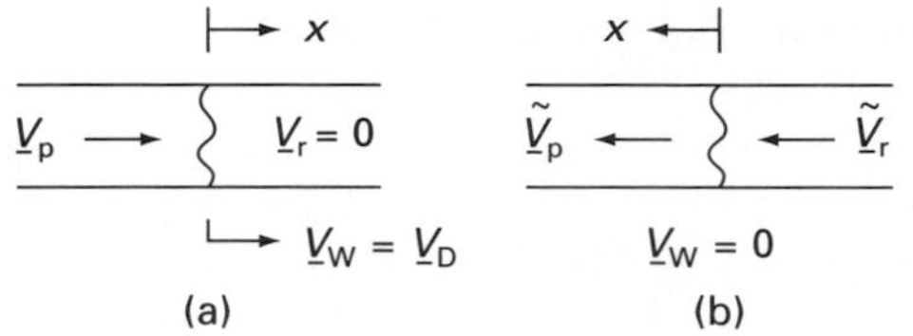

FIGURE 8.8
Coordinate transformation between traveling wave (*a*) and standing wave (*b*).

Equations of continuity, momentum, and energy across a detonation in the wave-fixed coordinate system are exactly the same as Eqs. 5.2 to 5.4, for the laminar flame, and in fact also describe the conditions across a normal shock wave. A flame is a rapid change in subsonic flow due to heat addition. A shock wave is an abrupt change from supersonic to subsonic flow without heat addition, and a detonation is an abrupt change from supersonic to subsonic flow with heat addition. The conservation equations across a detonation, shock, or flame formulated in wave-fixed coordinates are:

$$\rho_r \check{\underline{V}}_r = \rho_p \check{\underline{V}}_p \tag{8.3}$$

$$p_r + \rho_r \check{\underline{V}}_r^2 = p_p + \rho_p \check{\underline{V}}_p^2 \tag{8.4}$$

$$h_r + \frac{\check{\underline{V}}_r^2}{2} = h_p + \frac{\check{\underline{V}}_p^2}{2} \tag{8.5}$$

The problem for a detonation is generally stated as, Given p_1, T_1, ρ_1, and h_1 (the absolute enthalpy per unit mass of reactants), find $\check{\underline{V}}_r$, $\check{\underline{V}}_p$, p_p, T_p, and ρ_p. Using the equation of state $p = \rho RT$ and the enthalpy tables, it is apparent that, as with the laminar flame, we are again short one equation. For a detonation, this difficulty is overcome by asserting that the velocity of the products, $\check{\underline{V}}_p$, is sonic velocity (relative to the wave front and with respect to the hot gas). This is the so-called Chapman-Jouguet, or C − J, condition, and its justification is basically that it accurately predicts the observed detonation velocity $\check{\underline{V}}_r$. Physically, it is reasonable to view the heat addition as choking the flow, that is, driving the flow to $\check{\mathrm{Ma}}_p = 1$, where $\check{\mathrm{Ma}}$ is the Mach number. The solution to the problem can now be completed.

Across a detonation, the species composition changes, and it is assumed that the products are in chemical equilibrium. The speed of sound of the products is $a_p = \sqrt{(\partial p_p/\partial \rho_p)_s}$ and in general this is not equal to $\sqrt{\gamma_p R_p T_p}$ for a reacting mixture. Similarly, the isentropic exponent is not, in general for a reacting mixture, equal to the ratio of specific heats. The so-called Rankine-Hugoniot equations for a reacting mixture, which follow from Eqs. 8.3 to 8.5, are presented in detail by Kuo and can be solved by the STANJAN code. Examples of results from computations are given in Table 8.2.

It is instructive to proceed with solution of a Chapman-Jouguet detonation by assuming constant molecular weights and constant ratio of specific heats. For this case it is convenient to rewrite Eqs. 8.3 to 8.5 in terms of pressure, temperature, and Mach number. Then, the Mach number is given by

$$\check{\mathrm{Ma}}^2 = \frac{\check{\underline{V}}^2}{\gamma RT} = \frac{\rho \check{\underline{V}}^2}{\gamma p} \tag{8.6}$$

The continuity and momentum equations (Eqs. 8.3 and 8.4) become

$$\frac{p_r^2\,\check{\text{M}}\text{a}_r^2}{T_r} = \frac{p_p^2\,\check{\text{M}}\text{a}_p^2}{T_p} \tag{8.7}$$

or

$$p_r(1 + \gamma\,\check{\text{M}}\text{a}_r^2) = p_p(1 + \gamma\,\check{\text{M}}\text{a}_p^2) \tag{8.8}$$

Combining energy and continuity (Eqs. 8.3 and 8.5),

$$c_{p_p}T_p + \frac{\check{V}_p^2}{2} - c_{p_r}T_r - \frac{\check{V}_r^2}{2} = \frac{q}{\dot{m}} \tag{8.9}$$

where $q/\dot{m}$ is the heat of reaction per unit mass of reactants. Dividing by c_{p_r} and noting that $c_p = \gamma R/(\gamma - 1)$,

$$T_r\left(1 + \frac{\gamma - 1}{2}\check{\text{M}}\text{a}_r^2\right) + \frac{q}{c_{p_r}} = T_p\left(1 + \frac{\gamma - 1}{2}\check{\text{M}}\text{a}_p^2\right) \tag{8.10}$$

Now obtain the pressure and temperature jump in terms of the detonation Mach number, $\check{\text{M}}\text{a}_r = \check{\text{M}}\text{a}_D$, and using the Chapman-Jouguet condition that $\check{\text{M}}\text{a}_p = 1$. From Eq. 8.8,

$$\frac{p_p}{p_r} = \frac{1 + \gamma\,\check{\text{M}}\text{a}_r^2}{1 + \gamma} \tag{8.11}$$

From Eqs. 8.7 and 8.11,

$$\frac{T_p}{T_r} = \frac{(1 + \gamma\,\check{\text{M}}\text{a}_r^2)^2}{\check{\text{M}}\text{a}_r^2\,(1 + \gamma)^2} \tag{8.12}$$

Dividing Eq. 8.10 by T_r, substituting Eq. 8.12, and rearranging yields

$$\frac{(1 + \gamma\,\check{\text{M}}\text{a}_r^2)^2}{2\check{\text{M}}\text{a}_r^2\,(1 + \gamma)} - \left[1 + \frac{(\gamma - 1)}{2}\check{\text{M}}\text{a}_r^2\right] = \frac{q}{\dot{m}c_{p_r}}T_r \tag{8.13}$$

For $\check{\text{M}}\text{a}_r^2 \gg 1$, Eq. 8.13 may be simplified to

$$\check{\text{M}}\text{a}_r = \text{Ma}_D = \left[2(\gamma + 1)\left(\frac{q}{\dot{m}c_{p_r}T_r}\right)\right]^{1/2} \tag{8.14}$$

From Eq. 8.14, it is apparent that the detonation propagation rate depends primarily on the heat release per unit mass of reactants, which in turn depends on the heating value of the fuel and the air-fuel ratio. The effect of initial temperature and initial pressure is to influence the heat release. Preheating the initial gas mixture, which causes more final dissociation, decreases the detonation velocity; while increasing the initial pressure decreases dissociation and increases the detonation velocity. Thermodynamic equilibrium programs are used to calculate Chapman-Jouguet detonation properties, and some representative results are shown in Table 8.2. The effects of fuel-air ratio and initial pressure on a methane-air detonation are shown in Figure 8.9.

EXAMPLE 8.1. A large volume contains a stoichiometric mixture of methane and air at 1 atm and 25°C. Using the detonation velocity given in Table 8.1, calculate an approximate value for the pressure, temperature, and gas velocity immediately behind the detonation wave.

TABLE 8.2
Calculated detonation properties for several gaseous mixtures initially at 298 K and 1 atm [Soloukhin]

			Reactants				
Fuel (1 mole)	C_2H_2	C_2H_2	CO	H_2	H_2	CH_4	C_3H_8
O_2 (moles)	2.5	2.5	0.5	0.5	0.5	2	5
N_2 (moles)	0	9.32	0	0	1.88	7.52	18.8
			Detonation products (mole fractions)				
CO_2	0.0930	0.0880	0.4033	0	0	0.0696	0.0836
H_2O	0.0872	0.0615	0	0.5304	0.2943	0.1721	0.1384
O_2	0.1167	0.0221	0.1659	0.0486	0.0078	0.0098	0.0116
CO	0.3463	0.0660	0.3813	0	0	0.0235	0.0300
OH	0.1157	0.0146	0	0.1370	0.0183	0.0097	0.0099
O	0.1288	0.0056	0.0495	0.0386	0.0021	0.0012	0.0015
H	0.0746	0.0039	0	0.0811	0.0060	0.0017	0.0017
NO	0	0.0169	0	0	0.0078	0.0072	0.0085
H_2	0.0370	0.0062	0	0.1641	0.0317	0.0085	0.0072
N_2	0	0.7152	0	0	0.6319	0.6967	0.7076
			Detonation parameters				
V_D (m/s)	2425	1867	1799	2841	1971	1804	1801
Ma_D	7.36	5.41	5.24	5.28	4.84	5.11	5.31
T_p (K)	4214	3113	3525	3682	2949	2780	2823
p_p/p_r	33.87	19.13	13.98	18.85	15.62	17.20	18.27
γ_p	1.152	1.157	1.125	1.129	1.163	1.169	1.166
a_p (m/s)	1317	1027	977	1545	1092	999	994
M_p	23.3	28.4	34.5	14.5	23.9	27.0	27.7

Solution. Since stoichiometric methane-air contains 1 mole of CH_4 plus 9.52 moles of air, the molecular weight of the reactants is

$$\overline{M}_r = \frac{(1 \text{ kgmol } CH_4)(16 \text{ kg/kgmol } CH_4) + (9.52 \text{ kgmol air})(29 \text{ kg/kgmol air})}{10.52 \text{ kgmol}}$$

$$= 27.8 \text{ kg/kgmol}$$

From Table 8.1,

$$\underline{V}_D = \check{\underline{V}}_r = 1800 \text{ m/s}$$

The speed of sound is

$$a_r = (\gamma R_r T_r)^{1/2} = \left[(1.4)\frac{(8.314 \text{ kJ/kgmol}\cdot\text{K})}{27.8 \text{ kg/kgmol}}(298 \text{ K})\right]^{1/2}$$

$$= 353 \text{ m/s}$$

and the Mach number is

$$\check{\mathrm{Ma}}_r = 1800/353 = 5.10$$

Using Eqs. 8.11 and 8.12,

$$p_p = \frac{(1 \text{ atm})[1 + (1.4)(5.10)^2]}{1 + 1.4} = 15.6 \text{ atm}$$

$$T_p = \frac{(298 \text{ K})[1 + (1.4)(5.10)^2]^2}{(5.10)^2(1 + 1.4)^2} = 2784 \text{ K}$$

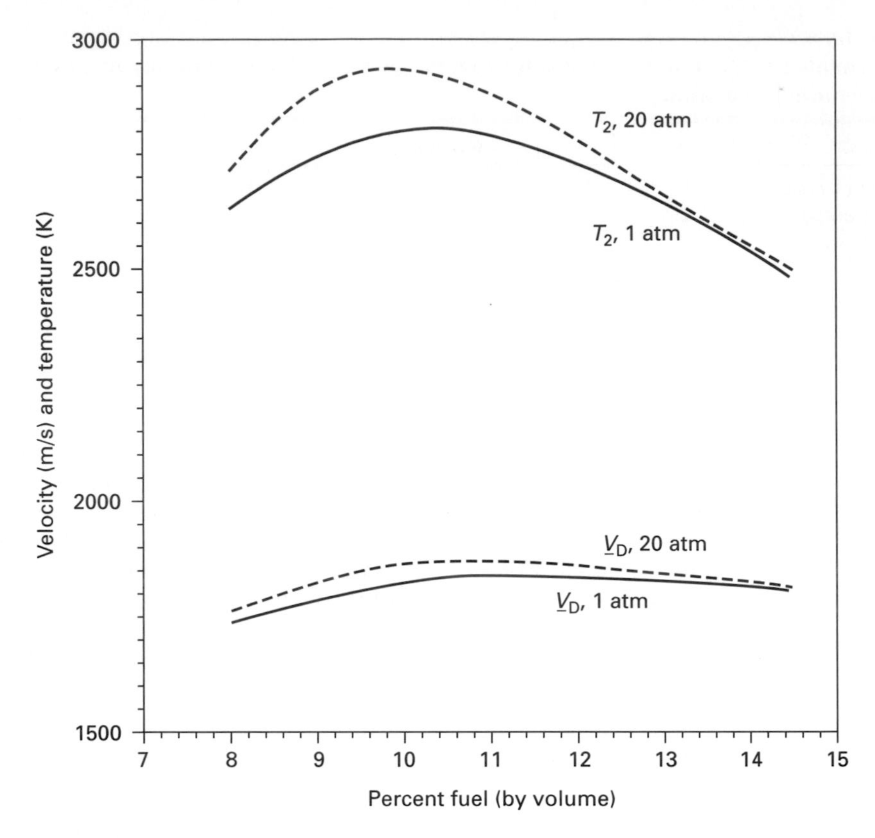

FIGURE 8.9
Chapman-Jouguet detonation velocity and products temperature for methane-air mixtures initially at 298 K and 1.20 atm between lean and rich limits. Stoichiometric is 9.51% methane.

The pressure p_p is lower than the exact calculation given in Table 8.2, and the temperature T_p is slightly higher. Note that if $\gamma_\pi = 1.169$ from Table 8.2 is used in the denominator of Eq. 8.11, then the correct value of p_p is predicted; however, this is fortuitous. Now use the Chapman-Jouguet condition and Eq. 8.2 to get the velocity behind the detonation in room-fixed coordinates:

$$\underline{\check{V}}_p = \underline{\check{V}}_r - a_p = 1800 \text{ m/s} - \left[\frac{(1.4)(8314 \text{ kg}\cdot\text{m}^2/\text{s}^2\cdot\text{kgmol}\cdot\text{K})(2784 \text{ K})}{27.8 \text{ kg/kgmol}}\right]^{1/2}$$

$$= 720 \text{ m/s}$$

Thus the Mach number behind the wave in room-fixed coordinates is

$$\text{Ma}_p = \frac{720 \text{ m/s}}{1080 \text{ m/s}} = 0.667$$

EXAMPLE 8.2. A large volume contains a stoichiometric mixture of methane and air at 1 atm and 25°C. Using the lower heating value of methane, calculate the detonation Mach number of this mixture and compare the results with Table 8.2.

Solution. The lower heating value of methane is 50,010 kJ/kg, and $f_s = 0.058$. Neglecting dissociation (although it is surely significant), the heat of reaction is estimated to be

$$q = \text{LHV}\left(\frac{f}{1+f}\right) = (50{,}010 \text{ kJ/kg})\left(\frac{0.058}{1+0.058}\right) = 2742 \text{ kJ/kg}$$

Using Eq. 8.14 with

$$\gamma = 1.4 \qquad c_{p_1} = 1.0 \text{ kJ/kg}\cdot\text{K}$$

we get

$$\check{\text{M}}\text{a}_r = \left[\frac{2(1.4+1)(2742 \text{ kJ/kg})}{(1.0 \text{ kJ/kg}\cdot\text{K})(298 \text{ K})}\right]^{1/2}$$
$$= 6.6$$

Using Eq. 8.13, $\check{\text{M}}\text{a}_r = 6.8$. However, Table 8.2, which is based on thermochemical calculations of reacting mixtures, gives $\check{\text{M}}\text{a}_r = \check{\text{M}}\text{a}_D = 5.1$. The error occurs because we have chosen a heat release which is too large due to dissociation of the products and because Eqs. 8.13 and 8.14 assume constant molecular weights and constant γ. The constant γ detonation model with heat release to completion overestimates the detonation Mach number but gives the correct trends.

8.4 MAINTAINED DETONATIONS

Flames can be fixed in position by flame holders or by special flow patterns such as swirl stabilization. Similarly, devices have been built to stabilize detonations behind standing normal shock waves. Three such designs are shown in Figure 8.10. A question arises whether these are true detonations, because three-dimensional transverse waves are not observed. However, the onset of detonation does force a readjustment of the normal shock to a new upstream position, and hence, they are considered detonations. Attempts have been made to develop chemical lasers using standing detonation waves.

The concept of a maintained detonation wave moving around an annulus which is continuously replenished with fresh combustibles has been investigated in England, the former Soviet Union, and the United States. Consider an annular chamber as shown in Figure 8.11. A gaseous fuel, such as methane or hydrogen, and either air or oxygen are introduced into manifolds which feed two rings of nozzles. The nozzles impinge, mix, and supply a detonable mixture to the combustion chamber. A detonation wave is initiated by a pulse of hot, high-pressure gas, which is introduced tangentially into the annulus. The detonation wave propagates around the annulus. The high pressure behind the wave front gradually decays, allowing fresh reactants to flow into the combustion chamber and exhaust products to flow out of the chamber.

Since detonations travel several thousand times faster than flames, the increase in heat release per unit volume would be enormous if a rotating detonation wave

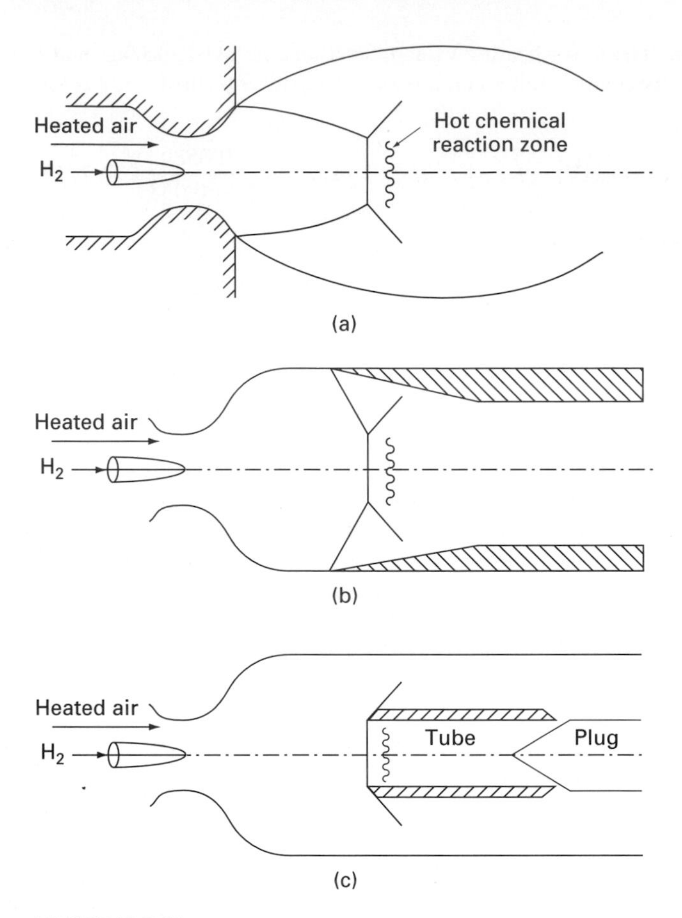

FIGURE 8.10
Standing detonation experiments: (*a*) behind a shock bottle of an overexpanded jet, (*b*) on a Mach stem of two oblique shocks, and (*c*) behind normal shock held on lip of tube by a choking plug [Strehlow, 1968].

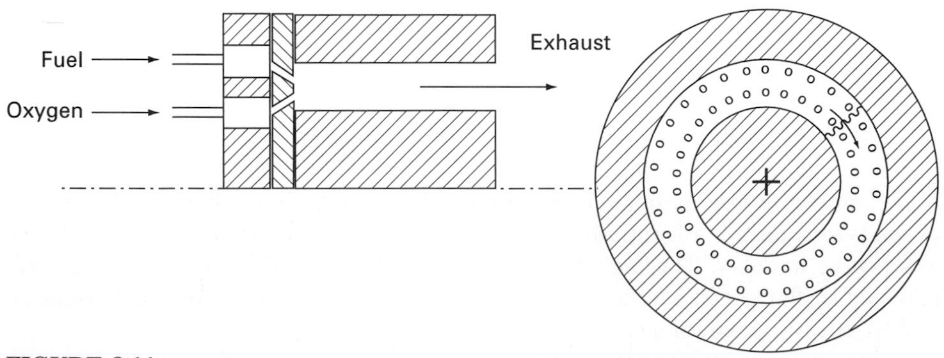

FIGURE 8.11
Concept for a rotating detonation wave combustor [Nicholls et al., © 1966 and reprinted with permission of AIAA].

could be sustained. Suggestions have been made to use a nozzle at the end of the combustion chamber and make a rotating detonation wave rocket engine. Others have suggested a rotating detonation wave gas turbine combustor. The high-frequency, high-velocity swirling flow of combustion products emitted from the annular chamber could perhaps be effectively mixed with dilution air in a short distance. Such a design could possibly reduce circumferential variability in temperature at the entry to the turbine, while also reducing the size of the combustor.

The rotating detonation wave combustor is not without problems, however. There is a tendency to form multiple waves, which make it impossible to supply enough fresh fuel, so that the combustion decays to a deflagration. Ignition must be done with care so as not to create two waves traveling in opposite directions. Combustor heat transfer will be extremely high, and special cooling requirements must be considered in the design. The required fuel and air flow rates are so great that the inlet pressure must be very large. This latter point as well as the first point regarding the formation of multiple waves suggests that liquid rather than gaseous fuel would perhaps be more appropriate for a rotating detonation wave combustor.

8.5 SUMMARY

In contrast to a flame, which is a constant-pressure process and has a low subsonic speed, the detonation mode of combustion produces a large increase in pressure and propagates at supersonic speeds with respect to the unburned gas (several thousand times faster than a flame). A propagating flame will undergo transition to a detonation given sufficient time and distance. Turbulence accelerates the transition. The one-dimensional equations of conservation of mass, momentum, and energy across the detonation, with the Chapman-Jouguet condition imposed on the burned gas, provide an accurate calculation of the detonation speed and average pressure jump when thermodynamic equilibrium calculations are included for determination of the heat release. However, in reality the detonation front has a highly three-dimensional nature consisting of a network of incident shocks, reflected shocks, and Mach stems, which move transversely and cause very transient high pressure and temperature spikes. Although detonations can be destructive when out of control, detonations have been controlled by standing detonations in nozzles and by a rotating detonation in an annular channel.

PROBLEMS

8.1. Use the information given in Table 8.2 to verify that Eq. 8.5 holds for a stoichiometric hydrogen-air detonation. Assume the contribution of O and H to the enthalpy of the products is negligible if these enthalpy data are not readily available. Use the absolute enthalpy form of Eq. 8.5.

8.2. Repeat Problem 8.1 using a stoichiometric propane-air mixture.

8.3. Use the detonation Mach number in Table 8.2 for stoichiometric hydrogen-air, to compare Eqs. 8.11 and 8.12 with p_p/p_r and T_p in Table 8.2.

8.4. Repeat Problem 8.3 using a stoichiometric propane-air mixture.

8.5. For a methane and air mixture at standard initial conditions, calculate the detonation velocity at stoichiometric and at the rich and lean limits using Eq. 8.13. Comment on the validity of Eq. 8.13.

8.6. Repeat Problem 8.5 for a stoichiometric mixture of propane and air.

8.7. A stoichiometric gasoline vapor and air mixture, which is initially at 300 K and 1 atm, is compressed isentropically to one-eighth the volume. Estimate the peak pressure and temperature which could be generated if a detonation occurred after compression.

8.8. For methane-air detonation at the lean detonation limit, estimate the peak pressure that could be generated. Assume that the reactants are at 20°C and 1 atm, and use the one-dimensional model. Comment on the validity of the one-dimensional model.

8.9. For a stoichiometric CH_4-air C-J detonation, use STANJAN to determine V_D, T_p, ρ_p. Then, from the STANJAN result for V_D, use the 1-D model to calculate p_p and compare with the STANJAN p_p. Is T_p different from the adiabatic flame temperature? If different, explain why. (When using STANJAN, call for C-J detonation with full—i.e., dissociated—products.)

8.10. What pressure is required to burst a 2-ft dia by $\frac{1}{8}$-in. wall steel pipe. Assume a yield stress for steel of 40,000 psi. If the initial pressure in the tube is 1 atm, what Mach number will generate this pressure based on (*a*) the pressure due to a normal shock wave, and (*b*) the pressure at the end of the reaction zone of a detonation? For a thin-wall tube, $S = pr_0/t$, where S is the stress, r_0 the tube radius, and t the thickness of the tube. Assume that $\gamma = 1.4$.

REFERENCES

Anderson, T. J., and Dabora, E. K., "Measurements of Normal Detonation Wave Structure Using Rayleigh Imaging," *Twenty-Fourth Symp. (Int.) on Combustion,* The Combustion Institute, Pittsburgh, pp. 1853–1860, 1992.

Bowen, J. R., Steffes, F., Loflin, T., and Ragland, K. W., "Heterogenous Detonation Supported by Fuel Fogs or Films," *Thirteenth Symp. (Int.) on Combustion,* The Combustion Institute, Pittsburgh, pp. 649–657, 1970.

Edwards, B. D., "Maintained Detonation Waves in an Annular Channel," *Sixteenth Symp. (Int.) on Combustion,* The Combustion Institute, Pittsburgh, pp. 1611–1618, 1976.

Gordon, G., and McBride, B. J., "Computer Program for Calculation of Complex Chemical Equilibrium Compositions, Rocket Performances, Incident and Reflected Shocks and Chapman-Jouguet Detonations," NASA SP-273, 1976.

Kailasanath, K., Oran, E. S., Boris, J. P., and Young, T. R., "Determination of Detonation of Cell Size and the Role of Transverse Waves in Two-Dimensional Detonations," *Combustion and Flame,* vol. 61, pp. 199–209, 1985. (This is a computational modeling paper.)

Knystautas, R., Guiraoc, C., Lee, J. H., and Sulmistras, A., "Measure of Cell Size in Hydrocarbon-Air Mixtures and Predictions of Critical Tube Diameter, Critical Initiation Energy, and Detonation Limits," *Prog. in Astronautics and Aeronautics,* vol. 94, pp. 23–37, 1984.

Kuo, K. K., *Principles of Combustion,* Wiley-Interscience, New York, 1986 (Chap. 4).

Lee, J. H. S., "Dynamic Parameters of Gaseous Detonations," *Ann. Rev. Fluid Mech.,* vol. 16, pp. 311–336, 1984, Annual Reviews Inc., Palo Alto, CA.

Nicholls, J. A., Cullen, R. E., and Ragland, K. W., "Feasibility Studies of a Rotating Detonation Wave Rocket Motor," *J. Spacecraft and Rockets,* vol. 3, pp. 893–898, 1966.

Nicholls, J. A., and Dabora, E. K., "Recent Results on Standing Detonation Waves," *Eighth Symp. (Int.) on Combustion,* pp. 644–655, The Combustion Institute, Pittsburgh, 1962.

Oppenheim, A. K., Urtiew, P. A., and Weinberg, F. J., "On the Use of Laser Light Sources in Schlieren and Interferometer Systems," *Proc. Royal Soc.,* A291, pp. 279–290, 1966.

Reynolds, W. C., "The Element Potential Method for Chemical Equilibrium Analysis. Implementation of the Interactive Program STANJAN," Mechanical Engineering Dept., Stanford University, 1986.

Shchelkin, K. I., and Troshin, Y. K., *Gas Dynamics of Combustion,* Izd. Akad. Nauk. USSR, Moscow, 1963. NASA Technical Translation NASA TT F-231, Washington, DC, 1964.

Soloukhin, R. I., *Shock Waves and Detonations in Gases,* Mono Book Corporation, Baltimore, 1966.

Strehlow, R. A., Liaugminas, R., Watson, R. H., and Eyman, J. R., "Transverse Wave Structure in Detonations," *Eleventh Symp. (Int.) on Combustion,* The Combustion Institute, Pittsburgh, pp. 683–692, 1967.

Strehlow, R. A., *Fundamentals of Combustion,* International Textbook Company, Scranton, PA, 1968. (See Chap. 9, Detonations.)

Taki, S., and Fujiwara, T., "Numerical Simulation of Triple Shock Behavior of Gaseous Detonation," *Eighteenth Symp. (Int.) on Combustion,* The Combustion Institute, Pittsburgh, pp. 1671–1681, 1981.

Williams, F. A., *Combustion Theory,* Addison Wesley, Reading, MA, 2nd ed., 1985.

PART III

Combustion of Liquid Fuels

Oil-fired furnaces, gas turbine combustors, and diesel engines involve spray combustion. Analogous to gaseous detonation, liquid fuels can also combust in a detonation mode. First, sprays are considered, and then spray combustion systems are discussed. The reader may wish to review the material on liquid fuels in Chapter 2 before proceeding.

CHAPTER 9

Spray Formation and Droplet Behavior

Oil-fired furnaces and boilers, diesel engines, and gas turbines utilize liquid fuel sprays in order to increase the fuel surface area and thus increase the vaporization and combustion rate. For example, breaking up a 3-mm sphere of liquid into 30-μm drops results in 1 million drops. The droplet mass burning rate is approximately proportional to diameter squared, and the increase in burning rate is 10,000 times if we assume that the large single droplet and the 1 million small droplets burn under the same ambient conditions. Thus the motivation for spray combustion is quite clear.

In order to understand spray combustion, let us first look at the overall problem and then go to a detailed study of the various parts of the spray mechanism. Upon injection of the liquid fuel into a combustion chamber, the liquid undergoes atomization, which causes the liquid to break up into a large number of droplets of various sizes and velocities. Depending on the density of the spray and the ambient conditions, some of the droplets may continue to shatter and some may recombine in droplet collisions. During this time, vaporization takes place. Fuel vapor produced by vaporization mixes with the surrounding gas, and either because of the high temperature of the ambient oxidizer or because of an existing flame front or other ignition source, combustion of the vapor-air mixture occurs. The hot products of combustion mix with the vapor and droplets. If enough time (or combustor length) is provided, the entire amount of fuel will be converted to combustion products. Carbon produced in the combustion process may either continue to oxidize to produce final gaseous products or may agglomerate to form exhaust particulates.

To diagnose this complex set of phenomena, let us first consider a steady-state spray combustion and then the more difficult case of unsteady (diesel) combustion. Suppose that we start at "station 1" near the fuel injector nozzle and ask what information is needed in order to understand and quantify the process and to predict the state of affairs downstream at some "station 2." The spray can be divided into three regions: the spray formation region, the vaporization region, and the combustion region.

At the end of the spray formation region, one would like to know the droplet size, velocity, and number distributions, the air velocity and temperature, and the droplet temperatures. In some sprays the breakup region will overlap the vaporization region. To follow the process through the vaporization region, a model is needed for air motion including turbulence and the interaction of air and droplet momentum. To follow droplet motion, droplet drag coefficients and droplet vaporization models are needed. Understanding the ignition process, both for the free vapor and the droplet boundary layer, is necessary in order that the onset of burning can be established. Then droplet burning rate relationships are needed. If emissions are to be predicted, models for the reaction kinetics are required. Radiation from carbon particles formed in the diffusion flames is an important component in the energy balance and in the diffusion flame temperature prediction. The convective heat losses may also play a role and are tied to the general problem of prediction of mixing, recirculation of products, and turbulence.

In diesel-type combustion, three additional factors enter into the combustion phenomena. First, the ignition delay (the time between start of fuel injection and first combustion) depends on the rate of heating, formation of a combustible mixture, and the mixture chemical kinetics. Second, the high pressures encountered may cause the droplets to approach their critical point, thus causing droplet breakup and a shift in the vaporization and burning rates. Third, in some engines (and some oil burners as well) residual fuels are used which may break down (crack) in the liquid droplet phase, causing different burning rates and the formation of residual carbon shells. In addition to these combustion effects, diesel engines have a very dense spray in which droplet interactions and local cooling by vaporization are important. For small engines and for cold starting conditions in all engines, the spray typically hits the piston surface, causing droplets to wet the surface, which changes the vaporization and mixing mechanism. Finally, combustion in an enclosure such as an engine cylinder causes combustion-induced motion because of the expansion of products in various parts of the cylinder volume.

9.1 SPRAY FORMATION

To start the topic of spray formation, let us first consider the different types of spray nozzles. Simple pressure nozzles either with 4–10 small holes or a pintle are used predominantly in diesels. Various types of air or steam atomization nozzles are used in burners and furnaces. Swirl-type nozzles are often found where less forward penetration is desired. The fuel is made to swirl by use of tangential inlets. The fuel comes out as a conical sheet. Each of these types of nozzles have different penetration and droplet size distributions caused by the differences in breakup mechanism.

Before discussing atomizer design, the basic phenomena of spray breakup and techniques for quantifying the resulting droplet sizes are presented. The precise mechanism of spray breakup varies with the injection pressure and type of atomizer. The resulting droplet size distributions for steady-flow injectors have been well formulated empirically, but as yet no validated theory for predicting droplet size has been formulated for practical atomizers. In qualitative terms, the breakup mechanism may be characterized by a set of six steps:

1. Stretching of fuel into sheets or streams
2. Appearance of ripples and protuberances
3. Formation of ligaments or holes in sheets
4. Collapse of ligaments or holes in sheets
5. Further breakup due to vibration of droplets
6. Agglomeration or shedding from large drops

For simple orifices the behavior of the jet may be characterized by three dimensionless groups:

Jet Reynolds number:

$$\mathrm{Re}_j = \left(\frac{\rho V}{\mu}\right)_l d_j \tag{9.1}$$

Jet Weber number:

$$\mathrm{We}_j = \frac{\rho_g V_j^2 d_j}{\sigma} \tag{9.2}$$

Ohnesorge number:

$$\mathrm{Oh} = \frac{\mu_l}{\sqrt{\rho_l \sigma d_j}} = \frac{[(\rho_l/\rho_g)\ \mathrm{We}_j]^{1/2}}{\mathrm{Re}_j} \tag{9.3}$$

The parameter d_j is the diameter of the undistributed jet, and σ is the surface tension of the liquid (in contact with the surrounding gas). Note that the jet Weber number is sometimes defined using ρ_l in place of ρ_g; we shall call this We_{jl}.

Criteria for spray formation are given by a plot of log Oh versus log Re_j, which is divided into three zones as shown in Figure 9.1. Zone I is named for Rayleigh, who did a wave instability analysis of jets showing that breakup is due to the effects of surface tension forces on the jet. The theory predicts that maximum instability (the breakup point) takes place where the jet disturbance wavelength λ is about $4.6d_j$. In zone II helicoidal waves in the jet are observed prior to breakup. Breakup reflects the beginnings of the influence of the ambient air; this is also called the wind-induced region. In zone III the jet is disrupted into droplets very close to the orifice. Here, breakup is due to the effects of the ambient air combined with the effects of flow turbulence.

For a swirl-type nozzle (Figure 9.2), the conical sheet breaks up first by formation of holes, and then the lace pattern breaks up into droplets. Droplets which are formed near the nozzle exit may undergo further breakup due to aerodynamic forces. Large droplets deform into a bag shape and then burst. For smaller droplets, the drop may vibrate and then separate into two or more parts. For such drops under conditions of sudden change in relative velocity, there is a time delay period followed by shedding of very small droplets from the sides of the existing

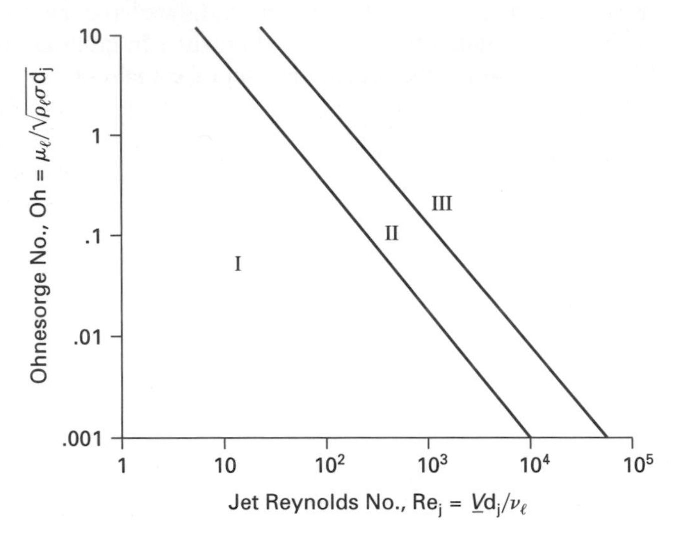

FIGURE 9.1
Stages of atomization with simple orifice [Barett and Hibbard].

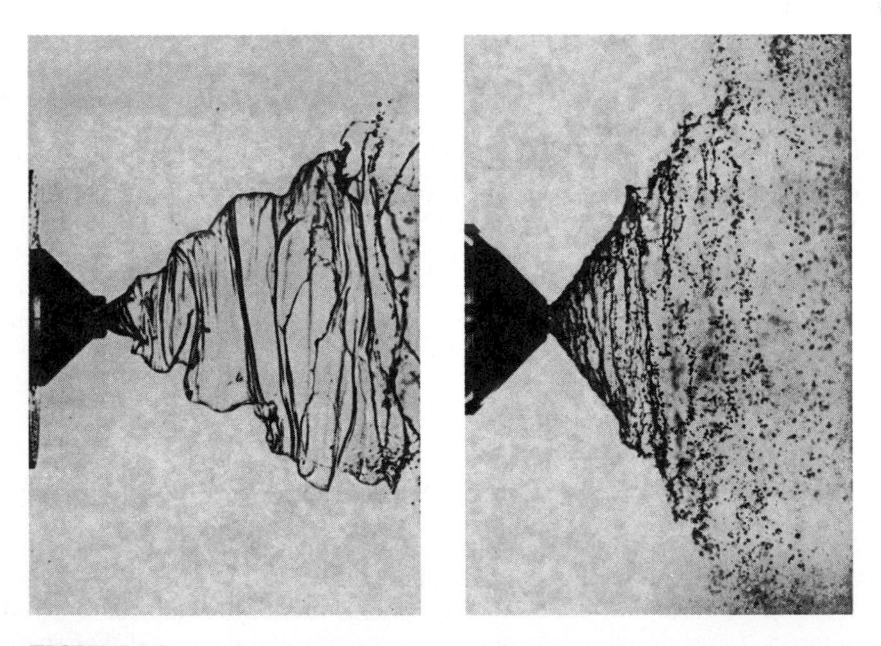

FIGURE 9.2
Thin conical jet: disturbances grow until the sheet disintegrates into droplets [Van Dyke, courtesy of H. E. Fiedler, professor Tech. Univ. of Berlin].

droplets. The shape of the shedding droplets at this point is elliptical at the forward surface and rather flat at the rear.

Observations from experiments with falling droplets of liquid CO_2 in helium at high pressures and temperatures (well above the liquid critical point) show that as the ambient pressure and temperature are increased, droplets at first break up and then for even higher pressures and temperatures they seem to totally disintegrate. This phenomenon indicates that the droplet surface mixture has reached its thermodynamic pseudo-critical point and that the surface tension has become negligible.

For noncritical conditions, the onset of droplet shattering due to aerodynamic forces is related to the droplet Weber number:

$$\mathrm{We}_d = \frac{\rho \underline{V}^2 d}{\sigma} \tag{9.4}$$

where $\underline{V}$ is the relative velocity of the ambient gas and d is the droplet diameter. For Weber numbers greater than about 12, one expects breakup. For large Weber numbers the mechanism of shedding is a boundary layer shearing phenomenon related to the Reynolds number. Shedding was observed in shock tube studies to take place for

$$\frac{\mathrm{We}_d}{\mathrm{Re}_d^{1/2}} > 0.7 \tag{9.5}$$

where
$$\mathrm{Re}_d = \frac{\underline{V}d}{\nu_g}$$

However, more recent studies of droplets injected into a moving cross stream cast some doubt on the applicability of this formula to some diesel engine conditions. For such cases, larger drops appear to fall into the very high Weber number region (over 100) and break up by a catastrophic mechanism. In this region aerodynamic forces flatten the drop into a sheet, which then breaks into parts due to Rayleigh-Taylor instability. The sheet fragments break up at the edges into ligaments which in turn break up into very small micron-sized droplets as shown schematically in Figure 9.3.

The various forces which can cause breakup of a droplet by a vibrational mechanism are illustrated in the *Taylor's analogy breakup model* (TAB model). In this heuristic model a droplet of spherical radius a is caused to oscillate either from an initial disturbance or by aerodynamic forces which continue after time zero. The perturbation of the radius is called x, and if it exceeds a the droplet is assumed to break apart. The perturbation is calculated from a simple spring-mass-damper model:

$$\rho_l a^3 \frac{d^2x}{dt^2} = \frac{2}{3}\rho_g a^2 \underline{V}^2 - 8\sigma x - 5\mu_l a \frac{dx}{dt} \tag{9.6a}$$

For many cases the damping and restoring forces due to viscosity and surface tension, respectively, can be small relative to the aerodynamic force. Then, taking

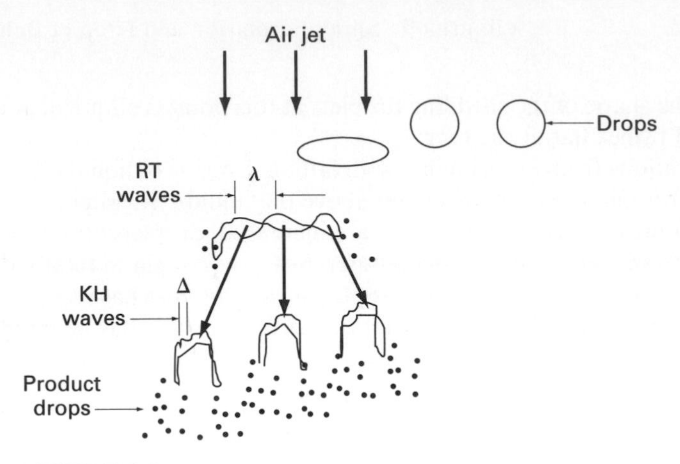

FIGURE 9.3
Breakup mechanism in the catastrophic breakup regime. RT, Rayleigh-Taylor instabilities; KH, Kelvin-Helmholtz waves. [Hwang et al., reprinted from *Atomization and Sprays,* vol. 16, pp. 353–376, 1966, by permission of Begell House, Inc.].

$dx/dt = 0$, the criterion for breakup ($x = a$) gives

$$a = \left[\frac{\frac{1}{3}(\rho_g/\rho_l)\underline{V}^2}{a}\right]t^2 \tag{9.6b}$$

and the minimum time delay for aerodynamic breakup is

$$\Delta t_b = 1.732\left(\frac{a}{\underline{V}}\right)\left(\frac{\rho_l}{\rho_g}\right)^{1/2} \tag{9.7}$$

if it is assumed that $\underline{V}$ and a are constant. In practice $\underline{V}$ changes rapidly due to drag forces and a decreases as the droplet vaporizes. Additional modeling can account for the changing shape on the aerodynamic term, and for the fact that the droplet can take on a dumbbell and/or an oblate spheroid shape. The number of droplets and their size distribution upon breakup is not well established, but the concept is nevertheless used in some CFD models of sprays.

EXAMPLE 9.1. Consider a 15-μm droplet which is traveling at a velocity relative to the surrounding air of 200 m/s. Assume a fuel density of 850 kg/m^3 and surface tension of $\sigma = 0.031$ N/m. The air is compressed with density 25 kg/m^3, 6.2 MPa pressure and 864 K temperature.

Calculate the droplet Weber number and Reynolds number, and predict if the drop will break up. If it will, then calculate the time required before breakup, Δt_b, from Eq. 9.7. Is the term $8\sigma x$ in Eq. 9.6*a* truly negligible for this case?

Solution

$$\text{We}_d = \frac{\rho_g \underline{V}^2 d}{\sigma} = \frac{(25)(200)^2(15 \times 10^{-6})}{3 \times 10^{-2}} = 500$$

$$\text{Re}_d = \frac{\underline{V}d}{\nu_g} = \frac{(200)(15 \times 10^{-6})}{1.63 \times 10^{-6}} = 1840$$

The droplet will break up, since $\mathrm{We}_d > 12$ and $\mathrm{We}_d/\mathrm{Re}_d^{1/2} = 12 > 0.7$. We should expect a catastrophic type of breakup for this large Weber number.

$$\Delta t_b = 1.732\left(\frac{15 \times 10^{-6}}{200}\right)\left(\frac{850}{25}\right)^{1/2} = 0.76\ \mu\mathrm{s}$$

At the given velocity the drop will travel about 10 diameters relative to the air during this time.

Now compare $\frac{2}{3}\rho_g a^2 V^2$ with $8\sigma x$ for $x = a = 15\ \mu$m. The first term $= \frac{2}{3}(25) \times (15 \times 10^{-6})^2(200)^2 = 1.5 \times 10^{-4}$ N, and the second term $= 8(3 \times 10^{-2}) \times (15 \times 10^{-6}) = 360 \times 10^{-8}$ N. Their ratio is $360 \times 10^{-8}/1.5 \times 10^{-4} = 0.0240$. Thus, neglecting the second term in Eq. 9.6a is a very good approximation for this case.

Spray breakup for single-hole nozzles with high-pressure injection such as used in diesel engines is the least understood of the various breakup phenomena. This is because the spray is in a transient state and is very thick. It is possible that high-pressure hole nozzles cause ligament breakup (zone III) but that the resulting rapidly moving droplets then further break up by shedding. Models based on this assumption have recently been used in 3-D computer models, but have not been validated. Detailed measurements are very difficult, and most data give only the variation of spray length (penetration distance) and spray angle with time.

From data for diesel sprays it appears that length-to-diameter ratio of the injector tip hole and its precise shape are important parameters in determining the breakup characteristics. For some cases the hole geometry may lead to cavitation within the injector, which may then have an effect on breakup. Much less is known about the effects of the shape of the rate-of-injection curve. It is logical to assume that if the injection pressure rises very rapidly, the droplet breakup will be more rapid and effective than if the pressure rises along a ramp with time. During this period the spray may go from zone II to zone III of Figure 9.1 as the exit velocity increases with time. For typical injections which give a triangular or trapezoidal rate of injection the spray consists of a conical outer shape near the orifice with a liquid core along the axis surrounded by disintegrating ligaments and droplets. Many small droplets may be formed very close to the nozzle and then agglomerate by collisional processes to produce larger droplets downstream. The liquid core itself can extend from 10 to 30 mm downstream of the injection tip, and for the typical ramp-shaped injection rate the liquid core exists even for very high injection pressures. The core length is a weak function of hole length to diameter ratio (L/d_j) for the range of ratios from 3 to 10, but the minimum core length occurs for holes with $L/d_j = 4$. For diesels, values of d_j range from 0.1 mm to 0.4 mm depending on the injection pressure used.

9.2 SIZE DISTRIBUTIONS

Since the theory of droplet breakup has not been adequate to predict the size distribution of droplets in practical sprays, it has been necessary to measure the distribution experimentally and then fit the resulting data with empirical functions.

When measuring size distributions, two basic types of measurement may be made: spatial or temporal. A *spatial distribution* is obtained by counting droplets in a given volume at a given instant. For example, using a pulse light source and camera to photograph a given volume of the spray and then counting the number of droplets in each size category yields a spatial number distribution. A *temporal distribution* is obtained by counting all droplets passing through a given surface. For a steady spray the distribution obtained is constant when averaged over a time sufficient to obtain a statistically valid sample.

To see the difference between spatial and temporal distributions, imagine the following analogy. Count the length of vehicles on a section of interstate highway by (1) taking a photograph of a section from the highway at different times, and (2) counting all vehicles passing through a tollgate. Suppose the tollgate data show equal numbers of vehicles over 20 and under 20 feet long. The photographic method will show the same equal length distribution if all vehicles are traveling at the same speed. However, if the short vehicles are traveling much faster than the long vehicles, the photos will show more long vehicles. In order to compare the temporal (tollgate) data and the spatial (photographic) data, the velocity distribution as well as the number distribution is needed.

Various photographic techniques of obtaining spatial distributions in sprays have been used. All methods except holography suffer from a lack of depth of field and poor definition of very small droplets. Holography gives great promise for more precise measurement; however, better techniques of quantifying the data (droplet counting) are required. One optical instrument currently available commercially for steady-state sprays uses Fraunhofer diffraction. In this method a laser beam passes through the spray and is focused on a detector. The diffraction pattern is measured and, after some numerically difficult analysis, produces the drop size distribution. A less accurate but easier method is to use the data to compute the constants in an assumed distribution function.

A variety of temporal methods have been used to measure steady spray droplet distributions. Freezing of droplets by use of liquid nitrogen with subsequent screening of the solid droplets has been used. Droplets have been collected on slides coated with soot and on absorbing paper. Impingement on charged wires has been attempted, but the data are difficult to interpret. Sprays of wax have been used to measure size by allowing the wax to solidify. Various methods have been used to try to separate out the sizes. In the cascade method the droplets are made to flow around a disk placed perpendicular to the flow path. Small droplets will change course and flow outward radially, while larger droplets travel more nearly along their original line of flight and impact the plate. A measure of the amount of liquid collected at various radial positions gives the distribution.

Droplet size and velocity distributions can be obtained simultaneously by use of an analyzer based on the same concepts as a laser Doppler velocimeter. In this method two laser beams of different wavelengths are intersected to form a small sampling volume. When a droplet passes through the volume, the light scattered forms an interference fringe pattern that appears to move at the Doppler difference frequency. The temporal frequency is a function of the particle velocity. The spatial frequency of the scattered interference fringe pattern is dependent on the particle diameter. Three detectors are typically used; each detector produces a similar Doppler burst signal, but with a phase shift between them. Data using such an analyzer have been reported for dilute sprays where the droplets are probably spherical. The application to thick sprays presents problems which have not yet been properly addressed, and thus the data are limited to the outer fringes of the spray.

Drop size distribution measurements are typically plotted as a histogram such as Figure 9.4, where ΔN_i is the fraction of droplets counted in size interval Δd_i. As Δd_i is made smaller, the histogram takes the form of a differential number distribution $dN_i/d(d)$ versus d. One may also plot the fraction with a size less than any given diameter to give a cumulative number distribution CNF, the fraction of all drops less than size d, and the cumulative volume distribution CVF of all drops less than size d. The histogram is replaced by a continuous smooth curve, and the summation is replaced by integration to give

$$\text{CVF} = \int_0^d d^3 \frac{dN}{d(d)}\, d(d) \tag{9.8}$$

Various average sizes have been defined to characterize the spray in a simple way. One may calculate the average diameter, the size which gives the average surface area, or the size which gives the average volume. The average volume is given by

$$\bar{V} = \sum_i \frac{\pi}{6} d_i^3\, \Delta N_i \tag{9.9}$$

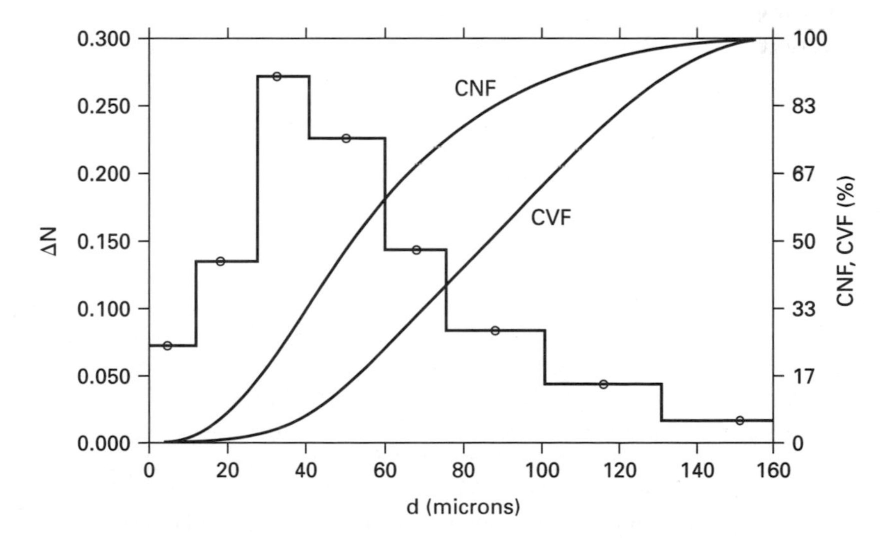

FIGURE 9.4
Droplet size distributions from Example 9.2.

The cumulative volume fraction of all drop sizes less than size d_k is

$$\text{CVF}_k = \frac{\sum_{i=1}^{k} (\Delta N_i d_i^3)}{\sum_{i=1}^{\infty} (\Delta N_i d_i^3)} \tag{9.10}$$

The most probable size is the size with the greatest number of counts. The mean size based on number is

$$\bar{d}_1 = \sum_{i=1}^{\infty} (\Delta N_i) d_i \tag{9.11}$$

The mean size based on surface area (the diameter which gives the average surface area) and that based on volume (the diameter which gives the average volume) are, respectively,

$$\bar{d}_2 = \left[\sum_{i=1}^{\infty} (\Delta N_i) d_i^2 \right]^{1/2} = \text{AMD} \tag{9.12}$$

$$\bar{d}_3 = \left[\sum_{i=1}^{\infty} (\Delta N_i) d_i^3 \right]^{1/3} = \text{VMD} \tag{9.13}$$

A number of spray models use the Sauter mean diameter (SMD), which is

$$\bar{d}_{32} = \frac{\sum_{i=1}^{\infty} (\Delta N_i) d_i^3}{\sum_{i=1}^{\infty} (\Delta N_i) d_i^2} = \text{SMD} \tag{9.14}$$

Figure 9.5 can be used to estimate the cumulative volume fraction of liquid less than a given drop size for a specified SMD. Note that the mass mean diameter (MMD) equals the volume mean diameter (VMD) if all droplets have the same density.

EXAMPLE 9.2. The following drop size data have been obtained experimentally:

Size bin (microns)	Count
0–10	80
11–25	150
26–40	300
41–60	250
61–75	160
76–100	95
100–130	50
131–170	20
171–220	0
0–220	1105

Determine the size distribution, and the cumulative number and volume fraction distributions. Determine the most probable diameter, the median and mean diameters, the volume median and mean diameters, and the Sauter mean diameter.

Solution

Column	1	2	3	4	5	6	7
	$d_i(\mu m)$	ΔN_i	$d_i\,\Delta N_i$	CNF (%)	$d_i^2\,\Delta N_i$	$d_i^3\,\Delta N_i$	CVF (%)
	5	0.072	0.36	0.7	1.8	9	0.003
	18	0.136	2.44	5.8	44.0	791	0.3
	33	0.272	8.96	24.3	295.7	9,757	3.8
	50.5	0.226	11.42	47.8	576.9	29,132	14.4
	68	0.145	9.85	68.0	669.6	45,530	31.0
	88	0.086	7.56	83.6	665.7	58,586	52.2
	115.5	0.045	5.23	94.4	603.6	69,721	77.6
	150.5	0.018	2.74	100	410.0	61,700	100
	195.5	0.000	0	100	0	0	100
Sum	—	1.000	48.5	—	3267	275,226	—

The above table yields:

Most probable diameter	33 μm	(from column 2)
Mean diameter	48 μm	(from sum 3)
Median diameter	52 μm	(from column 4)
Area mean diameter	57 μm	(from sum 5)
Volume mean diameter	65 μm	(from sum 6)
Volume median diameter	85 μm	(from column 7)
Sauter mean diameter	84 μm	(from sum 6/sum 5)

Various equations have been used to describe the drop size distribution functions. A general function often used dates to the early work of Nukiyama and Tanasawa (1939); the parameters a, b, c, and q are empirical constants.

$$\frac{d(\text{VF})}{d(d)} = ad^b \exp(-cd^q) \tag{9.15}$$

where VF is the volume fraction, or if the liquid density is constant, it also represents the mass fraction. Many other functions have been selected, but they are all lacking in any physical basis because the mechanisms of breakup are not understood.

The cumulative volume distribution may be represented by the Rosin-Rammler distribution, which was originally developed for pulverized powders:

$$\text{CVF}_i = 1 - \exp\left[-\left(\frac{d_i}{d_0}\right)^q\right] \tag{9.16}$$

where d_0 is a reference diameter which gives $\text{CVF}_i = 1 - \exp(-1)$, or $\text{CVF}_i =$ 63.2%, and where q may be obtained from the slope of the log-log plot of d_i/d_0 versus $\ln(1 - \text{CVF}_i)^{-1}$. For Example 9.2 it can be deduced from column 7 that $d_0 = 100\ \mu$m, and it can be shown by plotting that $q = 2.8$. A chart for graphically estimating cumulative volume fraction is shown in Figure 9.5. A correlation of data from several kinds of pressure atomizing nozzles was used to create this chart.

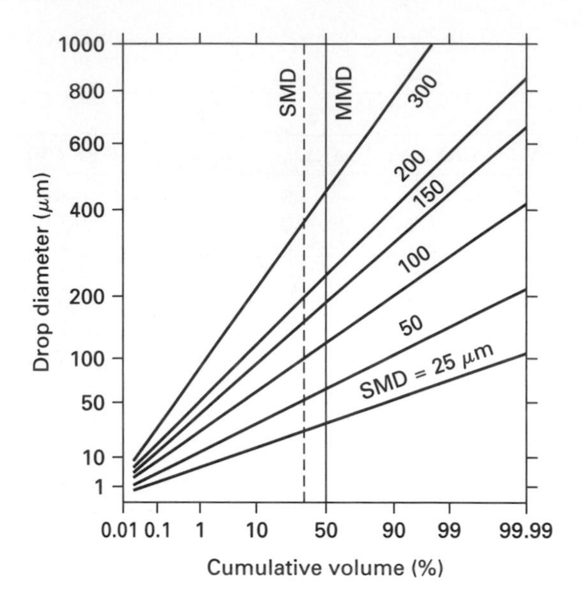

FIGURE 9.5
Chart for estimating cumulative volume fraction below a given droplet size [Simmons, *J. Engr. for Power*, © 1977, by permission of ASME].

9.3 FUEL INJECTORS

Having discussed the general mechanisms of spray breakup and the nature of the droplet distributions, we shall now return to discussion of injectors. Before looking at specific injector designs, a discussion of the general criteria which govern injector design will be given. For simple on-off operation in furnaces and other stationary burners, the injector must be relatively inexpensive and free of maintenance problems. For transportation engines the criteria are more demanding, since good atomization and dispersion are required over a wide range of fuel flows. In the case of gas turbines the combustor exit gas temperature must be uniform so as not to overheat portions of the turbine. In both diesels and turbines, excessive smoke and unburned hydrocarbons can result from poor injector performance. The degree of atomization required depends primarily upon the time available for the vaporization-mixing process. In a diesel engine this time can be as short as a few milliseconds and thus very small droplets are required. In all cases air utilization is important, and thus the injector should provide both penetration and dispersion. Typically, however, these parameters are related so that a high penetration is achieved only by loss of dispersion. For example, the diesel spray penetrates well through very dense air, but has a narrow cone angle and a very dense liquid core.

Steady-Flow Injectors

The simplest injector is the plain orifice with a hole of length L, and diameter d_0. The spray cone angle lies between 5° and 15° and is affected more by the fluid viscosity and surface tension than by d_0 or L/d_0. For low-viscosity liquids, the Sauter mean diameter (SMD) is influenced mostly by the jet velocity, being inversely proportional to the velocity. The use of such simple nozzles is limited because of their narrow cone angle and typically poor dispersion.

A considerable improvement is achieved over the plain orifice by the "simplex" or swirl atomizer shown in Figure 9.6. The fluid is caused to swirl by tangential slots or other similar means. The high velocity causes an air core vortex so that the fluid forms a hollow cone as it emerges from the orifice. The spray angle can be quite large—up to 90°. Friction losses are a consideration in designing swirl atomizers. The ratio of d_s/d_0 (see Figure 9.6) should be about 3.3 to obtain the best discharge coefficient. The L/d_0 of the final orifice should be as small as possible, but practical considerations limit it to the 0.2–0.5 range.

Figure 9.7 shows a measured mass distribution and relative number of drops versus drops diameter obtained from a small pressure-swirl type atomizer typically used in gas turbine combustors. The cross-sectional average SMD for this atomizer is about 45 μm. At 50 mm downstream from the exit the spray is 10 cm in diameter and the SMD varies from 80 μm at the edge to 10 μm at the axis. The volume fraction peaks at about 4 cm from the axis. Data of this type, which include cross-sectional information, are now reliably obtained by use of laser-diffraction and phase-Doppler methods.

The Sauter mean diameter for a pressure-swirl atomizer depends on liquid surface tension and viscosity, the mass flow of the liquid, and the pressure drop across the atomizer. The correlation by Radcliffe (see Lefebvre) is given by

$$\text{SMD} = 7.3\sigma^{0.6}\nu^{0.2}\dot{m}^{0.25}\,\Delta p^{-0.4} \tag{9.17}$$

where SMD is in microns and all other units are SI units.

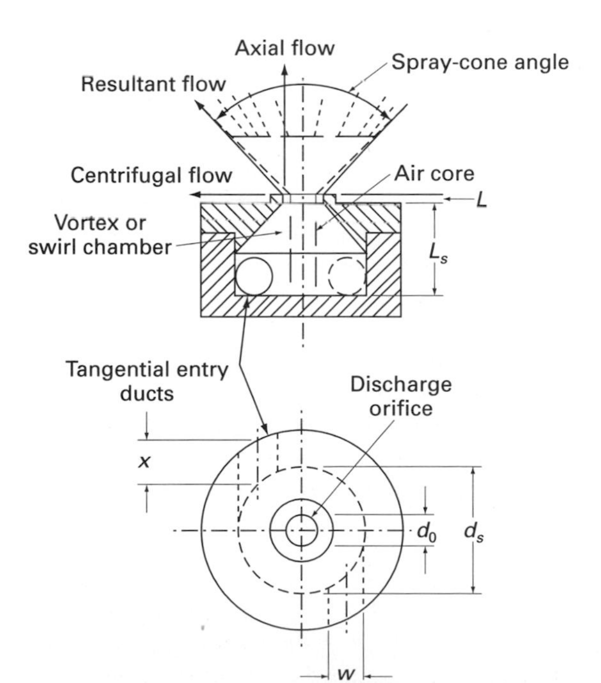

FIGURE 9.6
Simplex pressure-swirl atomizer.

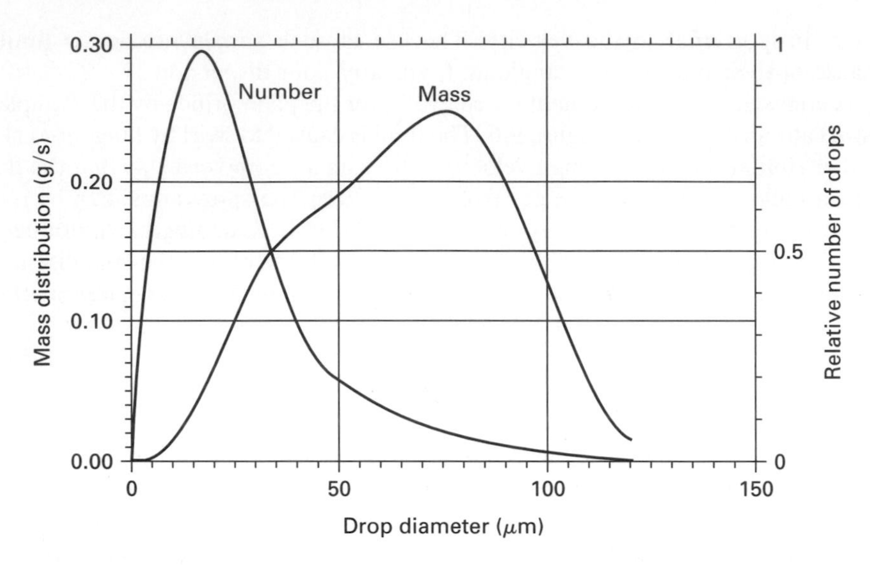

FIGURE 9.7
Measured drop size distribution for pressure-swirl atomizer with 80° cone angle; pressure difference across atomizer was 689 kPa (100 psi); aircraft gas turbine test fuel type II [Derived from data of Dodge and Schwalb, by permission of ASME].

The flow rate of the swirl atomizer is proportional to the square root of the injection pressure differential. For a 20:1 range of flow rates, as might be required in an aircraft jet turbine, the higher flow rate would require a pressure of about 400 atm to be sure that satisfactory operation would take place at the lowest flow rate. Such pressures represent the lower end of diesel injection pressures but are much too high for the large steady flows of a gas turbine or a furnace. To solve the problem of fuel turndown in aircraft and industrial gas turbines and in oil-fired furnaces, various forms of air-blast atomizers have been designed.

The prefilming pintle and prefilming double swirl designs shown in Figures 9.8*a* and *b* are typical of current practice. The additional air creates good mixing and thus reduces soot formation (at the expense of NO_x formation). The SMD of such atomizers increases with increasing liquid viscosity and surface tension and with decreasing air-to-liquid ratio. Increasing the air-to-liquid ratio above 5 has little effect. Liquid density has only a minor effect. The mechanism for air-blast atomization changes as the liquid viscosity is increased. For low viscosity the liquid film undergoes a wave type instability and breakup near the nozzle. For higher viscosity, liquid ligaments form, which break up farther downstream, thus giving larger drop sizes. The following empirical formula was derived from data obtained from the prefilming air-blast atomizer shown in Figure 9.8a [Lefebvre]:

$$\text{SMD} = 3.33 \times 10^{-3}\left(\frac{\sigma \rho_l d_p}{\rho_a^2 V_a^2}\right)^{0.5}\left(1 + \frac{\dot{m}_l}{\dot{m}_a}\right) + 13. \times 10^{-3}\left(\frac{\mu_l^2}{\sigma \rho_l}\right)^{0.425} d_p^{0.575}\left(1 + \frac{\dot{m}_l}{\dot{m}_a}\right)^2 \tag{9.18}$$

where SMD is in meters and all quantities are in SI units, and where d_p is the diameter of the pintle. The value of the pintle diameter was not changed in the experiments.

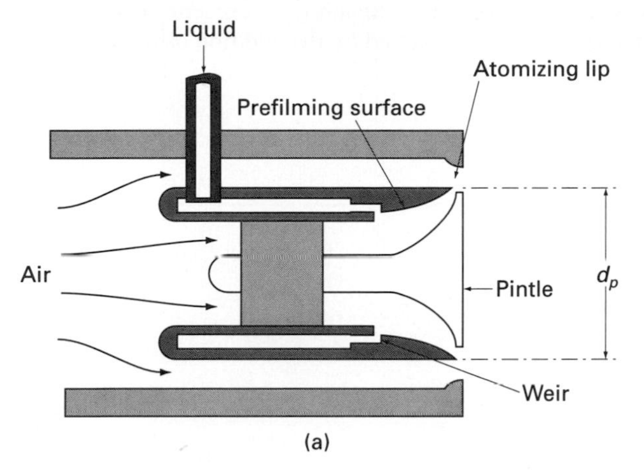

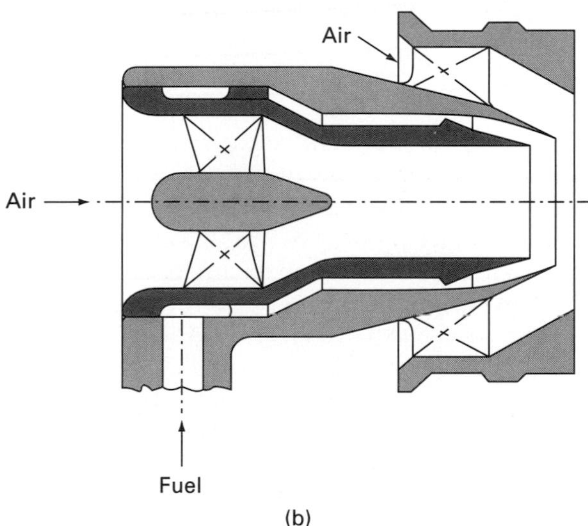

FIGURE 9.8
Prefilming types of air-blast atomizers: (*a*) pintle-type center body with pintle diameter of 36 mm [Rizkalla and Lefebvre, *J. Engr. Power*, © 1975, by permission of ASME]; (*b*) Parker-Hannifin double swirl type atomizer.

EXAMPLE 9.3. Compute and plot the SMD for a kerosene spray from a prefilming air-blast nozzle over a range of air-flow to liquid fuel flow ratios from 1 to 10. The air is at standard temperature and pressure, and let the velocity be a constant value of 50, 75, 100, 125 m/s. The pintle diameter is 36 mm.

Solution. The kerosene properties are:

$$\mu_l = 0.00129 \text{ N}\cdot\text{s/m}^2 \qquad \sigma_l = 0.0275 \text{ N/m} \qquad \rho_l = 784 \text{ kg/m}^3$$

The air density is 1.177 kg/m³. For SMD in μm units, Eq. 9.18 becomes

$$\text{SMD} = A\left(1 + \frac{\dot{m}_l}{\dot{m}_a}\right) + B\left(1 + \frac{\dot{m}_l}{\dot{m}_a}\right)^2$$

where

$$A = 2493/\underline{V}_a \text{ and } B = 0.44.$$

The plot below shows that the atomization quality starts to decline when the air/liquid ratio falls below about 2. When the air/liquid ratio exceeds about 3, only slight improvement in atomization quality is gained by the addition of more air.

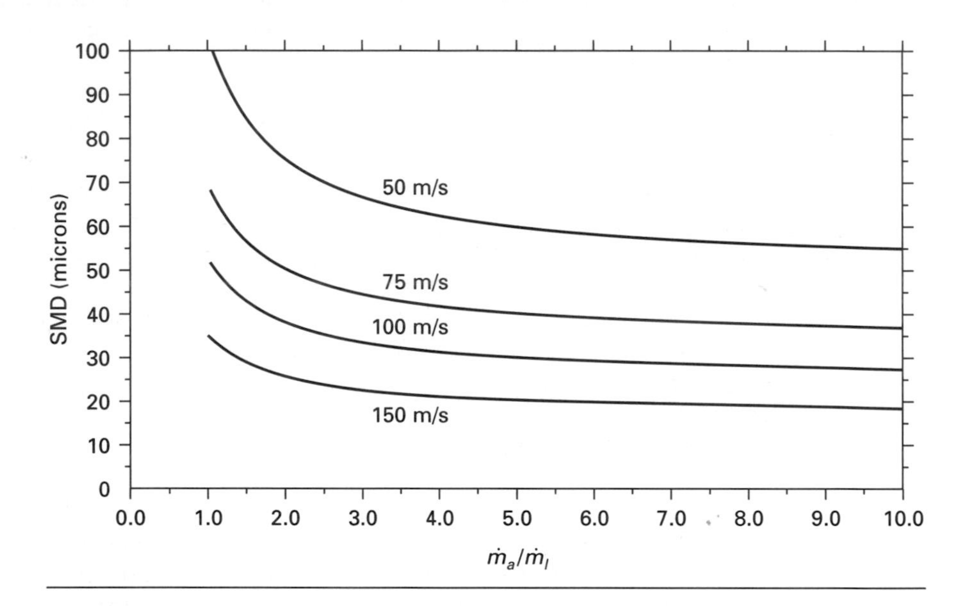

Intermittent Injectors

Injectors for direct injection into the intake port or cylinder of an I.C. engine operate only once per combustion cycle. Although direct injection into the cylinder may be used in future spark-ignition engines (especially in two-stroke engines), its current use is primarily in diesels. The typical diesel injection system uses high pressures (300 to 1400 atm) to cause atomization. The use of air-blast atomization, although used in the original work of Dr. Diesel, is not thought to be economical in today's high-pressure engines. To achieve better atomization and control, the present trend for heavy-duty engines is the use of electronic fuel injection with very high (1400–2000 atm) injection pressures. In this section we will discuss a few common types of diesel injection systems. The reader interested in more detail should see the handbook *Diesel Fuel Injection* [Bosch]. There are basically three types of injector systems for diesels: individual pump systems, distributor systems, and common-rail systems.

With an individual pump system, each cylinder is provided with an individual metering and compression pump. A low-pressure transfer pump is used to bring the fuel from the tank and through several stages of filtration to the sump of the high-pressure pumps. Figure 9.9 shows the pump elements and plunger assemblies and their operation. Fuel enters through A and A′ and is compressed by the plunger until the helix on the plunger uncovers the spill port A′. As the plunger rises, the pressure causes the delivery valve C to rise, and fuel flows to the injector until the pressure is suddenly lowered by the escape of fuel through the spill port. Rotation of the plunger by a rack changes the time of spill port opening and thus the amount of fuel injected. The fuel that leaves the pump travels to an injector, which contains a spring-loaded needle valve. The high pressure lifts the needle, causing the fuel to flow through a number of small holes in the injector tip. Typically, the pump is designed to give a constant beginning of injection crank angle and a variable ending. The number of crank degrees of injection is determined by the load.

The rapid start and stop of the flow in the line connecting the pump and injector causes pressure waves in the line which can cause changes in the start of injection and reopening of the needle after injection has stopped. Such afterinjections can cause serious smoke problems. The General Motors unit injector solves this problem by combining the individual pumps with the injectors. The distributor pump system both pumps and meters the fuel to the set of injectors, thus saving on cost. A single plunger is both continuously rotated and reciprocated.

With the common-rail system, the pump does not meter, but only supplies a constant pressure to a common pipe (rail), which carries the fuel to the injectors. Metering is accomplished by mechanical (or solenoid) opening of the injectors. The tubes leading to each injector must be closely matched to be sure each injector delivers the same amount of fuel. The Cummins PT (pressure-time) system is a type of common-rail system. In this system all metering and distributing take place at the low pressure produced by the gear pump with its pressure regulator. Fuel is metered in the injector, which bypasses about 80% of the fuel (thus cooling the injector). At the proper time the injector plunger compresses and discharges the fuel.

A number of electronic fuel injectors use variations of the unit injector concept, but with accumulators and hydraulic amplifiers. One design recently developed by the Nippondenso Corporation uses a high-pressure common rail pressure. Figure 9.10 shows the basic design. The common-rail pressure is about 100 MPa. When not injecting, fuel flows through the three-way valve to the leak—that is, it recirculates. For injection, the three-way valve first meters fuel to the needle, causing it to lift against the spring and hydraulic piston. The fuel above the piston can flow to the leak. The three-way valve then allows fuel to flow to the top of the hydraulic piston while closing the path to the leak. The pressure is amplified by the piston, causing higher-pressure injection (200 MPa) and positive needle closure. Note that the very first fuel injected is already at a fairly high pressure in this system. Another design recently developed by Caterpillar Inc. uses a separate electronically controlled high-pressure hydraulic circuit to drive the plunger, which causes the injection. Both the Caterpillar and the Nippondenso designs allow multiple injection events to be used.

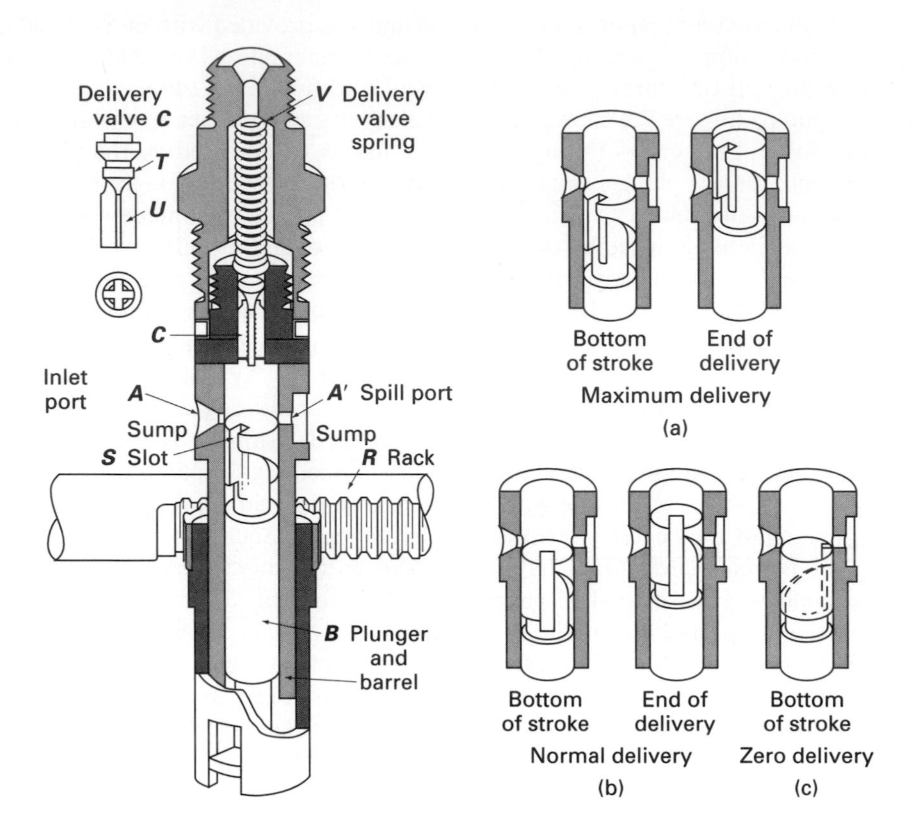

FIGURE 9.9
Diagram of pump elements of the individual pump system. Fuel under a pressure of 40 psi enters a sump, which is connected through ports A and A′ to the plunger-and-barrel assembly B. When the plunger compresses the fuel, the delivery valve C opens and fuel flows through the discharge tubing to the nozzle. The pump plunger is lifted by a cam on a camshaft driven by the engine. When the plunger is at the bottom of its stroke, ports A and A′ are uncovered. Fuel enters the barrel under pressure from the transfer pump. When the plunger rises, the ports are sealed and the compressed fuel lifts the delivery valve C and begins the injection period. Fuel is injected only during the high-velocity portion of the plunger stroke. As the plunger continues to rise, the spill port or bypass port A′ is uncovered by the helical relief on the plunger. At this point, the high-pressure oil above the plunger escapes through slot S and through the port A′ into the sump, while the delivery valve snaps shut, with consequent end of the injection period. The position of the helical groove in relation to the spill port A′ is changed by rotating the plunger with the rack or control rod R. By moving the rack, the quantity of fuel injected can be varied from zero to that demanded at full load. (*a*) Effective travel at full load before the spill port is uncovered by the helix; (*b*) a shorter effective travel, say, half load; (*c*) slot in plunger is in line with the spill port and no compression (or delivery) of fuel is obtained. Part (*c*) is the “stop” position for

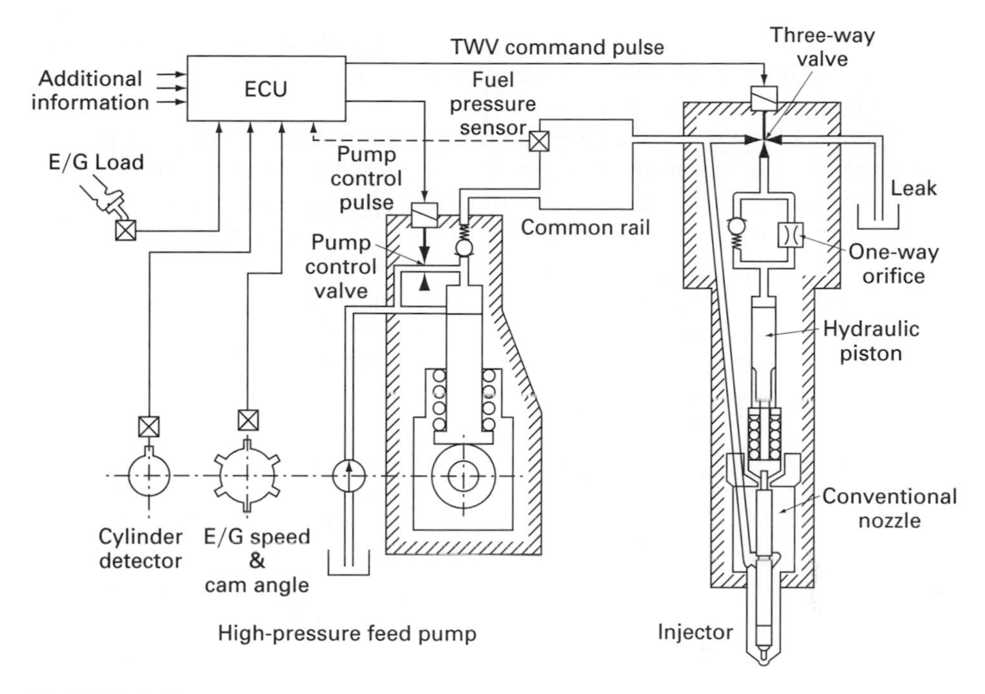

FIGURE 9.10
Schematic diagram of Nippondenso fuel-injection system. [Courtesy of Nippondenso, Inc.].

Nozzles for Intermittent Injectors

Diesel fuel injectors arc typically of the hole type in which the needle is inwardly opening. Figure 9.11 shows such multiple-hole diesel nozzles. The drawing in

---

FIGURE 9.9 (Continued)
shutting down the engine. Note that the overall travel or displacement of the plunger is constant at all speeds and loads but the effective travel is controlled by the helix (and spill port) in proportion to the load (displacement metering). The delivery value C allows a high pressure to be maintained in the delivery tubing and also stops the injection abruptly. Notice that when the pressure rises above the plunger, the delivery valve is forced upward, but flow cannot begin until the relief piston T leaves the passage (with flow passing through the flutes U). When the pressure falls, the delivery valve is closed by the spring V, with the relief piston retreating into the housing. This change in volume drops the pressure quickly in the delivery line, and the spring-loaded nozzle snaps shut. Thus the relief piston helps to obtain a quick cutoff of the nozzle, with less possibility of afterinjections. The valve action prevents the delivery line from being entirely relieved of pressure and so enables the pump on the next injection stroke to quickly increase the pressure in the line to a value sufficient to open the nozzle [Obert, courtesy of E. F. Obert estate].

Figure 9.11*a* shows a sac volume below the needle, which causes dribble at the end of injection. Elimination of this volume as shown in Figure 9.11*b* reduces exhaust smoke greatly.

Figure 9.12 shows standard and throttling pintle nozzles, which are typical of single-hole nozzles. Such single-hole nozzles are used in some designs of direct-injection stratified-charge engines. The throttling pintle tends to prevent weak injection at the start of injection and shuts faster to prevent dribble at the end of injection. Outwardly opening pintle nozzles are quite often used for lower-pressure injection (70 atm). Single-hole nozzles depend upon their wide conical spray to obtain a good fuel distribution. The use of fuel injection either at a single point, or several points, or in each port has become quite common in S.I. engines and has replaced the carburetor. The popularity of manifold or port injection is due to increased use of computers to provide precise fuel control.

Droplet size distribution for intermittent high-pressure single-hole nozzles is hard to obtain and thus considered to be not very accurate. A major problem is that the spray is very thick so that optical techniques are difficult to use. Table 9.1 gives three older correlations for Sauter mean diameter typical of the lower injection pressures used prior to the 1990s in heavy-duty diesels and still used in some smaller diesels. Note that the positive exponent on air density of the last two formulas may be the result of agglomeration effects. These formulas give trends, but should not be applied rigorously to spray modeling.

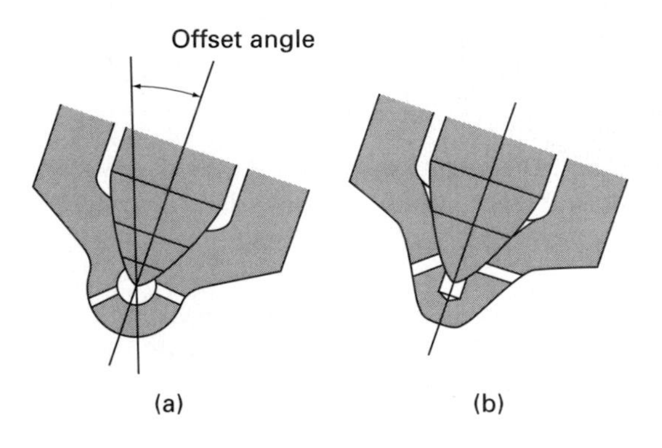

FIGURE 9.11
Multihole nozzle for diesel: (*a*) nozzle with sac volume, (*b*) sacless nozzle.

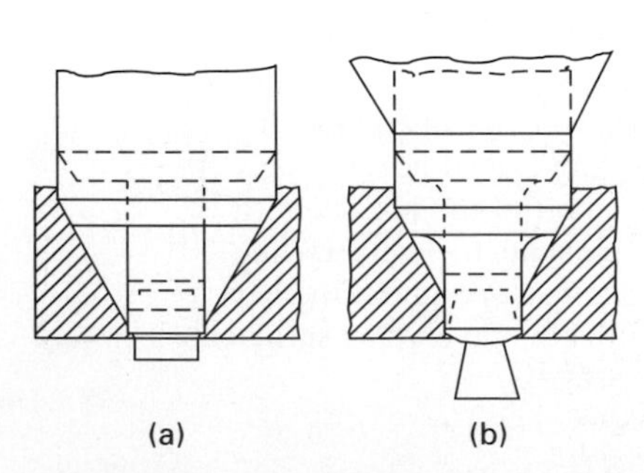

FIGURE 9.12
Pintle nozzle: (*a*) standard, (*b*) throttling.

TABLE 9.1
Equation of Sauter mean diameter for diesel spray

Source	SMD (microns)
Knight [1955]	$1.605 \times 10^6 (\Delta p)^{-0.458}(\dot{m})^{0.209}(\nu)^{0.215}\left(\frac{A_{orf}}{A(t)_{eff}}\right)^{0.916}$
Hiroyasu and Kadota [1974]	$2.33 \times 10^3(\Delta p)^{-0.135}(\rho_a)^{0.121}(V)^{0.131}$
El-Kotb [1982]	$3.08 \times 10^6 \nu^{0.385} \sigma^{0.737} \rho_l^{0.737} \rho_a^{0.06} \Delta p^{-0.54}$

Δp	Pressure difference	(Pa)
ρ_a	Density of air	(kg/m^3)
ρ_l	Density of fuel	(kg/m^3)
μ	Viscosity of fuel	(Pa·s)
ν	Kinematic viscosity of fuel	(m^2/s)
σ	Surface tension	(N/m)
V	Injection fuel volume	(m^3/stroke)
$\dot{m}$	Injection rate	(kg/s)
A_{orf}	Area of nozzle holes	(m^2)
$A(t)_{eff}$	Effective area of nozzle holes at Δp	(m^2)

EXAMPLE 9.4. Using the following data for the single hole of a multihole diesel injector, calculate the SMD from the formulas of Table 9.1:

$$d_j = 0.0003 \text{ m (0.3 mm)}$$

$$\Delta p = 35.5 \text{ MPa (based on a peak injection pressure and an air pressure of 6.2 MPa)}$$

$$\rho_a = 25 \text{ kg/m}^3$$

$$\rho_l = 850 \text{ kg/m}^3$$

$$\nu_l = 2.82 \times 10^{-6} \text{m}^2\text{/s}$$

$$\sigma = 0.03 \text{ N/m}$$

$$V = 10^{-8} \text{ m}^3 \text{ (based on 2 ms duration)}$$

$$\dot{m} = 8.68 \times 10^{-3} \text{ kg/s}$$

$$A_{orf} = 7.1 \times 10^{-8} \text{ m}^2$$

$$A_{eff} = 3.5 \times 10^{-8} \text{ m}^2 \text{ (includes effect of needle motion)}$$

For Knight's formula:

$$\text{SMD} = 1.605 \times 10^6(35.5 \times 10^6)^{-0.458}(8.68 \times 10^{-3})^{0.209}(2.82 \times 10^{-6})^{0.215}(2)^{0.916}$$
$$= 25.7 \ \mu\text{m}$$

For Hiroyasu and Kadota:

$$\text{SMD} = 2.33 \times 10^3(35.5 \times 10^6)^{-0.135}(25)^{0.121}(10^{-8})^{0.131} = 29.4 \ \mu\text{m}$$

For El-Kotb:

$$\text{SMD} = 2.08 \times 10^6(2.82 \times 10^{-6})^{0.385}(3 \times 10^{-2})^{0.737}(850)^{0.737}(25)^{0.06}(35.5 \times 10^6)^{-0.54}$$
$$= 24.8 \ \mu\text{m}$$

Although the three formulas give very similar results for these data, the trends are not the same for all variables. For example, Hiroyasu and Kadota do not show an effect of viscosity, while Knight does not show an effect of air density. Typically it has been found that the increase in SMD with air density is due to increased droplet agglomeration with increased air density. Of course, agglomeration effects increase as the location of the SMD measurement is moved downstream, away from the injector orifice.

More recently (1989), Hiroyasu et al. have given two formulas for SMD based on data obtained from Fraunhofer diffraction measurements applied to an intermittent injector with pressures of 3.5 to 90 MPa. Data for lower injection velocities (where the breakup mechanism may lie close to region II of Figure 9.1) were found to be fit by Eq. 9.19*a*, and for higher velocities (zone III) by Eq. 9.19*b*. The researchers suggest that use of the larger value of SMD, as given by these two equations, will give the correct value.

$$\mathrm{SMD} = 4.12\,\mathrm{Re}_{jl}^{0.12}\,\mathrm{We}_{jl}^{-0.75}\left(\frac{\mu_l}{\mu_a}\right)^{0.54}\left(\frac{\rho_l}{\rho_a}\right)^{0.18} d_j \tag{9.19a}$$

$$\mathrm{SMD} = 0.38\,\mathrm{Re}_{jl}^{0.25}\,\mathrm{We}_{jl}^{-0.32}\left(\frac{\mu_l}{\mu_a}\right)^{0.37}\left(\frac{\rho_l}{\rho_a}\right)^{-0.47} d_j \tag{9.19b}$$

It should be noted that the data showed significant scatter, and that the optical method has questionable accuracy for thick sprays and sprays with nonspherical droplets.

EXAMPLE 9.5. Consider a nozzle hole with an 0.2-mm diameter and a flow rate of diesel fuel of 0.0053125 kg/s. The fuel density is 850 kg/m^3, and the spray is into compressed air at 25 kg/m^3 density and 6.2 MPa pressure. Use Eq. 9.19 to calculate SMD.

Solution. First calculate Re_{jl}, We_{jl}, and Oh:

$$\mathrm{Re}_{jl} = \left[\frac{(850)(0.2 \times 10^{-3})}{24 \times 10^{-3}}\right]\underline{V}_j$$

$$\underline{V}_j = \frac{(0.0053125)}{(850)(\pi d_j^2/4)} = 199 \text{ m/s}$$

$$\mathrm{Re}_{jl} = 14{,}096$$

Estimating $\sigma = 3.0 \times 10^{-2}$ N/m from Appendix Table A.6.6,

$$\mathrm{We}_{jl} = \frac{(850)(199)^2(0.0002)}{(3.0 \times 10^{-2})} = 0.2244 \times 10^6$$

$$\mathrm{Oh} = \frac{[(0.2244 \times 10^6)]^{1/2}}{14{,}096} = 0.0336$$

From the ideal gas law, $T_a = 864$ K and thus $\mu_a = 4.084 \times 10^{-5}$ kg/m · s.

$$\frac{\mu_l}{\mu_a} = \frac{2.4 \times 10^{-3}}{4.084 \times 10^{-5}} = 58.8$$

$$\frac{\rho_l}{\rho_a} = \frac{850}{25} = 34$$

Thus

$$\frac{\text{SMD}}{d_j} = (4.12)(14{,}096)^{0.12}(224{,}400)^{-0.75}(58.8)^{0.54}(34)^{0.18} = 0.0214$$

$$\frac{\text{SMD}}{d_j} = 0.38(14{,}096)^{.25}(224{,}400)^{-0.32}(58.8)^{0.37}(34)^{-0.47} = 0.0691$$

Selecting the larger value as the correct answer,

$$\text{SMD} = (0.0691)(200) = 14\ \mu\text{m}$$

Note that Oh and Re_j when applied to Figure 9.1 show that this spray is in zone III, which agrees with the selection of Eq. 9.19*b*.

9.4 SPRAY DYNAMICS

Practical steady-flow fuel spray flames are of two basic types:

1. Those atomized by high-pressure air (or steam), which have such high momentum that the entrained air is sufficient for combustion
2. Those atomized by high liquid pressure which have relatively low momentum such that the dimensions of the flame are primarily determined by the surrounding airflow

The fuel sprays used in direct-injection reciprocating engines are unsteady, so that their length varies with time. Although some of these sprays, such as those used in direct-injected spark-ignited engines, are of much lower pressure and momentum than the typical high-momentum diesel spray, all of the in-cylinder engine sprays tend to influence the air motion significantly.

Given a spray droplet size and velocity distribution near the spray nozzle, relationships for droplet motion, vaporization, and agglomeration may be modeled. For some sprays such as the air atomization sprays and the very-high-pressure (1400 atm) diesel sprays, the spray momentum is large and the droplet size is very small. For such sprays, the droplets may vaporize quickly and the spray may be approximated as a gas jet. The gas jet problem for a steady jet is well worked out if the jet and air are traveling in the same direction. For a jet in cross flow, only approximate models and empirical formulations are available. Calculations using 3-D numerical codes exist for transient liquid sprays and gas jets, but have not yet been validated.

As a simple example, consider a gas jet in a stagnant surrounding gas. As gas leaves the nozzle, a boundary layer grows along the outer portion of the jet stream. The inner region that is unaffected by the boundary layer gradually diminishes until the boundary layer has filled the entire jet region. After a transition region, the profiles become fully developed. The behavior is very similar to the opposite problem of the entry region in pipe flow. The unaffected core region is about 4 to 5 nozzle diameters long, and the transition region is about 10 diameters long. Figure 9.13 shows the various regions.

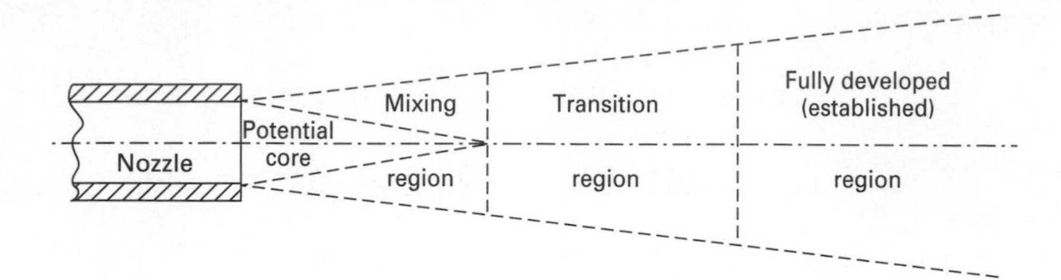

FIGURE 9.13
Regions of steady gas jet.

Figure 9.14 shows the normalized velocity, temperature, and concentration distributions as a function of r/x^*, where r is the radial distance at a given axial distance x^*. The distance x^* is the axial distance from the nozzle exit, x, plus the distance from the nozzle exit to the apparent point source origin of the jet, x_0. Figure 9.15 shows these quantities and constant-velocity lines. The normalizing values are the axial values ($r = 0$) at the particular value of x^*. The profiles are of the form $\exp[-C(r/x^*)^2]$. The constant C is 82–92 for velocity and 54–57 for concentration and temperature.

As the jet interacts with the surrounding air, it exchanges momentum and entrains the air. Figure 9.16 shows the patterns of streamlines for a circular, turbulent jet. The air flows perpendicular to the axis and then turns sharply as it enters the jet region. For an air density ρ_a and jet density ρ_0 from a nozzle orifice diameter, d_0, the ratio of entrained mass flow rate $\dot{m}_e$ to nozzle flow rate $\dot{m}_0$ is

$$\frac{\dot{m}_e}{\dot{m}_0} = 0.32\left(\frac{\rho_a}{\rho_0}\right)^{0.5}\left(\frac{x^*}{d_0}\right) - 1 \tag{9.20}$$

A transient gas jet injected into stationary air becomes fully developed after some distance from the injector. This fully developed region is self-similar, and thus if jet length and gas concentration are plotted in terms of dimensionless distance $x^*/(\rho_i/\rho_a)^{0.5}d_i$ and dimensionless time $t\underline{V}_i/(\rho_i/\rho_a)^{0.5}d_i$, they should collapse to a single set of curves. Here $\underline{V}_i$, ρ_i, and d_i are the jet velocity, density, and

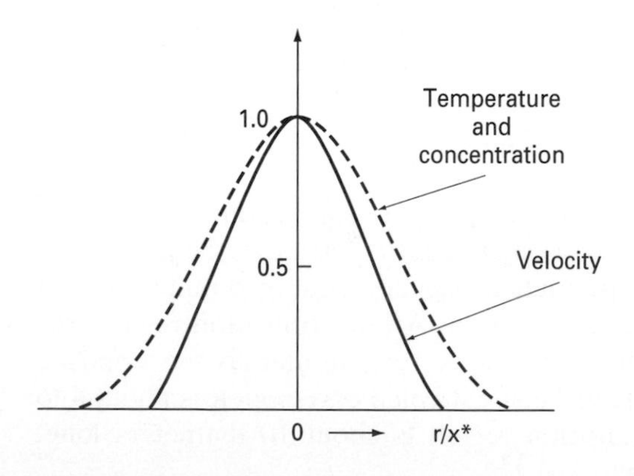

FIGURE 9.14
Normalized profiles of velocity, temperature of difference, and concentration for free jet.

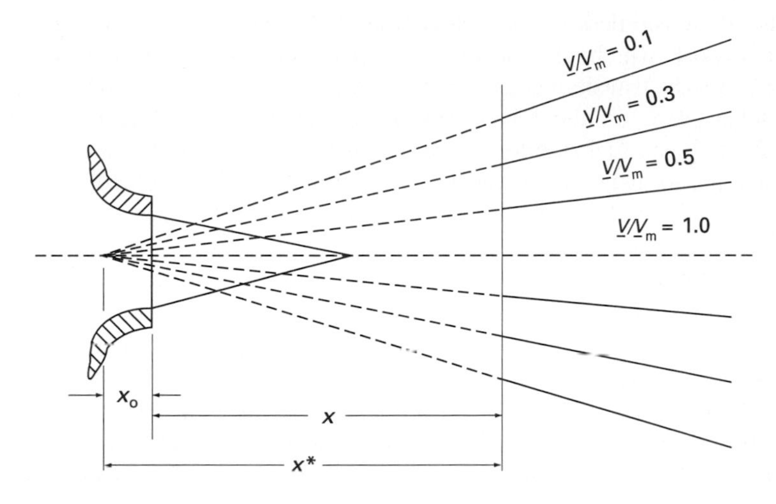

FIGURE 9.15
Lines of constant relative velocity from apparent origin at distance to behind exit.

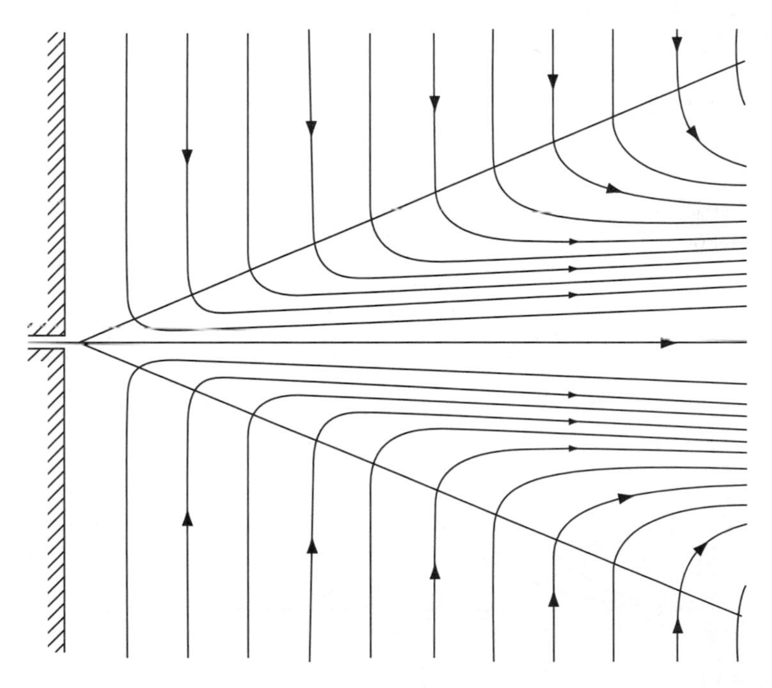

FIGURE 9.16
Streamlines in a circular, turbulent free jet, showing air entrainment.

diameter at the orifice exit. Abraham et al. have used CFD calculations to show this is true for dimensionless times greater than 100. Figure 9.17 shows a schematic from their work which gives some jet fuel concentration lines versus time as obtained from the transient calculations. The maximum in the flammable concentration takes place at about $t = 1000(\rho_i/\rho_a)^{0.5} d_i/\underline{V}_i$. The basic reason for this behavior is that the fuel rate for $t > 0$ is constant, but the entrained air flow rate is proportional to distance x from the orifice for the fully developed region. Such transient jet behavior is important for the design of direct-injected natural gas–fueled reciprocating engines.

For an impulsive liquid spray, the entrainment of air varies with time and position along the spray. Figures 9.18*a* and *b* show the progression of the spray and the pattern of entrainment of gas. From Figure 9.18*a*, note that air is at first entrained into the spray, and then the flow goes outward after a length l_c, as designated on the spray outline shown for $t = 1.05$ ms. Other work has shown that a recirculation pattern exists within the spray at the tip region. The mass of entrained gas per mass of fuel injected, m_a/m_f, up to a given distance x and at a given time can be calculated from

$$1 + \frac{m_a}{m_f} = \alpha_\epsilon \left(\frac{x}{d_0}\right)\left(\frac{\rho_a}{\rho_f}\right)^{1/2} \tag{9.21}$$

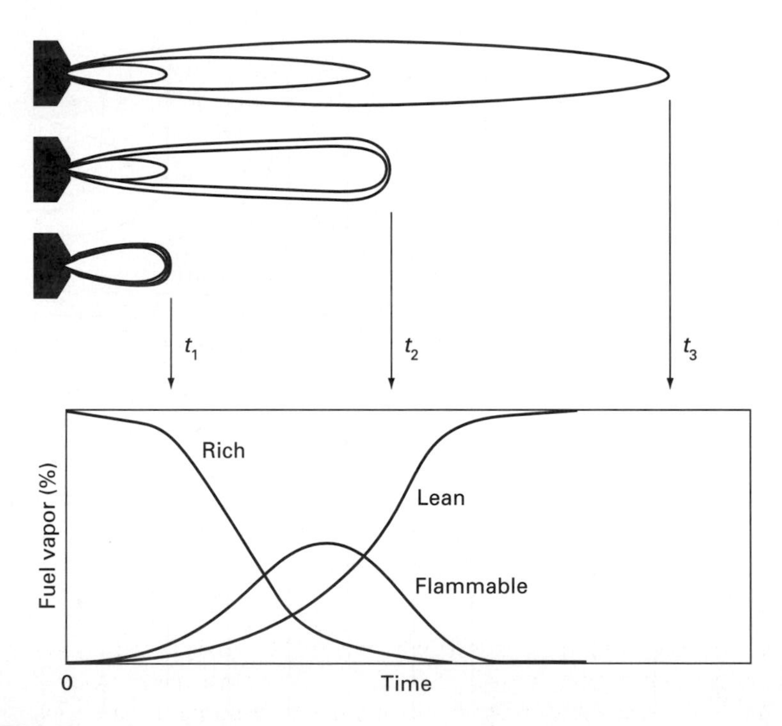

FIGURE 9.17
At top, schematic of the development of a gas jet. At bottom, fuel vapor profiles are shown versus time for each of the three envelope regions shown in the schematic [Abraham et al., reprinted with permission from SAE paper 940895 © 1994, Society of Automotive Engineers, Inc.].

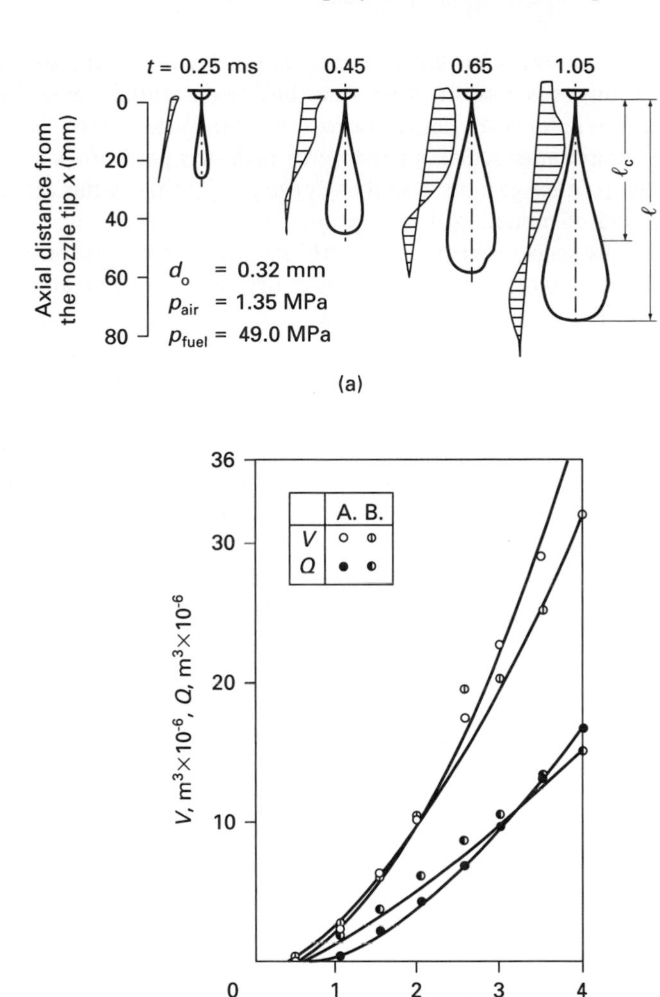

FIGURE 9.18
(*a*) Axial distribution of flow velocity for a liquid jet.
(*b*) Time history of entrainment gas quantity Q and overall volume V. Single-hole (0.32 mm) nozzle with fuel at 22.6 MPa and ambient air at 1.35 MPa [Ha et al., reprinted with permission from SAE paper 841078 © 1984, Society of Automotive Engineers, Inc.].

where α_ϵ is the entrainment coefficient, which is approximately constant after about 1 ms and has a value of 0.25 ± 0.05. The formula is based on data taken for injection of fuel into a pressurized bomb at ambient temperature and thus does not include the effects of vaporization or air motion other than that induced by the spray.

For many practical problems the jet does not flow into stationary air, but rather into a flowing airstream. For example, the flow may be perpendicular to the jet axis. For such cross flows, the jet bends until its axis is parallel to the stream direction. In the cross-flow region, the flow around the jet causes vortices to form behind the jet. The jet cross section is no longer circular, but becomes kidney-shaped. The behavior of burning gas jets in cross flow has been studied experimentally at atmospheric pressure; however, less is known about the behavior of such jets at high pressures and temperatures. Liquid sprays in high-temperature air are found to behave similarly to dense gas jets, and thus dense gas jet theory has often been used as a crude method of modeling liquid sprays.

For thin sprays such as used in gas turbines, the droplet vaporization time is important and can be worked out using single-droplet trajectory and vaporization equations. Typically, the size distribution is divided into a histogram and an initial velocity is assigned to each size category. Drag coefficients are used to determine exchange of momentum between the droplets and air. If the effect of this increase in air momentum may be ignored, the model is much easier to work out, since the air motion can then be determined independently. Note that in addition to momentum exchange, the droplets exchange heat, primarily by taking energy from the air, to provide the latent heat of vaporization needed to vaporize the liquid. The reduction in air temperature due to vaporization would be small if the entire amount of air were utilized. For most hydrocarbons $h_{fg} = 150$ Btu/lb$_m$ and $f = 0.06$, so that 9 Btu would be used to vaporize the fuel in one pound of air. This means that if the entire pound of air supplied this energy, it would experience a temperature drop of only 35°F. Similarly, the vapor pressure for the homogeneous mixture is small and would not play an important role in slowing down the droplet mass transfer process. For most sprays, however, the fuel/air ratio within the spray is much higher than the average value for the combustor. Thus, the effects of local cooling and local fuel vapor pressure are often significant for the spray zone. The complexity of flow patterns and spray-flow interactions in practical combustors has prompted the application of multidimensional modeling techniques to such problems. The initial work on such modeling was divided into two simpler cases: flow without sprays, and sprays with axisymmetric external flows.

The case of flows without sprays is undoubtedly most difficult for the case of reciprocating engines. Even for this case, great strides have been made, and it is now possible to model the flow with reasonable accuracy. Problems still exist, however, in obtaining accurate initial values for the start of the calculations, although it is now possible to model the flow through moving poppet valves. Similarly, existing models for turbulence are crude and require additional work. Despite these problems, much useful information can now be obtained about the effects of combustor geometry on flow patterns and turbulent intensity.

The modeling of sprays in simple flow geometries has also made good progress, but has many unknown and unevaluated factors. Among these factors is the atomization process, which, as previously discussed, is still in an empirical state. Detailed knowledge of the droplet size and velocity is missing, particularly for thick sprays. As was pointed out previously in this chapter, it is very difficult to measure droplet sizes in the region close to the injector. Distribution data taken downstream of the injector have been altered by droplet coalescence and breakup. The process of breakup dominates in thin sprays, but in thicker sprays the two processes are competitive. In such sprays it may be that this competitive process is more important in determining the spray behavior than the initial breakup mechanism.

Droplet vaporization for pure fuels can be reasonably well modeled, as we shall see in the next section. However, models for commercial fuels and for fuels near their thermodynamic critical point are not avaliable. Models for binary mixtures are available, but are not validated.

With so many factors not yet measured, it is very difficult to evaluate the accuracy of current spray models even for simple flows. Nevertheless, modeling has continued into the domain of sprays in three-dimensional flows such as exist in practical combustors. Thus, at the present time, extensive efforts are needed to experimentally evaluate the results of such computations.

The following section presents the basic concepts of droplet dynamics and vaporization, which are the basic building blocks of the spray models. Before going into details, however, a heuristic description of the behavior of thick sprays is given where the droplet and air motions are highly coupled.

Diesel Spray Dynamics

Consider a typical diesel spray. As breakup takes place for the first liquid out of the nozzle, the droplets encounter the undisturbed flow field. For the moment, let us consider stagnant surroundings. The first droplets then quickly give up their momentum to the air and quickly slow down. The next droplets formed at the nozzle now see air set in motion by the preceding droplets and thus do not slow down as quickly. In this way it is possible to build up the entrained airflow and to allow the following droplets to penetrate farther downstream. Droplets formed later pass droplets formed earlier. The unsteady air motion induced by the exchange of momentum with the droplets sweeps fresh air into the spray and transports vapor toward the tip of the spray.

Of course, sprays in real combustors often encounter cross flows. We have already commented on the complexity of this situation for gas jets. For high-pressure liquid sprays, the spray is not deflected very much by the cross flow; however, small droplets can be swept sideways out of the spray. These droplets may vaporize rapidly in the surrounding hot air and thus form a vapor cloud downstream of the spray axis.

In order to quantify the behavior just described, it is necessary to invoke the very complex two-phase, three-dimensional, transient equations of conservation of mass, momentum, and energy including effects of turbulence. However, such computations are still in the research stage, as can be judged from our previous discussion. Thus, a few empirical observations concerning the gross behavior of transient diesel-type sprays are given.

Empirical observations of thick sprays formed by pressure atomization have been carried out for more than 50 years. The basic measurements have been spray cone angle and tip penetration distance. There are several dozen formulas for penetration of diesel sprays into stagnant air. For a short time, t_b, during the early development of the spray, the tip moves linearly with time; after that the spray length is proportional to the square root of time. For the initial linear portion

($t \leq t_b$), the penetration distance L is given [Arai et al.] by:

$$\frac{L}{L_b} = 0.0349\left(\frac{\rho_a}{\rho_l}\right)^{1/2}\left(\frac{t}{d_0}\right)\left(\frac{\Delta p}{\rho_l}\right)^{1/2} \tag{9.22a}$$

where
$$t_b = 28.65\left(\frac{\rho_l}{\rho_a}\right)^{1/2}\left[\frac{d_0}{(\Delta p/\rho_l)^{1/2}}\right]$$

and
$$L_b = 15.8 d_0 \sqrt{\frac{\rho_l}{\rho_a}}$$

For $t \geq t_b$ the penetration distance is proportional to the $\frac{1}{4}$ power of the pressure difference and to the square root of the hole diameter:

$$L = 2.95\left[d_0 t\left(\frac{\Delta p}{\rho_a}\right)^{1/2}\right]^{1/2} \tag{9.22b}$$

It should be noted that these formulas give a discontinuity in slope at t_b, the slope of the linear portion being twice that of the other at t_b.

EXAMPLE 9.6. Calculate the penetration distance and time for a diesel fuel injector with the following data given: six-hole nozzle, each hole 0.2 mm diameter; 75 mm^3 of fuel injected over a duration of 0.002 s; air density 25 kg/m^3; pressure 6.2 MPa; nominal fuel density 850 kg/m^3; density ratio 34. Thus,

$$L_b = 15.8(0.2)(850/25)^{1/2} = 18.43 \text{ mm } (0.72 \text{ in.})$$

Solution. To calculate t_b, first find Δp. The average fuel mass flow rate is $(75)(850)(10^{-9})/0.002 = 0.0319$ kg/s, or $\dot{m}_f = 0.00531$ kg/s per hole. Using a flow coefficient of 0.7 obtained from experiment, the injection velocity is

$$\underline{V}_f = \frac{\dot{m}_f}{(\rho_f)(0.7A)} = \frac{0.0053 \text{ kg/s}}{(850 \text{ kg/m}^3)(0.7)(0.0314 \times 10^{-6} \text{ m}^2)} = 284 \text{ m/s}$$

and the pressure drop is

$$\Delta p = \frac{\rho_f \underline{V}_f^2}{2} = \frac{(850 \text{ kg/m}^3)(284 \text{ m/s})^2}{2} = 34.3 \times 10^6 \text{ Pa} = 34.3 \text{ MPa}$$

This is the average Δp; note the formula is written for a constant Δp. Thus the average injection pressure is 40.5 MPa (6080 psi), and

$$t_b = 28.65(34)^{1/2}\left[\frac{0.0002}{(34.3 \times 10^6/250)^{1/2}}\right]$$
$$= (1.66)(10^{-4}) = 0.166 \text{ ms } (8.3\% \text{ of the duration})$$

From Eq. 9.22*b*, the spray penetration is

$$L = 2.95[(0.0002)(0.002)(34.3 \times 10^6/25)^{1/2}]^{1/2} = 0.0639 \text{ m}$$
$$= 63.9 \text{ mm } (2.61 \text{ in.})$$

Thus the spray tip will be about 64 mm from the nozzle at the end of the injection period.

A correction factor for the effect of a cross-flow velocity is given by Arai et al. The cross flow is taken as a solid-body rotation with the spray traveling radially outward from the center of the rotation. The penetration with cross flow, L_f, is then given by

$$\frac{L_f}{L} = \frac{1}{1 + 2\pi\omega L/\underline{V}_l} \tag{9.23}$$

where ω = rotational speed of air and $\underline{V}_l$ = velocity of the liquid at the orifice.

EXAMPLE 9.7. Consider the previous example, with $L = 64$ mm and $\underline{V}_l = 200$ m/s. For an engine with 1800 rpm speed and a swirl rate of 3, find the penetration of the spray. Note that cavitation can change the density of the discharge fluid and thus the velocity is uncertain; however, this effect is neglected here.

Solution. The value of ω is $(3)(1800)/(60) = 90\ \text{s}^{-1}$. Using Eq. 9.23,

$$\frac{L_f}{L} = \frac{1}{1 + 2\pi(90)(64)/200{,}000} = 0.847$$

and

$$L_f = 54\ \text{mm}$$

Thus the swirl causes the fuel to be distributed in a volume of about two-thirds that of the nonswirl case, but improves fuel-air mixing within that smaller volume.

In small-bore diesel engines and larger-bore diesels during cold start, the spray impinges on the piston so that fuel droplets bounce off the surface or are swept along the piston surface by the airflow in a kind of wall jet layer. Various models for such fuel spreading have been developed, but no validated universal model has been established, although CFD models are currently under development.

The total cone angle θ (expressed in degrees), of the spray is probably a function of the hole geometry, but may be approximated by

$$\theta = 0.05\left(\frac{\Delta p\, d_0^2}{\rho_a \nu_a^2}\right)^{1/4} \tag{9.24}$$

Cone angle taken from measurements is sometimes used to approximately calculate the amount of entrained air in the spray.

EXAMPLE 9.8. Given the data of Example 9.5, calculate the total cone angle θ of the spray. Assume a flow coefficient of 0.7.

Solution. From the ideal gas law, $T_a = 864$ K, and from Appendix B, $\mu_a = 4.084 \times 10^{-5}$ kg/m·s so that $\nu_a = 1.6336 \times 10^{-6}$ m²/s. From Eq. 9.24,

$$\theta = 0.05\left[\frac{(34.3)(10^6)(0.0002)^2}{(0.667)(10^{-10})}\right]^{1/4}$$

$$= (0.05)(3.788)(10^2) = 18.9°$$

Single-Droplet Dynamics

For steady flow with the Reynolds number less than 1, the Stokes drag equation applies. The total force **F** of the fluid on the sphere of diameter d is

$$\mathbf{F} = \frac{\pi d^3 \mathbf{g}\rho}{6} + \pi d\, \mu \underline{\mathbf{V}} + 2\pi d\, \mu \underline{\mathbf{V}} \tag{9.25}$$

The first term is due to the buoyant force and can often be neglected, the second term is due to pressure drag, and the third term is due to viscous drag. The total pressure drag and viscous drag, $\pi d\, \mu \underline{\mathbf{V}}$, may be expressed by use of a drag coefficient C_D:

$$F = \left(\frac{\pi d^2}{4}\right)\left(\frac{\rho V^2}{2}\right) C_D \tag{9.26}$$

where $C_D = 24/\mathrm{Re}_d$ for Stokes flow.

For $\mathrm{Re}_d > 1$, empirical drag coefficients are used. For solid spheres,

$$C_D = \frac{24}{\mathrm{Re}_d}\left(1 + \frac{1}{6}\mathrm{Re}_d^{2/3}\right) \qquad \mathrm{Re}_d < 1000 \tag{9.27a}$$

$$C_D = 0.424 \qquad \mathrm{Re}_d > 1000 \tag{9.27b}$$

The deviation from Stokes's law can be quite sizable. For $\mathrm{Re}_d = 80$, $C_D = 1.23$, while Stokes's law gives 0.30.

Factors which can affect the drag for droplets include acceleration, internal circulation, vaporization, burning, nonspherical shape, and vibration. Various attempts have been made to correct for these effects. Ingebo directly measured the drag from photography of a spray. These results have been adjusted [Bartok and Sarofim] to give

$$C_D = 27\ \mathrm{Re}_d^{-0.84} \qquad \text{for} \qquad \mathrm{Re}_d < 80 \tag{9.28a}$$

and

$$C_D = 0.271\ \mathrm{Re}_d^{-0.217} \qquad \text{for} \qquad \mathrm{Re}_d > 80 \tag{9.28b}$$

This gives $C_D = 0.68$ and 0.70, respectively, for $\mathrm{Re}_d = 80$. Equations 9.28*a* and *b* give a 56% lower value than Eq. 9.27*a* at a Reynolds number of 80.

From work on rocket engines in the 1960s, it is known that at high Reynolds number the droplets distort and their drag coefficient approaches that of a disk. This has again been verified more recently by Liu et al., who have suggested use of the TAB model (see Eq. 9.6*a*) to make the transition from a sphere (Eq. 9.27) to a disk drag coefficient using the equation

$$C_D = C_{D,\mathrm{sphere}}\left(1 + \frac{2.632\, x}{a}\right) \tag{9.29}$$

The TAB model gives values of x/a, and for $x/a < 1$ there is no breakup and Eq. 9.29 is used directly.

For Stokes's law the force vector is $\mathbf{F} = 3\pi d\, \mu \underline{\mathbf{V}}$, so that $F_i = 3\pi d\, \mu \underline{V}_i$ with $i = 1, 2, 3$ gives the vector components of the force, where $\underline{V}_i$ = relative velocity component in direction i. Thus, the air velocity component in the i direction ($\underline{V}_{ai}$) influences only the i component of the drag force. For non-Stokes cases,

$$\mathbf{F} = \left(\frac{\pi d^2}{4}\right)\left(\frac{\rho \underline{\mathbf{V}}^2}{2}\right)(C_D)\left(\frac{\underline{\mathbf{V}}}{|\underline{\mathbf{V}}|}\right) \tag{9.30}$$

where

$$\underline{\mathbf{V}}^2 = \sum_{i=1}^{3} (\underline{V}_a - \underline{V}_d)_i^2$$

Since Eqs. 9.28a and b are of the form $C_D = C[\underline{V}d/\nu]^n$, it follows that $\mathbf{F}$ is proportional to $|\underline{\mathbf{V}}|^{1+n}\underline{\mathbf{V}}$. This means that for $n \neq -1$, F_i depends on all three components of the relatively velocity. In physical terms, an air velocity perpendicular to the direction of droplet travel can change the drag force in the direction of travel.

9.5 VAPORIZATION OF SINGLE DROPLETS

Calculation of spray behavior requires calculation of droplet vaporization. The vaporization rate of the spray can, in principle, be calculated by following the history of each droplet in the spray. Calculation of the air motion and composition would, of course, also be necessary in order to provide the boundary conditions for the droplet calculations. The simplest case is that of a very dilute spray such that each drop is in an abundant supply of air. By further assuming that the ambient air has constant temperature, pressure, and composition, we begin to approach the conditions most often studied experimentally. In the laboratory many studies have used suspended stationary droplets in free convection or flowing air, or have simulated droplets by wetted spheres. Experiments have also been conducted with falling droplets and droplets at microgravity (by using a free-falling bomb containing the gas and droplet).

The theoretical problem of droplet vaporization is one of a sphere with a boundary layer in which vapor is diffusing outward from the surface. The simplest theoretical problem is the case of a wetted solid sphere surrounded by an infinite supply of hot gas (air) at conditions of zero gravity and no bulk gas flow (air stationary with respect to droplet). This idealized situation gives a spherically symmetric boundary layer. If steady state is assumed, then the liquid surface temperature is such that the heat transfer to it is just equal to the energy needed to vaporize the liquid. The energy conservation equation gives the liquid temperature, and the conservation of mass and diffusion flux equations give the rate of vaporization. The problem is one of combined mass and heat transfer.

In practical situations a number of complicating situations arise. First, the effects of free and forced convection are important. Small droplets may be moving with almost zero velocity relative to the flow velocity of the air, but they will be influenced by the turbulent eddies which are typically 2000–3000 μm in size compared with the 20- to 200-μm droplets. In general, the droplet surface is moving because of vaporization, and for rapid vaporization this effect is significant. Second, the assumption of steady state is not realistic over a large portion of the droplet lifetime. For unsteady state, some of the energy reaching the surface goes into heating the droplet liquid, and heat transfer within the droplet is important. Third, the effects of high pressure (such as in diesel engines) cause changes in the properties and may cause the droplet to approach a thermodynamic critical state where the latent heat goes to zero. Fourth, for the practical case of high ambient temperature the properties in the boundary layer are functions of temperature and composition, and at high pressures they are nonideal.

For unsteady vaporization, the droplet surface boundary condition for the gas phase depends on the heat transfer in the liquid. For a solid sphere of diameter d and thermal conductivity k_s, suddenly heated by convection from a gas with a convection coefficient $\tilde{h}$, if the *Biot number* is less than 0.1, or

$$\mathrm{Bi} = \frac{\tilde{h}d}{k_s} \leq 0.1 \tag{9.31}$$

then the temperature gradients in the sphere are negligible and the temperature is a function only of time. For larger Biot numbers, finite rates of conduction in the sphere must be considered, and for droplets in flowing air the situation is complicated by the effects of internal mixing, which tends to augment the rate of heat transfer. The effects of mixing are to effectively increase k_s. The mixing is caused by the drag at the surface, which pushes surface liquid to the aft end of the droplet. The liquid then circulates back through the droplet, causing a double vortex (Figure 9.19). For very small droplets, the effect of mixing is less, and the gas side convection coefficient may be approximated by conduction. Thus, for small droplets, internal temperature gradients may be important in some cases. In the extreme case of very fast vaporization and small internal mixing effects, the heat conduction to the liquid may only penetrate a very short distance, so that the center core stays at a nearly constant temperature. This extreme case gives rise to the so-called onion skin model, in which the center core temperature stays low and heated liquid is removed as a thin shell.

In addition to the effect on the energy balance, heat transfer to the liquid also affects the behavior of liquid mixtures. In one extreme, the liquid mixture will be distilled as it vaporizes, so that less volatile components are left behind in the liquid state as the vaporization goes on. In the other extreme, the mixture vaporizes uniformly and the composition of the remaining liquid remains constant.

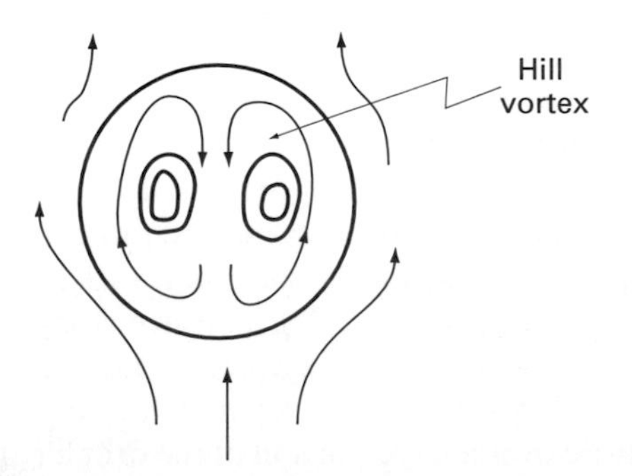

FIGURE 9.19
Schematic of droplet internal motion.

Modeling of the vaporizing fuel drop has been investigated by many; see Faeth's review. The formulation chosen here treats vaporization, including the effects of convective mass and heat transfer, by means of boundary layer film coefficients. First let us consider a model for unsteady droplet vaporization and then consider the steady-state solution.

Unsteady Vaporization

Consider an energy balance on a single droplet assuming a uniform liquid temperature which changes with time. The time rate of energy storage in the drop equals the heat flux to the drop minus the enthalpy carried away by the vapor:

$$\frac{dm_l h_l}{dt} = \tilde{h} A_0 (T_\infty - T_l) + \dot{m}_l h_v \tag{9.32}$$

Expanding,

$$m_l c_l \frac{dT_l}{dt} = \tilde{h} A_0 \,(T_\infty - T_l) + \dot{m}_l (h_v - h_l) \tag{9.33}$$

where T_∞ = ambient gas temperature and $h_v - h_l = h_{fg}$. It is assumed here that the ambient pressure is low enough so that none of the ambient gas is absorbed in the liquid. For a sphere,

$$A_0 = \pi d^2 \tag{9.34}$$

and

$$m_l = \frac{\rho_l \pi d^3}{6} \tag{9.35}$$

Differentiating Eq. 9.35 gives

$$\dot{m}_l = \frac{dm_l}{dt} = \left(\frac{\rho_l \pi d^2}{2}\right)\frac{d(d)}{dt} + \left(\frac{\pi d^3}{6}\right)\frac{d\rho_l}{dT_l}\frac{dT_l}{dt} \tag{9.36}$$

The second term in Eq. 9.36 represents the effect of thermal expansion as the droplet heats up. For typical conditions, $\dot{m}_l < 0$, $dT_l/dt > 0$, and $d\rho_l/dT < 0$. Thus, the droplet diameter can increase with time if the thermal expansion due to heating overcomes the decrease due to vaporization.

The convective heat transfer coefficient $\tilde{h}$ is obtained from a *Nusselt number* correlation. Ranz and Marshall found that for low rates of vaporization,

$$Nu = \frac{\tilde{h} d}{k_f} = 2 + 0.6\,\mathrm{Re}_d^{1/2}\,\mathrm{Pr}^{1/3} \tag{9.37}$$

For high rates of vaporization, $\tilde{h}$ must be corrected both for the effect of superheating of the vapor as it moves away from the surface and for the blowing effect of the vapor motion on the boundary layer. The value of $\tilde{h}$ corrected for the superheating effect is [Bird, Stewart, and Lightfoot]:

$$\tilde{h}^* = \tilde{h} Z \tag{9.38}$$

where

$$Z = \frac{z}{e^z - 1}$$

$$z = \frac{-\dot{m}_l c_{p_v}}{\tilde{h} A_0}$$

The value of Z is a function of $\dot{m}_l$, and thus an equation for the mass transfer rate is needed. It can be seen, however, that for small vaporization rates, Z will be about unity, while for high rates of vaporization Z will be less than unity and $\tilde{h}^* < \tilde{h}$. This is intuitively correct, since the streaming vapor leaves the surface and must be heated in the boundary layer until it reaches the ambient temperature (T_∞) at the edge of the boundary layer. The energy absorbed by this superheating subtracts from the energy which can reach the liquid surface.

The transfer of vapor mass from the liquid surface is influenced by convection and by the molecular driving force of the concentration gradient (see Appendix *E*). The vapor concentration at the liquid surface is determined by the vapor pressure of the liquid for the ideal low-pressure case. The difference in vapor partial pressure between the surface and the ambient ($p_{v_0} - p_{v_\infty}$) is then a driving force for mass transfer similar to $T_l - T_\infty$ for the heat transfer:

$$\dot{m}_l = \frac{-\tilde{h}_D A_0 (p_{v_0} - p_{v_\infty})}{R_v T_m} \tag{9.39}$$

The mass transfer film coefficient, $\tilde{h}_D$, is obtained by analogy with Eq. 9.37 using the *Sherwood number* (Nusselt number for mass transfer):

$$\text{Sh} = \frac{\tilde{h}_D d}{D_{AB}} = 2 + 0.6\,\text{Re}_d^{1/2}\,\text{Sc}^{1/3} \tag{9.40}$$

where $\text{Sc} = \mu/\rho D_{AB}$ = *Schmidt number*
$\tilde{h}_D$ = mass transfer coefficient
D_{AB} = binary diffusion coefficient for the vapor into the ambient gas

Equation 9.40 assumes low vaporization rates. For such low rates the effect of the bulk flow rate of the vapor is not important. For higher rates of mass transfer we must express the mass transfer in a different way. The corrected mass transfer coefficient, $\tilde{h}_D^*$, is given [Bird, Stewart, and Lightfoot] by:

$$\frac{\tilde{h}_D^*}{\tilde{h}_D} = \ln\left(\frac{p - p_{v_\infty}}{p - p_{v_0}}\right) \Big/ (p_{v_0} - p_{v_\infty})/p \tag{9.41}$$

Substituting Eqs. 9.41 and 9.39 into Eq. 9.38, the vapor flow rate $\dot{m}_v$ becomes

$$\dot{m}_v = -\dot{m}_l = \left(\frac{\pi d D_{AB} p}{R_v T_m}\right) \ln\left(\frac{p - p_{v_\infty}}{p - p_{v_0}}\right)(2 + 0.6\,\text{Re}_d^{1/2}\,\text{Sc}^{1/3}) \tag{9.42}$$

Note that if $p_{v_0} \to p$ (boiling point), then $\dot{m}_v \to \infty$. Thus, boiling by convective heat transfer is impossible except as a limiting case. In practice, the mean properties in the film are approximated by the film temperature, T_m:

$$T_m = \frac{T_\infty + T_l}{2} \tag{9.43}$$

The properties D_{AB}, μ_a, μ_v, k_a, k_v, c_{p_a}, and c_{p_v} are evaluated at T_m. The properties p_{v_0}, ρ_l, c_{p_l}, and σ are evaluated at T_l. Because the film has both concentration and temperature gradients, property equations contain the approximation that the average concentration at the average temperature gives the correct average film property. Using a linear concentration assuming $p_{v_\infty} = 0$, average values in terms of $x_v = p_{v_0}/p$ are given by

$$\overline{M} = \left(1 - \frac{x_v}{2}\right)M_a + \frac{x_v}{2}M_v \tag{9.44a}$$

$$\rho_m = \frac{p\overline{M}}{RT_m} \tag{9.44b}$$

$$k_m = \left(1 - \frac{x_v}{2}\right)k_a + \frac{x_v}{2}k_v \tag{9.44c}$$

$$\mu_m = \left(1 - \frac{x_v}{2}\right)\mu_a + \frac{x_v}{2}\mu_v \tag{9.44d}$$

$$c_{p_m} = \left(1 - \frac{x_v}{2}\right)\frac{M_a}{\overline{M}}c_{p_a} + \frac{x_v M_v}{2\overline{M}}c_{p_v} \tag{9.44e}$$

Equations 9.33 and 9.36 for dT_l/dt and dm_l/dt are coupled and nonlinear, but may be solved numerically. Figure 9.20 shows a set of results from such calculations for seven different single component fuels.

Steady-State Vaporization

For steady-state droplet temperature and pure air surroundings, Eq. 9.32 becomes

$$\tilde{h}^* A_0(T_\infty - T_l) = -\dot{m}_l h_{fg} \tag{9.45}$$

Substitution for $\tilde{h}^*$ gives

$$T_l = T_\infty - \left(\frac{h_{fg}}{c_{p_v}}\right)(e^z - 1) = T_s$$

where

$$z = \frac{c_{p_v} p D_{AB} \mathrm{Sh/Nu}}{R_v T_m k_m}\left[\ln\left(\frac{1}{1 - x_v}\right)\right] \tag{9.46}$$

The ratio of Sherwood to Nusselt numbers is unity for zero relative velocity or for Sc = Pr. The Reynolds number, and thus the velocity, will approximately cancel out for the case of larger Reynolds numbers. Thus, T_s is essentially independent of velocity. Because D_{AB} is inversely proportional to gas pressure, the product pD_{AB} is a function only of temperature. Generally, however, increasing pressure increases

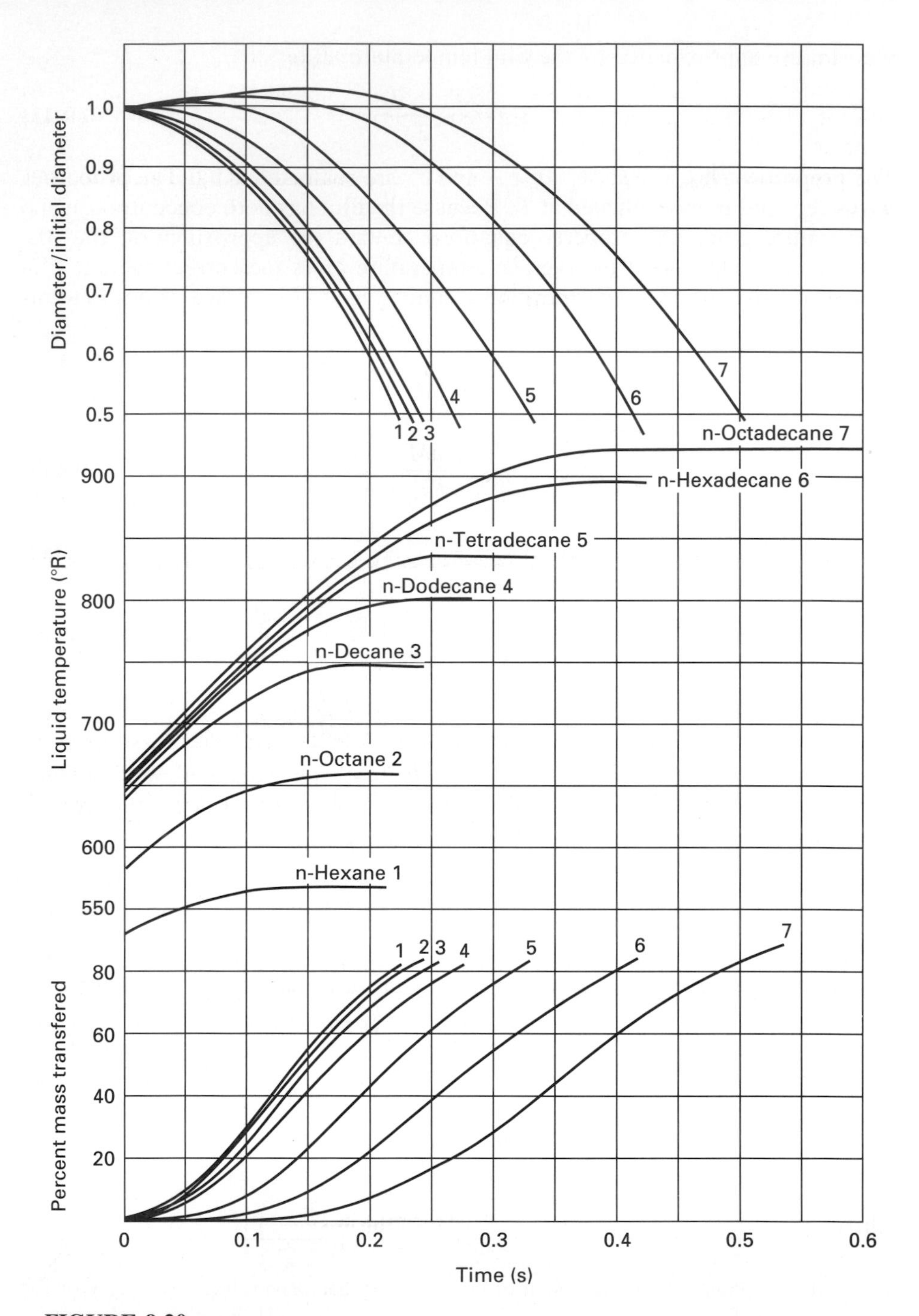

FIGURE 9.20
Radius, temperature, and percent mass transfered calculated for droplets in air at a pressure of 1 atm and temperature of 1360°R; air velocity 8.5 ft/s, initial drop diameter 500 μm [Priem et al.].

the steady-state temperature, because the mole fraction of the vapor, x_v, is decreased by an increase in total pressure. The effect of droplet radius is exactly like that of velocity since the radius only enters into the Reynolds number and thus cancels out for Sc = Pr.

For steady state, the mass transfer Eq. 9.42 with Re = 0 and with $p_{v_\infty} = 0$ gives

$$\dot{m}_l = -\pi \rho_l \beta d \tag{9.47}$$

where the vaporization constant β is

$$\beta = \frac{2pD_{AB}\ln[1/(1 - x_v)]}{R_v T_m \rho_l} \tag{9.48}$$

Equating Eqs. 9.36 and 9.47,

$$\frac{d(d^2)}{dt} = -4\beta \tag{9.49}$$

Upon integration, the vaporization time for a droplet of initial diameter d_0 is given by

$$t_v = \frac{(d_0)^2}{4\beta} \tag{9.50}$$

EXAMPLE 9.9. Calculate the steady-state droplet temperature and droplet vaporization time, t_v, for a 0.004-in. (101.6-μm) dodecane droplet in pure air at a pressure of 14.7 psia and an air temperature of 1260°R. The relative velocity is constant at 8.5 ft/s.

Solution. The basic equation to be solved from Eq. 9.46 is

$$T_\infty - T_l - \left(\frac{h_{fg}}{c_{p_v}}\right)(e^z - 1) = \text{residual}$$

where residual = 0 for the solution. To solve, we plot residual versus T_l. When residual crosses zero, we use the method of secants to improve the solution. Looking at Figure 9.20, we see the solution should be below 800°R. We thus start with T_l = 780, T_l = 790, and T_l = 800°R. Using Table A.8 of Appendix A, calculate the properties at T_l and T_m for the fuel, and using Appendix B, obtain the air properties. The residual for T_l = 780°R is 161 and for 790°R is 43. We thus guess T_l =800° R. The values and calculations for 780°R, 790°R, and 800°R follow.

		Properties at T_l		
	Units	**At 780°R**	**At 790°R**	**At 800°R**
ρ_l	$lb_m/in.^3$	0.0231	0.02295	0.02277
h_{fg}	Btu/lb_m	122.12	120.62	119.08
c_{pl}	Btu/lbm · °R	0.654	0.659	0.6646
p_v	$lb_f/in.^2$	3.085	3.68	4.368
x_v	—	0.21	0.25	0.297

		Properties at T_m		
		At 1020°R	**At 1025°R**	**At 1030°R**
c_{pv}	Btu/lbm · °R	0.654	0.656	0.659
k_v	Btu/in. · s · °R	4.453×10^{-7}	4.483×10^{-7}	4.512×10^{-7}
μ_v	lb_m/in. · s	4.430×10^{-7}	4.450×10^{-7}	4.471×10^{-7}
D_{AB}	in.^{2}/s	0.02576	0.0260	0.02622
μ_a	lb_m/in. · s	1.64×10^{-6}	1.64×10^{-6}	1.642×10^{-6}
k_a	Btu/in. · s · °R	5.84×10^{-7}	5.84×10^{-7}	5.87×10^{-7}
c_{pa}	Btu/lb_m · °R	0.250	0.250	0.2495
			Mean properties	
$\overline{M}$	lb_m/lbmol	43.8	46.63	49.95
ρ_m	lb_m/in.^{3}	3.405×10^{-5}	3.61×10^{-5}	3.845×10^{-5}
k_m	Btu/in. · s · °R	5.694×10^{-7}	5.668×10^{-7}	5.673×10^{-7}
μ_m	lb_m/in. · s	1.511×10^{-6}	1.511×10^{-6}	1.464×10^{-6}
ν_m	in.^{2}/s	0.0444	0.0410	0.0381
c_{pm}	Btu/lb_m · °R	0.415	0.435	0.4567
			Dimensionless groups	
Re_m		9.189	9.95	10.712
Pr_m		1.10	1.16	1.179
Sc_m		1.724	1.577	1.4527
Nu		3.88	3.99	4.074
Sh		4.18	4.203	4.224
Sh/Nu		1.08	1.053	1.0367
z		0.997	1.20	1.458

$$\begin{aligned}\text{Residual} &= 1260 - 800 - 596 = -136 \text{ at } 800°\text{R} \\ &= 1260 - 790 - 424 = +43 \text{ at } 790°\text{R} \\ &= 1260 - 780 - 319 = +161 \text{ at } 780°\text{R}\end{aligned}$$

Using values at 790 and 800°R, we find the steady-state droplet temperature to be

$$T_l = 790 + 10(43/179) = 792.4°R$$

Take $\sigma = 1.7 \times 10^{-4}$ lb_f/in. and compute We_d at 800°R:

$$We_d = \frac{3.845 \times 10^{-5}(102)^2(0.004)}{1.7 \times 10^{-4}} = 9.4$$

Since $We_d < 12$, the drop is stable.

Note that a 13% increase in $\overline{V}$ would give $We_d = 12$, but such an increase in velocity would have essentially no effect on T_l because $\overline{V}$ appears in Sh/Nu. Using Eq. 9.48, $\beta = 8.19 \times 10^{-5}$ in./s, and the droplet vaporization time is:

$$t_v = \frac{(0.004)^2}{4\beta} = 0.049 \text{ s}$$

Some Observations on Droplet Vaporization

With the above background in droplet vaporization film theory, let us summarize some of the observations and calculations for single drops:

1. The radius of the drop may increase at first (see Figure 9.20) because the expansion due to heating is greater than the mass loss due to evaporation. This is less true for volatile fuels and may not be true at all for very volatile fuels.
2. A significant portion of the lifetime of the drop is spent in reaching the steady-state temperature. Significant vaporization occurs during this heat-up period. Thus, under most conditions typical in combustion applications, the unsteady-state portion cannot be ignored.
3. The steady-state temperature increases as the air temperature increases (Figure 9.21). The steady-state temperature is higher for the lower-volatility fuels, and for all fuels asymptotically reaches the temperature where the vapor pressure equals the total pressure.
4. In heated air a fuel that is considered very volatile at room temperature will not show markedly smaller vaporization times than low-volatility fuels. For example, Figure 9.22 shows that the lifetime of the 610°R (approximately) boiling temperature fuel, hexane, was not much shorter than that of *n*-octane (720°R), and *n*-octane had only about half of the droplet lifetime of a cetane (1060°R) droplet. At room temperature a drop of cetane would remain for a very long period of time, while a drop of hexane would disappear in a short time.
5. For pure distillate fuel droplets which have reached a steady-state temperature, a plot of drop diameter squared versus time is approximately linear. This is the so-called *d*-squared law for steady-state droplet vaporization (see Eq. 9.49).
6. Fuels are often complex mixtures, and adequate models for their vaporization are not available. The outer surface of the droplet is quickly depleted of its more volatile components. The question is, then, how fast new volatile components are brought to the surface layer by liquid-phase diffusion, which is slow, and by convection due to internal circulation. Experiments indicate that the distillation is limited by the internal transport processes and reaches a steady-state balance with a limited rate of distillation.
7. Laboratory experiments with freely moving streams of burning single drops have shown that the droplets of certain fuel mixtures and emulsions may suddenly break apart into many small fragments. Such disruptive burning may be caused by the center core of the droplet suddenly vaporizing. This can happen if the outer layer has been depleted of its high-volatility components while the center core still contains such high-volatility components.

The surface of a droplet in a high pressure and temperature environment absorbs ambient gas. The result is significant changes in properties which must be accounted for in models. Figure 9.23 shows the effect of such conditions on the vaporization constant β. These data [Kadota and Hiroyasu] were taken for large initial droplet sizes (1800 μm) suspended in a stagnant atmosphere. The data indicate the large effect of pressure.

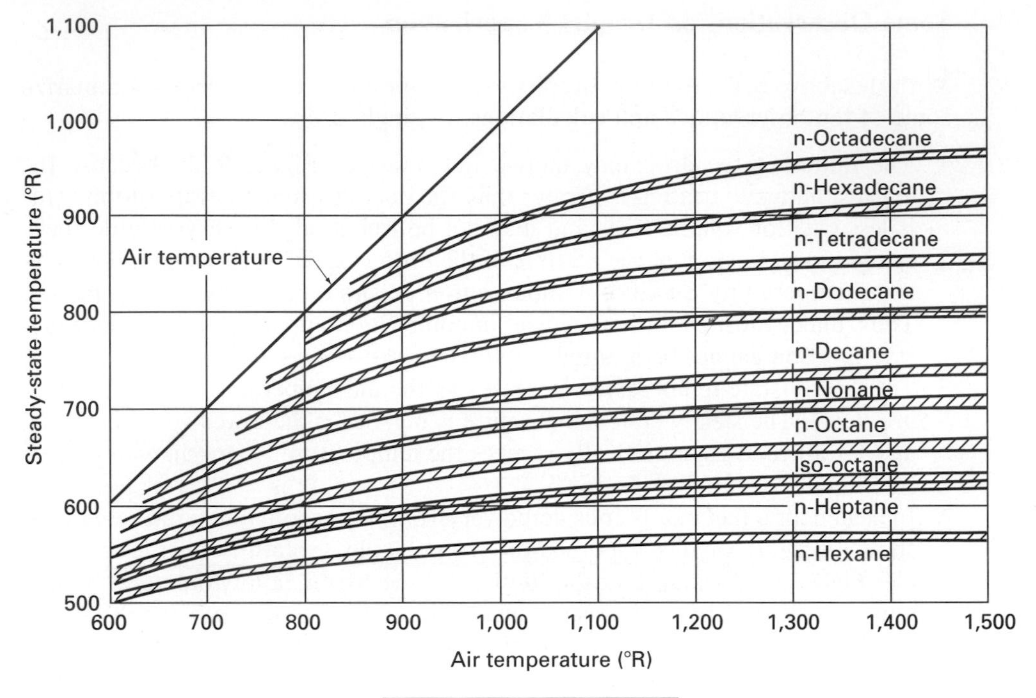

	Boiling temperature (°R)
n-Hexane	616
n-Heptane	669
Iso-octane	670
n-Octane	718
n-Nonane	736
n-Decane	805
n-Dodecane	881
n-Tetradecane	950
n-Hexadecane	1,007
n-Octadecane	1,063

FIGURE 9.21
Measured steady-state temperature as a function of air temperature for various fuels, 1 atm pressure [Priem et al.].

The effects of high ambient pressure (several times the critical pressure of the droplet liquid) are modeled by calculating all of the properties using nonideal thermodynamics [Manrique; Curtis]. Under high pressure the ambient gas penetrates into the surface of the droplet, causing a surface layer mixture which has its own pseudo-critical point. Computation of the mixture properties at the surface requires use of equations of state to calculate fugacities. As the droplet heats, the surface mixture may approach its pseudo-critical point while the pure fluid at the droplet interior remains below its critical point. When this happens, the energy required to change the phase of the droplet surface goes to zero, but more important, the surface tension becomes zero and the droplet distorts as would a dense gas

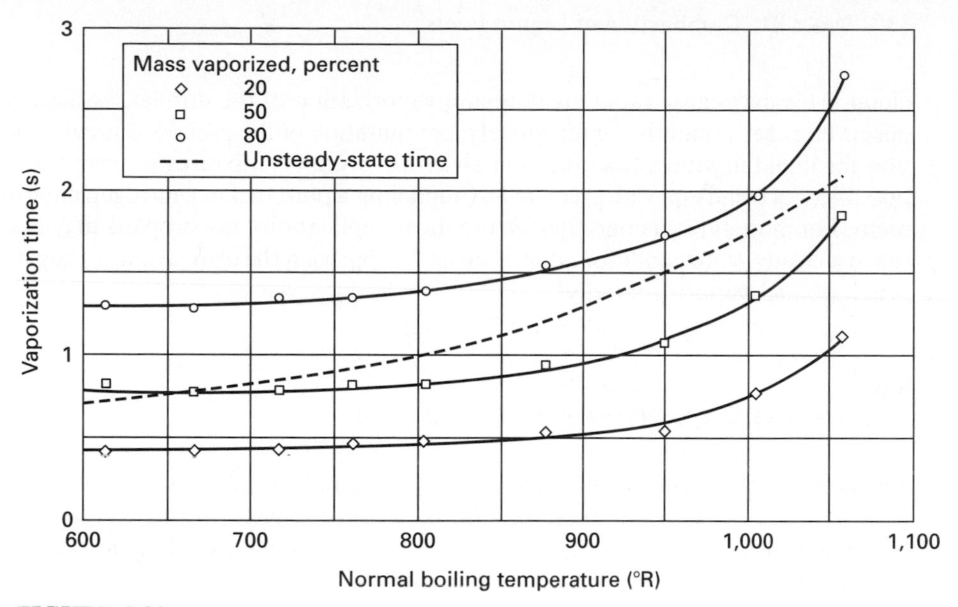

FIGURE 9.22
Measured vaporization times for fuels listed in Figure 9.21: 1 atm pressure, air temperature, 1430°R: drop diameter, 1500 to 1600 μm; relative velocity 6.67 ft/s [Priem et al.].

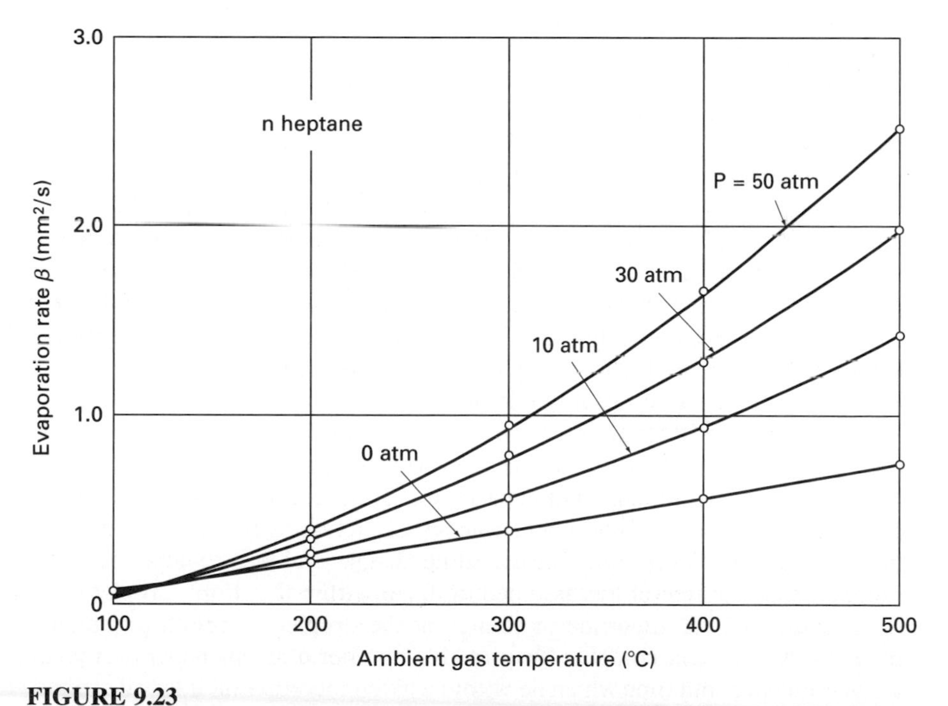

FIGURE 9.23
Vaporization rate β for large heptane droplets in a stagnant atmosphere of nitrogen vs. pressure and temperature [Kadota and Hiroyasu, reprinted with permission from the Japanese Society of Mechanical Engineers].

cloud. This may cause rapid breakup and vaporization of the droplet, as has been observed experimentally. Unfortunately, computation of the pseudo-critical condition for liquid mixtures (real fuels) in air or air-products mixtures is currently not possible, but trends may be predicted by modeling a pure fuel in a nitrogen environment. For more typical conditions in practical applications, the droplets may never reach a steady-state condition. As the droplets approach the critical state, they may break up and vaporize rapidly.

9.6
SPRAY MODELS FOR CFD PROGRAMS

The application of computational fluid dynamics (CFD) to spray combustion allows the fundamentals of single-droplet behavior discussed in this chapter to be combined with fluid mechanics to predict fuel preparation effects in practical spray combustors. Many of the CFD codes available were developed in the private sector so that the details of these codes are typically not available. The code KIVA, now available as KIVA II, was developed by the Los Alamos National Laboratory and is thus public. It has been upgraded and used at numerous other laboratories, in particular for engine applications at the University of Wisconsin–Madison and at the French Petroleum Institute (IFP). The basic program is described in the 1989 publication by Amsden et al., and its applications to engines is reviewed by Kong and Reitz in their 1995 review paper.

For sprays the basic conservation equations of mass, momentum, and energy for the fluid must be modified to include additional terms which account for two-phase effects. The continuity equation for gas-phase species includes a source term due to vaporization of droplets. The momentum equation includes a term for the rate of momentum gain per unit volume due to the spray. The energy conservation equation includes a source term for the energy exchange involved in droplet vaporization. As discussed in Chapter 5, the current status of turbulence modeling for practical CFD codes uses ensemble-averaged equations in which turbulence transport properties are computed from a turbulence model such as the k-ϵ model. For sprays, the turbulent kinetic energy k and dissipation rate ϵ equations each contain an additional term due to spray interactions.

The spray itself could in theory be modeled by following the behavior of each droplet, but the complexity of this approach for practical computations is prohibitive given current computer capabilities. Thus, the spray is described in terms of a droplet distribution function $\tilde{f}$, which is a function of 11 independent variables: three droplet position components, three droplet velocity components, droplet radius, droplet temperature (assumed uniform within the drop), droplet distortion from sphericity, the time rate of change of the droplet distortion parameter, and time. From $\tilde{f}$ one can compute the probable number of droplets per unit volume at a given position and time which lie within a given incremental interval around each of the other seven independent variables. The time evolution of $\tilde{f}$ is computed from the so-called "spray equation," which accounts for changes in $\tilde{f}$ due to each of the 11 independent variables plus changes due to droplet collisions and breakup.

The spray model considers the drop interactions with turbulence and walls, and calculates the changes of the independent variables (size, velocity, temperature, etc.) due to momentum change, evaporation, etc. Solution of the spray equation for $\tilde{f}$ then allows calculation of the source terms in the gas-phase equations of change for mass, momentum, and energy and the spray terms in the turbulence model equations.

It is not our purpose here to discuss the details of the CFD equations and models; that is an appropriate subject of a more advanced text. However, it is useful to understand the basic approaches and their limitations. The discussion just given shows the basic approach in very broad outline. Previous portions of the chapter give some of the details concerning droplet behavior. It should be understood that due to the probabilistic approach and the finite grid size of the calculations, many limitations are present. Because of the grid size (currently of the order of 1–2 mm) many droplets are contained within a given grid volume element. The subgrid modeling assumptions which determine behavior within each grid element are thus very important. For example, one may need to assume that newly formed fuel vapor within an element is uniformly distributed within the element, but this may affect the accuracy of the calculated droplet vaporization rate. The limitations on grid size also affect the modeling of heat transfer, momentum exchange, and droplet phenomena at solid surfaces. Thus, subgrid models for these interactions at the walls are required.

Which of the many empirically based assumptions are most important depends strongly on the application. For example, a spray in a relatively cold gas will not vaporize rapidly, and thus distillation of fuel components may be important so that the most volatile portions of the fuel control the mixture available for combustion. For a spray of very small droplets into high-temperature air, the vaporization will be rapid, so that nearly all droplets are vaporized before combustion. In this later case the turbulent mixing rate of the vapor and air becomes a determining parameter for combustion. Since various aspects of the spray model emerge as the dominant ones for different applications, it may be possible to eventually sort out the accuracy of various parts of the models. Currently, however, the lack of detailed experimental data makes this a difficult task. Overall measurements, such as spray penetration versus time for transient sprays, typically do not provide the sensitivity required to properly validate the models. However, the use of laser sheet diagnostics is improving the situation so that a better understanding of the overall phenomena is becoming possible [Dec]. Neverthless, the step from CFD modeling to design is still quite precarious.

9.7 SUMMARY

Spray combustion is widely used because it gives a practical method of rapidly vaporizing and mixing liquid fuels with air. Atomizers come in a large variety of designs and flow rates. Although various breakup regimes have been identified, modeling of the breakup mechanisms remains difficult, so that engineers typically must depend on empirical formulations. It is agreed, however, that the important dimensionless parameters for breakup are the jet Weber number $\rho_g \underline{V}_j^2 d_j/\sigma$ and the jet Reynolds number $\underline{V}_j d_j/v_l$. Droplets may continue to break up and in dense sprays coalesce due to collisions. Droplet shattering due to aerodynamics can be predicted from the critical droplet Weber number. Droplets may also break up by a boundary layer action, which strips very small droplets from the parent droplet surface.

Size distributions for sprays are expressed by various empirical distribution functions. Average sizes defined from these distributions or measured directly are often used to characterize the injector. The droplet size which characterizes volume to surface area, the Sauter mean diameter, is the most typically used parameter for such characterization, and empirical formulas which express Sauter mean values in terms of fuel and injection parameters are available for many of the common injector designs.

The computation of spray dynamics in terms of the equations of fluid mechanics and droplet ballistics is now possible by use of CFD codes. However, the application of such codes to intermittent, thick sprays still presents many unsolved problems, including questions concerning appropriate drag coefficient correlations and turbulent mixing models.

For combustion applications the mechanism of spray vaporization and mixing is of primary importance. Application of heat and mass transfer theory allows models to predict vaporization histories for single, undeformed droplets of pure fuels. For most applications the unsteady heating of the droplet must be taken into account. During unsteady heating the droplet temperature rises until it comes to steady state, where all of the heat reaching the surface is converted into the latent heat of vaporization. Once at the steady-state temperature, the square of droplet diameter decreases approximately linearly with time. For high-pressure combustion environments, droplet vaporization models need to include the effects of ambient gas absorption on the liquid surface properties. At such conditions the surface can approach a mixture critical point where surface tension and latent heat both approach zero.

Vaporization modeling for real fuels, which are complex liquid mixtures, is not yet possible. It is known, however, that the more volatile components are selectively removed. In some cases depletion of such volatile components at the surface causes the remaining volatile amounts in the interior to vaporize, causing the droplet to be shattered by the expanding vapor bubble.

PROBLEMS

9.1. Consider the following very simple distribution:

$\frac{1}{4}$ of the droplets have a diameter of 10 μm.
$\frac{1}{2}$ of the droplets have a diameter of 20 μm.
$\frac{1}{4}$ of the droplets have a diameter of 30 μm.

Show that $\overline{d}_1 = 20$, $\overline{d}_2 = 21.1$ $\overline{d}_3 = 22.2$ $\overline{d}_{32} = \text{SMD} = 24.4$ μm.

9.2. Using the chart of Figure 9.5 make a table of cumulative volume fraction versus drop diameter for SMDs of 50 μm and 25 μm.

9.3. For the prefilming air-blast nozzle shown in Figure 9.8, use the following data and assume the correlation of Eq. 9.18 holds. Compute the SMD for liquids A and B. The diameter of the pintle is 1 cm.

Property	Fuel A	Fuel B
$\sigma \times 10^3$ [kg/s^2]	27	74
ρ_l [kg/m^3]	784	1000
$\mu_l \times 10^3$ [kg/m · s]	1.0	53
p[kPa]	100	100
T_a[K]	295	295
$\underline{V}_a$[m/s]	100	100
$\dot{m}_a/\dot{m}_l$	3	3

If the air velocity is doubled, but keeping $\dot{m}_a$ constant, what is the effect on the SMD for each fuel? Discuss how the mechanisms of atomization differ for these two fuels.

9.4. Assume that the shape of a diesel spray can be modeled as a cone of angle θ with a hemisphere at the tip—it will look like a single scoop ice cream cone with half of the ice cream inside the cone. Using the angle and penetration formulas, find an expression for the spray volume versus time. The spray is in air at 1.35 atm pressure and 20°C temperature. The injector orifice diameter is 0.32 mm, the liquid mass flow rate is constant at 0.014 kg/s, and the injection pressure is 22.6 MPa. The spray is injected into a large volume of air, and after 4 ms the injection ends. The fuel is dodecane.

Plot the average fuel-air ratio in the spray volume for $t \leq 4$ ms. Assume that all of the fuel remains in the spray volume, and neglect vaporization. What effect would droplet vaporization have on the validity of the method? Compare with Figure 9.18.

9.5. A result similar to Eq. 9.49 can be obtained for Pr = Sc and $\mathrm{Re}_d \neq 0$. Show that for this case,

$$t_v = \frac{1}{2\beta} \int_0^{d_0} \frac{s\,ds}{1 + bs^{1/2}}$$

and define b, showing that it is independent of d.

9.6. Consider a droplet of 100 μm diameter moving in 1000 K air at 50 atm pressure.
(*a*) What droplet relative velocity will give an Re_d of 10? At this velocity, how far would the droplet move in 10 CA° at 2000 rpm engine speed?
(*b*) Determine the drag force F_x with and without a cross flow of velocity equal to the initial velocity.

$$\underline{V}_{dx} = \underline{V}_0 \qquad \underline{V}_{ax} = 0$$

$$\underline{V}_{dy} = 0 \qquad \underline{V}_{ay} = \underline{V}_0$$

$$\mathrm{Re}_d = \frac{d|\underline{\mathbf{V}}|}{\nu} = 10 \qquad \text{where} \qquad |\underline{\mathbf{V}}| = \sqrt{2}\underline{V}_0$$

9.7. Equation 9.48 gives an expression for the vaporization constant β when Re $= 0$ and $p_{v\infty} = 0$. Find the corrrect expression for the case Re $= 0$ and $p_{v\infty} > 0$. This is the case of fuel vapor mixed with the ambient air. What happens when $x_{v\infty} > x_{v0}$? Explain.

9.8. Consider a droplet of dodecane which has been partially vaporized and transported to a region where it is moving at the local bulk velocity and is vaporizing at steady state. (*a*) The temperature of the air is 1360°R and the pressure is 1 atm. Compute the vaporization constant and vaporization time of a 30-μm-dia droplet, assuming its steady-state temperature is 800°R as read from Figure 9.21. See Appendix Table A.8 for fuel properties. (*b*) The turbulent eddies in the region where the droplet in part *a* is located have an average size of 3000 μm and a characteristic turning velocity of 3.0 m/s. If the 30-μm droplet travels around the outer edge of the eddy with zero relative velocity, how many times must it go around the circular 3000-μm path before it has vaporized? What size droplet at the conditions of part *a* will travel just one eddy revolution in its lifetime?

9.9. For a simplex nozzle the Sauter mean diameter of the spray is given by

$$\text{SMD} = 2.25\sigma^{0.25}\mu_l^{0.25}\dot{m}_l^{0.25}\,\Delta p_l^{-0.5}\,\rho_a^{-0.25}$$

(*a*) Rewrite this equation in terms of σ, μ_l, d_0 (the orifice diameter), Δp_l, and ρ_a. That is, derive a relation between $\dot{m}_l$ and d_0, and substitute. Here $\dot{m}_l$ is the mass flow rate of liquid.

(*b*) Using the above expression for SMD, if the SMD is decreased by a factor of 2 while everything is held constant except the pressure drop, how much does the fuel pump power increase.

9.10. Calculate the final mixture temperatures for liquid dodecane droplets at an initial temperature of 600°R which vaporize in air initially at 1300°R, for a pressure of 1 atm and final equivalence ratios of 0.5, 1.0, 1.5. Neglect heat losses from the system.

REFERENCES

Abraham, J., Magi, V., MacInness, J., and Bracco, F. V., "Gas versus Spray Injection: Which Mixes Faster?" SAE paper no. 940895, SP-1028, 1994.

Amsden, A. A., O'Rourke, P. J., and Butler, T. D., "Kiva-II—A Computer Program for Chemically Reactive Flows with Sprays," Los Alamos National Laboratory, LA-11560–MS, 1989.

Arai, M., Tabata, M., and Hiroyasu, H., "Disintegration Process and Spray Characterization of Fuel Jet Injected by a Diesel Nozzle," SAE, paper no. 840275, 1984.

Bachalo, W. D., Houser, M. J., and Smith, J. N., "Evolutionary Behavior of Sprays Produced by Pressure Atomizers," AIAA paper no. 86–0296, 1986.

Barett, H. C., and Hibbard, R. R. (eds.), "Basic Considerations in the Combustion of Hydrocarbon Fuels with Air," NACA report 1300, NASA Lewis Flight Propulsion Laboratory, Cleveland, 1959.

Bartok, W. and Sarofim, A. F., *Fossil Fuel Combustion: A Source Book,* Wiley Interscience, New York, 1991.

Bird, R. B., Stewart, W. E., and Lightfoot, E. N., *Transport Phenomena,* Wiley, New York, 1991.

Bosch, *Diesel Fuel Injection,* R. Bosch GmbH, 7000 Stuttgart, Germany, 1944.

Bracco, F. V., "Modeling of Engine Sprays," SAE paper no. 850394, 1985.

Curtis, E. W., and Farrell, P. V., "A Numerical Study of High-Pressure Droplet Vaporation," *Combustion and Flame,* vol. 90, pp. 85–102, 1992.

Dec, J. E., "A Conceptual Model of DI Diesel Combustion Based on Laser-Sheet Imaging," SAE paper no. 970873, 1997.

Dodge, L., and Schwalb, J. A., "Fuel spray Evolution: Comparison of Experiment and CFD Simulation of Non-evaporating Spray," ASME paper no. 88-GT-27, June 1988.

El-Kotb, M. M., "Fuel Atomization for Spray Modeling," *Prog. Energy Comb. Sci.,* vol. 8, p. 61, 1982.

Faeth, G. M., "The Current Status of Droplet and Liquid Combustion," *Prog. Energy Comb. Sci.,* vol. 3, pp. 191–224, 1977.

Givler, S. D., and Abraham, J., "Supercritical Droplet Vaporization and Combustion Studies," *Prog. Energy Comb. Sci.,* vo. 22, pp. 1–28, 1996.

Glassey, S. F., Stockner, A. R., and Flinn, M. A., "HEUI—A New Direction for Diesel Engine Fuel Systems," SAE paper no. 930270, 1993.

Gosman, A. D., "Multidimensional Modeling of Cold Flows and Turbulence in Reciprocating Engines," SAE, paper no. 850344, 1985.

Ha, J., Iida, N., Sato, G. T., Hayashi, A. and Tanabe, H., "Experimental Investigation of the Entrainment into a Diesel Spray," SAE paper no. 841078, SAE SP-581, 1984.

Heywood, J., *I.C. Engine Fundamentals,* McGraw-Hill, New York, 1988.

Hiroyasu, H., Arai, M., and Tabati, M., "Empirical Equations for the Sauter Mean Diameter of a Diesel Spray," SAE paper no. 890464, 1989.

Hiroyasu, H., and Kadota, T., "Fuel Droplet Size Distribution in a Diesel Combustion Chamber," SAE paper no. 740715, 1974.

Hwang, S. S., Liu, Z., and Reitz, R. D., "Breakup Mechanisms and Drag Coefficients of High Speed Vaporizing Drops," *Atomization and Sprays,* vol. 6, pp 353–376, 1996.

Ingebo, R. D., "Vaporization Rates and Drag Coefficients for Isooctane Sprays in Turbulent Air Streams," NACA TN 3265, 1954.

Kadota, T., and Hiroyasu, H., "Evaporation of a Single Droplet at Elevated Pressures and Temperatures," *Bul. JSME,* vol. 19, no. 138, pp. 1515–1521, 1976.

Knight, B. E., "Communication on the Performance of a Type of Swirl Atomizer," *Proc. I. Mech Eng.,* vol. 104 1955.

Kong, S.-C., and Reitz, R. D., "Spray Combustion Processes in Internal Combustion Engines," in K. K. Kuo (ed), *Recent Advances in Spray Combustion* (AIAA Series), 1995.

Lefebvre, A. H., *Atomization and Sprays,* Hemisphere, New York, 1989.

Liu, A. B., Mather, D., and Reitz, R. D., "Modeling the Effects of Drop Drag and Breakup on Fuel Sprays," SAE paper no. 930072, 1993.

Manrique, J. A., "Theory of Droplet Vaporization in the Region of the Thermodynamic Critical Point," NASA CR-72558, June 1969.

Nukiyama, S., and Tanasawa, Y., "Experiments on Atomization of Liquids in an Airstream," *Trans. Soc, Mech. Engr. Japan,* vol. 5, pp. 68–75, 1939.

Obert, E. F., *Internal Combustion Engines and Air Pollution,* Intext, New York, 1973.

O'Rourke, P. J., and Amsden, A. A., "The TAB Method for Numerical Calculation of Spray Droplet Breakup," SAE paper no. 872089, 1987.

Priem, R. J., Borman, G. L., El-Wakil, M. M., Uyehara, O. A., and Myers, P. S., "Experimental and Calculated Histories of Vaporizing Fuel Drops," NACA TN 3988, 1957.

Ranger, A. A., and Nicholls, J. A., "Aerodynamic Shattering of Liquid Drops," *AIAA. J.,* vol. 7, no. 2, pp. 285–290, 1969.

Ranz, W. E. and Marshall, Jr., W. R., "Evaporation from Droplets," *Chem Eng. Progr.,* vol. 48, pp. 141–146 and pp. 173–180, 1952.

Reitz R., D., "Modeling Atomization Processes in High-Pressure Vaporizing Sprays," *Atomization and Spray Technology,* vol. 3, pp. 309–337, 1987.

Reitz, R. D., and Bracco, F. V., "Mechanisms of Breakup of Round Liquid Jets," *Encyclopedia of Fluid Mechanics,* Chap. 10, pp. 233–249, Gulf Publishing, Houston, 1986.

Rizkella, A. A., and Lefebvre, A. H., "Influence of Liquid Properties on Airblast Atomizer Spray Characteristics," *J. Engr. Power,* pp. 173–179, April 1975.

Simmons, H. C., "The Correlation of Drop-Size Distribution in Fuel Nozzle Sprays," *J. Eng. for Power,* vol. 99, pp. 309–319, 1977.

Van Dyke, M., *An Album of Fluid Motion,* Parabolic Press, Stanford, CA, 1982.

Zhao, F.-Q., Lai, M.-C., and Harrington, D. L., "The Spray Characteristics of Automotive Port Fuel Injection—A Critical Review," SAE paper no. 950506, 1995.

CHAPTER 10

Oil-Fired Furnace Combustion

Number 2 fuel oil (distillate oil) and no. 6 fuel oil (residual oil) are used in furnaces and boilers. Distillate fuel oil is used in residential, commercial, and industrial furnaces and boilers, while residual fuel oil is used in power plants and large industrial boilers. Due to the need to optimize crude oil refineries for gasoline, no. 3, 4, and 5 fuel oil grades are rarely available. Due to high cost, distillate fuel oil is rarely used in large burners.

Fuel oils are atomized by means of a spray nozzle in order to achieve rapid combustion. The exception is kerosene, which vaporizes rapidly as a warmed liquid, so that small space heaters are made with a wick or a so-called combustion pot. Number 2 fuel oil is sprayed at ambient temperature (even in cold climates), but no. 6 fuel oil must be heated to 100°C to ensure proper pumping and atomization.

In this chapter the system aspects of oil-fired furnaces and boilers are considered. Then burner combustion of sprays is discussed, and a simple plug flow model of droplet combustion is developed. Finally, emissions from oil-fired furnaces and boilers are examined.

10.1 OIL-FIRED SYSTEMS

A schematic diagram of an oil-fired furnace is shown in Figure 10.1. Fuel oil is pumped through a spray nozzle, the fuel droplets mix in a burner with air from a blower, and the mixture burns out in the combustion chamber. The fuel and air flow requirements for a given heat output are calculated from an energy balance around the combustion chamber and heat exchanger according to Eqs. 6.1–6.5. The pump and blower auxiliary power are obtained from Eq. 6.6.

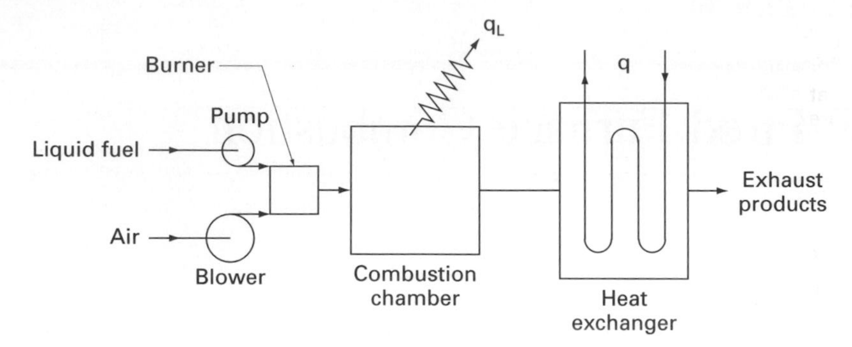

FIGURE 10.1
Schematic of oil-fired furnace, where q is the heat transfer from the combustion products to the heat exchanger tubes.

In small systems the combustion chamber is refractory-lined to promote combustion. In large systems the burners fire directly into a large volume which contains the heat exchanger surfaces. Fire-tube boilers (combustion products pass through tubes immersed in water) are used for small applications, and water-tube boilers (water in tubes is exposed to combustion products) are used for large systems.

As with gas-fired furnaces and boilers, the highest efficiency is obtained by completely burning the fuel with as little excess air as possible. This requires fine atomization of the fuel and rapid penetration of the air into the spray. Older domestic distillate oil-fired furnaces and boilers require 40–60% excess air and have efficiencies of 75–80%. Older commercial oil burners require 30% excess air, while industrial oil burners may require 15% excess air. Utility boilers using residual fuel oil require as little as 3% excess air. Attempts to reduce the excess air with inadequate mixing result in increased smoke emissions. Newer systems have improved mixing, lower excess air, and reduced soot emissions.

Atomization of fuel oil is typically accomplished by one of three methods—single fluid atomizers, twin fluid atomizers, or rotary cup atomizers. The burner design should include a fuel filter and fuel pump, fuel atomizer, and inlet air blower. The burner should provide a stable flame with low excess air, and low emissions, and be durable and reliable.

The single-fluid pressure jet atomizer is used widely for residential, commercial, and small industrial furnaces burning no. 2 fuel oil. In this type of application the burner is referred to as a high-pressure gun-type burner (Figure 10.2). Fuel oil is pressurized to 7 atm and forced through a nozzle which atomizes the liquid into small droplets typically with a mass mean diameter of 40 μm and 10% larger than 100 μm. Air from a low-pressure (about 25 cm H_2O gauge) blower flows around the fuel nozzle and mixes with the spray. In order to obtain good mixing of the fuel droplets and air, guide vanes are mounted near the spray nozzle.

The liquid jet from the fuel nozzle breaks up into a spray due to the high axial velocity of the liquid relative to the surrounding air and due to tangential velocity, which is imparted to the fuel jet by means of swirl. As shown in Figure 10.3, the

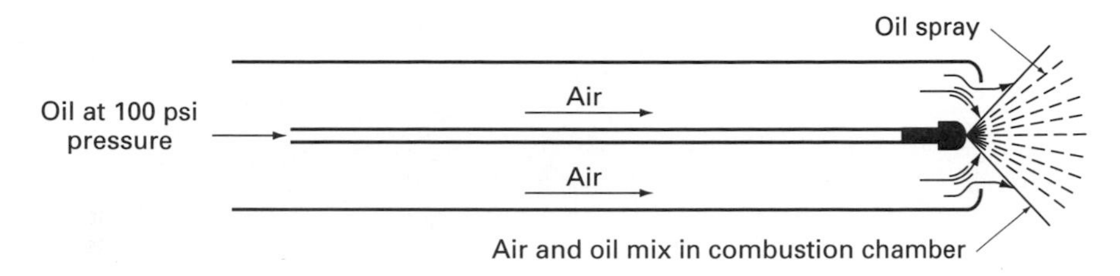

FIGURE 10.2
High-pressure gun-type oil burner.

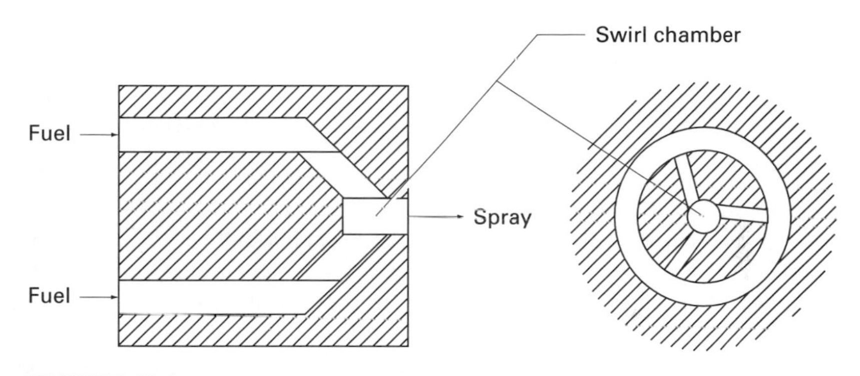

FIGURE 10.3
Pressure atomization nozzle for distillate fuel oil.

fuel oil flows tangentially into a small swirl chamber at the tip of the nozzle before exiting the nozzle. The angle of the external spray cone can be varied from 30 to 90° by varying the angle of the interior cone, which changes the ratio of the axial to rotational velocity of the jet. An 80° hollow-cone nozzle is typical of domestic use. Electrodes located near the upstream edge of the spray provide a continuous source of ignition.

Residential oil burners typically use 0.5 to 3 gal/h. Commercial and small industrial burners use nozzles which provide up to 35 gal/h per nozzle. Fuel drop size is distributed between 50 and 300 μm. Turndown cannot be achieved by reducing the oil pressure, because this increases the drop size, which causes incomplete burnout in small furnaces. Control is achieved by shutting off the oil pressure to the nozzle. Residential units have only one nozzle, and thus on-off control is used.

In recent years small oil burner designs have improved the mixing and reduced the excess air by adding swirl to the air and by adding a flame retention device (see Figure 10.4). Air swirl is achieved with guide vanes mounted to the inside of the burner can. The flame retention device is a metal cone which is mounted a short distance from the spray nozzle. The gap between the nozzle and the cone, and slots and holes in the cone, allow for air penetration into the spray. The cone acts as an air shield which stabilizes the flame, creating a more compact and intense flame. Swirl creates more mixing beyond the flame retention cone. In this way the excess air can be reduced and the efficiency increased.

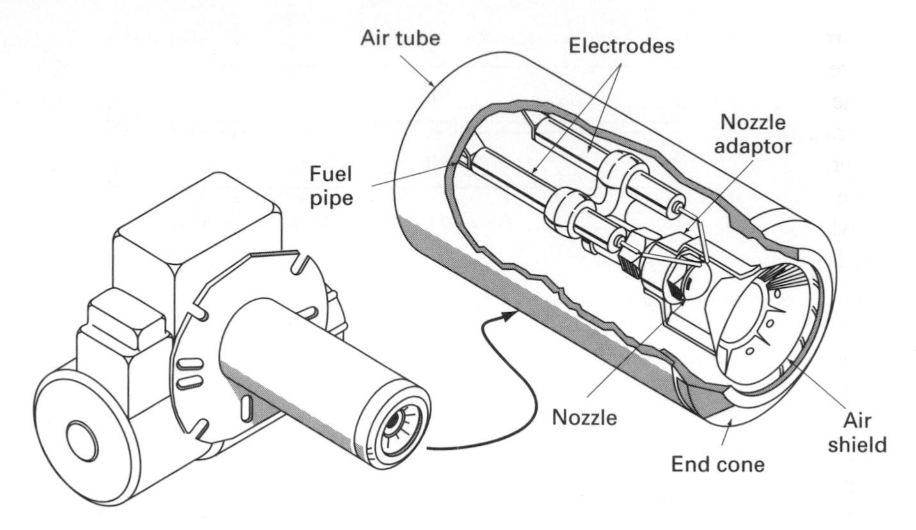

FIGURE 10.4
Residential oil burner with flame retention air shield [Offen].

The single-hole spray burner cannot be fired below 0.5 gal/min, which represents 70,000 Btu/h, because it is not practical to reduce the hole size. There are numerous applications where a smaller firing rate is needed. Prevaporizing the fuel in a so-called pot burner tends to yield a sooty flame. The Babbington burner, which operates on a different principle, can burn cleanly at much lower flow rates. This burner uses an "inside-out" nozzle, shown in Figure 10.5. The liquid fuel flows over the outside of a hollow sphere and drains off into a fuel return tube.

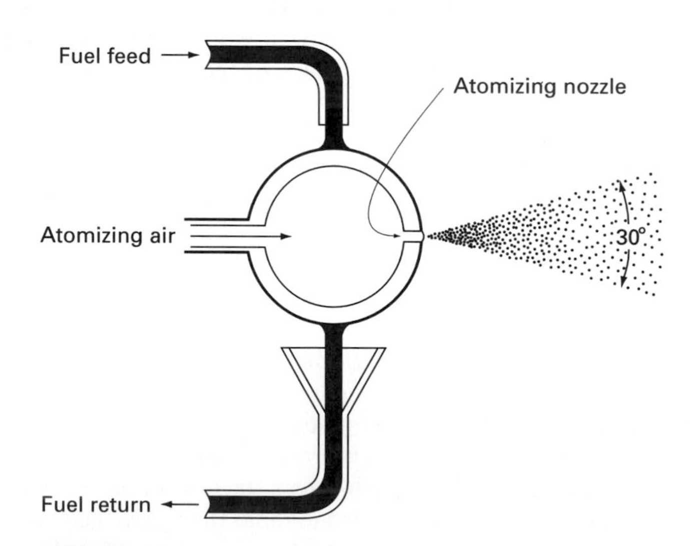

FIGURE 10.5
The Babbington atomization principle [courtesy of R. S. Babbington].

Low-pressure air (10 psi) flows into the sphere and through a small hole which atomizes a thin film of fuel. The atomization air is about 1% of the combustion air, and the mass mean drop size of no. 2 fuel oil is about 15 μm with less than 1% greater than 70 μm. This small drop size facilitates good mixing and low emissions. The spray is formed by rupture of the fuel film, and the drop size depends on the surface tension but not the viscosity. Turndown is achieved by reducing the fuel flow, which reduces the thickness of the fuel film on the sphere. Fuel flow rates from 1.0 gal/h down to 0.5 gal/week have been demonstrated in field tests for the U.S. Army.

Utility and large industrial burners operating with heavy fuel oil use single-fluid or twin-fluid atomizers. In a single-fluid pressure jet atomizer for heavy fuel oil, the pressure is typically about 30 atm with fuel rates up to 1250 gal/h per nozzle. Liquid enters a swirl chamber with a tangential component of velocity, and passes through a discharge orifice as an annular film which then disintegrates into a spray with additional assistance from a central core air jet (Figure 10.6*a*). The volumetric size distribution for heavy fuel oil is shown in Figure 10.6*b*. The mass mean diameter is 155 μm. The spill flow pressure jet atomizer shown in Figure 10.7 is a modification of Fig 10.6*a* to allow for turndown and shutoff; however, the central air core is sacrificed. A return flow path for the oil is provided inside the nozzle so that turndown of up to 3 to 1 is achieved by applying a back pressure to the return flow.

Twin fluid atomizers use a jet of steam or air impinging on the fuel jet. A steam atomizing nozzle of the so-called Y-jet type is shown in Figure 10.8. Typically, the steam pressure is only 1 to 3 atm above the oil pressure. A two-phase mixture is formed in the exit port which expands out of the port to form a spray. The flow rate of steam should not exceed 10% of the flow rate of oil. Of course, this requires makeup boiler water and reduces the boiler efficiency slightly. If steam is not available, compressed air can be used. The drop size distribution is similar to the single fluid atomizers. The main advantage is that the fuel pressure is reduced.

The rotary cup atomizer was introduced as a means of further reducing the required oil pressure. As shown in Figure 10.9, oil is fed to a conical cup through a tube inside a hollow rotating shaft. The shaft and cup rotate at 3500 rpm. Oil flows along the inside of the cup in a sheet and is thrown by centrifugal force into the surrounding air stream. The drop sizes formed are considerably larger than for the other two types mentioned, which tends to cause smoking. Thus, rotary cup atomizers have been used less in recent years.

10.2 SPRAY COMBUSTION IN FURNACES

The process by which a liquid fuel spray burns depends on the number density of the spray, the degree of turbulent mixing, and the fuel volatility. If the number density is low, the mixing is high, and the fuel volatility is relatively low, then the spray burns as individual droplets surrounded by individual flames. If the spray is dense, the mixing relatively low, and the fuel volatility high, then the droplets

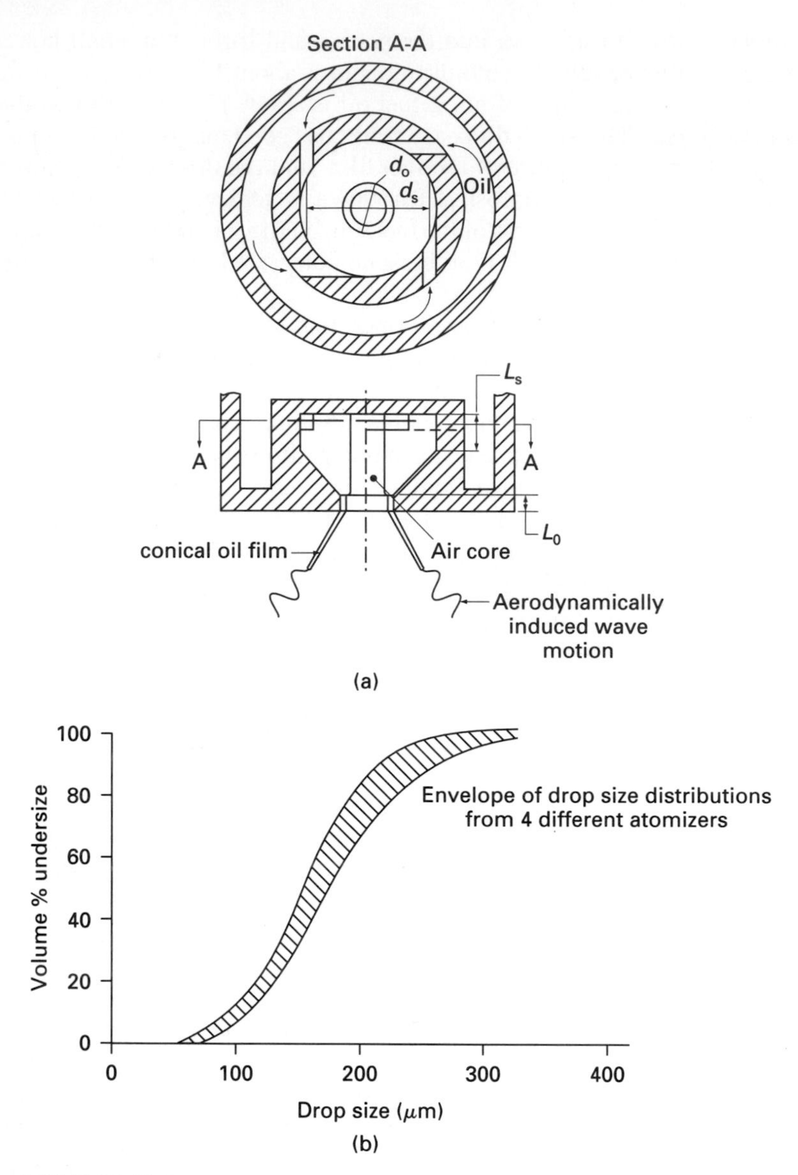

FIGURE 10.6
Swirl pressure jet atomizer for heavy fuel oil: (*a*) atomizer, (*b*) drop size distribution [Lawn, by permission of Academic Press, Inc.].

vaporize but the flame occurs at the outer edge of the spray. This is termed *external combustion.* The local fuel-air ratio is too high for combustion except at the outer edge of the spray. An intermediate case, which is termed *group combustion,* is when a group of drops within the spray burns in the manner of a large single drop. The external combustion flame tends to exhibit a blue flame, whereas the droplet flame tends to be yellow, indicating rich combustion with production of soot.

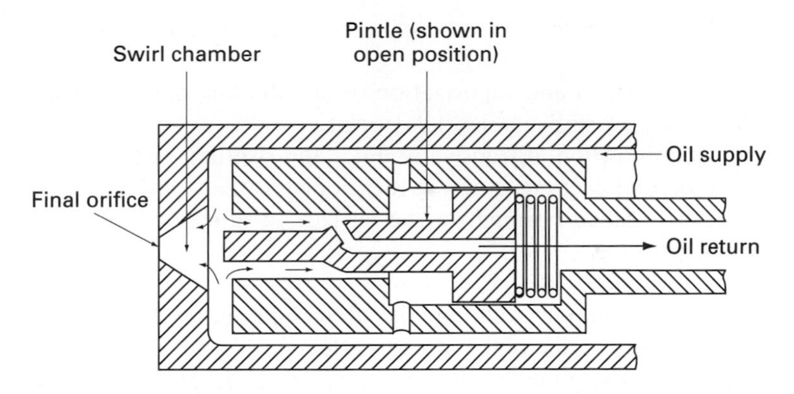

FIGURE 10.7
Spill pressure jet atomizer with shut-off [Lawn, by permission of Academic Press, Inc.].

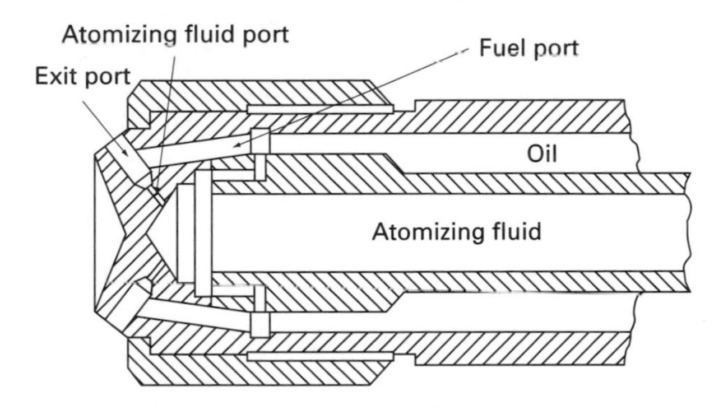

FIGURE 10.8
Y-jet atomizer [Lawn, by permission of Academic Press, Inc.].

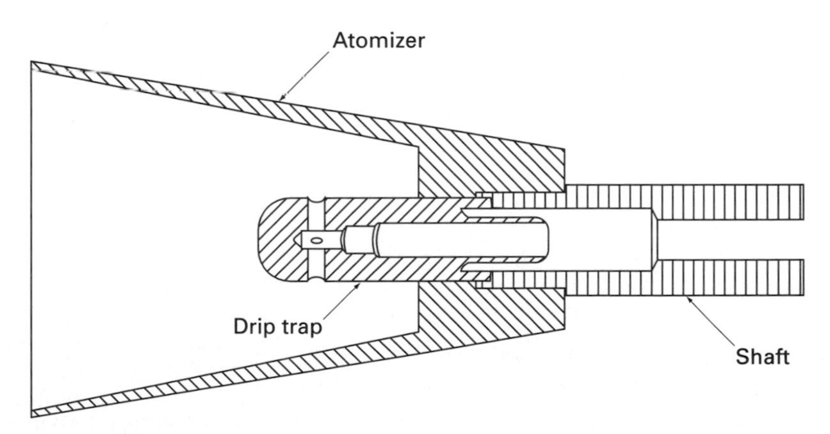

FIGURE 10.9
Rotary cap atomizer, showing arrangement for preventing drip of fuel oil into atomizer after shutdown of burner.

The spray processes for fuel oil combustion can be summarized as follows:

1. Heating of the droplet and vaporization of low–boiling point components
2. Ignition of volatiles surrounding the droplet
3. Thermal decomposition, disruptive boiling, and swelling of the droplet
4. Continued thermal decomposition of the droplet as the volatile flame continues
5. Carbonaceous residue burning on the surface at about one-tenth the initial droplet burning rate

Fuel oils which contain water exhibit enhanced boiling and swelling prior to ignition. Extensive disruption of the droplets effectively results in finer atomization by ejection of smaller satellite droplets from the parent droplet. The satellite drops may ignite before the parent drop.

The fluid dynamic flow of the air dominates the shape of the combustion zone. As noted above, most small oil fired burners use a flame retention cone (or end cone) (Figure 10.4), which acts as a partial baffle that stabilizes the initial flame zone, provides mixing of air into the droplets, and imparts a small amount of swirl. The droplets move on a trajectory in this cone region and then flow away from the cone in a more nearly uniform flow, all the while burning primarily as a collection of individual droplets. The droplet vaporization rate is proportional to the droplet diameter (see Eq. 9.47). The droplet vaporization rate constant was obtained theoretically in Chapter 9, and has been measured under furnace conditions for different fuels. Data of this sort are shown in Figure 10.10 for several light hydrocarbons.

In larger nozzle-type burners, the spray is injected axially with high momentum and the air is blown in with varying amounts of swirl. Combustion air comes from a windbox through movable registers which facilitate adjusting the flame length and shape. A slotted turbulator attached to the fuel nozzle causes mixing of the air and spray and provides flame stabilization (Figure 10.11*a*). With low swirl,

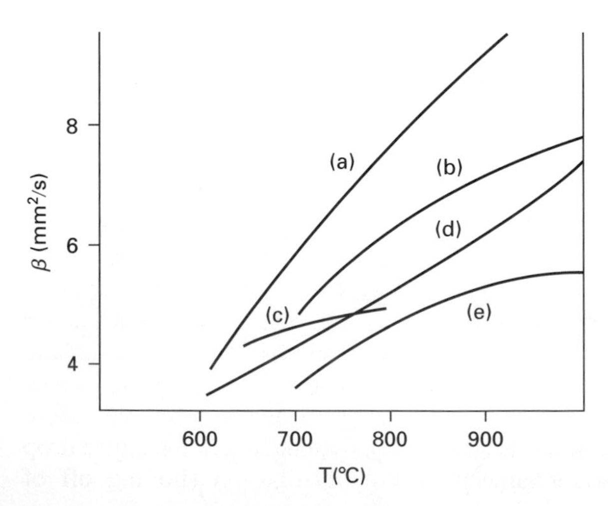

FIGURE 10.10
Droplet burning rate constants [*see* Eq. 9.47] for fuels burning in air at elevated temperatures: (*a*) cetane, (*b*) light diesel oil, (*c*) heptane, (*d*) aviation kerosene, and (*e*) benzene. Heavy fuel oils are similar to the light diesel oils up to 700°C, but measurements at higher temperatures are not possible because of droplet swelling [Williams, 1990, by permission of author].

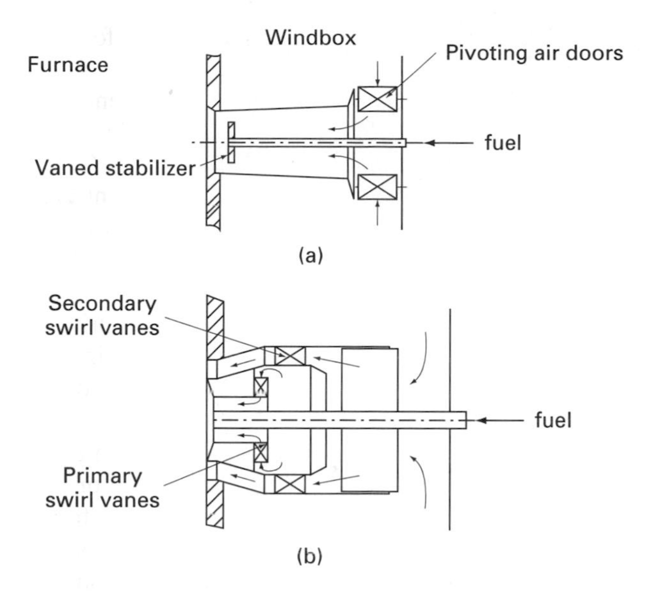

FIGURE 10.11
Industrial burners showing airflow path: (*a*) swirl air with turbulator, (*b*) dual swirl burner [Lawn, by permission of Academic Press, Inc.].

the flame is long and narrow as shown in Chapter 6, Figure 6.9. Gases within the combustion chamber near the flame are entrained by the momentum of the jet, causing a recirculation and mixing at the flame boundary. The droplets travel in a straight line and burn out as a combination of individual droplet flames and an external flame. If a divergent fuel spray is used and the air is blown in with swirl (Figure 10.11*b*), then a short, brushlike flame, which is stabilized on the surface of refractory nozzle of the burner (called the quarl), is produced. With high swirl there is a closed internal recirculation zone which promotes a short, high-intensity flame. An intermediate case of a combination jet flame and swirl flame where the jet penetrated through the recirculation zone is also shown in Figure 6.9. The different types of flames are used for different applications. Long jet flames are used in corner-fired boilers and with industrial processes where radiant heat transfer is used. Swirl flames are used in wall-fired boilers and industrial heaters. Temperature contours near the burner of Figure 10.11*b* are shown in Figure 10.12. The measurements were made with a suction pyrometer. There was no evidence of a closed recirculation zone in this case.

Large industrial and utility burners typically burn residual fuel oil, and the droplets burn quite differently and more slowly than distillate droplets. When a residual oil droplet is subjected to a high-temperature gas, say, 1300°C, as in Figure 10.13, evaporation starts at about 200°C with slight swelling until the onset of boiling, which is accompanied by distortion and swelling. Droplet expansions alternate with contractions in rapid succession, the overall diameter increasing and reaching about twice the initial diameter. Puffs of vapor are ejected at this stage, with considerable force and over distances up to 10 diameters from the parent droplet. Sometimes satellites are expelled, and in exceptional cases the entire drop disintegrates. Generally, volatilization proceeds by distillation (boiling off of

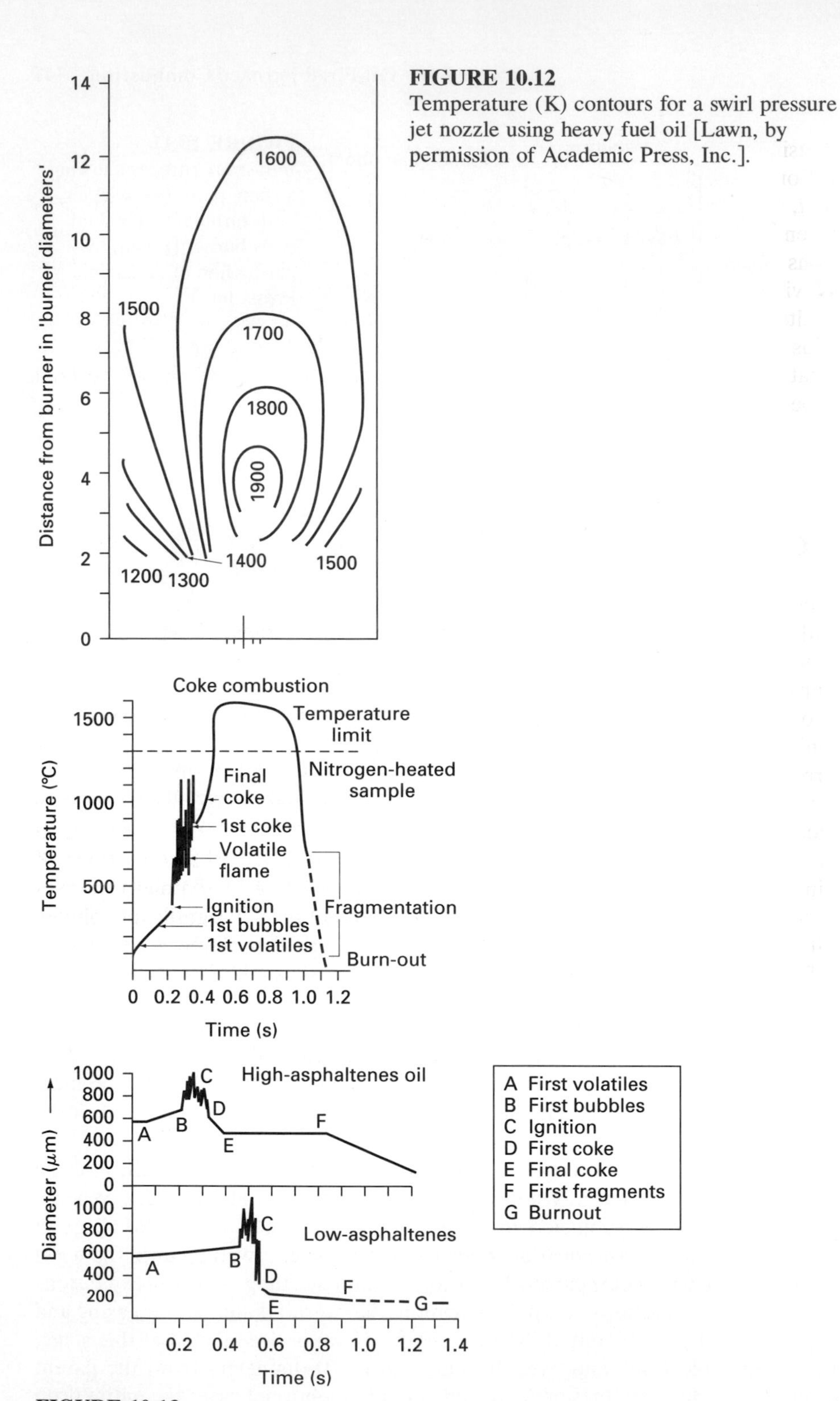

FIGURE 10.12
Temperature (K) contours for a swirl pressure jet nozzle using heavy fuel oil [Lawn, by permission of Academic Press, Inc.].

FIGURE 10.13
Temperature and size history of a single residual fuel oil drop in 1300°C air [Lightman and Street, by permission of *J. Institute of Energy,* © 1983].

increasingly higher boiling point compounds) and by pyrolysis (breaking of molecular bonds to form lower-molecular weight compounds). As the volatiles are released, 0.02- to 0.2-μm soot particles are formed, which can agglomerate into filaments up to several thousand micrometers long before burning out in leaner regions of the flame. Toward the end of the volatile loss phase, the droplets become very viscous and solidify into porous coke particles. Fuels which are high in asphaltenes tend to form more coke. The coke from a residual oil spray is a very porous, carbonaceous particle which burns out with decreasing density until fragmentation occurs toward the end of burnout (Figure 10.13). Solid particle burnout will be considered in more detail in Chapter 14.

10.3 PLUG FLOW MODEL OF DISTILLATE SPRAY COMBUSTION

The combustion of an oil burner spray typically involves a complex three-dimensional flow of droplets and air with turbulent mixing, droplet vaporization, and combustion. However, it is instructive to simplify the problem by assuming a one-dimensional, constant-pressure, constant-area stream tube containing a uniform flow of monodispersed droplets in air. Monodispersed means uniformly sized drops of uniform spacing. The droplets and gas have the same velocity. This case is referred to as *plug flow,* and it applies to a low swirl flow.

Vaporization and ignition start at $x = 0$ as indicated in Figure 10.14. The object is to find the temperature as a function of distance, the length of the reaction zone, and the overall combustion intensity. It is reasonable to assume that the mixing and chemical reaction time are short compared with the droplet vaporization time. The spray is dilute, so that the spray does not occupy significant volume, but it does add mass. At $x = 0$ the number of drops per unit volume, n_0' at a specified fuel/air ratio is determined as follows. In a volume $A\ dx$, the fuel/air ratio is

$$f = \frac{(n_0' \rho_l \pi d_0^3/6)A\ dx}{\rho_0 A\ dx - (n_0' \rho_l \pi d_0^3/6)A\ dx} \tag{10.1}$$

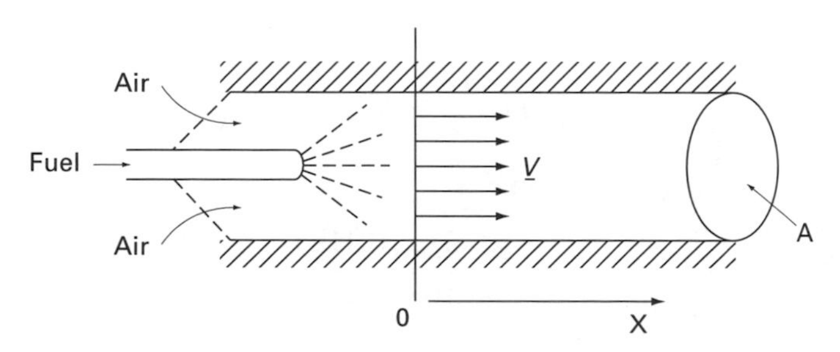

FIGURE 10.14
Diagram for plug flow model of spray combustion.

where the subscript 0 indicates the position $x = 0$. Solving for n'_0,

$$n'_0 = \frac{f}{1+f} \cdot \frac{\rho_0}{\rho_l} \cdot \frac{6}{\pi d_0^3} \tag{10.2}$$

Moving downstream, the temperature increases, the density decreases, and the velocity increases. Conservation of mass requires that

$$\rho_0 \underline{V}_0 A \doteq \rho \underline{V} A \tag{10.3}$$

The number of drops remains the same, but due to the expansion of the gas, the number per unit volume decreases:

$$n'_0 \underline{V}_0 A = n' \underline{V} A \tag{10.4}$$

and from Eq. 10.3,

$$n' = \frac{n'_0 \rho}{\rho_0} \tag{10.5}$$

From Chapter 9, a droplet at steady-state conditions vaporizes according to

$$\dot{m}_l = -\pi d \rho_l \beta \tag{9.47}$$

Integrating Eq. 9.36 with Eq. 9.47, it follows that

$$d = \sqrt{d_0^2 - 4\beta t} \tag{10.6}$$

and thus the droplet vaporization time is

$$t_v = \frac{d_0^2}{4\beta} \tag{9.50}$$

where β is the vaporization rate constant.

The temperature distribution is obtained by integrating the energy equation across a strip dx using the coordinate system shown in Figure 10.14. Assuming that the energy transport by conduction and diffusion are small compared with energy transport by convection for this extended reaction zone, the energy balance is

$$\rho \underline{V} A c_p \frac{dT}{dx} dx = q''' A \, dx \tag{10.7}$$

where q''' is the heat release rate per unit volume due to combustion. Canceling the volume element $A\,dx$ and noting that $\underline{V} = dx/dt$, the energy equation becomes

$$\rho c_p \frac{dT}{dt} = q''' \tag{10.8}$$

The heat release rate per unit volume is

$$q''' = n' (\dot{m})_{\text{drop}} (\text{LHV}) \tag{10.9}$$

Substituting Eqs. 10.2, 10.5, 9.47, and 10.9 into Eq. 10.8,

$$\frac{dT}{dt} = \frac{6f}{1+f} \cdot \frac{\beta(\text{LHV})}{\bar{c}_p d_0^3} \cdot d \tag{10.10}$$

Integrating Eq. 10.10 with Eq. 10.6,

$$T - T_0 = \frac{f}{1+f} \cdot \frac{(\text{LHV})}{\bar{c}_p}\left[1 - \left(\frac{d}{d_0}\right)^3\right] \tag{10.11}$$

Since the adiabatic flame temperature (see Chapter 3) is

$$T_{af} = T_0 + \frac{f}{1+f} \cdot \frac{\text{LHV}}{\bar{c}_p} \tag{3.66}$$

we get

$$T = T_0 + (T_{af} - T_0)\left[1 - \left(\frac{d}{d_0}\right)^3\right] \tag{10.12}$$

The reaction zone length is given by

$$L_v = \int_0^{t_b} \underline{V}\, dt$$

From Eqs. 10.3, and 10.11 and the ideal gas law,

$$L_v = \underline{V}_0 \int_0^{d_0^2/4\beta} \left[1 + \left(\frac{T_{af}}{T_0} - 1\right)\left(1 - \left(\frac{d}{d_0}\right)^3\right)\right] dt$$

Integration yields

$$L_v = \frac{\underline{V}_0 d_0^2}{4\beta}\left(\frac{2}{5} + \frac{3}{5}\frac{T_{af}}{T_0}\right) \tag{10.13}$$

Finally, the combustion intensity, which is the heat release per unit volume, can be estimated by assuming that the mixing and chemical reactions occur much faster than the vaporization. Then,

$$I = \frac{\rho_0 \underline{V}_0 \,\text{LHV}(f)}{L_v(1+f)} \tag{10.14}$$

or

$$I = \frac{\rho_0 \underline{V}_0 \bar{c}_p (T_{af} - T_0)}{L_v} \tag{10.15}$$

These relationships are explored in the example that follows.

EXAMPLE 10.1. A monodispersed spray moves in a plug flow of air with a fuel/air weight ratio of 0.077, initial droplet diameter of 100 μm, flame temperature of 2100 K, burning rate constant of 0.25 mm^2/s, initial velocity of 1 m/s, and an initial air temperature of 500 K. The droplet density is 900 kg/m^3, and the pressure is 1 atm. Find the initial droplet number density, the vaporization time, the length of the reaction zone, and the combustion intensity. Refer to Figure 10.14.

Solution. From Eq. 10.2, the initial droplet number density is,

$$n_0' = \left(\frac{0.077}{1.077}\right)\left(\frac{0.706 \text{ kg/m}^3}{900 \text{ kg/m}^3}\right)\left(\frac{6}{\pi(10^{-2}\text{cm})^3}\right) = 109/\text{cm}^3$$

From Eq. 9.50, the vaporization time is,

$$t_v = \frac{(100 \times 10^{-3}\text{mm})^2}{4(0.25 \text{ mm}^2/\text{s})} = 0.01 \text{ s} = 10 \text{ ms}$$

Evaluating Eq. 10.13,

$$L_v = \frac{(1 \text{ m/s})(0.1^2 \text{ mm}^2)}{(4)(0.25 \text{ mm}^2/\text{s})}[0.4 + (0.6)(2100/500)]$$

$$= 0.029 \text{ m} = 2.9 \text{ cm}$$

From Eq. 10.15 and taking $\rho_0 = 1.177(300/500) = 0.706 \text{ kg/m}^3$, and taking $\overline{c}_p$ for air at 1300 K from Appendix B,

$$I = \frac{(0.706 \text{ kg/m}^3)(1 \text{ m/s})(1.195 \text{ kJ/kg}\cdot\text{K})(2100 - 500 \text{ K})}{0.029 \text{ m}}$$

$$= 46.5 \text{ MW/m}^3$$

This calculation assumes that the mixing and chemical reaction rates are very rapid compared with the vaporization rate, so that the true reaction zone is somewhat longer than 2.9 cm. Using these relationships, the velocity and temperature profiles versus distance along the reaction zone may be determined (see Problem 10.5). Of course, when swirl is used, the flow field becomes more complex.

10.4 EMISSIONS FROM OIL FURNACES

Particulate and nitrogen oxides are the emissions of primary concern from distillate fuel oil furnaces, while sulfur oxides must also be considered when burning residual fuel oil. Fuel oil contains a small but significant percentage of noncombustible ash. Number 6 fuel oil contains up to 1.5% ash, while no. 2 fuel oil contains 0.1% ash. Representative composition of ash from no. 6 fuel oil is given in Table 10.1. Particulate emissions from residual fuel oil burners are due to three sources: ash, soot, and sulfates. Approximately 5% of the sulfur in the oil is converted to sulfur

TABLE 10.1
Representative ash from residual fuel oil

Compound	Weight %
SiO_2	1.7
Al_2O_3	0.3
Fe_2O_3	3.8
CaO	1.7
MgO	1.1
NiO	1.9
V_2O_5	7.9
Na_2O	31.8
SO_3	42.3

trioxide, while the remainder goes to sulfur dioxide. The SO_3 readily goes to H_2SO_4 and calcium, magnesium, and sodium sulfate. Vanadium pentoxide, which is present in some oils to a significant extent, acts as a catalyst to convert SO_2 to SO_3, and hence increased vanadium means increased particulate emissions. Sulfate compounds increase opacity (light scattering) and particulate emissions. Also, in the upper part of a stack the flue gas temperature can drop below the dewpoint of sulfuric acid. If this happens, the acid condenses on the fly ash, making it sticky and prone to deposition on walls of the stack. Periodically this material comes loose in flakes. This is the so-called smut problem.

Both sodium and vanadium in fuel oil may form sticky ash compounds having low melting point temperatures. These compounds lead to fouling of heat exchanger surfaces. Frequent soot blowing is needed to clean these surfaces. Fuel oil additives such as alumina, dolomite, and magnesia have been found effective in reducing fouling and corrosion. Additives either produce higher melting point ash deposits which do not fuse together, or form refractory sulfates which are more easily removed in soot blowing. Other fuel oil additives may reduce smoke. Organometallic compounds of manganese, iron, and cobalt have a catalytic influence on soot oxidation. Additives often increase the emissions of particulates; however, they turn sulfuric acid mist into a dry powder. Whereas soot can be controlled with good burner design, acid mist is controlled by additives.

The combustion products contain smoke, which consists of ash particles and soot. Hydrocarbon fuels such as fuel oil have a tendency to produce soot in fuel-rich zones of the combustion zone. During devolatilization at high temperature, the fuel oil tends to crack or split into compounds of lower molecular weight and then, in fuel-rich regions, to polymerize or build up again with the elimination of hydrogen (see Section 4.7). Fresh soot has the empirical formula C_8H. During the polymerization process, fuel ions nucleate to an atomic mass of about 10^4 and then form crystallites and finally soot spheres. The final stage of soot formation is aggregation of the spheres into filaments. Soot is identified by bright white-yellow radiation (continuum radiation). Hence, soot effectively transfers heat to the boiler walls. However, soot is more difficult to burn than gaseous volatile species. Thus, to avoid soot emissions, especially in smaller furnaces and boilers, it is important to bring air directly into contact with the devolatilizing spray.

Distillate oil burners exhibit a decrease in particulate emissions with increasing excess air due to reduction in soot. The *Bacharach smoke number* is used to characterize particulate emissions. In this method a small volume of flue gas is manually pulled through a filter and the darkness of the spot on the filter is matched by eye to a scale of 0 to 10. In practice the excess air is decreased (to increase the efficiency) until the smoke number exceeds 2. Design improvements such as the flame retention head allow reduced excess air. However, the flame retention head has higher NO_x emissions than the standard burner configuration due to the hotter flame zone.

Nitrogen oxide is formed from fuel-bound nitrogen and nitrogen in the combustion air. Number 6 fuel oil contains 0.1 to 0.5% nitrogen, while no. 2 fuel oil contains approximately 0.01% nitrogen. When fuel is burned, 10 to 60% of the fuel nitrogen is oxidized to NO. This fraction depends on the amount of oxygen available after the fuel molecules decompose. If the combustion zone is fuel-rich, the fuel molecules crack and much of the nitrogen forms N_2. Reduced excess air also helps to lower thermal NO and SO_3 formation. In utility boilers using residual fuel oil burners with reduced airflow, overfire air ports above the burner reduce NO emissions without excessive soot formation. Low NO_x oil burners for utilities have been developed which internally stage the air (Figure 10.15, which is similar to Figure 6.8 except that the fuel nozzle is surrounded by a flame stabilizer disk which improves flame stability and turndown). Combustion air is regulated by dual air zones with multistage swirl vanes. The flow rate and degree of swirl influence the mixing of the air into the fuel-rich core of the flame. The extended flame zone reduces the peak temperatures. Also, the low NO_x radially stratified burner, mentioned at the end of Chapter 6, can be modified for use with oil. With the radially stratified burner, the high swirl eliminates the need for the stabilizer disk in Figure 10.15.

Emission factors for generic residential, commercial, industrial, and utility burners are given in Table 10.2. These values can vary widely between models and depending on how well the units are maintained and adjusted. The federal emission standards for new large sources (greater than 250×10^6 Btu/h), are shown in

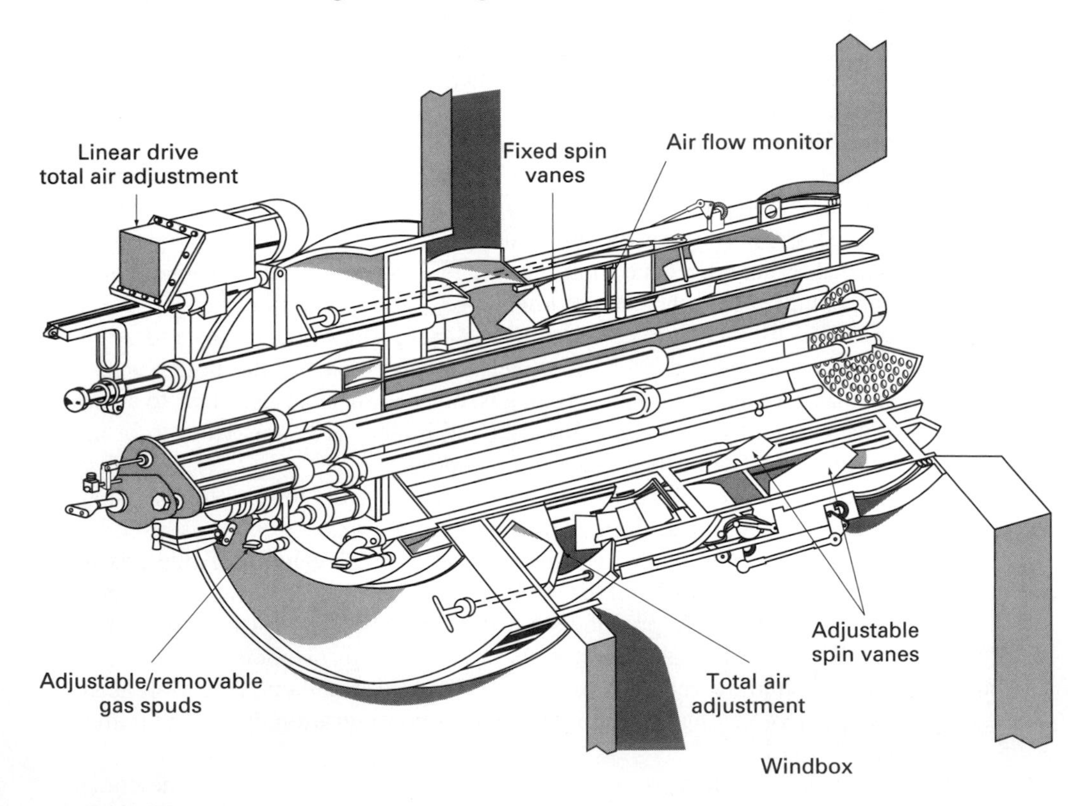

FIGURE 10.15
Low NO_x combination oil or gas burner [Stultz and Kitto, by permission of Babcock and Wilcox].

TABLE 10.2
Uncontrolled emission factors for fuel oil combustion in units of kg/10³ L of oil [EPA]

Type	Particulates	SO_2	NO_x as NO_2
Residential (D)	0.3	17S	2.2
Commercial and industrial (D)	0.24	17S	2.4
Industrial and utility (R)	1.25S + 0.38	19S	$2.75 + 50N^2$

D means distillate fuel oil; R means residual fuel oil no. 6.

S means multiply by percent sulfur in oil (3% means S = 3).

N means multiply by percent nitrogen in oil (0.2% means N = 0.2).

Table 10.3. Frequently, these standards can be met without external controls such as scrubbers or electrostatic precipitators; however, in instances where the local regulations are more strict than federal regulations or where fuel properties require it, external emission control equipment is used.

EXAMPLE 10.2. A large industrial burner proposes to use residual fuel oil containing 3% sulfur and 0.3% nitrogen. The higher heating value is 42,000 kJ/L. What degree of emissions control would be required to meet the US emission standards?

Solution. Note that 42,000 kJ/L = 42×10^6kJ/10^3L. Uncontrolled emissions are, from Table 10.2:

Particulates:

$$\frac{[1.25(3) + 0.38 \text{ kg}/10^3\text{L}](1000 \text{ g/kg})}{42 \times 10^6\text{kJ}/10^3 \text{ L}} = 98 \text{ g}/10^6 \text{ kJ}$$

Sulfur dioxide:

$$\frac{19(3)(1000)}{42} = 1357 \text{ g}/10^6 \text{ kJ}$$

Nitrogen dioxide:

$$\frac{[2.75 + 50(0.3)^2]\ 1000}{42} = 173 \text{ g}/10^6 \text{ kJ}$$

TABLE 10.3
Federal emission standards for large new residual fuel oil combustion sources in units of g/10⁶ kJ [EPA]

Emissions	Utility	Industrial
Particulates*	13	43
SO_2†	344	344
NO_x as NO_2	130	172

*Also require limit of 20% opacity.

†Also require 90% reduction but not lower than 86 g/10⁶ kJ.

Using Table 10.3, the required efficiencies of the emission controls are:

Particulates:

$$\frac{98 - 13}{98} = 87\%$$

Sulfur dioxide:

90%, which gives 135.7 g/10^6 kJ

Nitrogen dioxide:

$$\frac{173 - 130}{173} = 23\%$$

Particulate control would require a baghouse or electrostatic precipitator, SO_2 control would require a scrubber, and NO_x control would require combustion modification.

10.5 SUMMARY

Small distillate oil-fired burners use a single-fluid swirl atomizer pressurized to 7 atm. Improved mixing and reduced excess air are achieved with a flame retention head. Large industrial and utility burners use twin-fluid atomizers, often of the Y-jet type, to spray residual fuel oil at pressures to 50 atm with about 3% steam. The flame is stabilized on the refractory burner nozzle, and the shape of the flame is determined by the amount of air swirl used. The flame burns primarily as a collection of individual droplet flames, and for distillate fuel oil the burning rate is limited by the vaporization rate. The droplet vaporization rate is proportional to the droplet diameter, and the vaporization time is proportional to the square of the initial droplet diameter. For residual fuel oil the individual droplets burn in a more complicated manner, more soot is formed, and the burnout of the coke which is formed is the rate-limiting step.

Although the combustion of a burner spray involves a three-dimensional, two-phase flow with turbulent mixing, heat transfer, and chemical reaction, it is instructive to develop a simplified plug flow model of a vaporizing, reacting, monodispersed spray. In the plug flow model the droplet number density, droplet diameter, gas temperature, and gas velocity are determined as a function of time and distance.

As with gas-fired furnaces and boilers, the highest efficiency is obtained by completely burning the fuel with as little excess air as possible. The emissions of concern for well-adjusted oil-fired systems are particulates, sulfur dioxide, and nitrogen oxides. Particulates consist of fly ash, soot, and sulfates. Residual fuel oil contains up to 1.5% ash, 5% sulfur, and 0.5% nitrogen, while distillate fuel oil is one to two orders of magnitude lower in these percentages. Soot can be reduced by improved atomization and mixing. Nitrogen oxide emissions can be reduced by means of staging the burner air, and for utility-size systems flue gas recirculation is used. Control of particulate and sulfur dioxide emissions requires external control equipment when residual fuel oil is used.

PROBLEMS

10.1. Calculate the auxiliary power needed for a burner rated at 180,000,000 Btu/h input (based on the HHV) using no. 6 fuel oil pressurized to 1000 psig and sprayed through an atomizing nozzle. The air pressure drop across the burner is 5 in. H_2O and the inlet air is 300°F. Assume 10% excess air is used. The efficiency of the pump is 90%, and the efficiency of the blower is 70%. Calculate the power required by the pump and by the blower in Btu/h and horsepower.

10.2. An industrial boiler produces 150,000 lb_m/h of saturated steam at 650 psia using residual fuel oil (no. 6). The ambient air temperature is 0°F, which is also the temperature of the oil storage tank. There is no air preheater, but the oil is preheated to 212°F. The oil is pressurized to 750 psig. There is a 2-in. water pressure drop across the air ducts and windbox. An economizer heats the feedwater to 280°F. The stack temperature is 380°F. Assume that heat loss due to incomplete combustion is 0.5%, and that radiation losses are 5% of the total heat input. The excess air is measured to be 5%. The efficiency of the pump is 90%, and the efficiency of the blower is 70%. Calculate the boiler efficiency. Use properties of air for the products. Use the approach developed in Chapter 6, and data from Chapters 2 and 3 and the Appendixes. What recommendations can you make to attempt to improve the boiler efficiency?

10.3. A commercial oil-fired furnace is rated at 300,000 Btu/h input and uses no. 2 fuel oil. An atomizing nozzle is used with a pressure of 75 psig and a discharge coefficient of 0.8. Find the fuel flow rate (gal/h) and the diameter of the nozzle (in.).

10.4. Approximately how long does it take 50- and 300-μm distillate fuel oil droplets to burn out in a furnace? Should this result depend on the flame temperature, excess air, or turbulence? Explain.

10.5. For Example 10.1, calculate and plot the velocity and the temperature versus distance along the reaction zone. This can be done either analytically or using a numerical equation solver.

10.6. What is the maximum percent sulfur in residual fuel oil which can be burned without exceeding the federal emission standards?

REFERENCES

Babbington, R. S., McLean, Virginia, personal communication, 1993.

Burkhardt, C. H., *Domestic and Commercial Oil Burners*, McGraw-Hill, New York, 1969.

EPA, "Compilation of Air Pollution Emission Factors, Vol. I Stationary Point Sources and Area Sources," U.S. Environmental Protection Agency, EPA-AP-42, 1985.

Hall, R. E., Wasser, J. H. and Berkau, E. E., "A Study of Air Pollutant Emissions from Residential Heating Systems," U.S. Environmental Protection Agency, EPA 650/2-74-003, 1974.

Lawn, C. J. (ed), *Principles of Combustion Engineering for Boilers,* Academic Press, New York, 1987.

Lightman, P., and Street, P. J., "Single Drop Behaviour of Heavy Fuel Oils and Fuel Oil Fractions," *J. Inst Energy,* vol. 56, pp. 3–11, 1983.

Offen, G. R., et al., "Control of Particulate Matter from Oil Burners and Boilers," U.S. Environmental Protection Agency, EPA 450/3-76-005, 1976.

Sayre, A. N., Dugue, J., Weber, R., Domnick, J., and Lindenthal, A., "Characterization of Semi-Industrial-Scale Fuel-Oil Sprays Issued from a Y-Jet Atomiser," *J. Inst. Energy,* vol. 67, pp. 70–77, 1994.

Stultz, S. C., and Kitto, J. B.(eds.), *Steam: Its Generation and Use,* Babcock and Wilcox Co., Barberton, OH, 40th ed.,1992.

Williams, A., *Combustion of Liquid Fuel Sprays,* Butterworth, London, 1990.

Williams, A., "Fundamentals of Oil Combustion," in Chigier, N. (ed.), *Energy and Combustion Science,* Pergamon Press, London, 1979.

CHAPTER 11

Gas Turbine Spray Combustion

Gas turbine engines are used to produce thrust for aircraft and power for stationary applications. As shown schematically in Figure 11.1, the basic components consist of a rotating compressor, a combustor, and a turbine which drives the compressor plus a load such as an electric generator. In aircraft applications the combustion products are expanded through the turbine and exhaust at high velocity to produce thrust. In industrial applications the turbine generates shaft power, and the exhaust can be used in a heat recovery boiler for process heat or to drive a steam turbine in a combined cycle.

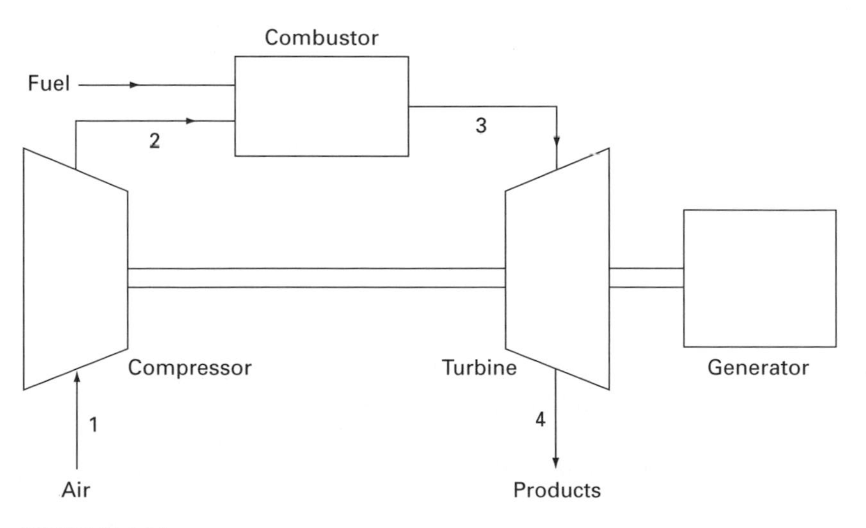

FIGURE 11.1
Schematic diagram of open-cycle gas turbine system.

Gas turbines have high output per unit volume compared with piston engines and hence are well suited for aircraft applications. However, the efficiency is relatively low because the exhaust temperature is high. Stationary source applications which utilize the exhaust gas from the turbine realize increased efficiency. Gas turbines primarily use distillate liquid fuels or natural gas in order to minimize damage to the turbine blades. Due to the need to conserve these premium fuels, there is recent interest in using residual fuel oils, coal, and biomass for industrial and utility gas turbines; this introduces additional concerns regarding erosion, corrosion, and deposition with respect to the turbine blades and seals.

First, the operating parameters for gas turbine combustors are considered, and then combustor design, combustion processes, heat transfer considerations, and emissions are examined in more detail.

11.1 GAS TURBINE OPERATING PARAMETERS

Gas turbine combustors operate at pressures of 3 atm for small simple engines to as high as 40 atm for advanced engines. Aircraft engines operate at compression ratios of 20/1 to 40/1, while stationary gas turbine combustors operate at 10 to 15 atm. The combustor inlet temperature depends on the compressor pressure ratio and ranges from 200 to 500°C. Combustor outlet temperatures are set by the metallurgical requirements of the turbine blades and range from 1300 to 1700°C for aircraft turbines and 1000 to 1500°C for stationary turbines. The highest cycle efficiency is achieved with the highest feasible turbine inlet temperature. The cycle efficiency for new advanced stationary turbines is about 35%. Since the stoichiometric flame temperature of gas turbine fuels is 2000°C or more, 100 to 150% excess air is used to achieve the desired turbine inlet temperature. Some of this excess air is used to cool the combustor liner to a suitable operating temperature of around 800°C. An example of design parameters for a large industrial gas turbine, which can be used as a stand alone simple-cycle system or in a cogeneration combined-cycle system, is given in Table 11.1.

To further consider the relationship between combustion pressure, temperature, and cycle efficiency, consider the ideal gas turbine or Brayton cycle shown in the p-V and T-s diagrams of Figure 11.2. Compression (1–2) and expansion through the turbine (3–4) are isentropic. Pressure drop through the combustor is usually about 3% but will be neglected here. The net work output of the cycle is

$$\dot{W}_{\text{net}} = \dot{m}[(h_3 - h_4) - (h_2 - h_1)] \tag{11.1}$$

For an ideal gas with constant specific heat, it follows that $p_2/p_1 = p_3/p_4 = p_r$, and $T_2/T_1 = T_3/T_4 = T_r$, and Eq. 11.1 becomes

$$\dot{W}_{\text{net}} = \dot{m}c_p\left[T_3\left(1 - \frac{1}{T_r}\right) - T_1(T_r - 1)\right] \tag{11.2}$$

By inspection of Eq. 11.2, there is an optimum temperature ratio for maximum power at fixed turbine inlet temperature. Differentiating Eq. 11.2 with respect to

TABLE 11.1
Heavy-duty industrial gas turbine design parameters

Simple cycle gas turbine system:	
Power output	136 MW
Heat rate	10,390 Btu/kWh
Combined cycle system with gas turbine and steam turbine:	
Power output	200 MW
Heat rate	6828 Btu/kWh
Compressor:	
Number of stages	18
Overall pressure ratio	13.5
Air flow rate	633 lb_m/s
Turbine:	
Number of stages	3
Inlet temperature	1260°C
Outlet temperature	593°C
Combustors:	
Number of chambers	14
Number of fuel nozzles per chamber	6

Source: Brandt, article courtesy of *Mechanical Engineering* magazine vol. 109, No. 7, July 1987, pp. 28–36, © *Mechanical Engineering* magazine (ASME Int.).

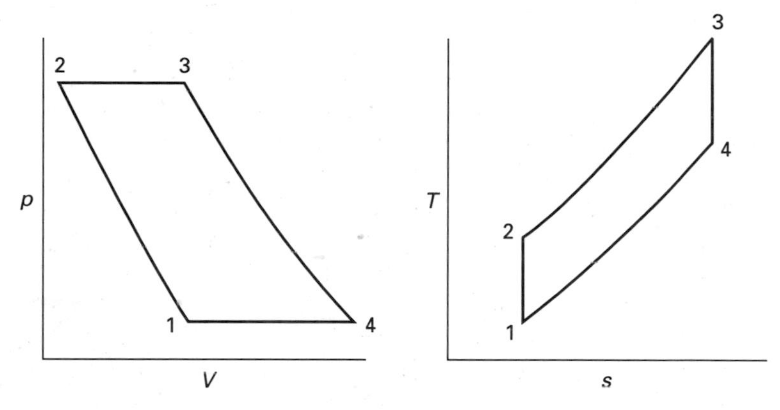

FIGURE 11.2
Ideal Brayton cycle pressure-volume and temperature-entropy diagrams.

T_r and setting it equal to zero yields

$$T_3 = \frac{T_2^2}{T_1} \tag{11.3}$$

or

$$(T_r)_{\text{opt}} = \left(\frac{T_3}{T_1}\right)^{1/2} \tag{11.4}$$

For isentropic compression the pressure ratio for maximum power is

$$(p_r)_{\text{opt}} = [(T_r)_{\text{opt}}]^{\gamma/(\gamma-1)} = \left(\frac{T_3}{T_1}\right)^{\gamma/2(\gamma-1)} \tag{11.5}$$

For example, for inlet air at 300 K and 1 atm the combustor pressure for maximum power is 11 atm if the combustor temperature is 1200 K, and 23 atm if the combustor temperature is 1800 K. The turbine efficiency, on the other hand, continues to increase with increasing compression ratio.

The gas turbine cycle efficiency is the net power output divided by the net energy output of the combustor, or

$$\eta = \frac{(h_3 - h_4) - (h_2 - h_1)}{h_3 - h_2}$$

$$= \frac{h_3(1 - h_4/h_3) - h_2(1 - h_1/h_2)}{h_3 - h_2}$$

$$= 1 - \frac{T_4}{T_3}$$

or

$$\eta = 1 - \frac{1}{p_r^{(\gamma-1)/\gamma}} \tag{11.6}$$

The *heat rate* (HR) is defined as the combustion energy input from the fuel divided by the net power output, which is the inverse of the cycle efficiency. In power engineering it is customary to use Btu/h input and kW output, and thus,

$$\text{HR} = \frac{3413}{\eta} \quad \text{(Btu/kWh)} \tag{11.7}$$

EXAMPLE 11.1. For compressor inlet air at 300 K and 1 atm and for a combustor outlet temperature of 1260°C, find the combustor pressure for maximum work. At this optimum pressure find the ideal compressor outlet temperature, the ideal cycle efficiency, and ideal heat rate. Assume a simple, ideal air cycle gas turbine and refer to Figure 11.2.

Solution. From Eq. 11.5,

$$p_2 = p_1\left(\frac{T_3}{T_1}\right)^{\gamma/2(\gamma-1)}$$

Since,

$$p_2 = p_3 \quad \text{and} \quad \gamma = 1.4$$

we find

$$p_3 = (1)\left(\frac{1260 + 273}{300}\right)^{1.75} = 17.4 \text{ atm}$$

For an isentropic compression from 1 to 17.4 atm the temperature rises to

$$T_2 = 300(17.4)^{(1.4-1)/1.4} = 679 \text{ K}$$

Using Eqs. 11.6 and 11.7,

$$\eta = 1 - 17.4^{-(1.4-1)/1.4} = 0.558$$

$$\text{HR} = \frac{3413}{0.558} = 6114 \text{ Btu/kWh}$$

Comparing these results with Table 11.1, we see that the pressure is too high and the heat rate is too low. This is because the compressor and turbine efficiency and combustor pressure loss have not been taken into account. Nevertheless the basic idea is correct. For a presentation of these equations for the nonideal gas turbine cycle see the book by El-Wakil.

11.2 COMBUSTOR DESIGN

Aircraft gas turbines must typically operate over a wide load range. The mass flow of air through the combustor increases as the load (and thus turbine rpm) increases. However, since the pressure also increases, the velocity through the combustor remains relatively constant with changes in load. Fuel flow rates may vary as much as 40 : 1; however, the range of fuel-air ratio varies less than 3 : 1. In addition to operating at high pressure and controlled temperature, the combustor should be designed to provide rapid and reliable ignition, operate over a wide range of mixtures without danger of blowoff, have minimum loss in pressure, have uniform exit temperature, be small in size and durable, and have low emissions. Industrial gas turbines have similar but less stringent requirements.

A conventional gas turbine combustor consists of (Figure 11.3) an inlet diffuser section, a fuel injector, an air swirler, a primary combustion zone, an intermediate combustion zone, a dilution zone, and a liner with holes and slots. The flow pattern through the combustor is shown in Figure 11.4. The diffuser section reduces the flow velocity and splits the flow between the primary combustion zone (15 to 20%) and the liner air (80 to 85%). The reduced velocity in the primary zone reduces the pressure loss due to heat addition and also reduces the chance for

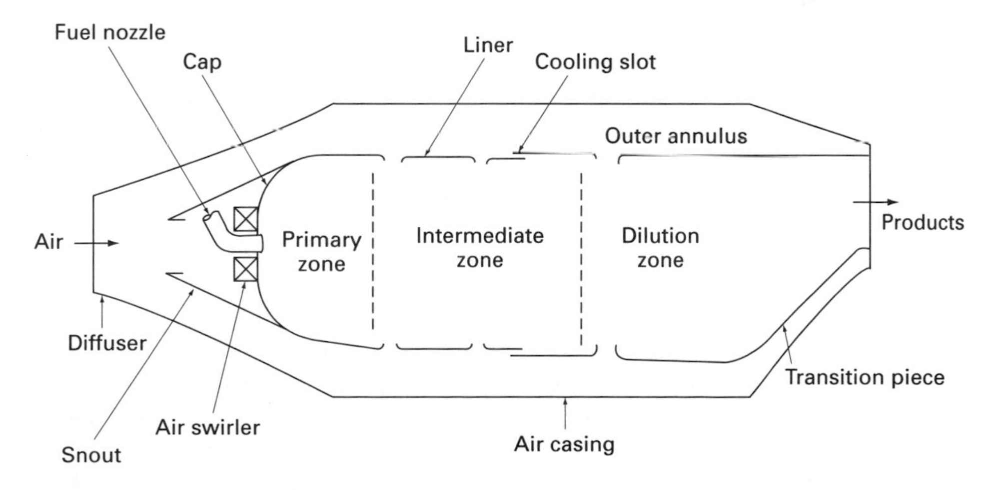

FIGURE 11.3
Main components of a gas turbine combustor [Lefebvre (1983), from *Gas Turbine Combustion,* p. 13, reproduced with permission of Taylor and Francis, Inc., Washington, D.C. All rights reserved.].

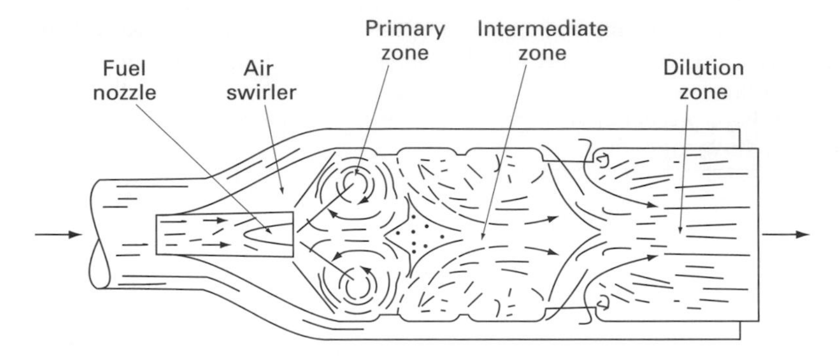

FIGURE 11.4
Flow pattern in combustor created by swirl vanes and radial jets (20% of air added in primary zone, 30% in intermediate zone, and 50% in dilution zone).

blowoff of the flame. The liner mixes air into the intermediate combustion zone and the dilution section to reduce the temperature to acceptable limits for the turbine blades. The primary combustion zone operates near stiochiometric, and although effective, it tends to form relatively high NO_x. The need to reduced emissions from turbine combustors has stimulated combustor redesign, and these new developments are discussed in Section 11.5.

Most fuel injectors used in gas turbine combustors are of the pressure-swirl or air-blast types (see Chapter 9). Typical Sauter mean diameters are 30 to 60 microns, and fuel nozzle pressure drops are in the range of 10 to 20 atm. Aircraft gas turbines use distillate fuels such as Jet A, while industrial gas turbines use kerosene, fuel oil, or natural gas.

There are three basic configurations of gas turbine engines: annular, cannular, and silo (Figure 11.5). In the *annular* configuration there is one annular combustion chamber mounted concentrically inside an annular casing. Older aircraft and heavy duty industrial gas turbines use the *cannular* design where up to 18 tubular combustion 'cans' are mounted in an annular plenum. Some industrial gas turbines use the single *silo* combustor design.

The aircraft gas turbine propulsion system shown in Figure 11.6 is a turbofan engine which generates 61,500 lb_f thrust. The high pressure turbine drives the fan and the compressor; the low pressure turbine and the fan provide the thrust. The annular combustor, which is very compact and lightweight, delivers a nearly uniformly heated gas to power the turbine. The heavy duty industrial engine of cannular design is similar to the one shown in Figure 11.6 except that there is no fan and low pressure compressor, and no nozzle. The cannular combustion chambers tend to be longer and thus have a longer residence time. An electrical generator is connected to the turbine shaft, and the engine runs at constant speed as the load varies. The gas turbine engine with a single silo combustor, which can be used for harder-to-burn fuels, is shown in Figure 11.7.

In either configuration, the combustion process is similar. Fuel is sprayed through an atomizing nozzle into the primary combustion zone. A spark plug is

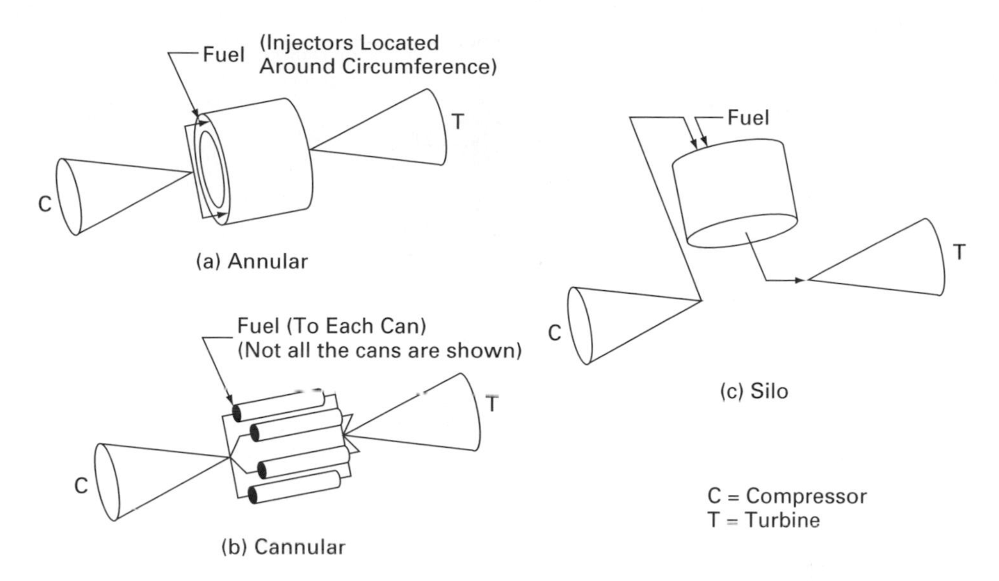

FIGURE 11.5
Gas turbine types: (*a*) annular combustor, (*b*) cannular combustors, and (*c*) silo combustor.

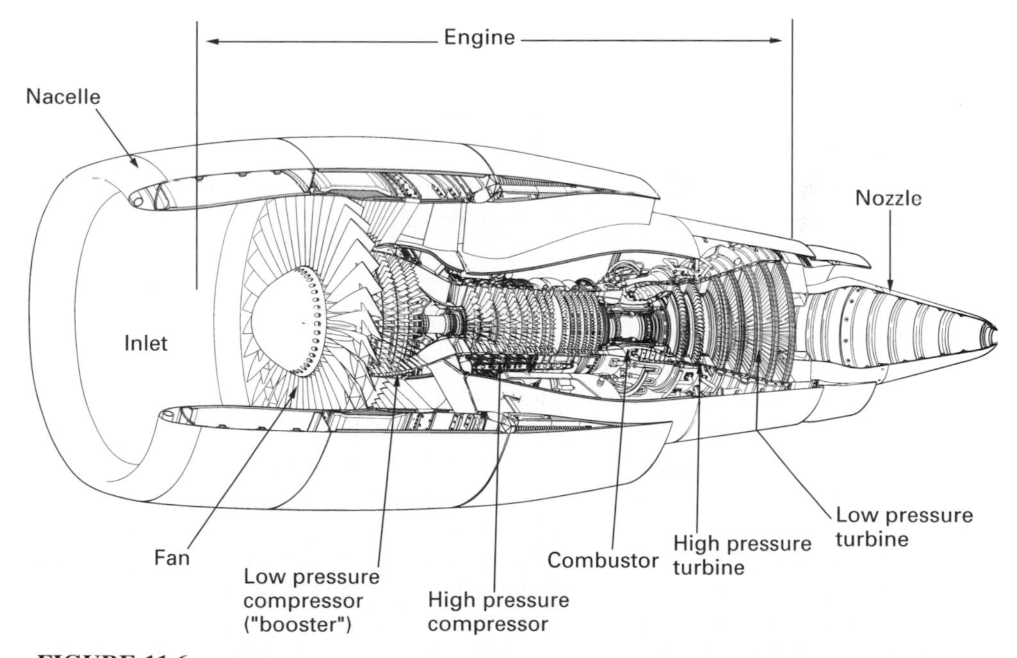

FIGURE 11.6
Aircraft gas turbine engine with annular combustion chamber. This engine has a mass flow of 1790 lb_m/s, pressure ratio of 31.9, overall length of 13.4 ft, fan diameter of 7.75 ft, weight of 9750 lb_m, and takeoff thrust of 61,500 lb_f [Courtesy of General Electric Co.].

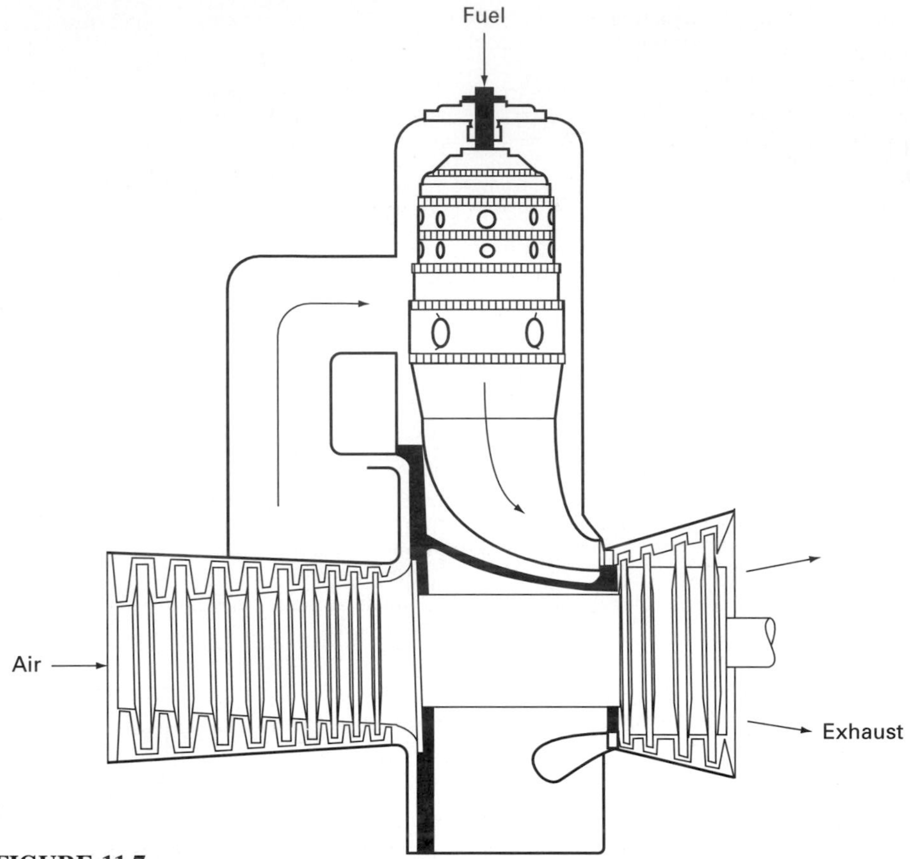

FIGURE 11.7
Industrial gas turbine with silo combustor [Lefebvre (1983), from *Gas Turbine Combustion,* p. 30, reproduced with permission of Taylor and Francis, Inc., Washington, D.C. All rights reserved.].

used for starting ignition. Flame stabilization is provided by swirl imparted to the primary inlet air. Swirl vanes curving 45° to the flow and located around the fuel nozzle give a tangential velocity component to the primary air. The flow expands into the combustion chamber and mixes with the fuel spray. As shown in Figure 11.4, a recirculation pattern is set up by means of the vortex motion created by the inlet swirl because the pressure is lower in the center of the vortex and because of the increasing axial pressure gradient in the primary combustion zone due to addition of secondary air through the walls of the flame tube. This provides aerodynamic stabilization of the flame by back-mixing of hot combustion products. The primary combustion zone is maintained at near-stoichiometric conditions.

Secondary air flows between the casing and the liner, and is admitted gradually into the combustion chamber through holes and slots. In this way cooling of the liner is achieved, combustion is completed, and combustion products are cooled to the required turbine inlet temperature. Approximately 30% of the air flows through the liner into the intermediate combustion zone, and the remaining 50% is added in the dilution zone. The size and number of holes in the liner is a compromise between a large number of small holes to give fine scale mixing and a smaller number of large holes to give better penetration and to cool the center core before entry into the turbine. By overlapping sections of the liner, a thin slot is formed to provide film cooling of the liner wall. Liners for aircraft combustors are typically only 1 mm thick, and hence the design of the slots is critical. The flow pattern shown in Figure 11.4 is also applicable to annular combustors, which use a ring of fuel nozzles and air swirlers discharging into the annular space.

The size of the combustor depends on the fuel type, droplet size distribution, and other factors. The cross-sectional area may be roughly sized by noting that aircraft turbines have an average velocity of 25–40 m/s, whereas industrial combustors typically have velocities of 15–25 m/s. The length of the combustor also varies, but cannular combustors typically have a length-to-diameter ratio of 3 to 6 based on the liner and 2 to 4 based on the casing. Combustors designed for aircraft jet fuels are also satisfactory for natural gas. More space and better liner cooling are needed for no. 2 fuel oil, since the fuel oil droplet vaporization rate is lower and radiation heat transfer is higher. Inorganic compounds in the fuel must be carefully limited to avoid erosion of the turbine blades.

Ignition

Ignition and re-light are of obvious importance in aircraft gas turbines, where flameout without reliable re-light can be disastrous. Typically, the ignition is by a spark plug using short-duration pulses at a continuous rate of 60–250 per minute. Surface discharge igniters which utilize a thin film of semiconductor material to separate the electrodes at the firing end are also used. The electrical resistance of the semiconductor falls with increasing temperature, giving a rapid discharge. The surface igniter gives optimal performance when a thin layer of fuel is on the plug face.

The ignition process itself begins with the formation of the kernel as discussed in Chapter 5. The flame must now spread through the chamber, and for cannular chambers it must also spread from lighted to unlighted liners. Typically, good flame spread conditions are similar to the conditions for good stability: reduced primary zone velocity, increased pressure and temperature, and near stoichiometric fuel-air ratios. Short, large-flow-area interconnectors with low heat loss facilitate flame spread between liners.

In Chapter 5 it was shown that the minimum ignition energy could be expressed in terms of the quench distance. This type of reasoning can be extended for spray conditions to include the effects of vaporization rates and mixing. For mists of uniform-sized droplets in quiescent atmospheres the quench distance is proportional to the droplet diameter and to the square root of fuel density. The minimum ignition energy is proportional to the cube of quench distance, and thus changes in the spray droplet size have an important effect on ignition. The low vaporization rate of the large-size droplets does not provide the fuel-air mixture needed for ignition. Figure 11.8 shows the effect of drop size on minimum ignition energy for several fuels.

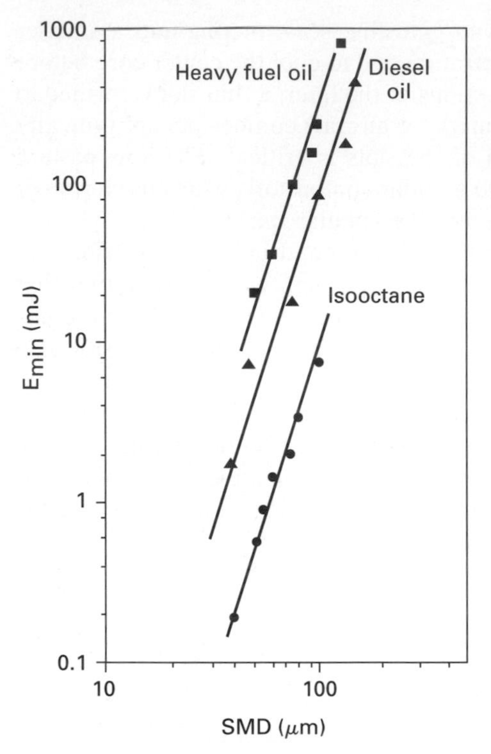

FIGURE 11.8
Effect of mean drop size on minimum ignition energy for conditions of 1 atm, 15 m/s and f = 0.65 [Reprinted by permission of Elsevier Science Inc. from "Ignition and Flame Quenching of Flowing Heterogeneous Fuel-Air Mixtures," by Ballal and Lefebvre, *Combustion and Flame,* vol 35, pp. 155–168, © 1979 by The Combustion Institute.].

Flame Stabilization

For the flame to be stable, the flame velocity must be equal and opposite to the reactant velocity at each point along the flame. For industrial burners, this presents much less of a problem than for gas turbine combustors, where the gas velocity is typically higher than the flame speed. Flame holding is achieved by creating high inlet swirl, which causes back mixing of hot products.

Experiments to determine flame stability for a given design are performed at constant inlet air temperature, pressure, and velocity. The fuel flow is varied until the lean and rich extinction limits are determined. The resulting data obtained by repeating the experiment at different airflow rates can be plotted to give a so-called stability loop diagram as illustrated in Figure 11.9. Interestingly, the shape of the curve is similar to the curve for blowout data in a well-stirred reactor (to be discussed in the next section). In general the blowout velocity and stability limits are extended by the following:

1. Reducing the main-stream velocity and turbulence intensity
2. Increasing the inlet temperature, pressure, and swirl
3. Keeping the primary combustion zone near stoichiometric
4. Using a higher volatility fuel
5. Decreasing the droplet size

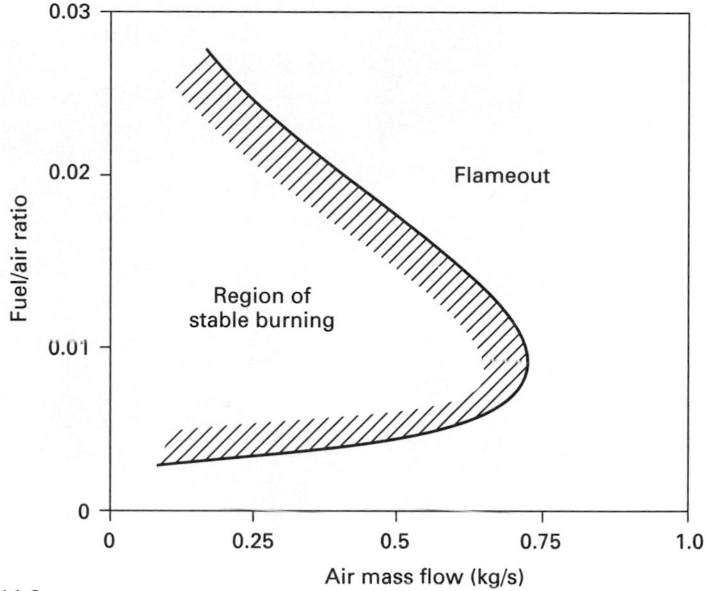

FIGURE 11.9
Combustion chamber stability plot for constant inlet pressure [Lefebvre (1983), from *Gas Turbine Combustion,* p. 180, reproduced with permission of Taylor and Francis, Inc., Washington, D.C. All rights reserved.].

Maximum stability is achieved by injecting primary air through a small number of large holes. The basic idea is to produce large-scale circulation patterns. Fuel effects are not important at high power conditions, but are important at idle conditions, where the flame stabilization becomes vaporization-limited. For pressure atomizers the inhomogeneity of the resulting mixture helps stability at the lean limit by providing zones of richer mixture. For prevaporized combustors or air-blast atomizers the mixture is nearly homogeneous, and well-stirred reactor correlations can be applied to predict the lean limit.

A Specific Combustor Design

Figures 11.10 and 11.11 show one of the 14 combustors for the heavy-duty industrial gas turbine indicated in Table 11.1. The 13° cant of the combustors allows for a more compact combustor and gas turbine design. The combustion system consists of a fuel nozzle assembly, a flow sleeve, a cap and liner assembly, and a transition piece. The combustion system incorporates cooling technology to permit firing temperatures of 1560 K. Ignition is achieved with a spark plug in 2 of the 14 chambers, with cross-firing connections to ignite the balance of the chambers. Successful ignition is sensed with four ultraviolet flame detectors located opposite the spark plugs.

The airflow paths are as follows. After leaving the diffuser area of the compressor, air flows past the transition pieces and forward toward the head end of the liner.

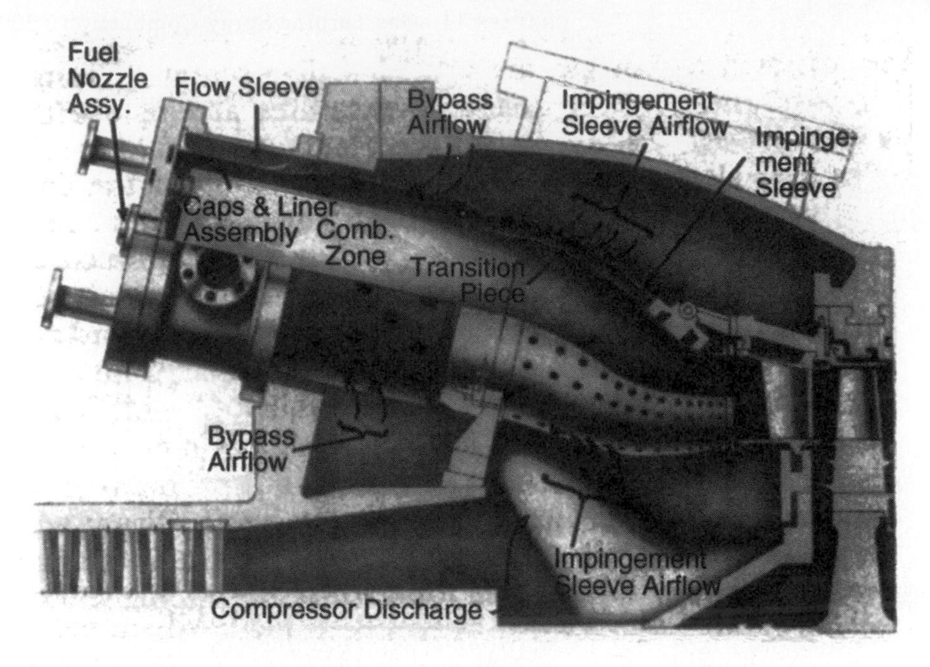

FIGURE 11.10
Gas turbine combustor showing components [Claeys et al., *J. Eng. Gas Turbines and Power,* © 1993, by permission of ASME].

The air splits between two parallel paths before entering the liner. Some of the air flows through the holes provided in the impingement sleeve of the transition piece for cooling; the balance of the air passes through an array of multiple holes in the flow sleeve. The air outside the liner passes through the cap of the liner, where the air and fuel react and flow downstream through the intermediate zone and transition piece before entering the turbine section.

The fuel nozzle assembly is designed to use either gas or distillate fuel and to transfer from one fuel to the other during operation. Six individual, dual-fuel nozzles are attached to an internally manifolded cover. Each fuel nozzle consists of a swirl tip, an atomizing air tip, and a distillate spray nozzle. Gaseous fuel is injected through metering holes in the swirler. Continuous atomizing air is used with the distillate fuel spray. Water or steam for NO_x control also enters through the combustion cover. The multinozzle design offers two significant advantages over single-nozzled designs. First, it allows more thorough mixing and control of the fuel and air in the reaction zone, resulting in a shorter flame length and improved combustor performance. Second, it produces significantly lower noise levels than single-nozzle combustors and thereby improves combustor component wear.

The cap and liner assembly, shown in Figure 11.11a, consists of a slot film-cooled, 0.35-m-dia, 0.76-m-long liner sleeve and a multinozzle cap assembly. The relatively short liner length is another benefit of the multinozzle design, which produces a shorter reaction zone than single-nozzle combustors. The shorter liner has less surface area to cool. Thus, in spite of the higher firing temperature, the liner operates at temperatures less than or equal to those of earlier designs. The cap is

FIGURE 11.11
Gas turbine combustor (*a*) cap and liner, (*b*) transition piece [Claeys et al., *J. Eng. Gas Turbines and Power,* © 1993, by permission of ASME].

film-cooled in the center and impingement-cooled in the region surrounding the fuel nozzles. The surfaces of the cap and liner exposed to hot gas are coated with a ceramic thermal barrier to further limit metal temperatures and reduce the effects of thermal gradients. The residence time in the combustor is about 20 ms.

The flow sleeve forms an annular air space about the liner sleeve through which spent cooling air from the transition piece is guided and the balance of combustion air is admitted by way of the array of holes. The flow sleeve also acts to maintain a sufficient flow velocity on the back side of the liner for liner cooling.

The transition piece (Figure 11.11b) directs the hot gases from the exit of the liner sleeve to the inlet of the turbine. The transition piece is surrounded by a sleeve in which an array of holes is provided. Air flows through the holes and impinges on the back side of the transition piece, thereby providing an effective impingement-type cooling of the surface. The shape of this impingement sleeve roughly parallels that of the transition piece with the gap between the two increasing from the aft end to the forward end to accommodate the increase in airflow from each of the consecutive rows of cooling holes. The use of impingement cooling allows more precise control of the cooling of the transition piece while having only a small, adverse effect on the combustion system's overall pressure drop. The internal surfaces of the transition piece have a ceramic thermal barrier coating applied to limit the maximum metal temperature.

11.3 COMBUSTION RATE

Aircraft gas turbines require that the combustion chamber be as small as possible, and thus the combustion rate should be as high as possible. Gas turbines used for stationary power need not meet such tight size constraints, although aircraft derivative engines are often used for stationary power generation. The combustor designer needs to know how the operating conditions, such as inlet pressure and temperature, control the combustion rate, and how to scale the size of the combustor for different fuels. The processes taking place in the primary combustion chamber include fuel atomization and vaporization, mixing, and chemical reaction. While relatively little is known about the detailed behavior occurring inside the combustor, gas turbine combustor technology has successfully advanced using intuitive design based on years of experience, extensive testing, careful use of empirical data, and computational fluid dynamics simulations. Due to the high-temperature, highly turbulent convective and radiant heat transfer, the droplet vaporization rate is very rapid. Mixing between the fuel vapor and gas in the primary combustion chamber occurs rapidly due to the swirling inlet flow, jets from the liner ports, and local turbulence on the scale of the drop spacing.

Because the vaporization and mixing rates are fast, and because the high swirl causes back mixing, it is instructive to model the primary zone of the combustor (Figure 11.3) as a *well-stirred reactor*. A well-stirred reactor, shown schematically in Figure 11.12, is a way to rapidly achieve the well-mixed state. Fuel and air flow through opposed jets into a spherical, insulated volume. Combustion products

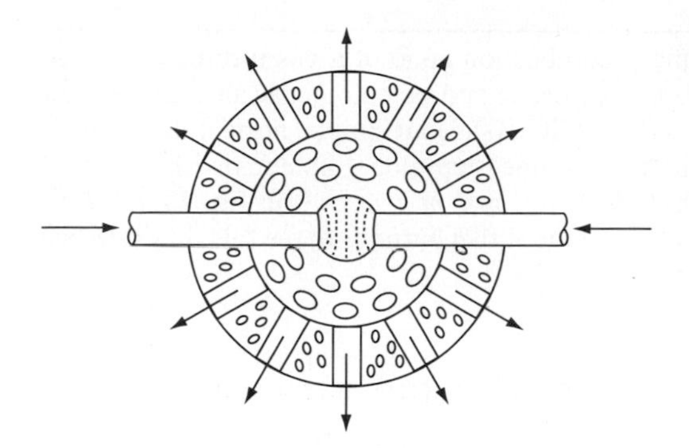

FIGURE 11.12
Physical model of well-stirred reactor.

flow outward through holes in the sphere. Because the mixing is intense, the temperature and species composition in the chamber is assumed uniform. The inlet mass flow rate is constant.

The species conservation equations for the well-stirred reactor are

$$\dot{m}(y_i - y_i^*) = \hat{r}_f M_f V \tag{11.8}$$

where y is the mass fraction, i indicates the species, V is the chamber volume, $\hat{r}_f$ is the molar reaction rate of the fuel per unit volume, and the asterisk designates the inlet conditions. Conservation of energy for the well-stirred reactor is expressed by

$$\dot{m} \sum_{i=1}^{I} (y_i h_i - y_i^* h_i^*) + q = 0 \tag{11.9}$$

where q is the heat loss through the walls and h is the absolute enthalpy. The nominal residence time can be obtained by,

$$\tau = \frac{\rho V}{\dot{m}} \tag{11.10}$$

where

$$\rho = \frac{p\overline{M}}{RT} \tag{11.11}$$

Equations 11.8 and 11.9 can involve many species and reaction steps. By way of simplification let us consider a one-step global chemical reaction of fuel and air going to products:

$$\dot{m}_f = \hat{r}_f M_f V \tag{11.12}$$

From Eq. 4.16, the molar reaction rate of the fuel is

$$\hat{r}_f = AT^n p^m \exp\left(\frac{-E}{RT}\right)(n_f)^a (n_{O_2})^b \tag{11.13}$$

Equations 11.8–11.13 provide a sufficient set of equations to determine the maximum heat release in a given volume, as well as the combustion efficiency. Let us explore this in Example 11.2.

EXAMPLE 11.2. Consider the primary combustion zone of a gas turbine as a well-stirred reactor with a volume of 900 cm^3. Kerosene and stoichiometric air at 298 K flow into the reactor, which is at 10 atm and 2000 K. To keep it simple, neglect dissociation and heat loss. Use LHV = 42.5 MJ/kg. Use one-step global kinetics with $A = 5 \times 10^{11}$, E = 30,000 cal/gmol, $a = 0.25$, $b = 1.5$, and $m = n = 0$ in Eq. 11.13. Take kerosene as $C_{12}H_{24}$. Find the fractional amount of fuel burned, η; the fuel flow rate, $\dot{m}_f$; and the residence time in the reactor, τ.

Solution. The reaction is

$$C_{12}H_{24} + 18(O_2 + 3.76N_2) \rightarrow 12\eta CO_2 + 12\eta H_2O + (1 - \eta)C_{12}H_{24} + 18\,(1 - \eta)O_2 + 67.7N_2$$

The stoichiometry is

$$\frac{\dot{m}_f}{\dot{m}_a} = \frac{1}{(18)(4.76)(29/168)} = 0.0676$$

or

$$\dot{m}_a = 14.8\,\dot{m}_f$$

Conservation of mass requires that

$$\dot{m}_a + \dot{m}_f = \dot{m}_p$$

Conservation of energy (using sensible enthalpies) is expressed by

$$\dot{m}_a h_{sa} + \dot{m}_f(h_{sf} + \eta\ \text{LHV}) = \dot{m}_p h_p$$

Since $h_{sa} = h_{sf} = 0$ in this case, the energy equation yields,

$$\eta = \left(\frac{\dot{m}_p h_p}{\dot{m}_f\,\text{LHV}}\right) = \frac{(1 + 14.8)(56.14/28)}{42.5} = 0.740$$

where h_p is taken from the N_2 tables of Appendix C for simplicity.

Evaluating the species in the reactor,

$$x_f = \frac{1 - \eta}{67.7 + 18 + 1 + (12 + 12 - 1 - 18)\eta} = 0.00288$$

$$n_f = \frac{x_f p}{\hat{R}T} = \frac{(0.00288)(10\text{ atm})}{(82.05\text{ cm}^3\cdot\text{atm/gmol}\cdot\text{K})(2000\text{ K})} = 1.76 \times 10^{-7}\text{ gmol/cm}^3$$

and

$$n_{O_2} = 18 n_f$$

For a stoichiometric reaction, the reaction rate from Eq. 11.13 is

$$\hat{r}_f = -5 \times 10^{11} \exp\left[\frac{-30{,}000\text{ cal/gmol}}{(1.987\text{ cal/gmol}\cdot\text{K})(2000\text{ K})}\right] \times (1.76 \times 10^{-7})^{0.25}\,(3.16 \times 10^{-6})^{1.5}\text{ gmol/cm}^3\cdot\text{s}$$

$$= -0.0303\text{ gmol/cm}^3\cdot\text{s}$$

Next, the species continuity (Eq. 11.8) is solved to obtain the mass flows:

$$\dot{m}_p y_f - \dot{m}_f = \hat{r}_f M_f V$$

or
$$15.8\dot{m}_f\left(\frac{x_f M_f}{M_p}\right) - \dot{m}_f = \hat{r}_f M_f V$$

Thus,

$$\dot{m}_f = \frac{\hat{r}_f M_f V}{(15.8 x_f M_f/\overline{M}_p) - 1}$$
$$= \frac{-(0.0303 \text{ gmol/cm}^3\cdot\text{s})(168 \text{ g/gmol})(900 \text{ cm}^3)}{(15.8)(0.00288)(168/29) - 1}$$
$$= 6220 \text{ g/s} = 6.22 \text{ kg/s}$$

It follows that

$$\dot{m}_a = 14.8\dot{m}_f = 92.1 \text{ kg/s}$$

and
$$\dot{m}_p = 92.1 + 6.2 = 98.3 \text{ kg/s}$$

Finally, the characteristic residence time is

$$\tau_{\text{res}} = \frac{m}{\dot{m}} = \frac{pV/RT}{\dot{m}_p}$$
$$= \frac{(10 \text{ atm})(900 \text{ cm}^3)}{\left[\dfrac{82.05 \text{ cm}^3\cdot\text{atm/gmol}\cdot\text{K}}{29 \text{ g/gmol}}\right](2000 \text{ K})(98{,}300 \text{ g/s})} = 16.2 \times 10^{-6} \text{ s}$$

and the characteristic chemical time is

$$\tau_{\text{chem}} = \frac{-n_f}{\hat{r}_f} = \frac{1.76 \times 10^{-7} \text{ gmol/cm}^3}{0.0303 \text{ gmol/cm}^3\cdot\text{s}} = 5.8 \times 10^{-6} \text{ s}$$

Thus the chemical time is less than the residence time, as it must be for significant heat release.

If Eqs. 11.8–11.13 are solved for the conditions of Example 11.2, but with the reactor temperature as a variable, then the reactor heat output rate ($\dot{m}_f\eta \times$ LHV) varies with temperature as shown in Figure 11.13. At low temperature the heat output is low because both $\dot{m}_f$ and η are low, whereas at high temperature η is close to unity but $\dot{m}_f$ is low because the fuel concentration n_f is low, which reduces the reaction rate, r_f. In the limit $\eta = 1$, T equals the flame temperature and $\dot{m}_f$ approaches zero. As seen in Figure 11.13 for a given fuel flow rate two temperature solutions are possible with two different heat outputs. If the fuel flow rate exceeds a certain value (6.25 kg/s in this example), then blowout occurs. The maximum heat output is very close to the blowout point.

This simplified well-stirred reactor calculation is very sensitive to the chemical reaction rate. A single global reaction rate is a large approximation. A more accurate and rigorous approach is to use a detailed kinetic mechanism such as GRI-Mech (Frenklach et al.). A computer program called ThermoChemical Calculator (TCC) has been developed to solve these types of kinetic calculations (see Problem 11.11).

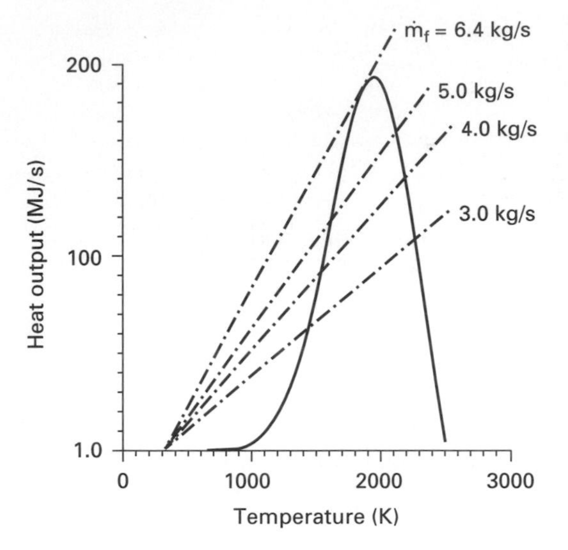

FIGURE 11.13
Output heat rate versus reactor temperature of a well-stirred reactor, following Example 11.2. Stoichiometric kerosene-air, $V = 900\ \text{cm}^3$, $p = 10$ atm.

11.4 LINER HEAT TRANSFER

The heat transfer to the combustion chamber liner comes from both convection and radiation. The primary radiation component is from the small soot particles which are formed in rich portions of the combustor. These particles are less than 1 micron in size and radiate approximately as black bodies. Because of their small size, their temperature is very close to the local gas temperature. The amount of soot formed by the combustion increases dramatically with pressure up to about 20 atm. Thus increasing pressure from 5 to 15 atm increases the radiation flux by about 4 times. Increasing the air-to-fuel ratio decreases the soot concentration and also lowers the temperature, thus greatly reducing the radiation flux. Doubling the air-to-fuel ratio can decrease the radiation by a factor of 3. The radiation flux is typically a function of distance from the injector, showing a maximum at 5–10 cm downstream of the injector. The location of the maximum moves downstream as the air-to-fuel ratio is decreased. The decrease in heat flux, following the maximum, is due to soot oxidation as well as decreasing temperature. The fraction of total heat transfer due to radiation can be as high as 50%, so it is important to be able to predict the radiation, but current correlations can only be considered as estimates. The reason for this difficulty is that to predict radiation one must be able to predict the soot production and oxidation rates, which depend on spray mixing and chemical kinetics, both of which are not well modeled at present.

Figure 11.14 shows the total heat flux to the liner wall as a function of liner diameter. As can be seen, the heat flux is very high, and thus efficient means of cooling the liner are required. Increasing pressure and inlet temperature increase the heat flux and cause a corresponding increase in liner temperature. Increased

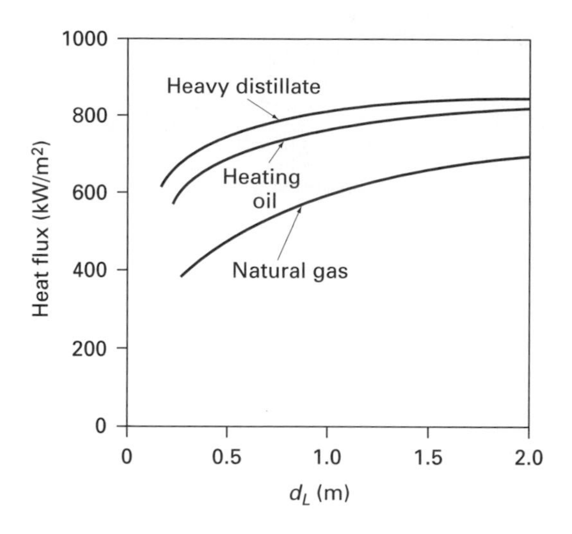

FIGURE 11.14
Typical heat flux to combustor liner wall versus liner diameter for different fuels [Lefebvre (1983), from *Gas Turbine Combustion,* p. 274, reproduced with permission of Taylor and Francis, Inc., Washington, D.C. All rights reserved.].

airflow rate decreases liner temperature because although both the cooling-side and combustion-side convection coefficients increase as $\dot{m}_a^{0.8}$, the convection component on the combustion side represents only half of the total flux. Thus a doubling of $\dot{m}_a^{0.8}$ will decrease the cooling-side thermal resistance by 2 but decrease the combustion side resistance by only $\frac{3}{4}$. Increasing the fraction of the inlet mass flow that gocs to the primary zone increases the convection coefficient in the primary zone and thus the liner temperature in that region if the air-to-fuel ratio is kept constant. The liner temperature reaches a maximum in the primary zone for mixtures 10% richer than the stoichiometric, which also corresponds to the maximum flame temperature for a homogeneous mixture.

High-performance combustor design depends on film cooling of the liner on the combustion side as well as cooling on the air side of the liner. The film cooling is provided by slots in the liner which introduce a wall jet along the inside of the liner. It is also possible to cool the wall by transpiration cooling, that is, by flowing the cooling air through a porous liner. Use of a liner with many small holes approximates transpiration cooling and is called *effusion cooling*. Figure 11.15 shows these methods in a conceptual manner.

Consider a slot of depth s, as shown in Figure 11.16, with mass flow rate per unit area of slot flow area $\rho_a \underline{V}_a$. We may define a slot Reynolds number $\text{Re}_s = s\underline{V}_a/\nu_a$. The coolant wall jet changes both the gas temperature and velocity in a region downstream of the slot. Further downstream the jet velocity decreases and the flow is better modeled as a turbulent boundary layer. The analysis requires a Nusselt number correlation, which gives the local convection coefficient in the region downstream of the slot. The effective temperature difference is taken as $T_{\text{wad}} - T_w$, where T_w is the wall surface temperature and T_{wad} is an adiabatic wall temperature calculated by using an effectiveness parameter η:

$$T_{\text{wad}} = T_g - \eta(T_g - T_a) \tag{11.14}$$

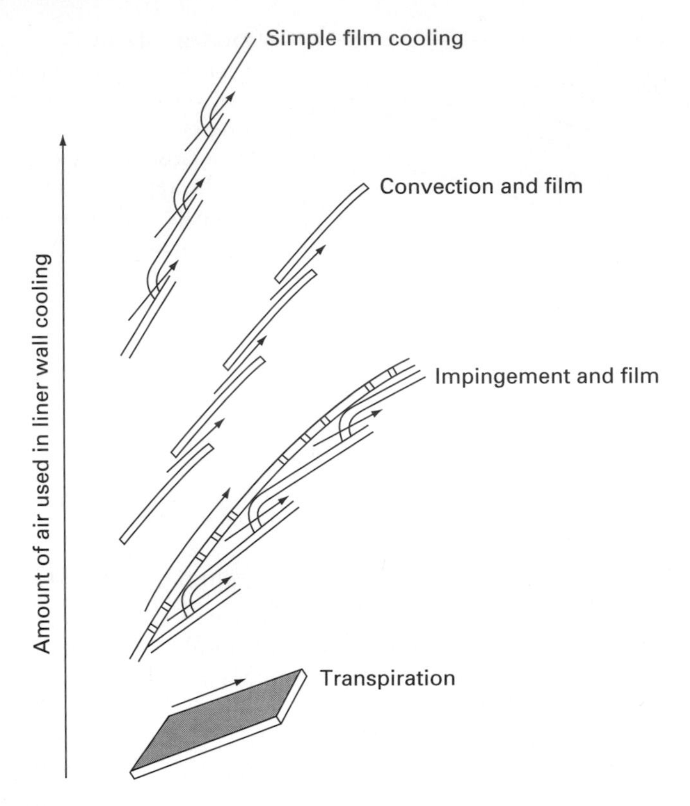

FIGURE 11.15
Types of combustor liner wall cooling designs [Lefebvre (1983), from *Gas Turbine Combustion,* p. 20, reproduced with permission of Taylor and Francis, Washington, D.C. All rights reserved.].

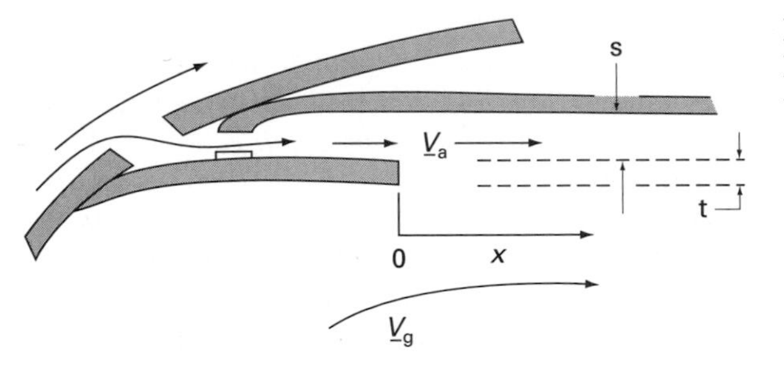

FIGURE 11.16
Notation for liner heat transfer equations.

The effectiveness is typically correlated in terms of the slot Reynolds number and the dimensionless flow parameter:

$$\tilde{W} = \frac{\rho_a \underline{V}_a}{\rho_g \underline{V}_g} \tag{11.15}$$

The effect of $\tilde{W}$, for $\tilde{W} < 1$, is to increase η as $\tilde{W}$ is increased. However, as $\tilde{W}$ is increased further the expected effectiveness increase due to increased coolant airflow is counteracted by increased turbulent mixing between the jets and the hot gas stream. Similarly, increasing the thickness of the slot lip, t, beyond $0.2s$ causes a waking action, which increases mixing and decreases the effectiveness. For clean slots with $0.50 \leq \tilde{W} \leq 1.3$, Ballal and Lefebvre (1972) give

$$\eta = 1 \qquad \text{for} \qquad \frac{x}{s} < 8 \tag{11.16a}$$

and
$$\eta = 1.10\tilde{W}^{0.65}\left[\frac{\mu_a}{\mu_g}\right]^{0.15}\left[\frac{x}{s}\right]^{-0.2}\left[\frac{t}{s}\right]^{-0.2} \qquad \text{for} \qquad \frac{x}{s} < 150 \tag{11.16b}$$

For this same range of $\tilde{W}$, the Nusselt number is given by

$$\mathrm{Nu}_x = \frac{\tilde{h}x}{k_a} = 0.069(\mathrm{Re}_x)^{0.7} \tag{11.17}$$

And the heat transfer to the liner is

$$q = \tilde{h}A(T_{\mathrm{wad}} - T_w) \tag{11.18}$$

Because a large portion of the heat transfer is by radiation, slot cooling can only partially reduce the air-side liner coolant load and the wall gas-side temperature. Transpiration or effusion cooling using multilayered walls with many interconnecting flow passages can be used to provide liner strength while also internally cooling the liner. In all cases the slots and holes are subject to clogging or partial blockage by soot. An alternative to these methods is to use a refractory liner backed by metal for strength. Refractory liners can operate with surface temperatures up to 1900 K (3000°F). Such high temperatures prevent carbon buildup and reduce wall quenching. The weight and volume of the refractory restrict their application to the industrial turbines. For aircraft, thin ceramic liner coatings, especially in downstream regions, have been attempted with modest success.

11.5 LOW-EMISSIONS COMBUSTORS

Over the last 50 years gas turbine combustors have evolved to achieve high stability and high combustion intensity (high heat release per unit volume) using swirl-stabilized diffusion flames. Today, in addition to high stability and intensity, gas turbine combustors must emit as low carbon monoxide, hydrocarbons, nitrogen oxides, and soot as is technically possible. The main factors controlling the emissions from gas turbine combustors using distillate liquid fuel or natural gas are: (1) the primary zone combustion temperature and equivalence ratio, (2) the degree of mixing in the primary zone, (3) the residence time, and (4) the combustor liner quenching characteristics.

Past practice was to maintain the primary combustion zone at an equivalence ratio of 0.7 to 0.9 (slightly lean). For leaner mixtures the CO levels are high because of the slow rate of oxidation and the relatively short residence times. Hydrocarbon emissions occur due to incomplete vaporization of the fuel spray. Both CO and HC emissions can be reduced by improved fuel atomization—for example, by use of well-designed air-blast spray nozzles. Air bleed through the liner in the intermediate combustion zone completes combustion of the HC and CO, while cooling the wall. Since mixing tends to be nonuniform, cool spots, which quench HC and CO burnout, and hot spots, which tend to generate NO_x, can occur.

Soot is formed in fuel-rich zones near the fuel droplets. Most of the soot is produced in the primary zone and oxidized in the intermediate zone. For high-temperature turbines soot is also consumed in the dilution zone. Injecting more air into the primary zone reduces soot emissions, but at the expense of increased CO and HC emissions. Sooting is more severe at high pressure, in part because soot is formed closer to the spray nozzles. Soot emissions are reduced by improved mixing, by water injection, and by increased residence time.

The basic idea for reducing NO_x emissions is to reduce the peak temperature during combustion and the time spent at high temperature. Modest reductions on NO_x emissions can be achieved by reducing the residence time at high temperature by improved mixing of the liner air, and by improved fuel injection. Large reductions in NO_x emissions can be achieved by: (1) water injection, (2) premixed lean burn combustion, and (3) staged combustion. Exhaust gas recirculation also reduces NO_x but requires additional compressor work, and hence is not used.

For utility gas turbines water injection at the rate of 0.5 to 1.5 times the fuel flow rate directly in the primary zone is effective in reducing NO_x emissions, while HC and CO emissions increase only slightly up to a water/fuel ratio of 1 and then increase rapidly after that (see Figure 11.17). Water addition increases the power by as much as 16%; however, the cycle efficiency drops up to 4% due to the lower temperature. Water injection is an added expense because the water needs to be demineralized so as not to foul the turbine. Water injection is used to reduce NO_x emissions to 100 ppm. Stricter standards, such as the California standard of 9 ppm, are met with premixed lean burn combustion.

With lean combustion the diluent is air rather than water. Reducing the fuel-air ratio to near the lean limit will reduce the NO_x because the flame temperature is

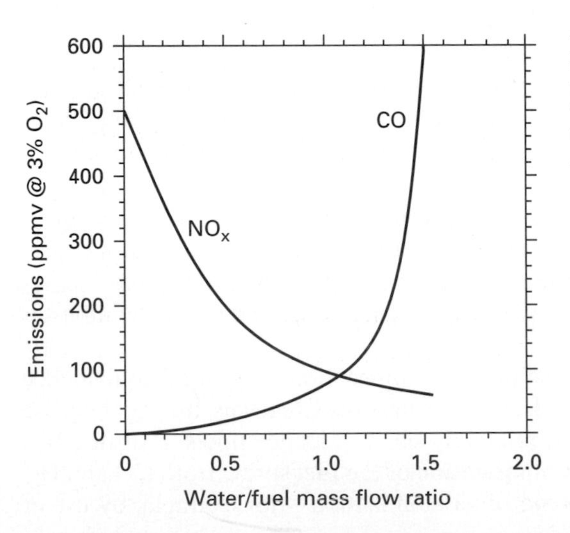

FIGURE 11.17
Effect of water injection on NO_x and CO emissions from large gas turbines [Bowman, by permission of The Combustion Institute].

significantly reduced. However, there are two design difficulties to be overcome with lean combustors. First, combustion stability must be assured; and second, turndown capability must be maintained, since a gas turbine must ignite, accelerate, and operate smoothly over a range of loads. The lowest NO_x levels are achieved with premixed gaseous fuel.

An example of a low NO_x heavy-duty gas turbine combustor is shown in Figure 11.18. The combustor is a single-stage, dual-mode combustor that can operate on either gaseous or liquid fuel. On gas the combustor operates in a diffusion flame mode at low loads, and in a premixed flame mode at high loads. Lowest NO_x levels are achieved in the lean, premixed mode. When using oil, operation is in the diffusion flame mode across the entire load range. There are five fuel nozzles (Figure 11.19).

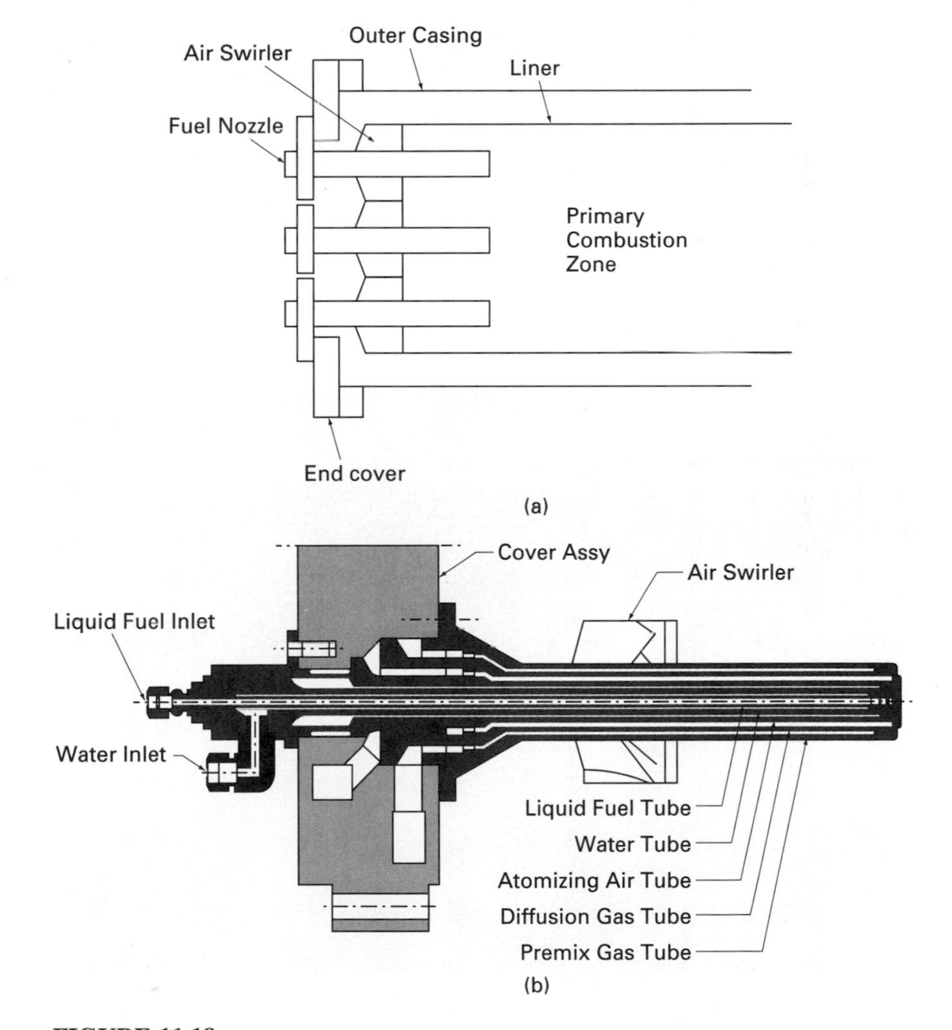

FIGURE 11.18
Schematic of low NO_x combustor for a gas turbine (*a*) combustor, (*b*) fuel nozzle [Davis (1996), Courtesy of General Electric Co.].

mounted in the end cap of the combustor, and each nozzle assembly has its own air swirler. The combustor operates with single burning zone formed by the combustor liner and the face of the end cap. Combustion is stabilized by the swirl and back mixing.

The fuel nozzle assembly, shown in Figure 11.18, has a central liquid fuel tube, surrounded by a water tube, an atomizing air tube, a gas tube for diffusion mode operation, and an outer tube in which gas and air are premixed. When operating with liquid fuels the atomizing air assists in producing fine droplets, while the water injection reduces the NO_x. With natural gas at low loads (<50% load) a diffusion flame is used to achieve stable burning. Once the combustor temperature exceeds 1200°C, the gas in the diffusion tube is shut off and the combustor is operated in the lean premixed flame mode. The combustor outlet temperature at full load is about 1300°C.

At 50% load in the diffusion mode the NO_x is 90 ppm at 15% O_2 and drops to 25 ppm at 15% O_2 in the lean premixed mode at that load, and remains at 25 ppm at full load. The CO is 75 ppm at 50% load in the diffusion mode and drops to 15 ppm at that load in the premixed mode, and remains at 15 ppm at full load. When using fuel oil the NO_x emissions decrease from 150 ppm to 40 ppm at 50% load with water injection, and CO drops from 150 ppm to 20 ppm; the NO_x and CO remain at 40 ppm and 20 ppm respectively at full load with water injection. The original combustor (Figure 11.10) produced 350 ppm NO_x and 5 ppm CO on liquid fuel, and 200 ppm NO_x and 5 ppm CO on natural gas at full load.

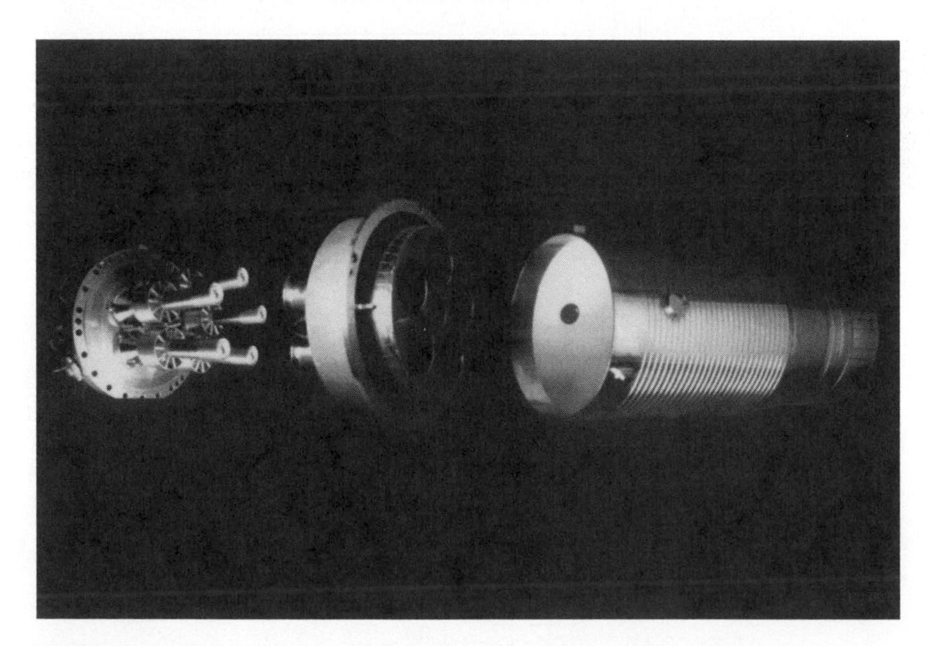

FIGURE 11.19
Low NO_x gas turbine combustor: left, fuel nozzles; center, nozzle outer body; right, combustor liner [Davis (1996), Courtesy of General Electric Co.].

With liquid fuels it is more difficult to design a lean premix burn combustor because the fuel must also be prevaporized. As with gaseous fuels, the goal is to supply the combustion zone with a perfectly homogeneous mixture of fuel and air, and to operate the combustion zone very close to the lean blowout limit in order to keep the flame temperature as low as possible. Fortunately, the flame temperature near the blowout limit is generally above the desired turbine inlet temperature. In addition to homogeneous premixing of the air and fuel, premature spontaneous ignition, flashback, and blowout must be avoided.

An alternative approach is a two-stage combustor such as Figure 11.20 where the first stage is operated rich followed by heat transfer cooling of the combustion products in the quench section by means of the atomizing air and the primary air flowing over the quench section, followed by lean burnout in the second stage. In the lean zone air jets rapidly bring the equivalence ratio to F = 0.5 to 0.7. The temperature is low enough to inhibit NO_x formation, but high enough to oxidize CO, HC, and soot which is formed in the primary zone. However, the mixing is not rapid enough for ultra low NO_x application.

From the above discussion it is apparent that operating performance and emissions from gas turbine combustors are affected by the design details of the front end of the combustor and the subsequent admittance of air into various combustion and dilution zones. For example, minor changes in the liner air distribution can create sizable variations in such parameters as lean blowout and ignition, exit gas temperature profile, and nitric oxide emissions. Improved combustor design requires better knowledge of the flow characteristics within the combustor, and the processes of fuel injection, evaporation, mixing, combustion chemistry, and convective and radiative heat transfer. Over the years empirical and semiempirical methods involving an enormous amount of testing have been used to evaluate performance and emissions of combustor design changes. In recent years three-dimensional computational fluid dynamic modeling of gas turbine combustors has provided valuable insight into the complex interaction between the swirling flow,

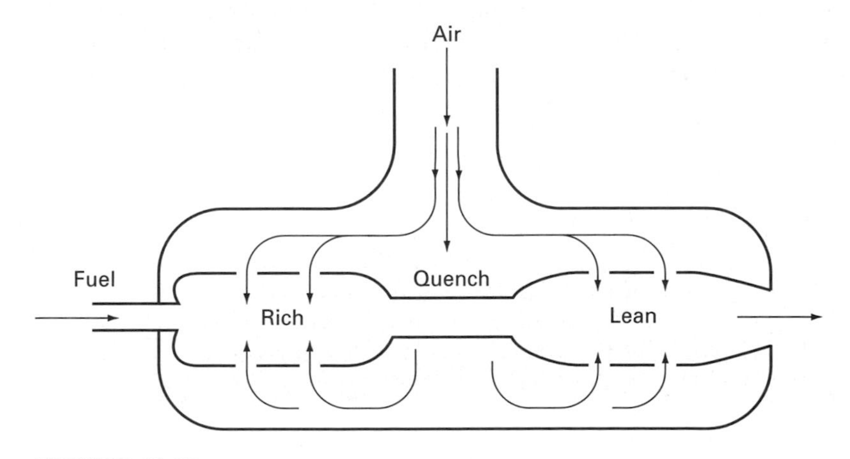

FIGURE 11.20
Two-stage, rich-quench-lean combustor for a low NO_x gas turbine.

spray injection, evaporation, mixing, turbulent combustion, and emissions. Nevertheless, further work is needed before the numerical models can be considered quantitatively accurate for engineering design. For further information on CFD modeling of gas turbine combustors, the papers by Rizk and Mongia and by Kandamby and Lockwood are suggested. Detailed chemical kinetic studies of pollutant formation at gas turbine conditions is also very important. The jet-stirred reactor is a useful laboratory device for investigation of this type of high-pressure, high-temperature, short-duration combustion. Kinetic modeling for short residence times, on the order of 10 ms, requires prompt reactions rather than the slower post flame reactions.

11.6 SUMMARY

Gas turbine engines are used to produce thrust for aircraft and power for stationary applications. The gas turbine cycle efficiency increases with increasing turbine inlet temperature (which equals the combustor outlet temperature), and there is an optimum pressure for a given turbine inlet temperature. Combustor pressures up to 25 atm and combustor outlet temperatures above 1500°C are used. A gas turbine combustor consists of an inlet diffuser section, a fuel injector, an air swirler, a primary combustion zone, an intermediate combustion zone, a dilution zone, and a liner with holes and slots for wall cooling. The primary combustor has high swirl for flame stabilization. Overall, the combustion is lean. Aircraft gas turbine combustors are designed to achieve high combustion intensity (heat release per unit volume). The well-stirred reactor was introduced as a simplified model for the primary combustion zone, and the equations were developed for a kinetically limited, one-step chemical reaction. Heat transfer to the combustor liner depends on convection from the high-velocity products, radiation from the soot, and film and jet cooling from the secondary and dilution air. Mandated reductions in emissions have forced the redesign of gas turbine combustors. A 95% reduction in NO_x emissions has been achieved in industrial gas turbines burning natural gas by means of dry, premixed, lean-burn combustors. Aircraft gas turbines burning distillate liquid fuel use combustor staging to achieve a 75% reduction in NO_x emissions.

PROBLEMS

11.1. For the gas turbine cited in Table 11.1, determine the following:
(*a*) The simple cycle and combined cycle thermal efficiency
(*b*) The fuel feed rate per nozzle (gal/h) assuming no. 2 fuel oil
(*c*) The combustor inlet temperature assuming no heat loss
(*d*) Overall excess air
(*e*) Velocity at combustor outlet
(*f*) Water injection rate to meet the specified turbine inlet temperature

11.2. Calculate and tabulate the ideal gas turbine power per unit mass flow versus combustor pressure for inlet air at 300 K and 1 atm and combustor temperatures of 1200 K and 1800 K. Let the combustor pressure vary from 1 atm to 35 atm. Assume $c_p = 1.0$ kJ/kg·K, $\gamma = 1.4$. Also calculate the efficiency and heat rate.

11.3. Since CO oxidation is the last step in the reaction of hydrocarbons, consider the reaction of a lean fuel-air mixture to form H_2O, CO_2, CO, O_2, N_2. If the combustion efficiency is 98%, what is the mole fraction of CO in the products? Use $C_{12}H_{24}$ for kerosene and take the fuel/air equivalence ratio $F = 0.5$.

11.4. For Problem 11.3 and assuming the combustor pressure is 10 atm and the temperature is 1000°C, how long does it take for the remaining CO to react?

11.5. A well-stirred reactor with a volume of 900 cm^3 operates at 10 atm pressure and 2000 K. Methane at 25°C flows in at 11.2 gmol/s, and air at 200°C flows in at 106.6 gmol/s. For a one-step reaction of methane to CO_2 and H_2O, what percent of the methane reacts before exiting the reactor?

11.6. For the conditions of Example 11.2, verify the heat output rate at a reactor temperature of 1600 K, and determine the combustion efficiency and mass flow rate of fuel at this temperature.

11.7. Find the maximum heat output rate for the conditions of Example 11.2, except that the air is preheated to 600 K. What is the reactor temperature for this case?

11.8. Find the maximum heat output rate for the conditions of Example 11.2 except that 140% stoichiometric air is used. What is the reactor temperature for this case?

11.9. Repeat Example 11.2 using 20 atm pressure, and compare with Figure 11.14.

11.10. Using the following data, calculate the heat flux versus x for the wall downstream of a slot cooler. Assume the wall temperature is constant.

$s = 0.002$ m
$t = 0.0004$ m
$\mu_a = 3.89 \times 10^{-5}$ kg/m·s at 880 K
$k_a = 0.0553$ W/(m·K)
$\rho_a \underline{V}_a = 486$ kg/m^2·s
$\tilde{W} = 1.2$
$T_g = 2280$ K
$T_a = 880$ K; $P_a = 3040$ kPa
$T_w = 1280$ K
q (radiation) $= 4.46 \times 10^5$ W/m^2

11.11. Repeat Problem 11.5 for a well-stirred reactor but use the detailed kinetics of GRI-Mech 1.2 in conjunction with the ThermoCalculator program, which may be accessed on the World Wide Web at http//www.me.berkeley.edu/gri_mech/.

REFERENCES

Ballal, D. B. and Lefebvre, A. H., "A Proposed Method for Calculating Film-Cooled Wall Temperatures in Gas Turbine Combustion Chambers," ASME Paper 72-WA/HT-24, 1972.

Ballal, D. R., and Lefebvre, A. H., "Film-Cooling Effectiveness in the Near Slot Region," *J. Heat Transfer,* vol. 95, pp. 265–266, 1973.

Ballal, D. R., and Lefebvre, A. H., "Ignition and Flame Quenching of Flowing Heterogeneous Fuel-Air Mixtures," *Comb. and Flame,* vol. 35, pp. 155–168, 1979.

Ballal, D. R., and Lefebvre, A. H., "General Model of Spark Ignition for Gaseous and Liquid Fuel/Air Mixtures," *Eighteenth Symp. (Int.) on Combustion,* The Combustion Institute, Pittsburgh, pp. 1737–1746, 1981.

Bowman, C. T., "Control of Combustion-Generated Nitrogen oxide Emissions: Technology Driven by Regulation," *Twenty-fourth Symp. (Int.) on Combustion,* The Combustion Institute, Pittsburgh, pp. 859–878, 1992.

Brandt, D. E., "Heavy Duty Turbopower: The MS7001F," *Mech. Eng.,* vol. 109, pp. 28–36, July 1987.

Burton, D. C., "Combustion Turbines at the Crossroads," *Mech. Eng.,* vol. 104, pp. 79–85, April 1982.

Claeys, J. P., Elward, K. M., Mick, W. J., and Symonds, R. A., "Combustion System Performance and Field Tests of the MS7001F Gas Turbine," *J. Eng. Gas Turbines and Power,* vol. 115, pp. 537–546, 1993.

Correa, S. M., "A Review of NO_x Formation under Gas Turbine Combustion Conditions," *Comb. Sci. Technol.,* vol. 87, pp. 329–362, 1992.

Davis, L. B., "Dry Low NO_x Combustion Systems for GE Heavy-Duty Gas Turbines," report GER-3568F, GE Power Systems, Schenectady, NY, 1996.

Davis, L. B., and Washam, R. M., "Development of a Dry Low NO_x Combustor," ASME paper no. 89-GT-255, Gas Turbine and Aeroengine Congress and Exposition, Toronto, 6 pp., 1989.

El-Wakil, M. M., *Powerplant Technology,* McGraw-Hill, New York, 1984, Chap. 8.

Frenklach, M., Wang, H., Goldenberg, M., Smith, G. P., Golden, D. M., Bowman, C. T., Hanson, R. K., Gardiner, W. C., and Lissianski, V., "GRI-Mech—An Optimized Detailed Chemical Reaction Mechanism for Methane Combustion," Gas Research Institute topical report GRI-95/0058, 1995. This mechanism is accessible on the World Wide Web (http//www.me.berkeley.edu/gri_mech/).

General Electric Co., "High Bypass Turbofan Engine, CF6-80C2," GE Aircraft Engines, Cincinnati, OH.

Glarborg, P., et al., "PSR: A Fortran Program for Modeling Well-Stirred Reactors," Sandia report SAND 86-8209, Livermore, CA, 1986.

Gupta, A. K., and Lilley, D. G., "Combustion and Environmental Challenges for Gas Turbines in the 1990's," *J. Propulsion and Power,* vol. 10, pp. 137–147, 1994.

Kandamby, N. H., and Lockwood, F. C., "On the Aero-Thermal Characteristics of Gasifier-Fuelled Gas Turbine Combustors with Complex Geometry," *Twenty-fifth Symp. (Int.) on Combustion,* The Combustion Institute, Pittsburgh, pp. 251–259, 1994.

Lefebvre, A. H. (ed.), *Gas Turbine Combustor Design Problems,* Hemisphere, Washington, DC, 1980.

Lefebvre, A. H., *Gas Turbine Combustion,* Hemisphere, New York, 1983.

Lefebvre, A. H. "The Role of Fuel Preparation in Low-Emission Combustion," *J. of Engr. for Gas Turbines and Power,* vol. 117, pp. 617–654, October 1995.

Longwell, J. P., and Weiss, M. A., "High Temperature Reaction Rates in Hydrocarbon Combustion," *Ind. Eng. Chem.,* vol. 45, no. 8, pp. 1634–1643, 1955.

Michaud, M. G., Westmoreland, P. R., and Feitelberg, A. S., "Chemical Mechanisms of NO_x Formation for Gas Turbine Conditions," *Twenty-Fourth Symp. (Int.) on Combustion,* The Combustion Institute, Pittsburgh pp. 879–887, 1992.

Plee, S. L., and Mellor, A. M., "Characteristic Time Correlation for Lean Blowoff of Bluff-Body-Stabilized Flames," *Comb. and Flame,* vol. 35, pp. 61–80, 1979.

Rizk, N. K., and Mongia, H. C., "Three Dimensional Gas Turbine Combustion Emissions Modeling," *J. Eng. for Gas Turbines and Power,* vol. 115, pp. 603–611, 1993.

Sawyer, J. W. (ed.), *Sawyer's Gas Turbine Engineering Handbook,* vols. I, II and III, Gas Turbine Publications, Inc., 3rd ed., 1985.

Steele, R. C., Jarrett, A. C., Malte, P. C., Tonouchi, J. H., and Nicol, D. G., "Variables Affecting NO_x Formation in Lean-Premixed Combustion," vol. 119, pp. 102–107, 1997.

CHAPTER 12

Direct-Injection Engine Combustion

Homogeneous charge engines, as discussed in Chapter 7, are limited in efficiency by fuel octane number, the ability to propagate flames in lean mixtures, cycle-to-cycle variability, and the fuel trapped in crevices. These restrictions are partly overcome by direct injection of gasoline, but direct-injection stratified-charge spark-ignited engines have yet to be accepted as an alternative to homogeneous spark-ignited engines. The diesel engine, which directly injects high-cetane fuel into the cylinder and uses compression ignition, has become the dominant type for heavy-duty applications for on-road, off-road, marine, and industrial applications. But problems of weight, noise, odor, and cost have limited its application in automobiles in the United States. In Europe, however, diesel passenger cars have begun to penetrate the market in countries where tax policy causes diesel fuel to have a favorable price. In Germany, for example, the major auto manufacturer produces 30% of its cars with diesel engines. If diesel cars can meet future emissions standards, they offer a viable means of meeting future fuel economy goals such as the 80 miles per gallon goal in the United States. (In Europe, the so-called three-liter car, meaning 3 liters per 100 km, is a nearly identical goal.) Table 12.1 shows estimates of fuel consumption for various future power plants. The diesel is clearly a contender for future high fuel economy vehicles.

In this chapter introductory concepts of diesel combustion will be discussed, followed by a more detailed discussion of the events which make up the diesel combustion process. A major portion of the chapter is then devoted to emissions—how they are produced and how they are changed by design and operating parameters. The chapter concludes with a discussion of CFD modeling of the diesel combustion process. The reader may wish to review the engine terminology given at the end of Chapter 7 before continuing.

TABLE 12.1
Best thermal efficiency estimates for various power plants

Power plant type	Efficiency (%)
Spark-ignited, port-injected, stoichiometric	31.5
Direct-injected, spark-ignited, stoichiometric	33
Direct-injected, spark-ignited, lean, early injection	34.5
Indirect-injected diesel	35.5
Direct-injected, spark-ignited, lean, late injection	38
Gas turbine	38
High-speed, direct-injected diesel	43
Heavy-duty, direct-injected diesel (HDDI)	46
Fuel cell	52
Turbocompounded, HDDI diesel	54

12.1 INTRODUCTION TO DIESEL ENGINE COMBUSTION

In diesel engines, the fuel is injected either into a small prechamber attached to the main cylinder chamber (an *indirect-injection* engine, IDI) or directly into the cylinder chamber (a *direct-injection* or open-chamber engine, DI). Prechamber IDI engines have the advantage of less noise and more rapid combustion. Starting is aided by a high compression ratio (24–27) and a glow plug mounted in the prechamber. However, IDI engines have somewhat poorer fuel economy than DI engines. Figure 12.1 shows a typical IDI, a light-duty DI design, and three heavy-duty DI chamber designs. Note that the injector in larger engines is typically vertical and located at the bore axis. Because of fuel economy considerations, DI engines are gradually replacing IDI designs even in light-duty applications.

Historically, direct-injection engines have followed one of two design philosophies. High-swirl engines have a deep bowl in the piston, a low number of holes in the injector (four), and moderate injection pressures (13–340 atm). Low-swirl or quiescent combustion chamber engines have a shallow bowl in the piston, a larger number of holes in the injector (eight), and higher injection pressures (500–1400 atm). Smaller engines tend to be of the swirl type, while larger engines tend toward the quiescent type. There is probably a somewhat different optimum combination of design parameters for each load and speed, so that there is a preferred design for each application. However, no single design can give the best possible performance over a wide range of speeds and loads.

Diesel engines come in a very wide range of sizes, from the enormous ship engines to tiny industrial engines. Although the speed ranges from 50 rpm for very large engines to 5000 rpm for very small engines, the average piston speed does not vary as much. Table 12.2 shows examples of the wide range of sizes produced.

Diesel engines are favored for high efficiency. The improved efficiency is caused by the relatively high compression ratios (12 to 18 for DI engines), the low pumping losses due to lack of throttling, the overall lean mixtures (0.2 to 0.8 fuel/air equivalence ratio), and the fact that the crevice volumes contain air or

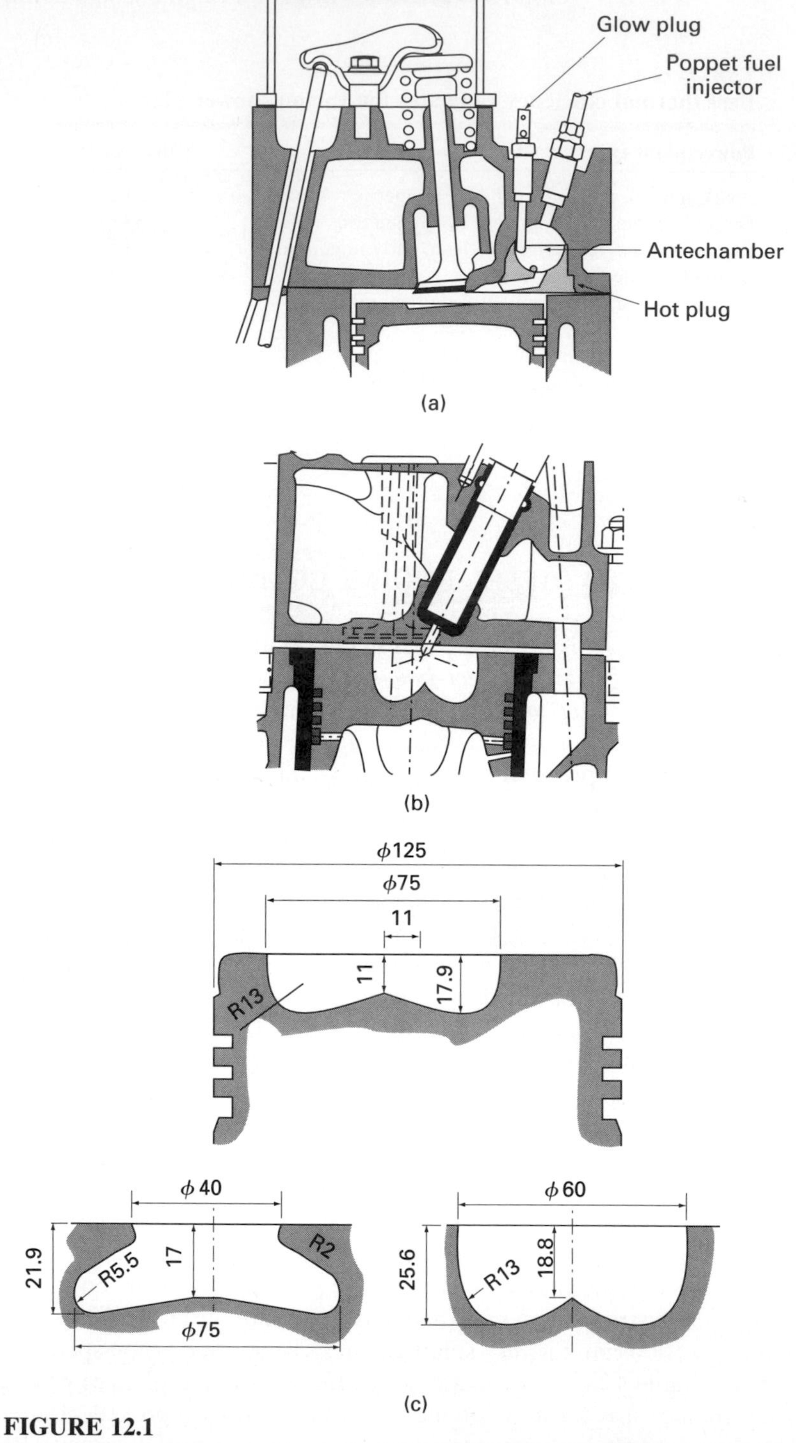

FIGURE 12.1
Diesel engine geometry: (*a*) prechamber engine for automobile applications, (*b*) deep bowl design for light-duty DI diesel engine, and (*c*) three chamber designs used in heavy-duty engines [Shimoda et al., reprinted with permission from SAE paper 850070 © 1985, Society of Automotive Engineers, Inc.].

TABLE 12.2
Typical size and output of diesel engines

Bore (mm)	45	80	127	280	400	840
Stroke (mm)	37	80	120	300	460	2900
Displacement (liter/cylinder)	0.06	0.402	1.77	18.5	57.82	1607
Number of cylinders	1	4L*	8V†	6–9L	6–9L	4–12L
Output/cylinder(kW)	0.7	10	40	325	550	3380
Rated speed (rpm)	3600	4800	2100	1000	514–520	55–76
BMEP (atm)	4	7.5	13	22	22.2	16.6

*L designates in-line cylinder arrangement.
†Designates V-shaped cylinder arrangement.

products rather than unburned fuel mixture. The power-to-volume ratio of the engine may be greatly increased by the use of turbocharging. (A *turbocharger* is a compressor which is driven by a turbine powered by the diesel exhaust.) By putting more air in the cylinder, the amount of fuel can also be increased, thus giving high brake mean effective pressures (BMEP) in the range 2.0 to 2.4 MPa. However, diesel engines are heavy, noisy, and expensive, and, as we shall discuss later, they have serious emission problems.

The design of combustion systems for diesel engines is still more of an art than a science, despite a long history of research. The reason for this is the very complex processes which take place in diesel combustion. At the start, air is drawn into the cylinder and compressed. For the DI engine, some amount of swirl may be introduced by the intake port shape. As the piston moves upward, the cylinder air is pushed into the bowl of the piston, giving the so called *squish flow.* Conservation of momentum demands that the swirl increase as the radius of the swirling gas is decreased. As the piston moves upward, the air at the piston surface must be at the velocity of the piston. The piston motion imparts a *squash velocity* to the air. The squash velocity is similar in magnitude to the swirl velocity for a *swirl ratio* (ratio of swirl revolutions per minute to engine revolutions per minute) of about unity. Such a swirl ratio is considered to be very low, and such chamber designs are often called "quiescent" even though the flows are quite turbulent and include complex large-scale flows. Induction swirl ratios of 3–4 are common and can result in piston bowl swirl ratios of 10–15 for small bowl-to-bore ratios. In the case of quiescent chambers the induction process produces tumbling vortex motions in the cylinder.

Although the amount of residual exhaust gas in the cylinder is low for four-stroke engines (a nominal value of 3–7%), it does raise the cylinder gas temperature slightly. However, the main increase in temperature is caused by the compression. For a naturally aspirated (NA) engine (without a turbocharger), a compression ratio of 16:1 will raise the cylinder gas temperature and pressure to about 830 K and 2.25 MPa at 20 crankdegrees before top dead center (TDC), which is a typical injection timing for best fuel economy at 2000 rpm. With turbocharging and no cooling between the compressor and engine (no intercooling), the absolute temperature will be about 12% higher than for the NA case. Fuel is injected at a time determined by best torque and emissions considerations.

Typical injection timing values range between 30° before TDC to TDC and depend on fuel cetane number and engine speed for a given engine design.

When the fuel is injected, some of it vaporizes and mixes with air to form a combustible mixture which autoignites due to the high temperature. Because of the high temperature the portion of the fuel which has been vaporized and mixed with air prior to ignition burns very rapidly. The mechanism of burning is uncertain—it may be by flame propagation, or by further autoignition induced by the rise in temperature due to the initial combustion. This initial burning of fuel-air mixture is called the *premixed burning period.* As the mixture around each spray plume burns, some of the entrained gas is hot products. This reduces the oxygen available for mixing with the fuel and raises the temperature within the spray. The ensuing combustion is rich, and its rate is controlled by turbulent mixing for all but very low pressure fuel injection or cold starting. In the case of low-pressure fuel injection in naturally aspirated engines, droplet vaporization may control the burning rate. For cold starting, the fuel vaporizes poorly and droplet burning is observed.

Because of the high temperatures, the premixed burning and mixing-controlled burning before peak pressure can result in the formation of large amounts of NO_x. Because of the rich burning at high pressure during the later turbulent diffusion burning, large amounts of particulates (mostly solid carbon) are produced. Because the crevice volumes are filled with air, the unburned hydrocarbon emissions are very low except during cold start. Carbon monoxide is low because the combustion products in the exhaust are lean.

Although the initial burning rate of the fuel is high, diesel combustion tends to have a rather long duration of burning which may extend to halfway down the stroke or even longer. Analysis of the burning process is hindered by the complexity of the transient injection process and the current lack of understanding of turbulent diffusion combustion. In the following sections each of the physical and chemical processes making up diesel combustion will be discussed in order of occurrence in the engine.

To gain an overall picture and to summarize the above introduction, see Figure 12.2, which shows a schematic of the interrelations between the combustion process events. Note the complex interactions between the spray, mixing, and combustion phenomena.

12.2 FUEL INJECTION

The primary purpose of the injector is to distribute and mix the fuel with the surrounding air. Because the engine load is controlled by the amount of fuel injected, the injector must handle a volume of fuel which ranges over an order of magnitude. The maximum fuel/air ratio is determined by the allowable smoke level and is typically at 0.8 equivalence ratio for an NA engine. Thus, for a naturally aspirated engine, the volume of fuel injected per cycle for each liter of displacement is about 80 mm^3. However, the trend is to use turbocharging with lower compression ratios and a maximum equivalence ratio of about 0.5. This raises the fuel volume to over 100 mm^3 per liter of displacement at full load.

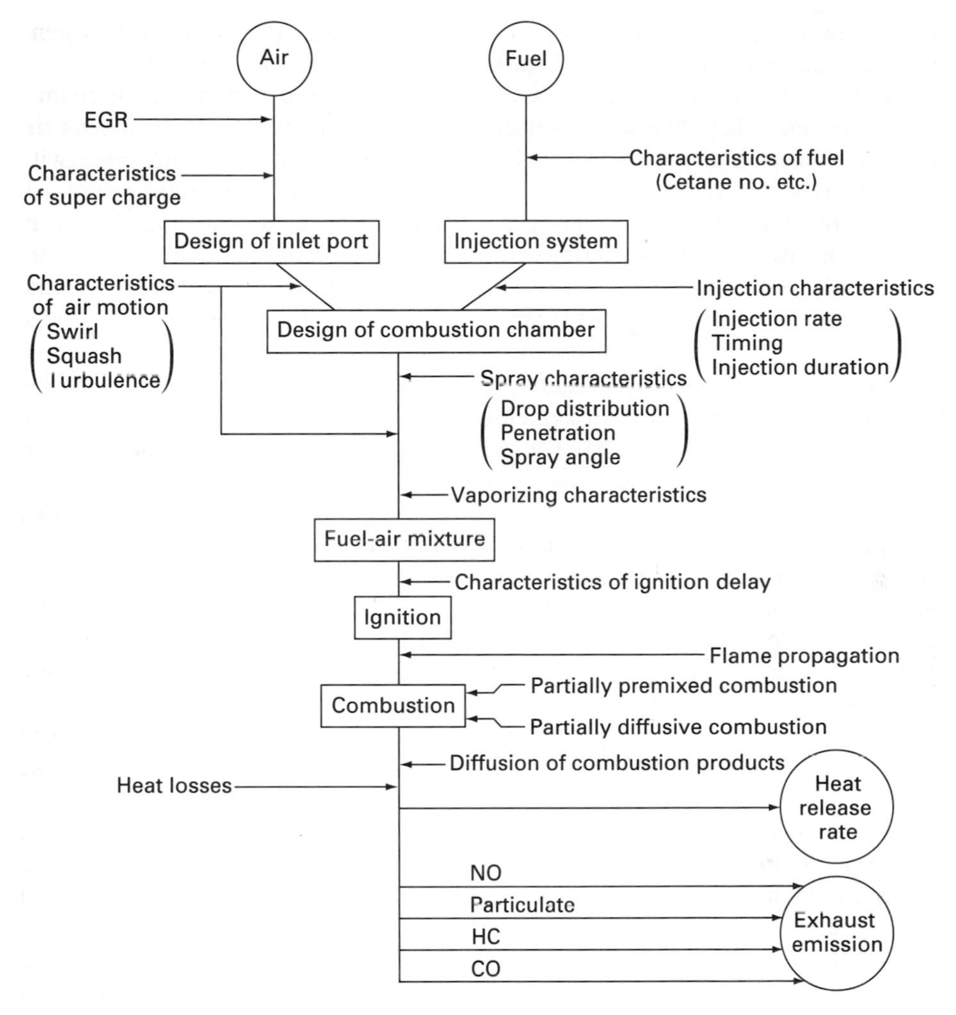

FIGURE 12.2
Diagram of diesel combustion processes [Hiroyasu, by permission of JSME].

For a fixed injection pressure, the amount of fuel injected is changed by changing the injection duration. The injection duration for a fixed pressure and amount of fuel per cylinder can be increased by decreasing the hole size or number of holes in the injector tip. Hole size is limited by the method of fabrication, which utilizes either small drills or electrodischarge machining. Holes of 0.2 mm diameter are routinely obtained with good reproducible quality. Holes down to 0.1 mm (0.004 in.) can be fabricated, but quality control is a problem. Even very small imperfections will cause quite significant changes in the spray pattern. In recent years the practice is to smooth the holes using electrochemical or abrasive flow techniques. Because discharge of the spray plume against the head or into the squish volume above the piston bowl can give unburned hydrocarbon and/or smoke emissions, each hole axis must be carefully set relative to the head. Even a 1° change in angle between spray axis and head or a few millimeters change in injection tip depth can cause serious problems. In order to provide minimum breakup length, the length-to-diameter ratio of the holes should be kept in the 3–5 range. Of course, each hole should have the same flow coefficient. Thus, testing of the injector by spraying into open air or into a see-through chamber containing nitrogen gas or kerosene is typically employed as a means of inspection. Measurement of the volume of fuel injected by each hole is also advisable. It is also important that each cylinder receive the same injection, especially from an emissions standpoint.

Recall, from Chapter 9, that the increased injection pressure decreases the average droplet size in the diesel spray proportional to about $(\Delta p)^{-1/2}$ while penetration increases as $(\Delta p)^{1/4}$. Thus, high injection pressure has the advantage of small droplets and rapid vaporization, but may cause the spray to impinge on the piston bowl, at least at some loads and for smaller engines. For a fixed amount of fuel injected per cycle and a fixed injection duration in crank-angle degrees, the injection pressure must increase as the square of the engine speed or the effective flow area of the nozzle must increase in proportion to the speed. Such control can be accomplished by adjustment of the pressure fed to the injector or can be automatically obtained by using a positive displacement system. (For a more detailed discussion of injector mechanisms, the reader may refer to Bosch's *Diesel Injector Handbook.*) With the introduction of electronically controlled injectors it should be possible to control both pressure and flow area. It is now also possible to control the shape of the rate-of-injection curve. For example, it is possible to have a pilot injection where a small amount of fuel is injected prior to injection of the rest of the fuel. Such flexibility, as will be seen in the next section, should allow designs for improved performance, emissions, and fuel tolerance.

The quality of diesel injection is often judged by gross factors such as droplet size distribution, penetration history, and spray angle. These parameters were discussed in Chapter 9. However, these gross parameters do not adequately deal with the more detailed behavior which influences combustion rate and emissions. Some examples of these details are the turbulence in the injector nozzle holes, effects of cavitation in the holes, effects of injection rate shape, the effects of coupling between spray parameters and droplet vaporization which affect mixing, effects of spatial distribution within the spray which affect the inner core and the outer fringe of the spray, the effects of spray impingement on the piston bowl surface, and the way the spray influences turbulent mixing rates, especially those which ensue after the end of injection. None of these details are currently well understood, and thus the following discussion is primarily heuristic. A more lengthly discussion is given by Arcoumanis et al.

Until recently, injection rate shaping was built into the injector design without much ability to change it during use. Conventional wisdom indicated that a gradual rise in rate at the start is more desirable than a rapid one. The rapid rate was observed to increase the rate-of-pressure rise and thus noise and to increase exhaust emissions of nitrogen oxides. On the other hand, the end of injection should be rapid and should not allow any poorly atomized fuel to be injected. Such rapid closing has been observed to decrease hydrocarbon and smoke emissions and to decrease the combustion duration, resulting in small improvements in fuel economy. As in most designs, achieving these goals should not compromise other aspects of performance. Thus, some nozzle tip designs which are sacless, and thus avoid escape of poorly atomized fuel, are not compatible with high-pressure injection and good spray uniformity.

The advent of electronic injection (see Chapter 9) has allowed more adventuresome changes in rate shape. A small amount of injection prior to main injection can improve ignition, as will be discussed in the next section, but in some designs can also increase particulate emissions. Recent research on *split injection,* where the injection process consists of two or more major injections with a delay between each of them, has shown promise as a means of reduction of particulates. In such practice the best result was found when the second injection event has a rapid initial rate. It is thought that the addition of spray momentum later in the cycle improves late mixing and thus allows more carbon particulates to oxidize before leaving the cylinder. Unfortunately, the delay between injections extends the end of combustion and adversely affects fuel economy unless other parameters such as chamber geometry can be used to speed up combustion and mitigate this effect. Such trade-offs will be discussed later in the section on emissions.

The spray formulas of Chapter 9 were developed for nonvaporizing cases. A comparison of penetration with and without vaporization and/or combustion from experiments carried out in a rapid compression machine and in a constant volume bomb shows decreases in penetration of only 10–20% due to vaporization; however, the important factor should be the overall effect on fuel distribution. One must also be careful to note that the spray momentum and droplet size distribution greatly confound this result. Recall that the liquid penetration is caused by newly formed drops overtaking and passing previously formed drops. Thus, the drops which are least vaporized lead the tip penetration. For very high-pressure (200 MPa) sprays with small nozzle holes (0.1 mm), the droplets are very small and thus the spray vaporizes rapidly, giving a gas-jet-like behavior. Laser sheet diagnostics indicate that at full load the spray becomes vaporized less than one inch from the nozzle. Although the jet will penetrate well, its ability to mix rapidly may be less than would be expected from previous experience with less vaporized sprays. However, the high momentum of such high-pressure sprays increases mixing greatly compared with lower pressure sprays. For heavy-duty engines, liquid impingement on the piston bowl surface should not be a problem for such sprays except during cold starting.

The effects of drop size on vaporization rate changes may be estimated from the theory of Chapter 9, but a more detailed understanding of the spatial distribution of droplets within the spray may be needed when considering ignition and emissions phenomena. As noted in Chapter 9, there is a narrow core along the spray axis which has a length approximately given by the breakup length (Eq. 9.22*a*). This core has a high momentum and shows some properties of a continuous liquid; it is thus often called the "intact core." Photographs of high-pressure sprays which are vaporizing have shown this core to persist while all the surrounding spray has vaporized. The possible influence of this core on particulate emission is unknown, but it may play an important role in mixing in small engines where liquid fuel impingement on the piston typically occurs. A second distribution phenomenon of interest is the irregular shape of clouds of fine droplets at the outer edge of the spray, giving a "Christmas tree" shape. Pulsed laser lighted photographs indicate that although the pattern is always similar, its details are random. Small droplets at the outer edge could lead to lean regions which do not combust, but could also provide the first bit of mixture to ignite. Phenomena of this sort, as well as other observed patterns within the spray, may be only curiosities, but they do also indicate a lack of detailed scientific understanding of the spray breakup mechanism.

In small high-speed diesels the liquid fuel impinges on the piston bowl surface because of the small bore. Swirl can be used to shorten the penetration and improve mixing, but typically at a cost in terms of reduced volumetric efficiency and increased heat transfer. Droplets reaching the surface may wet the surface or rebound. For many conditions, rebounding also results in droplet breakup. Detailed models involving all of these effects as well as heat transfer are currently being developed and evaluated for use in CFD codes.

12.3 IGNITION DELAY

Recall that the time (or crankdegrees) between the start of injection and the start of combustion is called the *ignition delay*. During the delay period, fuel is continuously injected and vaporized. For open-chamber engines, the effect of injection pressure, nozzle hole diameter, and number of holes seems to have only a minor influence on the delay period. This is particularly true for turbocharged engines. Thus, the delay is primarily a function of those parameters which affect the chemical reaction rate of the fuel-vapor mixtures. This fact is counterintuitive, since the liquid fuel must be disintegrated into droplets, vaporized, and mixed with air before significant reactions can take place. However, it appears that the physical delay due to these processes is not controlling. One explanation is that a small amount of appropriate fuel/air mixture is always quickly formed for all practical injection configurations. The effects of physical delay are more important in prechamber (IDI) engines, where fuel atomization and mixing are less rapid. Similarly, constant-volume combustion bombs, often used for experiments on ignition delay, typically have longer physical delay periods due to lower pressures, temperatures, and turbulence in the bomb than in engines.

The parameters that most affect ignition delay in open-chamber (DI) engines are temperature and pressure of the air during the delay period and the chemistry of the fuel. Recall (Chapter 2) that the cetane number rating is an attempt to rank the ignition delay of a given fuel relative to a reference fuel. The percentage of hexadecane that must be added to heptamethylnonane to produce 13° of ignition delay, at the same compression ratio which gives 13° ignition delay with ignition at TDC for the test fuel, determines the cetane number. The CFR engine speed is fixed at 900 rpm and the inlet air temperature at 339 K (150°F). Three important assumptions are inherent in this test. First, it is assumed that the reference fuel ignition chemistry is similar to that of the test fuel. This seems to be a good assumption for fuels derived from crude oil, but not so for fuels produced from vegetable oils (sunflower seeds or rapeseeds, for example). Second, it is assumed that increasing the compression ratio will continue to raise the air temperature in the prechamber of the standard (CFR) engine. This does not appear to be the case for low-cetane fuels (below 30 cetane number), where high compression ratios are required. Thus, rating of low-cetane fuels is unreliable, and an alternative test may be necessary. For example, one might raise the engine inlet temperature. Third, it is assumed that the low-speed prechamber engine ignition delay correlates well with higher-speed (2500 rpm) open-chamber engines. This seems to be the case, in the sense that ignition delay obtained in open-chamber engines using crude oil derived fuels give linear plots of ignition delay versus cetane number. However, cetane number (ignition delay) is not the only important combustion-related fuel parameter, and thus additional rating parameters seem necessary especially if fuel quality changes radically during the next century.

The major reason that long ignition delays are not tolerated is that a large amount of fuel vaporizes and mixes with air prior to ignition if the delay is long. When ignition takes place, this prepared fuel burns rapidly, giving a high rate of pressure rise, a high peak pressure, and a characteristic sharp knocking sound. Continued operation at such conditions can cause mechanical failure due to head gasket failures or to main bearing failures, and/or failure to meet noise control standards. Retarding the injection timing, recirculating some exhaust gas, or heating the inlet air may help, but may also produce undesirable changes in performance and/or emissions. Design changes to reduce these problems will be discussed later, but for now it is clear that ignition delay can be a limiting factor in diesel engine performance. The use of a spark plug to replace compression ignition requires the plug to be in a position where an ignitable fuel-air mixture is present for all operating conditions. Even if this is achieved, the plug life is typically found to be unacceptably short.

Formulas for prediction of ignition delay have been obtained from both engine and bomb tests. The independent variables are typically average cylinder gas pressure during the delay, average cylinder gas temperature during the delay, fuel cetane number, and engine fuel/air ratio. The most important variable is temperature; however, recent tests in a bomb have indicated that at temperatures above 800 K (1000°F) the effect of temperature is less significant. This indicates that if the chemical kinetic delay is very short, the physical delay may begin to play a role. However, engine experiments with a small amount of fuel burned prior to main injection have obtained delays smaller than 1 CA° at speeds of 2000 rpm (about 80 μs) indicating that the physical preparation time is quite small.

A typical correlation formula is of the form

$$\Delta t = C\left(\frac{p}{p_0}\right)^a (F)^b \exp\left(\frac{E}{RT}\right) \tag{12.1}$$

where the constant E depends on the fuel formulation, $-1.9 < a < -0.8$, $-1.9 < b < -1.6$, and p_0 is a reference pressure. The constant C depends on the engine. Inclusion of the equivalence ratio is not always necessary.

The various formulas such as Eq. 12.1 all give quite different delay values when applied to the same set of input data. It is thus quite clear that no single universal formula is available. A major reason for this lack of agreement is the inexact nature of determining the start of injection, the start of ignition, and the temperature in the region where the ignition first takes place. The mass-averaged temperature is used in these formulas, but it is quite clear that an error of at least 20°C in core temperature can occur due to neglecting the effects of the boundary layer in an engine with hot walls. For a bomb with cold walls, the error could be as high as 60°C. Taking $E/R = 4500$ K and T $= 900$ K, the error in the cold bomb determined value of delay would be about 27%. In addition to this error, the heterogeneous nature of the temperature field due to turbulent mixing and cooling caused by vaporization add to the uncertainty of the temperature. For many of the formulas the exact chemistry of the test fuel is unknown, so we must consider the value of E to be uncertain to within $\pm 10\%$; this could easily contribute another 30% error. The start of combustion is typically determined from cylinder pressure data, but this gives an uncertainty of about one crank-angle degree, or 20% for a 5° delay. Given these uncertainties the ignition delay formulas are useful only to predict trends unless one obtains the correlation data for a specific engine and fuel in a consistent way and then uses the correlation for that same combination.

EXAMPLE 12.1. The following ignition delay formula was obtained for a turbocharged diesel running on diesel fuel with cetane number of 45:

$$\Delta t = 0.075\left(\frac{1}{\overline{p}^{1.637}}\right)\left(\frac{1}{F^{0.445}}\right)\exp\left(\frac{3812}{\overline{T}}\right)$$

where Δt = ignition delay in ms
$\overline{p}$ = average p during the delay, MPa
$\overline{T}$ = average T during the delay, K
F = engine overall equivalence ratio

For this engine the compression ratio is 13.25 and the connecting rod to crank radius ratio is 4.25. We are given from data that $p = 3.13$ MPa and $T = 816$ K at $\theta =$ 20° BTDC for $F = 0.6$ and rpm $= 1500$.

Compute the ignition delay as a function of crank angle at start of injection (SOI). Assume the p and T are given by a polytropic function with exponent 1.35.

Solution. To obtain the tabulated values for $\overline{\theta} = 20°$, we substitute directly into the formula for Δt and obtain 1.555 ms. At 1500 rpm, $\Delta\theta = 9\Delta t = 14.0°$. We assume the SOI is at $\overline{\theta} + \Delta\theta/2 = 27°$. For other θ values, we calculate $V/V(20)$ from

$$\frac{1 + (\text{CR} - 1)x(\theta)}{1 + (\text{CR} - 1)x(\theta = 20)}$$

where x = fraction of stroke:

$$x = \frac{(1 - \cos\theta) + 4.25[1 - \sqrt{1 - (\sin\theta/4.25)^2}]}{2}$$

Then,

$$\frac{p}{p(20)} = \left[\frac{V(20)}{V}\right]^{1.35}, \quad \frac{T}{T(20)} = \left[\frac{V(20)}{V}\right]^{0.35}$$

The results are shown below.

$\bar{\theta}$	$V/V(20)$	$\bar{p}$ (MPa)	$\bar{T}$ (K)	$\Delta\theta$	SOI
0	0.688	5.186	930	3.5	1.75
5	0.708	4.986	921	3.8	6.9
10	0.764	4.502	897	5.0	12.5
15	0.865	3.807	858	8.0	19.0
20	1.000	3.130	816	14.0	27.0

Thus, for $\bar{\theta} = 15°$ and $\Delta\theta = 8°$, SOI $= 19°$ and start of combustion (SOC) is at 11° BTDC. This is close to MBT timing for these conditions.

Note that because of intercooling the effect of turbocharging is primarily to increase the compression pressure. We note that a 1-atm boost will decrease the delay by a factor of about $\frac{1}{3}$.

Before leaving this discussion of ignition delay it may be useful to comment on the chemical kinetics during the delay process. Experiments with pilot injection have shown that even very low-cetane fuels can be ignited in a warm engine if the pilot fuel is allowed to react for 15 to 20° before main injection. Such prereactions of the pilot fuel do not release much energy, indicating that the reactions have not yet produced significant CO_2. A similar effect was observed when ethanol was prevaporized and introduced as a lean mixture with the air (a process called fumigation). For low loads the ethanol, which has a very low cetane number, caused an increase in delay when the diesel fuel was injected. As load was increased, eventually a point was found where prereactions in the ethanol caused a decreased ignition delay for the diesel fuel. Again, the prereactions of the ethanol did not release much energy. Tests such as these indicate that a two-stage ignition process is possible given adequate time and high enough temperature. The existence of a two-stage reaction for the normal ignition of higher-cetane fuels is, however, not established and seems unlikely.

Modeling with CFD (discussed in Section 12.8) allows use of simplified ignition models such as the Shell model originally developed for knocking reactions in S.I. engines. This model contains 33 empirical constants which depend upon not only the chemistry but also the ability to model the temperature field accurately. Use of heated bomb data to set the constants is desirable.

12.4 COMBUSTION RATES

The amount of mixture which burns rapidly following ignition is determined by the fuel properties, the injection parameters, the flow pattern in the cylinder, the temperature and pressure of the cylinder gas, and the ignition delay. Following this rapid (premixed) burning, the combustion rate is controlled by mixing rates for most engines under fully warmed up conditions. Calculations based on spray models indicate that vaporization of droplets is not rate-controlling unless the engine is cold and/or the spray is very coarse. As discussed in Section 12.1, for modern turbocharged engines with high injection pressures the vaporization rate is unlikely to control the rate of combustion except under cold starting conditions.

Unfortunately, mixing models for sprays under the three-dimensional turbulent flow conditions encountered in engines have not been validated. It must be recognized that although vaporization is not rate-controlling, it does profoundly affect the fuel/air ratio distribution. Furthermore, the spray penetration determines the amount of fuel which interacts with the piston bowl surfaces. Given this modeling difficulty, it is not yet possible to predict diesel design configurations or even combustion trends from fundamental models based on first principles. Recently, however, the application of detailed CFD modeling has shown the ability to reproduce the burning rates obtained from pressure data. This is discussed later, in the section on CFD modeling. But these models have not been applied to a wide range of engines and conditions, so one cannot say yet that the CFD models are reliable for design purposes. As a result, the most common approach is to obtain the burning rate from analysis of cylinder pressure data. If enough data are taken for a given engine family (a series of engines of various sizes, but all with the same design philosophy), it is possible to correlate the burning rate data using empirical or quasi-empirical formulas. Attempts to obtain predictive models by semiempirical modeling (so-called phenomenological models, which use approximate equations containing many adjustable constants) have sometimes given good results when properly tuned using data, but have not shown the ability to be predictive without such empirical adjustments. Thus we shall give only a heuristic discussion of such models and focus most of our attention on pressure data analysis, which gives an apparent rate of burning.

Burning Rate Analysis

The rate of combustion can be determined from analysis of cylinder pressure if one assumes that the pressure is uniform so that the measured pressure is the pressure acting on the piston surface. The integral of $p\,dV$ thus gives the work done over the interval of integration. As a first step, consider the combustion to act as a uniform energy rate addition q_c, neglect all changes in composition, and assume constant specific heats. Then the energy equation for this *single-zone model* is

$$mc_v \frac{dT}{dt} + p\frac{dV}{dt} = q_c - q_L = q_{(\text{net})} \tag{12.2}$$

The quantity $q_{(\text{net})}$ is the energy rate addition due to combustion less the heat transfer rate q_L out through the chamber surfaces. From the ideal gas equation, neglecting the change in gas mass due to vaporization of liquid fuel,

$$mR\frac{dT}{dt} = p\frac{dV}{dt} + V\frac{dp}{dt}$$

and

$$R = c_p - c_v = c_v(\gamma - 1) \tag{12.3}$$

Thus,

$$q_{(\text{net})} = p\frac{dV}{dt} + \frac{p\dfrac{dV}{dt} + V\dfrac{dp}{dt}}{\gamma - 1} \tag{12.4}$$

Integrating with γ constant,

$$Q_{(\text{net})} = \int_{\theta_1}^{\theta_2} q_{(\text{net})}\left(\frac{dt}{d\theta}\right) d\theta = \int_{V_1}^{V_2} p\, dV + \frac{(pV)_2 - (pV)_1}{\gamma - 1} \tag{12.5}$$

The right side of Eq. 12.5 is evaluated from values of $V(\theta)$, and $p(\theta)$ obtained from the cylinder pressure data. It must be warned, however, that obtaining good pressure data is not easy. To obtain reasonable pressure data the crank-angle interval of sampling must be at least 0.5 crankdegree or, better, 0.1 crankdegree. Proper filtering of the signal and careful calibration of the transducer are required. Even then, it is typically necessary to ensemble-average the data over at least 100 cycles in order to remove noise and average random cyclic variations.

The simple heat release rate of Eq. 12.4 or the integrated value of Eq. 12.5 contains the confounding heat transfer q_L. Accurate evaluation of q_L is not possible from current theory. During the mixing-controlled combustion, the flame is quite luminous due to the formation of carbon particulate. These tiny particles, which are a fraction of a micron in size, each radiate as essentially black bodies. In older engines the resulting flux of radiation gave a peak value about equal to the peak convective flux. Modeling of this flux could be done in detail if the spatial distribution of carbon and its temperature were known. However, the distributions are not known, and thus only approximate representations can be given. Modern low-sooting engines have a much reduced radiation component. Recent data indicate that the peak radiation flux is only 30% of the total flux and that radiation accounts for only 15% of the total heat transfer to the combustion chamber surfaces. Figure 12.3 shows histories of total and radiation flux measured on the head of a fired turbocharged diesel. The convective flux measurement during combustion is confounded by the more difficult-to-measure radiation. In addition, measurements have shown that the total heat flux (radiation plus convection) is very nonuniformly distributed. At the present time only approximate correlations are available, and the most widely used is due to Woschni. This formula linearizes the radiation term and combines it with the convective term to give

$$q_L = \tilde{h}A(T - T_w) \tag{12.6}$$

where
T = mass average gas temperature, K
T_w = temperature of surface area A, K
$\tilde{h} = 0.82d^{-0.2}W^{0.8}p^{0.8}T^{-0.53}$, kW/m²·K
p = cylinder pressure, MPa
d = cylinder bore, m
$$W = 2.28C_m + 0.00324\left(\frac{V_d T_1}{p_1 V_1}\right)(p - p_0)$$

p_1, V_1, T_1 = reference values at intake valve closing
p_0 = motoring cylinder gas pressure, MPa
C_m = mean piston speed, m/s
V_d = displacement volume, m^3

In order to calibrate such a formula it is necessary to apply an energy balance between the start of injection and beyond the end of combustion. To be on the safe side it may be necessary to assume combustion ends at exhaust valve closing. The first law, with the internal energy of formation included in u and h_f, may be written as

$$ {}_1Q_2 = -\int_{V_1}^{V_2} p dV - (m_2 u_2 - m_1 u_1) + h_f(m_2 - m_1) \tag{12.7} $$

and ${}_1Q_2$ can also be calculated from the heat transfer model using

$$ {}_1Q_2 = C \int_{\theta_1}^{\theta_2} q_L \left(\frac{dt}{d\theta} \right) d\theta \tag{12.8} $$

Step-by-step integration of q_L using Eq. 12.6 with $T = pV/mR$ will give $C = 1$ in Eq. 12.8 if the equation for $\tilde{h}$ is correct. In practice C can be quite different from unity; for example, $C = 2$ is not uncommon. Of course, the factor C corrects the

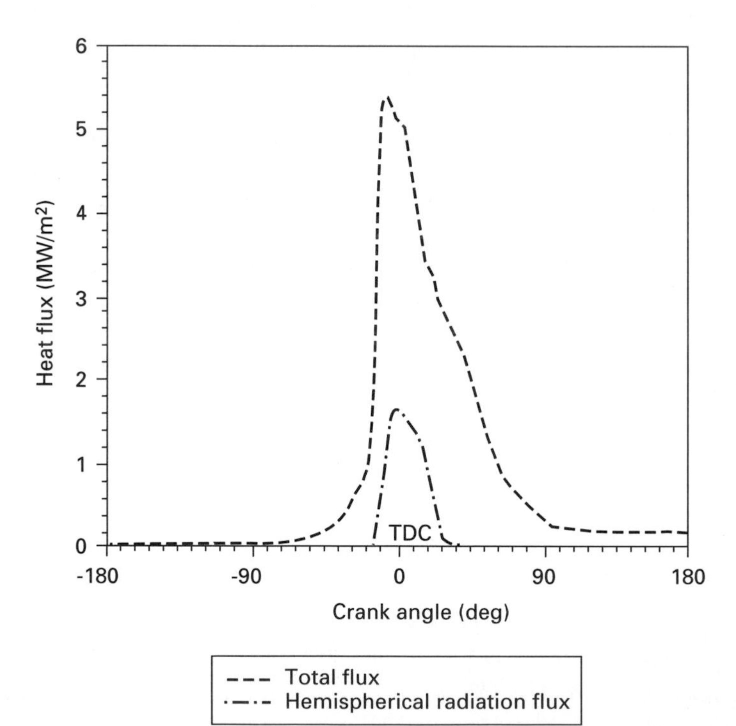

FIGURE 12.3
Total and radiant heat flux measured on the head of an open-chamber diesel of 2.33-L displacement at 1500 rpm, part load ($F = 0.4$) and 200-kPa intake pressure [Yan and Borman, reprinted with permission from SAE paper 891901 © 1989, Society of Automotive Engineers, Inc.].

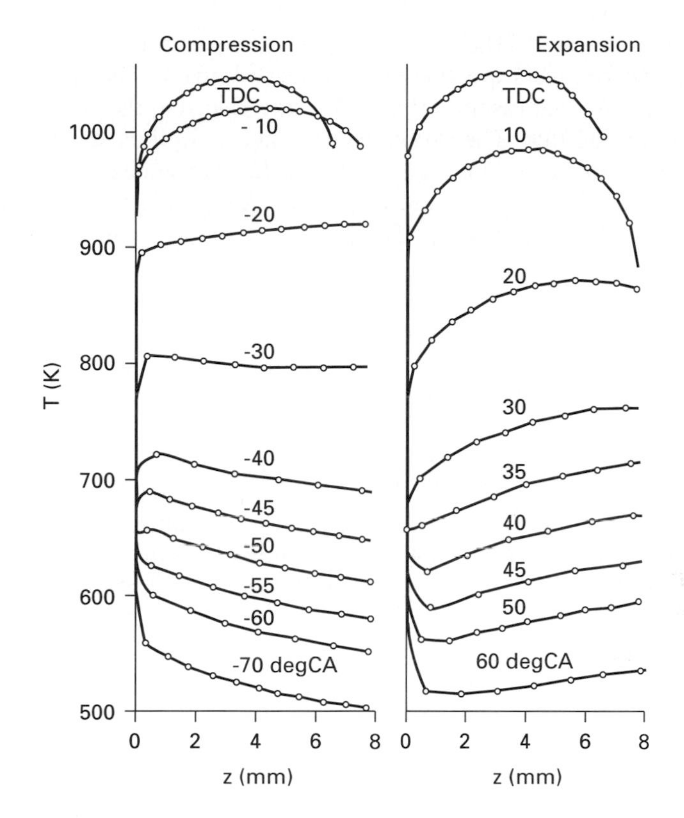

FIGURE 12.4
Temperature profiles in the gas calculated for a motored engine with heated surfaces; the wall temperature is 650 K.

total global value, ${}_1Q_2$, to balance the energy, but this does not mean that the shape of q_L versus θ is correct. In fact, it is known from experiment that local values of q_L can change sign during the period 55–65 crankdegrees after TDC even though $T > T_w$ during that period. The reason for this is shown by theory to be the expansion of the cool boundary layer gases to below the wall temperature, while the hot bulk gas, although also cooled by expansion, is still at a much higher temperature than the wall. Figure 12.4 shows temperature calculated for a motored engine using a multidimensional model for the gas and the law of the wall for the boundary layer profiles. One can see the minimum in the temperature profile for the 40 CA° curve during expansion. The empirical Eq. 12.6 is based on the mass average temperature and thus always predicts that the heat flux is from the gas to the wall when the mass-averaged gas temperature is higher than the wall temperature. The fact that both q_c and q_L are small at the same time and q_L is not well known means that it is very difficult to determine the exact end of burning.

To solve Eq. 12.2, the gas phase is modeled as air. Of course, the gas phase is a heterogeneous system of hot products, air, fuel vapor mixed with air, and liquid fuel. The question then becomes how to more accurately model the internal energy term in the first law rate equation. A simple one-zone model, although inaccurate, is often used. In this model the fuel vapor and liquid parts of the system are ignored and the air and products are taken as perfectly mixed. The model starts with air-residual mixture and assumes that fuel is introduced to the system at a rate just equal to the burning rate. The system equivalence ratio thus starts near zero and grows to the final overall engine value. Because hot rich zones are neglected in this model, the effects of dissociation are very small. A term must be introduced for the addition of the fuel enthalpy, h_f. The internal energy is for products and now includes the chemical energy:

$$m\frac{du}{dt} = -p\frac{dV}{dt} - q_L + \dot{m}_f(h_f - u) \qquad (12.9)$$

where $\dot{m}_f = dm/dt$. And neglecting dissociation,

$$\frac{du}{dt} = \left(\frac{\partial u}{\partial T}\right)\frac{dT}{dt} + \left(\frac{\partial u}{\partial F}\right)\frac{dF}{dt} \qquad (12.10)$$

Introduction of a heat transfer model, the ideal gas equation, and measured pressure data allows solution for $\dot{m}_f$ and T as functions of crank angle or time. In this model, $\dot{m}_f$ is an apparent fuel burning rate (AFBR) (see Problem 12.13).

Understanding the trends of burning rate in a turbocharged engine is confounded by the change in boost pressure with operating conditions. Even in an NA engine the conditions for ignition change because of heat transfer and residual fraction effects. For example, the ignition delay decreases with increasing load due to these effects. To avoid these confounding effects one can use a rapid compression machine (RCM) in which the conditions at start of injection are precisely controlled. Such studies have been widely conducted at the Tokyo Institute of Technology and at MIT. Figure 12.5 shows two ways of varying overall fuel-to-air ratio in an RCM. In Figure 12.5*a* the traditional method of increasing injection duration was used. The injection is the same for all three fuel amounts up to about 1 ms, and thus the AFBR is the same for each fueling rate up to about 1 ms after the start of combustion. In this case the premixed burning spike is also about 1 ms long. The end of premixed burning and start of mixing-controlled (diffusion) burning roughly corresponds to the first minimum in the AFBR curve, but overlapping of the two modes undoubtedly occurs after the maximum of the spike. Note that the increased fuel amount causes case A to have a higher diffusion burning rate, but that the duration of burning is not changed much for the three fueling rates. A similar observation has been made for a turbocharged engine [Woschni and Anisits]. Once injection has ended (at 3 ms for case A), the rate of mixing no longer benefits from the momentum of injection, and turbulent air motion (and piston motions in an engine) determines the mixing rate. Because the RCM has no expansion stroke, the combustion duration is much longer than in a heavy-duty engine operating at, say, 1660 rpm, where the duration is about 7 ms (about 70 CA°). The shape of the RCM generated AFBR is, however, very similar to that in an engine.

Figure 12.5*b* shows the effect of a constant duration injection with fueling level changed by increased injection rate. The case A is the same in both Figure 12.5*a* and Figure 12.5*b*. The premixed spike now increases with increased rate although the ignition delay is only slightly changed. It has been observed that very

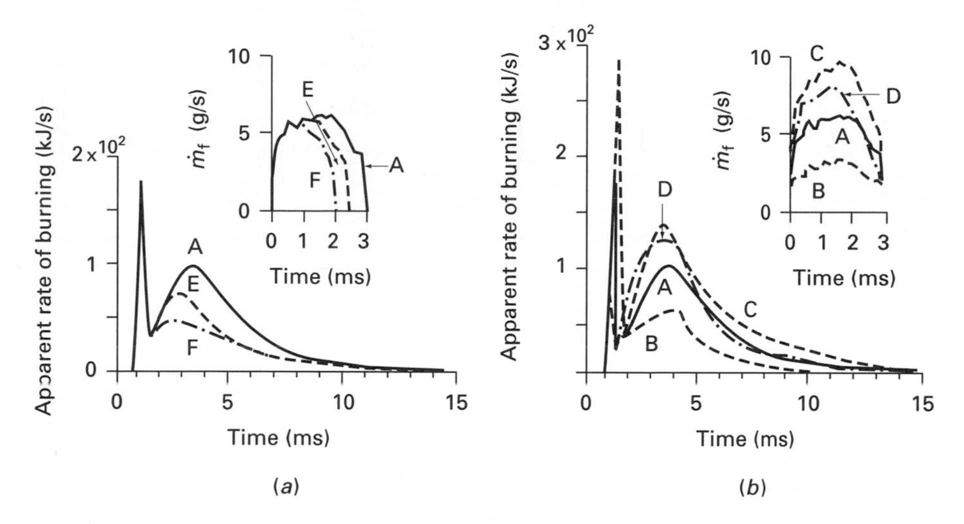

FIGURE 12.5
Heat release shapes obtained from a rapid compression machine with direct injection of various amounts of fuel: (*a*) effect of duration of injection, (*b*) effect of rate of injection [Kamimoto et al. (1984), reprinted with permission from SAE paper 841076 © 1984, Society of Automotive Engineers, Inc.].

high initial rates of injection shorten the ignition delay, but here the initial slope of the injection rate is not changed. Both the duration and diffusion burning rate are increased for the higher rate of fueling. The longer duration of case C compared with case A indicates that the increased rate of burning of C is not enough to offset the late addition of more fuel. Comparison of F and B shows that the decreased premixed burning of B shows up as more diffusion burning. We might expect case B to have a shorter duration because the spray momentum addition extends out longer for B than for F. This seems to be the correct conclusion, but the differences are small.

Figure 12.6 shows the effects of design and timing changes for a typical turbocharged truck engine, circa 1980. Figure 12.6*a* is for the standard 16:1 compression ratio engine. Note that the AFBR curves are normalized by dividing by total fuel burned. Thus, the uprated (more boost pressure and fuel) condition of Figure 12.6*b* shows essentially no change except for a slight increase in premixed burning due to the additional fuel. Figures 12.6*c* and *d* show the effects of reduced compression ratio. As expected, the ignition delay and the premixed spike increase with decreased compression ratio. The duration does not increase despite the lower expansion ratio, indicating that piston speed is dominating the later phases of burning rather than geometry effects. Typically, the lower-CR engines could be boosted more, but this was not done in the cases shown. Figures 12.6*e* and *f* show the effects of retarded timing. Note the large reduction in premixed burning due to retard. The figures do not clearly show the expected reduction in ignition delay, because of the figure scale; however, it is this reduction that causes the decreased premixed burning. The retard of premixed burning also reduces the peak

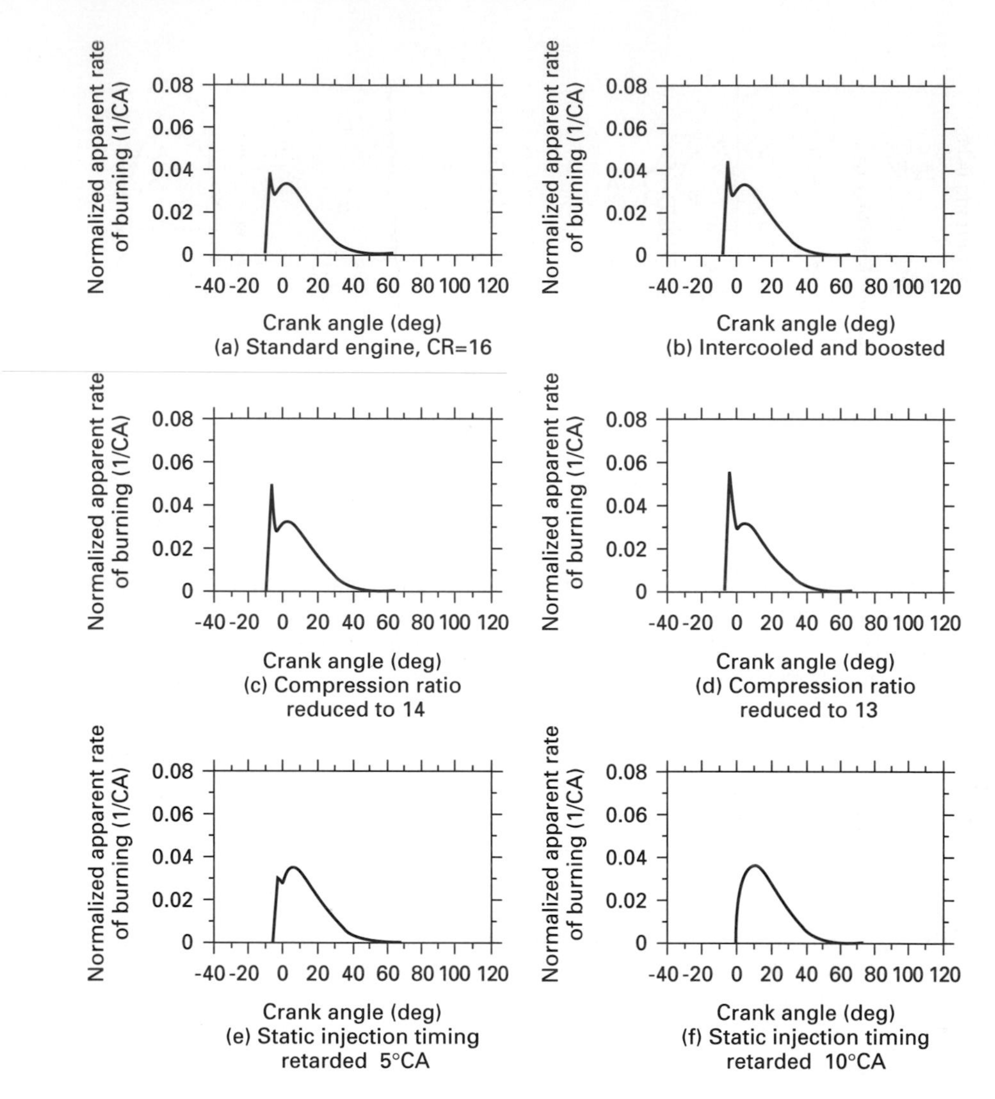

FIGURE 12.6
Normalized apparent fuel burning rate diagrams for a turbocharged truck diesel engine. The diagrams are normalized so that the integral of each is unity [Watson et al., reprinted with permission from SAE paper 800029 © 1980, Society of Automotive Engineers, Inc.].

temperature of the premixed burned products because the peak pressure is reduced and thus the products are not compressed as much after being formed.

Figure 12.7 shows the relationship between AFBR and heat transfer at two locations on the head of a single-cylinder boosted engine with a shallow bowl, low swirl design, and 16:1 compression ratio. The two surface thermocouple locations (STC 1 and STC 2) were on the head. STC 1 was above the combustion bowl, and STC 2 was to the outside of the bowl edge and thus covered by the piston at TDC.

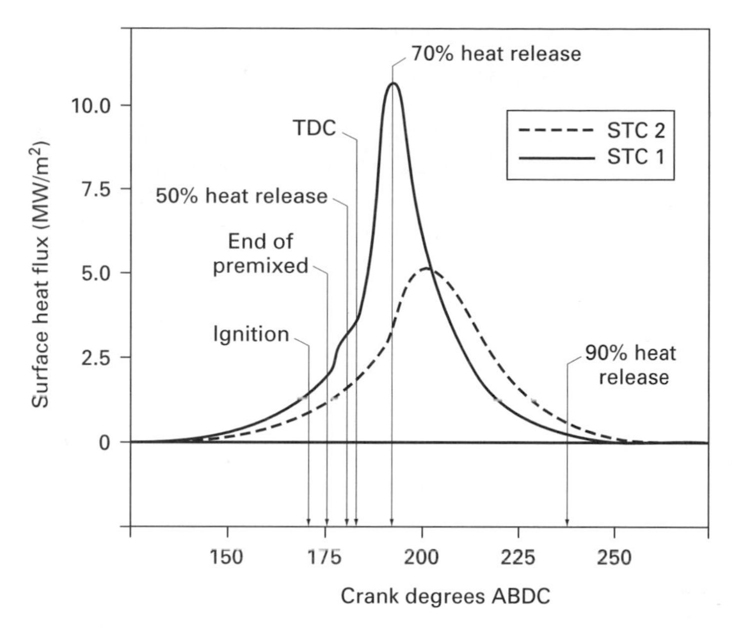

FIGURE 12.7
Surface heat flux at two locations on the head of a fired diesel engine. The engine conditions were: 1173 cm^3 displacement, 16:1 CR, shallow bowl, low swirl, 2000 rpm, 17° BTDC timing, 1.5 atm intake pressure, 6 × 0.178 mm injection nozzle with 15,000 psi injection pressure, equivalence ratio of 0.5, fuel cetane number 47 [Van Gerpen et al., reprinted with permission from SAE paper 850265 © 1985, Society of Automotive Engineers, Inc.].

The peak in the STC 1 location flux corresponds to the peak in mass-averaged gas temperature. About 30% of the fuel burned as premixed for the MBT timing case shown. The much lower peak flux at STC 2 can be accounted for by the effects of flame radiation, which could not reach that location. The STC 2 location did not experience the presence of hot products until the piston moved down and allowed outflow from the bowl. Also note that the heat flux for both locations was very low by the end of burning at about 60 CA° after top dead center (ATDC).

More complex thermodynamic models of heat release are possible, but require introduction of additional assumptions. For example, one may use a two-zone model where one zone is air and the other zone is products. Again the fuel vapor zone is neglected and fuel is introduced to the products zone at a rate equal to the burning rate. Now, however, the equivalence ratio of the products must be determined by assumption. It may be assumed that air from the air zone mixes with the products to allow any F versus θ curve desired. The final burning can be followed by mixing to give one final zone of lean products, or it may be assumed that complete mixing never takes place. It is clear that once one introduces two or more zones a fluid mixing model is required. This might lead to more and more zones and more and more unknown parameters. It would seem that this is not fruitful and one should take the great leap to a multidimensional model which solves the partial differential conservation equations. Although such a leap forward is now possible, it has only been recently that computers fast enough to solve the equations have been available.

One of the more successful of the older phenomenological models is that proposed by workers at Cummins Engine Company. In this model the spray is treated as a gas jet. To allow for the effects of cross flow, the gas jet theory was empirically modified. The model introduces the gas jet zone in which concentration profile shapes are assumed. An ignition delay formula determines the start of burning. At the start all regions which fall within prescribed lean and rich limits burn at an arbitrary rate. Products are lumped into one zone and mixing determines the rate of burning of the remaining fuel-air jet. Rather arbitrary assumptions are introduced for the mixing after injection has ended. The amount of NO produced in the products zone is calculated using Zeldovich kinetics.

More recently the model of Hiroyasu has offered a compromise between CFD modeling on supercomputers and a gas jet model such as provided by the Cummins model. The Hiroyasu model, which can be run on a personal computer, uses a crude grid and a spray droplet model. Because the grid is crude, the model depends strongly on semiempirical subgrid models. Such models can be tuned to a particular engine and then used for guidance in design, but are unlikely to predict correctly without tuning by use of engine-specific data.

Fitting of Burning Rate Curves

Use of the heat release model allows analysis and further understanding of the experimental data. However, such analysis does not allow predictive calculation of design change effects. Introduction of a burning rate curve into a detailed thermodynamic cycle calculation is needed to make such design studies. One may use a one- or two-zone model and introduce an arbitrary burning rate curve. Such analysis shows what might happen if such burning rates could be obtained.

Another approach is to introduce a functional form for the burning rate and to empirically adjust the constants in the function by using sets of heat release curves obtained from data. Once this is done the function is used in a predictive manner. A popular function used to fit heat release is the Wiebe function. Two Wiebe functions have been combined to produce a six-parameter function for fitting heat release data. The heat release rate HRR is represented [Miyamoto et al.] by

$$\begin{aligned}\text{HRR} = {} & C\left(\frac{Q_p}{\theta_p}\right)(M_p + 1)\left(\frac{\theta}{\theta_p}\right)^{M_p} \exp\left[-C\left(\frac{\theta}{\theta_p}\right)^{M_p+1}\right] \\ & + C\left(\frac{Q_d}{\theta_d}\right)(M_d + 1)\left(\frac{\theta}{\theta_d}\right)^{M_d} \exp\left[-C\left(\frac{\theta}{\theta_d}\right)^{M_d+1}\right]\end{aligned} \tag{12.11}$$

where C = constant
Q_p = integral under the premixed burning curve
θ_p = duration of premixed burning, CA°
Q_d = integral under the diffusion burning curve

θ_d = duration of diffusion burning, CA°
M_p, M_d = adjustable parameters

It was found that fixing C at 6.9 gave good fits in which M_p, M_d, and θ_p were essentially independent of operating conditions for a fixed engine design. The relationship between the functions is shown in Figure 12.8. Note that diffusion burning starts at $\theta = 0$ although radiation data indicate that diffusion burning actually starts somewhere near the peak of the premixed burning curve. It is not surprising that Q_p was found proportional to the amount of fuel injected during the delay period and that θ_d and Q_d were found to be proportional to each other. A different functional fit has been proposed by Watson et al., who claim that their formula fits a wide variety of engine conditions such as those shown in Figure 12.6.

The effects of the various heat release shapes can be explored without fitting to specific data. Figure 12.9 shows the results obtained by application of the single-zone model. The original heat release was first obtained from pressure data analysis and normalized to give the solid line in Figure 12.9*a*. The original curve was then simplified as two triangular shapes. Comparison of columns 1 and 2 shows that the simplification gave a reasonable approximation. Figure 12.9*b* shows a modified shape with a reduction of the peak of the premixed burning spike. This lowers the rate of pressure rise and peak pressure, but also decreases the indicated mean effective pressure (IMEP). In Figure 12.9*c* the modified curve is advanced by 7° to restore the IMEP while reducing the rate of change of pressure below the original value. The above exercise indicates the sensitivity of the engine to changes in heat release shape. Such a procedure could also be used to find the heat release shape that produces the best IMEP for a fixed maximum peak pressure. Unfortunately, such analysis does not indicate how to design the engine to obtain a given shape.

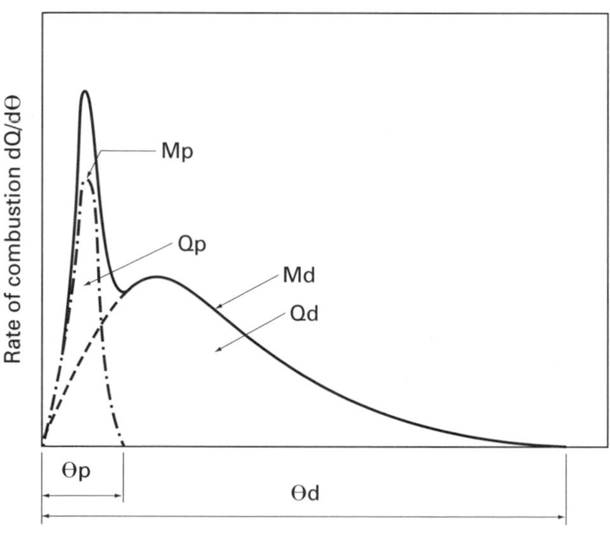

FIGURE 12.8
Description of rate of combustion in an engine by two combined Wiebe combustion functions [Miyamoto et al., reprinted with permission from SAE paper 850107 © 1985, Society of Automotive Engineers, Inc.].

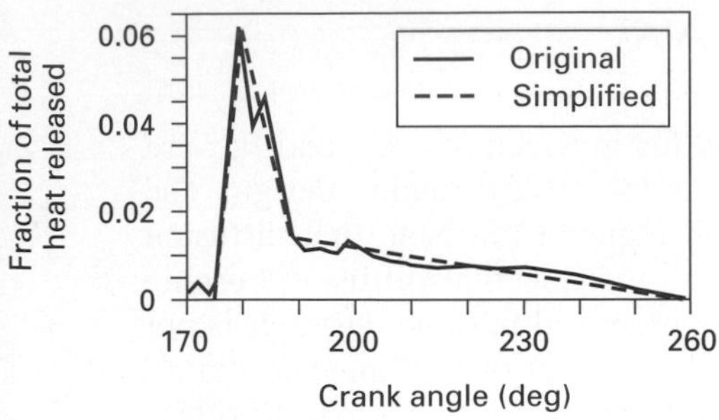

(a) Original experimental and simplified heat release versus crank angle at an engine speed of 3200 rpm

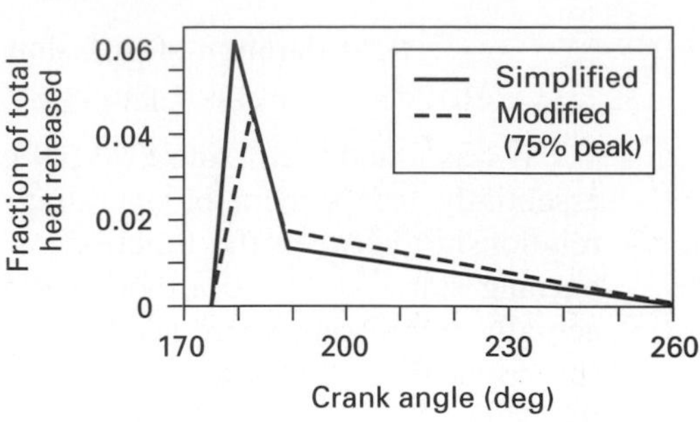

(b) Simplified heat release shape and modified heat release shape with 75% of simplified peak

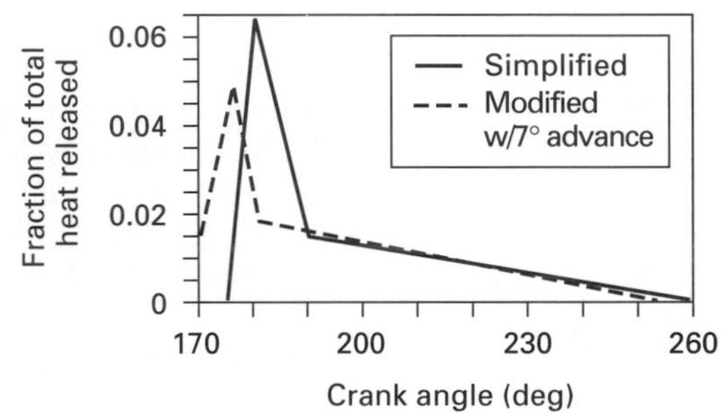

(c) Modified heat release shape and modified heat release shape advanced 7 deg

	(1)	(2)	(3)	(4)	(5)
Compression ratio	16.00	16.00	16.00	16.00	18.19
Peak cyl. temp., °R	3737.1	3750.5	3758.7	3764.1	3550.5
Peak cyl. pressure, psi	1130.0	1110.0	966.0	1112.0	1110.0
Max. pressure rise, psi/deg	88.38	81.03	50.82	70.64	56.19
Imep, psi	137.85	138.92	135.32	138.56	137.98
Heat transfer sum, Btu/cycle	0.423	0.421	0.413	0.432	0.41

Column	Heat release curve
1	Original Fig. (a)
2	Simplified Fig. (a)
3	Modified, 75% peak Fig. (b)
4	Modified, w/7 deg advance Fig. (c)
5	Modified, 75% peak Fig. (b), with increased compression ratio.

FIGURE 12.9
Effect of heat release shape modifications [Weber and Borman, reprinted with permission from SAE paper 670480 © 1967, Society of Automotive Engineers, Inc.].

12.5 CHAMBER GEOMETRY

Thus far the discussion of diesel combustion has centered on the spray characteristics. However, the shape of the combustion chamber and the fluid motion in it are also of importance. The characteristics of the flow at time of injection are determined by the inlet port and valve configuration, and the chamber shape. The major effect of the intake process is to determine the large-scale motions in the chamber at intake valve closing. As the piston moves upward, the gas is pushed into the piston bowl. If the gas is swirling, the swirl will increase due to conservation of angular momentum. Recall that the radial inward velocity of the flow caused by the piston motion is called the squish velocity. Although the direct effects of the squish velocity are probably small, the indirect effects on the motion in the bowl can be quite significant. The shape of the edge of the bowl is also important. A reentrant shape may set up vertical vortices in the bowl and can greatly influence the way the combustion products leave the bowl during the expansion stroke; this outflow is sometimes called *reverse squish.*

Swirl has a profound influence on mixing between the fuel plumes and the air during the injection period. After the end of injection the primary contribution of swirl is through its decay to turbulent motion. The swirl also tends to stabilize the flow, reducing cycle-to-cycle flow variations. Chambers with low swirl tend to have tumbling vortices which can shift from one cycle to the next. But the more stable swirling flow can also produce undesirable effects. The undesirable effects of high swirl are to cause the hot, less dense products to move to the center of the chamber, thus reducing mixing; to cause reduction in spray penetration, to produce lean mixtures which do not burn or only partially burn, thus causing unburned hydrocarbon and CO exhaust emissions; to increase heat transfer; and to increase the premixed burn fraction. Low-swirl engines with shallow, wide bowl combustion chambers and high injection pressures need to have nozzles with many holes (8–10) to compensate for the lack of tangential mixing that would be caused by swirl. Such chambers typically lack mixing during the later part of the combustion period. Unique designs of the chamber shape may provide a solution to these problems. Figure 12.10 shows three chamber shapes. The odd pocket shapes of the Hino Motors Micro Mixing System (HMMS) cause some of the swirl to be converted to vortices in each pocket during the expansion stroke. This system also seems to increase the higher-frequency components of the turbulent energy. The increased mixing after TDC results in improved performance and emissions as shown in Figure 12.11. A flat dish design gave more premixed burning and more late combustion than the HMMS design.

The HMMS chamber design influences the crank-angle history of the mixing rate, but cannot compensate for lack of mixing at low engine speeds. One way to solve the low-speed problem is to use a two-passage port with an adjustable vane in one port passage. By adjustment of the vane, more or less flow is directed through the port, which produces swirl. Thus, swirl can be increased at lower speeds, and performance at low speeds is improved. While such intake flow changes are easy to control, no method of dynamically adjusting the flow history during expansion has been found. The Hino design is a fixed geometry and thus can be optimized for only one condition of load and speed. As shown in the next section, some methods of fluid control are needed if diesels are to solve the problem of simultaneous reduction of particulates, nitric oxide, and fuel consumption.

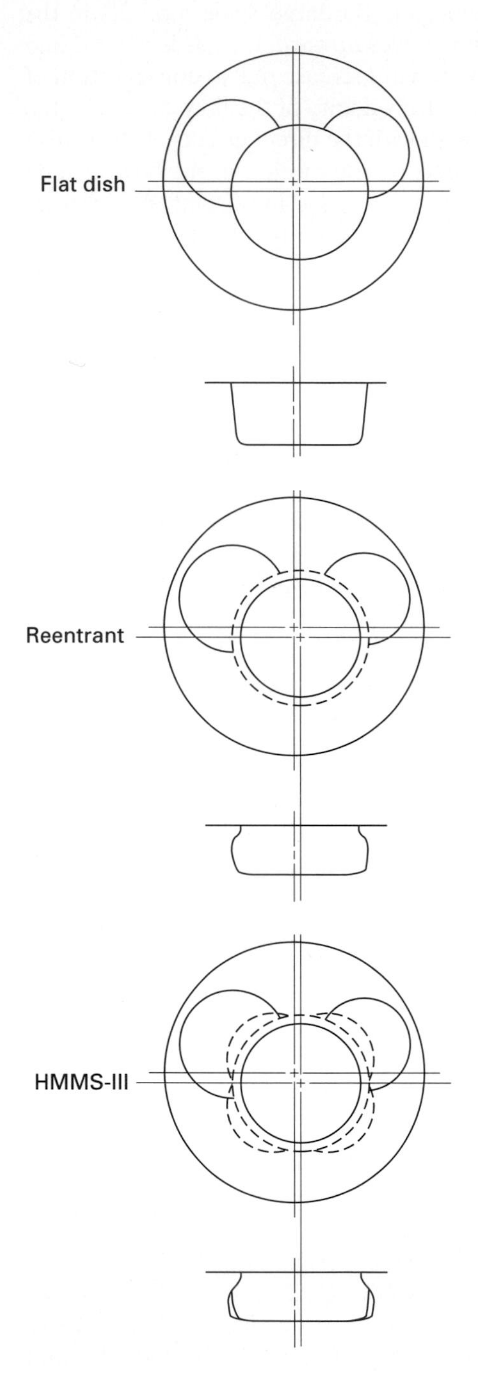

FIGURE 12.10
Comparison of combustion chamber configurations used in obtaining data of Figure 12.11 [Shimoda et al., reprinted with permission from SAE paper 850070 © 1985, Society of Automotive Engineers, Inc.].

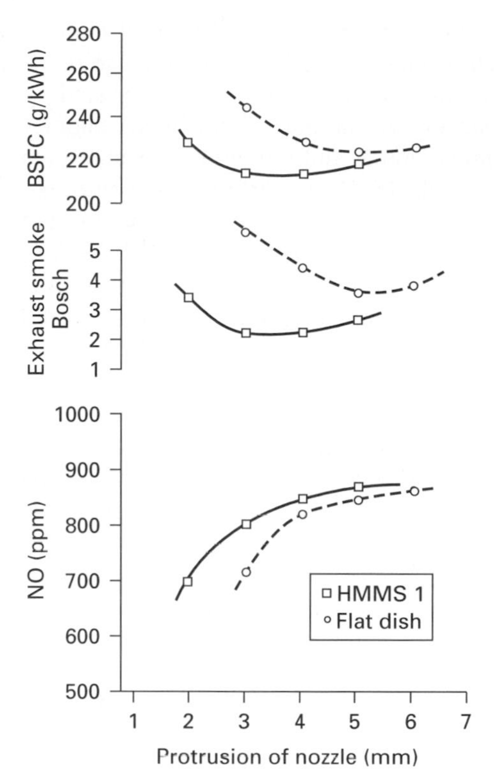

FIGURE 12.11
Comparison of engine performance between flat dish chamber and HMMS-III chamber shape [Shimoda et al., reprinted with permission from SAE paper 850070 ©1985, Society of Automotive Engineers, Inc.].

12.6 EMISSIONS

The two most troublesome emissions from diesel engines are soot (particulates) and nitrogen oxides. Typically, unburned hydrocarbons and CO are not a serious problem except at light loads. Hydrocarbons can arise from fuel absorbed in deposits and oil layers or from very lean mixtures. At light loads the fuel is less apt to impinge on surfaces; but because of poor fuel distribution, large amounts of excess air, and low exhaust temperature, lean fuel-air mixture regions may survive to escape into the exhaust. White smoke is often observed at low load conditions and is really a fuel particulate fog. The more typical black smoke sometimes observed during periods of rapid load increase, or for older engines at higher loads, is made up primarily of carbon particles. At higher loads the higher temperatures tend to oxidize even lean mixtures. In homogeneous charge automobile engines, unburned hydrocarbon emissions are high during startup, when the catalyst is too cold to be effective. Diesels provide reduced hydrocarbons in automobiles even though they have no catalyst system. This is only true, of course, if the diesel starts promptly, and thus starting aids such as glow plugs are often used in automotive diesel applications.

The problem of NO production in naturally aspirated (NA) diesel engines arises from the early rapid burning, which produces very-high-temperature products. Recall from Chapter 4 that if the reactants have a stoichiometric or slightly lean fuel-air ratio the amount of NO produced will be maximum. In NA engines with MBT injection timing, evidence from sampling and radiation measurements indicates the reactants are lean or stoichiometric during the premixed burning. However, recent data obtained by Dec and his co-workers [1997] for a boosted quiescent chamber with slightly retarded timing and moderately high injection pressure indicate that only a small initial portion of the premixed combustion is lean—the rest is rich. It is logical that for the shorter ignition delay of a boosted (turbocharged) engine, the amount of mixing prior to ignition will be limited. Similarly, in a quiescent chamber the mixture strength and distribution at the outer surface of the spray will depend on spray momentum rather than airflow effects. For swirl, the region downstream of the spray is likely to be a cloud of leaner mixture, while for a quiescent case the inward flow of entrained air could limit the lean mixture to a thin outer zone as reported by Dec. The author has observed that in several engines of the high-pressure injection type, the time and location of ignition is remarkably repeatable from cycle to cycle. This indicates that the very repeatable injection process rather than the cyclic variability of the airflow is governing the mixture preparation at the time of ignition.

Figure 12.12 shows a comparison of the NO history in the cylinder calculated from the Cummins model and from experiments for a heavy-duty NA engine with

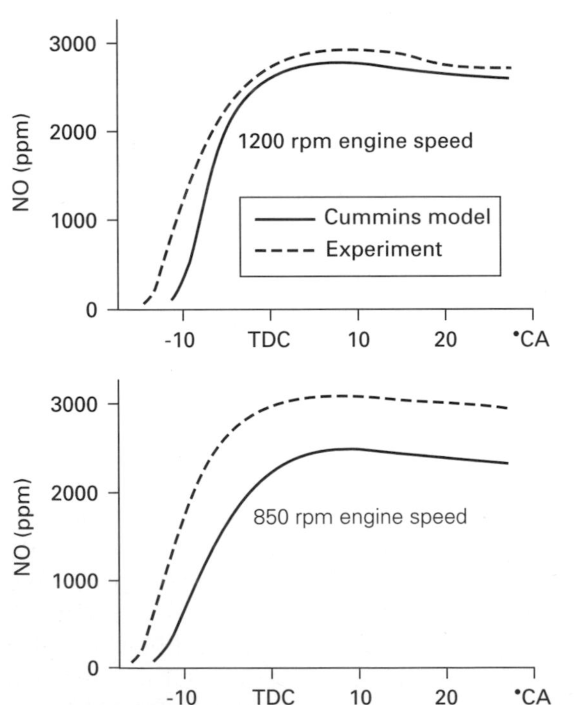

FIGURE 12.12
Comparisons of calculated and experimental nitric oxide histories for two speeds at constant equivalence ratio of 0.6 and injection at 27° before TDC [Voiculescu and Borman, reprinted with permission from SAE paper 780227 © 1978, Society of Automotive Engineers, Inc.].

direct injection and MBT timing. The experimental data were obtained by a method of total cylinder sampling. At a given crank angle, the cylinder contents were suddenly allowed to rapidly flow into a quenching tank, where the NO reactions were frozen. Measurement of the NO in the quenched gas for repeated experiments at different crank angles produced the NO history. These data show that NO is produced by the combustion prior to peak pressure. By the time of peak pressure the NO in these hot products is essentially at equilibrium. It would seem that NO could be modeled simply by calculating the amount of products at peak pressure using a heat release analysis and then calculating the NO by assuming the products are at their adiabatic flame temperature at each burning step and are unmixed. Unfortunately, such a model is not accurate, because the fuel/air ratio of the hot products significantly influences the equilibrium NO and this fuel/air ratio is unknown.

The problem of NO production in modern heavy-duty turbocharged diesels with high-pressure injection and retarded timing is more complex. Again, by use of total cylinder sampling, data show that in such engines a significant portion (40%) of the NO_x is produced after peak pressure, and that the premixed burning could at most produce less than half of the NO_x produced at peak pressure. One may be concerned that under such conditions the Zeldovich NO kinetics could be inadequate for the 40% produced after peak pressure. However, CFD calculations using only Zeldovich kinetics have been able to reproduce the experimental NO histories remarkable well. These model NO values are, however, very sensitive to the ignition and turbulence models so that the excellent agreement may be caused by fortuitous combinations of these submodels.

Two methods of reducing NO emissions are quite effective. The first method is to retard the injection timing. This causes some reduction in fuel economy, but is quite effective because it reduces the amount of premixed burning. The second method is to recirculate cooled exhaust gas so that product temperatures are lowered. These solutions when combined solve the NO problem, but unfortunately both methods increase the amount of exhaust particulate.

If the injection is progressively retarded from MBT at constant operating conditions, the particulate emissions will eventually rise, typically before the fuel economy begins to suffer significantly. A plot of specific particulate mass versus specific NO_x mass for the various injection timings produces what has come to be called the "particulate-NO_x trade-off curve." Figure 12.13 shows such a plot for a heavy-duty single-cylinder DI engine running at 1600 rpm and 75% load ($F = 0.55$) with the intake air pressure at 185 kPa. The first point on the right for each line represents a timing of 9° BTDC, and the following points are further retarded in 3° increments. The lines show four injection pressures each with the same injector tip configuration (125° spray included angle, 6 holes, 0.259 mm hole diameter). Because of the fixed holes, the increase in pressure requires a decrease in injection duration to keep the amount of fuel injected constant. The particulate decreases with injection pressure for a fixed NO_x value, but the amount of retard required to lower the NO_x increases with increasing pressure. This is probably due to the fact that the premixed burn fraction increased with increased injection pressure. The reduction in particulate with injection pressure was large between

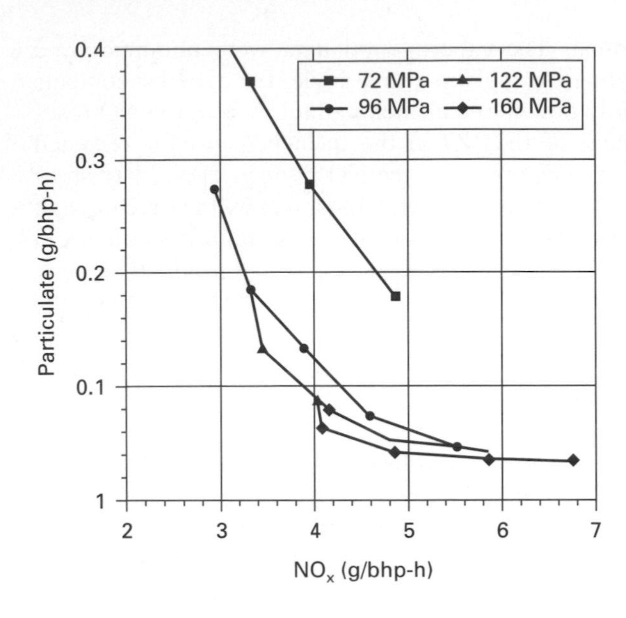

FIGURE 12.13
Effects of injection pressure and timing on particulate-NO_x trade-off. Each curve for a given pressure shows data points at 3° intervals of retard, starting with injection at 9° BTDC [Pierpont and Reitz, reprinted with permission from SAE paper 950604 ©1995, Society of Automotive Engineers, Inc.].

72 and 96 MPa but was relatively small thereafter. Other studies which have varied injection pressure, swirl, number of holes, and hole size indicate a similar trend; that is, the optimum trade-off is improved by increases in injection pressure, but the effect is much diminished after 130 MPa. This may be due to the very small droplets at very high pressure which rapidly vaporize, giving reduced mixing more similar to that of a gas jet.

At the present time no proven method of simultaneously reducing NO and particulates is known. We do know that a great portion of the particulate produced is oxidized in the cylinder. Only $\frac{1}{10}$ to $\frac{1}{20}$ of the soot formed escapes into the exhaust. The largest peak in soot production is just at the start of diffusion burning when the fuel spray is cut off from its air supply and surrounded by very hot products from the premixed burning. However, the high temperature and subsequent mixing with air cause this soot to rapidly oxidize. Furthermore, the time for the soot particles to agglomerate at this time is very short. Soot produced later, near the end of burning, is less apt to be oxidized because the temperature is much lower due to expansion. Soot remaining from the previous combustion is now also slower to oxidize, because of the lower temperature. Figure 12.14 shows the sequence of events in the soot formation process, and the appropriate times for each stage of the process are indicated in Table 12.3.

Modeling of the soot production-oxidation process is hampered by the present inability to predict soot production and to a lesser extent by the uncertainty of agglomeration, oxidation, and mixing models. It is clear, however, that neither very rapid mixing nor very slow mixing is desirable. Rapid mixing will cool the products too rapidly and thus quench oxidation. Slow mixing will prevent oxidation due to the low partial pressure of oxygen, and cooling by expansion will eventually prevent further oxidation. The optimum path of fuel/air ratio during combustion and rate of mixing of products with air is unknown. We can see, however, that both

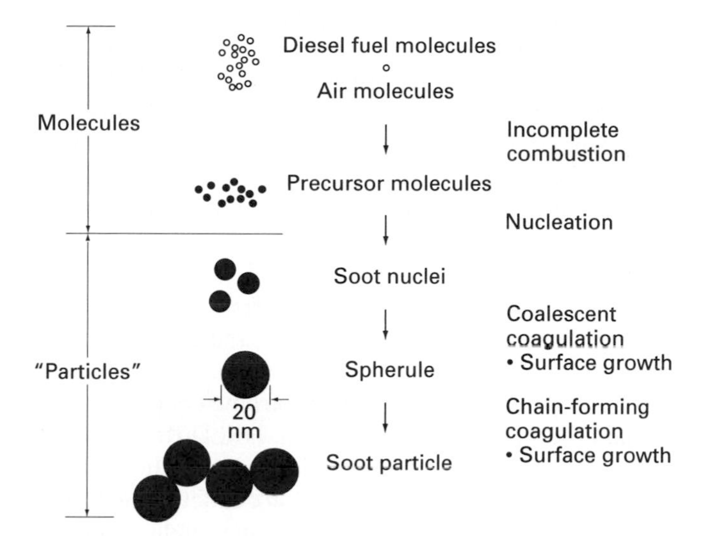

FIGURE 12.14
Sooting history in a small volume element of a diesel engine combustion chamber. Incomplete combustion of fuel molecules produces a supersaturation of condensable precursor molecules which are the building blocks of soot nuclei. Following nucleation, coalescent coagulation and surface growth by continuing condensation of precursor molecules proceed simultaneously, followed by chain-forming coagulation and surface growth (until precursor species are depleted or formation is terminated by cylinder expansion and cooling). Dehydrogenation and oxidation reactions occur simultaneously with these events.

TABLE 12.3
Summary of time constants for various aspects of diesel soot formation

Process	Approximate time constant or duration
Formation of precursors/nucleation	Few μs
Coalescent coagulation	0.5 ms after local nucleation
Spherule identity fixed	After coalescence ceases
Chain-forming coagulation	Few ms after coalescence ceases
Depletion of precursors	0.2 ms after nucleation
Nonsticking collisions	Few ms after nucleation
Oxidation of particles	4 ms
Combustion cycle complete	3–4 ms
Deposition of hydrocarbons	During expansion and exhaust

retarded injection and exhaust gas recirculation can produce more soot due to more diffusion burning, less oxygen, and lower oxidation temperatures.

Several kinds of experiments have been performed which augment mixing during the period after peak pressure. These experiments used either combustion in a small prechamber or mechanical compression in a small prechamber to create a jet of gas directed into the combustion chamber of the engine. These experiments all show reduction of soot without increases in NO. The techniques are not very practical, however, because of the additional equipment required and the added heat loss. Recently, the use of split injection has shown a similar effect. In split injection about half or more of the fuel is injected with retarded timing. The rest of the fuel is injected through the same injector after a significant delay (10 CA°). The combination of the split injection with EGR and retarded timing of the first injection has shown promising trends.

Before summarizing the effects of design parameters on diesel emissions it is necessary to explain the standards as set by U.S. law. Different standards apply to light-duty, heavy-duty, and off-road applications. In addition, the United States, European Community (EC), and Japanese laws are different in both test procedure and standards. This means that a heavy-duty engine might qualify in one part of the world but not in other places. Global standards have been developed for off-road applications, indicating some welcome movement toward global cooperation.

The evolution of U.S. limits for heavy-duty vehicles is shown in Table 12.4. Note that starting in 1983 a transient method was put in place. Both Japan and the EC still use combinations of steady-state modes. The steady-state mode method produces results that are easier to understand, but the transient method is claimed to be more realistic. The particulates are measured in the EPA procedure by dilution of exhaust and collection on a plastic-coated fiberglass filter at a

TABLE 12.4
History of U.S. EPA heavy-duty emission limits

	Emissions [g/(hp·h)]			
Year	**CO**	**HC + NO_x**		**Particulates**
1970	—	—		Smoke (4)
1974 (1)	40	16		(4)
1979	25	5		(4)
		HC	**NO_x**	(4)
1983 (2)	15.5	1.3	10.7	(4)
1988	15.5	1.3	6.0	0.6
1991	15.5	1.3	5.0	0.25
1993 (3)	15.5	1.3	5.0	0.1
1994	15.5	1.3	5.0	0.1
1994 (3)	15.5	1.3	2.5	0.07
1996 (3)	15.5	5.0	5.0	0.05
1998	15.5	4.0	4.0	0.1
1998 (3)	15.5	4.0	4.0	0.05

(1) 13 mode emissions; (2) transient cycle; (3) urban bus; (4) smoke measured by opacity rather than particulate mass.

temperature not to exceed 125°F. The collected mass includes condensed fuel, lubrication oil, and sulfates in addition to carbon particulate. The use of very-low-sulfur (0.05% by weight maximum) fuel starting in fall 1993 has greatly reduced the sulfates. Redesigning the engines for lower oil consumption has reduced the amount of lube oil, collected. Lube oil is thought to be from particulates of oil escaping from the piston-ring-liner lubrication system.

The ultra low emissions vehicle requirement of 0.05 g/mile of particulate for cars, based on the urban driving cycle, will be very difficult to meet and may limit the use of diesel-powered passenger cars in the United States. The U.S. EPA requirement of 0.10 g/mi is already very challenging.

The use of aftertreatment devices for diesels has been limited by the low temperature and lean fuel-air ratio of the exhaust. Recall that the three-way catalyst system requires products from a stoichiometric fuel-air ratio. Starting in about 1994 midsized diesel engines began to use oxidation catalytic converters, which oxidize the soluble organic fraction of the particulate and also some of the vapor-phase hydrocarbons. This lowers the mass of the particulates as defined by the emissions test procedure. Traps which collect particulates have also been developed. These traps must provide regeneration by being heated to combust the collected particulate. The high cost and durability of these traps has thus far limited their acceptance. Work is also progressing on catalyst systems which can destroy NO in lean-burn exhaust mixtures, but such systems are not yet in use.

Reformulation of diesel fuel by using lower aromatic content and higher-cetane fuels has shown the ability to help lower NO_x emissions. The beneficial effect of aromatics reduction seems to be caused by a reduction in peak flame temperature. The cetane effect decreases the premixed burning, which has a beneficial effect at all loads. Lower aromatics may or may not reduce particulate carbon (soot). The expected reduction is based on low-temperature diffusion flame experiments, but diesels can have very-high-temperature (2800 K) products which may break down the aromatic structures so that the anticipated reduction in soot may not take place. The influence of aromatics may also be confounded by the fuel hydrogen-to-carbon ratio. A recent engine fuels study [Rosenthal and Bendinsky] shows that increasing the H/C ratio decreases soot. This study also shows a small soot increase with increased fuel aromatics. Alternative fuels such as methanol and other higher-cetane fuels containing oxygen, such as DME (dimethyl ether), typically do not produce soot. Thus they could solve the particulate problem while allowing for more retard and EGR to lower NO_x. The current problem with these fuels is cost and availability.

Emission Trends with Operating Parameters

The above discussion indicates some of the combustion factors which determine performance and emissions in heavy-duty, open-chamber (DI) diesel engines with turbocharging and after cooling. Such engines are typically used in trucks. We now summarize these effects in a qualitative fashion. One must recognize that of these quantities only fuel economy (BSFC) is of concern to the user. The user also has many other concerns, however, such as reliability, starting, low maintenance, etc., which must be met. In Table 12.5, the effects of operating parameters are shown, and in each case the response is shown for an increase in the parameter.

It is clear that increased ambient temperature creates a problem for all factors except hydrocarbons. Recall that most turbocharged engines have an air-to-air intercooler so that increased ambient temperature does increase the intake temperature, thus decreasing the intake air density. Such a change lowers trapped mass as well as increasing compression temperature. The increase in temperature decreases ignition delay, but the higher products temperature overcomes this good effect and increases NO_x emissions. Similarly, the effect is to increase fuel/air ratio and diffusion burning fraction, thus increasing particulates, but improving particulate oxidation rates. Because oxidation rates are limited by mixing, the overall effect is to increase particulates. With onboard computers and modern controls it is possible to respond to ambient conditions by changes in engine fueling parameters, but as yet no perfect response is possible.

Understanding the parameters of engine speed and load is complicated by the engine governor system and its fueling system response. For example, in on-road vehicles engine torque is prevented from falling off with small decreases in speed resulting from increased load (due to a hill, for example); this reduces the need for downshifting. The speed-load operating factors are also affected by other system parameters such as the turbocharger performance map.

At low engine speeds, the turbocharger cannot provide adequate boost, and higher fuel/air ratios will result. At high speeds the pumping work increases, as do other loss factors. The engine system is typically designed to give optimum fuel economy at the moderate speeds at which it is usually operated. Temperature levels increase with speed due to reduced cooling and increased hot residuals; this increases NO_x. Hydrocarbons do not follow the expected decrease with temperature due to improved oxidation, because of the offsetting factor of less time to oxidize.

TABLE 12.5
Emission trends with operating parameters [Borman and Brown, by courtesy of CNR].

Operating parameter	BSFC	$BSNO_x$	BSHC	Particulates
Ambient humidity	→ (no effect)	↘ (decrease)	→ (no effect)	→ (no effect)
Ambient temperature	accelerating increase	↗ (increase)	↘ (decrease)	↗ (increase)
Ambient pressure	decelerating decrease	→ (no effect)	→ (no effect)	decelerating decrease
Engine speed	reaches an extremum	↗ (increase)	accelerating increase	reaches an extremum
Engine load	reaches an extremum	reaches an extremum	decelerating decrease	reaches an extremum
Fuel cetane number	→ (no effect)	decelerating decrease	decelerating decrease	→ (no effect)
Fuel sulfur content	→ (no effect)	→ (no effect)	→ (no effect)	↗ (increase)
Fuel aromatic content	→ (no effect)	↗ (increase)	accelerating increase	→ (no effect)
Fuel volatility	→ (no effect)	↗ (increase)	↗ (increase)	→ (no effect)
Coolant temperature	decelerating decrease	↗ (increase)	decelerating decrease	decelerating decrease
Time after start	decelerating decrease	decelerating increase	decelerating decrease	decelerating decrease

Symbol key: No Effect →; increase ↗; decrease ↘; accelerating increase; decelerating increase; decelerating decrease; accelerating decrease; reaches an extremum.

The optimum in specific particulates is, like fuel economy, partially a result of the system optimization, but at high speeds there is less time to mix and oxidize the soot.

Several factors change with load: the ratio of friction to output work decreases, fuel-to-air ratio increases, exhaust temperature goes up, and boost from the turbocharger increases. The net result is an optimum value at an intermediate load for all factors considered here except hydrocarbons, which continue to decrease with load primarily due to increased oxidation. The specific NO_x is typically decreasing with load because the premixed combustion decreases with load.

The effects of fuel properties, as discussed previously, are confounded by the fact that properties, are highly correlated. Thus increased aromatic content typically decreases cetane number unless a cetane improver is used. The primary effect of fuel volatility is to increase the rate of premixed burning and thus increase NO_x, and to create more lean regions which do not burn, thus increasing hydrocarbons. Increased coolant temperature reduces heat loss and increases temperature levels, especially during expansion. Countereffects can, however, be found due to effects of higher part temperatures on some fuel-injection systems.

The federal test procedure requires both a cold (room temperature) and a hot test. The test cycle is the same for both. At the cold start the engine is warming up with time; thus friction decreases and fuel vaporizes better as time goes on, but the temperature levels increase, causing increased NO_x but decreased particulate.

12.7 DIESEL DESIGN IMPROVEMENTS

The major losses in diesels other than the available energy destroyed by the combustion process are the lost available energy during exhaust blowdown, the heat transfer to the coolant, the mechanical friction, and unused exhaust energy. Increased turbocharging can reduce the exhaust blowdown loss by decreasing the cylinder-to-port pressure difference. No viable mechanism for replacing the blowdown throttling process by a reversible process has been found. Heat loss control and exhaust energy recovery are linked together because the main effect of reduced heat transfer to the coolant is to increase the exhaust energy. At the present time there is a major effort to find ways to insulate the engine parts, particularly the piston, head, and exhaust port, by use of ceramics or high-temperature metals. The use of increased part temperatures creates problems in decreased volumetric efficiency, lubrication difficulty, and serious thermal stress problems. However, increased temperature levels may decrease particulates through improved in-cylinder oxidation and can improve the fuel tolerance of the engine. The gains in efficiency from insulation are small unless the exhaust gas energy is recovered by use of turbocompounding. Turbocompounding is expensive and thus may be practical only for large systems or military applications.

Two major areas of design which influence both emissions and performance are fuel system design and engine geometry. The trends of fuel economy and emissions for these areas are shown in Tables 12.6 and 12.7.

Fuel system design factors have already been discussed, so Table 12.6 serves as a review. The most important design factor is high-pressure injection, which is the

TABLE 12.6
Emission trends with fuel system design [Borman and Brown, by courtesy of CNR].

Design parameter	BSFC	$BSNO_x$	BSHC	Particulates
High-presure injection	↘	↗	↘	↘
Retarded timing	↗	↘	↗	↗
Rate shaping	∪↗	∪↗	∪↗	∪↗
Pilot injection	↗	↳	→	→
Dual fuel injection	→	↘	→	↘
Sac volume	→	→	↗	∪↗
Number and size of holes	∪↗	∪↗	→	∪↗
Spray angle	∪↗	∪↗	→	∪↗
Eccentricity of nozzle	⤴	→	⤴	⤴
Tip projection	∪↗	→	∪↗	∪↗
Air assist	→	→	↘	↘
Impingement on pedestal or bowl	∪↗	∪↗	⤴	∪↗
Quality and uniformity of injection	→	→	↳	↳
Controllability	↘	↘	↘	↘

TABLE 12.7
Emission trends with engine geometry [Borman and Brown, by courtesy of CNR].

Design parameter	BSFC	$BSNO_x$	BSHC	Particulates
Compression ratio	∪↗	∪↗	↘	∪↗
Stroke/bore ratio	↳	→	↳	↳
Piston-to-head clearance	⤴	→	→	⤴
Crevice volumes	↗	→	→	↗
Piston crater shape	∪↗	∪↗	∪↗	∪↗
Swirl	∪↗	∪↗	∪↗	∪↗
Port geometry	∪↗	∪↗	∪↗	∪↗
Ring pack	∪↗	→	∪↗	∪↗
Valve size and number	↘	↳	→	→
Valve timing	∪↗	∪↗	∪↗	∪↗
Controllability	↳	↳	↳	↳

Symbol key: No Effect →; increase ↗; decrease ↘; accelerating increase ⤴; decelerating increase ↱; decelerating decrease ↳; accelerating decrease ⤵; reaches an extremum ∪↗.

major factor in reduction of particulates in modern engines. The effects of rate shaping and pilot injection are not yet proven for practical systems. Dual fuel injection has been suggested, and some injectors capable of injecting two different fuels in series have been built, but no system of this type is in use. An example would be injection of high cetane fuel for ignition followed by a low cetane fuel with good emissions reduction properties such as methanol. The trends shown are use of two fuels versus use of a single fuel. Users find the use of two fuels to be undesirable, so further development of such systems is unlikely. Sac volume reduction is desirable, but zero volume designs are sometimes incompatible with good injection quality due to the necessary flow pattern changes within the injector tip.

The next four parameters in Table 12.6 are the traditional design factors, and they must be selected for an optimum combination by experiment. Typically, the nozzle should be located along the cylinder axis for best injector performance, but small engines may require an inclined nozzle so as not to compromise valve size. Air-assisted injection for diesels and effects of using impingement to cause spray breakup or mixing are as yet unproved designs. The trends shown are thus speculative. The most desirable factor for a designer is control, especially in modern computer-controlled systems, which allow optimization for each operating condition. The problem here is the large amount of work required to find appropriate control algorithms.

The effects of engine geometry are far less well established in detail than are the fuel system parameters. Thus many parameters in Table 12.7 are simply shown as having an optimum. The first four factors, however, are relatively well understood. There is an optimum compression ratio for heavy-duty engines which is relatively low—about 16 for naturally aspirated engines and about 13 for turbocharged engines, where a lower optimum is prevented by starting considerations. Heat loss increases and increases in combustion duration with compression ratio increase are a major factor. Remember that an increased compression ratio also gives an increased expansion ratio. Thus the fluid mechanics of the combustion during the last part of burning are changed by the compression ratio. Variable compression ratio designs, including the Miller cycle, in which a late-closing intake value is used to limit compression at higher loads, have been attempted. Such designs allow control of peak pressure while still allowing higher compression ratios during starting and may prove practical if appropriate valve timing control mechanisms can be developed.

Large stroke-to-bore ratios have given improved fuel economy in large two-stroke engines due to chamber design effects and reduced heat transfer losses. However, four-stroke engines require adequate space for valves in the cylinder head, and thus most heavy-duty engines are about "square" (unity bore-to-stroke ratio). A larger stroke causes increased friction if the rpm is not reduced to keep the average piston speed constant as the stroke is increased.

Piston-to-head clearance should be as small as possible in order to obtain good air utilization and to minimize fuel getting into the crevice space. Crevice volumes should be minimized for similar reasons. Note that the mass of gas in crevices is higher than one might expect because the crevice gas is at the relatively low temperature of the crevice surface, but for most diesels of the heavy duty size the crevices contain air so that they do not contribute to the BSHC emissions.

The chamber shape and fluid mechanics are highly coupled. With modern high-pressure injection, the fluid mechanics has its greatest influence after the end of injection, when the spray momentum is no longer dominant. Swirl can, however, help to lean out the premixed burning and can improve air utilization and mixing in small engines, where spray impingement on the piston limits the utility of high-pressure injection. The major effect of bowl geometry other than to increase swirl is to influence the mixing as the products leave the bowl during the expansion after peak pressure; a reentrant shape aids mixing.

Increased valve flow area reduces the pumping work and heating of the trapped charge due to the filling process. Detailed cycle analysis and careful port design can be used to obtain optimum volumetric efficiency over a range of speeds and loads and to minimize pumping losses.

Control of valve timing and swirl can also help somewhat to improve performance, but large gains are not to be expected. Active control of geometry could be helpful to produce an optimum for each operating condition, but practical means for actuating such controls have not yet been devised.

An interesting comparison of 1991 and 1988 versions of the same engine model by Baumgard and Johnson has shown that without aftertreatment and using low-sulfur fuel for both engines, the volume of particulate at 75% load, 1900 rpm was reduced by a factor of 0.3, and at 25% load and 1900 rpm by a factor of 0.4 using the 1991 engine. The basic design changes were higher fuel injection pressure, higher turboboost pressure, and lower air intake temperature. Despite the lower particulate volume, the 1991 version produced a 35-fold increase in the number of particles at the 75% load condition and a 15-fold increase at 25% load. A large portion of this increase was due to a very large increase (30–60 times) in primary, nuclei-mode particles.

The above qualitative survey of design parameter effects has indicated some trends but does not help in the process of optimization. It is thus necessary at present to follow past practice, making experimental changes based on experience. This is a very costly process which could be greatly aided by predictive analysis tools. The following section thus discusses the current status of CFD analysis for diesel combustion. Because such models are currently in the process of development and are still primarily research tools, our discussion will be only heuristic in nature.

12.8 CFD MODELING OF DIESEL COMBUSTION

In the 1960s, thermodynamics, combined with the advent of digital computers, gave rise to detailed cycle analysis, which is zero-dimensional and can be used to improve valve timing, study thermal loading, and investigate the effects of heat release, turbocharging, and other systems design parameters [Benson and Whitehouse]. Zero-dimensional analysis is, however, unable to probe very far into the combustion process itself. The complexity of the transient effects of spray breakup, mixing, and inhomogeneous combustion require a detailed three-dimensional fluid dynamics model. Attempts to model such processes were limited by computation speed and computer memory size until the advent of supercomputers. Spatially resolved numerical methods and models developed in the 1970s began to appear in computer codes in the 1980s. By 1985 the generalized code KIVA, developed at Los Alamos National Lab, was available for public use. During the following 10 years the progress was remarkable, and several CFD codes for diesel design are now available commercially. We shall refer here primarily to the code KIVA II [Amsden et al., 1989] and to improvements in this code [Kong and Reitz, 1996] because details of the code are to be found in the open literature. The code KIVA III, made available in 1993, further improves the grid resolution. A more powerful KIVA code for computers using parallel processing is now available on an experimental basis.

The intent in the following review of the status of CFD models for diesel combustion is to motivate the reader to further study and to indicate how such programs may be helpful in design. As in all design work, models must be combined with past experience and experiments to be effective. The models for diesels form a hierarchy consisting of simple analysis models (heat release), zero-dimensional models (cycle analysis), quasi-one-dimensional models (gas dynamics in exhaust system), phenomenological models (coarse spatial resolution with highly empirical subgrid models which are typically system-specific), and CFD models (fully 3-D with semiempirical subgrid models). The engineer must learn to use all of these tools, both experimental and theoretical, in concert. There is no "perfect" model, and each must be used within the scope of its ability to predict. Sometimes the models can be used only to motivate experiments, sometimes to improve understanding or predict trends, and sometimes to make quantitative predictions that can be used directly in design. However, in no case do they yet substitute for good judgment and experience.

The following discussion of diesel CFD codes is divided into a few of the important issues, most of which involve subgrid models.

Grids and Grid-Element Equations

Because internal combustion engine ports, valves, and combustion chambers represent a complex geometry, the grid generation problem is complex and must conform to moving boundaries created by the moving piston and valves. In addition, some regions such as the port near the valve and around the injector tip region require fine grids to resolve the flow. Because engine boundary layers are very thin (less than 0.5 mm), subgrid models have been used near surfaces rather than trying to use grids fine enough to resolve the boundary layers. Given these important issues it is no wonder that grid generation and numerical techniques for restructuring grids efficiently are the subject of a large block of literature. As CFD programs begin to be used in application, the importance of quick and inexpensive grid generation is bound to increase.

A typical means of reducing both the grid generation and computational effort is to assume zonal symmetry at the time of injection and thereafter. Thus, for a nozzle located coaxially with the cylinder and having n equally spaced holes, one can calculate for a pie-shaped zone of included angle $360°/n$. Figure 12.15*a* shows a typical grid for the intake port and cylinder of a DI diesel. Figure 12.15*b* shows a zonal grid for the combustion chamber. The grids for ports and valves are used to generate intake flows. Because the flow field prior to injection is not symmetric, the zonal symmetry models represent a compromise between accuracy and cost.

The conservation equations of mass, momentum, and energy include turbulent diffusion terms. The mass conservation equation for a given species includes terms for sources due to Fick's law, chemical reactions, and spray introduction of fuel vapor. The momentum equation for the fluid mixture includes the turbulent kinetic energy, the turbulent viscous stress tensor, and the momentum gain due to the spray. The energy equation uses source terms to include chemical heat release,

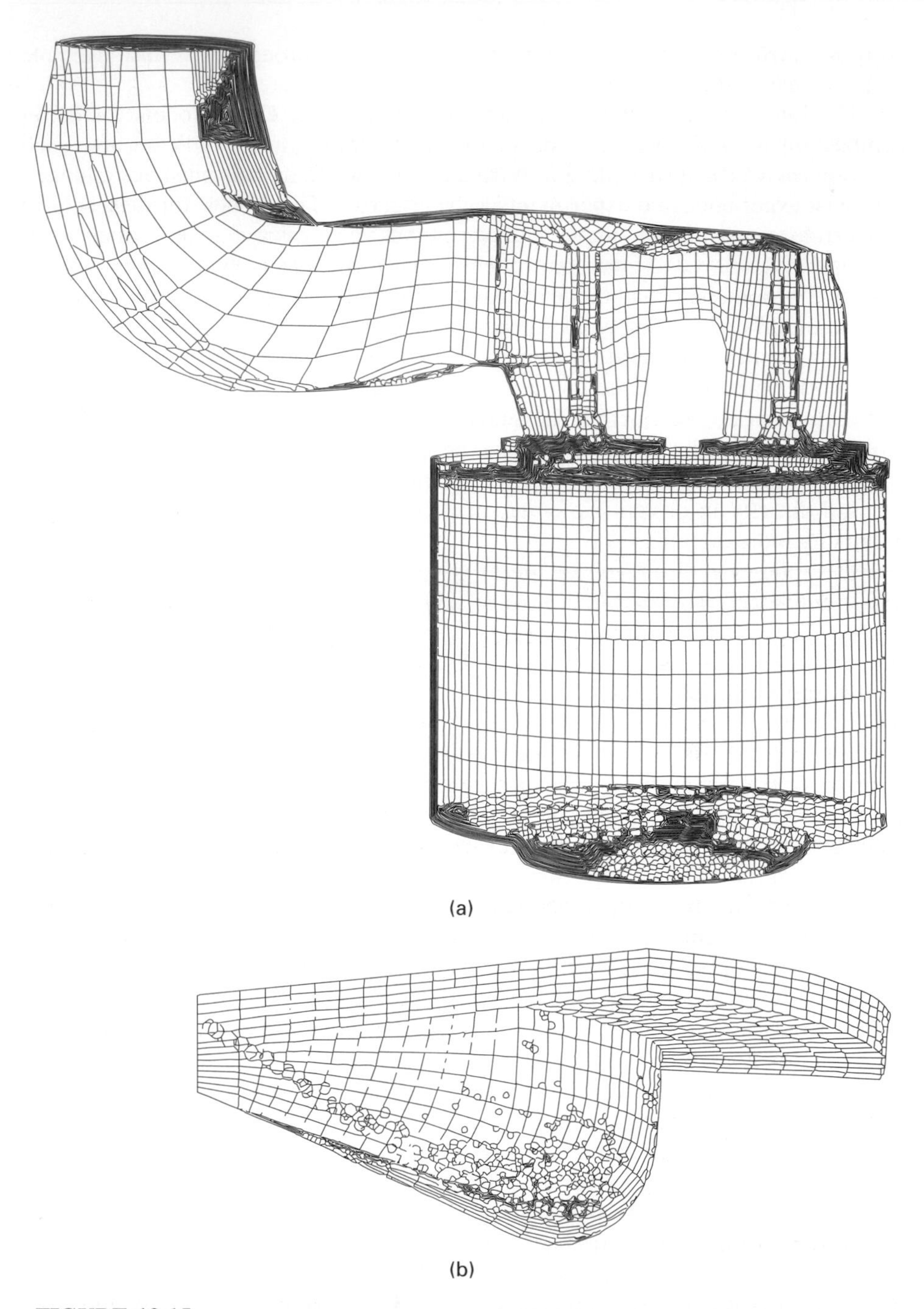

FIGURE 12.15
(*a*) Gridding for intake port, valves, and combustion chamber of a DI diesel with two intake valves; (*b*) perspective view of computational grid and fuel droplet distribution at 5° ATDC. The computational domain is one-sixth of the whole combustion chamber, since the injector tip has six equally spaced holes [Kong and Reitz, © 1996 and reprinted with permission of AIAA].

spray interactions, turbulent heat conduction, and turbulent enthalpy diffusion. The turbulence is modeled by the k–ϵ model in KIVA as discussed previously, but now must include source terms for interactions with the spray. Again, the turbulent diffusion coefficients are taken proportional to k^2/ϵ.

The law of the wall is used to introduce the nonslip condition at the wall and the gas-to-wall convective heat flux. Modifications to this model which have been shown necessary for homogeneous engines have also been included for diesels, but have not been validated for the diesel cases. Radiation has also been included, but not based on validated soot modeling. Because a large share of the cylinder heat transfer in diesels is to the piston bowl and because of spray impingement on the bowl, the heat flux model for the piston is at present in a state of research and not resolved. The flux to the head surface is seemingly less complex, but is known to be highly spatially variant. Thus even the head heat flux model is not sufficiently validated.

Spray Models

The spray models have been discussed in Chapter 9, and that discussion will not be repeated here. For diesels at fully warmed-up conditions, the droplets, which are very small (10 μm), vaporize rapidly. However, for starting or for small engines the liquid spray impinges on the piston. This can lead to instantaneous breakup as droplets rebound, vaporization of a thin liquid film on the surface, and sliding or bouncing back of droplets with distortion of their shape. The gas entrained by the moving droplets also enhances the convective effects on vaporization and heat transfer at the surface. The effects of surface deposits on these phenomena are not quantified, but it is known that liquid fuel can be trapped in the deposits, delaying its release as vapor.

For cold starting the effects of fuel volatility are important, and thus modeling of how various species vaporize from real fuels. One may, for example, ask how a volatile fuel fraction of high cetane number aids ignition. Multicomponent droplet vaporization models seem to be required, and such improvements to KIVA are now being introduced. However, diesel cold starting is quite ill behaved. For example, a cylinder may fire, then not fire again for a cycle or two before firing again. This suggests some mechanism of fuel carryover from one cycle to the next, such as might happen when fuel is stored in surface deposits or crevices. Thus droplet models may not be sufficient to explain starting behavior.

Ignition Models

Because low-temperature chemical kinetics plays a role in diesel ignition, the process should be modeled by mechanisms including effects of branching reactions and possibly two-stage ignition. To keep such calculations within the bounds set by current computing capacity, reduced mechanisms have been favored for use in CFD codes. The model used for diesel CFD ignition by Theobald and Cheng and more recently by Kong and Reitz is the so-called "Shell Model" of Halstead et al. This model has eight generic equations with eight rate constants. The equations for ignition of C_nH_{2m} fuel in air use a generic radical pool (R^*), an intermediate species (Q), and a branching agent (B) to produce CO, CO_2, and H_2O products (P):

$$C_nH_{2m} + O_2 \rightarrow 2R^*$$

$$R^* \rightarrow R^* + P + \text{(heat)}$$

$$R^* \rightarrow R^* + B$$

$$R^* \rightarrow R^* + Q$$

$$R^* + Q \rightarrow R^* + B$$

$$B \rightarrow 2R^*$$

$$R^* \rightarrow \text{(termination)}$$

$$2R^* \rightarrow \text{(termination)}$$

In order to use this model for a given fuel, one must fit the rate constants to combustion bomb data. Because the equations are not elementary, differences between bomb and engine conditions may influence the results. In particular, the sensitivity to local temperature means that the CFD code must be validated in conjunction with the ignition model. Errors in prediction should thus not all be attributed to the ignition model. The results of Kong and Reitz [1993] show quite good agreement with engine data, but the range of data and fuels is still too limited to judge the success. It may be that the model will always require some "tuning" for a specific engine even if CFD modeling of heated bomb data for the given fuel shows excellent validation over a range of initial pressures and temperatures. It is encouraging, however, that the model used in the UW version of KIVA not only predicts ignition delay trends correctly, but also closely approximates the geometric location in the engine where ignition is first observed by photography.

When split injection is used, the second injection is into higher-temperature gases which may contain radicals. Visualization diagnostics show that the ignition can take place instantly as soon as the fuel leaves the nozzle. Clearly the low-temperature kinetics models are inappropriate for this case. One might assume a zero ignition delay for the second injection.

Combustion

A characteristic time model based on earlier eddy breakup models (described in Chapter 5) has been tuned for diesel conditions and applied to KIVA. Recall that the characteristic time scale consists of a "laminar time scale" and a "turbulent time scale" modified by a factor $f(r)$. The value r depends on the ratio of products to total reactive species and thus goes from 0 (no reaction) to 1 (complete consumption of fuel). Thus, for diesel combustion the rate of premixed combustion is under conditions where r is nearly zero; that is, the local value of $f = (1 - e^{-r})/0.632 \approx 0$. Thus the time constant depends mostly on the laminar time constant during the premixed period. In spark-ignition engines this initial period is the kernel growth period, but in diesels it appears that the early combustion is dominated by chemical kinetics, not flame propagation. The laminar time constant is thus derived from a chemical rate expression. Once combustion starts, the pressure rises rapidly and many cells of mixture may autoignite. To take care of this, the Shell model is not used for cells with $T > 1000$ K and is replaced by the combustion model rate. Figure 12.16 shows comparisons of heat release rate (HRR) obtained from this CFD model and a single-zone analysis based on experimental pressure data. It is interesting to note that when the KIVA pressure history is used in the single-zone model, the HRR agrees with KIVA prediction. The ignition delay and premixed spike show excellent agreement, but the transition from premixed to mixing control is not as well modeled, resulting in a small error in peak pressure. The combustion model plays only one part in the HRR; the spray and mixing models play important roles too. Thus it is difficult to judge which model needs improvement or if any of the models are correct, because they have all been adjusted based on overall engine measurements.

The time constant combustion model is semiempirical and does not physically model the combustion process on a fundamental basis. Recently the use of flamelet models has been the source of more fundamentally based attempts to model diesel combustion. Recall from Chapter 5 that the flamelet models compute flame surface area per unit volume. The flame speed is computed from a library of strained laminar diffusion flame speeds. This concept is fairly clear for a classical diffusion flame where the fuel and air are initially separated. However, the diesel case is much more complicated and thus the models must introduce assumptions which are not fundamentally verified. In the diesel the oxidizer side of the flame may actually be hot products produced from an earlier combustion, which could be premixed or could involve droplets or puffs of fuel vapor embedded within the reaction zone of a premixed fuel vapor–air mixture. A challenge exists to sort out the various categories of burning that could exist within a given computational cell. It is thus not surprising that models [Dillies et al., 1993; Musculus and Rutland, 1995] differ in describing these categories. The Musculus and Rutland model first attempted to classify as diffusion burning all cases where mixing of fuel and oxygen is incomplete before combustion occurs and where a Damköhler number given by $\mathrm{Da} = Ae^{-E/RT}/(\epsilon/k)$ is larger than a selected critical value. All other cases of combustion were called "premixed burning." The premixed burning is modeled by a chemical rate-controlled combustion and the diffusion burning by the flamelet model. This approach gave an extended period of "premixed burning" of very rich regions, and an arbitrary rich limit of equivalence ratio 3 was used to cut off the premixed burning. Selection of the critical value of Da allows the initial premixed burning fraction to be varied, but values large enough to allow good agreement with experiment gave unstable computations. To fix this an additional transport equation was introduced to compute the amount of premixed fuel on the oxidizer side of the flamelet. Terms for chemical reaction as a sink and droplet vaporization as a source are included in this transport equation. Figure 12.17 shows a comparison of HRR computed with the model and the AHRR from experimental pressure data. The basic features of the AHRR are reproduced, but the premixed fraction is too small. Less intuitive are the details of the computations which show regions of premixed burning within the spray central regions.

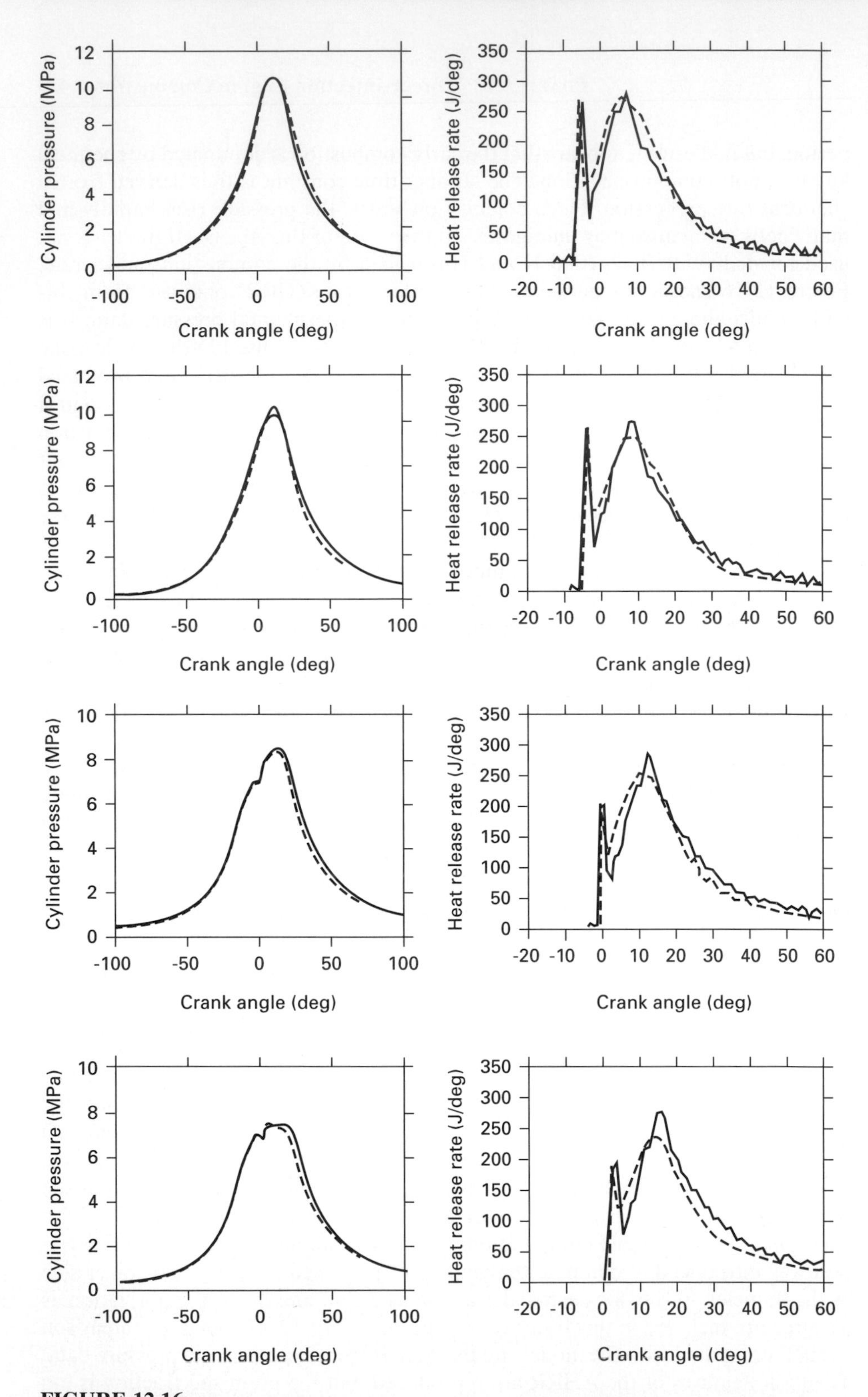

FIGURE 12.16
Comparison of predicted (dashed lines) and measured (solid lines) diesel engine data under different injection timing: (from top) 15°, 13°, 8°, and 5° BTDC [Kong and Reitz, *J. Eng. Gas Turbines and Power,* © 1993, by permission of ASME].

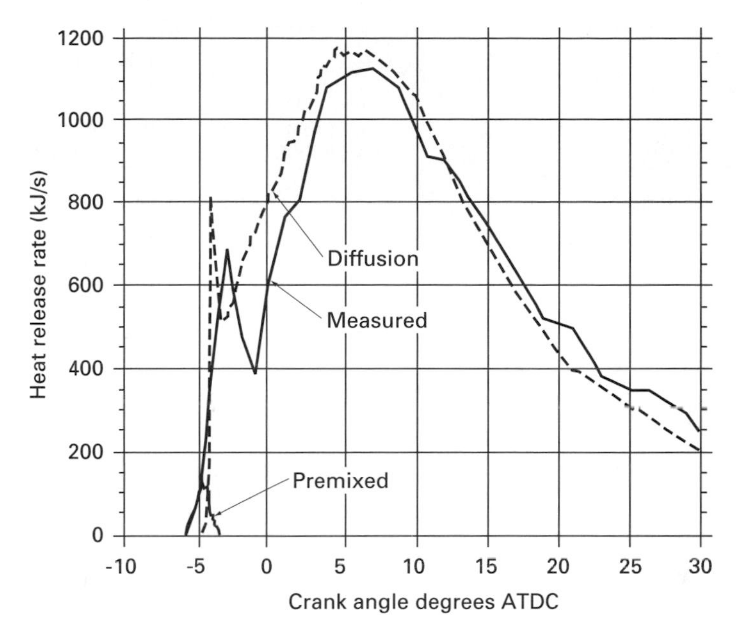

FIGURE 12.17
Heat release curves for a single cylinder TACOM-Labeco engine. The calculated values were obtained without the use of the premixed fuel transport equation and with a Damköhler number of 50 [Musculus and Rutland, © 1995, by permission of Gordon and Breach Publishers].

Emissions Models

The primary diesel exhaust emissions of NO_x and particulates are both very sensitive to temperature and mixing. The extended Zeldovich mechanism (Chapter 4) seems to model NO_x well in KIVA provided that the HRR and mixing are properly modeled. First attempts to model NO_x using the k–ϵ turbulence model in the UW version of KIVA II gave good trends with injection timing, etc., and also reproduced the total cylinder NO_x histories obtained by dumping experiments (Donahue et al.). However, the absolute values were off by a constant factor sometimes as large as 62. Many issues can play a part in such computations, but it is doubtful that the kinetic mechanism itself is a major source of error. Temperature seemed to be a major source of error, because the standard k–ϵ model predicted maximum local temperatures of about 2100 K at 10 crankdegrees after TDC. This is much lower than the 2700–2800 K values measured by optical means in similar engines. Introduction of a modified k–ϵ model called the RNG k–ϵ model by Han and Reitz produced a smaller turbulent viscosity, and higher temperatures resulted. Figure 12.18 shows the temperature contours. The dots or circles shown represent the fuel spray droplets. All of the spray droplets are projected on the plane shown, which is through the spray axis, and thus some drops fall outside the piston outline for the centerline plane. Note that the engine modeled has a smaller than conventional spray angle, so that the spray impinges on the bowl bottom rather than on

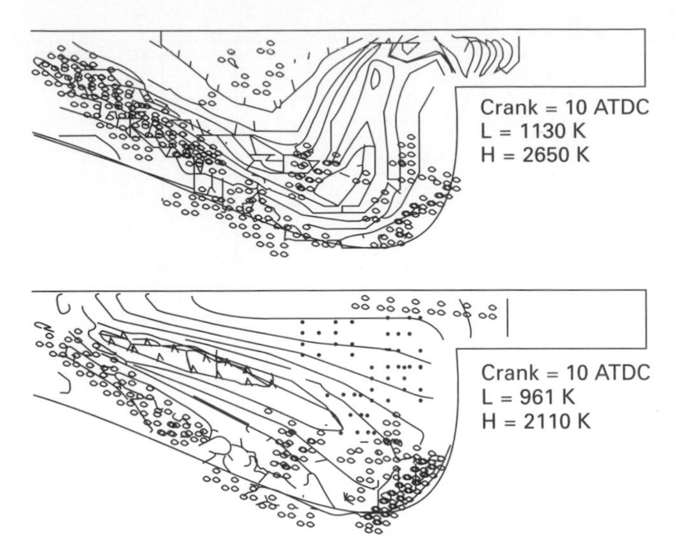

FIGURE 12.18
Temperature contours at TDC in the plane of the spray axis. Top: present RNG k-ϵ model. Bottom: previous standard k-ϵ model [Han and Reitz, © 1995, by permission of Gordon and Breach Publishers].

the bowl's vertical side as is more typical. The higher temperatures of the modified model caused the correction factor to go from 62 to 0.78. Figure 12.19 compares the shapes of NO_x curves for the two turbulence models, each corrected by a constant factor to make them agree with the experimental value as measured in the exhaust. The fact that significant NO is produced from products produced by diffusion (mixing-controlled) burning has caused some doubt about the validity of using only Zeldovich kinetics and not including other more complex reactions. However, the good agreement shown here and in many other comparisons indicates that a more complex kinetics model may not be needed.

The modeling of soot in the current UW version of KIVA II uses an empirical production equation and the Nagle and Strickland-Constable oxidation mechanism. More fundamentally based models show promise to work well, but are not yet fully validated. Figure 12.20 shows the soot history corresponding to the engine conditions of Figure 12.19. Note that the change in turbulence model produced higher soot production but also more oxidation. The shapes of these curves are more peaked than those obtained from previous experimental studies in a different engine; thus, although the predicted trends with injection timing as shown in Figure 12.21 are excellent, the shape of the soot production history could well be in error.

Modeling of emissions using the UW version of KIVA II has shown that NO_x agrees to within 15% of measured exhaust values and exhaust soot mass agrees to within 20%. Trends in heat release are predicted reasonably well, but quantitative agreement in shape is only fair as of this writing date. The agreement between measured and calculated trends with operating parameter changes is, however, very good and lends confidence to using the model to explain observed behavior. To illustrate such use the case of emissions behavior with split injection is considered further.

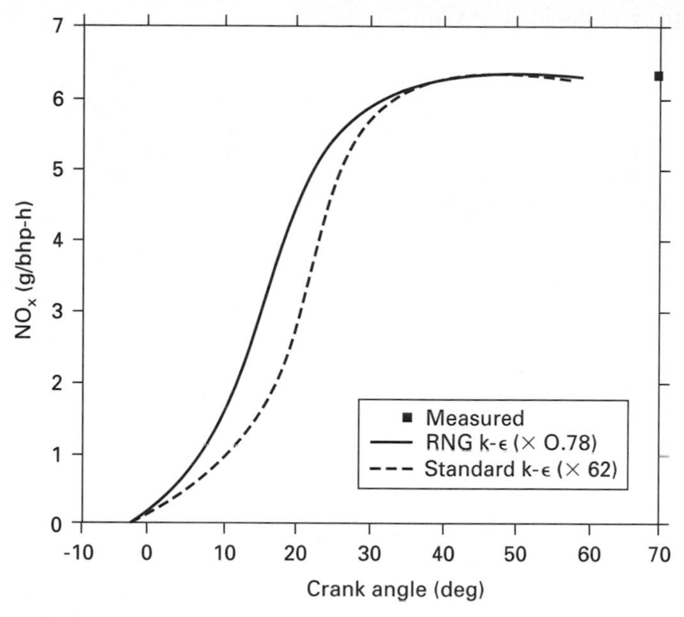

FIGURE 12.19
Effect of turbulence models on the total cylinder NO_x histories. Both use Zeldovich kinetics, but with correction factors required to match experimental data of 0.78 and 62 [Han and Reitz, © 1995, by permission of Gordon and Breach Publishers].

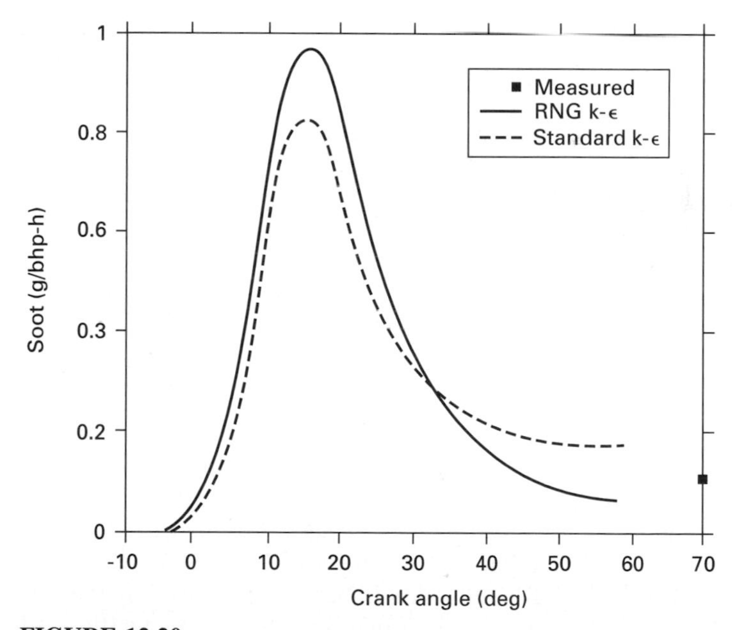

FIGURE 12.20
Soot histories computed from the empirical soot production model for two different turbulence models with oxidation. Measured value is from engine exhaust particulate mass on filter [Han and Reitz, © 1995, by permission of Gordon and Breach Publishers].

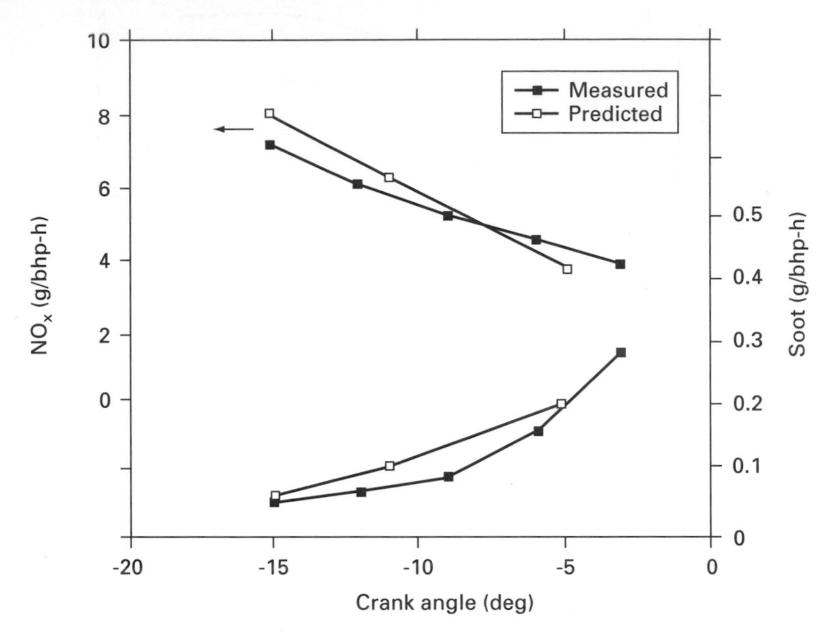

FIGURE 12.21
Comparison of computed and measured engine-out exhaust emissions for different injection timing values [Pierpont and Reitz, reprinted with permission from SAE paper 950604 © 1995, Society of Automotive Engineers, Inc.].

It has been shown experimentally that properly timed split injection can be used to improve the NO_x-soot trade-off condition. The primary design strategy is to lower NO_x by using retarded timing and EGR. This strategy is limited in conventional engines, because it tends to cause a rise in soot, as was shown for timing changes in Figure 12.13. However, split injection allows increased injection retard and EGR without the rise in soot.

Figure 12.22 shows the KIVA predicted histories of soot formation, oxidation, and net resulting soot for single and split injection each with injection starting at 10° BTDC. The split injection case injects 75% of the fuel in the first portion and then waits 8 crankdegrees before injecting the remaining 25%. The 28° crank-angle duration is held the same for both cases. Because the total fuel injected is the same for both cases, the rate of injection for the split case is 40% higher than for the single injection case. For single injection an increase in rate (decreased duration) can cause an undesirable rise in peak pressure, but with split injection such an increase is reduced by the dwell period. As shown in Figure 12.22, the total soot oxidized is a very high fraction of the soot formed, and about the same soot mass is oxidized in both cases. The reduction in net soot produced by the split injection is primarily the result of less soot formation. The reason for this reduced soot formation is shown schematically in Figure 12.23. In the single-injection case, soot formed in the rich regions of the spray accumulates in the spray tip region. This behavior has been observed experimentally by Kamimoto and Won; and also by Dec and Espey. Kamimoto also showed that the spray tip is a recirculation zone

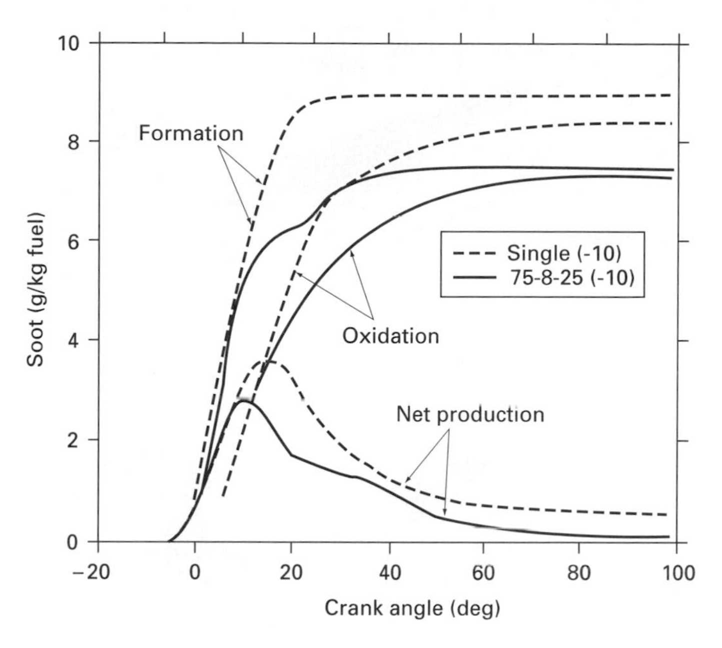

FIGURE 12.22
Computed soot histories for a diesel with single injection and split injection. Both have −10° ATDC start of injection and the same total mass and duration. The split is 75% mass, 8° dwell, 25% mass [Han et al., reprinted with permission from SAE paper 960633 © 1996, Society of Automotive Engineers, Inc.].

and that breaking up this zone using a jet or physical barrier will cause increased oxidation of the soot. In the split-injection case, the second injection enters a high-temperature fuel-lean zone, causing burning with soot reduction. The soot cloud created by the initial fuel jet is pushed upward out of the bowl. One may speculate that a lip on the bowl might increase oxidation of this cloud. It is likely that augmented mixing using a gas jet, which also reduces exhaust soot, has a similar effect to split injection by disrupting the soot formation process.

It is interesting to compare the above results from modeling with the conceptual model of Dec [1997], which is based on laser-sheet imaging in an optical-access diesel of 139.7 mm (5.5 in.) bore. Figure 12.24 shows a typical heat release diagram. Despite attempts to reproduce typical turbocharged engine conditions at start of injection the premixed burn fraction appears larger than in turbocharged engines with retarded timing. Figure 12.25 shows the conceptual model in schematic form. The schematic covers only the period from SOI to the initial stages of the "mixing-controlled" combustion. The boundaries, which are quite ragged for the real jet, are shown smooth for presentation purposes. The following is an abbreviated version of the discussion given by Dec.

The sketches up to 5° after start of injection (ASI) show the region of liquid-vapor-air as black with the outer line representing the edge of the vapor-air region. The liquid is confined to a region of 23 mm maximum length. Heat release, as

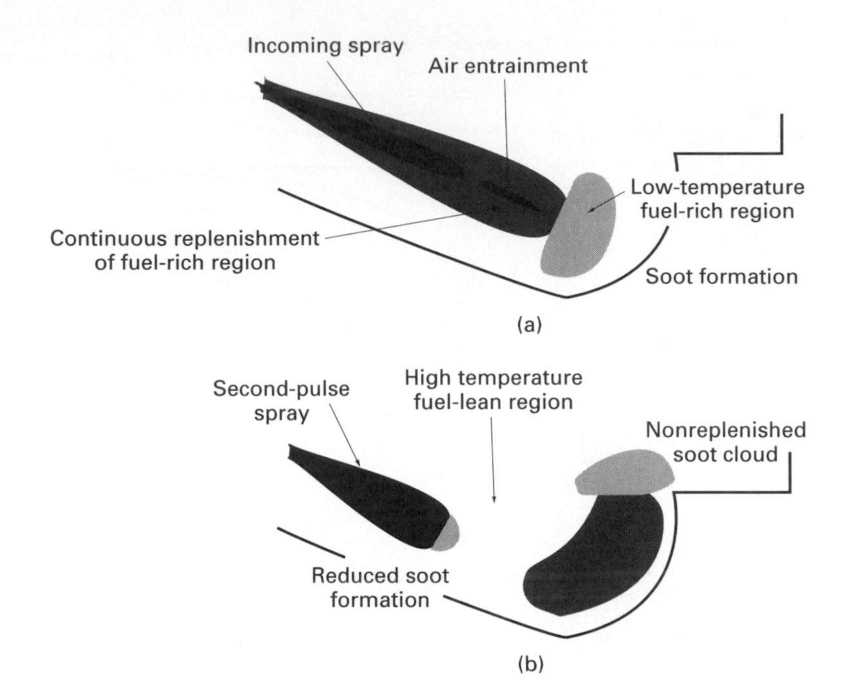

FIGURE 12.23
(*a*) Single-injection spray, and (*b*) start of the second injection of a split injection. The rich region formed by the first injection pulse is pushed upward in part *b* and is no longer replenished with fuel as it is in part *a* [Han et al., reprinted with permission from SAE paper 960633 © 1996, Society of Automotive Engineers, Inc.].

shown in Figure 12.24 starts at about 4° ASI. (The location of the start of combustion was not determined.) Although chemiluminescence was observed in the vapor region starting at 3° ASI, no polyaromatic hydrocarbons (PAH) were observed until about 5° ASI. Equivalence ratios ranged from 2–4 in this vapor region. Very small soot particles were observed throughout the forward region starting at 6° ASI. Note that this is about the peak of the premixed spike of Figure 12.24. A luminous diffusion flame appeared around the periphery of the jet starting from the end of the liquid-containing zone to the tip. In this model, the rich products of the premixed-zone combustion are burning as they mix with air in this thin region. The soot particles in this thin region are observed to be larger in size than in the core region. The mixing-controlled combustion continues to be confined to this thin region as time goes on. The higher soot concentration shown in the head zone at 8–10° ASI still contains only small particles, but as time goes on recirculation in the head zone allows agglomeration to produce larger particles. (The optical engine has no bowl-jet interaction within the viewing area. In real engines this bowl-side-surface is a region of mixing, but it is not included in the conceptual model.) Because of higher temperatures and perhaps radiation, the liquid-containing zone shortens. A rather abrupt falloff in soot is observed at about 27 mm from the

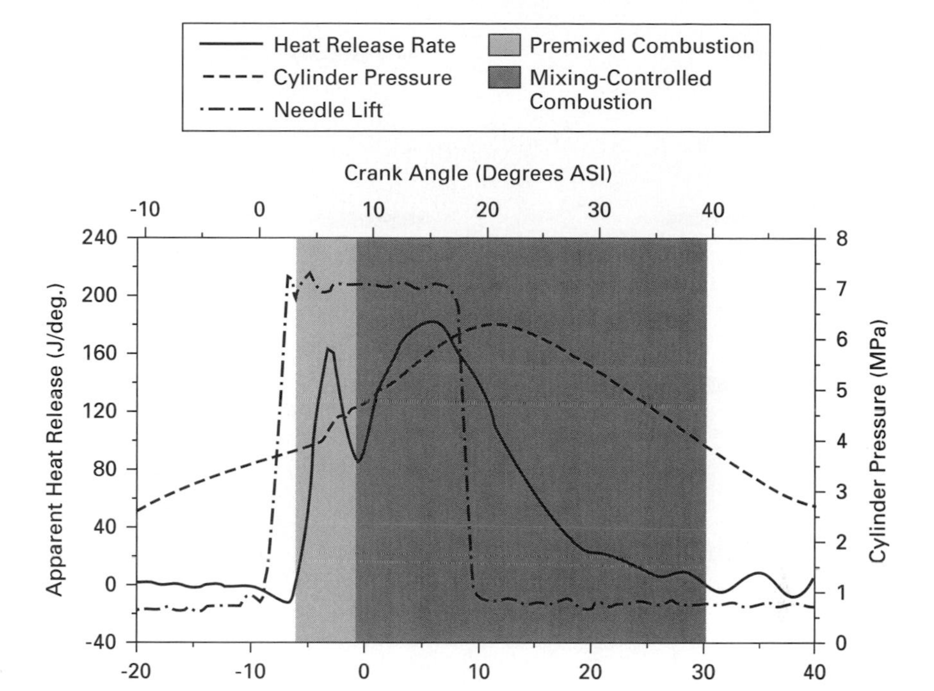

FIGURE 12.24
Apparent heat release rates cylinder pressure, and needle lift at F = 0.43, 1200 rpm, 147 kPa intake pressure, and 308 K intake temperature. The special optical-access engine has a bore of 139.7 mm (5.5 in.) and a stroke of 152.4 mm (6 in.) with CR = 11:1 for condition shown [Dec, reprinted with permission from SAE paper 970873 © 1997, Society of Automotive Engineers, Inc.].

injector. The thin outer zone of soot extends back to the liquid-containing zone for combustion with reference fuel, but could not be observed in this region for the low-sooting fuel. Dec speculates that the abrupt soot appearance at 27 mm may be a standing fuel-rich premixed flame front. However, it is difficult to estimate the flame speed of such a near-limit turbulent flame, so that the probability that it could be so anchored cannot be judged.

The conceptual model of Dec presents some difficulty when applied to what is known about NO production. It is unlikely that the model holds for an NA engine, because so much evidence shows that the premixed burning produces the bulk of the NO. But for turbocharged conditions, the data of Donahue et al. indicate that considerable NO is produced during the diffusion burning phase; but the premixed burning is nevertheless still responsible for more NO than could be envisioned in the Dec model. We may conclude that the conditions of the tests run by Dec may correctly portray the general trends, but that the typical conditions in real engines produce a higher fraction of the premixed combustion at near stoichiometric conditions than described by the Dec data.

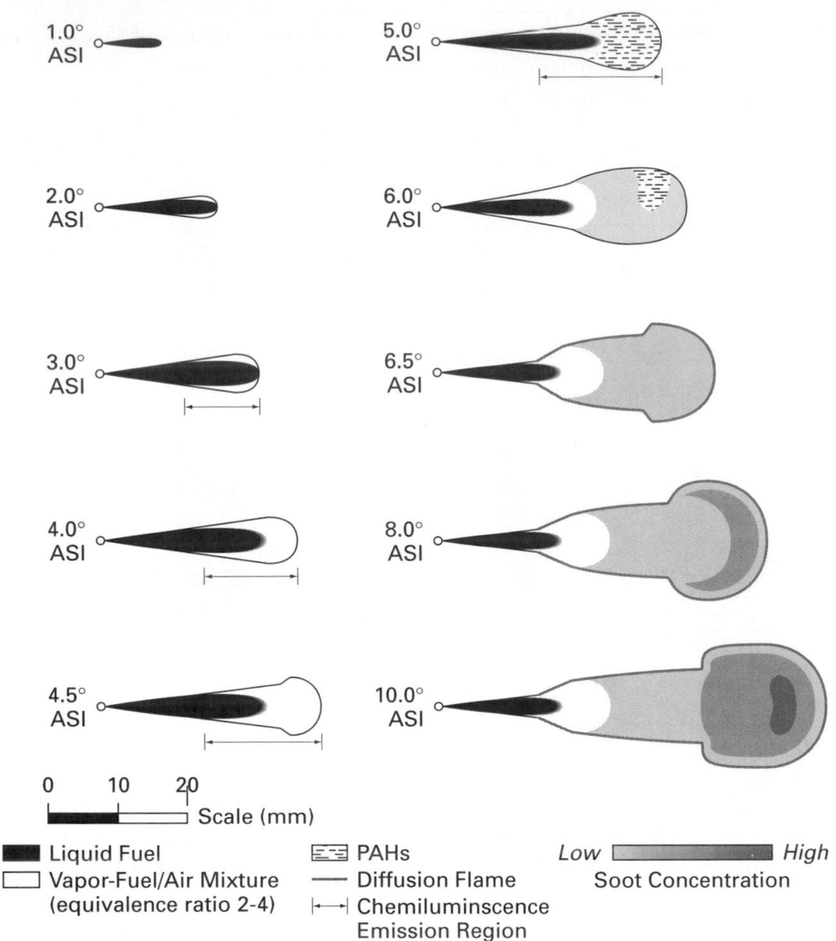

FIGURE 12.25
A temporal sequence of schematics based on diagnostics obtained from the optical-access engine, which shows the combustion history up to the early part of the mixing-controlled combustion [Dec, reprinted with permission from SAE paper 970873 © 1997, Society of Automotive Engineers, Inc.].

12.9 SUMMARY

Diesel engines offer the best fuel economy of any engine currently available and hold promise to be the short-term solution to the national goal of obtaining an 80 mile per gallon passenger car. Because of the move toward fuel economy, the chapter is devoted to direct-injection combustion technology with only brief mention of IDI engines. In DI engines fuel is directly injected into the cylinder, where it ignites by compression ignition. The rate of combustion is determined by the fuel mixing rate, and the engine load is determined by the amount of fuel injected. In this way the limitations of low compression ratio, crevice volume losses, throttling losses, lean flame speed limits, and cyclic variability inherent in S.I. engines are avoided. The disadvantages are an engine which is heavy, noisy, and expensive and which produces significant particulate and NO_x emissions.

To summarize the chapter, we return to Figure 12.2 and follow it from top to bottom. Starting at the top left of the diagram, the air induction design problem is to provide the engine with a large amount of trapped air with as low a pumping loss as possible. Use of a turbocharger followed, in larger engines, by an air-to-air heat exchanger significantly increases the air density. The use of EGR is accepted for smaller engines to help meet NO_x emissions standards. In the past, larger engines have avoided EGR, because of its complexity and possible adverse effects on durability. Large engines use high-pressure fuel injection to provide fuel-air mixing, but smaller engines suffer from overpenetration of the spray and thus typically use intake-induced swirl to improve mixing and reduce spray penetration. The swirl is imparted by the shape of the port. Because of the complexity of the spray mixing process the effects of intake-induced turbulence are not well understood, but clearly they are less important than in homogeneous charge engines, where they strongly influence the flame speed.

Looking to the top right of the Figure 12.2 diagram, we start with the effect of the fuel formulation. Fuel density and viscosity influence the injection process, but the main property is the ability of the fuel to ignite, which is indicated by the cetane number. High cetane number, very low sulfur content, low aromatic content, and high hydrogen-to-carbon ratio all favor lower emissions. Because fuel properties are highly correlated, it is difficult to determine the precise effect of a single property.

The injection system is at the heart of the combustion design process. High injection pressure (100–150 MPa) gives good atomization and mixing, which reduces particulate production. Electronic injectors allow rate shaping, which can be used to influence the rate of combustion and the rate of mixing. Retarding of the start of injection (SOI) from MBT lowers the production of NO_x, but is limited by the increase in particulates (soot) caused by too much retard. A curve of NO_x versus particulates for various SOI values shows the timing for the best NO_x-particulate trade-off. Use of split injection may offer a way to improve this trade-off value.

The spray characteristics have been discussed in Chapter 9; here it is important to note that good air utilization requires penetration of the spray and good air entrainment to provide mixing. Droplet vaporization is not a limiting parameter except for cold starting. The spray provides an ignitable mixture very quickly so that the ignition delay is primarily a function of cylinder temperature and pressure and is not influenced significantly by injection parameters. However, the injection does strongly influence the combustion, which in turn influences the temperature at SOI through the residual fraction and heat transfer from the surfaces.

The first portion of fuel that burns has been vaporized and premixed with air; it burns very rapidly. Turbocharged engines have a high cylinder pressure and temperature at SOI and thus a short ignition delay and a small amount of premixed burning. Naturally aspirated (NA) engines have a longer delay and a larger premixed burn, which causes a rapid cylinder pressure rise—so-called harsh combustion. The premixed combustion phase is very short, but because it occurs at or before TDC, its products are very hot and produce substantial amounts of NO_x, especially for NA engines. At these high temperatures the Zeldovich mechanism dominates the NO_x formation process. In modern turbocharged engines with low NO_x at least one-half of the NO_x is produced during the mixing-controlled phase of combustion and the NO_x kinetics may be more complex.

The turbulent diffusion or mixing-controlled combustion occurs over 80–85% of the combustion history. Flamelet models have been exercised using CFD codes, but the details of the processes are not understood well enough to judge the model validity.

Combustion rate is typically obtained from analysis of the experimentally obtained pressure versus crank angle diagrams by use of simple thermodynamic models (Eqs. 12.2–12.10). Such heat release rate diagrams are used to show the relationships between burning rate and other design parameters, but do not directly reveal the mechanisms involved. To obtain a detailed knowledge of the mechanisms will require continued use of in situ optical diagnostics and development of improved CFD codes.

PROBLEMS

12.1. Using the data of Table 12.2, calculate the average piston speed for each engine. Plot average piston speed versus displacement.

12.2. Consider a piston with an on-axis cylindrical bowl of diameter 60 mm and depth 30 mm. The bore and stroke are each 120 mm. Calculate the volume at bottom dead center (BDC) and the compression ratio. Assume the squish area clearance between piston and head is 1 mm.
Answer: $V_1 = 1453\ \text{cm}^3$, CR $= 15.1$.

12.3. For the data of Problem 12.2, the swirl ratio is measured in steady-flow bench tests to be 3. (*Swirl ratio* is the average rotation rate of the gas in rpm divided by the rpm of the engine.) Compute the swirl ratio at TDC by using angular momentum conservation and neglecting friction. Assume solid-body rotation of the fluid.

12.4. Calculate the time average squish velocity (the radial inward velocity caused by the flow into the bowl) at the bowl edge and compare with the average piston velocity and tangential swirl velocity (use data of Problem 12.3 for swirl). Assume 2000 rpm.

12.5. Calculate the discharge velocity and mass flow rate for a 0.2-mm injector hole with a Δp of 8000 psi. Use a flow coefficient of 0.7 and properties of dodecane.
Answer: $\dot{m}_f = 0.00631$ kg/s.

12.6. Using the data of problems 12.2 and 12.5, calculate the injection duration for a four-hole nozzle, $F = 0.8$ and volumetric efficiency of 95%, naturally aspirated with an inlet air density of 1.17 kg/m^3. Use dodecane fuel properties and use $f_s = 0.067$. Find the duration of the spray in milliseconds and crankdegrees for an engine speed of 2000 rpm. To do this, first find the trapped air mass.
Answer: $m_a = 1.62$ g, $\Delta t = 3.4$ ms, $\Delta\theta = 41$ CA°.

12.7. Using Eq. 9.22 for penetration with $t_b = 0.14$ ms and the data of Problems 12.2, 12.5, and 12.6, calculate the distance of penetration at the end of injection. Assume the spray starts at 20° before TDC. Estimate the crank angle when the spray strikes the side of the bowl.

12.8. Using a swirl ratio of 12 in the bowl and Eq. 9.23, calculate the penetration for Problem 12.7 and answer if the spray hits the bowl with the addition of swirl. Swirl ratio is the ratio of angular rotation of the gas to rpm of the engine.

12.9. Assume a shallow bowl piston with no swirl and an injector-to-bowl distance of 50 mm as measured along the spray plume axis. The engine bore and stroke are each 120 mm. The inlet pressure is 2 atm and inlet temperature is 120°F. The volumetric efficiency is 90%. The compression ratio is 16/1. The injector has eight holes, each of 0.2 mm diameter, and the injection pressure is constant at 8000 psi. The flow coefficient for each hole is 0.7. The fuel is dodecane. Calculate the volume of the eight spray plumes just when they have reached the piston bowl. Assume the piston is at TDC. (Use a cone shape and Eq. 9.22.) If the overall engine equivalence ratio is 0.8 at this point in the spray process, what is the average F in each spray plume? Use a polytropic compression of $n = 1.36$.
Answer: $F = 2.3$.

12.10. Calculate the temperature depression in each spray plume for Problem 12.9 by assuming that all the fuel is vaporized in each plume and all of the fuel injected is in the plumes. Assume the fuel is dodecane which enters at 540°R. Recall that $f_s = 0.067$ for this fuel.

12.11. Show that ${}_1Q_2$ is zero in Eq. 12.7 if the process follows $pV^\gamma =$ constant.

12.12. Assume the air in the cylinder of a 100-mm-bore, 100-mm-stroke engine with a 200-mm connecting rod is compressed from 15 psi and 540°R at BDC to 10° before TDC isentropically. Calculate Q_{net} versus θ using Eq. 12.5 and taking $p = p_1 + C\theta$ for θ between 10° before TDC and TDC. Take p at TDC to be 1000 psi. The compression ratio is 16:1. p_1 = pressure at 10° before TDC. *Hint:* $V(\text{TDC}) =$ 52,360 mm^3;

$$V(\theta) = V(\text{TDC}) + \left(\frac{\pi\, 100^2}{4}\right)\left[250 - 50\cos\theta - \sqrt{200^2 - (50\sin\theta)^2}\right]$$

12.13. Use Eqs. 12.9 and 12.10 and assume that $\partial R/\partial T = \partial R/\partial p = 0$. Write the differential equations for $\dot{m}$ and $\dot{T}$, which can be solved by use of a numerical step-by-step solver if the mass at SOI, p, $\dot{p}$, V, and $\dot{V}$ are given from data.

12.14. Using Eq. 12.1 with $C = 0.0197$ ms, $a = -1$, $b = -1.75$, and $E = 4500$ K, compare the ignition delay for the following cases. In each case the engine compression ratio is 16:1 and injection starts at 20 CA° before TDC. Assume the volumetric efficiency is 100% for each case and the fuel mass added is the same for each case. The equivalence ratio is $F = 0.8$. Calculate the ignition delay in crank-angle degrees for 2500 rpm. Neglect residual fraction effects.
(*a*) Naturally aspirated: inlet air at 1 atm, 300 K
(*b*) Turbocharged: inlet air at 2 atm, 380 K
(*c*) Turbocharged with intercooler: inlet air at 2 atm, 312 K

12.15. Near the top of p. 394 it is said that the diesel injection pattern can be observed by injection of diesel fuel into kerosene. How can this be? What phenomena would cause the fuel jet to be visible using ordinary light such as from a strobe?

12.16. Explain the effect of humidity on NO_x and the effect of ambient pressure on particulates, as shown in Table 12.5.

12.17. Construct a table similar to Table 12.5 for the effects of exhaust gas recirculation and oxygen enrichment of inlet air, and discuss the reason for these trends.

12.18. When fuel is suddenly increased in turbocharged diesels, a puff of smoke is typically emitted. Explain why this happens in qualitative terms.

12.19. At the start of this chapter it is stated that homogeneous charge engines are limited in efficiency by:
(*a*) Fuel octane number
(*b*) Flame propagation in lean mixtures
(*c*) Cycle-to-cyle variations
(*d*) Fuel trapped in crevice volumes
Explain qualitatively how each of these items limits homogeneous S.I. engines, and then explain how the diesel engine overcomes these difficulties.

12.20. In the heat release analysis (Problem 12.13) the cylinder contents go from air and residual at SOI to final products along a path of $F(\theta)$ which rises monotonically from zero to F_e, the overall engine value, which is always less than unity. In real engine combustion, burning takes place over a wide range of F values at any instant, including very rich regions where F may be greater than 3. In the model, the lean assumption is consistent with the assumption of uniformity. Explain why this is so. Because these rich, hot regions are neglected, dissociation in the model is very small and can be neglected. Discuss how this would affect the accuracy of the model heat release. Will the model under- or overpredict the correct heat release rate? Explain your answer.

12.21 In the heat release analysis using Eq. 12.9 (the results of Problem 12.13) the term $dp/d\theta$ is obtained from $p(\theta)$ data. During rapid combustion which takes place during the initial period of burning (especially in NA engines), pressure waves are produced. Because these waves repeat from cycle-to-cycle, the ensemble-averaged $p(\theta)$ (taken over 50 or 100 cycles) will show these waves. If they are included in the analysis, the resulting heat release will have very large waves. Is that realistic? If not, explain why it is incorrect, and indicate how to overcome this problem. If you had a CFD model which correctly predicted these waves and thus gave $p(\theta, x, y, z)$, how would you process the pressure calculated by the CFD program to produce the uniform $p(\theta)$ curve needed in the apparent heat release analysis?

REFERENCES

Abraham, J., and Magi, V., "Modeling Radiant Heat Loss Characteristics in a Diesel Engine," SAE paper no. 970888, also in SAE SP-1246, 1997.

Amann, C. A., Stivender, D., Plee, S., and MacDonald, J., "Some Rudiments of Diesel Particulate Emissions," SAE paper no. 800251, in SAE P-86, 1980.

Amsden, A. A., "KIVA III: A KIVA Program with Block Structured Mesh for Complex Geometries," Los Alamos rept. LA-12503-MS, 1993.

Amsden, A. A., O'Rourke, P. J., and Butler, T. D., "Kiva-II: A Computer Program for Chemically Reactive Flows with Sprays," Los Alamos, LA-11560-MS, 1989.

Arcoumanis, C., Gavaises, M., and French, B., "Effect of Fuel Injection Processes on the Structure of Diesel Sprays," SAE paper no. 970799, also in SAE SP 1219, 1997.

Baumgard, K. J., and Johnson, J. H., "The Effect of Fuel and Engine Design on Diesel Exhaust Particulate Size Distributions," SAE paper no. 960131, 1996.

Benson, R. S., and Whitehouse, N. D., *Internal Combustion Engines,* Vols. 1 and 2, Pergamon Press, Oxford, England, 1979.

Borman, G. L., and Brown, W. L. Jr., "Pathways to Emissions Reduction in Diesel Engines," *2nd Int. Engine Combustion Workshop,* C.N.R., Capri, Italy, 1992.

Bosch, *Diesel Fuel Injection,* ISBN 1-56091-542-0, 1994 (part of the Bosch Technical Instruction Series). Robert Bosch GmbH, Abt. KH/VDT, 7000 Stuttgart 1, Posffach 50, Germany.

Cartellieri, W. P., and Herzog, P. L., "Swirl Supported or Quiescent Combustion for 1990's Heavy-Duty D.I. Diesel Engines—An Analysis," SAE paper no. 880342, 1988.

Chiu, W., Shahed, S., and Lyn, W. T., "A Transient Spray Mixing Model for Diesel Combustion," SAE paper no. 760128, 1976.

Dec, J. E., "A Conceptual Model of D.I. Diesel Combustion Based on Laser-Sheet Imaging," SAE paper no. 970873, 1997.

Dec, J. E., and Espey, C., "Ignition and Early Soot Formation in a D.I. Diesel Engine Using Multiple 2-D Imaging Diagnostics," SAE paper no. 950456, 1995.

Dillies, B., Marx, K., Dec, J. and Espey, C., "Diesel Engine Combustion Modeling using the Coherent Flame Model in Kiva-II," SAE paper no. 930074, 1993.

Donahue, R. J., Borman, G. L., and Bower, G. R., "Cylinder-Averaged Histories of Nitrogen Oxide in a D.I. Diesel with Simulated Turbocharging," SAE paper no. 942046, 1994.

Ferguson, C. R., *Internal Combustion Engines,* Wiley, New York, 1986.

Fusco, A., Knox-Kelecy, A. L., and Foster, D. E., "Application of a Phenomenological Soot Model to Diesel Engine Combustion," *1994 COMODIA,* JSAE, Yokohama, Japan, August 1994.

Gosman, A. D., and Johns, R., "Computer Analysis of Fuel-Air Mixing in Direct Injection Engines," SAE paper no. 800091, 1980.

Halstead, M., Kirsh, L., and Quinn, C., "The Autoignition of Hydrocarbon Fuels at High Temperatures and Pressures—Fitting a Mathematical Model," *Combustion and Flame,* vol. 30, pp. 45–60, 1977.

Han, Z., and Reitz, R. D., "Turbulence Modeling of Internal Combustion Engines Using RNG k-ϵ Models," *Combustion Science and Technology,* vol. 106, pp. 267–295, 1995.

Han, Z., Uludogan, A., Hampson, G. I., and Reitz, R. D., "Mechanism of Soot and NO_x Emission Reduction Using Multiple-Injection in a Diesel Engine," SAE paper no. 960633, 1996.

Heywood, J. B., *I.C. Engine Fundamentals,* McGraw-Hill, New York, 1988, Chap. 10.

Hiroyasu, H. "Diesel Combustion and Its Modeling," COMODIA, JSME Symp., attn T. Nakajima, Tokyo, 1985.

Hunter, C. E., Gardner, T. P., and Zakrajsek, C. E., "Simultaneous Optimization of Diesel Engine Parameters for Low Emissions Using Taguchi Methods," SAE paper no. 902075, 1990.

Kamimoto, T., Chang, Y. J., and Kobayashi, H., "Rate of Heat Release and Its Prediction for a Diesel Flame in a Rapid Compression Machine," SAE paper no. 841076, in SP-581, 1984.

Kamimoto, T., and Kobayashi, H., "Combustion Processes in Diesel Engines," *Prog. Energy and Comb. Sci.,* vol. 17, pp. 163–189, 1991.

Kamimoto, T., and Won, Y.-H., "Soot Formation and Related Techniques for Its Reduction in Diesel Engines," *Second Int. Conf. Fluid Mech., Comb., Emissions and Reliability in Reciprocating Engines,* September 14–19, 1992, Capri, Italy. (Organized by Istito Motori, Naples, Italy.)

Kong, S.-C. and Reitz, R. D., "Multidimensional Modeling of Diesel Ignition and Combustion Using a Multistep Kinetics Model," ASME paper no. 93-ICE-22, *J. Eng. Gas Turbines and Power,* vol. 115, pp. 781–789, 1993.

Kong, S.-C., and Reitz, R. D., "Spray Combustion Processes in I.C. Engines," in *Recent Advances in Spray Combustion,* K. Kuo (ed.), Chapter 16, vol. 2, pp. 395–424, 1996; part of the series, *Progress in Aeronautics and Astronautics,* vol. 171, AIAA, Reston, VA.

Kosaka, H., Nishigaki, T., Kamimoto, T., Sano, T., Matsutani, A., and Harada, S., "Simultaneous 2-D Imaging of OH Radicals and Soot in a Diesel Flame by Laser Sheet Techniques," SAE paper no. 960834, 1996.

Miyamoto, N., Chikahisa, T., Murayama, T., and Sawyer, R., "Description and Analysis of Diesel Engine Rate of Combustion and Performance Using Wiebe's Functions," SAE paper no 850107, 1985.

Musculus, M. P., and Rutland, C. J., "Coherent Flamelet Modeling of Diesel Engine Combustion," *Comb. Sci. Tech.,* vol. 104, nos. 4–6, p. 295, 1995.

Naber, J. D., and Siebers, D. L., "Effects of Gas Density and Vaporization on Penetration and Dispersion of Diesel Sprays," SAE paper no. 960034, 1996.

Pierpont, D. A., and Reitz, R. D., "Effects of Injection Pressure and Nozzle Geometry on D.I. Diesel Emissions and Performance," SAE paper no. 950604, 1995.

Rosenthal, M. L., and Bendinsky, T., "The Effects of Fuel Properties and Chemistry on the Emissions and Heat Release of Low-Emission Heavy Duty Diesel Engines," SAE paper no. 932800, 1993.

Senda, J., Kobayashi, M. Iwashita, S., and Fujimoto, H., "Modeling of a Diesel Spray Impinging on a Flat Wall," SAE paper no. 941894, 1994.

Shimoda, M., Shigemori, M., and Tsuruoka, S., "Effect of Combustion Chamber Configuration in In-Cylinder Air Motion and Combustion Characteristics of D.I. Engine," SAE paper no. 850070, 1985.

Sihling, K., and Woschni, G., "Experimental Investigation of the Instantaneous Heat Transfer in the Cylinder of a High Speed Diesel Engine," SAE paper no. 790833, in SP-449, 1979.

Theobald, M. A., and Cheng, W. K., "A Numerical Study of Diesel Ignition," ASME paper no. 87-FE-2, 1987.

Van Gerpen, J. H., Huang, C. W., and Borman, G. L., "The Effects of Swirl and Injection Parameters on Diesel Combustion and Heat Transfer," SAE paper no. 850265, 1985.

Voiculescu, I., and Borman, G., "An Experimental Study of Diesel Engine Cylinder-Averaged NO_x Histories," SAE paper no. 780227, 1978.

Watson, N., Pilley, A. D., and Marzouk, M., "A Combustion Correlation for Diesel Engine Simulation," SAE paper no. 800029, 1980.

Weber, H., and Borman, G., "Parametric Studies Using a Mathematically Simulated Diesel Engine Cycle," SAE paper no. 670480, 1967.

Woschni, G., and Anisits, F., "Experimental Investigation and Mathematical Presentation of Rate of Heat Release in Diesel Engines: Dependence on Engine Operating Conditions," SAE paper no. 740086, 1974.

Yan, J., and Borman, G. L., "A New Instrument for Radiation Flux Measurement in Diesel Engines," SAE paper no. 891901, 1989.

Yoshikawa, S., Furusawa, R., Arai, M., and Hiroyasu, H., "Optimizing Spray Behavior to Improve Engine Performance and to Reduce Exhaust Emissions in a Small D.I. Diesel Engine," SAE paper no. 890463, 1989.

CHAPTER 13

Detonation of Liquid-Gaseous Mixtures

In contrast to gaseous detonation waves, which were discovered over 100 years ago, it was not recognized until the 1960s that a detonation wave can propagate through a two-phase mixture of liquid fuel and gaseous oxidizer. The fuel can be in the form of liquid droplets or a liquid film on the walls of a tube. This chapter will first consider spray detonations and then discuss liquid layer detonations. Two-phase detonations can occur with very nonvolatile fuels, as well as with more volatile fuels, where the detonation velocity is higher. While these types of detonations are generally to be avoided because of the high transient pressures involved, which can be very destructive, there may yet be some practical applications developed.

Detonation waves also can occur when combustible dust is suspended in air. An example of a dust detonation is a grain elevator explosion where combustion starts, due to some ignition source such as anaerobic heating, propagates rapidly through the dust suspension, and changes from a flame to a detonation in a manner similar to a gaseous detonation. The dust must be less than 10 μm to remain suspended and to have enough surface area for rapid reaction.

Phenomena related to a spray detonation may possibly occur in a diesel engine. As the fuel is injected into the cylinder at very high pressure, the leading droplets of the spray may possibly undergo stripping or collisions with droplets to form a microspray. The high pressure and temperature and rapid mixing may possibly be conducive to initiation of local pressure waves in the cylinder. Since the distances and times involved are short, this should not be viewed with alarm, but rather as another (speculative) view of what goes on in the cylinder during combustion.

Large liquid-fueled rocket motors have occasionally experienced wavelike pressure excursions which have been well documented. This type of combustion instability can be very destructive and can rupture the thin-walled rocket motor. While some combustion instability is acoustic in nature and leads to failure due to high heat transfer, there is another failure mode which is detonationlike with accompanying pressure waves. The cure for acoustic instabilities is appropriately designed liners, while elimination of detonationlike instabilities requires suitably placed baffles.

Fuel sprayed in air by fast-flying aircraft or drones and ignited by a small bomb has been proposed and tested as a way to rapidly consume all of the oxygen over a battlefield during a war.

Liquid fuel film detonations have been known to occur in long, partially filled pipelines. In one case in Texas where some cleaning and welding was being done on a pipeline which was partially filled with oil, combustion started at the open end; developed into a detonation; ruptured the tube in one section, thereby relieving the pressure and slowing the detonation; and then was reestablished further down the tube. The tube was ruptured in various locations for a distance of several miles before the detonation wave hit a valve. Another potential problem is in catapults with oil seals where high-pressure air in contact with a layer of oil on the walls could produce a flame started by frictional heating, and the flame could change into a detonation.

The rotating detonation combustor, which was mentioned in Chapter 8 as a possible way to utilize the incredible combustion intensity of a detonation wave, could perhaps be more easily implemented using liquid fuel rather than gaseous fuel. Since the density of liquid fuels is typically 500 times greater than gaseous fuel, it is easier to feed fresh liquid fuel ahead of the detonation wave. Since heterogeneous detonations have an extended reaction zone compared with gaseous detonations, there would be less tendency to form multiple waves in the annulus, which was one factor which plagued the gaseous rotating detonation combustor.

13.1 DETONATION OF LIQUID FUEL SPRAYS

Since the ratio of the volume occupied by the liquid droplets to the volume of the gas is less than 10^{-3} for near stoichiometric mixtures of interest, the sprays of interest are dilute, and shock waves can easily propagate through the mixture. Because the reaction zone of a gaseous detonation is completed in a few microseconds, one is tempted to assume that fuel from droplets cannot enter into an exothermic chemical reaction rapidly enough to allow a self-sustaining detonation. At least, one would think that there is a certain drop size above which a detonation wave would not be sustained. However, a careful look at the behavior of the fuel droplets behind the leading shock front is required to understand the nature of this type of combustion.

In an analogous fashion to a gaseous detonation, a spray detonation may be viewed as a normal shock wave, followed by a dynamic breakup of the droplets and then rapid combustion of the vaporizing microspray, which, due to the rapid expansion of product gases, drives the leading shock front. The existence of secondary, transverse shock waves, as in gaseous detonations, should not be ruled out.

Droplet Breakup

Before discussing the nature of the spray detonations further, let us consider the breakup of droplets in the convective flow behind a normal shock wave. Imagine a dilute spray of uniform-size drops in a quiescent gas at 1 atm pressure and room temperature. A shock wave passes over the spray, subjecting the drops to a high-pressure, high-temperature, and high-velocity flow. The dynamic pressure $(\frac{1}{2}\rho \underline{V}^2)$ of the flow tends to distort the drop by stretching it transverse to the flow, since the pressure is high on the front and lower at 90° from the stagnation point. This flattening effect is opposed by inertia and viscous stresses within the drop. For high-speed flow behind a shock wave, surface tension forces are relatively insignificant. The droplet flattens, and an internal circulation develops within the drop. The liquid in the drop flows to the edge of the flattened drop and is stripped from the surface (see Figure 13.1), thus forming a microspray in the wake of the parent drop. The microspray droplets are of very small size, probably less than 1% of the diameter of the parent drop.

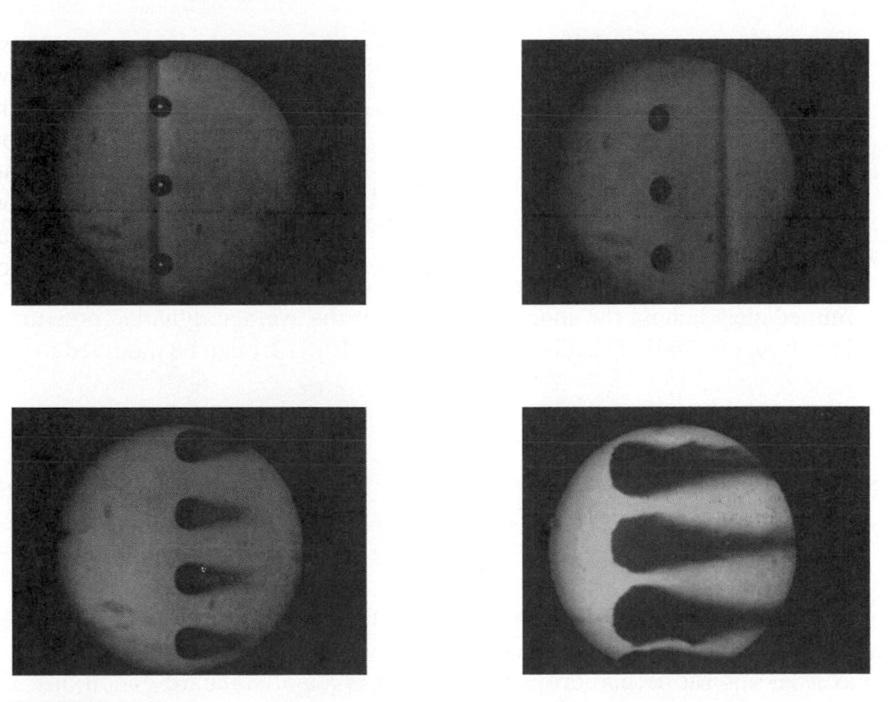

FIGURE 13.1
Shadow photographs of 750 μm water drops breaking up due to Mach number = 2.7 shock wave in atmospheric air: (i) undisturbed; (ii) 2.6 μs, (iii) 4.4 μs, and (iv) 14.4 μs after passage of the shock front [Dabora et al., by permission of The Combustion Institute].

The droplet breakup time t_b for nonburning drops due to convective flow behind a shock wave is given by

$$t_b = \frac{K_1 d \rho_l^{1/2}}{\sqrt{\rho_2 \underline{V}_2^2/2}} \tag{13.1}$$

where d = initial drop diameter
$\underline{V}_2$ = convective velocity behind the shock with respect to laboratory coordinates
K_1 = constant that depends weakly on dynamic pressure and may be taken as 7
ρ_2 = gas (air) density behind the shock
ρ_l = density of the liquid drop

The conditions behind the shock are obtained from the conservation of mass, momentum, and energy in the wave-fixed coordinates given by Eqs. 8.7–8.10. Setting $q = 0$ in Eq. 8.10 and substituting Eqs. 8.10 and 8.8 into Eq. 8.7 yields

$$\check{\mathrm{Ma}}_2 = \frac{(\gamma - 1)\check{\mathrm{Ma}}_1^2 + 2}{2\gamma\,\check{\mathrm{Ma}}_1 - (\gamma - 1)} \tag{13.2}$$

Equation 13.2 shows that for $\check{\mathrm{Ma}}_1$ greater than 1, $\check{\mathrm{Ma}}_2$ is always less than 1. Once $\check{\mathrm{Ma}}_2$ is obtained, p_2, T_2, and ρ_2 may be obtained, and Eq. 13.1 may be evaluated. Recall that $\check{\underline{V}}_2 = \check{\mathrm{Ma}}_2\sqrt{\gamma R T_2}$ and that, from Eq. 8.2, $\underline{V}_2 = \underline{V}_D - \check{\underline{V}}_2$. Also, note that in the lab coordinates Ma_2 can be either subsonic or supersonic depending on the propagation velocity $\underline{V}_D$.

In a detonation wave the dynamic pressure after chemical reaction is less than that immediately behind the shock wave, so that the average dynamic pressure is equal to about one-half of that in Eq. 13.1. Thus, Eq. 13.1 can be modified to read

$$t_b = 10d\left(\frac{\rho_l}{\rho_1}\right)^{1/2} \frac{(\rho_1/\rho_2)^{1/2}}{\underline{V}_2} \tag{13.3}$$

A nondimensional breakup time is given by

$$\tau_b \equiv t_b \frac{\underline{V}_s}{d} = 10\left(\frac{\rho_l}{\rho_1}\right)^{1/2} \frac{(\rho_1/\rho_2)^{1/2}}{1 - \rho_1/\rho_2} \tag{13.4}$$

In going from Eq. 13.3 to Eq. 13.4, the continuity equation across the shock front has been used. Equation 13.4 gives $\tau_b = 162$ at Ma = 3, and decreases to $\tau_b = 120$ as Ma $\rightarrow \infty$ for decane drops initially in oxygen at standard conditions. For example, a 3-mm drop in a Ma = 3 detonation in standard oxygen will take 0.5 ms to break up, and the breakup distance is (3)(330 m/s)(0.0005 s) = 0.5 m.

Spray Detonations

Experiments have been conducted in polydisperse sprays and sprays with precisely controlled droplet size in which the size was varied from 2 μm to 2600 μm diameter. In each case a self-sustaining detonation was achieved, and the propagation velocity was 2 to 35% below the detonation velocity of an equivalent gaseous case. These experiments were done with decane and diethylcyclohexane dispersed in standard oxygen. These fuels were chosen because they have very low vapor pressure at room temperature and hence preclude initial fuel vapor. The experiments were conducted in long tubes, and ignition was achieved by a pulse from a small shock tube located at the top of a larger tube. Experimental detonation velocities of diethylcyclohexane in standard oxygen are compared with the ideal gaseous detonation in Figure 13.2. Spray detonation of large-diameter drops in sprays in air require a more volatile fuel than decane.

Modeling of spray detonations indicates that for droplets less than 10 μm, vaporization alone is sufficiently rapid to allow detonation in oxygen. For droplets between 10 μm and 1000 μm, stripping of droplets to form microspray is sufficiently rapid to support detonation. For droplets greater than 1000 μm, an additional mechanism comes into play. Experiments have shown that this additional mechanism is local explosions about the parent drop.

A schlieren photograph of the shock front and reaction zone of a single stream of 2600-μm diethylcyclohexane drops in a vertical, 5 cm by 5 cm square tube filled with standard oxygen is shown in Figure 13.3. The schlieren photograph shows density gradients but is not sensitive to the light of combustion. Three drops are in the field of view. The leading shock front, which is quite planar, is traveling at 1020 m/s or Ma = 3.0 to the left. Behind the front, the second drop is seen to be forming a microspray in the wake of the drop. Since the convective flow is supersonic with respect to the drop, a bow shock and wake shocks are visible. Self-luminous photographs show that this wake is not yet burning, although there is combustion near the stagnation point of the drop. Disintegration of the third drop has proceeded further and has, in fact, generated a local explosion around the drop

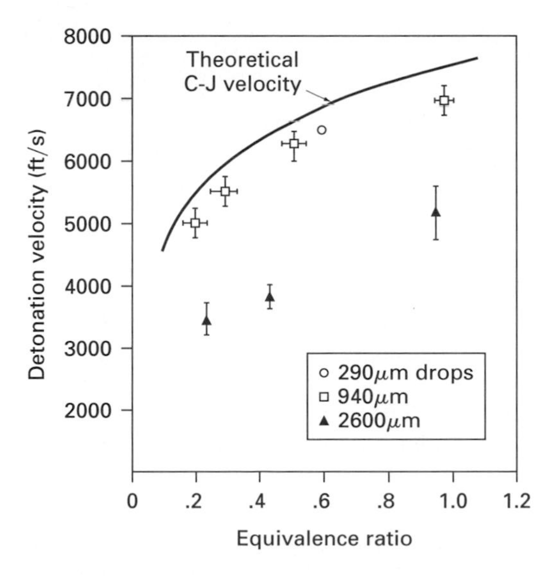

FIGURE 13.2
Comparison of experimental detonation velocity with the ideal C-J velocity for decane in standard oxygen [Ragland et al., reprinted with permission from *Physics of Fluids,* vol. 11, No. 11, pp. 2377–2388, © 1968, American Institute of Physics].

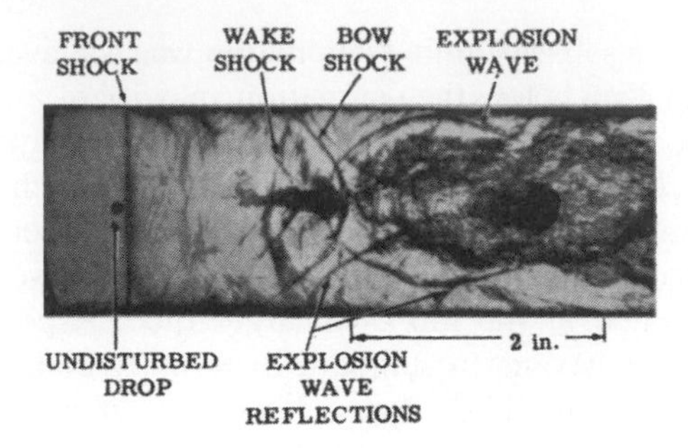

FIGURE 13.3
Spark-source (0.2-μs exposure) photograph of the detonation of single stream of 2.6-mm drops showing details of the phenomenon [Ragland et al., reprinted with permission from *Physics of Fluids*, vol. 11, No. 11, pp. 2377–2388, © 1968, American Institute of Physics].

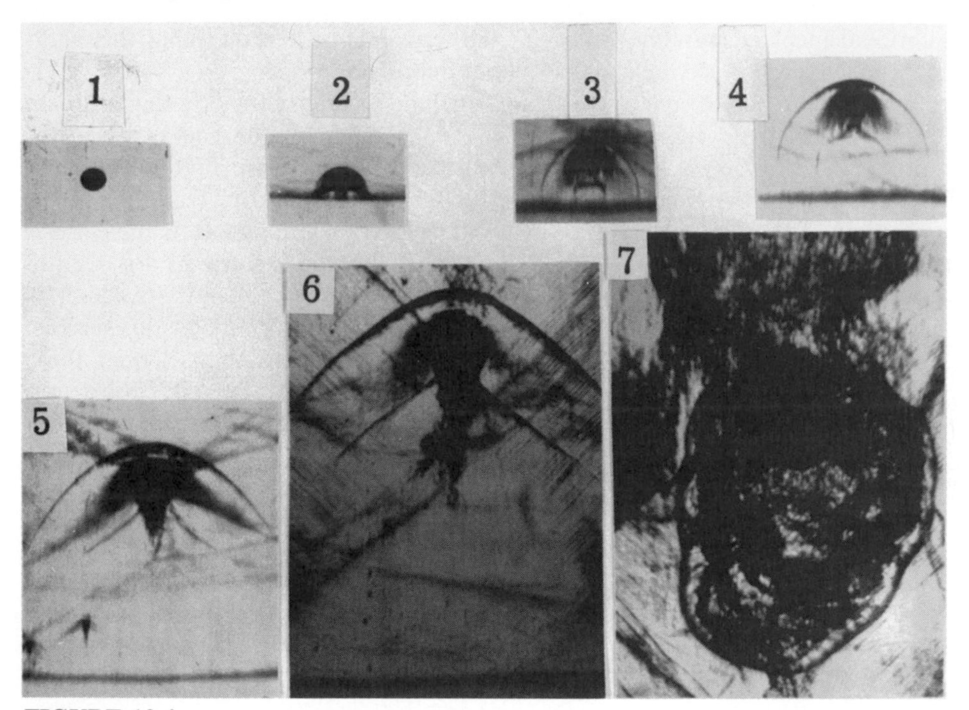

FIGURE 13.4
Schlieren photographs of 2.6-mm droplets within the detonation reaction zone: (1) undisturbed, (2) 3 μs after contact with the initial shock front, (3) 6 μs, (4) 11 μs, (5) 23 μs, (6) 40 μs, (7) 80 μs [Ragland et al., reprinted with permission from *Physics of Fluids*, vol. II, No. 11, pp. 2377–2388, © 1968, American Institute of Physics].

due to the buildup of the vaporized microspray. This local explosion of the sheared droplet propagates outward and further stimulates combustion of other droplets. Details of this process are shown in Figure 13.4. The time for complete disintegration of the 2600-μm diameter drop at these conditions is about 0.5 ms. Even though the reaction zone extends more than 30 cm behind the leading shock front, the combustion is coupled to the shock and drives the entire process at a steady-state velocity. A similar phenomenon occurs with smaller drops, although it is not known if the local explosions occur.

One-dimensional equations of conservation of mass, momentum, and energy may be applied to spray detonations as was done for gaseous detonations in Chapter 8. In addition, because of the extended reaction zone, there is heat transfer to the walls and momentum loss due to accelerating the drops and drag on the walls within the extended reaction zone. Again, the Chapman-Jouguet condition that the velocity at the end of the reaction zone, $\check{\underline{V}}_p$, is sonic with respect to the shock front is invoked. The notation is: 1 refers to upstream air, 2 is air just after the shock, and p refers to combustion products at the end of the reaction zone. The liquid is outside the control volume. The following results have been obtained for the pressure and temperature changes across the detonation wave:

$$\frac{p_p}{p_1} = \frac{\gamma(\underline{V}_D^2/a_1^2)(1 + f)Z}{1 + \gamma} \tag{13.5}$$

$$\frac{T_p}{T_1} = \frac{\gamma^2 M_3(\underline{V}_D^2/a_1^2)Z^2}{(1 + \gamma)^2 M_1} \tag{13.6}$$

where
$$Z = 1 + \frac{1}{\gamma(\underline{V}_D^2/a_1^2)(1 + f)} + \frac{C_H A_R \underline{V}_2^2}{(1 + f)A_s \underline{V}_D \check{\underline{V}}_2}$$

The subscripts are as indicated in Figure 13.5, A_R is the surface area of the reaction zone in contact with the tube walls, A_s is the frontal area of the leading shock wave, and f is the fuel/air ratio. The term $\check{\underline{V}}_2$ is the velocity just behind the shock front in a reference system with respect to the shock, which may be calculated or obtained from shock tables as a function of the Mach number. C_H is the heat

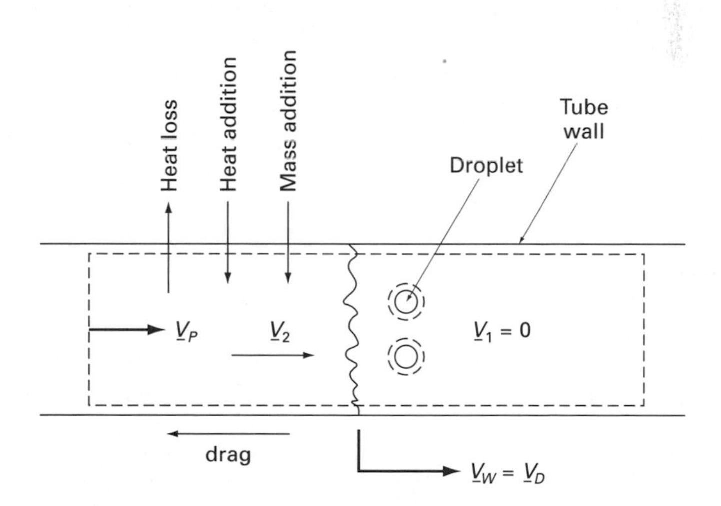

FIGURE 13.5
Control volume for analysis of heterogeneous detonation. Velocities are with respect to stationary walls.

transfer coefficient, which has been measured to be about 2.5×10^{-3} for spray detonations, where C_H is dimensionless and is given by

$$C_H = \frac{\int_2^3 q_w \, dx}{\rho_2 \check{\underline{V}}_2 (h_2 + 1/2\, \check{\underline{V}}_2^2 - h_w)} \tag{13.7}$$

The propagation velocity of a spray detonation is less than the equivalent gaseous detonation velocity $\underline{V}_{D0}$ and is given by

$$\underline{V}_D = \frac{\underline{V}_{D0}}{\left[1 + \dfrac{2\gamma^2 C_H A_R (\underline{V}_2)^2}{\underline{V}_D \check{\underline{V}}_2 (1 + f) A_s}\right]^{1/2}} \tag{13.8}$$

The drag coefficient does not appear in Eq. 13.8, because the Reynolds analogy that $C_D = 2C_H$ has been used. For drops less than 1000 μm in size A_R may be estimated from Eq. 13.4, since droplet breakup is the main rate-limiting step. Although the structure of spray detonations is complicated, one-dimensional theory with frictional and heat transfer losses in a reaction zone controlled by the dynamic breakup of the spray can predict the thermodynamic state and propagation velocity with reasonable accuracy.

13.2 DETONATION OF LIQUID FUEL FILMS

Consider a long tube which contains air or oxygen gas and in which the walls of the tube are coated with a nonvolatile fuel. A strong ignition source, such as a pulse from a small shock tube, will cause an accelerating shock followed by combustion of the fuel on the wall. After a certain transition distance, a steady-state propagation velocity will be reached. The process is similar to that of spray detonations, but there is no spray, only nonvolatile fuel on the walls. Behind the shock front, the fuel layer on the wall vaporizes and perhaps is stripped off the wall.

Experiments of film detonations show that the fuel burns in a thin region along the walls behind the shock front. For example, in a 5 cm by 5 cm square tube in which two walls were coated with diethylcyclohexane, the propagation velocity reached a steady value of 1370 m/s after 2 m of travel. A self-luminous photograph of the combustion is shown in Figure 13.6. The combustion starts at the wall and spreads inward. Radiating combustion products do not fill the center of the test section until about 0.5 ms or 0.6 m behind the point of ignition in the figure.

Spark schlieren photographs in Figure 13.7 show the shock structure and the highly turbulent nature of the reaction zone. In this case fuel is on one wall only and the propagation velocity is 1065 m/s. Even though the heat is released mainly near the wall, the leading shock front is surprisingly planar. However, it is evident that the leading shock front is being perturbed by various pressure disturbances as it propagates. The dark zone next to the wall is the burning layer of fuel vapor and associated turbulence. Evidence from pressure transducers and streak schlieren photographs indicates that local explosions occur within the boundary layer behind

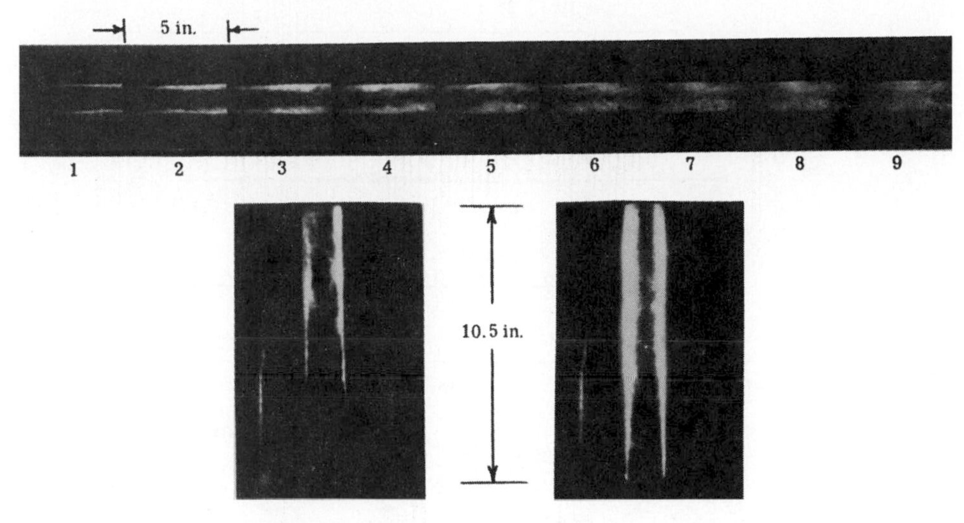

FIGURE 13.6
Self-luminous photographs of film detonation with fuel on two walls, 1-μs intervals [Ragland and Nicholls, by permission of AIAA].

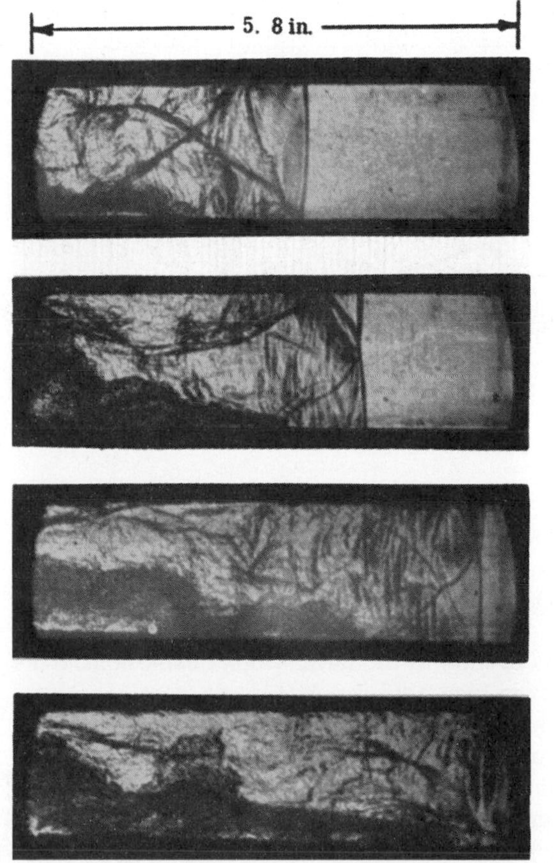

FIGURE 13.7
Spark schlieren photographs of film detonation with fuel on one wall [Ragland and Nicholls, by permission of AIAA].

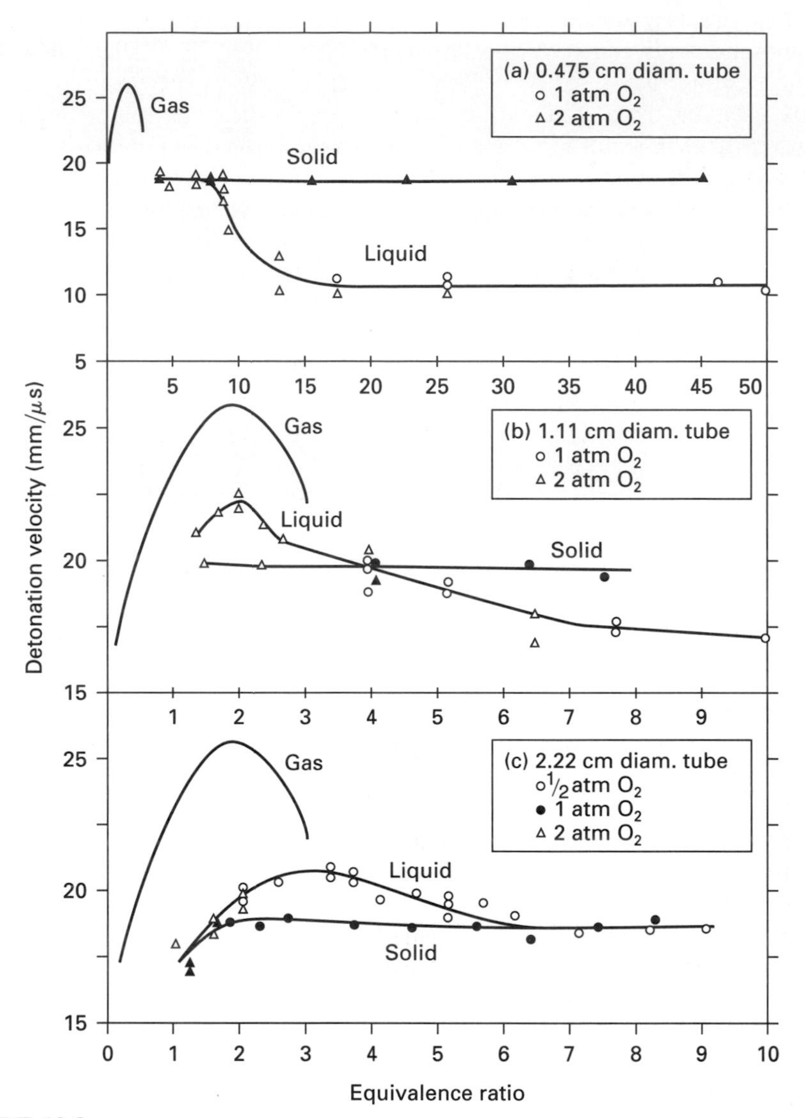

FIGURE 13.8
Experimental detonation velocity of decane and cetane layers in oxygen, and comparison with the ideal C-J velocity [Reprinted by permission of Elsevier Science Inc. from "Ignition Delay Measured in Two-Phase Detonations," by Ragland and Garcia, *Combustion and Flame,* vol. 18, pp. 53–58, © 1972 by The Combustion Institute.].

the shock in a somewhat analogous fashion to the large droplet explosions. This issuggested by the extraneous waves in Figure 13.7.

The detonation velocities of cetane liquid layers in various-size circular tubes filled with oxygen are shown in Figure 13.8. Also, the hexadecane was frozen to a solid by chilling the tube walls, and this represents the solid curve in Figure 13.8. The gaseous curve was obtained from computer calculations. The equivalence ratio was changed by varying the thickness of the fuel layer. An L/d of at least 120 was required to reach steady state, after which the velocity was constant to within ±2%. The minimum equivalence ratios were not the lean limit but rather the smallest amount of liquid that could be spread on the tube in this experiment. There is no rich limit; rather, the propagation velocity for liquid layers decreases to a plateau. At rich mixtures in small tubes the solid velocity curve is actually higher than the liquid curve. The solid curve is independent of equivalence ratio. These curves show the different rates at which the fuel enters the reaction for various thickness fuel layers.

The two-phase detonation velocities shown in Figure 13.8 are all lower than the corresponding gas-phase detonation. The equations developed for spray detonations, Eqs. 13.4–13.6, apply also to the film detonations. Of course, the heat transfer and drag coefficients will be different. In fact, the self-sustaining nature of the film detonation is facilitated by the fact that the shear stress at the fuel layer is greatly reduced as the layer vaporizes and burns. Similarly, the heat transfer to the walls is reduced by the presence of the fuel layer. Turbulent boundary layer heat transfer analysis shows that the vaporization rate is indeed rapid enough to support the observed propagation velocities and reaction zone thickness. In addition, the secondary shock waves catch up to the leading shock front and reinforce it.

13.3 SUMMARY

A detonation wave can propagate through a nonvolatile fuel spray in air or oxygen in a manner similar to a gaseous detonation, except that the reaction zone is more extended and the propagation velocity is lower. Large droplets participate in the reaction zone because the droplets break up by boundary layer stripping in the convective flow behind the leading shock front. Furthermore, a detonation wave can propagate down a tube containing a layer of nonvolatile liquid fuel on the walls. Vaporization and combustion in the boundary layer are coupled with and drive the leading shock front to a steady-state velocity. These types of processes indicate that combustion of liquid fuels, as well as gaseous fuels, can sometimes get out of control.

PROBLEMS

13.1. What are the breakup time and breakup distance for a 0.1-mm water drop subjected to a Mach number = 2.5 shock wave propagating through air initially at standard conditions?

13.2. What is the breakup time for a single 50-μm cetane (hexadecane) drop injected into high-pressure air at 50 m/s? The air is at 35 atm and 800 K.

13.3. Estimate the pressure and temperature at the end of the reaction zone for a stoichiometric mixture of cetane drops in standard oxygen due to a fully developed detonation which is propagating at 2100 m/s. Neglect effect of heat loss and friction.

13.4. For the conditions of Problem 13.3 with 0.5-mm drops in a 10-cm-dia tube, find the ratio of the spray detonation propagation velocity to the gaseous detonation velocity. Assume that the length of the reaction zone is 33 mm.

13.5. What is the thickness of a liquid layer of decane which will yield a stoichiometric mixture with (*a*) standard oxygen and (*b*) standard air in a 1-cm-dia tube?

REFERENCES

Borisov, A. A., Kogarko, S. M., and Lyubimov, A. V., "Ignition of Fuel Films behind Shock Waves in Air and Oxygen," *Eleventh Symp. (Int.) on Combustion,* The Combustion Institute, Pittsburgh, 1967.

Bowen, J. R., Ragland, K. W., Steffes, F. J., and Loflin, T. G., "Heterogeneous Detonations Supported by Fuel Fogs or Films," *Thirteenth Symp. (Int.) on Combustion,* The Combustion Institute, Pittsburgh, pp. 1131–1139, 1971.

Cramer, F. B., "The Onset of Detonation in a Droplet Field," *Ninth Symp. (Int.) on Combustion,* The Combustion Institute, Pittsburgh, pp. 482–487, 1963.

Dabora, E. K., Ragland, K. W., and Nicholls, J. A., "Drop Size Effects in Spray Detonations," *Twelfth Symp. (Int.) on Combustion,* The Combustion Institute, Pittsburgh, pp. 19–26, 1969.

Gordeev, V. E., and Komov, V. F., *Energetics,* vol. 24, pp. 12–15, 1964.

Gordeev, V. E., Komov, V. F., and Troshin, Y. K., "Concerning Detonation Combustion in Heterogeneous Systems," *Proceedings of the Academy of Science, USSR*, vol. 160, 4 (Physical Chemistry), 1965.

Komov, V. F., and Troshin, Y. K., "Detonation Properties in Certain Heterogeneous Systems," *Proceedings of the Academy of Science, USSR*, vol. 175, pp. 109–112, 1967.

Lin, Z. C., Nicholls, J. A., Tang, M. J., Kaufmann, C. W., and Sichel, M., "Vapor Pressure and Sensitization Effects in Detonation of a Decane Spray," *Twentieth Symp. (Int.) on Combustion,* The Combustion Institute, Pittsburgh, pp. 1709–1716, 1984.

Pinaev, A. V., and Sobbotin, V. A., "Reaction Zone Structure in Detonation of Gas-Film Type Systems," *Fizika Goreniya i Vzryva,* vol. 18, no. 5, pp. 103–111, 1982.

Ragland, K. W., Dabora, E. K., and Nicholls, J. A., "Observed Structure of Spray Detonations," *Physics of Fluids,* vol. 11, no. 11, pp. 2377–2388, 1968.

Ragland, K. W., and Garcia, C. F., "Ignition Delay Measure in Two-Phase Detonations," *Combustion and Flame,* vol. 18, pp. 53–58, 1972.

Ragland, K. W., and Nicholls, J. A., "Two-Phase Detonation of a Liquid Layer," *AIAA J.,* vol. 7, no. 5, pp. 859–863, 1969.

Webber, W. T., "Spray Combustion in the Presence of a Travelling Wave," *Eighth Symp. (Int.) on Combustion,* The Combustion Institute, Pittsburgh, pp. 1129–1140, 1962.

PART IV

Combustion of Solid Fuels

The three main types of combustion systems for firing solid fuels are fixed-bed firing, suspension firing, and fluidized-bed burning. Before considering solid fuel combustion systems, the behavior of a single coal particle in a hot, flowing gas stream is discussed. Although finely pulverized solid fuels can also support a detonation wave under certain conditions, this will not be included. The reader may wish to review the material on solid fuels in Chapter 2 before proceeding.

CHAPTER 14

Solid Fuel Combustion Mechanisms

The combustion of individual solid fuel particles such as coal and wood is considered in this chapter. The fuel size ranges from pulverized fuels to larger sizes such as crushed, chipped, or shredded fuels, or logs. Investigation of the behavior of individual solid fuel particles provides insight into the design and performance of furnaces and boilers which are discussed in following chapters.

When a solid fuel particle is exposed to a hot flowing gas stream, it undergoes three stages of mass loss—drying, devolatilization, and char combustion. The relative significance of each of these three processes is indicated by the proximate analysis of the fuel. For example, coal has relatively little water, fewer volatiles, and more fixed carbon (char) as compared with wood. For pulverized fuel particles, drying, devolatilization, and char burn occur sequentially, and the char burn period lasts much longer than the devolatilization and drying stages. For larger particles, drying, devolatilization, and char burn occur simultaneously.

14.1 DRYING OF SOLID FUELS

Moisture in solid fuels can exist in two forms—as free water within the pores of the fuel, and as bound water which is adsorbed to the interior surface structure of the fuel. Wood, being rather porous, has both free and bound water. Green wood contains about 45% water on an as-received basis, of which half is bound water. Lignites contain up to 40% water, while bituminous coals have relatively small pores containing only a few percent water, mostly bound water.

Consider a pulverized coal or wood particle which is inserted into a furnace. Upon entry into the gas stream, heat is convected and radiated to the particle surface and conducted into the particle. For a pulverized particle (say, 100 μm in size), the water is vaporized and forced out of the particle rapidly before volatiles are released. The drying time of a small pulverized particle is the time required to heat up the particle to the vaporization point and drive off the water. An energy balance on the small particle says that the time rate of change of energy within the particle equals the heat rate to vaporize the water plus the net heat transfer rate to the particle by convection and radiation:

$$\frac{d}{dt}(m_w u_w + m_{df} u_{df}) = -\dot{m}_w h_{fg} + q \tag{14.1}$$

where h_{fg} = latent heat of vaporization per unit mass of water
w = water
df = dry fuel

The heat transfer rate to the particle, q, depends on the background furnace temperature T_b, which is assumed to be equal to the surrounding gas temperature. Gray body radiation with emissivity ε and a view factor of 1, and a convective film coefficient $\tilde{h}^*$, are used (recall from Chapter 9 that $\tilde{h}^*$ is the convective heat transfer coefficient corrected for outgassing):

$$q = \epsilon\sigma A_p(T_b^4 - T_p^4) + \tilde{h}^* A_p(T_g - T_p) \tag{14.2}$$

Integrating Eqs. 14.1 and 14.2 from the initial temperature to the boiling point of water, using the relation $du = c\,dT$, and noting that the boiling point temperature is much lower than the background furnace temperature, we obtain the particle drying time:

$$t_{\text{dry}} \approx \frac{(m_{wi}c_w + m_{df}c_{df})(373 - T_i) + m_{wi}h_{fg}}{\epsilon\sigma A_p(T_b^4 - T_p^4) + \tilde{h}A_p(T_g - T_p)} \tag{14.3}$$

where m_{wi} is the initial mass of water in the particle. The differential heat of wetting (i.e., the heat to release the absorbed water from the surface) is about 70 kJ/kg dry solid and is neglected compared with the heat of vaporization of water of 2400 kJ/kg of water. For evaluation of the convective heat transfer coefficient, the particle film temperature is used:

$$T_m = \frac{T_p + T_g}{2} \tag{14.4}$$

along with a Nusselt number correlation.

EXAMPLE 14.1. A 100-μm oak particle which contains 40% moisture (dry basis) is inserted into a 1500-K furnace. The oak particle was initially at 300 K and had a dry density of 690 kg/m^3. Find the drying time.

Solution. The thermal conductivity of the gas (air) obtained from Appendix B using a film temperature of 900 K is k_g = 0.0625 W/m·K. Using a Nusselt number of 2.0,

$$\tilde{h} = \frac{2.0k_g}{d} = \frac{(2.0)(0.0625\ \text{W/m}\cdot\text{K})}{100 \times 10^{-6}\ \text{m}} = 1250\ \text{W/m}^2\cdot\text{K}$$

The specific heat of dry wood from Table A.3 in Appendix A is c_{df} = 1 kJ/kg·K, and for water c_w = 4.18 kJ/kg·K. The heat of vaporization is 2400 kJ/kg. The emissivity of wood is assumed to be $\epsilon = 0.90$. The masses of dry wood and water are

$$m_{df} = \frac{\pi d_p^3}{6} \rho_p = \frac{(\pi)(100 \times 10^{-6} \text{ m})^3 (690 \text{ kg/m}^3)}{6} = 3.61 \times 10^{-10} \text{ kg}$$

$$m_w = (0.40)(3.61 \times 10^{-10} \text{ kg}) = 1.44 \times 10^{-10} \text{ kg}$$

Evaluating Eq. 14.3 for the drying time,

$$t_{\text{dry}} = \frac{a + b + c}{d + e}$$

where

$$a = (1.44 \times 10^{-10})(4.18)(373 - 300)$$
$$b = (3.61 \times 10^{-10})(1)(373 - 300)$$
$$c = (2400)(1.44 \times 10^{-10})$$
$$d = (0.9)(5.67 \times 10^{-8})[\pi(100 \times 10^{-6})^2](1500^4 - 336^4)$$
$$e = \frac{1250}{5} [\pi(100 \times 10^{-6})^2](1500 - 336)$$

so that

$$t_{\text{dry}} = 2.4 \times 10^{-5} \text{ kJ/W} = 24 \text{ ms}$$

While 24 ms may seem like a relatively short time, if, for example, we were trying to retrofit an oil-fired furnace to burn sawdust, the distance traveled while drying before ignition would be 24 cm if the velocity were 10 m/s. This could create difficulties in trying to stabilize the flame, and thus the fuel should be dried before being fed to the burner.

For relatively large fuel particles such as stoker coal or wood chips in a convective flow, Eq. 14.1 is not valid. Because of temperature gradients within the particle, moisture is evolved from inside the particle while volatiles are being driven off near the outer shell of the particle. Due to the high pressure in the fuel pores during devolatilization of the outer layer of the particle, some of the moisture is forced toward the center of the particle until the pressure builds up throughout the particle. Hence, drying of large solid fuel particles initially involves inward migration of the water vapor as well as the outward flow. A pyrolysis layer starts at the outer edge of the particle and gradually moves inward, releasing volatiles and forming char. The moisture release reduces the heat and mass transfer to the particle surface so that the mass loss of the particle (burning rate) is reduced. As the moisture and volatile release is reduced, the char surface begins to react. For example, consider a 10-mm pine cube suspended from an electronic balance into a 1100-K airstream. The mass divided by the initial mass was measured versus time (Figure 14.1) for initial moisture contents of 0%, 15%, and 200% (moisture content on a dry basis). The normalized mass versus time curves were differentiated to yield the normalized burning rates, which are reduced by the fuel moisture. Gas measurements of H_2O, CO_2, and CO indicate that pyrolysis products (indicated by CO_2) are released while the fuel is drying. Drying and pyrolysis are complete at 120 s for the 200% moisture case, at which point some of the char has burned (as indicated by the mass remaining), and the remaining char burn generates CO and CO_2.

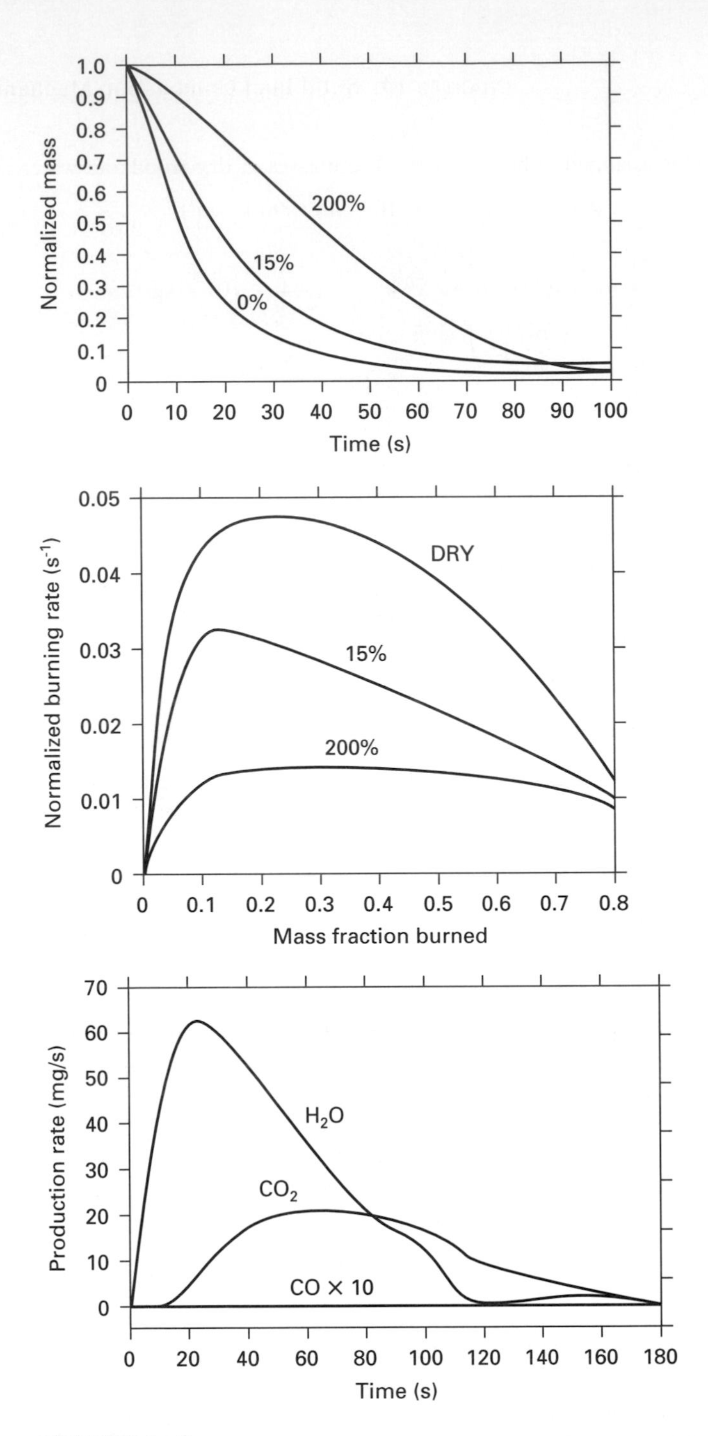

FIGURE 14.1
Effect of moisture (% dry basis) on 10-mm pine cubes burned in air at 1100 K and Re = 120: (*a*) the normalized mass, (*b*) normalized burning rate (time derivative of mass per initial mass), and (*c*) CO, CO_2, and H_2O products, 200% moisture [Simmons and Ragland, © 1986, by permission of Gordon and Breach Publishers].

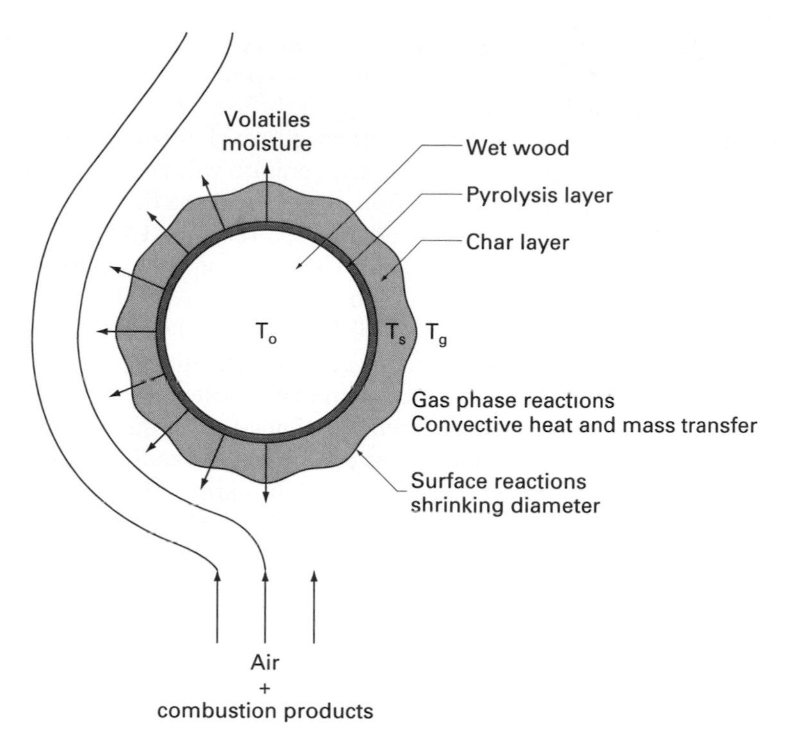

FIGURE 14.2
Cross section of reacting log showing char, pyrolysis, and undisturbed wood regions.

When a log burns in a campfire, a fireplace, or a fixed bed of logs, the drying, pyrolysis, and char burn occur simultaneously until most of the mass is consumed, at which point only char remains. Figure 14.2 indicates the three zones for a partially reacted log. The outside char layer is black and porous. The pyrolysis zone is a thin brown layer, and the interior portion of the log is white, moist wood with a temperature less than 100°C even though the outside of the log is very hot. The moisture impedes the inward conduction of heat, and some moisture is retained during almost the entire burn time. For short logs, some moisture escapes out the ends of the log during combustion, while for long logs some moisture remains until 99% of the initial mass of the log is consumed.

14.2 DEVOLATILIZATION OF SOLID FUELS

When the drying of a small fuel particle or a zone within a large particle is completed, the temperature rises and the solid fuel begins to decompose, releasing volatiles. Since the volatiles flow out of the solid through the pores, external oxygen cannot penetrate into the particle, and hence the devolatilization is referred to as the *pyrolysis* stage. The rate of devolatilization and the pyrolysis products depend on the temperature and the type of fuel. The pyrolysis products then ignite and form an attached flame around the particle as oxygen diffuses into the products. The flame in turn heats the particle, causing enhanced devolatilization. While water vapor is flowing out of the pores, the flame temperature will be low and the flame weak. Once all the water vapor is driven from the particle, the flame will be hotter.

For lignite coals, pyrolysis begins at 300–400°C releasing CO and CO_2. Ignition of the volatiles occurs at 400–600°C. Carbon monoxide, carbon dioxide, chemically formed water, hydrocarbon vapors, tars, and hydrogen are produced rapidly as the temperature reaches 700–900°C. Above 900°C pyrolysis is essentially complete, and the char (fixed carbon) and ash remain.

The devolatilization of bituminous coal proceeds differently from lignite because bituminous coal contains less oxygen than lignite. First the bituminous coal becomes plastic and some of the bituminous coals swell markedly. Pressure within the particle builds up, and tars are squeezed out of the particle. Fracture of the coal particle into several pieces may occur due to the internal pressure. Meanwhile, pyrolysis proceeds, releasing carbon monoxide, hydrocarbons, and soot, which burn as an attached diffusion flame around the particle. The reaction rate of the released volatiles can be calculated approximately using the global hydrocarbon reaction rate of Section 4.4.

For wood, the hemicellulose pyrolyzes at 225–325°C, the cellulose at 325–375°C, and the lignin at 300–500°C. Certain extractives such as terpenes, which amount to only a few percent of the wood, escape at less than 225°C. Various hydrocarbon vapors, liquids, and tars and water are formed and quickly break down under combustion conditions so that the pyrolysis products in a combustion environment may be considered to be short-chained hydrocarbons, carbon dioxide, carbon monoxide, hydrogen, and water vapor. As with coal, the exact composition is a function of the heating rate. The pyrolysis products burn as a diffusion flame around the particle if sufficient oxygen is present.

If the object is to gasify the solid fuel but not to complete combustion, such as for a gasifier to run a reciprocating engine, then only enough oxygen is provided to generate exothermic reaction for drying and pyrolysis of the particles. The volatile flame is not established. The products are short-chained hydrocarbons, carbon monoxide, hydrogen, carbon dioxide, water vapor, nitrogen, and tar. The tar, char, and ash particles and any vaporized inorganic compounds must be separated from the gaseous pyrolysis products before using them in an engine.

The rate of devolatilization of a solid fuel may be represented approximately as a first-order reaction with an Arrhenius rate constant:

$$\frac{dm_v}{dt} = -m_v k_{\text{pyr}} \tag{14.5}$$

where $k_{\text{pyr}} = -k_{0,\text{pyr}} \exp(-E_{\text{pyr}}/\hat{R}T_p)$, and where pyr refers to pyrolysis and $m_v = m_p - m_c - m_a$ (i.e., the mass of the volatiles equals the mass of the dry particle minus the mass of the char and ash). Note that the pyrolysis rate is independent of the particle size as long as the particle temperature is constant. Generally, the heat-up time is short compared with the pyrolysis time for pulverized fuels. For large particles the transient heating of the particle must be considered.

The activation energy and preexponential factor must be determined experimentally for specific combustion conditions and fuel type. Representative values of the preexponential factor and the activation energy are shown in Table 14.1. The mass of the char can be determined from the proximate analysis; however, under high heating rates such as experienced in pulverized coal flames, the volatile yield is higher and the char yield is lower than suggested by the proximate analysis. Volatile yields of 50% for pulverized coal and 90% for pulverized wood are typical.

If the particle temperature is constant during devolatilization, then Eq. 14.5 may be integrated to obtain the mass loss as a function of time during pyrolysis:

$$\ln\left[\frac{m_p - m_c - m_a}{m_{pi} - m_c - m_a}\right] = -k_{\mathrm{pyr}}\, t \tag{14.6}$$

Equation 14.6 implies that a single chemical reaction converts the solid fuel to pyrolysis products. In reality, solid fuels such as coal and wood are complex compounds which undergo many reactions when they are heated. Some of these reactions are endothermic and some are exothermic, and they proceed at different rates. Nevertheless, the net heat of devolatilization is thought to be near zero, and Eq. 14.5 is a useful approximate global pyrolysis rate.

EXAMPLE 14.2. A bituminous coal particle has a temperature of 1500 K. Find the time required to devolatilize 90% of the volatile mass.

Solution. From Table 14.1, $k_{0,\mathrm{pyr}} = 700\ \mathrm{s}^{-1}$ and $E_{\mathrm{pyr}} = 11.8$ kcal/gmol. Thus,

$$k_{\mathrm{pyr}} = 700\ \mathrm{s}^{-1} \exp\left[\frac{-11{,}800\ \mathrm{cal/gmol}}{(1.987\ \mathrm{cal/gmol\cdot K})(1500\ \mathrm{K})}\right] = 13.35\ \mathrm{s}^{-1}$$

From Eq. 14.6,

$$t_{\mathrm{pyr}} = \frac{-\ln(0.10)}{13.35\ \mathrm{s}^{-1}} = 0.17\ \mathrm{s}$$

For larger fuel particles, considerable time is required to heat the particles to pyrolysis temperatures after they are inserted into the combustion environment, and the pyrolysis process gradually penetrates into the particle. For example,

TABLE 14.1
Representative pyrolysis parameters for several solid fuels

Fuel	$k_{0,\mathrm{pyr}}$ (s^{-1})	E_{pyr} (kcal/gmol)
Lignite	280	11.3
Bituminous coal	700	11.8
Wood	7×10^7	31.0

experimental data obtained by inserting a single coal particle into a hot gas stream are shown in Figures 14.3 and 14.4. The coal particle is attached to a fine quartz rod which is suspended from an electronic balance. In this case a class C bituminous coal particle of 5.3 mm diameter (100 mg) was suspended in a 1100-K airstream (Figure 14.3). The ignition delay was a few seconds; then a volatile flame ignited and remained attached to the particle for 30 s. Upon extinction of the volatile flame, 55% of the mass was lost and the rate of mass loss decreased markedly. The proximate analysis would suggest that only 40% of the mass should be lost at the end of devolatilization. It is typical of solid fuels undergoing high heating rates that the volatile yield is greater than indicated by the proximate analysis. Figure 14.4 shows data for a 5.3-mm-dia bituminous particle inserted into a 1200-K stream of air flowing at 2 m/s, the particle requires 22 s to pyrolyze, while a 10-mm particle requires 62 s to pyrolyze. Similarly, a 10-mm cube of dry pine shown in Figure 14.1 requires 25 s to pyrolyze. The 10-mm bituminous coal particle initially weighed 675 mg and pyrolyzed 55% of this weight, while the pine cube weighed 400 mg and pyrolyzed 90% of the initial weight. Hence, the average pyrolysis rate was 6.0 mg/s for the 10-mm coal particle and 14.4 mg/s for the 10-mm pine particle. This is in agreement with the general statement that wood is more reactive than coal.

Ignition of solid fuels can occur either by ignition of the fixed carbon (char) on the surface of the fuel, or by ignition of the volatiles in the boundary layer around the particle. Which mechanism actually occurs first depends on the rate of convective and radiative heat transfer to the particle. If the radiative heat transfer is high

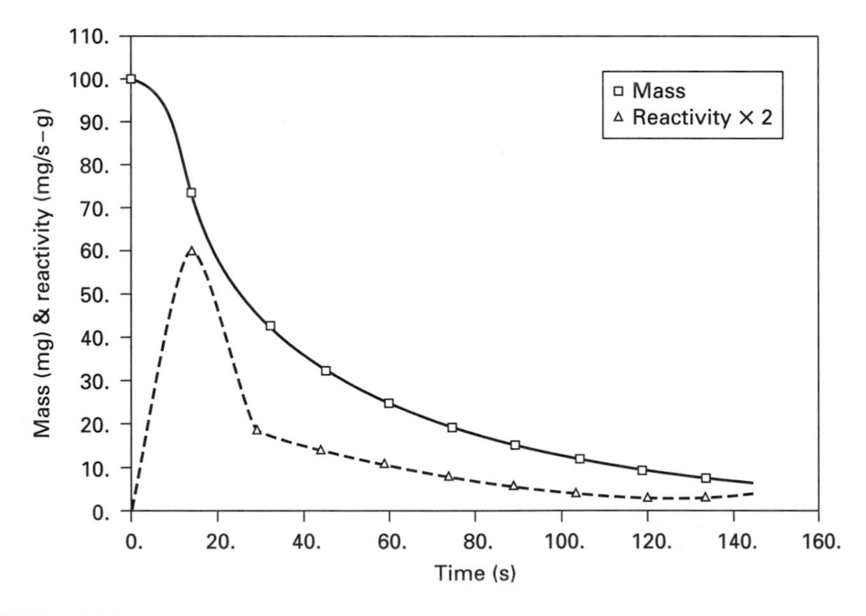

FIGURE 14.3
Typical curves of transient mass and reactivity for bituminous coal particles of 5.3 mm with 1100-K air at 2 m/s [Reprinted by permission of Elsevier Science Inc. from "Combustion of Millimeter Sized Coal Particles in Convective Flow," by Ragland and Yang, *Combustion and Flame,* vol. 60, pp. 285–297, © 1985 by The Combustion Institute.].

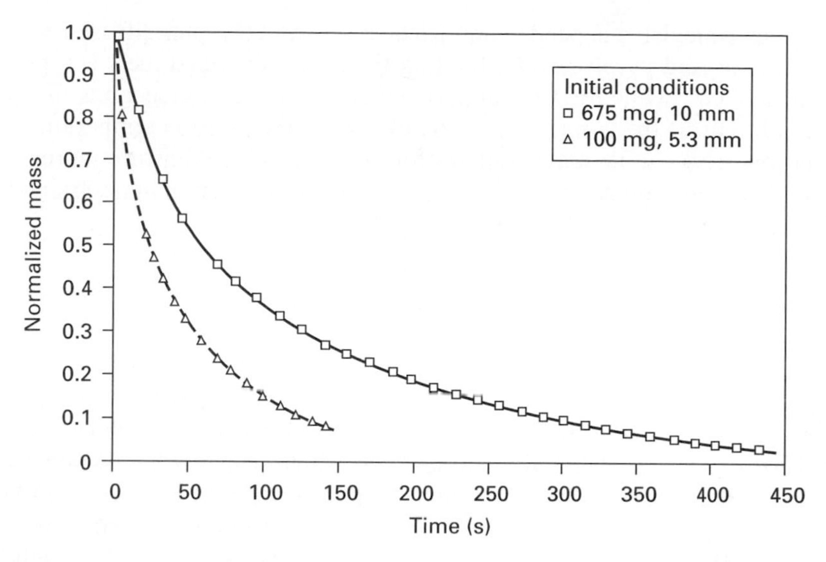

FIGURE 14.4
Effect of particle size on transient mass for bituminous coal particles with 1200-K air at 2 m/s [Reprinted by permission of Elsevier Science Inc. from "Combustion of Millimeter Sized Coal Particles in Convective Flow," by Ragland and Yang, *Combustion and Flame,* vol. 60, pp. 285–297, © 1985 by The Combustion Institute.].

so that the surface quickly heats up to the ignition temperature of the carbon, or if the rate of convective heating is high so that the surface rapidly heats but the volatiles are swept away before a combustible mixture can accumulate, the ignition will occur first at the surface. On the other hand, if surface heating is low, then the volatiles may ignite first, since they have a lower ignition temperature than carbon.

Ignition temperatures of various fuels are shown in Table 14.2. Note that char has a much lower ignition temperature than graphite. This implies that char is not pure carbon, and in fact char contains some hydrogen as well as carbon. Pyrolysis under the usual combustion conditions does not drive off all of the hydrogen. Note that wood char ignites at a lower temperature than coal char. Similarly, volatiles from wood ignite at a lower temperature than volatiles from coal.

TABLE 14.2
Typical ignition temperature for selected solid fuels

	Ignition temperature	
Fuel	**(°F)**	**(°C)**
Graphite	1500	820
Bituminous coal char	765	410
Bituminous coal volatiles	660	350
Wood char	650	340
White pine volatiles	500	260
Paper	450	230

Ignition time delay depends upon particle size and thermal diffusivity as well as heating rate and pyrolysis rate. Ignition time for pulverized fuels is typically a few milliseconds, while for 10-mm particles it can be several seconds under furnace conditions. If the temperature is barely above the ignition temperature, then the ignition delay can be many minutes for large particles. Moisture increases the ignition delay. Ignition delay can be an important consideration in designing burners for pulverized fuels and for coal water slurries.

14.3 CHAR COMBUSTION

The final step in the solid fuel combustion process is the char combustion. When devolatilization is completed, the char and ash remain. Char is highly porous. The porosity of wood char is about 0.9 (90% voids), while coal char tends to have a porosity of about 0.7, although this can vary widely. The internal surface area is on the order of 100 m^2/g for coal char and 10,000 m^2/g for wood char. When no volatiles are escaping from the char and since the char is porous, oxygen can diffuse through the external boundary layer and into the char particles. The burning rate of the char depends on both the chemical rate of the carbon-oxygen reaction at the surfaces and the rate of boundary layer and internal diffusion of oxygen. The surface reaction generates primarily CO. The CO then reacts outside the particle to form CO_2. The surface reactions typically raise the temperature of the char 100–200°C above the external gas temperature.

The burning rate of the char depends on the oxygen concentration, gas temperature, Reynolds number, and char size and porosity. For engineering purposes it is appropriate to use a global reaction rate to determine the burning rate of a char particle. The alternative is to use intrinsic (elementary) reaction rates; however, this requires detailed calculations in the boundary layer and within the pore structure of the char. The global reaction rate is formulated in terms of rate of reaction of the char mass per unit external surface area and per unit oxygen concentration outside the particle boundary layer. As with all global reactions, the results should be verified experimentally over the range of application.

The carbon char reacts with oxygen at the surface to form carbon monoxide and carbon dioxide, but generally carbon monoxide is the main product:

$$C + \tfrac{1}{2}O_2 \rightarrow CO \qquad (a)$$

The carbon surface also reacts with carbon dioxide and water vapor according to the following reduction reactions:

$$C + CO_2 \rightarrow 2CO \qquad (b)$$

$$C + H_2O \rightarrow CO + H_2 \qquad (c)$$

Reduction reactions (*b*) and (*c*) are generally much slower than the oxidation reaction (*a*), and for combustion usually only reaction (*a*) need be considered. Where the oxygen is depleted, then these reduction reactions are important.

For a global reaction rate of order n with respect to oxygen, the char burning rate is given by

$$\frac{dm_c}{dt} = -i\left(\frac{M_c}{M_{O_2}}\right)A_p k_c(\rho_{O_2}(s))^n \tag{14.7}$$

where i is the stoichiometric ratio of moles of carbon per mole of oxygen (which is 2 for reaction (a)), A_p is the external particle surface area, k_c is the kinetic rate constant, $\rho_{O_2}(s)$ is the oxygen partial density at the surface of the particle, and n is the order of reaction. The oxygen concentration at the particle surface is not known; however, it can be eliminated by equating the oxygen consumed by the char to the diffusion of oxygen across the particle boundary layer. Since a simplification occurs when the reaction order is 1, we consider only this case. Then,

$$A_p k_c \rho_{O_2}(s) = A_p \tilde{h}_D(\rho_{O_2}(\infty) - \rho_{O_2}(s)) \tag{14.8}$$

Solving for the oxygen density at the surface,

$$\rho_{O_2}(s) = \left(\frac{\tilde{h}_D}{k_c + \tilde{h}_D}\right)(\rho_{O_2}(\infty)) \tag{14.9}$$

Hence, Eq. 14.7 becomes

$$\frac{dm_c}{dt} = -\tfrac{12}{16}A_p k_e \rho_{O_2}(\infty) \tag{14.10}$$

where the effective rate constant k_e includes both kinetics and diffusion:

$$k_e = \frac{\tilde{h}_D k_c}{\tilde{h}_D + k_c} \tag{14.11}$$

The kinetic rate constant is calculated from the Arrhenius relation

$$k_c = k_{c,0}\exp\left(\frac{-\hat{E}_c}{\hat{R}T_p}\right) \tag{14.12}$$

Sometimes global char kinetic reaction rate constants are based on oxygen pressure rather than oxygen concentration:

$$k_p = k_{p,0}\exp\left(\frac{-\hat{E}_c}{\hat{R}T_p}\right) \tag{14.13}$$

Global kinetic parameters for several coal chars are given in Table 14.3. Note that k_p has units of g/(cm^2 · s · atm O_2), whereas k_c has units of g/(cm^2 · s · g/cm^3) or cm/s. Values of k_c may be calculated from k_p values, from

$$k_c = \frac{k_p T_g \hat{R}}{M_{O_2}} \tag{14.14}$$

The mass transfer coefficient $\tilde{h}_D$ (cm/s) is obtained from the Sherwood number ($\mathrm{Sh} = \tilde{h}_D d/D_{AB}$). For very low flow around the particle, Sh = 2, according to the following reasoning. Consider a spherical char particle surrounded by stagnant air in a zero gravity field. Diffusion of oxygen to the surface is governed by the

TABLE 14.3
Representative global coal char oxidation rate constants [Smoot and Smith, by permission of Plenum Publishing Corp.]

Coal type	$k_{p,0}$ [g/(cm²·s·atm O₂)]	$\hat{E}$ (cal/gmol)
Anthracite	20.4	19,000
Bituminous (high volatile A)	66	20,360
Bituminous (high volatile C)	60	17,150
Subbituminous (class C)	145	19,970

diffusion equation, which is analogous to the heat conduction equation (note: r is radius here, not reaction rate):

$$\frac{d}{dr}\left(r^2 \frac{d\rho_{O_2}}{dr}\right) = 0 \tag{14.15}$$

Integrating twice,

$$\rho_{O_2} = \frac{-a_1}{r} + a_2 \tag{14.16}$$

The boundary conditions are as follows: for r large, $\rho_{O_2} = \rho_{O_2}(\infty)$; and at the particle surface ($r = d/2$), it is assumed that all of the oxygen is consumed ($\rho_{O_2}(s) = 0$). Hence, Eq. 14.16 becomes

$$\rho_{O_2} = \rho_{O_2}(\infty)\left(\frac{1 - d}{2r}\right) \tag{14.17}$$

Differentiating Eq. 14.17, the flux of oxygen to the surface is

$$D_{AB}A_p \frac{d\rho_{O_2}}{dr}\Big|_{r=d/2} = \frac{2D_{AB}A_p\rho_{O_2}(\infty)}{d} \tag{14.18}$$

Also, by definition of the mass transfer coefficient, the oxygen flow to the surface is given by

$$D_{AB}A_p \frac{d\rho_{O_2}}{dr}\Big|_{r=d/2} = \tilde{h}_D A_p(\rho_{O_2}(\infty) - \rho_{O_2}(s)) \tag{14.19}$$

Equating Eqs. 14.18 and 14.19, with zero oxygen at the surface,

$$\frac{\tilde{h}_D d}{D_{AB}} = \mathrm{Sh} = 2 \tag{14.20}$$

When the particle Reynolds number is not much less than 1, then the Ranz-Marshall correlation (see Eq. 9.40), may be used:

$$\mathrm{Sh} = (2 + 0.6\,\mathrm{Re}_d^{1/2}\,\mathrm{Sc}^{1/3})\phi \tag{14.21}$$

where the Schmidt number Sc is typically 0.73. Also, a mass transfer correction factor ϕ is introduced due to rapid outward flow of combustion products.

ϕ is not well known for char but may vary from 0.6 to 1 depending on the moisture and volatile release rate, and is about 0.9 for char.

Char Burnout

From Eq. 14.10, assuming carbon monoxide is formed at the surface and the oxygen density far from the particle surface is used (but the ∞ notation is dropped for convenience), the global char burnout rate is

$$\frac{dm_c}{dt} = -\pi d^2 (\tfrac{12}{16}) k_e \rho_{O_2} \tag{14.22}$$

As limiting cases, consider that the char burns out at constant diameter (with decreasing density) or at constant density (with decreasing diameter). For constant-diameter burnout, Eq. 14.22 may be integrated directly. For constant density note that $d = (6m/\pi\rho)^{1/3}$. Under conditions where diffusion is rate-limiting ($k_c \gg \tilde{h}_D$) associated with high temperature and large particles, and using Eq. 14.22,

$$\frac{dm_c}{dt} = -2\pi \left(\frac{6m_c}{\pi\rho_c}\right)^{1/3} \left(\frac{12}{16}\right) D_{AB} \rho_{O_2} \tag{14.23}$$

Under kinetic control ($\tilde{h}_D \gg k_c$) associated with low temperature and small particles, the constant-density char burnout becomes

$$\frac{dm_c}{dt} = -\pi \left(\frac{6m_c}{\pi\rho_c}\right)^{2/3} \left(\frac{12}{16}\right) k_c \rho_{O_2} \tag{14.24}$$

The char burnout time is obtained by integrating Eqs. 14.22 to 14.24 from the initial char mass to zero, and the following special cases are obtained:

Constant-diameter model:

$$t_c = \frac{\rho_{ci} d_i}{4.5 k_e \rho_{O_2}} \tag{14.25}$$

Constant density, diffusion control:

$$t_c = \frac{\rho_c d_i^2}{6 D_{AB} \rho_{O_2}} \tag{14.26}$$

Constant density, kinetic control:

$$t_c = \frac{\rho_c d_i}{1.5 k_c \rho_{O_2}} \tag{14.27}$$

When evaluating the binary diffusion coefficient, it is suggested to use the molecular diffusion coefficient for oxygen into nitrogen, which is given in Appendix B. Turbulence increases diffusion, and to account for turbulence it is suggested to increase the Sherwood number by using an appropriate root-mean-square turbulent velocity in the Reynolds number of Eq. 14.21.

A summary of total burn time versus coal particle diameter is given in Figure 14.5. These data cover many types of experiments with coals under a variety of combustion conditions. Also, ignition delay and total burn time data for several sizes of Douglas fir bark with 25% excess air and peak temperatures of 1400–1500 K are shown in Table 14.4. As seen by comparing Table 14.4 with Figure 14.5, a particle of bark will burn out roughly twice as fast as a coal particle of the same size. This is because bark has more volatiles and less char than coal.

Solutions of Eq. 14.22 for the constant-diameter mode and Eq. 14.23 for the constant-density mode are shown in Figure 14.6 assuming $k_p = 0.1$ g/(cm$^2 \cdot$ s $\cdot$ atm O_2) and also $k_c \rightarrow \infty$, which implies diffusion control of the burning rate. Also implied in this figure is that the particle Reynolds number is much less than 1. Figure 14.6 shows the approximate residence time required to burn out a 100-micron suspended char particle under four different assumptions.

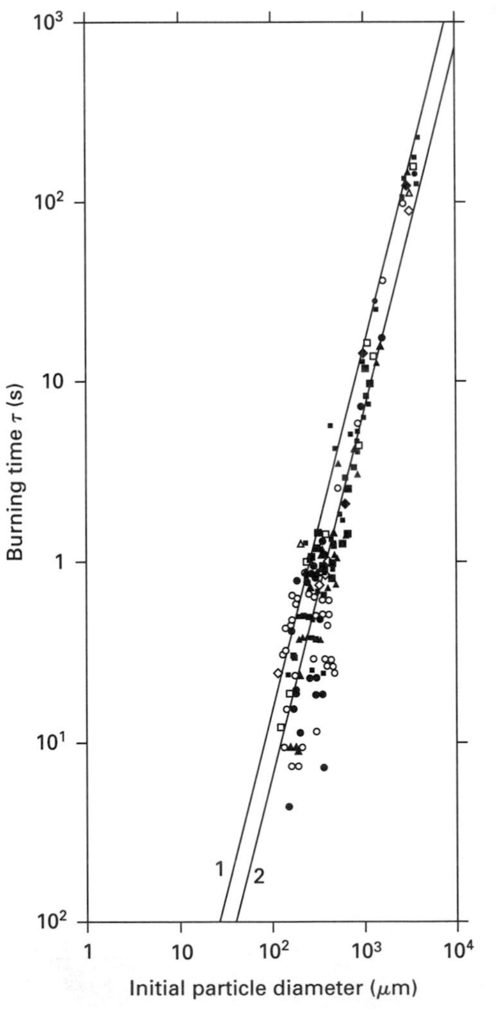

FIGURE 14.5
Single-particle burn time in air at 1 atm and 1500 K vs. particle size. Curves are calculated from Eq. 14.27; curve 1, $\rho = 2$ g/cm^3; curve 2, $\rho = 1$ g/cm^3 [Essenhigh, by permission of National Academy Press, originally from Mulcahy and Smith].

TABLE 14.4
Typical ignition delay and total burn times for Douglas fir bark particles

Particle size (μm)	Moisture (%)	Ignition delay (ms)	Burnout time (ms)
36	10	0	30
300	10	5	540
300	28	30	570
612	10	20	655

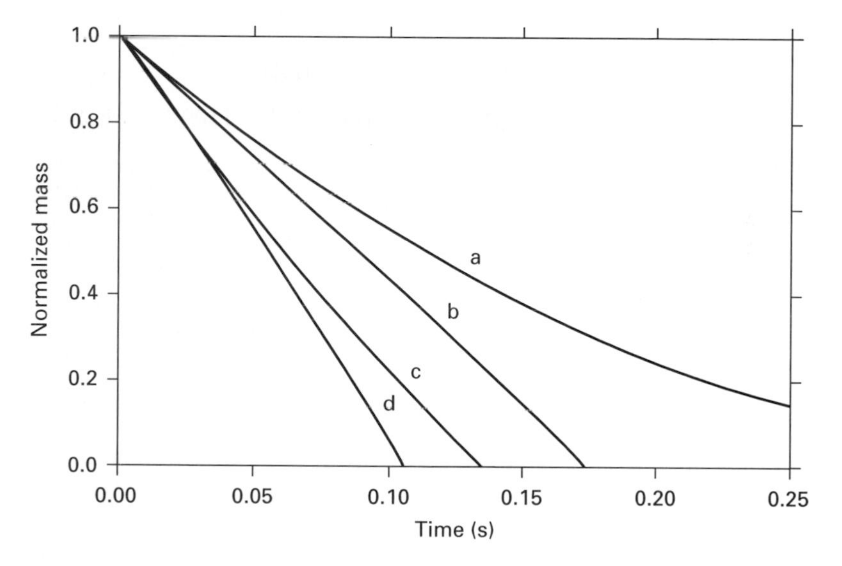

FIGURE 14.6
Burnout of 100-μm dia coal char particles at 1600 K, 10% oxygen, 1 atm pressure, initial char density 0.8 g/cm: (*a*) ρ = constant, $k_p = 0.1$ g/cm$^2 \cdot$s$\cdot$atm; (*b*) d = constant, $k_p = 0.1$; (*c*) ρ = constant, $k_e = \tilde{h}_D$; (*d*) d = constant, $k_e = \tilde{h}_D$.

Char Surface Temperature

When the surface of the char particle is in an oxidizing environment, the particle surface temperature is typically hotter than the surrounding gas temperature. A steady-state energy balance equates heat generation at the surface to heat loss by convection and radiation, neglecting heat loss by conduction:

$$\frac{dm_c}{dt}H_c = \tilde{h}A_p(T_p - T_g) + \sigma\epsilon A_p(T_p^4 - T_b^4) \tag{14.28}$$

The particle temperature is coupled to the burning rate. The heat of reaction of the char is based on reaction (*a*). For a small char particle the Nusselt number is 2. The particle temperature is determined for a specific case in the following example.

EXAMPLE 14.3. Determine the char particle temperature due to combustion of C to CO at the char surface for the following conditions: oxygen partial pressure 0.1 atm, particle size 100 μm, gas temperature 1700 K, background temperature 1500 K, particle emissivity 0.9, and char reaction rate constant of 0.071 g/(cm$^2\cdot$s$\cdot$atm O_2).

Solution. Consider Eq. 14.28. For the C-CO reaction, the heat of reaction is 9.2 kJ/g. For Nu = 2, and using a gas conductivity for air obtained from Appendix B,

$$\tilde{h} = \frac{2\tilde{k}_g}{d} = \frac{(2)(0.105\ \text{W/m}\cdot\text{K})}{100 \times 10^{-6}\ \text{m}} = 2.1\ \text{kW/m}^2\cdot\text{K}$$

Equating Eq. 14.22 and 14.28 and dividing by A_p yields

$$\begin{aligned}
&\tfrac{12}{16}[0.071\ \text{g/(cm}^2\cdot\text{s}\cdot\text{atm O}_2)](0.1\ \text{atm O}_2)(9.2\ \text{kJ/g}) \\
&\quad = [2.1 \times 10^{-4}\ \text{kW/(cm}^2\cdot\text{K)}](T_p - 1700\ \text{K}) \\
&\qquad + [5.67 \times 10^{-15}\ \text{kW/(cm}^2\cdot\text{K}^4)](0.9)(T_p^4 - 1500^4) \qquad \text{kJ/s}\cdot\text{cm}^2
\end{aligned}$$

Solving implicitly, T_p = 1800 K, which is 100 K above the gas temperature.

Ash Formation

The type and extent of the mineral matter in the solid fuel can influence the reaction rate. Mineral matter in biomass is less than a few percent, while in coal the mineral matter can range from a few percent to 50% or more for very low-grade coal. As the char burns, the minerals, which are dispersed as ions and submicron particles in the fuel, are converted to a layer of ash on the char surface. In high-temperature pulverized coal combustion, the ash tends to form hollow glassy spheres called *cenospheres*. At lower temperatures the ash tends to remain softer. The ash layer can have a significant effect on the heat capacity, radiative heat transfer, and catalytic surface reactions, as well as result in increased diffusive resistance to oxygen, especially late in the char burn stage. In combustion systems, the ash which is formed from the mineral matter can slag on radiant heat transfer surfaces and foul convective heat transfer surfaces if the particle temperature is too high. Slagging and fouling will be discussed in Chapter 16. The size and composition of the particulate emissions are influenced by the nature of the mineral matter and time-temperature history of the combustion.

14.4 SUMMARY

Solid fuel particles (pulverized, crushed, chipped, or cut) in a combustion environment undergo drying, devolatilization, and char burn. The rate of these processes depends on fuel type, fuel moisture content, size, and heat and mass transfer to the particle. For small particles, drying, devolatilization, and char burn occur in series, while for large particles these processes occur simultaneously. For small particles drying is the fastest step, and char burn takes much longer than devolatilization. For large fuel segments such as logs, char burn is the rate-limiting step, and char burn is slowed by evolution of the moisture. Devolatilization is a kinetic process which is often modeled as a first-order reaction. The volatiles contain H_2, CO, CO_2, H_2O, hydrocarbon gases, and tars which mix with oxygen and burn out in the gas phase. Char is a porous carbon matrix with a small amount of hydrogen interspersed with inorganic compounds. Char combustion involves diffusion of oxygen to the surface of the char and chemical reaction at the surface. An effective global rate constant was formulated which includes both diffusion and chemical kinetics (Eq. 14.11). Constant-diameter and constant-density char burnout models were given, see Eq. 14.22 and Eqs. 14.23 (diffusion control) and 14.24 (kinetic control), respectively. Due to the surface reaction the char surface is hotter than the surrounding gas. As the char burns, the tiny inorganic particles build up on the surface of the char, and ash particulates are formed.

PROBLEMS

14.1. A 100-μm bituminous coal particle initially at 300 K and containing 5% moisture (as-received) is inserted into a furnace at 1500 K. Find the time to reach the ignition temperature, assuming that the Nusselt number is 2. Neglect any pyrolysis which may occur.

14.2. Find the time required for (*a*) a 100-μm lignite particle and (*b*) a 500-μm lignite particle to reach 99% devolatilization assuming the particle is uniformly at (1) 1200 K and (2) 1500 K. Neglect the particle heat-up and drying time.

14.3. Find the time required to release 90% of the volatiles from a wood chip which is abruptly brought to a uniform temperature of (*a*) 900 K and (*b*) 1200 K.

14.4. A dry lignite coal particle, which consists of 50% volatiles, 40% char, and 10% ash, is abruptly brought to a uniform temperature of 1500 K. Calculate and plot the particle mass divided by the initial mass versus time during devolatilization.

14.5. A coal char with an apparent activation energy of 17.15 kcal/gmol and a preexponential constant ($k_{p,0}$) of 60 g/(cm$^2 \cdot$ s $\cdot$ atm O_2) is burned in a 1450-K and 1-atm gas stream containing 10% oxygen by volume. Assuming the particle temperature is 1450 K and the slip velocity between the particle and the gas is zero, calculate the char burn time for 10-, 100- , and 1000-μm particles. Consider the two limiting cases of constant-density and constant-diameter burning. Make a table of your results. Use a char specific gravity of 0.8.

14.6. Compare the burn time data of Table 14.4 for the 612- and 300-μm Douglas fir particles with calculated char burn times, assuming that diffusion is the rate-limiting step.

14.7. A 1-mm-dia particle of bituminous A coal char burns in a gas stream containing 5% oxygen at (*a*) 1500 K and (*b*) 1700 K. The slip velocity between the particle and the gas is 0.15 m/s. What is the char burn time, assuming constant char density? Assume ρ_c = 0.8 g/cm^3 and the particle temperature equals the gas temperature.

14.8. Derive Eq. 14.26 for the char burnout time. State all assumptions made.

14.9. For a 100-μm bituminous C char particle (see Table 14.3), find the oxygen partial pressure at the char surface when the gas temperature is 2000 K. Assume that the particle temperature is 100 K greater than the gas temperature, that there is no relative velocity between the particle and the gas, and that the oxygen partial pressure far from the particle is 0.1 atm. Repeat for a gas temperature of 1500 K.

14.10. A 50-μm-dia bituminous A char particle burns in product gas at 1 atm and 1600 K. As the particle burns to form carbon monoxide, the heat release is 2200 cal/g . The burning rate constant is 0.05 g/(cm$^2 \cdot$ s $\cdot$ atm O_2). Calculate the steady-state surface temperature of the particle for oxygen particle pressures of 0.06 atm, 0.13 atm, and 0.21 atm. Account for convection and radiation heat loss from the particle. Assume that the particle emissivity is 0.8, and use a Nusselt number of 2.

14.11. Consider two subbituminous C char particles in quiescent air at 1 atm and 2000 K. One is a 50-μm particle, and the other is a 200-μm particle. Compare the initial burning rates of the two particles.

14.12. For a lignite char particle in a 1500-K gas stream at 1 atm with a gas velocity of 1 m/s relative to the particle, what is the particle diameter for which the kinetic rate constant equals the diffusion reaction rate constant? Assume k_c = 24.4 cm/s and reaction order $n = 1$.

14.13. A wood char has a surface area (external plus internal) of 1000 m^2/g. If this were in the form of a solid sheet with a density of 2 g/cm^3, how thick would the sheet be?

REFERENCES

DiBlasi, C., "Modeling and Simulation of Combustion Processes of Charring and Non-Charring Solid Fuels," *Prog. Energy Comb. Sci.*, vol. 19, pp. 71–104, 1993.

Essenhigh, R. H., "Fundamentals of Coal Combustion," Chap. 19 in M. A. Elliott (ed.), *Chemistry of Coal Utilization,* Wiley-Interscience, New York, 1981.

Howard, J. B., "Fundamentals of Coal Pyrolysis and Hydro-pyrolysis," Chap. 12 in M. A. Elliott (ed.), *Chemistry of Coal Utilization,* Wiley-Interscience, New York, 1981.

Lyczkowski, R. W., and Chao, Y. T., "Comparison of Stefan Model with Two-Phase Model of Coal Drying," *Int. J. Heat Mass Transfer,* vol. 27, pp. 1157–1169, 1984.

Mulcahy, M. F. R. and Smith, I. W., "Kinetics of Combustion of Pulverized Coal," *Rev. Pure Appl. Chem.,* vol. 19, pp. 81–108, 1969.

Overend, R. P., Milne, T. A., and Mudge, L. K. (eds.), *Fundamentals of Thermochemical Biomass Conversion,* Elsevier, London, 1985.

Ragland, K. W., and Yang, J. T., "Combustion of Millimeter Sized Coal Particles in Convective Flow," *Combustion and Flame,* vol. 60, pp. 285–297, 1985.

Simmons, W. W., and Ragland, K. W., "Burning Rate of Millimeter Sized Wood Particles in a Furnace," *Comb. Sci. Tech.,* vol. 46, pp. 1–15, 1986.

Smith, I. W., "Combustion Rates of Coal Chars: A Review," *Nineteenth Symp. (Int.) on Combustion,* The Combustion Institute, Pittsburgh, pp. 1045–1066, 1982.

Smoot, L. D., and Smith, P. J., *Coal Combustion and Gasification,* Plenum Press, New York, 1985.

CHAPTER 15

Fixed-Bed Combustion

The traditional campfire is an example of fixed-bed combustion with natural convection of combustion air. The simple natural-draft pile burning method has low combustion intensity (heat output per unit volume) and suffers from poor control of the heat output and emissions. In order to increase the combustion intensity, improve control of heat output, and reduce emissions, forced convection and grates are used. Fans are used to force the air up through the fuel bed (updraft), or in other designs the air flows down through the fuel bed (downdraft).

In addition to the combustion itself, fuel handling and feeding, ash fouling and slagging, and gaseous and particulate emissions must be considered in any solid fuel combustor design. The nature of the combustion system dictates the required fuel preparation, and fixed-bed systems require the least fuel size reduction compared with suspension burning and fluidized beds. For biomass and refuse fuels this is an important consideration, because these fuels are difficult to pulverize compared with coal. Solid fuel handling and feeding are the focus of much effort compared with gas or liquid fuels. Considerations for reduced fouling and slagging and reduced emissions have led to many different types of combustion system designs. Even with the best of designs, however, extensive soot-blowing jets and external emission control equipment are needed when burning most solid fuels.

Several types of fixed-bed combustors for solid fuels are considered, including a representative wood stove, the Dutch oven, the spreader stoker, and several innovative downdraft systems. Generally, a particular type of furnace or boiler is designed for a particular type of solid fuel with a particular range of heating values, ash and moisture contents, and fuel sizes. These factors should be kept in mind when examining the various types of fixed-bed combustors, as well as suspension burners and fluidized-bed combustors.

15.1 WOOD STOVES AND DUTCH OVENS

The majority of residential wood stoves rely on natural circulation of air; there is no fan. The stoves typically burn logs or sticks and are for residential space heating. Although there are a great many designs currently available, a generic wood stove design which has several important features will be considered. The Dutch oven is typically implemented on a larger scale than the wood stove and uses a fan. The Dutch oven is used as a small industrial process heat source.

A typical residential stove for logs is shown in Figure 15.1. The logs rest on pipes which preheat the air and direct the air upward beneath the logs toward the secondary chamber, where the volatiles are more completely burned. An internal baffle helps to regulate the flow through the stove and create turbulence, which leads to mixing and more complete combustion. A bypass gate is used for startup. A catalytic converter, consisting of a ceramic honeycomb impregnated with a thin layer of noble metal, is sometimes used to further reduce organic emissions. The lower part of the firebox is insulated to increase the flame temperature, which increases combustion efficiency.

A large mass of noncombustible material such as stone, brick, or a water tank immediately adjacent to the stove may be used to store heat. In general, the smallest stove that will provide enough heat should be used. A large stove, burning fuel at the same rate as a small stove, will have more heat loss and produce more condensed organic compounds such as creosote.

The best fuel is hardwood which has been air-dried to reduce the moisture to 20% or less. Air-dried softwoods are slightly more difficult to burn cleanly due to their higher resin content. Logs of 10–15 cm diameter or larger are best. The larger pieces limit the devolatilization rate so that the combustibles can be completely burned by the available air supply within the stove. Wood which has too much

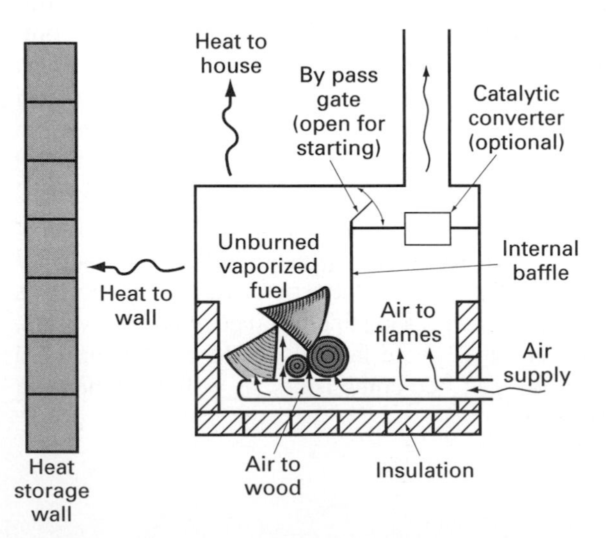

FIGURE 15.1
Representative residential wood stove.

moisture, such as freshly cut wood, will burn with too cool a flame, resulting in unburned organic matter and lower heat output. Wood which is too dry (kiln- or oven-dried) and wood which is too small in size will burn too fast, resulting in insufficient air and increased organic emissions and soot. Coal, treated wood, and refuse are not recommended in this type of system.

Design improvements for wood stoves include preheated secondary air, a somewhat hemispherically shaped combustion chamber, a catalytic flame shield, and an automatic thermostatically controlled damper. With these improvements it is possible to average less than 3 g/h of particulate emissions. For example, the state of Oregon has set a standard of 4 g/h for residential catalytic wood stoves and 9 g/h for noncatalytic stoves. This is a 75% reduction in emissions over conventional stoves. These levels can be reached with good design practice.

The rate of wood burning is related to the air supply. Adding new wood requires opening the dampers. For wood char the dampers may be reduced. Small, frequent additions of wood are preferred to many pieces all at once. For overnight operation when a large charge of wood is desired, charred wood should be accumulated within the stove over a period of several hours and then the damper turned down.

The Dutch oven type of furnace or boiler has traditionally been used to burn shredded wood, bark, sawdust, and other types of biomass for industrial process heat. The basic system is shown in Figure 15.2. Fuel falls by gravity into a conical pile on a grate in a refractory-lined chamber. Air is blown by a fan up through the grate, and fuel-rich combustion occurs in the pile. The incomplete combustion products flow around the arch to a second chamber, where secondary air from wall jets completes the combustion. Fuel particles tend to remain in the primary chamber if velocities are low in that section. Refractory walls above the fuel pile radiate

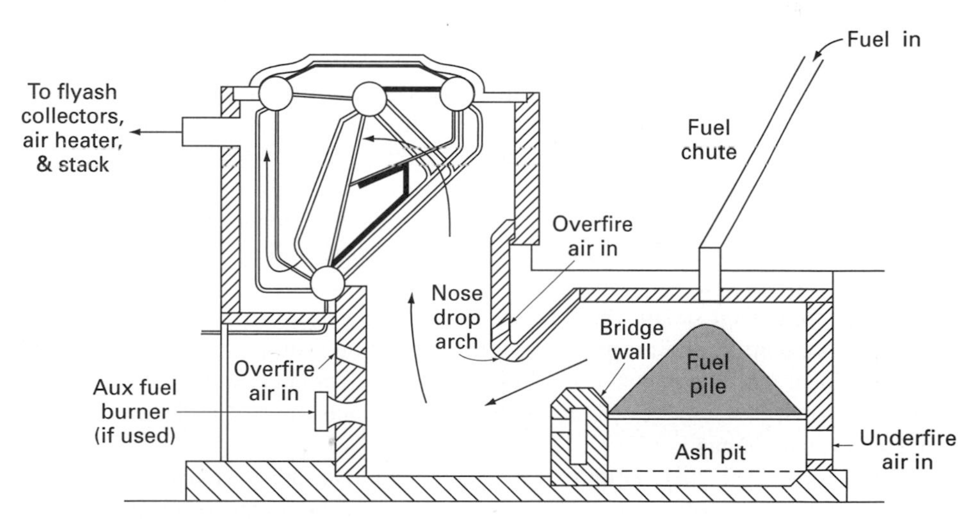

FIGURE 15.2
Dutch oven furnace and boiler.

heat, which assists in drying the upper layers of the fuel pile so that fuels with up to 50% moisture can be used. Combustion products then flow to a products-air or products-water heat exchanger.

Controlling the airflow rate in a Dutch oven is not easy, because the pressure drop across the bed depends on the height, size, and density of the fuel, which is continually changing. Wet fuel will cause a greater pressure drop and hence lower airflow rate for a given plenum pressure. Primary air is typically about 50% of the total air. The difficulty of controlling the air distribution, the slow response, and excessive emissions have made the Dutch oven less attractive for large-scale applications than other systems such as stokers and fluidized beds. However, a wide variety of solid fuels are successfully burned in Dutch ovens at low cost.

15.2 STOKER-FIRED BOILERS

A stoker type of furnace or boiler is an example of shallow-bed, updraft combustion. Shallow fixed-bed combustion systems require a continuous fuel feed system, which is referred to as a *stoker*. Air flows up through the grate and through the bed of ash, char, and fuel. The pressure drop across the bed tends to remain constant, because the bed depth is constant, and hence control of combustion air is easier than in a deep variable bed. Since the bed is thin, the pressure drop is less and the blower costs are reduced. Overfire air jets located in the walls above the bed are used to complete the combustion.

A great many types of stokers have been built over the years around the world. There are three main types of stokers based on the way the coal is fed onto the grate: overfeed, underfeed, and crossfeed stokers. Crushed coal up to 4 cm in size is used. Wood chips, peat, and refuse are also burned in stoker-fired boilers.

In the overfeed stoker, the flow of fuel and air is countercurrent. Fuel is fed onto the top of the bed and moves downward as it is consumed. Air flows up through the layers of ash, char, and fresh fuel. Volatile gases burn above the bed, and some fine fuel particles also burn above the bed. Overfire air is supplied to complete the combustion. Ash is removed by dumping, shaking, vibrating, or a continuously moving grate. Most overfeed systems use a spreader stoker, shown in Figure 15.3, where the fuel is mechanically thrown onto the grate. The bed is usually 10–20 cm deep. Fresh fuel is heated by the upward-moving gasses and by radiation from the flames above the bed.

In the underfeed stoker, the flow of fuel and air is upward. The evolved moisture, volatile matter, and air pass through the burning fuel layer. The bed is up to 1 m deep near the center. Fuel is fed by a ram feeder from below. There is less fines carryover than in the overfeed stoker. However, since the bed is deeper, the response to load change is slower.

In the crossfeed stoker (Figure 15.4), the fuel moves across a sloping vibrating grate, while the air flows upward through the grate. The underfire air is zoned to account for the varying amount of fuel. A refractory-lined arch above the initial part of the bed is often used for moist fuels such as refuse in order to ensure ignition. This type of system is often used for hard-to-feed fuels such as unprocessed refuse.

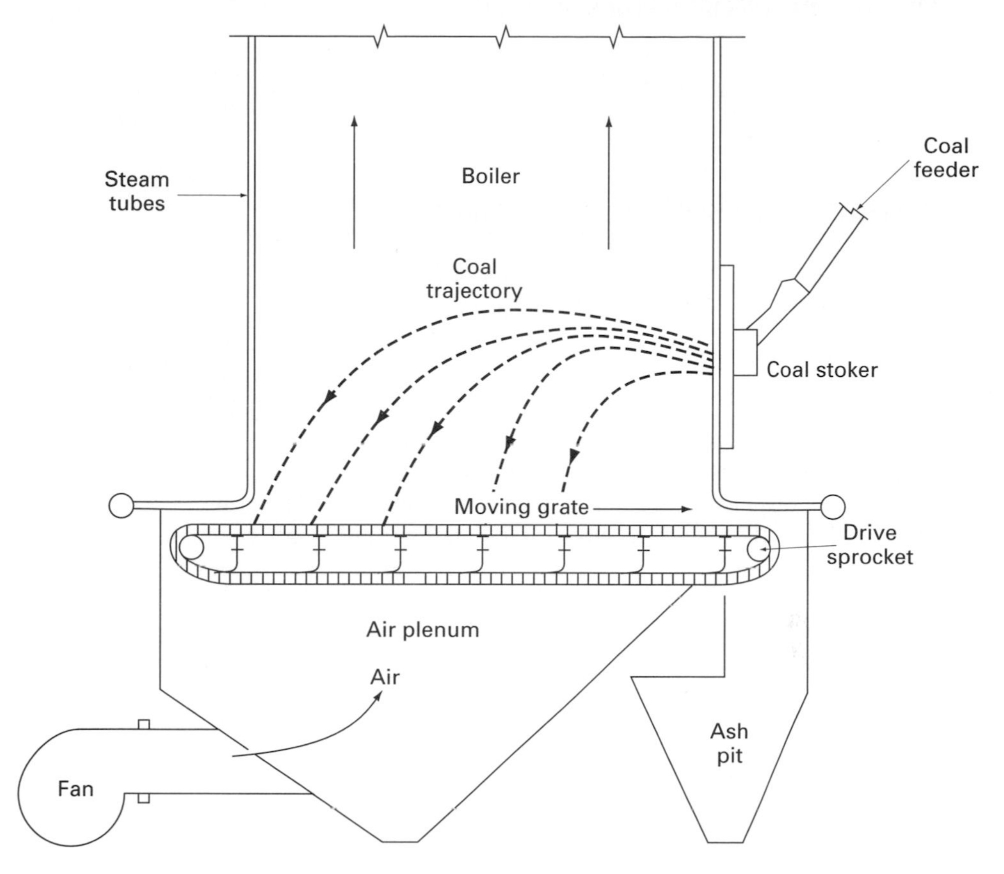

FIGURE 15.3
Spreader stoker boiler with traveling grate.

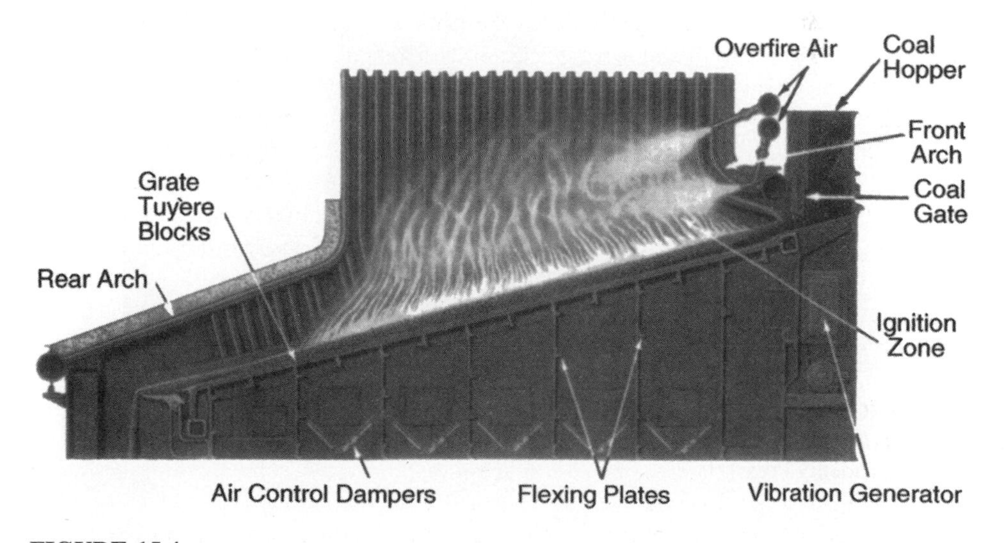

FIGURE 15.4
Front hopper fed water-cooled vibrating-grate stoker [Ceely and Daman, courtesy of National Academy Press].

Because the traveling-grate spreader stoker boiler is widely used, let us consider the combustion processes, combustion efficiency, and emissions in more detail.

Combustion in a Spreader Stoker Boiler

The spreader stoker boiler with travelling grate is shown in Figures 15.3 and 15.5. The stoker flings the coal toward the far end of the grate. The larger fuel particles ($\sim$ 25 mm) tend to go to the rear, while the midsize particles drop out in the middle section of the grate. The fine particles ($<$ 1 mm) are carried upward by the upward motion of the gases. Hence, the fines burn in suspension rather than on the grate. A representative coal size distribution for a spreader stoker is shown in Figure 15.6 Generally, a size distribution associated with the lower portion of the curve will result in less suspension burning. However, if a more rapid response to load changes is needed, then the size distribution associated with the higher portion of the curve is suggested. Washed stoker coal can be purchased with reduces the percentage of fines.

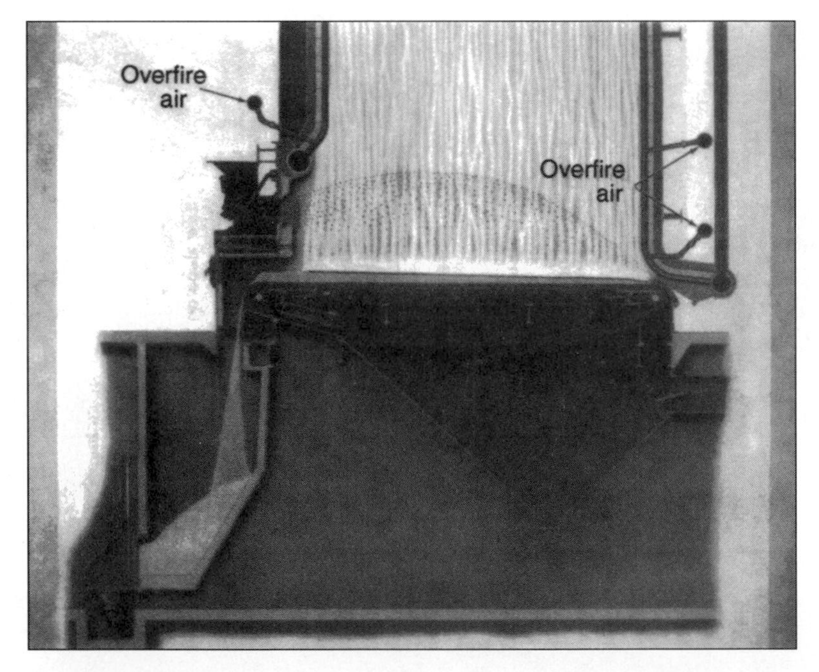

FIGURE 15.5
Traveling-grate spreader stoker [Ceely and Daman, courtesy of National Academy Press].

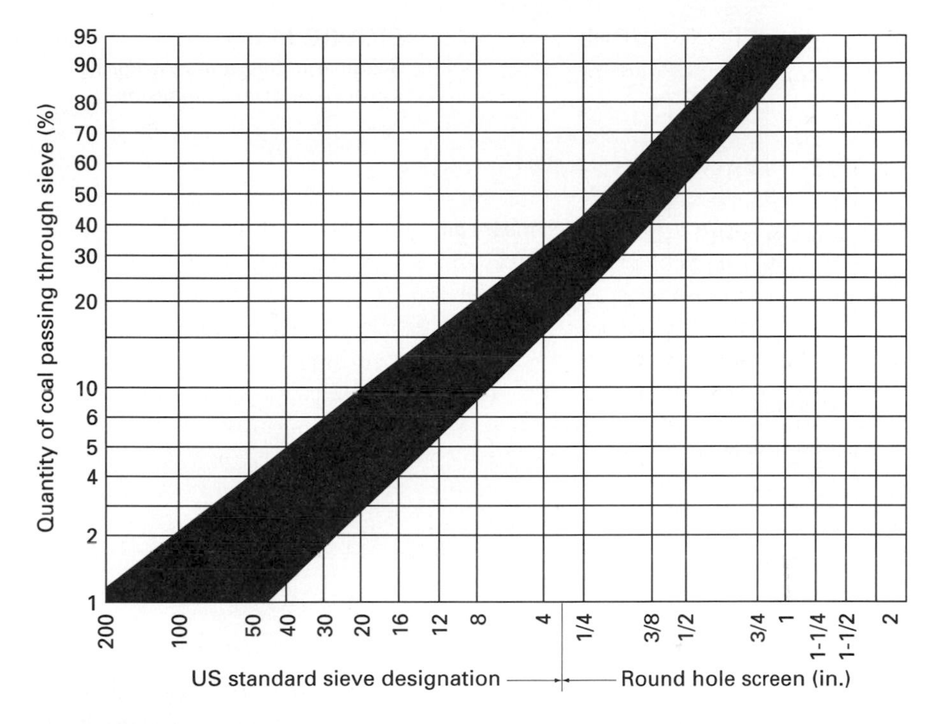

FIGURE 15.6
Typical size distribution of coal for spreader stokers.

The stoker feeding mechanism indicated in Figure 15.3 works as follows. The coal slides down the hopper onto a spill plate. A ram moves forward to push the coal over the spill plate and onto rotary blades, which fling the coal into the boiler. The feed rate is adjusted by the speed of the ram feeder, and the trajectory of the fuel particles is adjusted by setting the position of the spill plate and the speed of the rotor.

The endless grate moves forward at a rate of 1–6 m/h. The grate speed is adjusted so that the coal burns out before it reaches the edge of the grate, and the ash dumps into the ash pit below. When the coal feed rate is adjusted, the grate speed is also adjusted to control the thickness of the bed to a thickness of 10–20 cm. If the coal size or coal properties such as moisture or coal type change, causing the burning rate to change, the grate speed is adjusted accordingly. The grate is made of steel and has holes approximately 6 mm in diameter for the underfire (primary) air.

The fuel bed is shown schematically in Figure 15.7. At the top of the bed, fresh fuel receives heat from the convective flow of combustion products, which dries the top layer of fuel. Further heating of the fuel in the next layer down drives out the volatile matter consisting primarily of hydrocarbon vapors and tars, carbon monoxide, and hydrogen. There is little oxygen in the gas at this point, because it has been consumed in the bed below. The next layer down is the char layer. Initially, the char has very nearly the same size as the raw fuel except for certain so-called caking

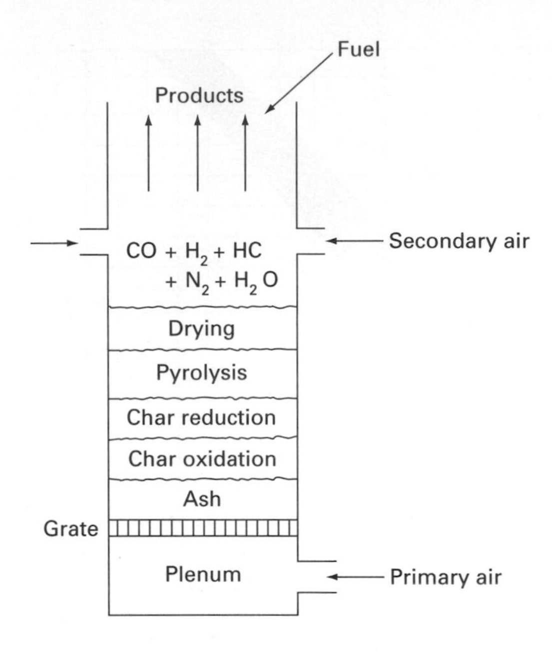

FIGURE 15.7
Schematic diagram showing layers of top-feed updraft combustor.

coals which swell during devolatization. Caking coals may fuse together as they swell, thus blocking the gas flow. Hence, strongly swelling coals should not be used in fixed beds unless there is some means to break up the cakes such as an underbed ram feeder.

The char which is in the lower part of the bed consists of a reducing zone and an oxidizing zone. In the oxidizing zone oxygen molecules penetrate to the surface of the char and form carbon monoxide and carbon dioxide:

$$C + \tfrac{1}{2}O_2 \rightarrow CO \tag{a}$$

$$C + O_2 \rightarrow CO_2 \tag{b}$$

and simultaneously the gas-phase oxidation of carbon monoxide occurs:

$$CO + \tfrac{1}{2}O_2 \rightarrow CO_2 \tag{c}$$

The oxygen is typically consumed within 5–10 cm above the grate, and above this there is a reducing zone in which carbon dioxide and water vapor are converted to carbon monoxide and hydrogen at the surface of the char:

$$C + CO_2 \rightarrow 2CO \tag{d}$$

$$C + H_2O \rightarrow H_2 + CO \tag{e}$$

Reactions (*d*) and (*e*) depend on the surface area, so that the thicker the bed, the more CO and H_2 are formed. Increasing the velocity of the air through the bed increases the burning rate of the char and reduces the amount of CO and H_2 formed but generally does not eliminate the reducing zone. Typical gaseous carbon species profiles are shown in Figure 15.8.

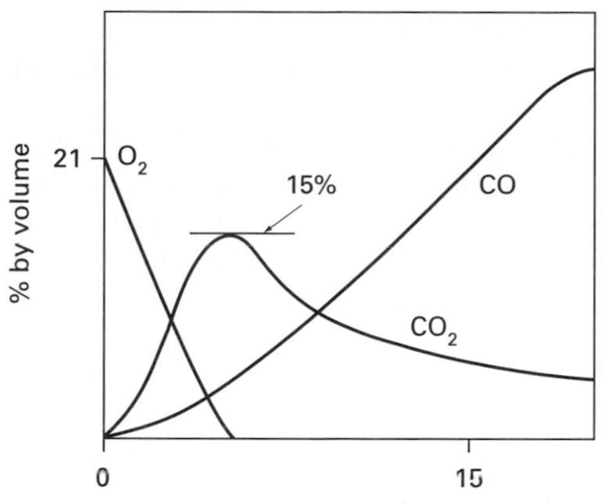

FIGURE 15.8
Typical gas analysis in char layer. [Thring, by permission of John Wiley & Sons].

The char first exposed to the air burns out and the ash is mostly retained on the grate, thereby insulating the grate from the highest temperatures. The temperature is highest in the char oxidation zone. If the ash temperature exceeds the ash softening temperature (about 1300°C for coal and 1200°C for wood ash), clinkers (agglomerated ash) will form. Thus, low–ash-fusion-temperature fuels must be used with caution in a fixed-bed combustor.

The undergrate airflow rate should be kept low enough so as not to blow ash, char, and fuel out of the bed and yet high enough for rapid burning. Increasing the undergrate air will increase the burning rate of the char and hence the heat output, but will also increase the particulate carryover. Overfire air jets above the bed, called overfire air, are used to complete combustion of the volatiles and carbon particles. From 40 to 60% of the heat release occurs above the bed where volatiles from the bed burn along with fuel fines which do not reach the bed. Overfire air jets which penetrate into the fireball and promote good mixing are vital to complete combustion.

The spreader stoker can achieve high heat release per unit area of grate, can use a wide range of solid fuels, and can follow changes in heat load requirements rapidly. Some problems are excessive smoke at low load, high particulate emissions at high loads, a tendency to burn too many fuel fines in suspension with resulting high carbon carryover, and difficulty in achieving good mixing with overfire air jets.

Combustion Efficiency, Furnace Efficiency, and Plant Efficiency

Combustion efficiency refers to the completeness of combustion. Thus, the combustion efficiency is 1 minus the quotient of heat loss due to incomplete combustion divided by the heat input to the boiler. In practice, carbon monoxide and unburned hydrocarbon emissions typically represent a negligible heat loss. However, combustibles in the bottom ash and fly ash can be significant. Field tests have shown that combustibles in the bottom ash of spreader stokers with traveling grate are 0–2% when operating properly. Combustible levels in the fly ash typically are higher in spreader stokers than in overfeed stokers and underfeed stokers. Proper design requires that the fuel remain on the grate through char burnout, which means that the underfire airflow rate must not be too high. In addition, carbon fines burning in suspension must have sufficient residence time in the furnace to burn out. Carbon fines burnout can be improved by careful design of the overfire air jets. Increasing overall excess air does not necessarily improve carbon burnout, and may in fact make it worse if the temperature is too low. Fly ash reinjection into the furnace from cyclone collectors located after the heat exchangers generally improves the carbon burnout, but at the expense of increased fine ash emissions.

Generally, solid fuel furnaces and boilers are run at slightly negative pressure to avoid leaks to the boiler room, and thus both a forced draft fan and an induced draft fan are used. As noted previously, heat exchangers are usually an integral part of the combustion chamber. The air heater preheats the air to about 200°C, which improves the combustion efficiency as well as the furnace efficiency by decreasing the exhaust temperature. To avoid corrosion the exhaust temperature should remain above the dewpoint of sulfuric acid (~160°C), if sulfur is present in the fuel.

Furnace efficiency is defined (in Chapter 6) as the useful heat out divided by the heat input. Referring to Figure 15.9

$$\eta = \frac{q}{\dot{m}_f \, \mathrm{HHV} + \dot{W}_a} \tag{6.7}$$

where $\dot{W}_a$ is the power input due to the fans. If the fuel flow rate is known, then the boiler efficiency may be determined directly from Eq. 6.7. Frequently the fuel flow rate is not measured, and then the indirect method given by Eq. 6.8 may be used. The indirect method is also useful for investigating how to improve the efficiency. In Chapter 6 we noted that the efficiency can be increased by reducing the excess air and reducing the extraneous heat loss. The excess air should be reduced to the point where the CO, combustibles in the fly ash, and stack opacity just begin to

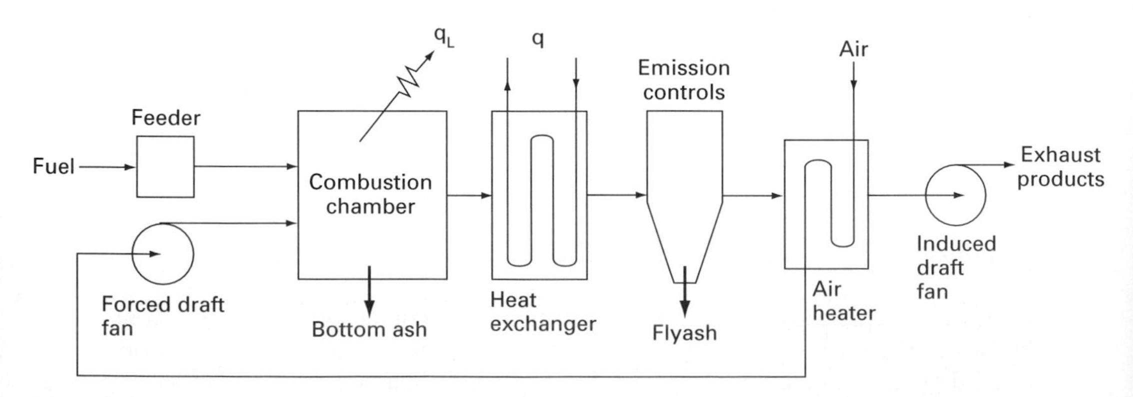

FIGURE 15.9
Schematic of solid fuel fired boiler system.

increase. Conversely, reducing heat loss due to combustibles in the ash may require increasing the excess air. Radiation heat loss from the walls of the boiler is an additional term which is needed in the indirect method of boiler efficiency determination, and is generally less than 1% for large boilers.

Excess air should be reduced as low as possible without increasing particulate emissions, and in current practice this means 25–35% at the furnace assuming minimal leakage around the edges of the grate, etc. Too much underfire air can cause ash and char particles to lift off the bed. On the other hand, too little air can cause clinkering of the ash on the bed. When the air is preheated, the amount of underfire air must be reduced slightly to avoid liftoff of bed particles.

Overfire air is necessary to burn out the volatiles, entrained fuel particles, and soot. Overfire air jets must have sufficient velocity to penetrate into the fireball and promote good mixing. Too much overfire air will cool the flame temperature, decrease the residence time, and entrain larger fuel particles. However, too little overfire air will not allow proper burnout of the particulates. Recent trends have been toward increased overfire air and decreased underfire air. Design of the overfire air system to ensure good penetration and mixing is done using computational fluid dynamics. Overfire air ranges from 25 to 50% of the total air depending on fuel type and furnace design.

Fly ash reinjection is used if the carbon carryover is high. In some extreme situations combustion efficiency can drop by as much as 7% if fly ash reinjection is not used. Separate nozzles just above the grate near the hottest part of the flame are used for reinjection. Even so, not all of the char may burn out on the second pass.

Plant efficiency refers to the overall thermodynamic efficiency of converting the energy in the fuel (based on the HHV) to useful power and/or useful heat. Thus the plant efficiency depends on the combustion efficiency and boiler efficiency, as well as the thermodynamic cycle efficiency and the power plant auxiliaries.

15.3 EMISSIONS FROM SPREADER STOKERS

Emissions of particulates, sulfur dioxide, and nitrogen oxides are of primary concern when burning coal in traveling-grate spreader stokers. Hydrocarbons and carbon monoxide emissions are usually low. Emission factors from field measurements of many units, which were obtained by the U.S. Environmental Protection Agency, are given in Table 15.1. High uncontrolled particulate emissions are due to the ash content in the fuel as well as combustibles in the fly ash, and sulfate and nitrate compounds which may be formed during combustion. In addition, particulate emissions depend on furnace design and operating conditions such as load (or grate heat release rate), air preheat, excess air, location and amount of overfire air, and fuel particle size. In spreader stokers, up to 80% of the ash is typically retained as bottom ash, as opposed to suspension burning and fluidized beds, which have little bottom ash because the ash is blown upward out of the combustor. Generally, higher heat release per unit area (higher load) results in higher particulate

TABLE 15.1
Emission factors for uncontrolled emissions from spreader stokers firing bituminous and subbituminous coal [EPA]

Pollutant	Emission factor (g/kg coal)
Particulates	30
Nitrogen oxides (as NO_2)	7
Sulfur dioxide	19.5 S
Carbon monoxide	2.5

S = weight percentage of sulfur in coal (5% sulfur means S = 5).

emissions but there are no clear relationships between all the variables. Best operation is obtained by experienced operators experimenting with the controls and equipment.

Sulfur in coal is emitted primarily as sulfur dioxide. Approximately 5% of the sulfur combines with the ash to form sulfate or forms sulfuric acid. Enhancement of sulfur dioxide capture by adding pulverized limestone with the coal or by forming pulverized coal and limestone into briquettes has not proved to be an efficient method. However, injection of pulverized limestone in the ductwork downstream of the superheater tubes and air heater has resulted in up to 50% conversion of sulfur dioxide to calcium sulfate, which is then removed by particulate control equipment.

Nitrogen oxide emissions consist of NO formed from fuel-bound nitrogen and atmospheric nitrogen. Only a few percent of the NO_x emissions are in the form of nitrogen dioxide and nitrate compounds. Nitrogen oxide emissions have been observed to decrease with decreasing excess air, as shown in Figure 15.10. Note that although the emissions are mostly NO, for regulatory purposes they are reported as NO_2 by multiplying the NO emissions by the ratio of the molecular weights (46/30). This is because NO is transformed to NO_2 in the atmosphere and NO_2 has health and environmental effects, whereas NO does not.

Flue gas recirculation has proved to be a successful way to reduce NO emissions in stoker boilers by reducing the excess air. A fraction of the exhaust gas is ducted back into the underfire air plenum and into the overfire air nozzles. For example, when a certain flow rate of air which yields 10% excess oxygen is replaced by the same amount of flow consisting 65% air and 35% flue gas, the excess O_2 level is reduced to 4%. This serves to reduce the NO formation and also to increase the efficiency. Additionally there is the possibility of decreasing the temperature in the char zone to avoid slagging of the ash. By increasing the overfire air without increasing the oxygen, the penetration and turbulent mixing produced by the overfire air is increased, thereby reducing combustibles in the fly ash in some cases.

Emission standards are set at the federal level for large sources ($> 250 \times 10^6$ Btu/h) and by state government for smaller sources. The emission standards, shown in Table 15.2, apply to all types of solid fuel–fired furnaces and boilers, not just stokers. In addition to the above standards, large new furnaces and boilers must

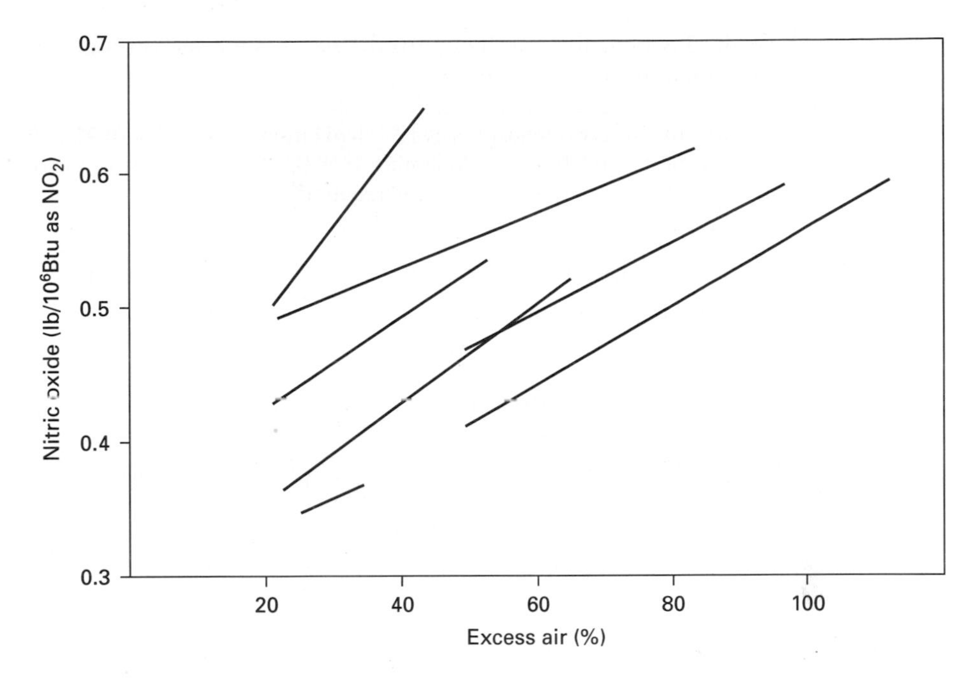

FIGURE 15.10
Nitric oxide emissions vs. excess air under full load conditions for seven spreader stokers [Langsjoen et al.].

TABLE 15.2
Federal new source emission standards for large coal-fired furnaces and boilers [EPA]

Pollutant	Emission standard
Particulates*	13 g/10^6 kJ
Sulfur dioxide	70–90% removal†
Nitrogen oxides as NO_2	260 g/10^6 kJ

*Also require 20% opacity.

†Require at least 70% removal up to emission rate of 560 g/10^6 kJ, then increased removal efficiency to maximum of 90%.

meet an opacity standard of 20% based on 6-min averages. Exceedances of 27% opacity for 6 min/h are allowed for soot blowing. *Opacity* refers to the light transmissivity of the stack gases, which is a measure of visibility degradation. The opacity standard may require more stringent control of particulates than the particulate standard, especially for fine particles.

Meeting the particulate and opacity standards typically requires the use of cyclones and baghouses or electrostatic precipitators. Meeting the sulfur dioxide emission standards requires a scrubber or the use of very-low-sulfur coal. The nitrogen oxide standard is typically met by controlling excess air and through the use of flue gas recirculation.

EXAMPLE 15.1. A large stoker spreader burns 12 T/h of bituminous coal with 9% ash and a heating value of 12,000 Btu/lb$_m$. Assuming that the typical emission factor applies, how much particulate emission control is required?

Solution. The heat input rate is

$$(12 \text{ T/h})(2000 \text{ lb}_m/\text{T})(12{,}000 \text{ Btu/lb}_m)(1.054 \text{ kJ/Btu}) = 303 \times 10^6 \text{ kJ/h}$$

Using Table 15.1, the particulate emission rate is

$$(30 \text{ g/kg})(12 \text{ T coal/h})(2000 \times 0.454 \text{ kg/T}) = 3.27 \times 10^5 \text{ g/h}$$

or

$$\frac{3.27 \times 10^5 \text{ g/h}}{303 \times 10^6 \text{ kJ/h}} = 1079 \text{ g}/10^6 \text{ kJ}$$

The percent control needed to meet the particulate standard of 13 g/10^6 kJ is

$$\eta_{\text{control}} = \left(\frac{1079 - 13}{1079}\right)100 = 98.8\%$$

This degree of control will require a fabric filter baghouse or an electrostatic precipitator.

15.4 MODELLING FIXED-BED COMBUSTION

Consider the top-feed updraft fixed bed of fuel shown in Figure 15.7. A bed of stoker-sized coal particles is supported by a grate. Air flows upward through the grate and the fuel bed. The coal flows slowly downward at a velocity $\underline{V}_s$ as the coal burns out in the lower layer of the bed. To simplify the situation let us use only a bed of char so that the pyrolysis and drying layers may be neglected. A one-dimensional model of the fuel bed is used to represent a vertical section out of a large fuel bed. Conservation of mass and energy for the solid and gas phases are needed. The product gas consists of nitrogen, oxygen, carbon monoxide, and carbon dioxide.

Oxygen reacts with char to produce CO in the lower portion of the bed. The CO reacts rapidly in the gas to form CO_2. The CO_2 in turn is reduced by the char. The latter reaction causes CO buildup when the oxygen is depleted. The rate constants

TABLE 15.3
Rate constants and heat of reaction in char bed

Reaction	k_{CO} (g/cm$^2\cdot$s$\cdot$atm)	$\hat{E}$ (kcal/mol)	H (MJ/kg)
1. $C + \frac{1}{2}O_2 \rightarrow CO$	(See section 14.3)	(See Section 14.3)	9.2 MJ/kg C
2. $CO + \frac{1}{2}O_2 \rightarrow CO_2$	(See section 4.4)	40	10.1 MJ/kg CO
3. $C + CO_2 \rightarrow 2CO$	2×10^5	50	−14.4 MJ/kg C
4. $C + H_2O \rightarrow H_2 + CO$	4×10^5	50	−10.9 MJ/kg C

and heat of reaction for the heterogeneous reactions are given in Table 15.3. The gas-phase reaction of CO is assumed to be instantaneous compared with the heterogeneous reactions in the following discussion. Also, in the following, water is not present, but it is included in the table for completeness.

Consider a one-dimensional slice dz thick through the bed of cross-sectional area A. From conservation of mass for the char, it follows that the net rate at which the char particles subside toward the grate equals the rate at which the char is consumed:

$$\frac{d}{dz}(\rho_s \underline{V}_s A_s)\, dz = -\frac{dm_c}{dt} \tag{15.1}$$

The char moves down because the char is shrinking as it burns. Introducing the bed void fraction ϵ, $A_s = A(1 - \epsilon)$. Assuming the void fraction remains constant, dividing by A, and introducing the particle area per unit volume A_v, the conservation of mass expression for the char becomes

$$(1 - 2\epsilon)\frac{d}{dz}(\rho_s \underline{V}_s) = r_1 + r_3 \tag{15.2}$$

where $r_1 = (12/16)k_{e1}\rho_{O_2}A_v$ and $r_3 = (12/44)k_{e3}\rho_{CO_2}A_v$, and where the subscripts 1 and 3 refer to reactions listed in Table 15.3.

The mean particle surface area per unit bed volume is

$$\overline{A}_v = \frac{6\overline{d}^2(1 - \epsilon)}{\overline{d}^3} = \frac{6(1 - \epsilon)}{\overline{d}} \tag{15.3}$$

where $\overline{d}$ is the mean diameter of the particles. If we assume that the char burns according to a constant-density shrinking-core model, then ρ_s is known and $\underline{V}_s$ is obtained from Eq. 15.2 once the temperature and species densities are known from the energy and species continuity equations.

In order to keep the analysis relatively simple, assume that the gas and particle temperatures are equal, and that energy transport by conduction, radiation, and diffusion is small compared with convective energy transport. Also, the sensible energy flux of the solids is small compared with that of the gas. The simplified energy equation is

$$\frac{d}{dz}(\rho_g \underline{V}_g h_g) = r_1 H_1 + r_2 H_2 + r_3 H_3 \tag{15.4}$$

where the heats of reaction are given in Table 15.3, r_1 and r_3 are defined above, and r_2 is given by Eq. 4.14. Note that H_3 is an endothermic reaction and hence has a negative heat of reaction. To allow for different temperatures between the gas and the solid, an energy equation for both phases must be written and convective heat transfer between the phases must be included using a Nusselt number appropriate to fixed beds [Wakao]. The effect of heat conduction and radiation should be included in a full analysis.

Species continuity equations for CO_2, O_2, and CO with the reactions shown in Table 15.3 are needed to obtain the individual densities:

$$\frac{d}{dz}(\rho_{CO_2}\underline{V}_g) = \frac{r_2 M_{CO_2}}{M_{CO}} - \frac{r_3 M_{CO_2}}{M_C} \tag{15.5}$$

$$\frac{d}{dz}(\rho_{O_2}\underline{V}_g) = -\frac{r_1 M_{O_2}}{M_C} - \frac{0.5 r_2 M_{O_2}}{M_{CO}} \tag{15.6}$$

$$\frac{d}{dz}(\rho_{CO}\underline{V}_g) = \frac{(r_1 + 2r_3) M_{CO}}{M_C} - r_2 \tag{15.7}$$

The solution procedure is to start at the grate, where the temperature and species densities are known, and numerically integrate the energy and species continuity equations. The overall mass flux of gas is obtained from

$$\rho_g \underline{V}_g = \rho_{CO_2}\underline{V}_g + \rho_{O_2}\underline{V}_g + \rho_{CO}\underline{V}_g + \rho_{N_2}\underline{V}_g \tag{15.8}$$

Since the gas density may be obtained from the equation of state $\rho_g = p/RT$, Eq. 15.8 determines $\underline{V}_g$. Once the temperature and species profiles are obtained, the downward velocity of the char $\underline{V}_s$ may be determined from Eq. 15.2. The heat input rate per unit area, which equals the heat output rate per unit area, is obtained from $(1 - \epsilon)\rho_c \underline{V}_s(\text{LHV})$.

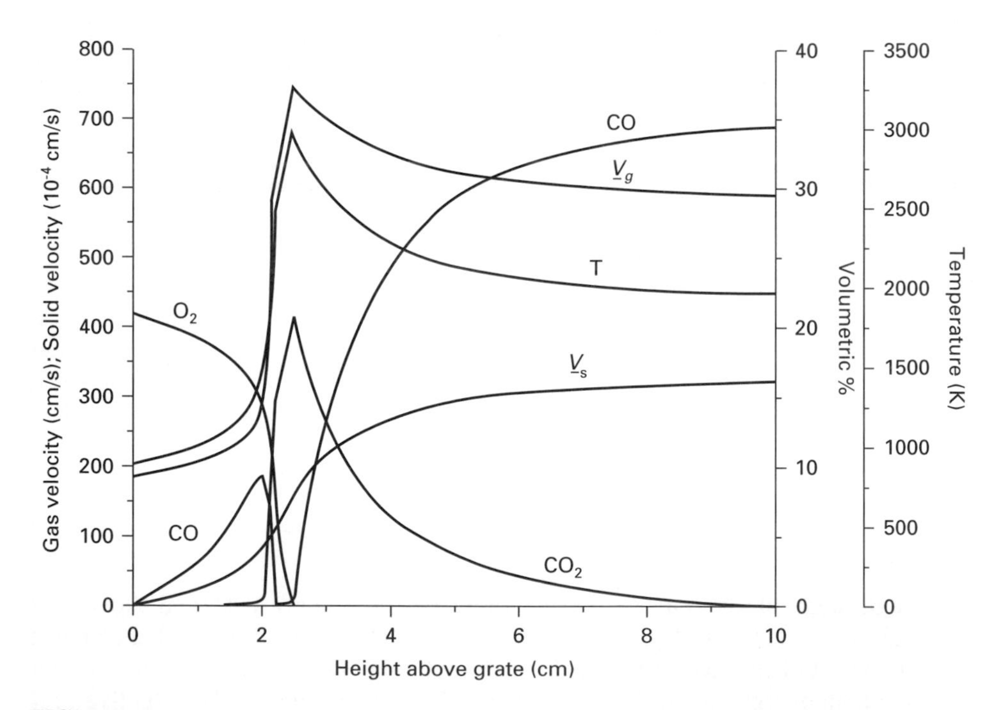

FIGURE 15.11
Updraft packed-bed combustion of coal char. Solution of Eqs. 15.1–15.8 with d (initial) = 1.0 cm, d (final) = 0.1 cm, $\rho_c = 0.8$ g/cm^3, $\epsilon = 0.4$, and $\underline{V}_g = 2$ m/s at 800 K.

Representative solutions for the set of equations 15.1–15.8 for T, x_{O_2}, x_{CO}, x_{CO_2}, and $\underline{V}_g$ are shown in Figure 15.11. The bed height is 10 cm, and the constant-density shrinking-sphere model is used. The temperature versus distance from the grate rises slowly at first, peaks rapidly, and then decreases at the point where the oxygen is consumed. The CO increases at first, then is converted to CO_2 as the temperature increases, and then rises again when the oxygen is depleted. The downward char velocity increases to 0.32 mm/s at the top of the bed, which represents a heat release rate of 2.5 MW/m^2 due to the char, when the overfire air is added. Note that the char-CO_2 reducing reaction adds to the solid velocity and thus the overall heat release rate, but to a lesser extent than the char-O_2 oxidation reaction. Increasing the airflow rate moves the peak temperature away from the grate and increases the heat release rate. Increasing the airflow rate too much may increase emissions of carbon fines. When heat conduction and radiation are included the temperature profile would be more rounded; and also when the inlet air preheat is below the ignition temperature, it can be shown that increasing the airflow rate too much will blow out the combustion [Goldman et al.].

15.5 REFUSE- AND BIOMASS-FIRED BOILERS

Municipal and industrial refuse and hogged wood waste are also burned in spreader stokers. This fuel is fed pneumatically into the furnace above the coal stokers. In addition to the feed air, more overfire air is needed because refuse and wood

TABLE 15.4
Example of industrial boiler using spreader stoker [Courtesy of Oscar Mayer Corp.]

Boiler	
Design steaming capacity	125,000 lb_m/h
Design pressure	425 psi
Final steam temperature	600°F
Boiler heating surface	13,865 ft^2
Waterwall heating surface	2540 ft^2
Economizer heating surface	8100 ft^2
Furnace volume	8150 ft^3
Stoker	
Number of feeders	6
Grate type	Continuous front discharge
Grate length (shaft to shaft)	18 ft $7\frac{1}{2}$ in.
Grate width	17 ft 8 in.
Effective grate area	248 ft^2
Recommended coal sizing	$1\frac{1}{4}$ in. to 10 mesh
Overfire air	
Upper rear wall	16 jets 1 in. dia located 10 ft above grate, 30° below horizontal
Lower rear wall	16 jets 1 in. dia located 18 in. above grate
Upper front wall	16 jets 1 in. dia 10 ft above grate, 35° below horizontal
Lower front wall	20 jets $\frac{7}{8}$ in. dia 1 ft above grate

contain more volatiles and less char than coal. Flue gas driers are often used to dry the wet fuel prior to the boiler. Fuel with greater than 45% moisture (as-received) cannot be burned in a conventional spreader stoker.

An example of a specific design for a boiler using a spreader stoker with a traveling grate is shown in Table 15.4 and Figure 15.12. This boiler is equipped to burn coal and refuse-derived fuel (RDF), as well as natural gas. Full load is 125,000 lb_m/h of superheated steam per hour. The vertical walls consist of waterfilled tubes for raising steam to the upper drums. Convective and superheat sections, consisting of large arrays of tubes exposed to the convective flow of combustion products, are located behind baffles before the economizer. The econ-

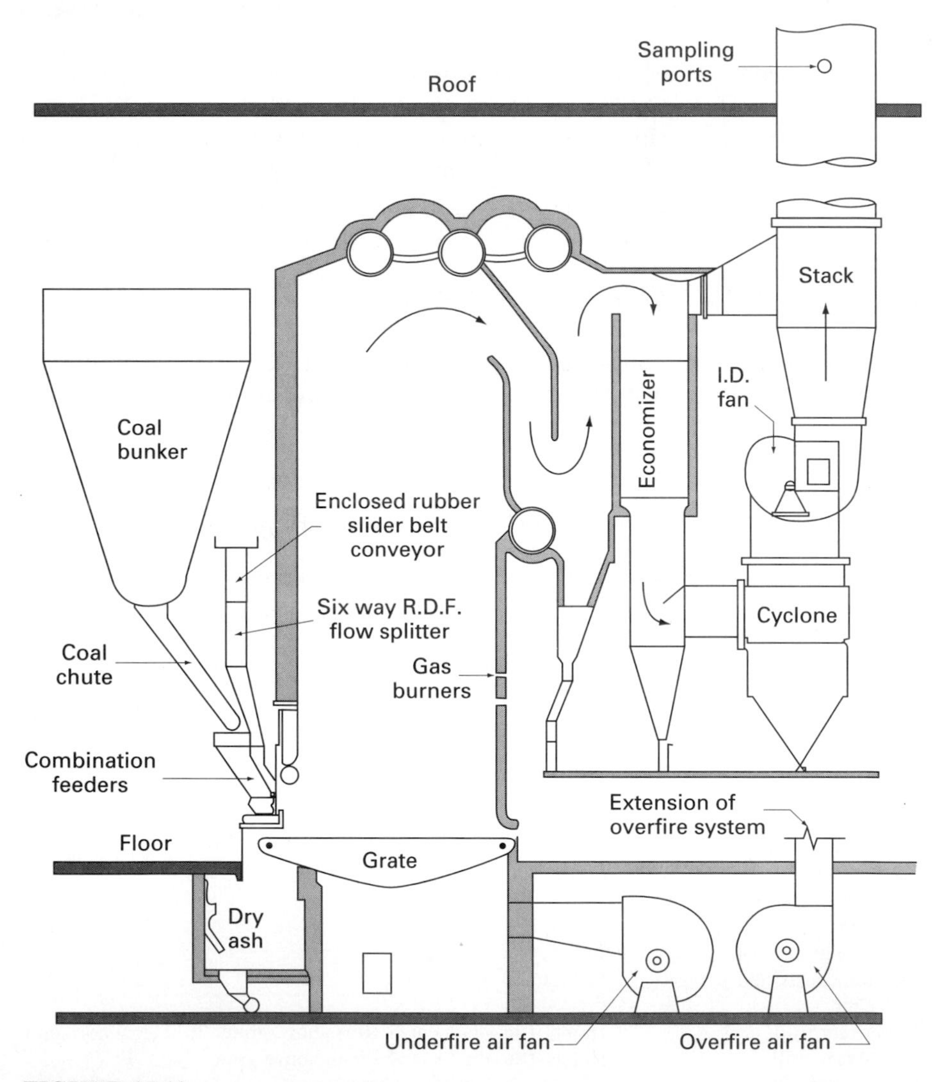

FIGURE 15.12
Traveling-grate spreader stoker boiler. Design parameters are given in Table 15.4.

omizer preheats the boiler feedwater before it flows to the waterwalls. The combustion products are then sent to a cyclone collector to control particulate emissions before flowing up the stack. There are three fans for this system: a low-pressure (~5 in. H_2O) forced draft fan for undergrate air, a high-pressure (~25 in. H_2O) forced draft fan for overfire, and an induced draft fan after the cyclone to maintain a few inches of water negative pressure in the furnace to prevent leaks into the powerhouse. Soot blowers, which are jets of steam, are used to clean the boiler tubes. Some units are also equipped with air preheaters which preheat the underfire air up to 450°F.

A traveling-grate spreader stoker designed specifically to burn high-moisture wood waste and wood chips is shown in Figure 15.13. The wood fuel with a top size

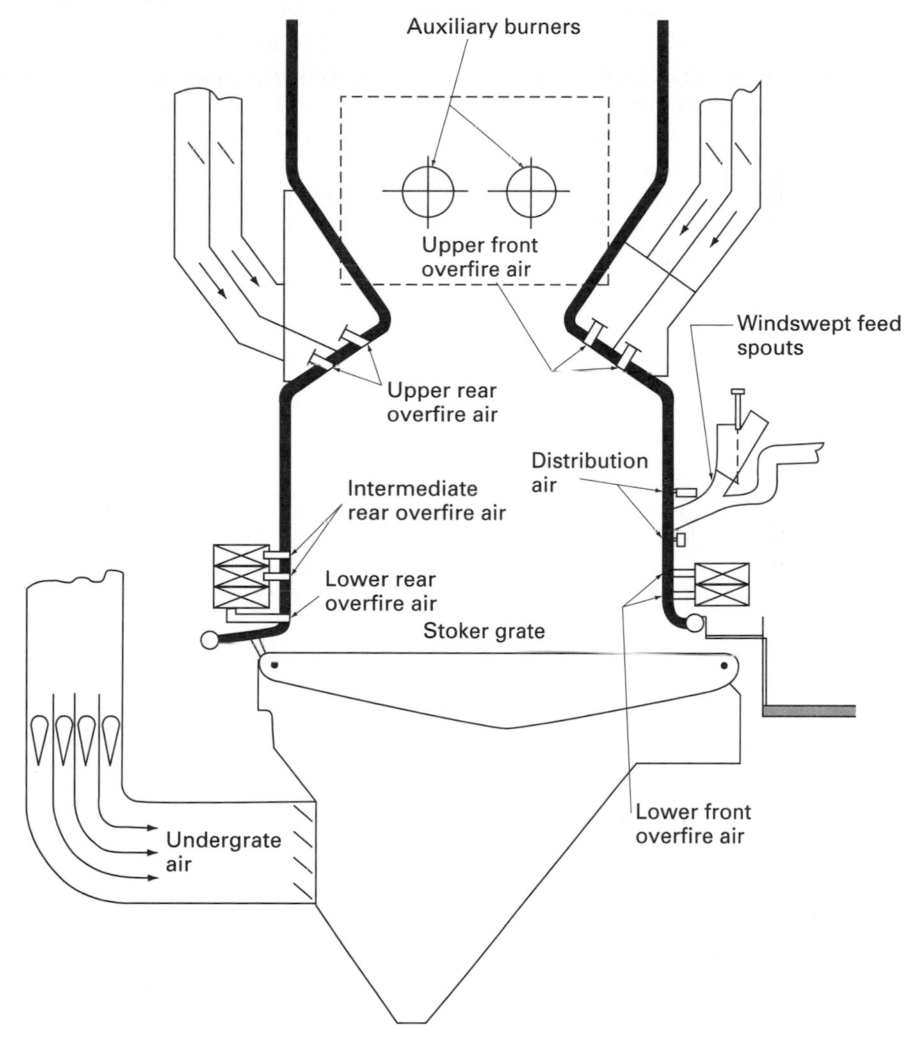

FIGURE 15.13
Spreader stoker designed for high-moisture fuels.

of 75 mm enters through the windswept spouts and is directed toward the rear of the grate. Arches above the grate reflect radiation back into the fireball and also provide a good location for overfire air jets, which promote good mixing. The radiation arches help to maintain sufficient temperature for complete combustion of wet wood.

Biomass specifically grown for energy such as short-rotation woody crops are a form of "closed-loop" biomass. No net carbon dioxide is added to the atmosphere, because the carbon in the wood comes entirely from the atmosphere. Genetically engineered trees, such as hybrid poplar, can grow 5–10 times faster than natural aspen trees. Hybrid poplar trees, which are harvested after 5–7 years, yield about 35 dry tons/acre. A 100-MW steam power plant requires 80,000–100,000 acres of dedicated land to sustain the power plant, which is about 2% of the land within a 50-mile radius. Typically the harvested wood is chipped before transport to the power plant; however, a new concept is to cut the tree only once at the base and transport and burn the trees whole. This technique saves the appreciable cost of chipping, allows for long-term storrage and drying without loss of volatiles, and has advantages for combustion.

The whole trees are transported from the drying dome to the boiler on a conveyor, and tree sections are fed with a ram feeder into a deep fixed-bed combustor. Preheated air flows up through a water-cooled grate which supports the tree sections (Figure 15.14). Since the fuel is large, the undergrate air velocity can be

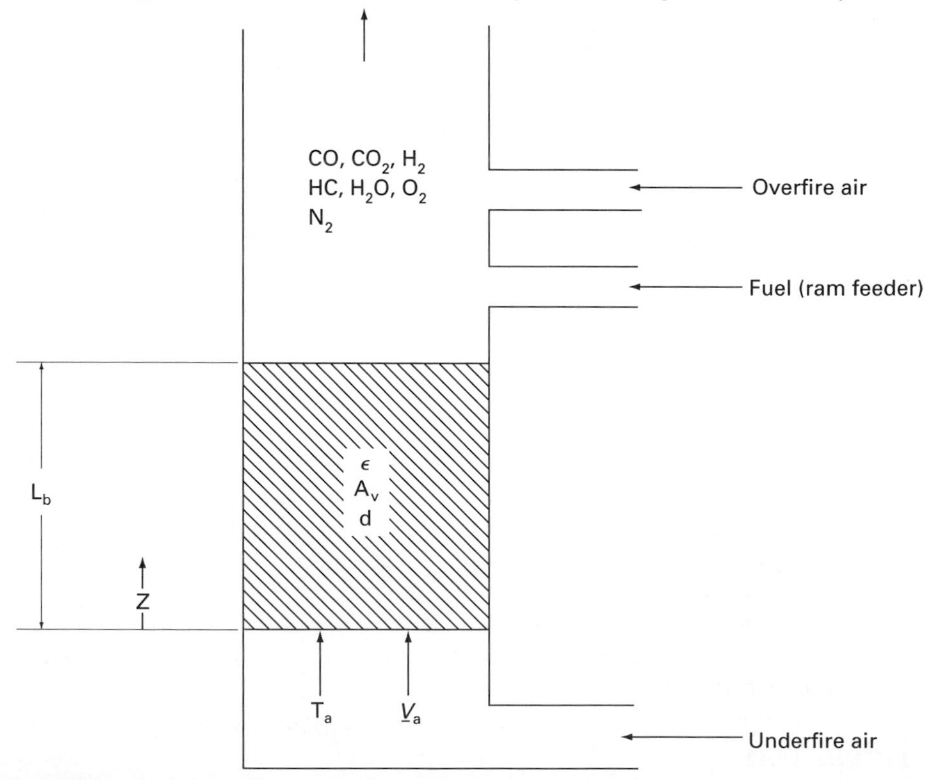

FIGURE 15.14
Schematic of deep fixed-bed updraft combustor.

high without blowing the fuel off the grate. The fuel next to the grate is char, and above this are tree sections with an external char layer and a core of undisturbed wood (see Figure 14.2), and at the top of the bed is the raw wood. Combustion takes place in the lower portion of the fuel bed, and the remainder of the fuel bed acts as a gasifier. Combustion is completed with overbed air. Heat release rates of over 3 million Btu/h · ft^2 have been obtained in a 12-ft-deep fuel bed, which is much higher than a conventional stoker-fired fixed-bed system using coal or wood chips. For a 100-MW (electrical) power plant which has a plant efficiency of 33%, the grate is about 28 ft long by 14 ft wide. Hence, the plan area is compact compared with the more conventional systems. New boiler designs using higher temperature and pressure steam may be able to improve the power plant efficiency to 45%. The use of higher-temperature steam in biomass-fired boilers than in coal-fired boilers may be feasible because the inorganic compounds in woody biomass are less corrosive than those in coal.

The deep fixed-bed combustor for the whole-tree system can be modeled using the equations described in Section 15.4 and also including the volatiles and the moisture. Some results are shown in Figure 15.15 for a maximum fuel diameter of 8 in. with 23% moisture as-received and 75° F inlet air temperature. The predicted gaseous species profiles in the fuel bed with an inlet air velocity of 12 ft/s are shown. The first 25% of the bed is an oxidizing region and the upper 75% is a reducing region. The hydrocarbons, carbon monoxide, and hydrogen which are formed in the reducing region burn out in the overfire air region. The fuel is devolatilizing in the upper 98% of the bed, and pure char exists only in the lowest 2% of the bed. The predicted gas and solid surface temperatures are shown in Figure 15.15. The solid surface temperature just above the grate is high because the high oxygen concentration is reacting with pure char. As the char surface reaction decreases and as the volatiles and moisture escape through the surface of the fuel, the surface temperature decreases rapidly, but then rises further above the grate due to mixing with the gaseous combustion products. At 3 ft above the grate the oxygen is consumed and the char reducing reactions decrease the temperatures. Overfire air is needed to complete combustion, and the predicted overfire air to underfire air ratio is 0.85.

Several different options are available to the designer and operator to meet the goal of producing a given heat release rate to meet a given load. Increasing the underfire air velocity causes a higher burning rate as the oxygen penetrates further into the bed and the convective heat and mass transfer to the fuel increases, thus increasing the heat output. Increasing the bed height lengthens the gasification zone and also increases the heat release rate, provided more overfire air is used. Increasing the underfire air preheat decreases the fuel burning rate for a fixed underfire air velocity, because less air mass flow is delivered to the fuel bed. Although the chemical reaction rates tend to be increased by increased temperature, the reactions in the bed are more limited by mass transfer than kinetics. Increasing the fuel moisture content decreases the devolatilization rate because the temperatures are lowered, thus lowering the heat release rate, all else remaining constant.

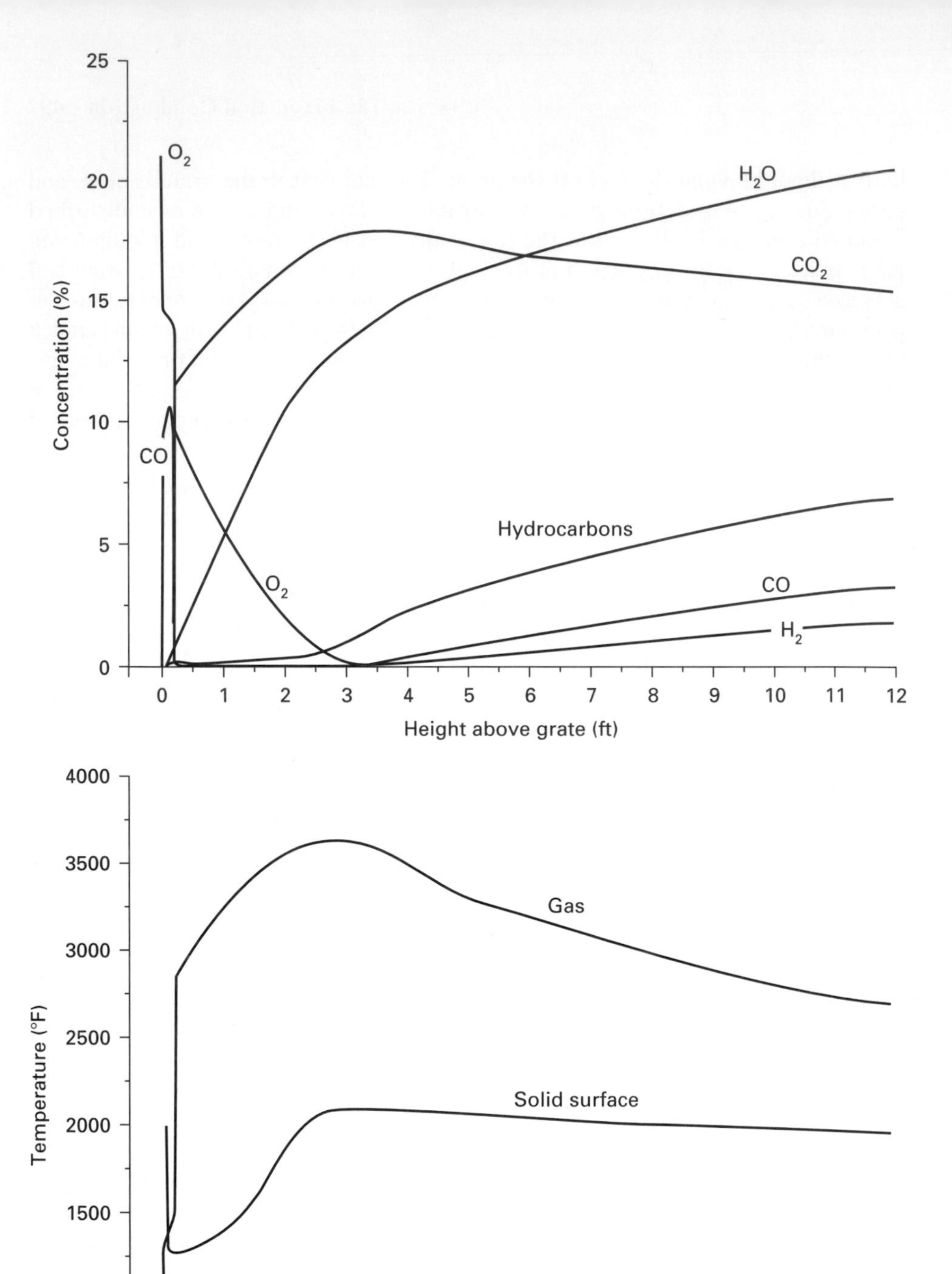

FIGURE 15.15
Model calculations of deep fixed-bed combustor using hybrid poplar: d (initial) = 8 in., $\rho_c = 0.8$ g/cm^3, $\epsilon = 0.4$, and $\underline{V}_g = 12$ ft/s at 750°F underfire air, initial fuel moisture = 23% (top) gaseous concentrations; (bottom) solid surface and gas temperatures [Reprinted with permission from Bryden and Ragland in *Energy and Fuels*, vol. 10, pp. 269–275, © 1996, American Chemical Society].

15.6 DOWNDRAFT SYSTEMS

Deep-bed combustion systems also have been developed for small applications such as residential and commercial users. Deep beds have the advantage of allowing intermittent fuel feed. Both downdraft and updraft designs have been developed, but downdraft systems have been shown to produce less particulate emissions. The design of a smokeless coal-burning residential furnace was achieved in the late 1940s, about the time when liquid fuels replaced coal for residential furnaces in the United States.

Figure 15.16 shows a cross-sectional view of a downdraft furnace called the *Rayburn smoke eater*. Primary air enters from the side and flows around a refractory wall and down through the fuel bed to a bottom rear arch where secondary air is mixed with the incompletely burned products. Startup is achieved by igniting a layer of coal on the grate by some means such as kindling, and then slowly increasing the depth of the bed. After a good fire and draft have been established, charging with fuel can proceed intermittently without disturbing the combustion. The combustion propagates upward to some equilibrium point; and as the coal burns, shrinking in size, it moves downward by gravity. The ash layer on the grate is removed by vibrating the grate. Combustion is completed by secondary air. A careful adjustment of primary air and secondary air must be made for proper operation without smoke and combustible emissions.

The unit operates in two levels of output: full output when the fan is operating, and low when the fan is off. The fan may be operated thermostatically, and it may be off for a day without extinguishing the fire. Crushed coal of low sulfur content and low to moderate swelling characteristics is most suitable. Some test results for the Rayburn smoke eater using a bituminous coal with a higher heating value of 13,030 Btu/lb_m are shown in Figure 15.17. The charging time was once every four hours. Unfortunately, the particulate emissions were not reported.

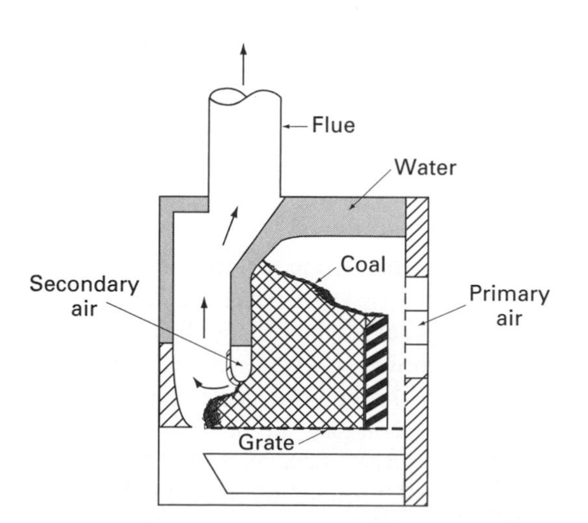

FIGURE 15.16
The Rayburn smoke eater [Ceely and Daman, courtesy of National Academy Press].

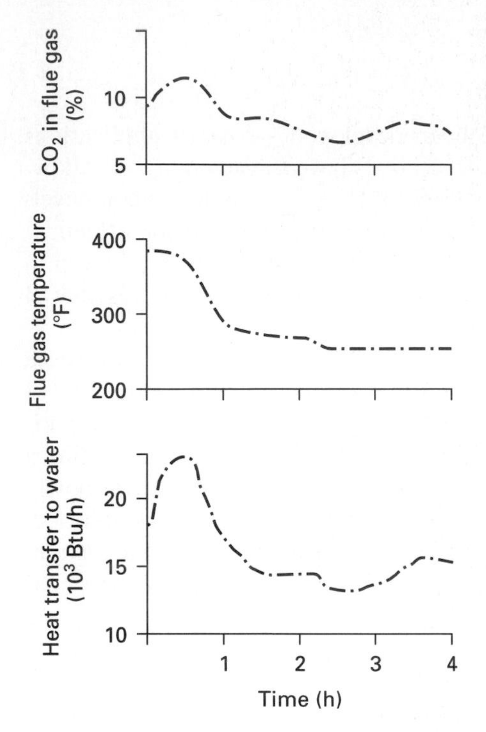

FIGURE 15.17
Tests of the Rayburn smoke eater [Ceely and Daman, courtesy of National Academy Press].

Downdraft combustion is subject to possible difficulty with ash removal and air blockage if the temperature exceeds the ash fusion temperature and clinkers (large ash particles) are formed. In addition to coal, briquettes of peat, sod, and brown coal have been successfully burned in downdraft furnaces. Renewed interest in solid fuels may spur the development of this type of combustion system because of the possibility for low particulate emissions.

The top-feed downdraft fixed-bed combustor is shown schematically in Figure 15.18. Five zones within the bed may be denoted. The highest temperature occurs somewhere in the center of the bed. Heat is transferred upward against the gas flow by conduction and radiation, thereby devolatizing and drying the top layer of fuel. The volatiles and tars, which are driven out of the solid fuel by heating, ignite and burn readily, since there is plenty of oxygen. Depending on the airflow rate and the temperature, the char will be consumed by oxidation and reduction or simply by reduction.

Due to the intense heat within the bed, the hydrocarbon vapors and tars which are emitted during devolatilization are cracked as they pass over the char so that the products are CO, H_2, and CH_4 and a small percentage of higher hydrocarbons. Carbon dioxide is reduced to CO. The solid bed moves slowly downward as the char shrinks. Ash is retained on the grate or blown through the grate. The partially burned combustion products are mixed with secondary air below the grate to complete combustion. The combustion products then flow to a heat exchanger. If the secondary air is not used and if minimal primary air is used, then this

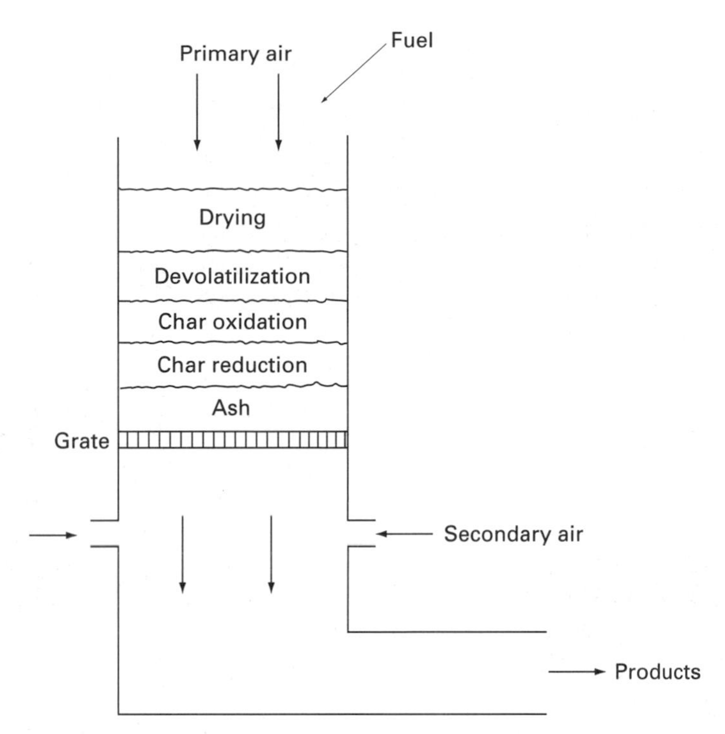

FIGURE 15.18
Schematic diagram of top-feed downdraft combustor.

configuration becomes a downdraft gasifier in which the object is to produce as much CO, H_2, and CH_4 as possible.

If the airflow rate is low so that heat transfer upward by radiation and conduction exceeds heat transfer downward by convection, then the combustion zone will tend to stabilize at the top of the bed. On the other hand, if the inlet airflow rate is high, convective heat transfer will drive the reaction zone to the bottom of the bed, and the devolatilization and char layers will be fairly thin. Since the char burning rate increases with increasing Reynolds number, higher airflow will result in higher heat output. However, even higher airflow will cause the temperature to decrease, thus causing lower heat output.

The downdraft combustion of a solid fuel bed is somewhat analogous to a laminar premixed gaseous flame in that flame propagation is due to heat conduction upstream against the flow. Continuing this analogy, flashback is when the flame moves upward to the top of the bed, and blowoff occurs at extreme airflow rates, which blow out the fire in the bed.

The main difficulty with downdraft systems is that when they operate with the proper airflow to produce high heat output per unit volume, the temperature of the char and ash is too high and slagging of the ash occurs. To avoid excessive temperature, the downdraft combustor must be run very lean, which means that for furnace applications the furnace efficiency is poor because of the high excess air.

In a gas turbine combustor, lean combustion is desirable, and this is a possible application for a pressurized downdraft combustor using solid fuels. Experiments have shown that lean pressurized downdraft combustion works for wood chips, but not for coal due to slagging of the ash. However, hot gas cleanup of the particulates and alkali in the combustion products is required prior to the gas turbine, which is a difficult task. The other alternative is to operate the downdraft system as a gasifier with relatively low airflow. The problem here has been channeling of the airflow in the fuel bed, which makes it diffucult to control the stoichiometry. Hot gas cleanup is still necessary for engine applications, and a water-cooled grate is required in either case.

15.7 SUMMARY

Fixed-bed combustion is used widely for coal, biomass, and waste fuels. Wood stoves employ fixed-bed combustion, and a well-designed stove burns cleanly. The overfeed, updraft, traveling-grate, spreader stoker boiler is used extensively throughout world for generating steam for heating and steam turbines. A well-designed spreader stoker boiler is durable and reliable. Overfire air wall jets are used to complete the combustion, and this limits the scale-up of fixed-bed combustors to large size utility boilers because good mixing cannot be achieved beyond a certain size. Up to 80% of the ash is retained as bottom ash. Flue gas recirculation is used to control NO_x emissions, and sulfur dioxide is controlled by limiting the amount of sulfur in the fuel. Hydrocarbon and carbon monoxide emissions are generally low.

When analyzing fixed bed combustion of relatively small particles such as crushed coal or woodchips, it is useful to visualize the combustion as occurring in layers. For an overfeed, updraft system the layers are (starting with the grate): ash, char, pyrolysis, drying, and volatile burnout with overfire air. A one-dimensional model of the reacting fixed bed of coal char was formulated and solved using conservation of mass and energy, and species continuity equations for CO, CO_2, and O_2. For a deep fixed bed of large sized fuel such as logs or tree sections the modeling is more complex but the basic idea is the same. Devolatilization and char burn take place simultaneously throughout most of the bed, and pure char combustion takes place only near the grate. Several novel fixed-bed downdraft combustors have been investigated but have not gained acceptance; when the air flow rate is low the reaction zone is extended, and when the air flow rate is high the reaction zone is thin.

PROBLEMS

15.1. What is the volume (liters) of wood chips required to heat 5 L of water from 20°C to 100°C? Assume the wood has 20% moisture as-received and assume 50% thermal efficiency based on the higher heating value. Repeat using dry bituminous crushed coal.

15.2. A wood stove, which has a heat output of 20,000 Btu/h, is found to emit 9 g/h of particulates. Using emission factors and other data given in previous chapters, how does this emission rate compare with a furnace running on (*a*) natural gas and (*b*) no. 2 fuel oil?

15.3. A wood stove has a heat output of 35,000 Btu/h and an efficiency of 60%. How many pounds of wood is needed in a 4-h period, assuming oak with 20% moisture (dry basis)? What is the volume occupied by this charge of wood?

15.4. Estimate the required ratio of underfire air to overfire air for top-feed updraft fixed-bed combustion under the assumption that a stoichiometric amount of underfire air is used to burn the char to completion while stoichiometric overfire air is used for the volatiles. Use lignite coal with the following analysis:

Proximate analysis (as-received)	
Moisture	35% (wt)
Volatile matter	24%
Fixed carbon	29%
Ash	12%
Ultimate analysis (dry, ash-free)	
C	75% (wt)
H	6%
O	19%
HHV (dry, ash-free)	12,570 Btu/lb_m

15.5. For Problem 15.4 calculate the maximum possible temperature in the bed assuming no radiation or conductive heat loss within the bed or from the bed. Assume air and coal enter at 77°F. Assume the char is carbon for this calculation. (*a*) Work the problem assuming no dissociation using Appendix C. (*b*) Run on STANJAN including dissociation and compare the results.

15.6. For top-feed downdraft fixed-bed combustion, estimate the required primary air to reach stoichiometric at the end of devolatilization. Assume complete combustion at that point. Also estimate the temperature at that point assuming no heat loss. Use the coal of Problem 15.4. The air and coal enter at 77°F.

15.7. The coal of Problem 15.4 is burned to completion on a traveling-grate spreader stoker with 20% excess air. If flue gas recirculation is used at the rate of 35%, what is the excess O_2% in the the flue gas, and how many moles of products per 100 g of coal at the combustor outlet are formed?

15.8. Refuse-derived fuel (RDF) is burned to completion in a spreader stoker boiler at the rate of 10,000 kg/h with 30% excess air. Stack gases are emitted at 200°C and 1 atm pressure. For the as-received fuel analysis below find the following:
(*a*) Volume % of CO_2, H_2O, O_2, and N_2 in stack gas.
(*b*) Concentration (ppm) of SO_2, HCl, and NO (from fuel N only) in stack gas.

(*c*) Volumetric flow rate (m^3/h) of stack gases.
(*d*) Uncontrolled emission rate of SO_2, HCl, and NO in g/s.
(*e*) Assuming that all the lead is vaporized and then recondenses on the fly ash, that 65% of the ash is fly ash, and that 85% of the fly ash is controlled, what are the emissions of lead (g/s)?

RDF analysis	**% (by weight)**
Moisture	20.90
Carbon	33.80
Hydrogen	4.50
Nitrogen	0.42
Chlorine	0.31
Sulfur	0.21
Ash	13.63
Oxygen	26.20
Lead	0.03
Higher heading value	13,440 kJ/kg

15.9. Show that for spherical particles, $A_v = 6(1 - \epsilon)/d$ and $\text{Re} = 6\ \rho_g \underline{V}_g / A_v \mu$. Define each term.

REFERENCES

"ASME Power Test Code for Steam Generating Units," PTC, 4.1, 1973, American Society of Mechanical Engineers, New York, 1973.

Bryden, K. M., and Ragland, K. W., "Numerical Modeling of a Deep, Fixed Bed Combustor," *Energy and Fuels,* vol. 10, pp. 269–275, 1996.

Ceely, F. J., and Daman, E. L., "Combustion Process Technology," chap. 20 in M. A. Elliott (ed.), *Chemistry of Coal Utilization,* Wiley-Interscience, New York, 1981.

EPA, "Compilation of Air Pollutant Emission Factors, Vol. 1: Stationary Point and Area Sources," AP-42, U.S. Environmental Protection Agency, 1985.

Goldman, J., Xieu, D., Oko, A., Milne, R., and Essenhigh, R. H., "A Comparison of Prediction and Experiment in the Gasification of Anthracite in Air and Oxygen-Enriched Steam Mixtures," *Twentieth Symp. (Int.) on Combustion,* The Combustion Institute, Pittsburgh, pp. 1365–1372, 1984.

Kammen, D. M., "Cookstoves for the Developing World," *Scientific American,* pp. 72–75, July, 1995.

Kowalczyk, J. F., and Tombleson, B. J., "Oregon's Woodstove Certification Program," *J. Air Poll. Cont. Assoc.,* vol. 6, pp. 619–625, 1985.

Langsjoen, P. L., Burlingame, J. O., and Gabrielson, J. E., "Emissions and Efficiency Performance of Industrial Coal Stoker Fired Boilers," EPA-600/7-81-11a, 1981.

McGowin, C., "Proceedings: Strategic Benefits of Biomass and Waste Fuels," Electric Power Research Institute report TR-103146, Palo Alto, CA, 1993.

Ostlie, L. D., Schaller, B. J., Ragland, K. W., Bryden, K. M., and Wiltsee, G. A., "Whole Tree Energy™ Design," vols. 1–3, EPRI report TR-101564, December 1993.

Purnomo, Aerts, D. J., and Ragland, K. W., "Pressurized Downdraft Combustion of Woodchips," *Twenty-Third Symp. (Int.) on Combustion,* The Combustion Institute, Pittsburgh, pp. 1025–1032, 1990.

Reed, T. B., and Das, D., "Handbook of Biomass Downdraft Gasifier Engine Systems," Solar Energy Research Inst. report SERI/SP-271-3022, 1988.

Singer, J. G. (ed.), *Combustion: Fossil Power Systems,* Combustion Engineering Inc., 1981.

Smoot, L. D. (ed.), *Fundamentals of Coal Combustion for Clean and Efficient Use,* Elsevier Science, 1993.

Thring, M.W., *The Science of Flames and Furnaces,* John Wiley & Sons, 2nd Ed. 1962.

Wakao, N., and Kaguei, S., *Heat and Mass Transfer in Packed Beds,* Gordon & Breach, New York, 1982.

CHAPTER 16

Suspension Burning

Suspension-burning furnaces burn pulverized fuel particles which are blown through burner nozzles into a furnace volume which is large enough to allow burnout of the fuel char. In addition to pulverized coal, pulverized biomass and shredded refuse-derived fuel (RDF) are also burned in suspension-fired furnaces and boilers. Coal is relatively easy to pulverize, since it is brittle, whereas biomass, being fibrous, is more difficult to pulverize. Much of the electricity in the world today is generated by pulverized coal–fired steam power plants.

A suspension-burning steam power plant is shown schematically in Figure 16.1. As-received coal is fed from bunkers or bins into pulverizers. From the pulverizers the face powder–sized fuel is piped with air to burners, where the fuel

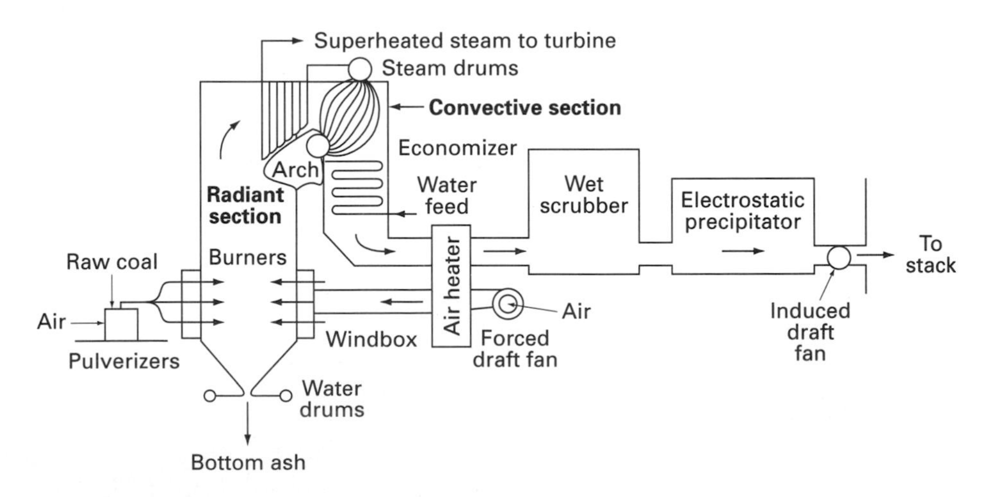

FIGURE 16.1
Utility-scale pulverized coal combustion system.

is mixed with preheated air from the *windbox* (air plenum). The burners stabilize the volatile flame, and the char burns out in the radiant section of the furnace. The furnace walls are made of alloy steel tubes with narrow steel strips welded between the tubes to form gastight walls. The incoming water is pumped to high pressure, preheated in the economizer, and sent to the lower steam drum and through downcomers to headers at the bottom of the furnace. The water is converted to a steam-water mixture in the waterwall tubes of the radiant section, and the wet steam rises to the upper drum. In the upper drum the two-phase mixture is separated; the steam flows to the superheater tubes in the convective section, while the water drains to the lower drum and rises as wet steam in the convective boiler tubes. The superheated steam flows to a steam turbine generator. For supercritical steam boilers (for steam pressures above 218 atm), steam drums are not used. On the gas side, the flue gases flow past the convective tubes, the economizer tubes, and the air heater to a wet scrubber to control sulfur dioxide emissions and then to an electrostatic precipitator or fabric filter baghouse for particulate emissions control (fly ash and unburned carbon). Induced draft fans before the stack and forced draft inlet air fans overcome the system pressure drop and maintain a slightly negative pressure in the furnace as insurance against leaks.

The advantages of suspension burning over fixed-bed combustion are that the system can be scaled up to very large sizes and is more responsive to required load swings. Suspension-burning systems are more complex than fixed-bed systems due to the pulverizers and more responsive air controls. Most coal-fired industrial steam generators with a capacity greater than 100,000 kg steam/h and large coal-fired electric utility units utilize suspension burning rather than fixed-bed stokers. Single boilers with a capacity of 4,500,000 kg steam/h burning more than 500 tons of coal/h are now in operation.

Operating steam pressures in these large units range from 150 to 300 atm, and the maximum steam temperature is 540–590°C. A higher steam temperature would increase the plant efficiency, but the steam temperature has been limited primarily by corrosion of the boiler tubes due to sulfates and chlorides in the hot combustion products. While the design inlet temperature to gas turbines has increased steadily in recent years and now exceeds 1200°C, the inlet steam temperature to utility steam turbines has remained in the 540–590°C range for the last 35 years. The plant efficiency for modern power plants with dual reheat and multiple feedwater heaters is 33–35%. As new boiler tube and steam turbine materials are utilized, it is to be expected that the steam temperatures will gradually rise and take full advantage of the combustion temperatures which are produced, and thereby push the power plant efficiency up to 45%.

16.1 PULVERIZED COAL–BURNING SYSTEMS

The pulverizer is typically a type of ball mill, the larger of which can handle up to 100 tons coal/h. Air heated to about 340°C or more is blown through the pulverizer to dry the coal particles and convey them to the burner located in the furnace wall.

The coal emerges from the pulverizer at 50–100°C. Carbon monoxide is monitored in the pulverizer to prevent explosion. The conveying lines should have velocities greater than 15 m/s to avoid settling of the pulverized coal, and right angles should be avoided to minimize erosion of the tubes.

Pulverized coal is ground to face powder size to ensure rapid combustion. The mass mean size is about 50 μm, which is roughly the diameter of a human hair. Essentially all of the particles are less than 300 μm. The size distribution follows the Rosin-Rammler distribution:

$$y_i = 1 - \exp\left[-\left(\frac{d_i}{d_0}\right)^q\right] \tag{16.1}$$

where y_i is the mass fraction less than size d_i. The parameters d_0 and q are obtained from experiment. Note that when $d_i = d_0$, $y_i = 0.632$, so that d_0 is the size such that 63.2% of the particles are smaller than d_0. A typical size distribution curve is shown in Figure 16.2. For this figure $d_0 = 60$ μm and $q = 1.4$.

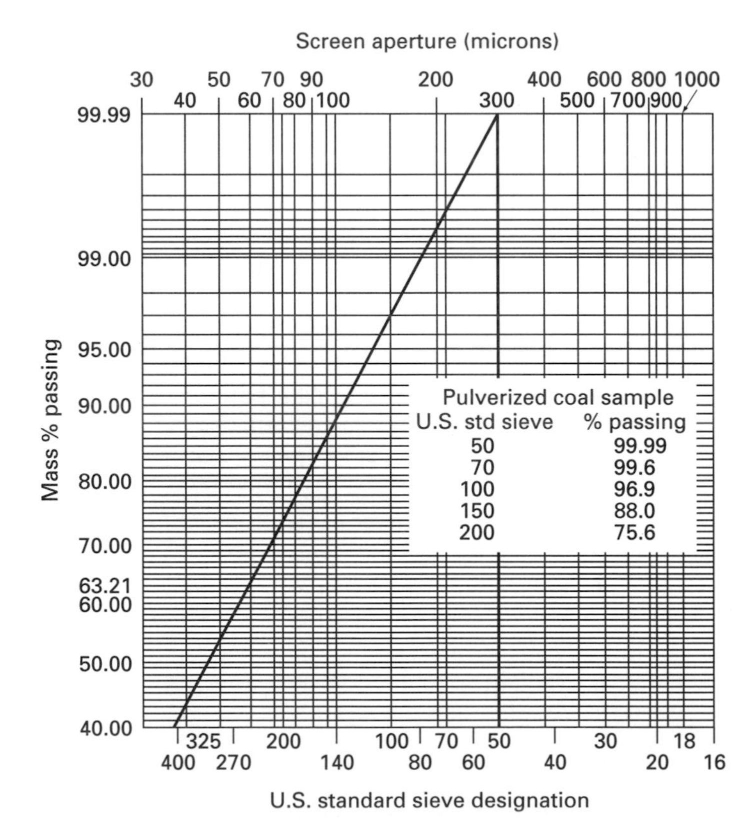

FIGURE 16.2
Rosin-Rammler chart for plotting pulverized coal sample sieve analyses; the table is the data from which the straight line curve was obtained [Field et al.].

From the pulverizer the coal-air mixture is piped to the burners, as represented in Figure 16.3. The conveying air is called *primary air,* and this is about 20% of the required combustion air. Typically the coal and primary air exit the burner nozzle at 75°C and 25 m/s. The *secondary* or main *air* supply is preheated to 300°C and exits the swirl vanes at about 40 m/s. The flame shape is controlled by the amount of secondary air swirl and the contour of the burner throat. The recirculation pattern which is set up inside and extends several throat diameters into the furnace provides a stabilized zone for ignition and combustion of the volatiles. Hence, air preheat and swirl are the two means of ensuring that ignition occurs in the throat of the burner and that the volatile flame is stable. The pulverized fuels should have a volatile content of at least 20% in order to maintain stability. This includes most solid fuels except anthracite. Nozzles with up to 165×10^6 kJ/h heat input per nozzle are used with pulverized coal.

The velocity of the primary coal plus air stream of the nozzle must exceed the speed of flame propagation so as to avoid flashback. The flame speed depends on fuel-air ratio, the amount of volatile matter and ash in the coal, the particle size distribution (fineness of grind), the air preheat, and tube diameter. Figure 16.4 is representative of this type of flame speed data. The maximum flame speed is comparable to a turbulent premixed gaseous hydrocarbon flame. Fuel-rich mixtures have the highest flame speeds, but if 20% of stoichiometric air is used, as in the primary air plus coal jet, the flame speed is lower and is near the rich limit. For low-volatile and/or high-ash coal the flame speed is low and the air from the secondary nozzle should be mixed in more slowly to avoid instability.

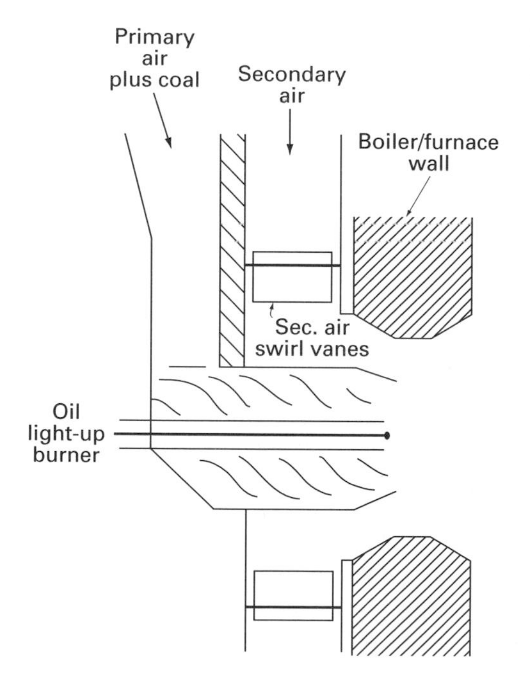

FIGURE 16.3
Pulverized coal nozzle burner.

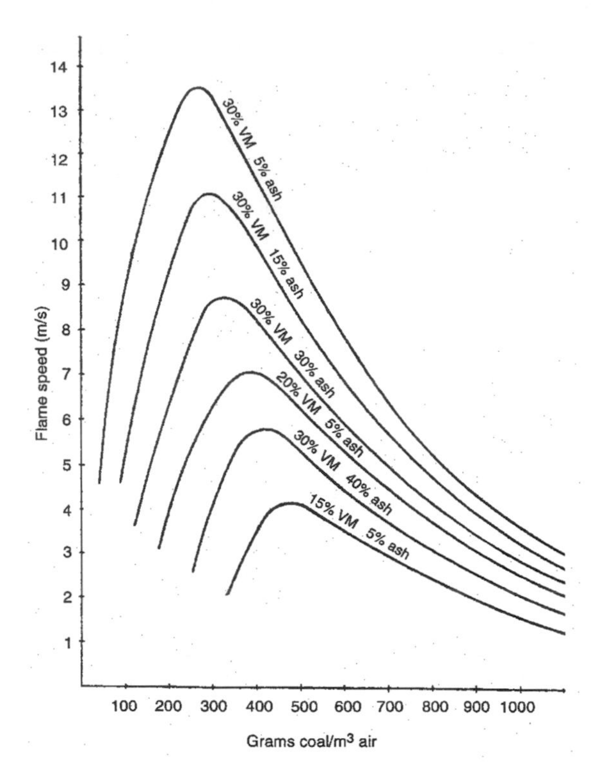

FIGURE 16.4
Typical flame speeds for mixtures of sub-bituminous pulverized coal and air [de Grey].

Location of Fuel and Air Nozzles

Numerous methods of locating and orienting the fuel and air nozzles have been tried, as illustrated in Figure 16.5. In the horizontal firing method the nozzles are located on the front wall—or on both the front and back walls, which is called opposed horizontal firing. Another approach for firing pulverized coal in large boilers is the so-called tangential firing method (Figures 16.5*c* and 16.6). The fuel nozzles, which do not have swirl, are located in the corners of the boiler and are directed along a line tangent to a circle in the center of the boiler. There is relatively little mixing in the primary jets, but the interaction of the jets sets up a large-scale vortex in the center of the boiler, which creates intense mixing. The nozzles in the corners consist of a vertical array of coal nozzles, and secondary air nozzles. The angle and velocity of the jets can be adjusted to optimize the fireball size and location in terms of heat release and ash slagging on the walls. The opposed inclined firing, and the U and double U designs shown in Figures 16.5*d*, *e*, and *f*, are used for harder-to-burn fuels such as coke, anthracite coal, and high-ash coal. The downward firing results in a greater residence time within the furnace.

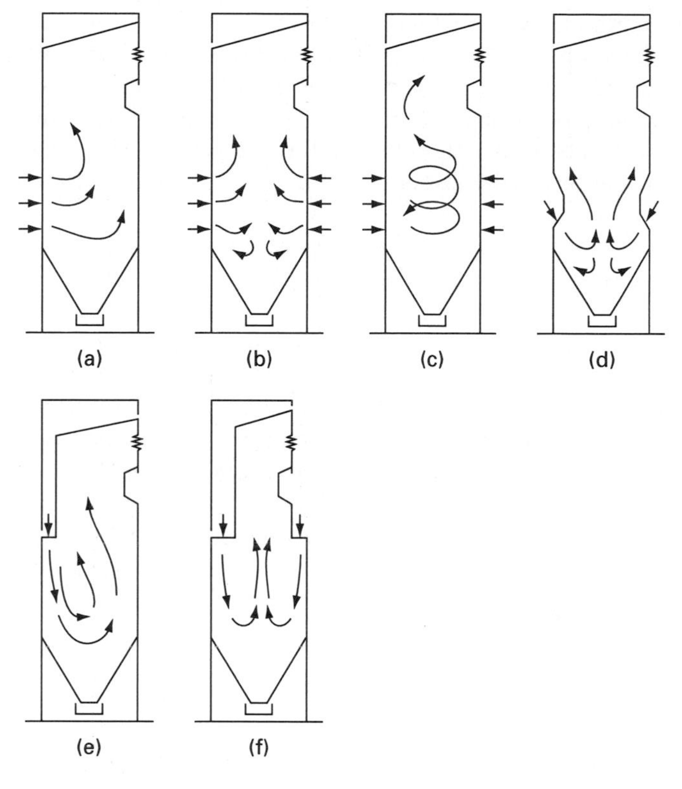

FIGURE 16.5
Dry bottom furnace and burner configurations: (*a*) horizontal (front and rear, (*b*) opposed horizontal, (*c*) tangential, (*d*) opposed inclined, (*e*) single U-flame, (*f*) double U-flame [Elliott, by permission of National Academy Press].

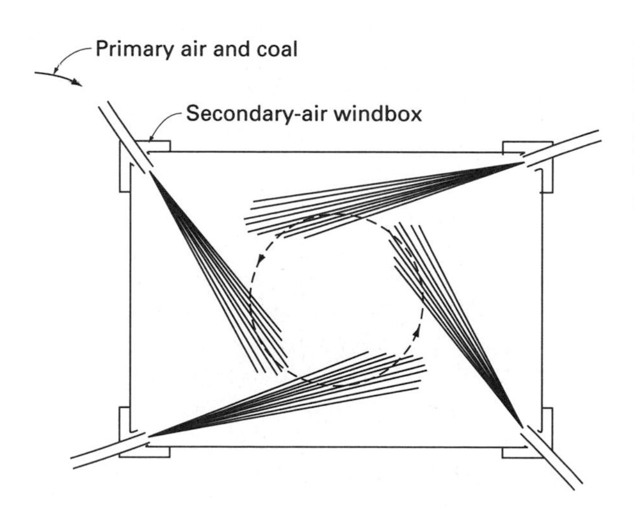

FIGURE 16.6
Schematic of cross section of corner-fired boiler with tangential injection. Burner tips are also tilting with numerous burner levels.

The firing methods shown in Figure 16.5 are examples of *dry bottom* furnaces, meaning that the temperatures are such that most of the ash does not agglomerate into molten globules which fall to the bottom of the furnace. The ash which does fall to the bottom, and this is only 10–20% of the total ash, remains nonmolten. This is accomplished by providing sufficient furnace volume with sufficient cooling surfaces so that the peak combustion temperatures are reduced by radiation heat transfer to the walls.

Wet bottom firing is another design approach which has been used. The high temperature necessary for ash melting is produced by locating the nozzles close together near the bottom of the furnace floor and lining nearby walls with refractory. In addition, various types of double chambers or baffled chambers have been built to provide two-stage combustion wherein the first stage is very hot, thereby slagging the ash. Up to 75% ash retention has been achieved in certain wet bottom designs with certain coals and operating conditions. Experience has shown that long-term operation (30 years or more) with a variety of coals and load conditions can best be achieved with dry bottom systems. Wet bottom systems are more prone to slagging and corrosion and have higher NO_x emissions. Dry bottom systems have higher uncontrolled fly ash emissions, but particulate control equipment is required in either case.

Furnace Design

A utility-sized furnace may be up to 50 m high and 10 m on a side (Figure 16.7). Combustion begins in the throat of the fuel nozzles and is completed in the radiant section. Peak temperatures near the nozzles may reach 1650°C, and the temperature drops gradually to 1100°C before entering the convection section due to radiant heat transfer to the walls. The radiant section must be large enough to provide sufficient residence time for complete burnout of the fuel particles. Also, the radiant section must extract enough heat to drop the temperature below the ash softening temperature prior to the convective section of the boiler to avoid fouling of the boiler tubes.

An arch in the upper part of the radiant section is used to provide more uniform flow around the upper corner of the combustion chamber, thereby increasing the heat transfer in the convective section. The convection section contains tubes to superheat the steam and provide reheat for multistage turbines. The convective section is designed to extract as much heat in as small a space as possible. Gas velocities in the convective section are restricted to about 20 m/s for coal-fired furnaces to minimize tube erosion due to fly ash. If the ash content of the fuel is high, or if the ash contains especially hard compounds such as quartz, then the gas velocity must be reduced by enlarging the convective section. Flue gas temperature at the inlet to the convective section generally is limited to 1100°C to minimize tube corrosion. Soot blowers, consisting of either steam or compressed air jets, are used regularly on the convective tubes as well as the economizer and air preheater tubes to maintain effective heat transfer and reduce flue gas pressure drop due to deposition of ash on the tube surfaces.

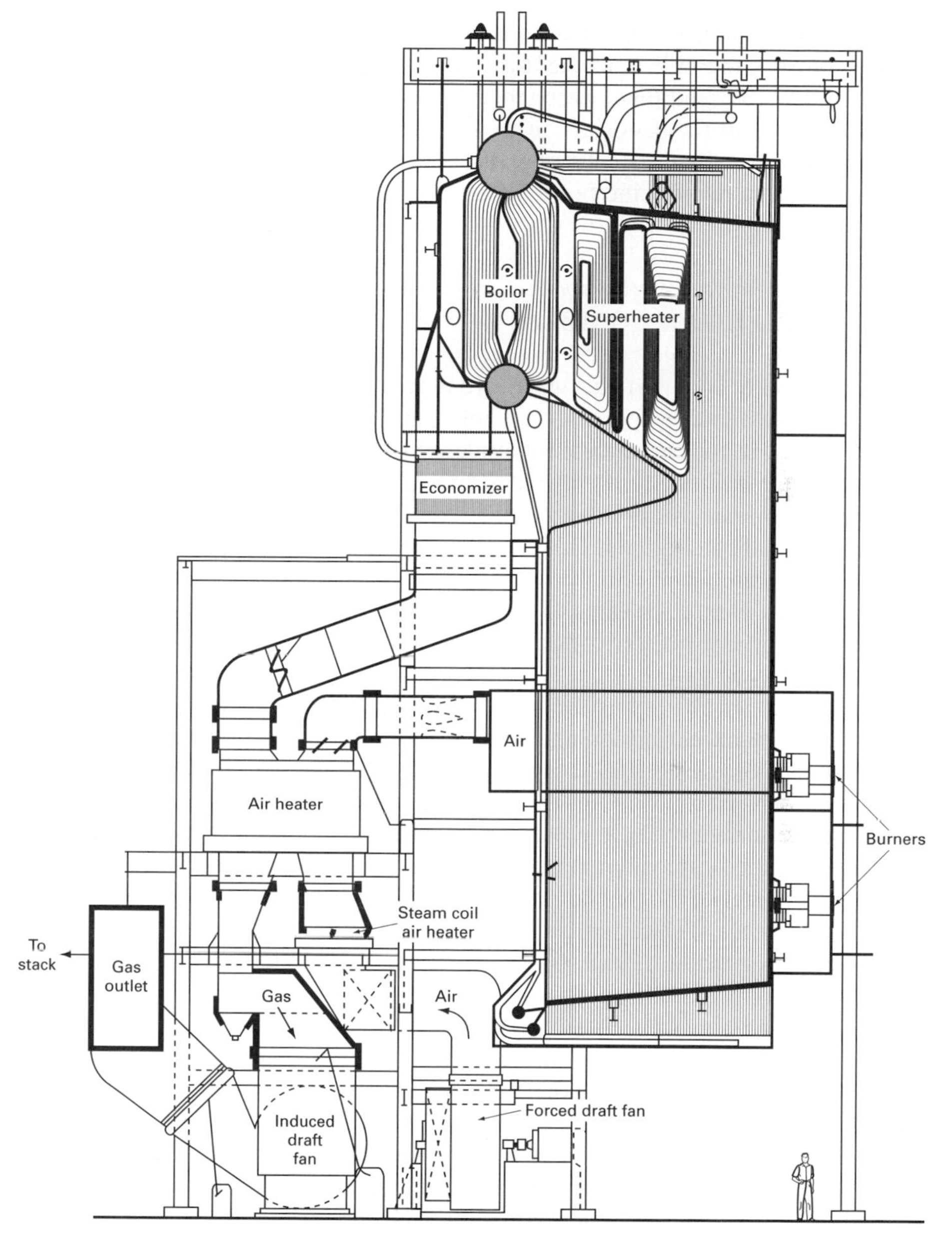

FIGURE 16.7
Pulverized coal furnace and associated equipment [Stultz and Kitto, compliments of Babcock and Wilcox].

The inorganic constituents of the fuel are an important consideration in pulverized fuel firing. For dry bottom furnaces the ash fusion temperature should be sufficiently high to avoid slagging in the furnace. Ash deposits can cause fouling of convective and superheat sections of the boiler. The presence of elements such as sodium, potassium, and iron in the ash can cause these deposits to sinter or fuse onto the boiler tubes and thus reduce the effectiveness of the soot blowers. Sulfate and chloride ions in the deposits can cause corrosion of the tube surfaces. When changing fuels, the effects of slagging and erosion and corrosion must be considered in relation to the specific boiler design.

16.2 PULVERIZED COAL COMBUSTION

Consider a cloud of coal particles which are blown through a burner into a furnace. Approximately 80% of the particles are between 10 and 100 μm in size. The swirl and back mixing of the burner cause a residence time near the burner where radiant heating from the burner quarl (a refractory-lined nozzle) ignites the particle cloud. As the particles are transported into the furnace away from the burner, the heat of combustion is radiated to the relatively cool waterwalls. As combustion proceeds, the gas temperature drops and the oxygen concentration decreases to a final value of a few percent. Hence the steps of drying, devolatilization, char combustion, and ash formation proceed under different conditions as the particle moves along its trajectory.

To be more specific, consider an example of a 100-μm coal particle which flows through the burner nozzle and encounters a 1400°C flame zone. In less than 1 ms the particle reaches 100°C and moisture is drawn off. (Some drying also occurs in the pulverizer.) By 1 ms the particle reaches 400°C and volatile gases and tars start to be forced out of the particle. With swelling bituminous coals, plastic flow occurs. After 10 ms the particle has reached 1000°C, devolatilization is complete, approximately 50% of the particle weight is released, and the porosity is established. The initial char particle size is equal to or greater than the original coal particle size at this point. For the first time oxygen reaches the particle surface and char burning begins.

Between 0.01 s and 1 s char burn continues and the particle surface temperature is several hundred °C hotter than the gas temperature due to surface reaction with oxygen. After 0.5 s, half of the char is consumed. Fissures are formed in the char particle, the porosity has increased, and the particle has begun to shrink in size. The mineral matter, which is dispersed throughout the char in nodules mostly less than 2 μm in size, becomes molten, and trace metals in the mineral matter volatilize. The many molten mineral matter nodules agglomerate into a few nodules, and toward the latter stages of char burnout the char may fragment into several pieces. The mineral matter, meanwhile, has grown large enough through agglomeration that vapor pressure overcomes surface tension effects, and the mineral matter nodules puff up into hollow glasslike spheres called *cenospheres*. These cenospheres are like miniature Christmas tree ornaments which vary in size from

0.1–50 μm and are perfectly spherical. Hence, fly ash from pulverized coal–fired boilers, and also residual fuel oil–fired boilers, is identified by its characteristically spherical cenospheres, as opposed to soil dust, which is irregular in shape, for example. After approximately 1 s, char combustion and cenosphere formation are complete and the combustion products begin to cool as they flow past the convective tubes.

The char burning rate is a key factor in determining the size of the boiler. As noted, char consists of porous carbon, mineral matter, and a small amount of organic matter. The gas temperature, particle temperature, and oxygen concentration change as the particles move through the furnace. Chemical reaction proceeds by reaction of oxygen, carbon dioxide, and water vapor with the surface of the char; however, the dominant reaction is with oxygen. Combustion of char particles depends on heat transfer and diffusion of reactant gases and products through the external boundary layer; diffusion and heat transfer within the pores of the char particle; and chemisorption of reactant gases, surface reaction, and desorption of the products. External and internal diffusion occur, and chemical kinetics is important. The surface accessibility and surface reactivity of an individual particle change with time.

In order to gain an understanding of the combustion process for suspension burning of solid fuels, let us consider the simplified case of char combustion in plug flow using a reaction rate based on the external surface area. Isothermal plug flow will be presented first, followed by nonisothermal plug flow.

Isothermal Plug Flow of Pulverized Coal Char

Consider an isothermal, uniform flow of uniformly dispersed char particles in a hot gas containing oxygen. The heat release rate due to combustion just equals the heat transfer to the walls of the combustor. For an effective global surface reaction rate constant k_e, and assuming the char goes to CO at the surface, the char particles burn according to the equation

$$\frac{dm_c}{dt} = -\tfrac{12}{16}\, k_e A_p p_{O_2} \tag{16.2}$$

where p_{O_2} is the oxygen partial pressure in the gas away from the particle surface and k_e has units of g/(cm$^2 \cdot$ s $\cdot$ atm O_2). For a spherical particle the area is related to the mass by

$$A_p = \pi\left(\frac{6m_c}{\rho_c \pi}\right)^{2/3} \tag{16.3}$$

The oxygen pressure decreases along the flow path due to combustion of the char:

$$p_{O_2} = p_{O_2}(i)\frac{m_{O_2}}{m_{O_2}(i)}$$

where i refers to initial. Using the excess air EA (or, equivalently, excess oxygen), and letting comb refer to that consumed by combustion and s refer to stoichiometric conditions,

$$p_{O_2} = p_{O_2}(i)\left[\frac{m_{O_2}(s)(1 + \text{EA}) - m_{O_2}(\text{comb})}{m_{O_2}(s)(1 + \text{EA})}\right]$$

Rearranging,

$$p_{O_2} = p_{O_2}(i)\left[1 - \frac{m_{O_2}(\text{comb})}{m_{O_2}(s)} + \text{EA}\right]\left(\frac{1}{1 + \text{EA}}\right)$$

or

$$p_{O_2} = p_{O_2}(i)\left(\frac{m_c}{m_c(i)} + \text{EA}\right)\left(\frac{1}{1 + \text{EA}}\right) \tag{16.4}$$

Substituting Eqs. 16.3 and 16.4 into Eq. 16.2, the char burning rate in isothermal plug flow becomes

$$\frac{dm_c}{dt} = -\frac{12}{16}k_e\pi\left(\frac{6m_c}{\rho_c\pi}\right)^{2/3} p_{O_2}(i)\left(\frac{m_c}{m_c(i)} + \text{EA}\right)\left(\frac{1}{1 + \text{EA}}\right) \tag{16.5}$$

These equations are solved for a reacting cloud of uniformly sized char particles in the following example. Note that stoichiometry and the excess air refer to the $C + O_2 \rightarrow CO_2$ reaction, while the surface reaction is for $C + \frac{1}{2}O_2 \rightarrow CO$.

EXAMPLE 16.1. For a uniform cloud of 50-μm char particles flowing in a stream of hot, isothermal gas containing oxygen, determine the particle mass divided by the initial mass versus time and the particle burnout time. Use an effective char kinetic rate constant of 0.5 g/(cm$^2 \cdot$ s $\cdot$ atm O_2), a char density of 0.8 g/cm^3, initial oxygen pressure of 0.1 atm, and excess oxygen of 20%. Repeat for 100-μm particles and compare the burnout times based on 99% burnout.

Solution. An equation solver routine is used to solve Eq. 16.5. As shown on the graph, the burnout time is 0.24 s for the 50-μm particle and 0.40 s for the 100-μm particle for these conditions.

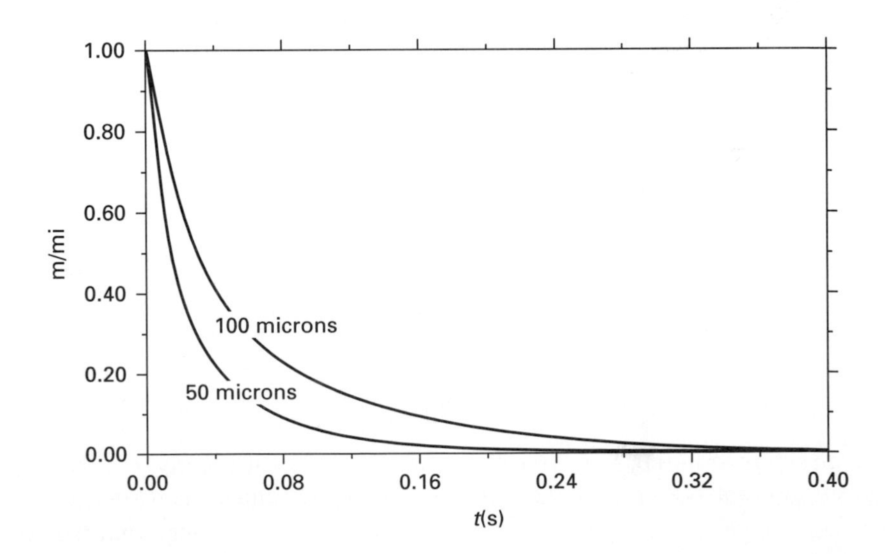

The effect of changing the excess air on the burnout of a cloud of particles in isothermal plug flow may be examined in a similar way. For example, using 5% initial excess air requires 65% more time to reach 95% burnout than does 20% initial excess air. By knowing the average particle velocity, the distance traveled may be determined. For a distribution of particle sizes, the initial rate of combustion is greater than for the monosize suspension, due to the fine particles, and there is a longer tail to the burnout curve due to the large particles which burn out with lower oxygen.

The heat input per unit flow area, HR″, depends on the number of char particles per unit volume n', the mass of each particle, the flow velocity, and the lower heating value:

$$\text{HR}'' = n' m_c \underline{V}(\text{LHV}) \tag{16.6}$$

The number of particles per unit volume is

$$n' = \frac{\text{Mass fuel per unit vol.}}{\text{Mass per particle}} = \frac{f\rho_{\text{gas}}}{m_c} \tag{16.7}$$

where f is the fuel/air weight ratio. The use of Eqs. 16.6 and 16.7 for isothermal plug flow of char particles is illustrated in Example 16.2.

EXAMPLE 16.2. Determine the heat input rate per unit flow area and the burnout distance for 50-μm char particles for the conditions of Example 16.1, and assuming a velocity of 10 m/s and a reaction temperature of 1600 K.

Solution. For the reaction of 1 mol of carbon char going to 1 mol of CO_2 in air with 20% excess air, $f = 0.0724$, and the lower heating value is $393.5/12 = 32.8$ MJ/kg (using an energy balance for the char reaction). From Appendix B the approximate gas density based on air is $\rho_{\text{gas}} = 0.000221$ g/cm^3. The particle mass is

$$m_c = (0.80 \text{ g/cm}^3)(\pi/6)(0.0050 \text{ cm})^3 = 5.23 \times 10^{-8} \text{ g/particle}$$

From Eq. 16.7,

$$n' = \frac{(0.0724)(2.21 \times 10^{-4} \text{ g/cm}^3)}{5.23 \times 10^{-8} \text{ g}} = 305 \text{ particles/cm}^3$$

Using Eq. 16.6, the heat input rate is,

$$\begin{aligned} \text{HR}'' &= (305 \text{ part/cm}^3)(5.23 \times 10^{-8} \text{ g/part.})(1000 \text{ cm/s})(32.8 \text{ kJ/g}) \\ &= 0.52 \text{ kW/cm}^2 = 5.2 \text{ MW/m}^2 \end{aligned}$$

For a 50-mm char particle the burnout time is 0.24 s, so that the travel distance is (0.24 s)(10 m/s) = 2.4 m. The energy density per unit volume is 2.1 MW/m^3. This illustrates the relatively high power density of pulverized char combustion.

Nonisothermal Plug Flow of Pulverized Char Suspension

Consider a uniform flow of a uniform mixture of coal char particles suspended in air. The particle and gas temperatures are identical, and there is no relative velocity between the gas and the particles. The particles are dilute enough that the volume occupied by the suspended particles is negligible. The flow is steady and pressure drop is negligible. Heat conduction in the gas phase is neglected compared with convection. As the particles flow along the streamlines, the char reacts with the oxygen due to kinetics and diffusion of oxygen from the gas to the solid to produce CO near the particles' surfaces, which then reacts rapidly in the gas phase to produce CO_2. Reactions of CO_2 with char are neglected compared with reactions with oxygen. Following Section 14.3, the rate of combustion of char per unit volume of gas is

$$r_c = \frac{12}{16} k_e \rho_{O_2} A_v \tag{16.8}$$

A_v is the fuel external surface area per unit volume of gas and is given by

$$A_v = (\pi d^2)(n') \tag{16.9}$$

As the char particle cloud moves along, the particle diameter and oxygen density decrease, and the temperature and velocity increase due to combustion.

The gas phase consists of O_2, CO_2, and N_2. Since the CO reacts rapidly compared with the char, the CO in the gas is negligible, and

$$r_{CO_2} = \frac{r_c M_{CO_2}}{M_c} \tag{16.10}$$

$$r_{O_2} = -\frac{r_c M_{O_2}}{M_c} \tag{16.11}$$

Conservation of energy, mass, and species continuity are used to obtain the temperature, velocity, and species profiles along the streamlines. The conservation of energy equation is

$$\frac{d}{dx}(\rho_g \underline{V} h_g) = r_c H_c - q_r \tag{16.12}$$

where q_r is heat loss to the walls of the combustor per unit volume due to radiation, r_c is governed by CO formation at the surface, and H_c is the heat of reaction for $C + O_2 \rightarrow CO_2$. The expression for conservation of mass for the gas phase is

$$\frac{d}{dx}(\rho_g \underline{V}) = r_c \tag{16.13}$$

The species continuity equations are needed to obtain the gas density:

$$\rho_g = \rho_{O_2} + \rho_{CO_2} + \rho_{N_2} \tag{16.14}$$

$$\frac{d(\rho_{O_2} \underline{V})}{dx} = r_{O_2} \tag{16.15}$$

$$\frac{d(\rho_{CO_2} \underline{V})}{dx} = r_{CO_2} \tag{16.16}$$

and

$$\rho_{N_2} = \frac{\rho_{N_2}(\text{initial}) \cdot T(\text{initial})}{T} \tag{16.17}$$

For the mixture,

$$h_g = \sum h_i y_i \tag{16.18}$$

where

$$y_i = \frac{\rho_i}{\rho_g} \tag{16.19}$$

The solution, which is implicit, proceeds as follows. For a given inlet p, T, $\underline{V}$, particle size d, and fuel-air ratio f, the above equations are solved incrementally for a small Δx by obtaining r_c from Eq. 16.8, T from Eq. 16.12, gas density from Eqs. 16.14–16.17, and velocity from Eq. 16.13. The new values of A_v, ρ_i and r_c are obtained and the steps repeated until char burnout. The particle diameter, for use in Eq. 16.9 and in obtaining $\tilde{h}_D$ (which is part of k_e), is obtained by knowing the time from $x/\underline{V}$ and using Eq. 16.8, as done in Section 14.3. The particle temperature, which may be different from the gas temperature, is also obtained from the relations in Section 14.3.

Results of applying detailed plug flow relations to a bituminous coal suspension including drying, particle heat-up, pyrolysis, and drying are shown in Figure 16.8 [see Smoot and Smith]. Particle heat-up requires about one-third of the distance in this example, pyrolysis proceeds rapidly, and excess oxygen is available at the combustor exit. The 85-μm particle burnout exceeds the available combustor length. Of course, in an actual furnace the flow is more complicated than the plug flow example, but nevertheless these types of calculations provide useful insight into why the furnace height must be so large. Full three-dimensional computational fluid dynamics calculations have been carried out extensively for boilers using similar relations for the fuel particles [Abbas, et al.]. One of the objectives of this CFD modeling is to assist in the design of the radiant waterwall tubes and the convective boiler tubes.

16.3 BEHAVIOR OF ASH

The ash is a major factor affecting the performance of boilers. Not only does ash deposit on heat transfer surfaces and in flue gas passages, it also leads to corrosion and erosion of heat transfer tubes. Deposition, erosion, and corrosion can seriously limit the efficiency and lifetime of a boiler.

Deposition of ash is called slagging and fouling. *Slagging* generally refers to the sticking of molten ash on heat transfer surfaces—usually the radiant tubes. *Fouling* refers to deposition of nonmolten ash on the convective tubes and surfaces by impaction and diffusion. The severity of the slagging and fouling depends on the nature of the mineral matter and the temperatures and velocities involved.

Mineral matter in coal varies widely but typically consists of aluminosilicates (clays), sulfides, carbonates, oxides, and chlorides. Commonly found compounds are shown in Table 16.1. During combustion the mineral compounds are converted to oxides. Seven major constituents constitute most coal ash: SiO_2, Al_2O_3, Fe_2O_3,

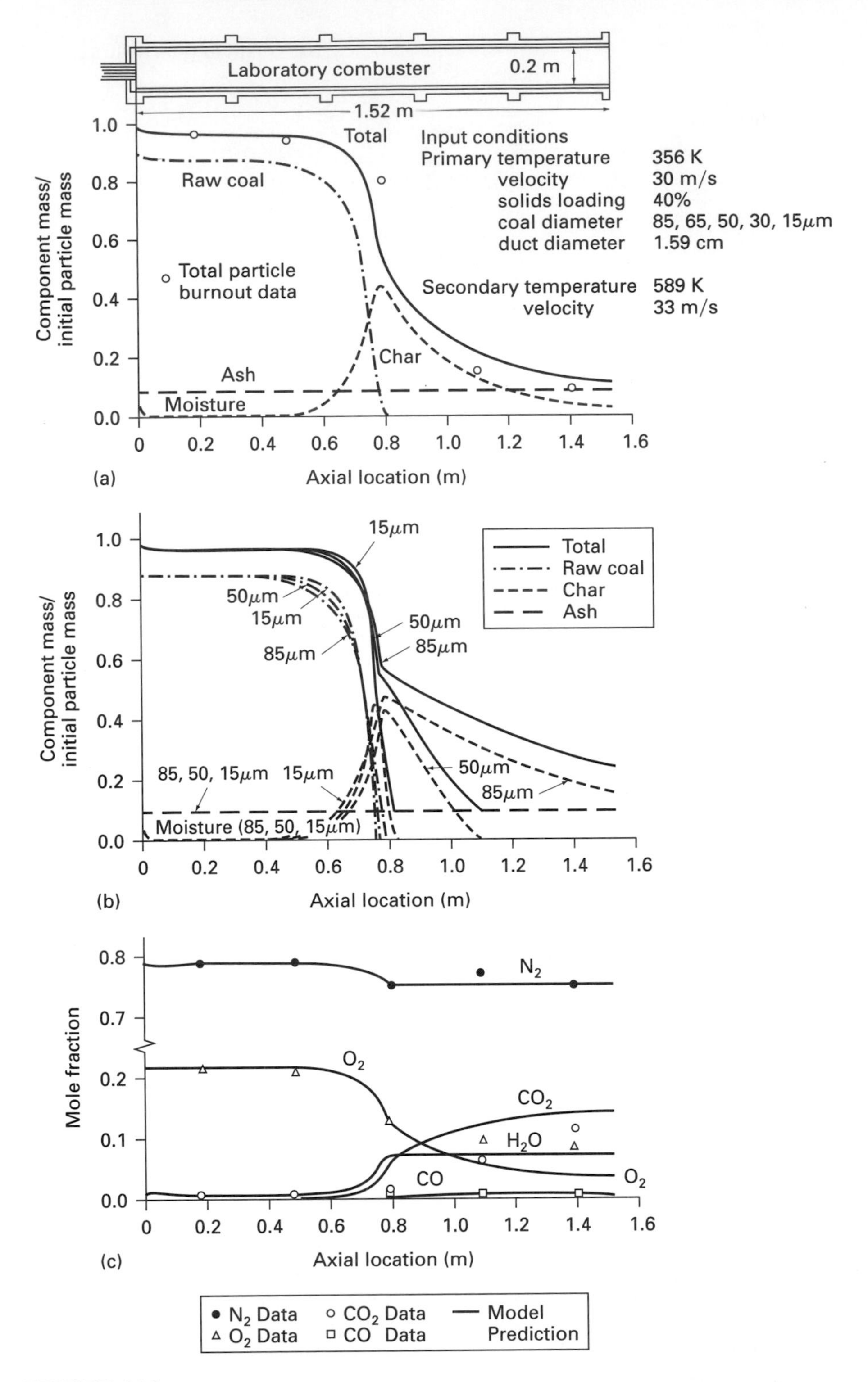

FIGURE 16.8
Nonisothermal plug flow of pulverized coal combustion, predictions and measurements: (*a*) coal particle burnout; (*b*) individual particle histories in a polydispersed cloud; (*c*) gas mole fractions [Smoot and Smith, *Coal Combustion and Gasification*, © 1985, with permission of Plenum Pub. Corp.].

TABLE 16.1
Common* Mineral Matter in Coal [Elliott, by permission of National Academy Press]

Type	Compound	Formula
Clays	Illite	$KAl_2(AlSi_3O_{10})(OH)_2$
	Kaolinite	$Al_2Si_4O_{10}(OH)_2$
Sulfides	Pyrite	FeS_2
Carbonates	Calcite	$CaCO_3$
	Dolomite	$CaMg(CO_3)_2$
	Siderite	$FeCO_3$
Oxides	Quartz	SiO_2
Chlorides	Halite	NaCl
	Sylvite	KCl

*Compounds vary widely; only a few common ones shown.

CaO, MgO, Na_2O, and K_2O. An example of a bituminous ash is shown in Table 16.2. Typically the ash constituents are reported as separate oxides, as in Table 16.2; however, in actuality more complex inorganic compounds are formed. Coal ash may begin to sinter at temperatures as low as 650°C. *Sintering* is chemical bonding of ash particles, which occurs without melting. At temperatures above 1000°C appreciable melting can occur. Above approximately 1400°C a highly viscous liquid slag is formed. Formation of the liquid phase can lead to slagging of boiler tubes particularly in the radiant section of the boiler. The cone ash fusion temperature test mentioned in Section 2.3 is a rough indicator of the slagging potential of a particular coal.

Two techniques are used to remove troublesome deposits of slag: (1) high-velocity jets of steam, air, or water are used to break the bond between the slag and the tube wall; and (2) the firing rate can be reduced to decrease furnace temperatures. Bonding of slag to tube walls is thought to be due to alkali compounds such as Na_2SO_4 and K_2SO_4. Sodium and potassium are volatilized in the flame zone, react rapidly with sulfur dioxide, and condense on the tubes, which are about 500°C, causing the tubes to become sticky. As the flue gas flows over the tubes,

TABLE 16.2
Representative composition of bituminous coal ash

Composition	Weight %
Silica, SiO_2	45.3
Alumina, Al_2O_3	19.5
Ferric oxide, Fe_2O_3	23.3
Calcia (lime), CaO	3.7
Magnesia, MgO	0.8
Potassium oxide, K_2O	2.1
Sodium oxide, Na_2O	0.6
Sulfur trioxide, SO_3	3.0
Other	1.7
	100.0

TABLE 16.3
Ash fouling index for bituminous coal [Elliott, by permission of National Academy Press]

Fouling index* R_f	Severity
< 0.2	Low
0.2–0.5	Medium
0.5–1.0	High
> 1.0	Severe

*$R_f = (m_B/m_A)(m_{Na_2O})$, where m_B is mass of basic compounds such as Fe_2O_3, CaO, MgO, K_2O, or Na_2O and m_A is mass of acidic compounds such as SiO_2, Al_2O_3, or TiO_2.

some of the larger fly ash particles are deposited by impaction on the upstream side of the tubes. Particles less than about 3 μm can be deposited on the back side of the tubes due to turbulent diffusion. Coals with relatively high sodium and potassium content tend to be more prone to fouling. A fouling index has been developed for bituminous coals, which is explained in Table 16.3. This provides a criterion for rating a coal as having potentially high or low slagging characteristics. Additives to the coal nozzles, such as alumina powder, are sometimes used to mitigate fouling.

16.4 EMISSIONS FROM PULVERIZED COAL FURNACES

Particulate matter, sulfur dioxide, and nitrogen oxides are the significant emissions from pulverized coal furnaces. Hydrocarbons and carbon monoxide are not emitted in significant amounts in a properly operating furnace. Each of the three main types of emissions will be discussed in this section.

Emission factors for dry bottom and wet bottom boilers and for cyclone-fired boilers (to be discussed in the next section) are given in Table 16.4. These emission factors give typical uncontrolled emissions without combustion modification and external control equipment. Emission standards which must be met are the same

TABLE 16.4
Emission factors for uncontrolled emissions from pulverized coal-fired boilers (g/kg coal input) [EPA]

Boiler type	Particulates	SO_2	NO_x as NO_2
Dry bottom	5 A*	19.5 S†	10.5
Wet bottom	3.5 A	19.5 S	17
Cyclone	1 A	19.5 S	18

*A = wt % ash in coal (10% ash means A = 10).
†S = wt % sulfur in coal (3% sulfur means S = 3).

as those given in Table 15.2. The amount of emission control required may be determined by using Tables 15.2 and 16.4, and knowing the heating value of the fuel.

Particulate emissions, referred to as fly ash, originate from the mineral matter in the coal. Some of the mineral matter vaporizes and recondenses onto the non-volatile ash in the cooler regions of the furnace, and a small fraction of the sulfur may combine with the ash. In dry bottom furnaces 80% or more of the ash is converted to fly as while the remainder falls out in the bottom ash hopper, and in the economizer and air heater hoppers. Generally carbon in the fly ash is very low since sufficient furnace volume is provided for good carbon burnout. About 40% of the fly ash tends to be less than 10 μm in size. Particulate matter less than 10 μm in size tends to remain suspended in the atmosphere and can penetrate into human lungs, causing adverse health effects. External particulate controls are required to reduce the particulate emissions.

Volatilization of the alkali-metal oxides (Na_2O and K_2O) in the ash becomes appreciable as the temperature rises above 1100°C. The other compounds shown in Table 16.2 do not begin to volatilize at temperatures normally found in dry bottom furnaces. As the temperature drops in the upper reaches of the furnace, the alkali vapors condense onto the surface of the fly ash particles. In addition, trace metals such as lead and zinc, which are found in coal ash in the parts per million range, are vaporized in the furnace and condense onto the particulates in the cooler sections of the furnace. Hence, most of the trace elements are captured and removed by the fine particulate control equipment.

Sulfur in the fuel is readily converted to sulfur dioxide during combustion. Approximately 4% of the sulfur combines with the calcium and magnesium in the fly ash to form calcium and magnesium sulfate. About 1–2% of the sulfur forms sulfuric acid, which condenses on the submicron particles and contributes to the plume opacity. External control equipment such as large scrubbers is required for new pulverized coal–fired boilers in order to control sulfur dioxide emissions. Low-sulfur coal, typically less than about 0.6% sulfur, can be used without scrubbers, but this coal tends to be more expensive.

Injection of dry pulverized limestone ($CaCO_3$) or hydrated lime ($Ca(OH)_2$) along with the pulverized coal has been tried as a means of controlling sulfur dioxide emissions without scrubbers. The extra particulate emissions are captured in a baghouse. Early results were disappointing: a Ca/S molar ratio of 2 resulted in only 20% control of SO_2, and boiler tube plugging was observed. The poor performance was attributed to the high flame temperatures, which deactivated the lime by making it less porous. With the advent of the multistage burners to control NO_x emissions, there has been renewed interest in dry limestone injection. If the flame temperatures are kept below 1260°C (2300°F), tests have shown that a Ca/S molar ratio of 2 will capture 50% of the SO_2.

Nitrogen oxide emissions, which are essentially all nitric oxide, are formed from fuel-bound nitrogen and atmospheric nitrogen. Studies of NO emissions from all types of steam generators indicated that those from tangentially fired units were about half the values of those from horizontally fired systems. This is due to the relatively low rate of mixing between the parallel streams of coal and secondary air emitted from the corner windboxes. Thus, ignition and partial devolatilization occur within an air-deficient primary combustion zone that exists from the fuel nozzle to a point within the furnace at which the stream is absorbed into the rotating fireball. The fireball itself contains the air required for complete combustion of the fuel, but because only a portion of the devolatilization occurs after the coal stream enters the fireball, the potential for fuel nitrogen oxide formation is limited.

Two significant modifications have been made to the design and operation of the air supply of tangentially fired boilers to extend the oxygen-deficient combustion zone. The first modification added air compartments within the windbox above the uppermost coal nozzle. These overfire air ports divert approximately 20% of the total combustion air to a burning zone above the windboxes, thus extending the fireball. The second modification was an operating procedure which moved the ignition point closer to the nozzle. This change extended the duration of the primary combustion zone so that a greater portion of the devolatilization took place before entering the fireball. The ignition point is controlled by varying the quantity and velocity of the air though the annulus of the fuel compartment by changing the position of the damper. These modifications resulted in a 20–30% reduction in NO emissions.

For horizontally fired furnaces the burners have been modified to reduce the NO_x by adding more secondary air and also tertiary air nozzles above the burner. In the first zone of the flame, which consists of primary air and coal plus a portion of the secondary air and is the start of devolatilization, the stoichiometry is maintained very rich. Supplemental secondary air, termed outer secondary air, is then added to the rich products to establish near stoichiometric conditions that maximize the decay of the volatilized fuel-nitrogen compounds. The tertiary air provides the oxygen necessary for complete combustion and provides an oxidizing zone around the rich primary zone.

In pulverized coal utility boilers the flame temperatures and residence times are such that NO_x formation is more from the fuel-bound nitrogen than from atmospheric nitrogen. Nitrogen-containing heterocyclic ring structures in the coal decompose to gaseous compounds such as HCN and NH_3 during pyrolysis (the so-called cyanide-amine pool). The extent to which these nitrogen compounds react to form NO depends on the local temperature and oxygen concentration. Hence, the flow field, the trajectories of the individual coal particles, and the turbulent mixing are vitally important. Since the chemistry can proceed only after molecular diffusion has taken place between reactants, the size of the reaction zone is closely related to the microscale of the turbulence. However, macromixing must first take place. In swirling pulverized coal flames the macromixing occurs between pyrolysis products, combustion air, and recirculated combustion products within the internal and external recirculation zones. The turbulent flame structure consists of macro-sized regions comprising fuel, oxidant, and combustion products. Within these macroscale regions are successively smaller eddies down to the microscale; the combustion reactions occur within the microscale eddies.

Measurements in swirling flames of high- and medium-volatile bituminous coals show that NO_x production and destruction take place in a region close to the burner. Char burnout as high as 80% and NO_x concentrations representative of the flue gas concentration levels are typically measured within two quarl diameters (refractory nozzle diameters) downstream of the burner outlet. Thus an understanding of the near-field fluid dynamics of swirl burners is key to achieving low NO_x emissions. For example, laboratory tests in a 2-MW burner with a type-0 flame (see Figure 6.9) with little or no swirl and using a high-volatile coal containing 1.6% nitrogen produced 470–520 ppm NO_x and 98% carbon burnout. In the same setup, when the swirl was increased to a type-2 flame (with a closed internal recirculation zone), the NO_x emissions were 900–1000 ppm and the carbon burnout was 99.6%. For a type-1 flame (a combination of type-0 and type-2 flames such that the fuel jet partially penetrates through the internal recirculation zone), the NO_x emissions were 200–400 ppm and the carbon burnout was 99.5% [Weber, 1992]. A number of low-NO_x burners designed to retrofit existing boilers now use type-1 flames. Computational fluid dynamic models have recently been developed and validated with detailed measurements such that confidence in the turbulent mixing and the chemistry simulation is to the point where CFD modeling can guide further improvements reducing emissions.

16.5 CYCLONE COMBUSTORS

Cyclone-fired boilers were first introduced in the early 1940s, and their use grew steadily until the early 1970s, when they were in use for approximately 9% of the electrical generating capacity of the United States. Cyclone combustors have the ability to use low–ash fusion temperature, high-ash fuels which are unsuitable for use in pulverized fuel boilers. A cyclone combustor is shown in Figure 16.9.

The cyclone combustor is basically a horizontal cylindrical prechamber in which crushed coal burns in a cyclonic motion due to air which is introduced tangentially into the cylinder. The combustor surfaces are covered with refractory so that gas temperatures of about 1650°C (3000°F) are achieved. This temperature is sufficient to melt the ash into a molten slag, which forms on the cylinder walls.

Coal is crushed so that 95% is less than 5 mm and the mass mean diameter is approximately 0.5 mm. The crushed coal is fed into the burner end of the cyclone. About 20% of the combustion air (primary air) also enters the burner tangentially and imparts swirl to the incoming coal. Secondary air with a velocity of 100 m/s is admitted tangentially at the roof of the main barrel of the cyclone and imparts a further centrifugal action. A small amount of tertiary air (5%) is admitted at the center of the burner. The burning coal particles are thrown to the cylinder wall and are held in the slag layer as they burn out. Combustion gases leave through the reentrant throat at the exit. Molten slag drains from the bottom rear through a small opening.

Cyclone combustors operate at much higher levels of swirl than swirl burners. Cyclone combustors typically operate at swirl numbers of 6 to 30, compared with swirl numbers of less than 2 for swirl burners. The *swirl number* is the ratio of the axial flow angular momentum to the axial flow of linear momentum times the radius. The strong tangential flow in the cyclone chamber induces secondary flows

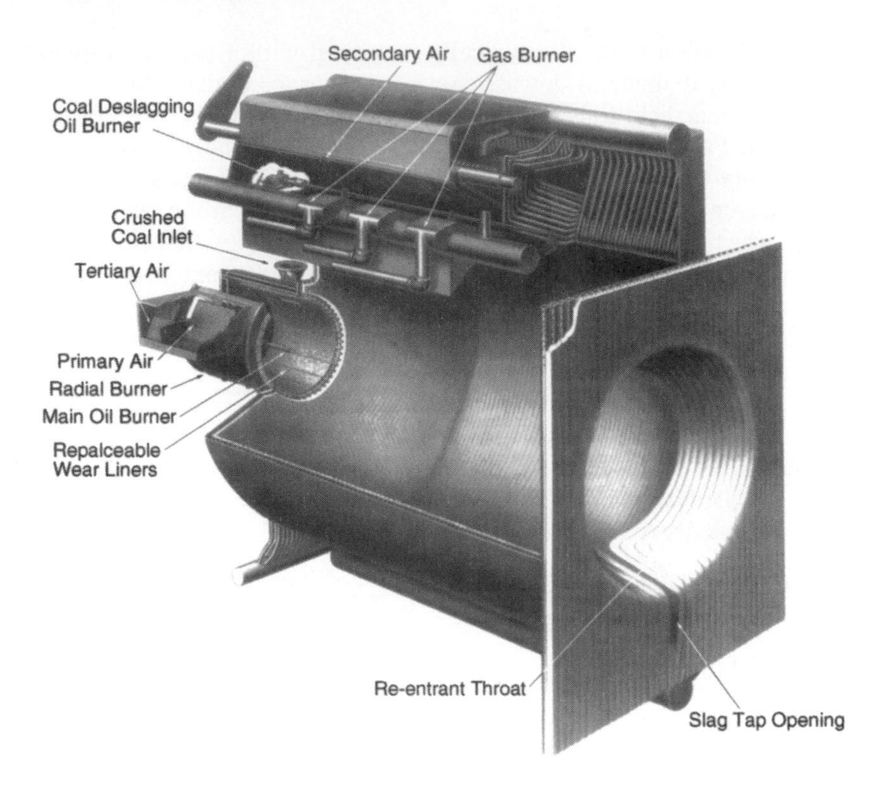

FIGURE 16.9
Cyclone combustor for crushed coal [Stultz and Kitto, compliments of Babcock and Wilcox].

in the axial and radial directions (see Figure 16.10). The flow near the wall spirals around the periphery of the chamber until it reaches the front wall, where it reverses at mid-radius and flows upstream to the back wall, and then flows forward near the centerline to the exit of the cyclone chamber. The shearing between the forward and reverse flows produces high levels of turbulent mass transfer, which facilitates the wide stability limits and high combustion efficiency of cyclone combustors.

The cyclone combustor can burn a wide range of coals from low-volatile bituminous to lignite as well as other solid fuels such as wood bark and petroleum coke, for example. Fuel oils and gaseous fuels may also be burned in cyclone combustors. The suitability of fuels is dependent on the moisture, volatile content, and viscosity of the ash. The hot gases from the cyclone combustor flow into a conventional waterwall boiler with associated convective section. Since 20–30% of the ash is converted to fly ash, instead of 80% or more for pulverized coal combustion, the convective boiler tubes may be spaced closer together. However, since some of the coal fines burn in suspension, the furnace volume must be large enough to allow for char burnout. High nitrogen oxide emissions have restricted the use of cyclone burners. However, recent developments have shown that staged combustion can control the formation of nitrogen oxide during combustion.

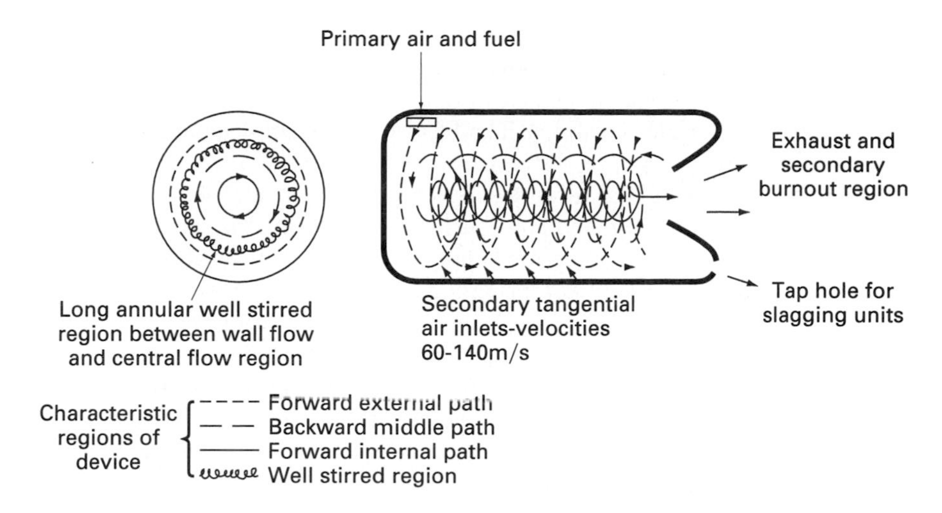

FIGURE 16.10
Flow pattern in a cyclone combustor [Reprinted from Syred et al., "Cyclone Combustors," in *Principles of Combustion Engineering for Boilers,* Lawn (ed.), © 1987, by permission of Academic Press Ltd.].

16.6 PULVERIZED BIOMASS AND REFUSE-FIRED BOILERS

Wood and bark are an order of magnitude more difficult to grind than coal because of the wood fibers, which are tough rather than brittle like coal. However, pulverized wood burners are economically attractive for retrofitting oil burners when clean waste sawdust is available. When designing a new boiler for wood, often a fixed-bed or fluid-bed system which does not require pulverizing the wood is chosen. Co-firing biomass with pulverized coal is being done to a limited extent, by replacing up to 3 to 5% of the coal with wood chips. The wood chips are added to the coal pulverizers with the coal. Higher replacement amounts degrade the performance of the pulverizer. Co-firing crushed coal and wood with a top size of 6 mm can be done in cyclone burners with replacements up to 5%.

The design problem is to determine how fine the wood must be pulverized for proper combustor performance. Since wood is more reactive than coal, the particle size for wood can be larger than for coal. The size distribution must also be taken into account. Both burner stability and particle burnout in the boiler must be considered. Research data on combustion performance are being collected for large power plants. Industrial size burners for sander dust and other finely pulverized biomass are commercially available.

Refuse-derived fuels (RDFs) are being co-fired with coal at various locations in the United States and Europe. When the design is done properly, operational problems have been overcome. There is still a question about long-term corrosion and erosion of boiler tubes. Typically, grates are installed above the ash pit at the boiler neck to maximize burnout of RDF and prevent clinkering in the ash pit.

Larger fans are needed to supply overfire and underfire air to the grates. The RDF is screened to minus 2 cm (sizes above 2 cm are removed) and pneumatically injected into the furnace via air-cooled nozzles located in the front wall between the two lowest coal nozzle elevations. The RDF feed system is independent of the primary fuel system.

RDF burns well with coal at an average replacement rate of 15% on a heat input basis. Typically an estimated 50% burns in suspension, the remainder on the dump grates. The boilers, which do not slag on coal, have shown little tendency to slag on RDF and coal. Some clinker formations are observed above the grate at the furnace throat, and are removed by lowering the grate and rodding the clinker from an access door. By slowing the RDF feed rate to the boiler and dumping the grates immediately after soot blowing, clinker formations can be minimized. In some cases stack outlet temperatures increase some 15°C, which results in a loss in efficiency of about 1.5%. When co-firing with RDF, emissions of volatile inorganic and organic compounds should be carefully monitored.

16.7 COAL-WATER FUEL

Pulverized coal can be mixed with water and made into a slurry which has the appearance of residual fuel oil. However, it burns more like a solid fuel than a liquid fuel and hence will be discussed in this chapter. The motivation is to retrofit furnaces and boilers which were designed for oil to burn a coal-water fuel (CWF).

The reason for using a slurry fuel rather than converting directly to pulverized coal is that the fuel handling and feeding of a slurry can be easier than for pulverized coal. The slurry fuel can be made at a remote site and transported by truck or pipeline to the furnace or boiler. Hence, there is no coal pile on site, which may be important for certain applications. Also, coal beneficiation (cleaning) by means of fine grinding and flotation techniques adds water to the coal. Using a slurry fuel eliminates the need to dry the beneficiated coal.

The CWF concept is to grind the coal to a top size of 75 μm or less, remove as much ash as possible using flotation techniques, and produce a pumpable slurry with as little water as possible. Current practice using bituminous coals results in a slurry with a coal loading of 70%, water 29%, and additives 1% by weight. The additives help to disperse the solids and stabilize them so they do not settle out during short-term storage. Lower-rank coal, such as lignite, often tends to be more hydroscopic and may require 50% water to make a slurry.

The CWF is sprayed into the combustion chamber through an atomizing nozzle. The Y-jet atomizer (Chapter 10) has been used successfully with tungsten carbide inserts for wear prevention. Also, straight shot nozzles and pintle nozzles have been used. The design problem is to generate a fine spray and yet not plug up the nozzle. The slurry must be composed of individual particles, because agglomerates will clog the nozzle. Sometimes other additives are used to control the viscosity in order to improve atomization. Without additives, the viscosity can be 10 to 100 times greater than no. 6 oil; however, with additives of 1% or less the viscosity can be made comparable to no. 6 oil.

Combustion of CWF is different from either oil or pulverized coal. Burner tests have shown that the air should be preheated to 260°C to achieve flame stability. CWF containing 50% water will sustain a flame; however, slurries containing 30% water burn much better. The ignition delay time is increased significantly due to the water. A higher–volatile matter coal has a more stable flame than a lower-volatile coal. Interestingly, char burnout occurs with a particle size associated more with the spray droplet size than with the individual coal particle size. During devolatilization of the droplets, the coal particles in the droplet fuse together and burn out as a porous agglomerate. Hence, residence time may be insufficient to burn out the carbon if the drop size is too large. This in turn relates to the need for special nozzles which develop a fine spray without clogging.

The basic decision to be made when considering the use of CWF fuel is whether the ease in on-site handling of the fuel overcomes the loss in efficiency due to the water. If highly beneficiated coal becomes available at reasonable cost, CWF fuel may find application in industrial furnaces and boilers, in selected utility boilers, and possibly as a gas turbine fuel or as an industrial diesel fuel. In converting boilers designed for oil to CWF, derating of the boiler output and/or modifications to the preheater, superheater, and economizer sections to minimize erosion are to be expected, as well as the addition of particulate control equipment such as a fabric filter baghouse or an electrostatic precipitator.

16.8 SUMMARY

Suspension burning of pulverized coal is used primarily by large electric utilities. Coal is typically pulverized in a mill to the point where 90% of the particles are less than 100 μm in size. Since the velocities are relatively high to achieve high heat output, the combustor tends to be very tall. Heat released by combustion is absorbed by the water walls and the convective section. Temperatures and velocities in the convective section need to be kept low enough to minimize fouling, slagging, and erosion of the convective tubes. Following a 100 μm coal particle, drying occurs in the first millisecond, devolatilization is complete in about 10 ms, followed by char burnout which requires about 1 s. Volatiles burn out near the burner nozzle. Char burnout needs to be completed before the convective section where the temperature drops rapidly. During the final stages of char burnout hollow cenospheres of molton ash are formed. An isothermal plug flow model of char burn, which was formulated, shows that the burnout time and heat release rate per unit volume depends on the char reactivity and the excess air. A non-isothermal plug flow model of char burn, which was formulated but not solved, gives more realistic profiles of particle burnout and mole fractions of product species. Full simulation requires CFD modeling.

Cyclone combustors were developed to burn coals with a low melting point and high ash content. Other fuels such as biomass- and refuse-derived fuels are sometimes burned in suspension if the size distribution is sufficiently fine, although these fuels are more typically burned in fixed-bed systems due to the cost of size reduction. Coal-water slurries have been investigated for possible retrofitting of oil burners. With slurry combustion the spray droplet size needs to be small because the coal particles in an individual droplet tend to agglomerate as the char is formed and burnout as an agglomerate rather than as individual particles.

Particulate, NO_x, and SO_x emissions are regulated in industrial and utility suspension burning systems. Most of the ash in the fuel ends up as flyash and must be controlled with external particulate control equipment. NO_x emissions are controlled primarily by design of the air flow mixing in the burner and with tertiary air mixing in the boiler. SO_x emissions are controlled by using low sulfur fuels or with external scrubbers.

PROBLEMS

16.1. A large pulverized coal–fired dry bottom furnace is used with a Rankine cycle steam turbine to provide 500 MW of electrical power. Subbituminous coal with a higher heating value of 9000 Btu/lb_m as-received, 10% ash, and 3% sulfur is used. An electrostatic precipitator removes 99% of the particulate, and a wet limestone scrubber removes 90% of the SO_2 from the exhaust. The overall system efficiency is 35%. Calculate the following:

(*a*) The coal feed rate (tons/h)

(*b*) The particulate emissions (tons/h and $lb_m/10^6$ Btu input)

(*c*) The sulfur dioxide emissions (tons/h and $lb_m/10^6$ Btu input)

(*d*) The annual fuel cost assuming full power every hour of the year and a coal cost of $30/ton

(*e*) The limestone ($CaCO_3$) feed rate (tons/h) if a Ca/S mole ratio of 3 is required

16.2. A bituminous coal is pulverized and burned in a furnace. The ultimate analysis is 5% H, 82% C, and 13% O on a dry ash-free basis. The higher heating value is 14,000 Btu/lb_m on a dry ash-free basis. The ash is 10% on a dry basis. Assume that the fuel moisture is vaporized in the pulverizer. The fuel and 20% of the air enter the furnace at 170°F. Secondary and tertiary air enter at 600°F. Overall 10% excess air is used. Assume that the coal burns completely near the fuel nozzles. Use the gas property data of Appendix C.

(*a*) What is the transfer per lb_m coal (as-received) to the walls in order to limit the peak flame temperature to 3000°R?

(*b*) What is the heat transfer per lb_m coal in order to drop the temperature from 3000 to 2000°R in the radiant section of the furnace?

(*c*) What is the heat lost out of the stack per lb_m coal if the stack gas temperature is 340°F and the ambient temperature is 77°F?

16.3. For a pulverized coal burner rated at 165×10^6 kJ/h, estimate the diameter of the primary air plus coal nozzle and the diameter of the primary plus secondary air nozzle. Use information in the chapter and state your assumptions. Use the conditions and data given in Problem 16.2.

16.4. Verify Example 16.1. A numerical solution is suggested.

16.5. Repeat Example 16.1 using an initial pressure of 10 atm. Discuss the effect of elevated pressure on the burnout time, and indicate the relevance for an industrial gas turbine combustor.

16.6. What is the distance between particles and the number of particles per cm^3 for isothermal plug flow of 50-μm char particles to burnout completely with 20% excess air at 1600 K and 1 atm?

16.7. Repeat Example 16.1 for an isothermal cloud of coal particles which contains 50% by weight of 50-μm particles and 50% of 100-μm particles. Discuss the difference between the burnout of a monosized cloud and the two-size cut cloud.

16.8. Repeat Example 16.1 for isothermal plug flow using a char reaction rate constant which is due to diffusional effects only and thus changes as the particle shrinks. Use the shrinking-sphere constant-density model. The gas temperature is 1600 K.

16.9. For the conditions of Example 16.1, determine the effect of different excess air levels on the char burnout curve.

16.10. For the conditions of Example 16.1, estimate the pyrolysis time for bituminous coal particles of (*a*) 50 μm diameter and (*b*) 100 μm diameter using the information given in Chapter 14. Assume the particle temperature is (1) 1000 K and (2) 1500 K. How does this compare with the char burnout time?

16.11. A bituminous coal containing 12% ash and 3% sulfur with a higher heating value of 13,000 Btu/lb_m, all on an as-received basis, is burned in a large horizontally fired dry bottom pulverized coal boiler. Using the emission factors of Table 16.4 and the emission standards of Table 15.2, find the percent control required for particulate and sulfur dioxide.

16.12. If all of the sulfur in a coal were converted to sulfur dioxide, what would be the emission factor for sulfur dioxide in Table 16.4? Why are the sulfur dioxide emission factors shown in Table 16.4 less than for complete conversion of sulfur to sulfur dioxide? Explain.

16.13. Look up the melting point of each of the compounds in Table 16.2. Next, check the melting points of oxides of Ag, As, Ba, Cd, Cr, Cu, Hg, Mn, Pb, Sn, and Zn. Explain the relevance of these data for cenosphere formation and health effects from fly ash.

16.14. An 850-MW electrical power plant is designed to burn residual fuel oil. Conversion to a coal-water slurry (CWF) fuel or to pulverized coal is under consideration. A bituminous coal with a higher heating value of 13,000 Btu/lb_m on a dry basis is proposed. The CWF would have 30% water. Because of the fly ash the boiler would be derated to 685 MW. Assume that the boiler efficiency is 80% for pulverized coal. The coal costs $50/ton and the coal-water slurry costs $56/ton. Estimate the boiler efficiency when operating on CWF. Compare the 20-year fuel cost for CWF vs. pulverized coal. Assume that the plant runs at full load every hour of the year.

16.15. For a Rosin-Rammler size distribution of char particles with $d_0 = 50\ \mu m$ and $q = 1.2$, find the diameters which divide the size distribution into 10 equally weighted size increments ($y = 0.05, 0.15, 0.25, \ldots, 0.95$).

REFERENCES

Abbas, T., Costen, P. G., and Lockwood, F. C., "Solid Fuel Utilization: From Coal to Biomass," *Twenty-Sixth Symp. (Int.) on Combustion,* The Combustion Institute, Pittsburgh, pp. 3041–3058, 1996.

Atal, A., and Levendis, Y. A., "Observations on the Combustion Behavior of Coal Water Fuels Impregnated with Calcium Magnesium Acetate," *Combustion and Flame,* vol. 93, pp. 61–89, 1993.

Breen, B. P., "Combustion in Large Boilers: Design and Operating Effects on Efficiency and Emissions," *Sixteenth Symp. (Int.) on Combustion,* The Combustion Institute, Pittsburgh, 1977.

de Grey, M. A., "Les Experiences Sur Les Poussieres de Houille et la Combustion du Charbon Pulverise," Revue de Metallurgie, vol. 19, pp. 645–655, 1922.

Elliott, M. A. (ed.), *Chemistry of Coal Utilization,* Wiley-Interscience, New York, 1981.

EPA, "Compilation of Air Pollutant Emission Factors, Vol. 1: Stationary Point and Area Sources," AP-42, U.S. Environmental Protection Agency, 1985.

Essenhigh, R. H., "Coal Combustion," Chap. 3 in Wen and Lee (eds.), *Coal Conversion Technology,* Addison-Wesley, Reading, MA, 1979.

Field, M. A., Gill, D. W., Morgan, B. B., and Hawksley, P. G. W., *Combustion of Pulverized Coal,* British Coal Utilization Research Association, Leatherhead England, 1967.

"Fossil-Fired Systems: A Century of Progress," *Power,* pp. 166–203, April 1982.

Lawn, C. J. (ed.), *Principles of Combustion Engineering for Boilers,* Academic Press, New York, 1987.

Makanski, J., "Few Options Emerge for Complying with CAA Phase II," *Power,* pp. 21–27, July 1994a.

Makanski, J., "SOX/NOX Control: Fine Tuning for Phase I Compliance," *Power,* pp. 15–28, March 1994b.

Reid, W. T., "The Relation of Mineral Composition to Slagging, Fouling and Erosion during and after Combustion," *Prog. Energy Combust. Sci.,* vol. 10, pp. 159–175, 1984.

Singer, J. G., *Combustion: Fossil Power Systems,* Combustion Engineering Inc., 1981.

Smart, J. P., and Weber, R., "Reduction of NO_x and Optimisation of Burnout with an Aerodynamically Air-Staged Burner and an Air-Staged Precombustor Burner," *J. Institute of Energy,* vol. LXII, no. 453, pp. 237–245, 1989.

Smoot, L. D. (ed.), *Fundamentals of Coal Combustion for Clean and Efficient Use,* Elsevier Science, 1993.

Smoot, L. D., and Pratt, D. T., *Pulverized Coal Combustion and Gasification,* Plenum Press, New York, 1983.

Stultz, S. C., and Kitto, J. B. (eds.), *Steam: Its Generation and Use,* Babcock and Wilcox Co., Barberton, OH, 40th ed., 1992.

Syred, N., Claypole, T. C., and MacGregor, S. A., "Cyclone Combustors," Chap. 5 in C. J. Lawn (ed.), *Principles of Combustion Engineering for Boilers,* Academic Press, New York, 1987.

Tomeczek, J., *Coal Combustion,* Krieger, 1994.

Wall, T. F., "The Combustion of Coal as Pulverized Fuel through Swirl Burners," Chap. 3 in C. J. Lawn (ed.), *Principles of Combustion Engineering for Boilers,* Academic Press, New York, 1987.

Weber, R., Dugue, J., Sayre, A., and Visser, M. B., "Quarl Zone Flow Field and Chemistry of Swirling Pulverized Coal Flames: Measurement and Computations," *Twenty-Fourth Symp. (Int.) on Combustion,* The Combustion Institute, Pittsburgh, pp. 1373–1380, 1992.

Weber, R., Peters, A. F., Breithaupt, P. P., and Visser, M. B., "Mathematical Modeling of Swirling Pulverized Coal Flames: What Can Engineers Expect from Modeling?", ASME *FACT*, Vol. 17, *Combustion Modeling,* book no. H00827, 1993.

CHAPTER 17

Fluidized-Bed Combustion

A *fluidized bed* is a bed of solid particles which are set into motion by blowing a gas stream upward through the bed at a sufficient velocity to locally suspend the particles, but yet not too great a velocity to blow the particles out of the bed. The bed appears like a boiling liquid and has other features of liquids such as exhibiting buoyancy and hydrostatic head. Hence it is called a fluidized bed. The main components of a *bubbling* fluidized bed are the air plenum, the air distributor, the bed, and freeboard as shown in Figure 17.1. The freeboard is used to disengage the particles which are thrown up above the bed. A second type of fluidized bed—the *circulating* fluidized bed—is discussed in Section 17.4.

Fluidized beds have been used by the oil industry for many years to catalytically crack crude oil to make gasoline, and they are used for many other applications such as metallurgical ore roasting, limestone calcination, and petrochemical

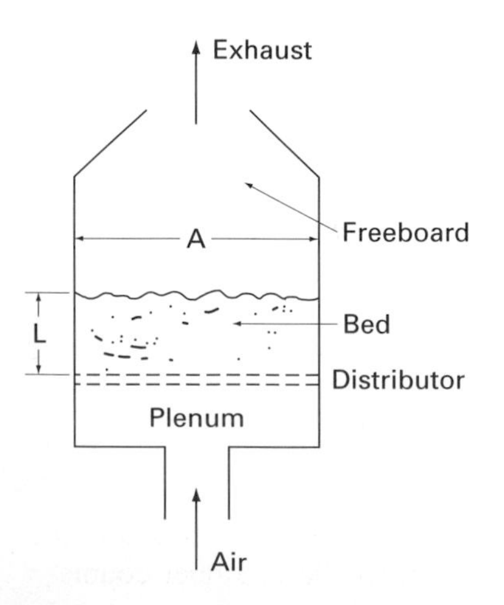

FIGURE 17.1
Schematic diagram of fluidized-bed.

production. Fluidized-bed combustion, however, is a relatively new development which has recently entered commercial markets.

While gaseous and liquid fuels may be burned in fluidized beds, the main application is for combustion of coal and other solid fuels. Furnaces and boilers utilize fluidized beds at atmospheric pressure. *Pressurized* fluidized-bed combustors are being investigated for gas turbine–steam turbine combined cycle systems (Section 17.5).

Fluidized-bed boilers overcome two fundamental limitations of conventional stoker-fired and pulverized coal boilers. Fluid-bed boilers control SO_x and NO_x emissions within the combustion chamber, thus eliminating the need for scrubbers and elaborate combustion modifications. Second, fuel flexibility is enhanced by allowing the use of a range of solid fuels with widely varying ash and moisture contents. High-moisture fuels can be burned in fluidized beds without danger of combustion instability or loss of ignition. High-ash fuels can be burned with much less danger of slagging, because the combustion temperature is lower due to in-bed heat transfer.

This chapter focuses on the fundamentals of fluidization and in-bed combustion processes related to bubbling fluidized beds, circulating fluidized beds, and pressurized fluidized beds.

17.1 FLUIDIZATION FUNDAMENTALS

For most combustion applications the particle size is greater than 1 mm, so this chapter omits the fluidization of powders. Typically the bed consists of solids of a variety of sizes and shapes. As a practical matter, the effective bed diameter can be determined from a screen analysis which separates the bed material into a series of size increments. A mean diameter is calculated from the screen analysis based on the mean particle surface area, which is appropriate for consideration of pressure drop and surface reaction rates. The mean specific surface area per unit volume is defined as

$$\bar{A}_p = \sum_i \left(\frac{\text{Surface area}}{\text{Volume}}\right)_i y_i \tag{17.1}$$

where y_i is the mass fraction of increment i. For an equivalent sphere, the surface/volume ratio is $6/d_p$, and thus,

$$\bar{A}_p = 6 \sum_i \left(\frac{y_i}{d_i}\right) = \frac{6}{\bar{d}} \tag{17.2}$$

where $\bar{d}$ is the surface mean diameter given by

$$\bar{d} = \frac{1}{\sum_i (y_i/d_i)} \tag{17.3}$$

Note that this is equivalent to the Sauter mean diameter, defined in Section 9.2 (see Problem 17.1), but is obtained from mass fractions rather than number counts.

The bed will occupy a certain total volume, and there will be a certain void space containing only gas within that volume. The void fraction ϵ is defined as

$$\epsilon = \frac{\text{Void volume}}{\text{Bed volume}} \tag{17.4}$$

The void volume fraction is approximately equal to the fractional cross-sectional area of the voids at any point in the bed. The effective local velocity through the bed is called the *interstitial velocity,*

$$\underline{V}_I = \frac{\dot{V}}{A\epsilon} \tag{17.5}$$

where $\dot{V}$ is the volume flow rate through the bed and A is the cross-sectional area of the bed as shown in Figure 17.1. The so-called *superficial velocity* $\underline{V}_s$ is simply the gas velocity if the bed particles were not present:

$$\underline{V}_s = \frac{\dot{V}}{A} \tag{17.6}$$

EXAMPLE 17.1. For the screen analysis below, find the mean particle diameter, the mean particle surface area per unit particle volume, and the mean particle surface area per unit volume of bed. The void fraction is measured to be 0.40.

Tyler mesh no.	Diameter (mm)	Weight on screen (kg)
8	2.36	0
10	1.65	60
14	1.17	80
20	0.83	40
35	0.42	20
48	0.29	0
Total		200

Solution. Determine the mass fractions in each size range and evaluate Eq. 17.3. Note that $d_i = 2.00$ below comes from $(2.36 + 1.65)/2$ above, for example.

d_i (mm)	y_i	y_i/d_i
2.00	0.30	0.150
1.41	0.40	0.284
1.00	0.20	0.20
0.62	0.10	0.161
0.35	0.0	0.0
Sum	1.00	0.795

Hence, from Eq. 17.3, the particle surface mean diameter is

$$\overline{d} = \frac{1}{0.795} = 1.26 \text{ mm}$$

The mean surface area per unit particle volume is

$$\overline{A}_p = \frac{6}{1.26 \times 10^{-3}} = 4762 \text{ m}^2/\text{m}^3$$

The mean particle surface area per unit bed volume is

$$\overline{A}_v = \overline{A}_p(1 - \epsilon) = 4762(1 - 0.4) = 2857 \text{ m}^{-1}$$

Pressure Drop across the Bed

Consider first a bed in which the flow rate is not sufficient to fluidize the bed, but rather it remains as a packed bed. The gas flows up through a tortuous path in the bed. In an analogous manner to pipe flow, the pressure drop across the bed divided by the depth of bed, L, is related to the interstitial velocity, an effective diameter of the tortuous path through the bed, and the density and viscosity of the gas:

$$\frac{\Delta p}{L} = \text{function}(\underline{V}_I, d_{\text{eff}}, \rho, \mu) \tag{17.7}$$

From dimensional analysis there are two dimensionless groups, which may be written as

$$\left(\frac{\Delta p}{L}\right)\left(\frac{d_{\text{eff}}}{\rho \underline{V}_I^2}\right) = \text{function}\left(\frac{\underline{V}_I d_{\text{eff}} \rho}{\mu}\right) \tag{17.8}$$

The effective diameter of the tortuous path may be obtained from the hydraulic diameter concept. The hydraulic diameter may be defined as 4 times the volume of fluid divided by the surface area wetted by the fluid. The volume of fluid is ϵLA, where A is the cross-sectional area of the bed. The wetted surface area is the total number of particles in the bed times the surface area of one particle. The total number of particles is $LA(1 - \epsilon)/(\pi \overline{d}^3/6)$. Putting all of this together,

$$d_{\text{eff}} = \frac{(4\epsilon LA)(\pi \overline{d}^3/6)}{LA(1 - \epsilon)(\pi \overline{d}^2)} = \frac{2}{3}\frac{\epsilon \overline{d}}{(1 - \epsilon)} \tag{17.9}$$

Following the practice with pipe flow, Eq. 17.8 is written as

$$f_{pm} = \text{function}(\text{Re}_{pm}) \tag{17.10}$$

where the porous media friction factor f_{pm} is defined as

$$f_{pm} = \left[\frac{\epsilon^3 \overline{d}}{\rho \underline{V}_s^3(1 - \epsilon)}\right]\left(\frac{\Delta p}{L}\right) \tag{17.11}$$

and the porous media Reynolds number is defined as

$$\text{Re}_{pm} = \frac{\overline{d}\,\underline{V}_s \rho}{(1 - \epsilon)\mu} \tag{17.12}$$

Experiments in packed beds have shown that Eq. 17.10 holds and has the functional form

$$f_{pm} = \frac{150}{\text{Re}_{pm}} + 1.75 \tag{17.13}$$

which is the so-called Ergun equation [Kunii and Levenspiel]. The first term represents laminar flow and the second term represents turbulent flow, which, because of the extreme roughness, is independent of Reynolds number.

Combining Eqs. 17.11–17.13, the pressure drop across a packed bed may be written as

$$\Delta p = \frac{150\underline{V}_s\mu(1-\epsilon)^2L}{\epsilon^3\overline{d}^2} + \frac{1.75\underline{V}_s^2(1-\epsilon)\rho L}{\epsilon^3\overline{d}} \tag{17.14}$$

Minimum Fluidization Velocity

The onset of fluidization occurs when the drag force on the particles in the bed due to the upward-flowing gas just equals the weight of the bed. This is equivalent to stating that the pressure drop across the bed times the area of the bed just equals the weight of the bed at the onset of fluidization:

$$\Delta p\, A = LA(1-\epsilon)(\rho_p - \rho)g \tag{17.15}$$

Substituting Eq. 17.14 into Eq. 17.15 and rearranging slightly, a quadratic equation is obtained for the minimum superficial fluidization velocity, $\underline{V}_{mf}$:

$$\frac{1.75\rho}{\overline{d}\epsilon^3}\underline{V}_{mf}^2 + \frac{150\mu(1-\epsilon)}{\overline{d}^2\epsilon^3}\underline{V}_{mf} - (\rho_p - \rho)g = 0 \tag{17.16}$$

The first term is negligible if Re_{pm} is less than 10, and the second term is negligible if Re_{pm} is greater than 1000.

As the flow rate is increased beyond the point of minimum fluidization, the pressure drop remains essentially constant. When the flow rate is increased further, eventually the particles will be blown out of the bed. The limiting case is the single-particle terminal velocity $\underline{V}_t$. This pressure-velocity behavior is summarized in Figure 17.2. Since $\underline{V}_t$ is at least 10 times greater than $\underline{V}_{mf}$ for uniform-size

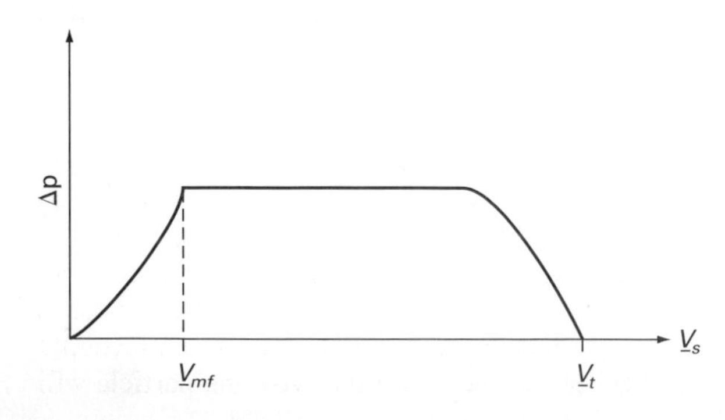

FIGURE 17.2
Pressure drop across bed vs. superficial velocity.

particles, there is a range of operating flow rates for fluidization. Of course, if the bed contains large particles and small particles, the small particles will be elutriated (blown out) before the large particles.

Single-Particle Terminal Velocity

The terminal velocity of a single particle is obtained by setting the weight equal to the drag. The drag is the drag coefficient C_D times the frontal area times the dynamic pressure. Thus, in equilibrium,

$$\left(\frac{\pi d^3}{6}\right)\rho_p g = C_D\left(\frac{\pi d^2}{4}\right)\left(\frac{\rho \underline{V}_t^2}{2}\right)$$

so that

$$\underline{V}_t = \left[\frac{4d\rho_p g}{3C_D\rho}\right]^{1/2} \tag{17.17}$$

For a spherical particle,

$$C_D = \frac{24}{\text{Re}_d} + \frac{4}{\text{Re}_d^{0.333}} \quad \text{for} \quad \text{Re}_d < 1000 \tag{17.18a}$$

and

$$C_D = 0.424 \quad \text{for} \quad 1000 < \text{Re}_d < 200{,}000 \tag{17.18b}$$

Bubbling Beds

At minimum fluidization the bed is homogeneous in the sense that the gas surrounds each particle, and the particles are set into motion. The entire bed behaves as a so-called *dense phase* fluid. As the superficial velocity is increased above minimum fluidization, *bubbles* are formed. These bubbles are essentially void of solids and contain primarily gas. The bubbles are referred to as the *dilute phase.* The initial size of the bubbles depends on the type of air distributor plate used, as indicated in Figure 17.3. A plate with a few large orifice inlets will have larger bubbles near the plate, while a plate with many small inlets will have many small bubbles near the plate. However, as the bubbles rise, they coalesce, growing in size, and the distribution between dense phase and dilute phase becomes independent of the inlet air distribution design.

The bubbles play a vital role in the behavior of the fluidized bed. In practice, bubbles cannot be avoided. As the superficial velocity is increased, the added gas flow goes into the bubbles (dilute phase) rather than into the dense phase. The dense phase remains essentially at minimum fluidization. As the bubbles rise due to buoyancy, they carry along nearby solid particles, thus providing the main mechanism for large-scale mixing of the bed. When the bubbles reach the top of the bed, they break through the surface and the upward momentum flings the particles above the bed. If the superficial velocity above the bed is less than the terminal velocity of a given particle, it will fall back to the bed; otherwise the particle will

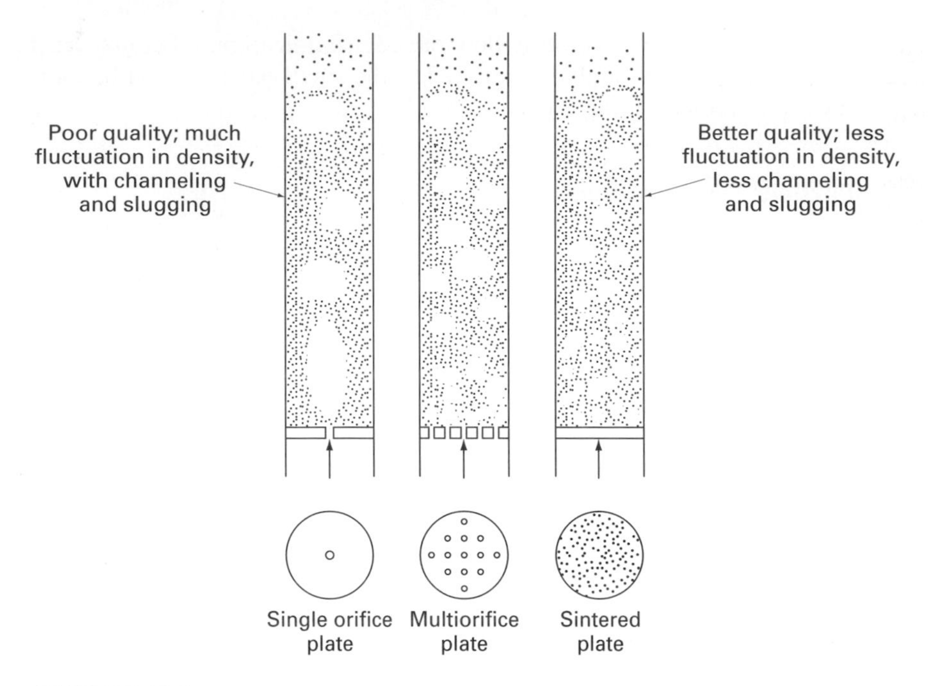

FIGURE 17.3
Quality of fluidization as influenced by type of gas distributor.

be entrained by the exhaust. In vigorously bubbling beds the particles may be flung upward for distances of several bed depths, and a so-called transport disengagement height or freeboard must be provided to allow the particles to fall back into the bed.

The rise rate of a single bubble of diameter d_B in a fluidized bed is given by

$$\underline{V}_B = K\sqrt{\frac{gd_B}{2}} \tag{17.19}$$

where K is a constant which depends on the size and shape of the particles and is generally about 0.9. For millimeter-size bed particles the bubble rising velocity is typically less than the interstitial velocity, and the bubbles coalesce due to lateral motion rather than being overtaken from below. Heat transfer tubes within the bed will tend to break up large bubbles should they form.

With deep beds operating at high superficial velocities without in-bed tubes, the bubbles may grow until they occupy the whole cross-sectional area of the bed. These bubbles carry a slug of particles ahead of them until instability occurs and the solids collapse back to the bed. This is the so-called *slugging* mode of operation, which can cause considerable vibration of the system and is generally undesirable. More fine particles will tend to be elutriated under slugging conditions.

In the case of fluidized-bed combustors, the fuel may be of a different size and density from the bed material. If the fuel is lighter and of larger size than the bed material, the fuel may tend to segregate or float on the top of the bed. In many instances this can be overcome by setting the flow rate so that the bed is bubbling vigorously. Of course, this will also tend to elutriate small particles from the bed.

Due to the bubbles, there are pressure fluctuations within the bed, and these pressure fluctuations can feed back across the inlet distributor plate, thus altering the flow rate. Experience has shown that a pressure drop across the inlet air distributor plate of about 12% of the pressure drop across the bed will isolate the inlet plenum sufficiently from the bed pressure fluctuations. Various inlet air distributor plate designs have been utilized to ensure good fluidization. The nozzle stand pipe type of distributor (Figure 17.4*a*) has been found to be well suited for fluidized-bed combustion. The air enters the bed from holes at the top of the nozzle, and the static bed forms an insulating layer between the hot fluidizing layer and the base plate. The nozzles are usually arranged on a pitch of 75–100 mm over the base plate and have a diameter of 12–15 mm and a height of 50–100 mm. The nozzle hole size is a compromise between having an excessive number of nozzles and preventing particle fallthrough. The bubble cap design is sometimes used to prevent bed particle fallthrough (Figure 17.4*b*).

Heat and Mass Transfer

Heat and mass transfer considerations play an important role in the design and operation of fluidized-bed combustors. Heat transfer from gas to particles is important in determining the rate of heat-up and devolatilization of solid fuel particles. Botterill reviewed the literature in this area and recommends the following expression for a particle Nusselt number for millimeter-sized particles up to 20 atm pressure:

$$\mathrm{Nu}_p = 0.055\ \mathrm{Re}_p^{0.77}\left(\frac{\rho_g}{\rho_{g0}}\right)^{0.2} \tag{17.20}$$

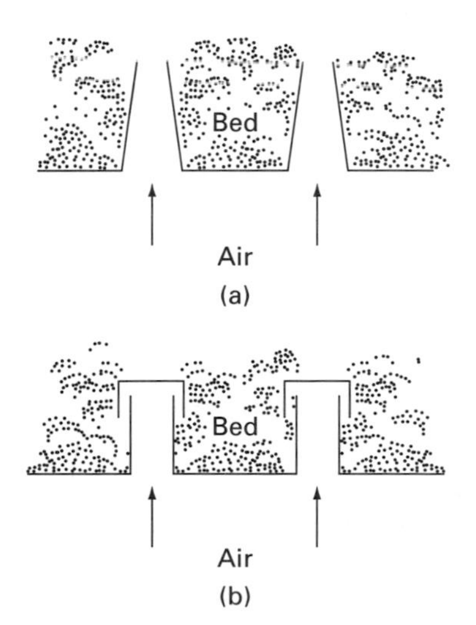

FIGURE 17.4
Air distributor plate for a bubbling fluidized bed: (*a*) nozzle stand pipe type, and (*b*) bubble cap type.

where ρ_{g0} is the gas density at atmospheric pressure and ρ_g is the gas density at the working pressure. The increase in heat transfer with increasing pressure is attributed to improvement in the quality of fluidization due to more gas passing through the dense phase.

The heat transfer to tubes immersed in a fluidized bed increases with decreasing bed particle size and with increasing tube size. The heat transfer coefficient is 5 to 10 times higher than for conventional gas-to-surface heat exchange surfaces, depending on the particle size. For example, in fluidized-bed combustors burning coal in millimeter-sized beds, heat transfer coefficients to water tubes of approximately 200–350 $W/m^2 \cdot K$ are observed.

Mass transfer of oxygen from the gas to the fuel particles sets the burning rate of the char. Part of the inlet air passes through the bubble phase and does not contact the particles. Mass transfer theories to account for the bubbles tend to be complex. A certain fraction of the oxygen, say, 20–40%, does not react with the char surface, but rather reacts with the volatiles in the bed or in the freeboard. Considering diffusion within the dense phase (excluding the bubbles), the oxygen diffusion equation, assuming a relatively quiescent gas, may be written (r is radius here), as

$$\frac{d}{dr}\left(\epsilon r^2 \frac{d\rho_{O_2}}{dr}\right) = 0 \tag{17.21}$$

The solution, following Chapter 14, is

$$\frac{\tilde{h}_D \bar{d}}{D_{AB}} = \text{Sh} = 2\epsilon \tag{17.22}$$

where ϵ, the void fraction, is that associated with minimum fluidization, and typically for combustion applications may be taken as 0.4. For significant flow through the dense phase, Eq. 17.22 may be modified, based on Eq. 14.21, to give

$$\text{Sh} = \left[2\epsilon + 0.6\left(\frac{\text{Re}_{mf}}{\epsilon}\right)^{0.5} \text{Sc}^{0.33}\right]\phi \tag{17.23}$$

Using Eq. 17.23, the theory of Chapter 14 may be used, as an approximation, for the burning rate of single particles in a fluidized bed with certain additional features discussed in the next section.

17.2 COMBUSTION IN A BUBBLING BED

In fluidized-bed combustion the bed temperature is maintained well below the melting point of the ash in the fuel to eliminate slagging of the ash. Frequently it is desired to capture SO_2 in the bed using an active bed such as limestone ($CaCO_3$), which is calcined in the bed to form calcium oxide (CaO). The optimum temperature for CaO reaction with SO_2 to form $CaSO_4$ at atmospheric pressure is

815–900°C (1500–1650°F). This temperature is low for combustion, compared with premixed flames, suspension burning, and grate burning of solid fuels. However, this is a perfectly adequate temperature for many applications. For example, current steam turbine temperatures do not exceed 540–590°C (1000–1100°F). In addition to facilitating sulfur dioxide capture and soft ash particles, bed temperatures of 815–900°C result in relatively low NO_x emissions, less volatilization of alkali compounds, and less erosion of in-bed boiler tube surfaces.

In order to take advantage of the high heat transfer rates associated with fluidized beds, boiler tubes are frequently immersed in the bed. As heat is removed more fuel must be added to maintain a constant bed temperature, but the bed fuel air ratio should not exceed stoichiometric. The question is what fraction of the fuel energy can be removed without lowering the bed temperature too much. To answer this question let us consider mass and energy balances on a control volume around the bed as shown in Figure 17.5. First let us neglect the effect of the bubbles and assume that combustion is complete within the bed, then allow for some combustion above the bed, and finally include the effect of the bubbles and combustion above the bed.

Neglect Bubbles and Assume Complete Combustion in Bed

Conservation of mass across the bed requires that

$$\dot{m}_a + \dot{m}_f = \dot{m}_p \tag{17.24}$$

The flow rate of products depends on density of the gaseous products, the superficial fluidization velocity, and the cross-sectional area of the bed:

$$\dot{m}_p = \rho_g \underline{V}_s A \tag{17.25}$$

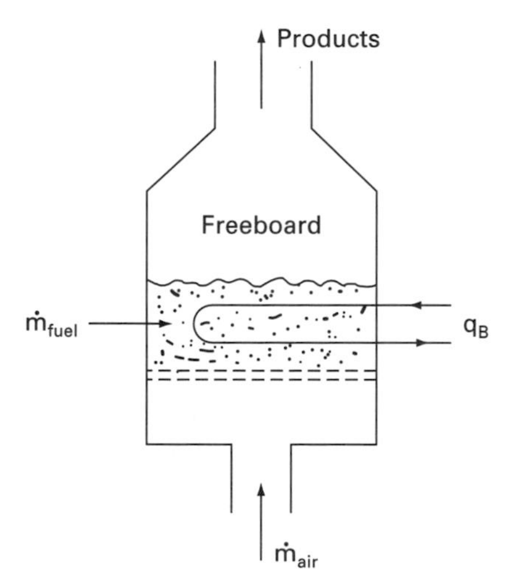

FIGURE 17.5
Schematic for mass and energy balance.

Combining Eqs. 17.24 and 17.25 and using $\dot{m}_a = \dot{m}_f/f$, the fuel feed rate becomes

$$\dot{m}_f = \left(\frac{f}{1+f}\right)(\rho_g \underline{V}_s A) \tag{17.26}$$

The energy equation across the bed with heat transfer from the bed to in-bed tubes, q_b, and assuming 100% heat release in the bed, is

$$\dot{m}_a h_a + \dot{m}_f h_f = \dot{m}_p h_p + q_B \tag{17.27}$$

Using sensible enthalpies and assuming complete combustion,

$$\dot{m}_a h_{sa} + \dot{m}_f(\text{LHV}) = \dot{m}_p h_{sp} + q_B \tag{17.28}$$

Dividing by $\dot{m}_f$ and introducing $r_B = q_B/\dot{m}_f$ LHV, which is the fraction of input heat extracted from the bed, the energy equation becomes

$$h_{sa} + f\,\text{LHV}(1 - r_B) = (1 + f)h_{sp} \tag{17.29}$$

Equation 17.29 shows the relationship between the heat extracted and the fuel-to-air ratio required to maintain a given bed temperature. Of course, the value of f should not exceed stoichiometric for Eq. 17.29 to be valid.

In order to extract the heat q_B from the bed with in-bed heat exchanger tubes, the tubes must have sufficient surface area A_t such that

$$q_B = \tilde{h} A_t\, \Delta T \tag{17.30}$$

where ΔT is the log mean temperature difference across the tubes:

$$\Delta T = \frac{(T_B - T_{\text{out}}) - (T_B - T_{\text{in}})}{\ln\left(\dfrac{T_B - T_{\text{out}}}{T_B - T_{\text{in}}}\right)} \tag{17.31}$$

where T_{in} and T_{out} are the temperatures into and out of the heat exchanger. Now, A_t can be related to the fraction of the bed volume occupied by the tubes, Φ; the bed depth L; the cross-sectional area of the bed A; and the tube diameter d_t:

$$A_t = \frac{4\Phi L A}{d_t} \tag{17.32}$$

For a given heat exchanger design, it is apparent from Eq. 17.32 that the fluidized bed must have sufficient depth in order to provide for a given heat removal.

EXAMPLE 17.2. A fluidized bed burns coal with a lower heating value of 25,000 kJ/kg. The bed temperature is maintained at 877°C, and the inlet air is at 127°C. Neglect the bubbles. Assume that the combustion is complete in the bed and neglect any ash in the fuel. Find the fuel/air ratio for bed heat removal fractions from 0 to 0.7.

Solution. Assuming the products to be air (for simplicity) and using Appendix B with a reference temperature of 300 K for compatibility with the lower heating value, Eq. 17.29 becomes

$$(401.3 - 300.4)\ \text{kJ/kg} + (25{,}000\ \text{kJ/kg})f(1 - r_B) = (1 + f)(1219.2 - 300.4)\ \text{kJ/kg}$$

Tabulating r_B versus f,

Heat removal ratio r_B	fuel/air ratio f
0	0.0340
0.1	0.0379
0.2	0.0429
0.3	0.0493
0.4	0.0581
0.5	0.0706
0.6	0.0901
0.7	> 0.1

This table gives the relation between r_B and f which is necessary to maintain a bed temperature of 877°C. Knowing f, the fuel feed rate is obtained from Eq. 17.26. As the heat extracted is increased, the fuel feed rate is increased accordingly. The equation does not hold for f greater than stoichiometric which is typically about 0.1.

EXAMPLE 17.3. Find the required bed height for a fluidized-bed combustor with the following design parameters:

$$\underline{V}_s = 3 \text{ m/s} \qquad d_t = 25.4 \text{ mm}$$

$$T_B = 877°\text{C} \qquad \Phi = 0.17$$

$$T_a = 127°\text{C} \qquad r_B = 0.5$$

$$\text{LHV} = 24{,}428 \text{ kJ/kg} \qquad p = 1 \text{ atm}$$

$$T_{\text{in}} = 127°\text{C} \qquad \tilde{h} = 250 \text{ W/m}^2\cdot\text{K}$$

$$T_{\text{out}} = 827°\text{C}$$

Solution. From the definition of r_B,

$$q_B = r_B \dot{m}_f(\text{LHV})$$

Substituting Eq. 17.26 for $\dot{m}_f$,

$$q_B = r_B\left(\frac{f}{1+f}\right)\rho_g \underline{V}_s A(\text{LHV})$$

Dividing by the area of the bed and using the ideal gas law,

$$\frac{q_B}{A} = r_B\left(\frac{f}{1+f}\right)\frac{p_g}{RT_B}\underline{V}_s(\text{LHV})$$

Following the method of Example 17.2, $f = 0.0725$, and assuming the molecular weight is 29.0,

$$\frac{q_B}{A} = (0.5)\left(\frac{0.0725}{1.0725}\right)\left[\frac{(101 \text{ kPa})(29 \text{ kg/kgmol})(3 \text{ m/s})(30{,}000 \text{ kJ/kg})}{(8.314 \text{ kPa}\cdot\text{m}^3/\text{kgmol}\cdot\text{K})(1150 \text{ K})}\right]$$

$$= 932 \text{ kW/m}^2$$

Substituting Eq. 17.30 into Eq. 17.32 and solving for the bed depth,

$$L = \frac{q_B}{A} \frac{d_t}{4\tilde{h}\Phi\, \Delta T}$$

where from Eq. 17.31,

$$\Delta T = \frac{(877 - 827) - (877 - 127)}{\ln\left(\dfrac{877 - 827}{877 - 127}\right)} = 258 \text{ K}$$

Substituting numerical values,

$$L = \left(932 \frac{\text{kW}}{\text{m}^2}\right)\left[\frac{0.0254 \text{ m}}{4(0.250 \text{ kW/m}^2\cdot\text{K})(258 \text{ K})(0.17)}\right] = 0.54 \text{ m}$$

The minimum bed depth to cover the tubes is 0.54 m, and allowing for clearance at the top and bottom of the bed, a bed height of 1 m is reasonable.

Neglect Bubbles but Include Some Combustion above the Bed

Let X equal the fraction of fuel burned in the bed, and $\dot{m}_{pB}$ equal the mass flow rate of combustion products in the bed, which includes some oxygen. Then conservation of mass in the bed requires

$$\dot{m}_a + X\dot{m}_f = \dot{m}_{pB} \tag{17.33}$$

Conservation of mass in the freeboard above the bed requires

$$\dot{m}_{pB} + (1 - X)\dot{m}_f = \dot{m}_p \tag{17.34}$$

where

$$\dot{m}_{pB} = \rho_g \underline{V}_s A$$

Conservation of energy in the bed requires

$$\dot{m}_a h_a + X\dot{m}_f h_f = \dot{m}_{pB} h_B + q_B \tag{17.35}$$

Relationships similar to those in Examples 17.2 and 17.3 can be developed for this case also.

Include Effect of Bubbles and Some Combustion above the Bed

To account for the bubbles, assume that a certain fraction of air, B, flows through the dense phase and $1 - B$ passes through the bed via the bubbles without reacting with fuel in the bed. Further assume that there is no heat and mass transfer between the bubbles and the dense phase. The conservation of mass expression for the dense phase in the bed is

$$(1 - B)\,\dot{m}_a + X\dot{m}_f = \dot{m}_{pD} \tag{17.36}$$

where

$$\dot{m}_{pD} = \rho_g \underline{V}_{mf} A$$

Conservation of mass in the freeboard requires that

$$\dot{m}_{pD} + B\dot{m}_a + (1 - X)\dot{m}_f = \dot{m}_p \tag{17.37}$$

The dense-phase bed energy equation is

$$(1 - B)\dot{m}_a h_a + X\dot{m}_f h_f = \dot{m}_{pD} h_B + q_B \tag{17.38}$$

Relationships similar to Examples 17.2 and 17.3 can be developed for this case. Advanced theories of fluidization account for mass and heat transfer between the dense phase and the bubbles [Howard].

Fuel Holdup in the Bed

As indicated in Example 17.2, when the in-bed heat load is increased, the fuel feed rate is increased to maintain the bed temperature. This means that the excess air and the burning rate of the fuel change. With a fluidized bed the fuel solids circulate in the bed until they are burned up, except for fines, which tend to be blown out of the bed. As the burning rate changes, the amount of fuel in the bed at any given instant, called the *fuel holdup,* also changes. The airflow must be great enough to fluidize the bed, and yet low enough to prevent excessive entrainment of fine particles from the bed.

The amount of fuel in the bed at any given time (the fuel holdup) may be approximated from the fuel feed rate and the total burn time of the individual particle:

$$m_f = \dot{m}_f t_b \tag{17.39}$$

Since the char burn is typically an order of magnitude longer than the volatile burn, the single-particle char burn time from Chapter 14 is suggested here. The shrinking-sphere model, modified by Eq. 17.23 to account for the bed material surrounding the fuel particle, may be used to obtain the burn time. However, the oxygen concentration will be lower in the dense phase than in the freeboard due to the bypassing effect of the bubbles. Furthermore, there are two additional complicating factors—coal particles tend to *fracture* into smaller sizes, and the particle surface regression rate is increased by *attrition* due to collisions with the bed material. Hence the apparent burning rate of a single particle in a fluidized bed is more rapid than that of a free burning particle.

Fracturing of coal particles occurs when the coal particles are fed into the high–heat transfer environment of the fluidized bed. During devolatilization, pressure builds up inside the particle, causing fracture planes and fragmentation of the parent particle. The fracture planes are especially evident in lignite coals, as shown in Figure 17.6. Fragmentation during devolatilization is called *primary fragmentation,* and typically one to three primary fragmentations occur per particle. As the char burns, the fracture planes deepen, and *secondary fragmentation* takes place, producing an additional 20–50% increase in the number of char particles. Also, as the char burns, the surface of the char is weakened, since the oxygen penetrates into the char; and as bed particles collide with the char surface, small pieces of char are

FIGURE 17.6
Microphotograph of lignite char particle after removal from fluidized-bed combustor. White mark is 100 μm.

broken off by a process of attrition. In addition, char particles can collide with in-bed tubes, causing further fragmentation.

The processes of fragmentation and attrition are shown schematically in Figure 17.7. Fragmentation alters the size distribution, and attrition increases the apparent particle burning rate. Experiments suggest that the attrition rate per unit bed volume can be calculated from an attrition rate constant k_{attr}, the difference between the superficial velocity and the minimum fluidization velocity, and the surface area per unit volume of the char in the bed:

$$\frac{\dot{m}_{attr}}{V} = \frac{k_{attr}\,(\underline{V} - \underline{V}_{mf})\,\rho_c\,\overline{A}_p}{6} \tag{17.40}$$

The attrition rate constant depends on the bed particle size and the type of char. Figure 17.8 shows that the attrition rate constant is 1×10^{-6} for coal char in 1-mm sand bed material. The effect of attrition is to increase the effective burning rate of char particles in a fluidized bed by a factor of 25–100% compared with char particles burning in air without attrition. The net result is that the fuel consumption rate $\dot{m}_f$ due to chemical reaction and attrition is such that typically the bed consists of 95% inerts and 5% coal.

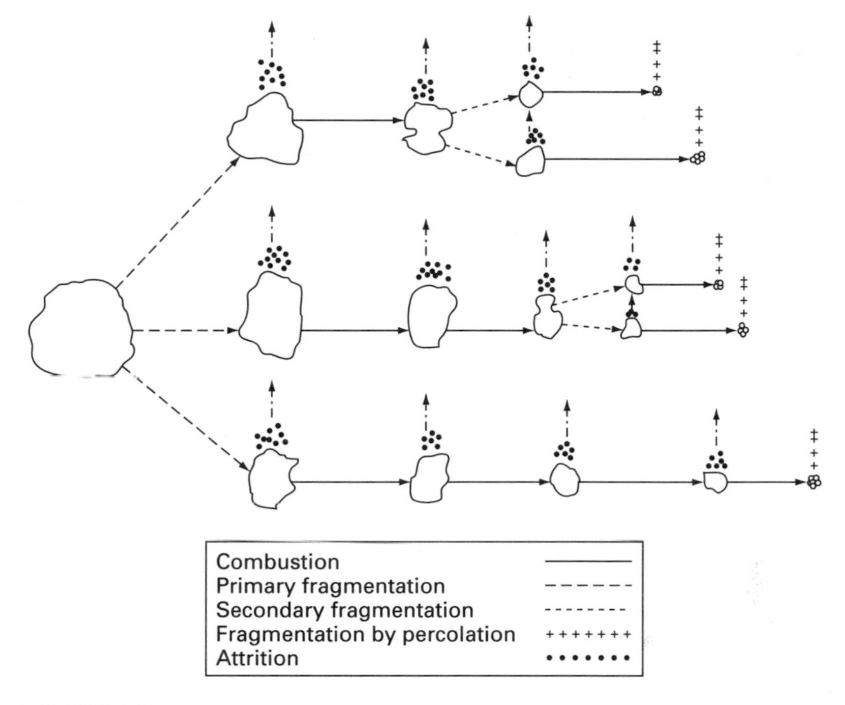

FIGURE 17.7
Schematic representation of coal particle fragmentation and attrition [Arena et al., reproduced with permission of the American Inst. of Chemical Engineers, © 1983 AICHE. All rights reserved.].

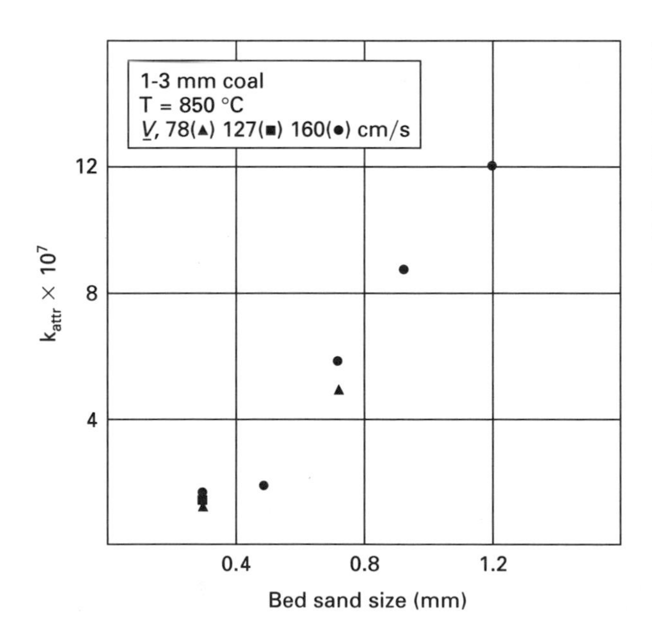

FIGURE 17.8
Attrition rate constant for coal char vs. bed sand size [Arena et al., reproduced with permission of the American Inst. of Chemical Engineers, © 1983 AICHE. All rights reserved.].

17.3 ATMOSPHERIC FLUIDIZED-BED COMBUSTION SYSTEMS

Fluidized-bed combustion systems are attractive because better control over combustion is realized so that peak flame temperatures are avoided. Slagging of the ash is eliminated, corrosion problems are reduced, nitrogen oxide emissions are reduced, and sulfur control with limestone addition to the bed is possible. Since in-bed heat transfer coefficients are high, fluidized-bed boilers are more compact than fixed-bed and suspension-burning systems. The disadvantages are that the pressure drop across the bed is greater, the ash carryover is higher, and in-bed tube erosion can be high. Crushed stoker-size coal is used. Fuel is fed from the top, or it can be fed under the bed by a ram or pneumatically. A wide range of fuels including high-ash, high-moisture fuels can be used.

A fluidized-bed boiler, shown schematically in Figure 17.9, contains the bed, in-bed boiler tubes, sufficient freeboard height to avoid excessive elutriation of bed material and unburned carbon, a convective steam raising section, and a superheater section. Coal and limestone are fed from above the bed. The bed temperature is maintained below 950°C to avoid agglomerating the ash and above 800°C to avoid poor combustion efficiency. The gases leave the bed at the bed temperature. Combustion of fines and volatiles continues in the freeboard. In-bed heat transfer allows control of excess air without increasing the bed temperature, and as noted previously, the lower the excess air, the higher the boiler efficiency, up to the point where combustion is incomplete.

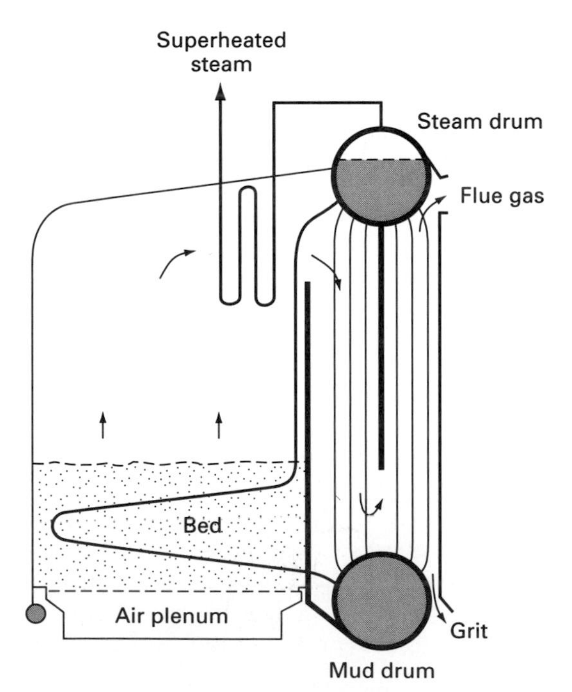

FIGURE 17.9
Schematic of bubbling fluidized-bed boiler [Howard].

With a low-grade, high-moisture fuel, the heat release may be only slightly in excess of that required to heat the fuel and air to a bed temperature required for efficient combustion. In this case immersed tubes are not used, and the bed temperature is controlled by supplying excess air to remove heat from the bed for subsequent recovery in convective tube banks. With high-grade fuels, combustion in a bed without cooling surfaces requires at least 150% excess air to maintain a 900°C bed.

For optimum efficiency over the turndown range of a boiler, the excess air level should remain constant as the air supply and fuel feed rate are changed to vary the boiler output. Since the bed temperature can be raised only 100–200°C without increasing emissions of SO_x and NO_x, and since load turndown of 2/1 or, better, 4/1 is desired, the main alternatives are to provide multiple compartments which can be shut off independently, or to provide a means of gradually exposing in-bed tubes. When the fluidizing air is reduced, bed expansion is reduced, which can provide a limited method of exposing bed tubes. With limestone feed the bed height can be raised by increasing the limestone feed, and conversely bed material can be drained off to lower the bed. Some fluidized-bed designs employ novel geometry to control circulation and thus heat transfer to the tubes, as a means of load control.

Atmospheric Emissions

Emissions of major significance are particulates, sulfur dioxide, and nitric oxide. Particulates include ash, carbon, and bed material. Essentially all of the ash in the fuel is blown out of the bed. A few percent of unburned carbon may also be blown out of the bed. The challenge is to design the system so that the unburned carbon carryover is low. In addition, the bed material (usually limestone) tends to elute fine particles due to attrition. Hence, particulate carryover is high and must be controlled by a fabric filter baghouse or related device. The emission standards noted in Table 15.2 also apply to fluidized-bed combustion of solid fuels.

Sulfur dioxide reacts with the limestone bed to form calcium sulfate. The efficiency of the sulfur dioxide removal depends on the calcium/sulfur molar ratio. As indicated in Figure 17.10, Ca/S ratios of 2 to 5 are required for 90% removal of SO_2 in an atmospheric fluidized bed operating at 900°C depending on the reactivity of the limestone. If the bed temperature increases, the SO_2 removal efficiency decreases. For a pressurized fluidized bed using dolomitic limestone (which contains $MgCO_3$ as well as $CaCO_3$), the Ca/S ratio is in the range 1.5–2 for 90% SO_2 removal.

Nitrogen oxide emissions are lower than with spreader stokers and pulverized combustors, since the combustion temperature is lower. As a first approximation we may assume that the thermal NO is negligible. However, the fuel-bound organic nitrogen is converted to NO, and as a starting point it can be assumed that all of the fuel N is converted to NO.

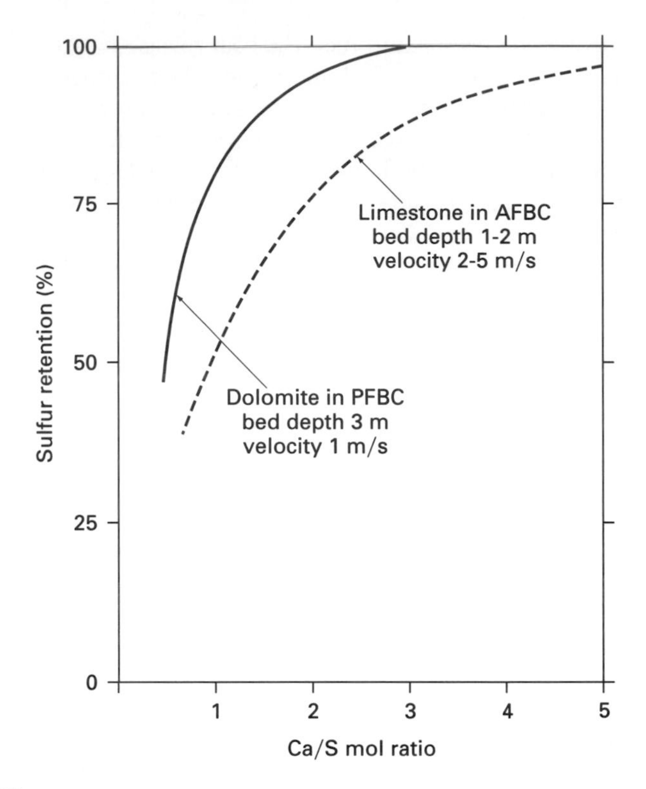

FIGURE 17.10
Variation of sulfur retention with Ca/S molar ratio for atmospheric (AFBC) and pressurized fluidized-bed combustion (PFBC) [McClung et al., in *Fluidized Bed Projects and Technology,* pp. 85–92, © 1994, by permission of ASME].

17.4 CIRCULATING FLUIDIZED BEDS

In the standard bubbling atmospheric fluidized-bed combustor, small, lightweight particles such as char near burnout tend to be blown out of the bed before combustion is complete. Also, a feed point is needed for approximately every 1 m^2 for good mixing of the coal, which is a limitation for large systems. Circulating fluidized beds have been developed to overcome the tendency for high carbon carryover of bubbling beds and to facilitate fuel feeding.

Fixed-bed, bubbling fluidized-bed, and circulating fluidized-bed combustors are schematically represented in Figure 17.11. For circulating fluidized beds, the velocity is increased beyond the particle entrainment velocity, so that the solids are transported up the full height of the chamber and returned in the downward leg of a cyclone separator. The pressure drop across the system is a function of the velocity and the particle loading. This type of fluidization is characterized by high turbulence, solids back mixing, and the absence of a defined bed level. Fuel is fed into the lower part of the combustion chamber, and primary air is introduced

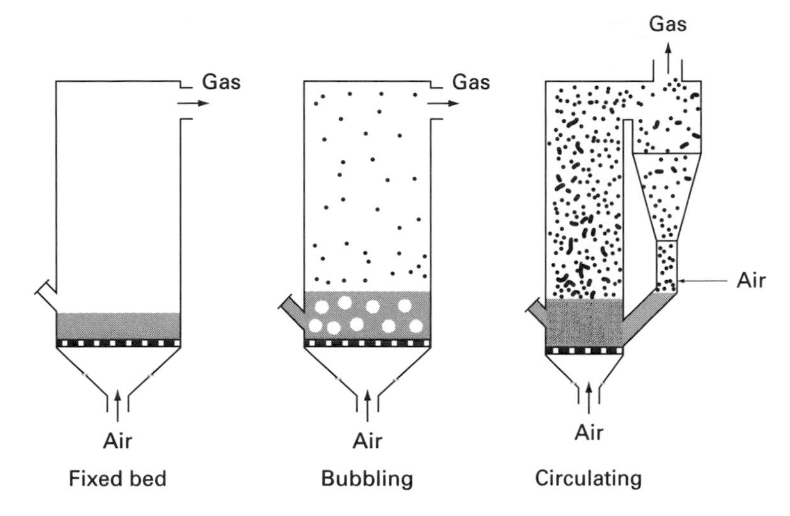

FIGURE 17.11
Fixed-bed, bubbling fluidized-bed, and circulating fluidized-bed combustors.

through the grid plate. Secondary air in the upper part of the chamber is used to ensure solids circulation. A small amount of air is usually introduced near the bottom of the cyclone downcomer to control the return rate of the solids. Mixing of fuel with bed material is rapid due to the high turbulence. Although there is no definite fixed bed depth, the density of the bed varies throughout the system, with the highest density near the grid plate.

In a typical circulating fluidized-bed system, heat from the cyclone exhaust is absorbed in a boiler bank, superheater, economizer, and air heater, as indicated in Figure 17.12. In the design shown there are no heat transfer tubes in the bed. Secondary air is used to ensure adequate air for fines combustion in the upper part of the combustion chamber. Velocities of 10 m/s in the main combustion chamber are typical. The combustion chamber may have waterwalls, although the lower portion near the inlet air grid is usually covered with refractory. The cyclone collectors, located at the outlet of the combustion chamber, are steel vessels lined with hard-faced refractory backed by lightweight insulating refractory. Char and bed particles continue to circulate until they are reduced to 5–10 μm, at which point they escape the cyclone collector. The fuel feed rate and airflow rates are adjusted depending on the steam load so that combustion takes place near 850°C. Turndown of 3:1 can be achieved by reducing the air and fuel flow rates. Load changes of 50% can be made within a few minutes.

Test results for a circulating fluidized-bed combustor running on a high-ash, high-sulfur, low-heating value coal are shown in Table 17.1. Long residence times result in good limestone utilization. Note that 100–300 μm limestone was used, and 90% removal of SO_2 was achieved with 0.85 calcium/sulfur molar ratio. The high ash content of the coal, which contains calcium, accounts for 44% SO_2

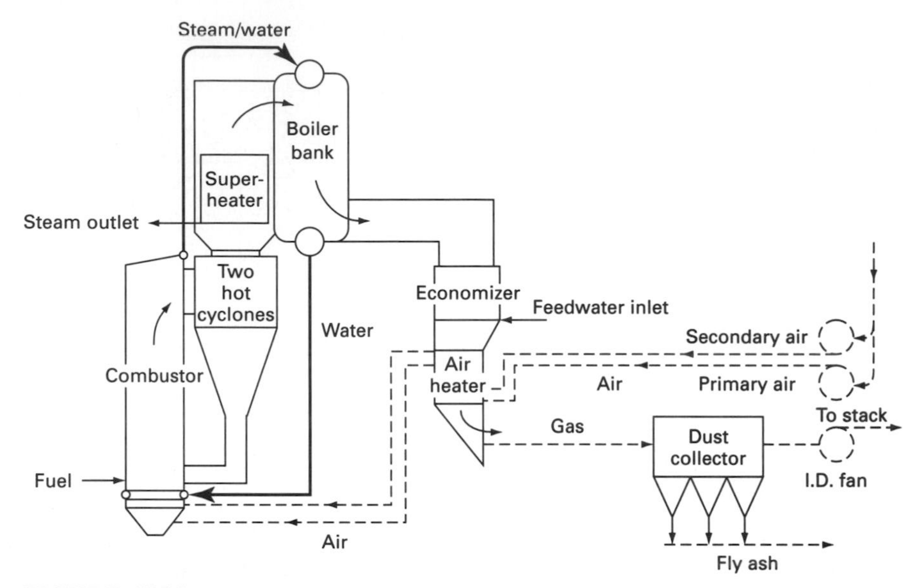

FIGURE 17.12
Circulating fluidized-bed combustion system.

TABLE 17.1
Test of high-ash coal in a circulating fluidized combustor

	Coal analysis	
Moisture		7.0 % wt
Volatile matter		23.6%
Ultimate analysis (%, dry)		
Carbon		38.8%
Hydrogen		2.66%
Sulfur		3.73%
Nitrogen		0.83%
Oxygen		9.58%
Ash		44.4%
HHV		17.9 MJ/kg (7730 Btu/lb$_m$)
	Operating conditions	
Coal size		< 25 mm
Limestone size (mean)		100–300 microns
Combustion temperature		850–890°C (1550–1643°F)
Excess air		20–29%
	Performance	
	No limestone	**With limestone**
SO_2 removal	44%	90%
Calcium/sulfur moles	—	0.85
NO_x	250 ppm	250 ppm
CO	50 ppm	100 ppm
Carbon burnup	98.6%	99.1%

Source: Maitland et al., in *Fluidized Bed Projects and Technology*, pp. 69–76, © 1994, by permission of ASME.

retention, and thus less limestone is needed than with lower-ash coals. Excess air, NO, CO, and carbon carryover are low.

In sum, circulating fluidized beds have the ability to burn a wide variety of fuels such as coal, peat, wood, and refuse while minimizing unburned carbon carryover. Test results are very encouraging, and this technology is rapidly gaining acceptance.

17.5 PRESSURIZED FLUIDIZED-BED COMBUSTION

The objective of pressurized fluidized-bed combustion is to directly power a gas turbine using coal or other solid fuels. When a combined-cycle gas turbine–steam turbine system is used, the cycle efficiency is greater than for the conventional steam cycle alone. The high-efficiency combined-cycle power plant is the main motivation for pressurized combustion of solid fuels. Both pressurized bubbling fluidized-bed combustors and pressurized circulating fluidized-bed combustors are being developed. In either case, particulate matter size and loading at the gas turbine inlet must be carefully controlled by separate hot gas cleanup equipment to prevent erosion of the turbine blades.

In addition to pressurized fluidized-bed combustion, pressurized fluidized-bed gasification is an option under development. In gasification a substoichiometric amount of air is used in the fluidized bed, and thus hot gas cleanup involves a smaller amount of gaseous products at a lower temperature. After cleanup the gasification products are mixed with air and burned in a separate combustor to power the gas turbine. The focus here is on combustion rather than gasification.

Since the oxygen throughput is higher in a pressurized bed than an atmospheric bed, the heat release rates in the bed are higher, and the volume of the system is smaller than an atmospheric pressure system. The bed temperature is maintained at 750–950°C for efficient sulfur removal and to minimize in-bed tube erosion. Pressurized fluidized-bed combustion and gasification have been under investigation in recent years, and several full-scale demonstration plants are now being built.

The thermodynamic cycle for a pressurized fluidized-bed combustion system can have several forms. Two main types are the basic air cycle (Figure 17.13) and the supercharged boiler cycle (Figure 17.14). In the air cycle, part of the air from the compressor fluidizes the bed and part is heated by the bed. The amount of air heated by the in-bed tubes can vary from zero to about twice minimum fluidizing air. An exhaust gas heat exchanger is used to drive a steam turbine in order to achieve high efficiency.

The supercharged boiler combined gas turbine–steam turbine cycle is shown in Figure 17.14. Bed temperature is maintained at 850–950°C by generating steam in tubes immersed in the bed. Feedwater is preheated with a heat exchanger using the gas turbine exhaust. The gas turbine compresses the fluidizing air and produces power.

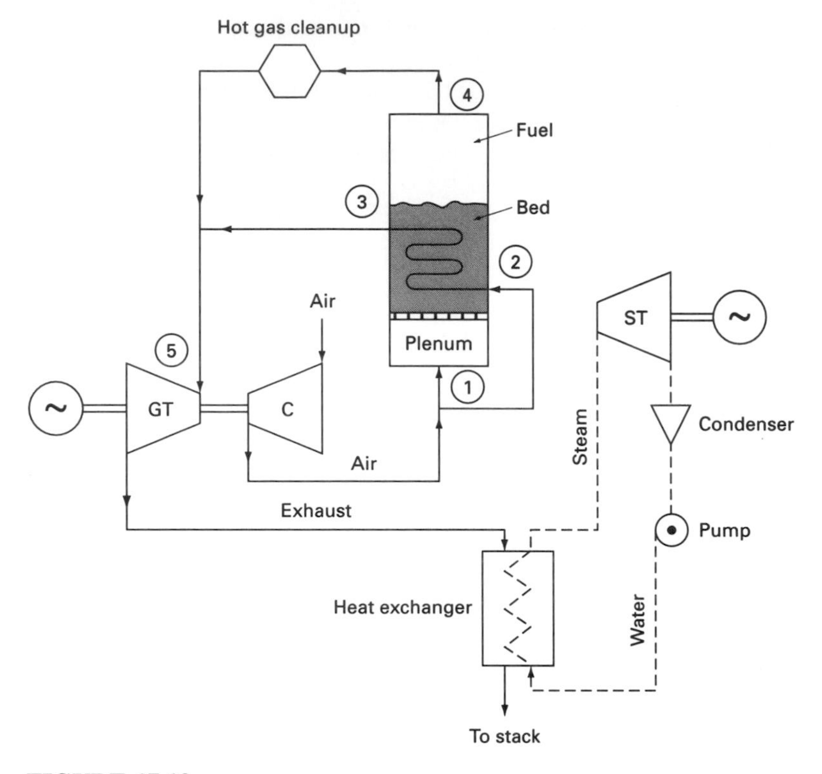

FIGURE 17.13
Air cycle pressurized fluidized bed combined cycle. GT, gas turbine; C, air compressor; ~, electrical generator; ST, steam turbine.

Optimization of these types of cycles will not be analyzed here, but because of the gas turbine, the best efficiency is not achieved at minimum excess air as in a conventional boiler. Typically, 100% excess air is desirable. High excess air is beneficial to combustion, reduces corrosion of turbine blades, and provides a margin of safety for turndown. On the other hand, the bed has to be larger for a given output (hence pressure vessel costs are increased), and the amount of gas to be cleaned increases.

The bed height in a pressurized fluidized bed tends to be large compared with an atmospheric fluidized-bed because of the volume required to remove sufficient heat in the bed. Consider the air cycle shown in Figure 17.13. The heat transfer surface A_t may be calculated from the convective heat transfer using the log mean temperature difference across the tubes:

$$\tilde{h}A_t\,\Delta T = \dot{m}(h_3 - h_2) \tag{17.41}$$

where

$$\Delta T = \frac{(T_4 - T_3) - (T_4 - T_2)}{\ln\dfrac{T_4 - T_3}{T_4 - T_2}} \tag{17.42}$$

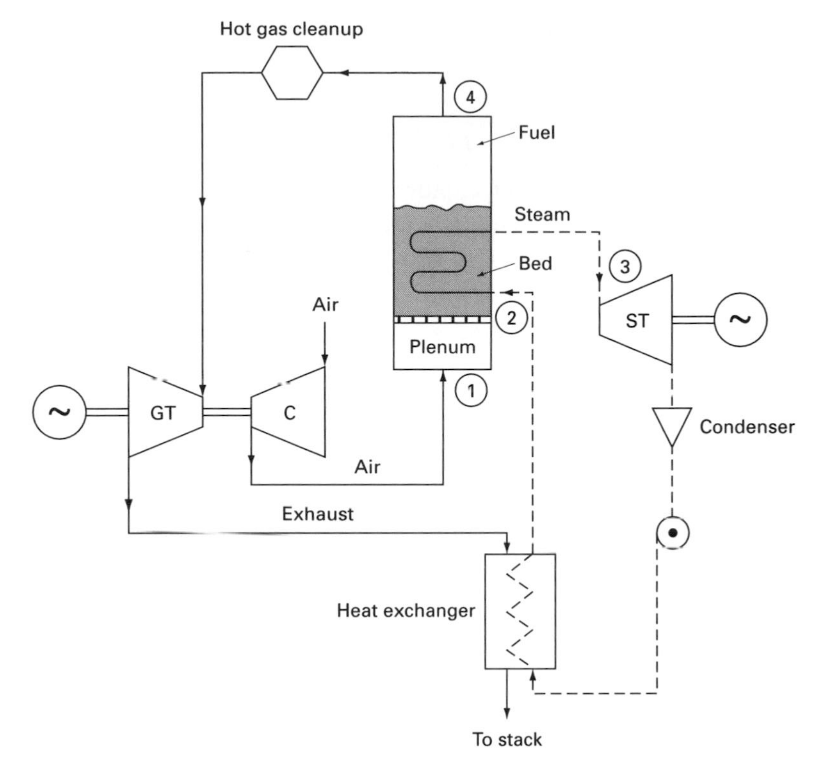

FIGURE 17.14
Supercharged boiler pressurized fluidized bed combined cycle. GT, gas turbine; C, air compressor; ~, electrical generator; ST, steam turbine.

Now, A_t can be related to the bed depth L, the bed cross-sectional area A, the tube diameter d_t, and the fraction of the bed volume occupied by the tubes, Φ, by Equation 17.32. Equations 17.32, 17.41, and 17.42 may be solved for L for a given geometry by using a mass and energy balance around the bed:

$$\dot{m}_1 + \dot{m}_f = \dot{m}_4 \tag{17.43}$$

$$\dot{m}_1 h_{s1} + \text{LHV}\ \dot{m}_f = \dot{m}_2(h_{s3} - h_{s2}) + \dot{m}_4 h_{s4} \tag{17.44}$$

The bubbles are neglected for pressurized systems because they are smaller and because the closely spaced in-bed tubes tend to break up the bubbles. Selecting the velocity through the bed sets $\dot{m}_4$:

$$\dot{m}_4 = \rho_4 \underline{V}_s A_b \tag{17.45}$$

Choosing the desired fuel-to-air ratio sets $\dot{m}_1$ in terms of $\dot{m}_4$ by means of Eq. 17.43:

$$\dot{m}_1 = \frac{\dot{m}_4}{1 + f} \tag{17.46}$$

Substituting these relations into Eq. 17.41 and solving for the required height of the bed to cover the tubes yields

$$L = \left(\frac{\rho_4 \underline{V}_s d_t}{4(1 + f)\Phi\tilde{h}\,\Delta T}\right)[\text{LHV}\, f + h_{s1} - (1 + f)h_{s4}] \tag{17.47}$$

Typically, T_4 is set to optimize SO_2 capture in the bed. T_3, which is needed for ΔT, depends on the mass flow through the tubes and the desired turbine inlet temperature T_5. Both are obtained from a mass and energy balance around the gas turbine inlet mixing station:

$$\dot{m}_2 + \dot{m}_4 = \dot{m}_5 \tag{17.48}$$

$$\dot{m}_2 h_3 + \dot{m}_4 h_4 = \dot{m}_5 h_5 \tag{17.49}$$

Combining Eqs. 17.41–17.49 yields

$$\dot{m}_2 = \dot{m}_4\left[\frac{h_{s1} + \text{LHV}\, f}{(1 + f)(h_{s5} - h_{s2})} - \frac{h_{s5}}{h_{s5} - h_{s2}}\right] \tag{17.50}$$

$$h_3 = h_5 - \frac{\dot{m}_4(h_4 - h_5)}{\dot{m}_2} \tag{17.51}$$

This analysis is explored further in the next example and in the problems at the end of the chapter. For the supercharged boiler setup shown in Figure 17.14 it can be seen that Eq. 17.47 holds also. However, the heat transfer coefficient and the temperature difference ΔT are considerably greater in the boiler than in the air heater, and therefore the bed height is reduced.

EXAMPLE 17.4. For the air cycle pressurized fluidized-bed system shown in Figure 17.13, find the bed height and ratio of flow through the bed tubes to flow through the bed ($\dot{m}_2/\dot{m}_1$) for the following design parameters:

$\underline{V}_s = 1$ m/s	$\Phi = 0.17$
$T_1 = T_2 = 600$ K	LHV $= 30{,}000$ kJ/kg
$T_4 = 1200$ K, $p_4 = 10$ atm	$f_s = 0.1$
$T_5 = 1150$ K	$f = 0.074$, i.e., 35% excess air
$d_t = 25.4$ mm	$\tilde{h} = 250$ W/m$^2\cdot$K

Solution. For simplicity assume the combustion products have the properties of nitrogen. Evaluating the properties and solving,

$$\rho_4 = \frac{p_4}{RT_4} = \frac{(10\text{ atm})(101.3\text{ kPa/atm})(28\text{ kg/kgmol})}{(8.314\text{ kJ/kgmol}\cdot\text{K})(1200\text{ K})} = 2.84\text{ kg/m}^3$$

From Appendix C, using nitrogen to represent flue gas for simplicity,

$$h_4 = 28.11\text{ MJ/kgmol} = 1004\text{ kJ/kg}$$

$$h_1 = h_2 = 8.89\text{ MJ/kgmol} = 317\text{ kJ/kg}$$

$$h_5 = 27.08\text{ MJ/kgmol} = 967\text{ kJ/kg}$$

Equations 17.50 and 17.51 give

$$\frac{\dot{m}_2}{\dot{m}_4} = \frac{317 \text{ kJ/kg} + (30{,}000 \text{ kJ/kg})(0.074)}{1.074(967 - 317 \text{ kJ/kg})} - \frac{967 \text{ kJ/kg}}{(967 - 317 \text{ kJ/kg})} = 2.15$$

$$h_3 = 27.08 \text{ MJ/kgmol} - \frac{28.11 - 27.08 \text{ MJ/kgmol}}{2.15}$$

$$= 26.60 \text{ MJ/kgmol}$$

so that from Appendix C, $T_3 = 1155$ K and

$$\Delta T = \frac{(1200 - 1155) - (1200 - 600)}{\ln \dfrac{1200 - 1155}{1200 - 600}} = 214 \text{ K}$$

Finally, from Eq. 17.47,

$$L = \frac{\left(2.84 \dfrac{\text{kg}}{\text{m}^3}\right)\left(1 \dfrac{\text{m}}{\text{s}}\right)(0.0254 \text{ m})\left(30{,}000 \dfrac{\text{kJ}}{\text{kg}}\right)(0.074) + 317 \dfrac{\text{kJ}}{\text{kg}} - (1.074)\left(1004 \dfrac{\text{kJ}}{\text{kg}}\right)}{4(1.074)(0.17)(0.250 \text{ kW/m}^2 \cdot \text{K})(214 \text{ K})}$$

$$= 2.7 \text{ m}$$

Note that this result for bed depth is sensitive to the log mean temperature difference, and in order to increase the temperature to the turbine, more tubes are required, which means a deeper bed. Also, a space of 0.5–1 m is required below the tubes to allow for adequate mixing. Thus the bed would be about 4 m deep, which is much deeper than an atmospheric fluidized bed.

The performance of a pressurized bed is improved in significant ways over an atmospheric bed. The fluidization tends to be smoother because the bubbles are smaller. As remarked earlier, the burning rate is higher because of the higher oxygen concentration. Combustion is complete because of the deep bed, which means a longer residence time in the bed with little combustion in the freeboard. The pressure drop across the bed is about 5%, which is greater than the pressure drop across a conventional gas turbine combustor, and additional pressure drop is required across the hot gas cleanup system. Thus, the compressor must be larger than that used with a conventional liquid- or gas-fired gas turbine. The heat transfer coefficient does not change due to increased pressure, all else being equal. However, in practice smaller bed particles can be used so that the heat transfer rate is actually increased.

The sulfur retention of limestone is similar to atmospheric pressure beds. Nitrogen oxide emissions are reduced in proportion to the square root of the pressure. Particulate and condensable vapors such as sodium, potassium, and chlorine are of concern regarding erosion deposition and corrosion of turbine blades, and special alkali vapor getters are required as part of the hot gas cleanup before the gas turbine. The degree of required hot gas cleanup is not known at this time; one criterion is suggested in Figure 17.15. High-efficiency cyclones may possibly be able to meet the suggested requirement of 8 ppm at 5 μm particle size, but

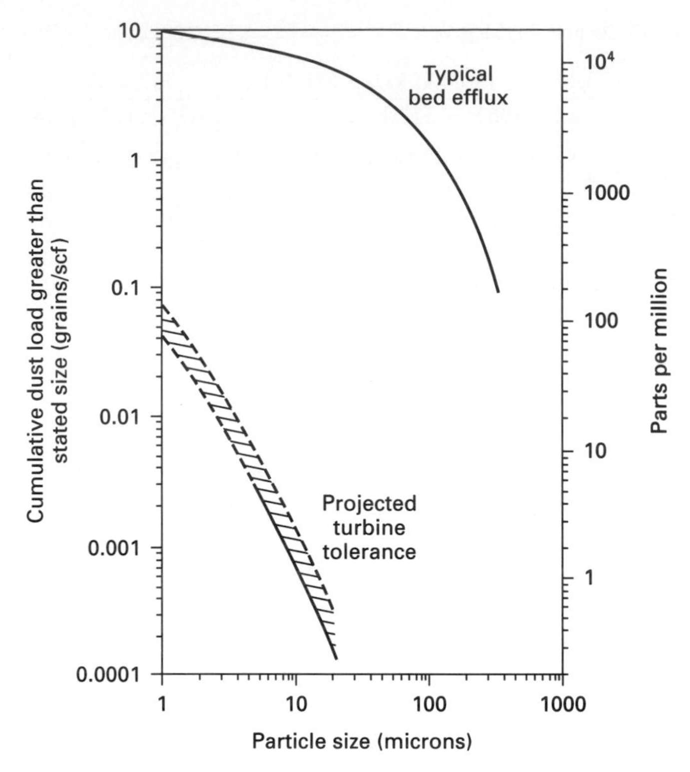

FIGURE 17.15
Uncontrolled fluidized-bed dust loading and projected turbine tolerance [McClung et al., in *Fluidized Bed Projects and Technology,* pp. 85–92, © 1994, by permission of ASME].

probably ceramic barrier filters are required. Hot gas cleanup is an area of ongoing research. Other challenging areas of engineering research and development include coal feeding at pressure, turbine load control, startup, and turbine blade performance.

17.6 SUMMARY

Fluidized-bed combustors are a relatively new technology which may gradually replace fixed-bed and suspension-burning combustors in a variety of applications because they are more compact, have lower NO_x and SO_x emissions, and can use a wider range of fuels. The disadvantages of fluidized-bed combustors are that the pressure drop is higher and boiler tube erosion is higher than in fixed bed and suspension burning. The superficial velocity in the bed must be greater than the minimum fluidization velocity and less than the single particle terminal velocity of the bed particles. In this regime air bubbles are formed in the bed, and in the dense phase (excluding the bubbles) the flow is at minimum fluidization. In a bubbling fluidized-bed combustor typically 5% of the bed is unburned or partially burned fuels and 95% is inert bed material. Stoker-sized coal and sand or crushed limestone bed material is used. Volatiles burn partially in the bed and partially above the bed. Char burns in the bed, and the burning rate is enhanced by the turbulence, fragmentation, and attrition due to char-bed particle collisions. Heat transfer to in-bed tubes is enhanced by the bed, and thus the required surface area of the boiler tubes is reduced compared to conventional systems. Circulating fluidized beds operate at a higher velocity and recycle the entrained bed material. Circulating beds overcome the tendency for high carbon carryover in bubbling beds and facilitate fuel feeding in large systems. Pressurized fluidized-bed combustion of solid fuels using both bubbling and circulating beds is under development for powering gas turbines in combined cycle gas turbine-steam turbine systems. Pressurized fluidized-bed systems are attractive provided the combustion products can be sufficiently cleaned of particulate and alkali vapor prior to entering the gas turbine.

PROBLEMS

17.1. Show that Eq. 17.3 for the surface mean diameter is equivalent to the Sauter mean diameter (SMD) of Chapter 9.

17.2. Calculate the minimum fluidization velocity for 0.2- and 2.0-mm-dia limestone particles which have a specific gravity of 2.0. Assume that the void fraction is 0.45 and that the air is at 1 atm and 900°C.

17.3. Calculate the terminal velocity for limestone particles (specific gravity of 2.0) with diameters of 0.2 mm and 2 mm. The fluid is air at 1 atm and 900°C. Repeat for char particles with a specific gravity of 0.3.

17.4. Calculate and plot the pressure drop versus superficial velocity across a 0.5-m-deep bed of crushed limestone. Assume the void fraction is 0.4 and the bed temperature is 1200 K and the pressure is 1 atm. The bed particle size is 2 mm. Let the velocity vary from zero to twice the minimum fluidization velocity. Use a limestone specific gravity of 2.0. Repeat for a pressure of 20 atm.

17.5. Repeat Example 17.2, but assume that LHV is 20,000 kJ/kg and that the coal contains 30% ash. The specific heat of the ash is 0.7 kJ/kg·K.

17.6. A fluidized-bed operates with a superficial velocity of 5 ft/s at 1 atm and 1550°F. Lignite coal is burned in the bed with 20% excess air. Assume that the fuel burns completely in the bed. Assume the coal and air enter at 77°F. Use lignite coal with the following analysis:

Proximate analysis (as-received)	
Moisture	35% (wt)
Volatile matter	24%
Fixed carbon	29%
Ash	12%
Ultimate analysis (dry, ash-free)	
C	75% (wt)
H	6%
O	19%
HHV (dry, ash-free)	12,570 Btu/lb_m

Find:

(*a*) Coal feed rate per unit area of bed ($lb_m/h \cdot ft^2$)
(*b*) Heat input rate per unit area of bed ($Btu/h \cdot ft^2$)
(*c*) Heat rate removed from bed per unit area of bed ($Btu/h \cdot ft$)
(*d*) Heat rate from product gases by external heat exchanger to drop the gas temperature to 340°F ($Btu/h \cdot ft^2$).

17.7. It is desired to build an atmospheric-pressure fluidized-bed boiler to deliver 200,000 lb_m/h of saturated steam at 1250 psi using coal with a lower heating value of 10,000 Btu/lb_m. The desired superficial velocity in the bed is 8 ft/s. The bed temperature is 1550°F. Estimate the coal feed rate, the cross-sectional area of the bed, and the excess air. Half of the steam is produced by tubes in the bed and half by tubes in the freeboard. Neglect pump and blower power. Consider two cases: (*a*) no bubbles, and (*b*) 30% of the air is retained in the bubbles and does not contact the solids. Assume the coal enters at 77°F and the air enters at 225°F. Assume overall thermal efficiency is 85% based on the LHV. Assume that the stoichiometric fuel/air ratio is 0.10, and all the combustion takes place in the bed. The inlet temperature of the feedwater is 77°F.

17.8. For a 0.8-m-deep atmospheric fluidized bed 10 m by 10 m across using limestone with a bulk density of 1200 kg/m^3, find the pressure drop across the bed and the required blower power. Assume the bed temperature is 1250 K and the superficial velocity is 1.3 m/s.

17.9. A 2500-ft^2 atmospheric fluidized bed burns 25,000 lb_m coal/h. The bed consists of 3 ft depth of crushed limestone with a bulk density of 90 lb_m/ft^3. If the burn time of a typical coal particle is 180 s, what is the approximate percent by weight of char in the bed?

17.10. Repeat Example 17.2 but with air preheat of 227°C.

17.11. Repeat Example 17.2 if 10% heat release occurs in the freeboard and the inlet air is preheated to 227°C.

17.12. A circulating fluidized bed burns 5% sulfur coal with an as-received higher heating value of 12,000 Btu/lb_m and an ash content of 10%. A calcium/sulfur molar ratio of 2 is used. The coal costs $24/ton, the limestone costs $15/ton, and ash disposal is \$2/ton. If the fluidized-bed has a heat input of 250×10^6 Btu/h, what are the full-load annual fuel cost, limestone cost, and ash disposal cost? Compare these costs with burning no. 2 oil, which costs \$6/10^6 Btu. Assume 90% sulfur capture. Limestone is $CaCO_3$ and goes to CaO in the bed and then to $CaSO_4$.

17.13. Make a plot of superficial velocity through a fluidized bed versus heat input (MW) per bed cross-sectional area, while holding the excess air constant at 10%. Let the velocity vary from 0.5 to 3.0 m/s. The coal has a higher heating value of 26,000 kJ/kg and a stoichiometric fuel/air ratio of 0.11. Do this for bed pressures of 1 atm and 10 atm, and an inlet air temperature of 300 K.

17.14. Explain why Eq. 17.47 holds for the supercharged boiler pressurized fluidized-bed combined cycle of Figure 17.14.

17.15. For the design parameters given in Example 17.3, but with a pressure of 20 atm and superficial velocity of 1 m/s, find the minimum bed height for the supercharged boiler pressurized fluidized-bed combined cycle shown in Figure 17.14.

REFERENCES

Arena, U., D'Amore, M., and Massimilla, L., "Carbon Attrition during the Fluidized Combustion of Coal," *AIChE J.*, vol. 29, pp. 40–49, 1983.

Basu, P., *Fluidized Bed Boilers: Design and Application,* Pergamon Press, New York, 1984.

Basu, P., and Large, J. F. (eds.), *Circulating Fluidized Bed Technology II,* Pergamon Press, New York, 1988.

Basu, P., Masayuki, H., and Hasatani, M., *Circulating Fluidized Bed Technology III,* Pergamon Press, New York, 1991.

Botterill, J. S. M., *Fluid Bed Heat Transfer,* Academic Press, New York, 1975.

Gamble, R. L., "The 100,000 lb/hr Fluidized Bed Steam Generation System for George Washington University," *American Power Conference,* pp. 295–301, 1979.

Hafer, D. R., and Bauer, D. A., "AEP's Program for Enhanced Environmental Performance of PFBC Plants," *American Power Conference,* vol. 55, pp. 127–132, 1993.

Howard, J. R. (ed.), *Fluidized Beds: Combustion and Application,* Applied Science Publishers Ltd., London, 1983.

Kelly, A. J., "Fuel Flexibility in a Circulating Fluidized Bed Combustion System," *Industrial Power Conference,* New Orleans, 1982.

Kunii, D., and Levenspiel, O., *Fluidization Engineering,* Wiley, New York, 1969.

Maitland, J. E., Skowyra, R. S., and Wilheim, B. W., "Design Considerations for Utility Size CFB Steam Generators," in T. J. Boyd and D. Mahr (eds.), *FACT,* vol. 19, *Fluidized Bed Projects and Technology,* ASME, pp. 69–76, 1994.

Makansi, J., and Schwieger, R., "Fluidized Bed Boilers," *Power,* pp. Sl–S16, August 1982.

McClung, J. D., Quandt, M. T., and Froelich, R. E., "Design and Operating Considerations for an Advanced PFBC Facility at Wilsonville, Alabama," in T. J. Boyd and D. Mahr (eds.), *FACT,* vol. 19, *Fluidized Bed Projects and Technology,* ASME, pp. 85–92, 1994.

Pecson, F. A., and Ragland, K. W., "Coal Fragmentation in a Fluidized Bed Boiler," *Twenty-second Symp. (Int.) on Combustion,* The Combustion Institute, Pittsburgh, pp. 259–265, 1988.

APPENDIX A

Properties of Fuels

Guide to Appendices

A Properties of Fuels
B Properties of Air
C Thermodynamic Properties of Combustion Products
D Viscosity and Thermal Conductivity of Gas Mixtures
E A Brief Introduction to Mass Transport By Diffusion in Gaseous Systems

TABLE A.1
Thermodynamic properties of alkane fuels

Formula	Name	M	sg	T_b	p_v	c_{pg}	c_{pl}	T_{ig}	HHV	LHV	h_{fg}	a/f	Octane		$\Delta\hat{h}°$	p_c	T_c
													Res.	Mot.			
CH_4	Methane	16.04	0.466	−161		2.21		537	55,536	50,048	510	17.2	120	120	−74.4	45.4	190
C_2H_6	Ethane	30.07	0.572	−89		1.75		472	51,902	47,511	489	16.1	115	99	−83.8	48.2	305
C_3H_8	Propane	44.10	0.585	−42	12.8	1.62	2.48	470	50,322	46,330	432	15.7	112	97	−104.7	41.9	370
C_4H_{10}	*n*-Butane	58.12	0.579	0	3.51	1.64	2.42	365	49,511	45,725	386	15.5	94	90	−146.6	37.5	425
C_4H_{10}	Isobutane[1]	58.12	0.557	−12	4.94	1.62	2.39	460	49,363	45,577	366	15.5	102	98	−153.5	36.0	408
C_5H_{12}	*n*-Pentane	72.15	0.626	36	1.06	1.62	2.32	284	49,003	45,343	357	15.3	62	63	−173.5	33.3	470
C_5H_{12}	Isopentane[2]	72.15	0.620	28	1.39	1.60	2.28	420	48,909	45,249	342	15.3	93	90	−178.5	33.5	460
C_6H_{14}	*n*-Hexane	86.18	0.659	69	0.337	1.62	2.27	233	48,674	45,099	335	15.2	25	26	−198.7	29.7	507
C_6H_{14}	Isohexane[3]	86.18	0.662	50	0.503	1.58	2.20	421	48,454	44,879	305	15.2	104	94	−207.4	30.9	500
C_7H_{16}	*n*-Heptane	100.20	0.684	99	0.110	1.61	2.24	215	48,438	44,925	317	15.2	0	0	−224.2	27.0	540
C_7H_{16}	Triptane[4]	100.20	0.690	81	0.229	1.60	2.13	412	48,270	44,757	289	15.2	112	101		29.2	531
C_8H_{18}	*n*-Octane	114.23	0.703	126	0.036	1.61	2.23	206	48,254	44,786	301	15.1	20	17	−250.1	24.6	569
C_8H_{18}	Isooctane[5]	114.23	0.692	114	0.117	1.59	2.09	418	48,119	44,651	283	15.1	100	100	−259.2	25.4	544
C_9H_{20}	*n*-Nonane	128.26	0.718	151	0.012	1.61	2.21		48,119	44,688	288	15.1			−274.7	22.6	595
$C_{10}H_{22}$	*n*-Decane	142.28	0.730	174	0.005	1.61	2.21		48,002	44,599	272	15.1			−300.9	20.9	618
$C_{10}H_{22}$	Isodecane[6]	142.28	0.768	161		1.61	2.21					15.1	113	92		24.8	623
$C_{11}H_{24}$	*n*-Undecane	156.31	0.740	196		1.60	2.21		47,903	44,524	265	15.0			−327.2	19.4	639
$C_{12}H_{26}$	*n*-Dodecane	170.33	0.749	216		1.60	2.21		47,838	44,574	256	15.0			−350.9	18.0	658
$C_{13}H_{28}$	*n*-Tridecane	184.35	0.756	236		1.60	2.21				246	15.0				17.0	676
$C_{14}H_{30}$	*n*-Tetradecane	198.38	0.763	253		1.60	2.21				239	15.0				14.2	693
$C_{15}H_{32}$	*n*-Pentadecane	212.45	0.768	271		1.60	2.21				232	15.0				15.0	707
$C_{16}H_{34}$	*n*-Hexadecane[7]	226.43	0.773	287		1.60	2.21		47,611	44,307	225	15.0			−456.1	13.9	722
$C_{17}H_{36}$	*n*-Heptadecane	240.46	0.778	302		1.60	2.21				221	15.0				12.8	733
$C_{18}H_{38}$	*n*-Octadecane	254.50	0.782	317		1.60	2.21		47,542	44,256	214	15.0				11.8	748

Definitions: M, molecular weight; sg, specific gravity (density of substances at 20°C relative to water at 4°C, for gases determined at boiling point of the liquefied gas); T_b, boiling point temperature, °C at 1 atm; p_v, vapor pressure at 38°C, atm; c_{pg}, specific heat of the gas at 25°C, kJ/kg · °C; c_{pl}, specific heat of the liquid at 25°C, kJ/kg · °C; T_{ig}, autoignition temperature, °C; HHV, higher heating value, kJ/kg, first three are gas, all others liquid; LHV, lower heating value, kJ/kg, first three are gas, all others liquid; h_{fg}, heat of vaporization at 1 atm and boiling point temperature, kJ/kg; a/f, stoichiometric air-to-fuel mass ratio; octane res., research octane no.; octane mot., motor octane no.; $\Delta\hat{h}°$, enthalpy of formation at 25°C, kJ/gmol, first three are gas, all others liquid; p_c, critical point pressure, atm; T_c, critical point temperature, K.

Notes: (1) 2-Methylpropane; (2) 2,2-methylbutane; (3) 2,3-dimethylbutane; (4) 2,2,3-trimethylbutane; (5) 2,2,4-trimethylpentane; (6) 2,2,3,3-tetramethylhexane; (7) cetane.

Sources: Lide, D. L. (ed.), *Handbook of Chemistry and Physics,* 74th ed., CRC Press, Boca Raton, FL, 1993; Bartok and Sarofim, *Fossil Fuel Combustion: A Source Book,* Wiley, New York, 1991. [By permission of CRC Press, © 1993 and John Wiley and Sons Inc., © 1991]

TABLE A.2

Thermodynamic properties of some other hydrocarbon fuels

Class	Formula	Name	M	sg	T_b	p_v	c_{pg}	c_{pl}	T_{ig}	HHV	LHV	h_{fg}	Octane a/f	Octane Res.	Octane Mot.	$\Delta\hat{h}^\circ$	p_c	T_c
Cycloalkane	C_5H_{10}	Cyclopentane	70.13	0.746	49	0.673	1.135	1.836	385	46,936	43,798	389	14.8	101	85	−76.4	44.5	512
	C_6H_{12}	Cyclohexane	84.16	0.779	81	0.224	1.214	1.813	270	46,573	43,435	358	14.8	84	78	−123.2	40.2	553
	C_7H_{14}	Cycloheptane	98.19	0.810	119	0.058	1.181	1.826		46,836	43,698	335	14.8	39	41	−118.1	37.6	604
	C_8H_{16}	Cyclooctane	112.2	0.835	149	0.021	1.173	1.838		46,943	43,808	309	14.8	71	58	−124.4	35.1	647
Aromatic	C_6H_6	Benzene	78.11	0.874	80	0.224	1.005	1.717	592	41,833	40,145	393	13.3		115	82.6	48.3	562
	C_7H_8	Toluene[1]	92.14	0.867	111	0.070	1.089	1.683	568	42,439	40,528	362	13.5	120	109	50.4	40.5	592
	C_8H_{10}	Ethylbenzene	106.17	0.867	136	0.025	1.173	1.721	460	42,996	40,923	339	13.7	111	98	29.9	35.5	617
	C_8H_{10}	*m*-Xylene[2]	106.17	0.864	139	0.022	1.164	1.692	563	42,873	40,800	342	13.7	118	115	17.3	34.9	617
Olefin	C_3H_6	Propylene	42.08	0.519	−47	15.401	1.482	2.450		48,472	45,334	437	14.8	102	85	20.0	45.4	365
	C_4H8	1-Butene	56.11	0.595	−6	4.286	1.487	2.240		48,073	44,937	390	14.8	99	80	−0.1	39.7	420
	C_5H_{10}	1-Pentene	70.13	0.641	30	1.293	1.524	2.178	298	47,766	44,528	358	14.8	91	77	−21.3	34.8	465
	C_6H_{12}	1-Hexene	84.16	0.673	63	0.408	1.533	2.144	272	47,550	44,312	335	14.8	76	63	−43.5	31.3	504
Diolefin	C_5H_8	Isoprene[3]	68.11	0.681	34	1.136	1.495	2.199		46,382	43,798	356	14.2	99	81	75.5	38.0	484
	C_6H_{10}	1,5-Hexadiene	82.15	0.688	59	0.483	1.390	2.136		46,796	43,582	312	14.3	71	38	84.1	33.9	507
Cycloolefin	C_5H_8	Cyclopentene	68.12	0.772	44		1.064	1.759		45,733	43,149		14.2	93	70	33.9	47.4	506
	C_6H_{10}	Cyclohexene	82.15							45,674	42,995					−5.0		
Alcohol	CH_4O	Methanol	32.04	0.791	65	0.310	1.370	2.531	385	22,663	19,915	1099	6.5	106	92	−201.5	79.8	513
	C_2H_6O	Ethanol	46.07	0.789	78	0.153	1.420	2.438	365	29,668	26,803	836	9.0	107	89	−235.1	60.6	514
	C_3H_8O	1-Propanol	60.10	0.803	97	0.061	1.424	2.395		33,632	30,709	690	10.5			−255.1	51.0	537
	$C_4H_{10}O$	1-Butanol	74.10	0.805	118	0.022		2.391		36,112	33,142	584	11.1			−274.9	43.6	563
	$C_5H_{12}O$	1-Pentanol	88.15	0.814	137			2.361		37,787	34,791	503	11.7			−298.9	38.6	588
	$C_6H_{14}O$	1-Hexanol	102.18	0.814	158			2.353		38,994	35,979	436	12.2			−317.8	40.0	611

Definitions: M, molecular weight; sg, specific gravity (density of substances at 20°C relative to water at 4°C, for gases determined at boiling point of the liquefied gas); T_b, boiling point temperature, °C at 1 atm; p_v, vapor pressure at 38°C, atm; c_{pg} specific heat of the gas at 25°C, kJ/kg · °C; c_{pl}, specific heat of the liquid at 25°C, kJ/kg · °C; T_{ig}, autoignition temperature, °C; HHV, higher heating value of liquid, kJ/kg; LHV, lower heating value of liquid, kJ/kg; h_{fg}, heat of vaporization at 1 atm and boiling point temperature, kJ/kg; a/f, stoichiometric air-to-fuel mass ratio; octane res., research octane no.; octane mot., motor octane no.; $\Delta\hat{h}^\circ$, enthalpy of formation of liquid at 25°C, kJ/gmol; p_c, critical point pressure, atm; T_c, critical point temperature, K.

Notes: (1) Methylbenzene; (2) 1,3-dimethylbenzene, (3) 2-methyl-1,3-butadiene.

Sources: Lide, D. L. (ed.), *Handbook of Chemistry and Physics,* 74th ed., CRC Press, Boca Raton, FL, 1993; Bartok and Sarofim, *Fossil Fuel Combustion: A Source Book,* Wiley, New York, 1991. [By permission of CRC Press, ©1993 and John Wiley and Sons Inc., © 1991]

TABLE A.3
Specific heats of selected fuels

Fuel	c_p (kJ/kg·K)
Methane (gas)	$-42.05 + 27.48\theta^{0.25} - 1.555\theta^{0.75} + 20.24\theta^{-0.5}$ (300 K < T < 2000 K)
Propane (gas)	$-0.0918 + 0.692\theta - 0.0357\theta^2 + 0.00072\theta^3$ (300 K < T < 1500 K)
Liquid petroleum fuels	$0.76 + 0.000335T$
Petroleum fuel vapor	$0.136 + 0.0012T(4 - \text{sg})$
Coal	1.3
Wood	1.3

Notes: $\theta = T\,(\text{K})/100$; sg is specific gravity of the liquid; T is in K. For dodecane and octane see Table A.8. For density multiply sg by 998 kg/m^3.

TABLE A.4
Specific gravity and bulk density of selected fuels

Fuel type	Specific gravity (at 293 K)
Kerosene	0.80
Gasoline	0.72–0.78
Diesel fuel (no. 2)	0.85
No. 2 fuel oil	0.86
No. 6 fuel oil	0.99
Lignite coal*	1.2
Bituminous coal*	1.4
Anthracite coal*	1.7
Wood, pine*	0.5
Wood, oak*	0.7
Pelletized wood	0.6
Pelletized municipal RDF†	0.5

*Dry.
† Refuse-derived fuel.

Fuel type	Bulk density (kg/m^3)
Stoker coal (bituminous)	780
Cordwood (dry)	580
Wood hog fuel	240
Wood chips (dry, 19 mm)	200
Municipal RDF*	180

*Refuse-derived fuel.

TABLE A.5
Transport properties of selected fuels

Fuel	Viscosity ($N \cdot s/m^2$)	Thermal conductivity ($W/m \cdot K$)
Methane (gas)*	1.078×10^{-5}	$-1.869 \times 10^{-3} + 8.727 \times 10^{-5}T + 1.179 \times 10^{-7}T^2 - 3.614 \times 10^{-11}T^3$
Propane (gas)*		$1.858 \times 10^{-3} - 4.698 \times 10^{-6}T + 2.177 \times 10^{-7}T^2 - 8.409 \times 10^{-11}T^3$
Gasoline	$5.0 \times 10^{-4\dagger}$	$0.117(1 - 0.00054T)\text{sg}^{\ddagger}$
No. 2 diesel	$2.4 \times 10^{-3\dagger}$	$0.117(1 - 0.00054T)\text{sg}^{\ddagger}$
Methanol	$4.6 \times 10^{-4\dagger}$	$-7.79 \times 10^{-3} + 4.167 \times 10^{-5}T + 1.214 \times 10^{-7}T^2 - 5.184 \times 10^{-11}T^{3\ddagger}$
Ethanol	$8.3 \times 10^{-4\dagger}$	$-7.79 \times 10^{-3} + 4.167 \times 10^{-5}T + 1.214 \times 10^{-7}T^2 - 5.184 \times 10^{-11}T^{3\ddagger}$
No. 2 fuel oil	$2.3 \times 10^{-3\dagger}$	$0.117(1 - 0.00054T)\text{sg}^{\ddagger}$
No. 6 fuel oil	$0.36^{\dagger}$	$0.117(1 - 0.00054T)\text{sg}^{\ddagger}$
Wood	—	$0.238 + \text{sg}(0.200 + 0.404\ \text{MC})$
Coal	—	0.25

Notes: sg is specific gravity of the liquid; T is in K; MC is moisture content. For dodecane and octane see Table A.8.

*Viscosity at 288 K, conductivity range 273–1270 K, T in K.

†Viscosity at 313 K, for liquid.

‡Conductivities for vapor.

TABLE A.6
Surface tension of several fuels and water (typical values)

Substance	Surface tension (N/m)*
Gasoline	2.3×10^{-2}
Kerosene	$2.5\text{–}3.0 \times 10^{-2}$
No. 2 fuel oil	$2.9\text{–}3.2 \times 10^{-2}$
n-Octane	2.18×10^{-2}
Water	7.2×10^{-6}

*20°C in contact with vapor.

TABLE A.7
Latent heat of vaporization of selected fuels

Fuel	Latent heat (kJ/kg)
Gasoline	339
Kerosene	291
Light diesel	267
Medium diesel	244
Heavy diesel	232

TABLE A.8
Properties of dodecane and octane

n-Dodecane properties	Equation	Range, °R
Density of liquid, $lb_m/in.^3$	$0.0322608 - 6.26 \times 10^{-6}T - 7 \times 10^{-9}T^2$	560–860
Heat of vaporization, Btu/lb_m	$115.88 + 0.164T - 2 \times 10^{-4}T^2$	660–860
Specific heat of liquid, $Btu/lb_m \cdot °R$	$0.2326 + 5.4 \times 10^{-4}T$	660–860
Specific heat of vapor, $Btu/lb_m \cdot °R$	$0.151085 + 4.9275 \times 10^{-4}T$	660–1060
Thermal conductivity, vapor, Btu/ in. · s · °R	$-1.61 \times 10^{-7} + 5.9441 \times 10^{-10}T$	660–1460
Vapor pressure, $lb_f/in.^2$	$\exp[12.12767 - 6738.9067/(T - 167.44)]$	
Viscosity of vapor, $lb_m/in. \cdot s$	$2.885 \times 10^{-8} + 4.0605 \times 10^{-10}T$	600–1060
Diffusion coefficient,* fuel-air, $in.^2 \cdot s$	$-2.316 \times 10^{-3} + 9.1269 \times 10^{-6}T + 1.803705 \times 10^{-8}T^2$	

*At 1 atm pressure.

n-Octane properties	Equation	Range, °R
Density of liquid, $lb_m/in.^3$	$0.02933 - 6.26 \times 10^{-6}T - 1.3148 \times 10^{-8}T^2$	580–850
Heat of vaporization, Btu/lb_m	$160.3 + 0.0815\,T - 1.75 \times 10^{-4}T^2$	460–860
Specific heat of liquid, $Btu/lb_m \cdot °R$	$0.1705 + 6.57 \times 10^{-4}T$	560–860
Specific heat of vapor, $Btu/lb_m \cdot °R$	$0.1165 + 5.45 \times 10^{-4}T$	500–900
Thermal conductivity, vapor, Btu/in. · s · °R	$-1.47 \times 10^{-7} + 6.2 \times 10^{-10}T$	500–950
Vapor pressure, $lb_f/in.^2$	$\exp[11.99684 - 5616.5277/(T - 114.869)]$	
Viscosity of vapor, $lb_m/in. \cdot s$	$4.5 \times 10^{-8} + 5.67 \times 10^{-10}T$	500–1000
Diffusion coefficient*, fuel-air, $in.^2 \cdot s$	$-2.8287 \times 10^{-3} + 1.17815 \times 10^{-5}T + 2.4718 \times 10^{-8}T^2$	550–840

*At 1 atm pressure.

Source: Priem, R. J., "Vaporization of Fuel Drops Including the Heating-Up Period," Ph.D. thesis, University of Wisconsin–Madison, 1955.

TABLE A.9
Specifications of unleaded gasoline used in U.S. emissions certification

Item	ASTM	Value
Octane, research, minimum	D2699	93
Sensitivity, minimum		7.5
Lead (organic), g/gal		0.00–0.05
Distillation range		
IBP,* °F	D86	75–95
10% point, °F	D86	120–135
50% point, °F	D86	200–230
90% point, °F	D86	300–325
EP point, °F	D86	415
Sulfur, weight %, max	D1266	0.01
Phosphorus, g/gas, max		0.005
Reid vapor pressure,†,‡ $lb_f/in.^2$	D323	8.7–9.2
Hydrocarbon composition		
Olefins, %, max	D1319	10
Aromatics, %, max	D1319	35
Saturates, %	D1319	Remainder

*For testing at altitudes above 4000 ft the specified range is 75–105.

†For testing which is unrelated to evaporative emissions control, the specified range is 8.0–9.2.

‡For testing at altitudes above 4000 ft the specified range is 7.9–9.2.

Source: CFR 40, Part 86, Subpart A, p. 482, July 1, 1985.

TABLE A.10
Atomic weights of elements and trace elements in fuels

Element	Symbol	Atomic Weight
Aluminum	Al	26.98
Arsenic	As	74.92
Bromine	Br	79.90
Cadmium	Cd	112.41
Calcium	Ca	40.08
Carbon	C	12.01
Chlorine	C1	35.45
Copper	Cu	63.55
Fluorine	F	19.00
Hydrogen	H	1.01
Iron	Fe	55.85
Lead	Pb	207.2
Magnesium	Mg	24.31
Manganese	Mn	54.94
Mercury	Hg	200.59
Nickel	Ni	58.71
Nitrogen	N	14.01
Oxygen	O	16.00
Phosphorus	P	30.97
Potassium	K	39.10
Silicon	Si	28.09
Sodium	Na	22.99
Sulfur	S	32.06
Tin	Sn	118.69
Vanadium	V	50.94
Zinc	Zn	65.38

APPENDIX B

Properties of Air (at 1 atm)

T(K)	u (kJ/kg)	h (kJ/kg)	ρ (kg/m^3)	c_p (kJ/kg·K)	$\mu \times 10^5$ (kg/m·s)	D_{AB} (cm^2/s)	k (W/m·K)	Pr	Sc
300	214.32	300.43	1.177	1.005	1.853	0.21	0.0261	0.711	0.749
400	286.42	401.26	0.882	1.013	2.294	0.34	0.0330	0.703	0.788
500	359.79	503.30	0.706	1.029	2.682	0.50	0.0395	0.699	0.760
600	435.03	607.27	0.588	1.051	3.030	0.68	0.0456	0.698	0.751
700	512.58	713.50	0.505	1.075	3.349	0.89	0.0513	0.702	0.745
800	592.53	822.15	0.441	1.099	3.643	1.11	0.0569	0.703	0.744
900	674.77	933.10	0.392	1.121	3.918	1.35	0.0625	0.703	0.740
1000	759.14	1046.17	0.353	1.141	4.177	1.61	0.0672	0.709	0.735
1100	845.38	1161.12	0.321	1.160	4.44	1.88	0.0732	0.704	0.736
1200	933.28	1277.72	0.294	1.177	4.69	2.17	0.0782	0.704	0.735
1300	1022.67	1395.81	0.271	1.195	4.93	2.47	0.0837	0.704	0.736
1400	1113.34	1515.18	0.252	1.212	5.17	2.79	0.0891	0.703	0.736
1500	1205.15	1635.69	0.235	1.230	5.40	3.13	0.0946	0.702	0.734
1600	1297.93	1757.18	0.221	1.248	5.63	3.49	0.100	0.703	0.730
1700	1391.62	1879.57	0.208	1.266	5.85	3.86	0.105	0.703	0.728
1800	1486.13	2002.78	0.196	1.286	6.07	4.26	0.111	0.703	0.727
1900	1581.33	2126.71	0.186	1.307	6.29	4.67	0.117	0.703	0.724
2000	1677.21	2251.29	0.176	1.331	6.50	5.10	0.124	0.698	0.724
2100	1773.69	2376.45	0.168	1.359	6.72	5.53	0.131	0.696	0.723
2200	1870.73	2502.19	0.160	1.392	6.93	5.97	0.139	0.694	0.725
2300	1968.28	2628.45	0.153	1.434	7.14	6.42	0.149	0.687	0.727
2400	2066.30	2755.17	0.147	1.487	7.35	6.88	0.161	0.679	0.727
2500	2164.76	2882.33	0.141	1.556	7.57	7.36	0.175	0.673	0.729

Note: For other pressures multiply ρ by the pressure in atmospheres; divide D_{AB} by the pressure in atmospheres; c_p, μ, k and Pr do not change with pressure. D_{AB} is the binary diffusion coefficient for O_2 into N_2 and Sc is based on this D_{AB}.

Source: Data from Keenan, Chao, and Kaye, *Gas Tables,* Wiley, New York, 1983; Kays and Crawford, *Convective Heat and Mass Transfer,* McGraw-Hill, New York, 1980, Table A-1; and Field, M. A., *Combustion of Pulverized Coal,* Cheney & Sons, London, 1967, App. Q.

APPENDIX C

Thermodynamic Properties of Combustion Products

The absolute enthalpy of a substance is its enthalpy relative to stable elements at the reference state (T_0, 1 atm), which is taken as zero. The absolute entropy of a substance is its entropy relative to 0 K, 1 atm, which is taken as zero. Absolute properties at a pressure of 1 atm are denoted by a degree sign. The reference temperature for these tables is $T_0 = 298$ K for SI units and 537°R for English units. At other temperatures and pressures absolute enthalpy is given by

$$h(T) = \Delta h^\circ(T_0) + h_{\text{sensible}}$$

where
$$h_{\text{sensible}} = \int_{T_0}^{T} c_p(T')\, dT'$$

The absolute entropy is given by

$$s(T, p) = s^\circ(T) - R \ln\left(\frac{p}{p_0}\right)$$

where
$$s^\circ(T) = \int_{T_0}^{T} \frac{c_p(T')\, dT'}{T'}$$

Normalized values of $g^\circ(T) = \Delta h^\circ(T) - Ts^\circ(T)$ are also tabulated to assist in equilibrium computations.

Tables in SI units are given first, followed by tables in English units. The most commonly used gases (CO_2, H_2O, N_2, O_2) are listed first in each table, followed by the others in alphabetical order.

Courtesy of Prof. Glen E. Myers who derived these data from D. R. Stull and H. Prophet, *JANAF Thermochemical Tables*, 2nd ed., NSRDS-NBS 37, National Bureau of Standards, Washington, DC, 1971.

Carbon dioxide (CO_2)

T (K)	$\hat{c}_p$ (kJ/kgmol·K)	$\hat{h}$ (MJ/kgmol)	$\hat{s}^\circ$ (kJ/kgmol·K)	$\hat{g}^\circ/\hat{R}T$
200	32.36	−3.41	199.87	−262.746
293	36.88	−0.19	213.09	−187.161
298	37.13	0.00	213.69	−184.449
400	41.33	4.01	225.22	−144.210
500	44.63	8.31	234.81	−120.904
600	47.32	12.92	243.20	−105.546
700	49.56	17.76	250.66	−94.712
800	51.43	22.82	257.41	−86.693
900	53.00	28.04	263.56	−80.542
1000	54.31	33.41	269.22	−75.693
1100	55.41	38.89	274.45	−71.784
1200	56.34	44.48	279.31	−68.577
1300	57.14	50.16	283.85	−65.907
1400	57.80	55.91	288.11	−63.657
1500	58.38	61.71	292.11	−61.739
1600	58.89	67.58	295.90	−60.091
1700	59.32	73.49	299.48	−58.662
1800	59.70	79.44	302.88	−57.416
1900	60.05	85.43	306.12	−56.322
2000	60.35	91.45	309.21	−55.356
2100	60.62	97.50	312.16	−54.499
2200	60.86	103.57	314.99	−53.737
2300	61.09	109.67	317.70	−53.054
2400	61.29	115.79	320.30	−52.443
2500	61.47	121.93	322.81	−51.892

$\Delta\hat{h}^\circ(298) = -393.52$ MJ/kgmol.

Water (H_2O, gas)

T (K)	$\hat{c}_p$ (kJ/kgmol·K)	$\hat{h}$ (MJ/kgmol)	$\hat{s}^\circ$ (kJ/kgmol·K)	$\hat{g}^\circ/\hat{R}T$
200	33.34	−3.28	175.38	−168.494
293	33.56	−0.17	188.19	−121.920
298	33.58	0.00	188.72	−120.252
400	34.25	3.45	198.67	−95.572
500	35.21	6.92	206.41	−81.333
600	36.30	10.50	212.93	−71.982
700	37.46	14.18	218.61	−65.407
800	38.69	17.99	223.69	−60.557
900	39.94	21.92	228.32	−56.849
1000	41.22	25.98	232.60	−53.937
1100	42.48	30.17	236.58	−51.598
1200	43.70	34.48	240.33	−49.689
1300	44.99	38.90	243.88	−48.107
1400	45.97	43.45	247.24	−46.780
1500	47.00	48.10	250.45	−45.657
1600	47.96	52.84	253.51	−44.697
1700	48.84	57.68	256.45	−43.872
1800	49.66	62.61	259.26	−43.158
1900	50.41	67.61	261.97	−42.536
2000	51.10	72.69	264.57	−41.993
2100	51.74	77.83	267.08	−41.516
2200	52.32	83.04	269.50	−41.095
2300	52.86	88.29	271.84	−40.724
2400	53.36	93.60	274.10	−40.395
2500	53.82	98.96	276.29	−40.103

$\Delta\hat{h}^\circ(298) = -241.83$ MJ/kgmol.

Nitrogen (N_2)

T (K)	$\hat{c}_p$ (kJ/kgmol · K)	$\hat{h}$ (MJ/kgmol)	$\hat{s}°$ (kJ/kgmol · K)	$\hat{g}°/\hat{R}T$
200	29.11	−2.86	179.88	−23.353
293	29.12	−0.15	191.04	−23.037
298	29.12	0.00	191.50	−23.033
400	29.25	2.97	200.07	−23.170
500	29.58	5.91	206.63	−23.430
600	30.00	8.89	212.07	−23.724
700	30.75	11.94	216.76	−24.019
800	31.43	15.05	220.91	−24.307
900	32.09	18.22	224.65	−24.584
1000	32.70	21.46	228.06	−24.848
1100	33.24	24.76	231.20	−25.101
1200	33.73	28.11	234.12	−25.341
1300	34.15	31.50	236.83	−25.570
1400	34.53	34.94	239.38	−25.789
1500	34.85	38.40	241.77	−25.999
1600	35.14	41.90	244.03	−26.200
1700	35.39	45.43	246.17	−26.393
1800	35.61	48.98	248.19	−26.579
1900	35.81	52.55	250.13	−26.757
2000	35.99	56.14	251.97	−26.929
2100	36.14	59.75	253.73	−27.095
2200	36.28	63.37	255.41	−27.255
2300	36.41	67.01	257.03	−27.410
2400	36.53	70.65	258.58	−27.560
2500	36.64	74.31	260.07	−27.705

$\Delta\hat{h}°(298) = 0.00$ MJ/kgmol.

Oxygen (O_2)

T (K)	$\hat{c}_p$ (kJ/kgmol · K)	$\hat{h}$ (MJ/kgmol)	$\hat{s}°$ (kJ/kgmol · K)	$\hat{g}°/\hat{R}T$
200	29.12	−2.87	193.38	−24.982
293	29.35	−0.15	204.56	−24.664
298	29.37	0.00	205.03	−24.660
400	30.11	3.03	213.76	−24.800
500	31.09	6.09	220.59	−25.067
600	32.09	9.25	226.35	−25.370
700	32.98	12.50	231.36	−25.679
800	33.74	15.84	235.81	−25.981
900	34.36	19.25	239.83	−26.273
1000	34.88	22.71	243.48	−26.553
1100	35.31	26.22	246.82	−26.819
1200	35.68	29.76	249.91	−27.074
1300	36.00	33.35	252.78	−27.317
1400	36.29	36.97	255.45	−27.549
1500	36.56	40.61	257.97	−27.771
1600	36.82	44.28	260.34	−27.983
1700	37.06	47.97	262.58	−28.187
1800	37.30	51.69	264.70	−28.383
1900	37.54	55.43	266.73	−28.571
2000	37.78	59.20	268.65	−28.752
2100	38.01	62.99	270.50	−28.927
2200	38.24	66.80	272.28	−29.096
2300	38.47	70.63	273.98	−29.259
2400	38.69	74.49	275.63	−29.418
2500	38.92	78.37	277.21	−29.570

$\Delta\hat{h}°(298) = 0.00$ MJ/kgmol.

Carbon monoxide (CO)

T (K)	$\hat{c}_p$ (kJ/kgmol·K)	$\hat{h}$ (MJ/kgmol)	$\hat{s}°$ (kJ/kgmol·K)	$\hat{g}°/\hat{R}T$
200	29.11	−2.86	185.92	−90.549
293	29.14	−0.15	197.08	−69.111
298	29.14	0.00	197.54	−68.347
400	29.34	2.97	206.12	−57.132
500	29.79	5.93	212.72	−50.746
600	30.44	8.94	218.20	−46.608
700	30.17	12.02	222.95	−43.741
800	31.90	15.18	227.16	−41.658
900	32.58	18.40	230.96	−40.091
1000	33.18	21.69	234.42	−38.881
1100	33.71	25.03	237.61	−37.927
1200	34.17	28.43	240.56	−37.163
1300	34.57	31.87	243.32	−36.543
1400	34.92	35.34	245.89	−36.034
1500	35.22	38.85	248.31	−35.613
1600	35.48	42.38	250.59	−35.262
1700	35.71	45.94	252.75	−34.969
1800	35.91	49.52	254.80	−34.722
1900	36.09	53.12	256.74	−34.514
2000	36.25	56.74	258.60	−34.338
2100	36.39	60.38	260.37	−34.188
2200	36.52	64.02	262.06	−34.062
2300	36.64	67.68	263.69	−33.956
2400	36.74	71.35	265.25	−33.867
2500	36.84	75.02	266.76	−33.792

$\Delta\hat{h}°(298) = -110.53$ MJ/kgmol.

Carbon (C, Gas)

T (K)	$\hat{c}_p$ (kJ/kgmol·K)	$\hat{h}$ (MJ/kgmol)	$\hat{s}°$ (kJ/kgmol·K)	$\hat{g}°/\hat{R}T$
200	20.91	−2.05	149.66	410.753
293	20.84	−0.10	157.66	274.349
298	20.84	0.00	157.99	269.432
400	20.82	2.12	164.11	195.891
500	20.81	4.20	168.76	152.707
600	20.80	6.28	172.55	123.835
700	20.79	8.36	175.75	103.151
800	20.79	10.44	178.53	87.593
900	20.79	12.52	180.98	75.458
1000	20.79	14.60	183.17	65.722
1100	20.79	16.68	185.15	57.733
1200	20.79	18.76	186.96	51.057
1300	20.80	20.84	188.63	45.392
1400	20.80	22.92	190.17	40.523
1500	20.82	25.00	191.60	36.291
1600	20.83	27.08	192.95	32.577
1700	20.85	29.17	194.21	29.291
1800	20.88	31.25	195.41	26.362
1900	20.91	33.34	196.54	23.734
2000	20.95	35.43	197.61	21.362
2100	21.00	37.53	198.63	19.210
2200	21.05	39.64	199.61	17.248
2300	21.11	41.74	200.55	15.452
2400	21.18	43.86	201.45	13.801
2500	21.24	45.98	202.31	12.278

$\Delta\hat{h}°(298) = 715.00$ MJ/kgmol.

Carbon (C, graphite)

T (K)	$\hat{c}_p$ (kJ/kgmol·K)	$\hat{h}$ (MJ/kgmol)	$\hat{s}°$ (kJ/kgmol·K)	$\hat{g}°/\hat{R}T$
200	5.03	−0.67	3.01	−0.765
293	8.35	−0.04	5.54	−0.684
298	8.53	0.00	5.69	−0.684
400	11.93	1.05	8.68	−0.730
500	14.63	2.38	11.65	−0.828
600	16.89	3.96	14.52	−0.952
700	18.58	5.74	17.26	−1.090
800	19.83	7.66	19.83	−1.233
900	20.79	9.70	22.22	−1.377
1000	21.54	11.82	24.45	−1.520
1100	22.19	14.00	26.53	−1.660
1200	22.72	16.25	28.49	−1.798
1300	23.12	18.54	30.33	−1.932
1400	23.45	20.87	32.05	−2.062
1500	23.72	23.23	33.68	−2.188
1600	23.94	25.61	35.22	−2.310
1700	24.12	28.02	36.67	−2.429
1800	24.28	30.44	38.06	−2.543
1900	24.42	32.87	39.38	−2.655
2000	24.54	35.32	40.63	−2.763
2100	24.65	37.78	41.83	−2.868
2200	24.74	40.25	42.98	−2.969
2300	24.84	42.73	44.08	−3.068
2400	24.92	45.22	45.14	−3.163
2500	25.00	47.71	46.16	−3.256

$\Delta\hat{h}°(298) = 0.00$ MJ/kgmol.

Methane (CH_4)

T (K)	$\hat{c}_p$ (kJ/kgmol·K)	$\hat{h}$ (MJ/kgmol)	$\hat{s}°$ (kJ/kgmol·K)	$\hat{g}°/\hat{R}T$
200	33.48	−3.37	172.47	−67.796
293	35.48	−0.18	185.58	−53.113
298	35.64	0.00	186.15	−52.593
400	40.50	3.86	197.25	−45.076
500	46.34	8.20	206.91	−40.924
600	52.23	13.13	215.88	−38.342
700	57.79	18.64	224.35	−36.647
800	62.93	24.67	232.41	−35.501
900	67.60	31.20	240.10	−34.714
1000	71.80	38.18	247.45	−34.175
1100	75.53	45.55	254.47	−33.812
1200	78.83	53.27	261.18	−33.579
1300	81.75	61.30	267.61	−33.442
1400	84.31	69.61	273.76	−33.379
1500	86.56	78.15	279.66	−33.373
1600	88.54	86.91	285.31	−33.411
1700	90.29	95.86	290.73	−33.483
1800	91.83	104.96	295.93	−33.583
1900	93.19	114.21	300.94	−33.705
2000	94.40	123.60	305.75	−33.844
2100	95.48	133.09	310.38	−33.997
2200	96.44	142.69	314.85	−34.161
2300	97.30	152.37	319.15	−34.333
2400	98.08	162.14	323.31	−34.513
2500	98.78	171.99	327.33	−34.697

$\Delta\hat{h}°(298) = -74.87$ MJ/kgmol.

Hydrogen (H_2)

T (K)	$\hat{c}_p$ (kJ/kgmol·K)	$\hat{h}$ (MJ/kgmol)	$\hat{s}°$ (kJ/kgmol·K)	$\hat{g}°/\hat{R}T$
200	27.27	−2.77	119.33	−16.018
293	28.80	−0.14	130.10	−15.707
298	28.84	0.00	130.57	−15.705
400	29.18	2.96	139.11	−15.841
500	29.26	5.88	145.63	−16.100
600	29.33	8.81	150.97	−16.391
700	29.44	11.75	155.50	−16.684
800	29.65	14.70	159.44	−16.966
900	29.91	17.68	162.95	−17.236
1000	30.20	20.69	166.11	−17.491
1100	30.54	23.72	169.01	−17.734
1200	30.92	26.79	171.68	−17.963
1300	31.34	29.91	174.17	−18.181
1400	31.80	33.06	176.51	−18.389
1500	32.43	36.27	178.72	−18.588
1600	32.73	39.52	180.82	−18.777
1700	33.14	42.81	182.82	−18.959
1800	33.54	46.15	184.72	−19.134
1900	33.92	49.52	186.55	−19.302
2000	34.29	52.93	188.30	−19.464
2100	34.64	56.38	189.98	−19.621
2200	34.97	59.86	191.60	−19.772
2300	35.29	63.37	193.16	−19.918
2400	35.59	66.91	194.67	−20.060
2500	35.88	70.49	196.13	−20.198

$\Delta\hat{h}°(298) = 0.00$ MJ/kgmol.

Hydrogen, monatomic (H)

T (K)	$\hat{c}_p$ (kJ/kgmol·K)	$\hat{h}$ (MJ/kgmol)	$\hat{s}°$ (kJ/kgmol·K)	$\hat{g}°/\hat{R}T$
200	20.79	−2.04	106.31	117.077
293	20.79	−0.10	114.28	75.648
298	20.79	0.00	114.61	74.152
400	20.79	2.12	120.72	51.663
500	20.79	4.20	125.36	38.369
600	20.79	6.28	129.15	29.422
700	20.79	8.35	132.35	22.971
800	20.79	10.43	135.13	18.089
900	20.79	12.51	137.57	14.257
1000	20.79	14.59	139.76	11.163
1100	20.79	16.67	141.75	8.609
1200	20.79	18.75	143.55	6.462
1300	20.79	20.82	145.22	4.628
1400	20.79	22.90	146.76	3.044
1500	20.79	24.98	148.19	1.658
1600	20.79	27.06	149.53	0.436
1700	20.79	29.14	150.80	−0.653
1800	20.79	31.22	151.98	−1.628
1900	20.79	33.30	153.11	−2.508
2000	20.79	35.38	154.17	−3.306
2100	20.79	37.46	155.18	−4.035
2200	20.79	39.53	156.16	−4.703
2300	20.79	41.61	157.08	−5.317
2400	20.79	43.69	157.96	−5.885
2500	20.79	45.77	158.81	−6.412

$\Delta\hat{h}°(298) = 217.99$ MJ/kgmol.

Nitrogen, monatomic (N)

T (K)	$\hat{c}_p$ (kJ/kgmol · K)	$\hat{h}$ (MJ/kgmol)	$\hat{s}°$ (kJ/kgmol · K)	$\hat{g}°/\hat{R}T$
200	20.79	−2.04	144.90	265.669
293	20.79	−0.10	152.86	175.550
298	20.79	0.00	153.19	172.301
400	20.79	2.12	159.30	123.639
500	20.79	4.20	163.94	95.021
600	20.79	6.28	167.73	75.859
700	20.79	8.35	170.94	62.111
800	20.79	10.43	173.71	51.756
900	20.79	12.51	176.16	43.668
1000	20.79	14.59	178.35	37.169
1100	20.79	16.67	180.33	31.829
1200	20.79	18.75	182.14	27.360
1300	20.79	20.82	183.80	23.562
1400	20.79	22.90	185.34	20.293
1500	20.79	24.98	186.78	17.448
1600	20.79	27.06	188.12	14.949
1700	20.79	29.14	189.38	12.734
1800	20.79	31.22	190.57	10.757
1900	20.79	33.30	191.69	8.981
2000	20.79	35.38	192.76	7.376
2100	20.79	37.46	193.77	5.918
2200	20.80	39.53	194.74	4.587
2300	20.80	41.61	195.66	3.366
2400	20.82	43.70	196.55	2.243
2500	20.83	45.78	197.40	1.206

$\Delta\hat{h}°(298) = 472.79$ MJ/kgmol.

Nitric oxide (NO)

T (K)	$\hat{c}_p$ (kJ/kgmol · K)	$\hat{h}$ (MJ/kgmol)	$\hat{s}°$ (kJ/kgmol · K)	$\hat{g}°/\hat{R}T$
200	30.42	−2.95	198.64	28.633
293	29.85	−0.15	210.19	11.703
298	29.84	0.00	210.65	11.087
400	29.94	3.04	219.43	1.672
500	30.49	6.06	226.16	−4.024
600	31.24	9.15	231.78	−7.944
700	32.03	12.31	236.66	−10.835
800	32.77	15.55	240.98	−13.072
900	33.42	18.86	244.88	−14.867
1000	33.99	22.23	248.43	−16.347
1100	34.47	25.65	251.70	−17.596
1200	34.88	29.12	254.71	−18.667
1300	35.23	32.63	257.52	−19.601
1400	35.53	36.17	260.14	−20.424
1500	35.78	39.73	262.60	−21.159
1600	36.00	43.32	264.92	−21.819
1700	36.20	46.93	267.11	−22.418
1800	36.37	50.56	269.18	−20.964
1900	36.51	54.20	271.15	−23.465
2000	36.65	57.86	273.03	−23.929
2100	36.77	61.53	274.82	−24.358
2200	36.87	65.22	276.53	−24.758
2300	36.97	68.91	278.17	−25.132
2400	37.06	72.61	279.75	−25.483
2500	37.14	76.32	281.26	−25.813

$\Delta\hat{h}°(298) = 90.29$ MJ/kgmol.

Nitrogen dioxide (NO_2)

T (K)	$\hat{c}_p$ (kJ/kgmol·K)	$\hat{h}$ (MJ/kgmol)	$\hat{s}°$ (kJ/kgmol·K)	$\hat{g}°/\hat{R}T$
200	34.38	−3.49	225.74	−9.350
293	36.82	−0.19	239.34	−15.284
298	36.97	0.00	239.92	−15.506
400	40.17	3.93	251.23	−19.084
500	43.21	8.10	260.53	−21.426
600	45.84	12.56	268.65	−23.160
700	47.99	17.25	275.88	−24.531
800	49.71	22.14	282.40	−25.662
900	51.08	27.18	288.34	−26.625
1000	52.17	32.34	293.78	−27.464
1100	53.04	37.61	298.80	−28.207
1200	53.75	42.95	303.44	−28.874
1300	54.33	48.35	307.77	−29.481
1400	54.81	53.81	311.81	−30.037
1500	55.20	59.31	315.61	−30.550
1600	55.53	64.85	319.18	−31.027
1700	55.81	70.42	322.56	−31.472
1800	56.06	76.01	325.75	−31.890
1900	56.26	81.63	328.79	−32.283
2000	56.44	87.26	331.68	−32.655
2100	56.60	92.91	334.44	−33.008
2200	56.74	98.58	337.08	−33.343
2300	56.85	104.26	339.60	−33.662
2400	56.96	109.95	342.02	−33.968
2500	57.05	115.65	344.35	−34.260

$\Delta\hat{h}°(298) = 33.10$ MJ/kgmol

Oxygen, monatomic (O)

T (K)	$\hat{c}_p$ (kJ/kgmol·K)	$\hat{h}$ (MJ/kgmol)	$\hat{s}°$ (kJ/kgmol·K)	$\hat{g}°/\hat{R}T$
200	22.74	−2.19	152.05	130.256
293	21.95	−0.11	160.61	82.879
298	21.91	0.00	160.95	81.168
400	21.48	2.21	167.32	55.469
500	21.26	4.34	172.09	40.290
600	21.13	6.46	175.95	30.085
700	21.04	8.57	179.20	22.735
800	20.98	10.67	182.01	17.178
900	20.95	12.77	184.48	12.820
1000	20.92	14.86	186.69	9.306
1100	20.89	16.95	188.68	6.407
1200	20.88	19.04	190.49	3.974
1300	20.87	21.13	192.16	1.897
1400	20.85	23.21	193.71	0.104
1500	20.84	25.30	195.15	−1.462
1600	20.84	27.38	196.49	−2.843
1700	20.83	29.46	197.76	−4.070
1800	20.83	31.55	198.95	−5.170
1900	20.83	33.63	200.07	−6.160
2000	20.83	35.71	201.14	−7.059
2100	20.83	37.80	202.16	−7.877
2200	20.83	39.88	203.13	−8.627
2300	20.84	41.96	204.05	−9.317
2400	20.84	44.04	204.94	−9.954
2500	20.85	46.13	205.79	−10.543

$\Delta\hat{h}°(298) = 249.19$ MJ/kgmol.

Hydroxyl (OH)

T (K)	$\hat{c}_p$ (kJ/kgmol · K)	$\hat{h}$ (MJ/kgmol)	$\hat{s}°$ (kJ/kgmol · K)	$\hat{g}°/\hat{R}T$
200	30.78	−2.97	171.48	1.318
293	30.01	−0.15	183.13	−5.896
298	29.99	0.00	183.59	−6.162
400	29.65	3.03	192.36	−10.357
500	29.52	5.99	198.95	−12.995
600	29.53	8.94	204.33	−14.873
700	29.66	11.90	208.89	−16.299
800	29.92	14.88	212.87	−17.433
900	30.26	17.89	216.41	−18.365
1000	30.68	20.93	219.62	−19.151
1100	31.12	24.02	222.57	−19.827
1200	31.59	27.16	225.30	−20.420
1300	32.05	30.34	227.84	−20.945
1400	32.49	33.57	230.23	−21.417
1500	32.92	36.84	232.49	−21.844
1600	33.32	40.15	234.63	−22.235
1700	33.69	43.50	236.66	−22.594
1800	34.05	46.89	238.59	−22.927
1900	34.37	50.31	240.44	−23.236
2000	34.67	53.76	242.22	−23.526
2100	34.95	57.24	243.91	−23.798
2200	35.21	60.75	245.54	−24.054
2300	35.45	64.28	247.12	−24.296
2400	35.67	67.84	248.63	−24.526
2500	35.84	71.42	250.09	−24.745

$\Delta\hat{h}°(298) = 39.46$ MJ/kgmol.

Carbon dioxide (CO_2)

T (°R)	$\hat{c}_p$ (Btu/lbmol·°R)	$\hat{h}$ (kBtu/lbmol)	$\hat{s}°$ (Btu/lbmol·°R)	$\hat{g}°/\hat{R}T$
400	7.972	−1.152	48.590	−238.907
530	8.823	−0.062	50.927	−186.551
537	8.868	0.000	51.038	−184.449
600	9.237	0.574	52.054	−167.723
800	10.242	2.528	54.855	−132.526
1000	11.032	4.659	57.228	−111.667
1200	11.670	6.931	59.297	−97.948
1400	12.192	9.319	61.136	−88.288
1600	12.620	11.803	62.793	−81.153
1800	12.971	14.362	64.301	−75.693
2000	13.261	16.987	65.683	−71.397
2200	13.502	19.664	66.959	−67.942
2400	13.703	22.385	68.142	−65.115
2600	13.869	25.142	69.245	−62.768
2800	14.013	27.931	70.279	−60.794
3000	14.135	30.747	71.250	−59.117
3200	14.240	33.584	72.166	−57.679
3400	14.334	36.441	73.032	−56.437
3600	14.414	39.316	73.854	−55.356
3800	14.486	42.207	74.635	−54.410
4000	14.549	45.111	75.380	−53.578
4200	14.607	48.026	76.090	−52.843
4400	14.658	50.952	76.772	−52.191
4600	14.706	53.889	77.424	−51.610
4800	14.749	56.835	78.051	−51.090

$\Delta\hat{h}°(537) = -169.184$ kBtu/lbmol.

Water (H_2O, gas)

T (°R)	$\hat{c}_p$ (Btu/lbmol·°R)	$\hat{h}$ (kBtu/lbmol)	$\hat{s}°$ (Btu/lbmol·°R)	$\hat{g}°/\hat{R}T$
400	7.970	−1.092	42.756	−153.790
530	8.016	−0.056	44.977	−121.545
537	8.020	0.000	45.076	−120.252
600	8.065	0.510	45.980	−109.983
800	8.276	2.142	48.322	−88.428
1000	8.551	3.824	50.196	−75.705
1200	8.854	5.564	51.780	−67.368
1400	9.174	7.366	53.169	−61.520
1600	9.507	9.235	54.415	−57.216
1800	9.844	11.169	55.555	−53.937
2000	10.178	13.173	56.609	−51.367
2200	10.507	15.241	57.595	−49.312
2400	10.828	17.371	58.521	−47.639
2600	11.089	19.561	59.398	−46.258
2800	11.355	21.806	60.229	−45.106
3000	11.597	24.102	61.021	−44.134
3200	11.819	26.444	61.777	−43.308
3400	12.021	28.828	62.500	−42.601
3600	12.206	31.250	63.192	−41.993
3800	12.374	33.709	63.857	−41.466
4000	12.526	36.199	64.495	−41.009
4200	12.666	38.718	65.109	−40.610
4400	12.794	41.264	65.702	−40.261
4600	12.912	43.836	66.273	−39.956
4800	13.020	46.429	66.825	−39.687

$\Delta\hat{h}°(537) = -103.967$ kBtu/lbmol.

Nitrogen (N_2)

T (°R)	$\hat{c}_p$ (Btu/lbmol·°R)	$\hat{h}$ (kBtu/lbmol)	$\hat{s}°$ (Btu/lbmol·°R)	$\hat{g}°/\hat{R}T$
400	6.952	−0.950	43.721	−23.213
530	6.956	−0.048	45.654	−23.036
537	6.956	0.000	45.739	−23.033
600	6.962	0.441	46.522	−23.057
800	7.016	1.838	48.526	−23.280
1000	7.113	3.250	50.102	−23.593
1200	7.282	4.692	51.415	−23.922
1400	7.472	6.169	52.552	−24.244
1600	7.648	7.681	53.561	−24.554
1800	7.810	9.226	54.471	−24.848
2000	7.953	10.803	55.301	−25.128
2200	8.079	12.407	56.066	−25.393
2400	8.188	14.034	56.773	−25.644
2600	8.282	15.681	57.432	−25.884
2800	8.363	17.345	58.049	−26.112
3000	8.433	19.025	58.628	−26.330
3200	8.495	20.718	59.175	−26.538
3400	8.548	22.422	59.692	−26.738
3600	8.595	24.136	60.182	−26.929
3800	8.636	25.860	60.647	−27.113
4000	8.673	27.592	61.091	−27.290
4200	8.707	29.329	61.515	−27.460
4400	8.737	31.073	61.921	−27.625
4600	8.763	32.824	62.310	−27.784
4800	8.787	34.580	62.684	−27.938

$\Delta\hat{h}°(537) = 0.000$ kBtu/lbmol.

Oxygen (O_2)

T (°R)	$\hat{c}_p$ (Btu/lbmol·°R)	$\hat{h}$ (kBtu/lbmol)	$\hat{s}°$ (Btu/lbmol·°R)	$\hat{g}°/\hat{R}T$
400	6.963	−0.954	46.945	−24.841
530	7.012	−0.049	48.885	−24.663
537	7.015	0.000	48.971	−24.660
600	7.066	0.446	49.764	−24.685
800	7.291	1.882	51.823	−24.912
1000	7.560	3.367	53.477	−25.234
1200	7.810	4.904	54.878	−25.577
1400	8.020	6.489	56.097	−25.915
1600	8.191	8.111	57.180	−26.241
1800	8.330	9.762	58.153	−26.553
2000	8.444	11.440	59.036	−26.848
2200	8.539	13.138	59.846	−27.129
2400	8.622	14.855	60.593	−27.396
2600	8.697	16.587	61.286	−27.649
2800	8.767	18.334	61.933	−27.890
3000	8.833	20.093	62.540	−28.120
3200	8.897	21.866	63.112	−28.340
3400	8.961	23.653	63.654	−28.551
3600	9.023	25.451	64.167	−28.752
3800	9.084	27.261	64.657	−28.946
4000	9.145	29.085	65.124	−29.133
4200	9.206	30.919	65.572	−29.313
4400	9.265	32.767	66.002	−29.486
4600	9.324	34.625	66.414	−29.654
4800	9.382	36.495	66.813	−29.816

$\Delta\hat{h}°(537) = 0.000$ kBtu/lbmol.

Carbon monoxide (CO)

T (°R)	$\hat{c}_p$ (Btu/lbmol·°R)	$\hat{h}$ (kBtu/lbmol)	$\hat{s}°$ (Btu/lbmol·°R)	$\hat{g}°/\hat{R}T$
400	6.952	−0.950	45.163	−83.761
530	6.959	−0.049	47.097	−68.939
537	6.960	0.000	47.182	−68.347
600	6.971	0.441	47.966	−63.665
800	7.050	1.841	49.975	−53.918
1000	7.198	3.265	51.562	−48.249
1200	7.386	4.723	52.890	−44.592
1400	7.581	6.220	54.043	−42.069
1600	7.764	7.754	55.067	−40.245
1800	7.926	9.323	55.991	−38.881
2000	8.065	10.923	56.833	−37.833
2200	8.185	12.548	57.608	−37.014
2400	8.286	14.196	58.324	−36.362
2600	8.373	15.862	58.991	−35.837
2800	8.447	17.545	59.615	−35.411
3000	8.512	19.240	60.200	−35.061
3200	8.567	20.948	60.751	−34.773
3400	8.616	22.667	61.271	−34.535
3600	8.658	24.393	61.766	−34.338
3800	8.696	26.131	62.234	−34.173
4000	8.729	27.872	62.681	−34.037
4200	8.759	29.621	63.108	−33.925
4400	8.786	31.375	63.516	−33.832
4600	8.810	33.135	63.907	−33.756
4800	8.832	34.899	64.282	−33.694

$\Delta\hat{h}°(537) = -47.519$ kBtu/lbmol.

Carbon (C, gas)

T (°R)	$\hat{c}_p$ (Btu/lbmol·°R)	$\hat{h}$ (kBtu/lbmol)	$\hat{s}°$ (Btu/lbmol·°R)	$\hat{g}°/\hat{R}T$
400	4.986	−0.680	36.291	367.855
530	4.978	−0.035	37.675	273.242
537	4.978	0.000	37.736	269.433
600	4.975	0.315	38.296	238.970
800	4.971	1.309	39.723	174.314
1000	4.968	2.304	40.831	135.393
1200	4.967	3.298	41.736	109.362
1400	4.967	4.291	42.502	90.709
1600	4.967	5.284	43.164	76.673
1800	4.966	6.278	43.750	65.722
2000	4.966	7.269	44.273	56.933
2200	4.967	8.263	44.746	49.720
2400	4.968	9.257	45.179	43.689
2600	4.970	10.251	45.576	38.571
2800	4.973	11.245	45.945	34.170
3000	4.979	12.241	46.288	30.344
3200	4.985	13.236	46.610	26.985
3400	4.994	14.235	46.912	24.013
3600	5.005	15.234	47.197	21.362
3800	5.017	16.236	47.469	18.983
4000	5.032	17.241	47.727	16.836
4200	5.048	18.249	47.973	14.886
4400	5.065	19.260	48.208	13.109
4600	5.083	20.274	48.433	11.481
4800	5.103	21.293	48.650	9.984

$\Delta\hat{h}°(537) = 307.396$ kBtu/lbmol.

Carbon (C, graphite)

T (°R)	$\hat{c}_p$ (Btu/lbmol·°R)	$\hat{h}$ (kBtu/lbmol)	$\hat{s}°$ (Btu/lbmol·°R)	$\hat{g}°/\hat{R}T$
400	1.390	−0.235	0.855	−0.727
530	2.004	−0.014	1.331	−0.684
537	2.037	0.000	1.358	−0.684
600	2.329	0.139	1.602	−0.690
800	3.151	0.690	2.389	−0.769
1000	3.810	1.390	3.167	−0.895
1200	4.316	2.205	3.909	−1.043
1400	4.677	3.105	4.603	−1.201
1600	4.944	4.070	5.246	−1.361
1800	5.146	5.080	5.840	−1.520
2000	5.316	6.127	6.391	−1.676
2200	5.450	7.203	6.904	−1.828
2400	5.551	8.304	7.384	−1.976
2600	5.631	9.422	7.831	−2.119
2800	5.695	10.555	8.251	−2.257
3000	5.747	11.700	8.645	−2.390
3200	5.791	12.854	9.018	−2.518
3400	5.829	14.017	9.370	−2.643
3600	5.861	15.185	9.705	−2.763
3800	5.890	16.359	10.022	−2.879
4000	5.915	17.541	10.325	−2.991
4200	5.939	18.726	10.614	−3.100
4400	5.960	19.916	10.891	−3.205
4600	5.980	21.110	11.156	−3.307
4800	5.999	22.308	11.411	−3.406

$\Delta\hat{h}°(537) = 0.000$ kBtu/lbmol.

Methane (CH_4)

T (°R)	$\hat{c}_p$ (Btu/lbmol·°R)	$\hat{h}$ (kBtu/lbmol)	$\hat{s}°$ (Btu/lbmol·°R)	$\hat{g}°/\hat{R}T$
400	8.068	−1.128	42.070	−63.128
530	8.483	−0.059	44.356	−52.996
537	8.512	0.000	44.460	−52.593
600	8.861	0.552	45.438	−49.434
800	10.277	2.459	48.166	−42.969
1000	11.854	4.672	50.628	−39.351
1200	13.371	7.196	52.924	−37.139
1400	14.768	10.012	55.091	−35.719
1600	16.028	13.094	57.147	−34.787
1800	17.149	16.414	59.101	−34.175
2000	18.133	19.945	60.960	−33.780
2200	18.992	23.659	62.729	−33.541
2400	19.738	27.534	64.414	−33.414
2600	20.384	31.547	66.020	−33.370
2800	20.944	35.681	67.552	−33.389
3000	21.431	39.920	69.014	−33.455
3200	21.855	44.249	70.411	−33.559
3400	22.224	48.659	71.747	−33.690
3600	22.547	53.136	73.027	−33.844
3800	22.832	57.674	74.254	−34.015
4000	23.082	62.267	75.431	−34.198
4200	23.304	66.905	76.563	−34.392
4400	23.502	71.586	77.652	−34.594
4600	23.677	76.304	78.700	−34.801
4800	23.834	81.056	79.712	−35.014

$\Delta\hat{h}°(537) = -32.189$ kBtu/lbmol.

Hydrogen (H_2)

T (°R)	$\hat{c}_p$ (Btu/lbmol·°R)	$\hat{h}$ (kBtu/lbmol)	$\hat{s}°$ (Btu/lbmol·°R)	$\hat{g}°/\hat{R}T$
400	6.646	−0.929	29.205	−15.876
530	6.880	−0.048	31.100	−15.707
537	6.887	0.000	31.187	−15.705
600	6.938	0.437	31.962	−15.728
800	6.982	1.830	33.962	−15.950
1000	6.997	3.228	35.520	−16.261
1200	7.020	4.629	36.798	−16.587
1400	7.070	6.038	37.883	−16.905
1600	7.136	7.459	38.831	−17.207
1800	7.214	8.893	39.675	−17.491
2000	7.305	10.345	40.440	−17.760
2200	7.406	11.815	41.141	−18.013
2400	7.519	13.308	41.790	−18.252
2600	7.664	14.824	42.397	−18.479
2800	7.788	16.367	42.969	−18.694
3000	7.882	17.933	43.509	−18.900
3200	7.990	19.521	44.021	−19.096
3400	8.093	21.129	44.509	−19.284
3600	8.190	22.757	44.974	−19.464
3800	8.282	24.404	45.419	−19.638
4000	8.370	26.070	45.846	−19.805
4200	8.453	27.751	46.257	−19.966
4400	8.531	29.450	46.652	−20.122
4600	8.605	31.164	47.033	−20.272
4800	8.674	32.891	47.400	−20.419

$\Delta\hat{h}°(537) = 0.000$ kBtu/lbmol.

Hydrogen, monatomic (H)

T (°R)	$\hat{c}_p$ (Btu/lbmol·°R)	$\hat{h}$ (kBtu/lbmol)	$\hat{s}°$ (Btu/lbmol·°R)	$\hat{g}°/\hat{R}T$
400	4.965	−0.679	25.932	104.068
530	4.965	−0.035	27.313	75.311
537	4.965	0.000	27.374	74.152
600	4.965	0.314	27.933	64.853
800	4.965	1.308	29.358	45.031
1000	4.965	2.301	30.465	33.010
1200	4.965	3.293	31.369	24.913
1400	4.965	4.286	32.135	19.069
1600	4.965	5.279	32.797	14.641
1800	4.965	6.272	33.382	11.163
2000	4.965	7.266	33.905	8.352
2200	4.965	8.259	34.378	6.030
2400	4.965	9.250	34.810	4.075
2600	4.965	10.244	35.208	2.406
2800	4.965	11.237	35.575	0.961
3000	4.965	12.231	35.918	−0.303
3200	4.965	13.223	36.239	−1.420
3400	4.965	14.215	36.539	−2.414
3600	4.965	15.209	36.823	−3.306
3800	4.965	16.202	37.091	−4.112
4000	4.965	17.195	37.347	−4.844
4200	4.965	18.187	37.588	−5.511
4400	4.965	19.180	37.820	−6.124
4600	4.965	20.174	38.041	−6.688
4800	4.965	21.167	38.252	−7.210

$\Delta\hat{h}°(537) = 93.717$ kBtu/lbmol.

Nitrogen, monatomic (N)

T (°R)	$\hat{c}_p$ (Btu/lbmol·°R)	$\hat{h}$ (kBtu/lbmol)	$\hat{s}°$ (Btu/lbmol·°R)	$\hat{g}°/\hat{R}T$
400	4.965	−0.679	35.149	237.337
530	4.965	−0.035	36.529	174.818
537	4.965	0.000	36.590	172.301
600	4.965	0.314	37.148	152.152
800	4.965	1.308	38.574	109.345
1000	4.965	2.301	39.681	83.534
1200	4.965	3.293	40.586	66.242
1400	4.965	4.286	41.351	53.831
1600	4.965	5.279	42.013	44.478
1800	4.965	6.272	42.598	37.169
2000	4.965	7.266	43.121	31.294
2200	4.965	8.259	43.594	26.464
2400	4.965	9.250	44.026	22.419
2600	4.965	10.244	44.424	18.982
2800	4.965	11.237	44.792	16.021
3000	4.965	12.231	45.134	13.444
3200	4.965	13.223	45.455	11.178
3400	4.966	14.215	45.755	9.170
3600	4.966	15.209	46.039	7.376
3800	4.967	16.202	46.308	5.764
4000	4.968	17.195	46.563	4.307
4200	4.970	18.189	46.805	2.982
4400	4.973	19.184	47.037	1.772
4600	4.977	20.178	47.258	0.663
4800	4.982	21.175	47.469	−0.358

$\Delta\hat{h}°(537) = 203.264$ kBtu/lbmol.

Nitric oxide (NO)

T (°R)	$\hat{c}_p$ (Btu/lbmol·°R)	$\hat{h}$ (kBtu/lbmol)	$\hat{s}°$ (Btu/lbmol·°R)	$\hat{g}°/\hat{R}T$
400	7.214	−0.978	48.239	23.345
530	7.129	−0.050	50.228	11.564
537	7.128	0.000	50.313	11.087
600	7.120	0.451	51.117	7.217
800	7.200	1.882	53.168	−1.155
1000	7.378	3.338	54.790	−6.362
1200	7.587	4.835	56.153	−9.958
1400	7.789	6.372	57.338	−12.619
1600	7.966	7.948	58.390	−14.685
1800	8.118	9.557	59.337	−16.347
2000	8.244	11.193	60.200	−17.723
2200	8.350	12.853	60.990	−18.885
2400	8.439	14.533	61.720	−19.886
2600	8.514	16.229	62.399	−20.761
2800	8.577	17.938	63.033	−21.534
3000	8.631	19.659	63.626	−22.224
3200	8.678	21.389	64.184	−22.847
3400	8.717	23.129	64.711	−23.412
3600	8.753	24.876	65.211	−23.929
3800	8.785	26.629	65.685	−24.404
4000	8.813	28.390	66.136	−24.843
4200	8.838	30.155	66.567	−25.251
4400	8.861	31.925	66.979	−25.632
4600	8.881	33.699	67.373	−25.988
4800	8.901	35.477	67.752	−26.323

$\Delta\hat{h}°(537) = 38.818$ kBtu/lbmol.

Nitrogen dioxide (NO_2)

T (°R)	$\hat{c}_p$ (Btu/lbmol·°R)	$\hat{h}$ (kBtu/lbmol)	$\hat{s}°$ (Btu/lbmol·°R)	$\hat{g}°/\hat{R}T$
400	8.330	−1.171	54.819	−11.166
530	8.802	−0.062	57.196	−15.334
537	8.831	0.000	57.305	−15.506
600	9.088	0.568	58.312	−16.946
800	9.924	2.470	61.037	−20.225
1000	10.682	4.533	63.334	−22.445
1200	11.302	6.734	65.338	−24.105
1400	11.789	9.044	67.118	−25.427
1600	12.167	11.441	68.718	−26.525
1800	12.460	13.905	70.168	−27.464
2000	12.689	16.421	71.494	−28.285
2200	12.872	18.979	72.711	−29.014
2400	13.017	21.567	73.838	−29.672
2600	13.134	24.184	74.885	−30.270
2800	13.230	26.820	75.862	−30.819
3000	13.310	29.474	76.778	−31.327
3200	13.377	32.143	77.639	−31.799
3400	13.433	34.824	78.452	−32.241
3600	13.481	37.516	79.221	−32.655
3800	13.522	40.216	79.951	−33.046
4000	13.557	42.924	80.645	−33.415
4200	13.588	45.639	81.307	−33.766
4400	13.615	48.359	81.940	−34.099
4600	13.638	51.084	82.545	−34.417
4800	13.661	53.815	83.126	−34.721

$\Delta\hat{h}°(537) = 14.228$ kBtu/lbmol.

Oxygen, monatomic (O)

T (°R)	$\hat{c}_p$ (Btu/lbmol·°R)	$\hat{h}$ (kBtu/lbmol)	$\hat{s}°$ (Btu/lbmol·°R)	$\hat{g}°/\hat{R}T$
400	5.381	−0.725	36.907	115.376
537	5.241	−0.036	38.379	82.494
537	5.233	0.000	38.442	81.168
600	5.190	0.331	39.030	70.539
800	5.103	1.359	40.506	47.895
1000	5.058	2.373	41.638	34.177
1200	5.031	3.382	42.557	24.947
1400	5.014	4.386	43.331	18.293
1600	5.003	5.390	44.000	13.258
1800	4.996	6.389	44.589	9.306
2000	4.990	7.387	45.115	6.116
2200	4.986	8.386	45.590	3.484
2400	4.983	9.381	46.024	1.271
2600	4.980	10.378	46.423	−0.617
2800	4.978	11.373	46.792	−2.250
3000	4.976	12.369	47.135	−3.676
3200	4.976	13.364	47.456	−4.935
3400	4.975	14.359	47.758	−6.055
3600	4.975	15.355	48.042	−7.059
3800	4.975	16.350	48.311	−7.964
4000	4.976	17.345	48.566	−8.786
4200	4.977	18.339	48.809	−9.535
4400	4.979	19.335	49.040	−10.221
4600	4.982	20.331	49.262	−10.853
4800	4.985	21.328	49.474	−11.437

$\Delta\hat{h}°(537) = 107.134$ kBtu/lbmol.

Hydroxyl (OH)

T (°R)	$\hat{c}_p$ (Btu/lbmol·°R)	$\hat{h}$ (kBtu/lbmol)	$\hat{s}°$ (Btu/lbmol·°R)	$\hat{g}°/\hat{R}T$
400	7.292	−0.986	41.761	−0.911
530	7.167	−0.050	43.765	−5.956
537	7.162	0.000	43.851	−6.162
600	7.125	0.452	44.656	−7.869
800	7.063	1.870	46.691	−11.655
1000	7.047	3.280	48.263	−14.108
1200	7.071	4.692	49.549	−15.862
1400	7.130	6.111	50.642	−17.201
1600	7.218	7.545	51.600	−18.269
1800	7.327	8.999	52.456	−19.151
2000	7.446	10.477	53.234	−19.897
2200	7.568	11.978	53.950	−20.542
2400	7.690	13.505	54.613	−21.108
2600	7.806	15.055	55.233	−21.612
2800	7.916	16.627	55.816	−22.065
3000	8.018	18.220	56.366	−22.478
3200	8.113	19.834	56.886	−22.855
3400	8.200	21.465	57.380	−23.203
3600	8.280	23.113	57.852	−23.526
3800	8.355	24.776	58.301	−23.827
4000	8.422	26.455	58.731	−24.109
4200	8.485	28.145	59.145	−24.374
4400	8.539	29.848	59.541	−24.625
4600	8.592	31.562	59.922	−24.862
4800	8.646	33.287	60.288	−25.087

$\Delta\hat{h}°(537) = 16.966$ kBtu/lbmol.

APPENDIX D

Viscosity and Thermal Conductivity of Gas Mixtures

Reid and Sherwood discuss various approximate methods for calculating thermal properties and recommend those procedures which they believe to be best. They recommend the formulas of Wilke for the computation of viscosities of mixtures. The viscosity of the mixture is given in terms of constituent viscosities by

$$\mu_{\text{mix}} = \sum_i \left[\frac{x_i \mu_i}{\Sigma_j \, x_j \phi_{ij}} \right]$$

where

$$\phi_{ij} = \frac{[1 + (\mu_i/\mu_j)^{1/2}(M_j/M_i)^{1/4}]^2}{2\sqrt{2}(1 + M_i/M_j)^{1/2}} \tag{D.1}$$

The viscosity is also a weak function of pressure, but the effect of pressure on the mixture viscosity has not been fully investigated. The combustion products are always at high temperature when they are at high pressure. Examination of the change of viscosity of nitrogen with pressure shows that the pressure effect is less important at high temperatures. For example, at 2700°R the viscosity of nitrogen increases only 0.7% as the pressure is increased from 1 atm to 100 atm. The same change in pressure at 540°R causes an 11% change in viscosity. Thus, the same pressure-temperature relationship which limits dissociation of the products also limits the effect of pressure on viscosity.

Calculation of the viscosity of the complete undissociated lean products of combustion of air and hydrocarbon fuel ($C_n H_{2n}$) at low pressures using Eq. D.1 shows that the viscosity of stoichiometric products of combustion is about 9% higher than that of air at room temperature and about 16% higher at 1500 K. The variation with equivalence ratio is approximately linear. Given the accuracy of most combustion models this warrants the assumption used in the text of using the viscosity of air as an approximation for the viscosity of combustion products.

EXAMPLE D.1. Calculate the viscosity of a 15 : 1 air-fuel ratio mixture of octane vapor and air at 300 K and 1 atm pressure.

Solution. From Appendix B,

$$\mu_1(\text{air}) = 1.853 \times 10^{-5}\ \text{N}\cdot\text{s/m}^2$$

From Table A.8 for octane vapor (C_8H_{18}),

$$\mu_2 = 35.12 \times 10^{-8}\ \text{lb}_\text{m}/\text{in.}\cdot\text{s} = 0.6272 \times 10^{-5}\ \text{N}\cdot\text{s/m}^2$$

and using

$$M_1 = 28.967 \qquad N_1 = 0.5178$$

$$M_2 = 114.22 \qquad N_2 = 0.00875$$

$$x_1 = 0.9834 \qquad N_1 + N_2 = 0.52655$$

$$x_2 = 0.0166$$

$$\sum_j x_j\phi_{1j} = x_1\phi_{11} + x_2\phi_{12} = \text{sum}(1)$$

$$\sum_j x_j\phi_{2j} = x_1\phi_{21} + x_2\phi_{22} = \text{sum}(2)$$

$$\phi_{11} = \phi_{22} = 1$$

we get

$$\phi_{12} = \frac{[1 + (\mu_1/\mu_2)^{1/2}(M_2/M_1)^{1/4}]^2}{2\sqrt{2}(1 + M_1/M_2)^{1/2}} = 3.698$$

$$\phi_{21} = \frac{[1 + (\mu_2/\mu_1)^{1/2}(M_1/M_2)^{1/4}]^2}{2\sqrt{2}(1 + M_2/M_1)^{1/2}} = 0.31743$$

$$\text{sum}(1) = 1.04479$$

$$\text{sum}(2) = 0.32876$$

and finally

$$\mu_{\text{mix}} = \frac{(0.9834)(1.853 \times 10^{-5})}{1.04479} + \frac{(0.0166)(0.6272 \times 10^{-5})}{0.32876}$$

$$= (0.941)(1.853 \times 10^{-5}) + (0.505)(0.6272 \times 10^{-5})$$

$$= 1.776 \times 10^{-5}\ \text{N}\cdot\text{s/m}^2$$

Note: For this binary mixture the result is fortuitously the same as that obtained from a mass-averaged value,

$$\mu_{\text{mix}} = (0.9375)(1.853 \times 10^{-5}) + (0.0625)(0.6272 \times 10^{-5})$$

$$= 1.776 \times 10^{-5}\ \text{N}\cdot\text{s/m}^2$$

The thermal conductivity of a mixture is much more sensitive to interaction effects than is the viscosity. The conductivity is an energy transport coefficient and is thus very closely tied to the structure of the molecules. In particular, reacting mixtures may behave in a very complex way. Unfortunately, the data for conductivities of pure substances are quite limited. The effect of pressure is not well established, although for many regions, the effects appear negligible. Reid and Sherwood suggest the use of the formula of Lindsay and Bromley for mixtures which contain polar gases. The formulas are quite similar to those for the viscosity of a mixture. In cases where the Prandtl number is known, it can be used in conjunction with viscosity and specific heat data to calculate the conductivity. The formulas for the mixture conductivity depend on knowledge of the viscosity and conductivity of the pure components. In addition, the Sutherland constants are to be approximated from the normal boiling points T_{bi} of the pure components.

$$k_{\text{mix}} = \sum_i \left[\frac{k_i}{\sum_j (A_{ij} x_j / x_i)} \right] \tag{D.2}$$

where $A_{ij} = 1 \quad \text{for} \quad i = j$

$$A_{ij} = \frac{1}{4}\left[1 + \sqrt{\left(\frac{\mu_i}{\mu_j}\right)\left(\frac{M_j}{M_i}\right)^{0.75}\left(\frac{1 + S_i/T}{1 + S_j/T}\right)}\right]^2 \left(\frac{1 + S_{ij}/T}{1 + S_j/T}\right) \qquad \text{for} \quad i \neq j$$

$$S_i \approx 1.5 T_{bi}$$

$$S_{ij} \approx 0.735 (S_i S_j)^{0.5}$$

The thermal conductivity values of stoichiometric combustion products of hydrocarbon fuels ($C_n H_{2n}$) and air parallel those of air as a function of temperature, and the value is about 4.5% lower than that of air at 600 K. Again, the variation with equivalence ratio is about linear on the lean side.

In view of the uncertainties and approximations involved in the above formulas and the limited range of data available, the use of viscosity and Prandtl number data for air is adequate for most calculations.

A FORTRAN program produced by Kee et al. can also be used to calculate the gas-phase transport properties. The program uses a simplified mixture rule from Mathur et al. The results are close to those computed from the Wilke formula given here.

REFERENCES

Hilsenrath, J., *Tables of Thermodynamic and Transport Properties,* Pergamon, London, 1960.

Kee, R. J., Warnatz, J., and Miller, J. A., "A FORTRAN Computer Code Package for the Evaluation of Gas-Phase Viscosities, Conductivities, and Diffusion Coefficients," SANDIA report, SAND83-8209 UC-32, March 1983.

Lindsay, A. L., and Bromley, L. A., "Thermal Conductivity of Gas Mixtures," *Ind. Eng. Chem.,* vol. 2, p. 1508, 1950.

Mathur, S., Tondon, V. K., and Saxena, S. C., "Thermal Conductivity of Binary, Ternary and Quaternary Mixtures of Rare Gases," *Mol. Phys.,* vol. 12, p. 569, 1967.
Reid, R. J. C., and Sherwood, T. K., *The Properties of Gases and Liquids,* McGraw-Hill, New York, 1958.
Vargaftik, N. B., Filippov, L. P., Tarzimanov, A. A., and Totskii, E. E., *Handbook of Thermal Conductivity of Liquids and Gases,* CRC Press, Boca Raton FL, 1994.
Wilke, C. R., "A Viscosity Equation for Gas Mixtures," *J. Chem. Phys.,* vol. 18, p. 517, 1950.

APPENDIX E

A Brief Introduction to Mass Transport by Diffusion in Gaseous Systems

The concept of mass diffusion first appears in this text during the presentation of the governing equations for premixed laminar flame propagation. While mass diffusion of species plays a secondary role in such flames, it can be the predominant mechanism governing laminar gaseous fuel diffusion flames and plays an important role in the vaporization of liquid fuels and the combustion of solid fuels. Readers who have not encountered diffusion theory previously may thus wish to read this appendix to obtain an elementary understanding of diffusion mechanisms. Those wishing a more comprehensive presentation should consult the appropriate chapters of the references listed at the end of this appendix.

This appendix is restricted to diffusion among species of ideal gases and begins with basic concepts for binary mixtures. Examples are then given to illustrate the simple cases of equimolal and unidirection diffusion. The appendix ends with a few comments concerning multicomponent diffusion in ideal gas mixtures.

BASIC DEFINITIONS

Each species in the binary mixture is traveling at a different velocity $\underline{\mathbf{V}}_i$, in fixed coordinates. The diffusion velocity $\tilde{\underline{\mathbf{V}}}_i$ relative to the mass average velocity $\underline{\mathbf{V}}$ is

$$\tilde{\underline{\mathbf{V}}}_i = \underline{\mathbf{V}}_i - \underline{\mathbf{V}} \tag{E.1}$$

where, for a binary mixture,

$$\underline{\mathbf{V}} = y_1 \underline{\mathbf{V}}_1 + y_2 \underline{\mathbf{V}}_2$$

The mass flux is

$$\begin{aligned} \dot{\mathbf{m}}'' &= \rho \underline{\mathbf{V}} = \rho y_1 \underline{\mathbf{V}}_1 + \rho y_2 \underline{\mathbf{V}}_2 \\ &= \rho_1 \underline{\mathbf{V}}_1 + \rho_2 \underline{\mathbf{V}}_2 \end{aligned} \tag{E.2}$$

The mass flux of species i with respect to $\underline{\mathbf{V}}$ then becomes

$$\tilde{\dot{\mathbf{m}}}_i'' = \rho_i \underline{\tilde{\mathbf{V}}}_i \tag{E.3}$$

It is left as an exercise for the reader to demonstrate that $\tilde{\dot{\mathbf{m}}}_1'' + \tilde{\dot{\mathbf{m}}}_2'' = 0$.

The molar flux is given by

$$\dot{\mathbf{N}}_i' = \frac{\dot{\mathbf{m}}_i''}{M_i} \tag{E.4}$$

and

$$\tilde{\dot{\mathbf{N}}}_i' = \frac{\tilde{\dot{\mathbf{m}}}_i''}{M_i} = \left(\frac{\rho_i}{M_i}\right)\underline{\tilde{\mathbf{V}}}_i \tag{E.5}$$

The term ρ_i/M_i is the molar concentration of species i; recall that the molar concentration of a species i is written as n_i (or in Chapter 4 as [A] for chemical species A).

FICK'S FIRST LAW OF DIFFUSION

Diffusion of species can take place due to gradients in concentration, pressure, and temperature. Species may also move relative to each other due to unequal external forces on each species. Only the diffusion due to concentration gradients, called *ordinary diffusion,* will be discussed here. Typically the other types are small, but note that in flames the steep temperature gradients can cause thermal diffusion to be significant although still small relative to ordinary diffusion. For ordinary diffusion in a binary mixture of components A and B, Fick's law of diffusion is given by

$$\tilde{\dot{\mathbf{m}}}_A'' = -\rho D_{AB} \nabla y_A \tag{E.6}$$

where ∇y_A is the gradient vector with components in x, y, z coordinates of

$$\left(\frac{\partial y_A}{\partial x}, \frac{\partial y_A}{\partial y}, \frac{\partial y_A}{\partial z}\right)$$

EXAMPLE E.1. Find the expression for Fick's law in terms of mass flux.

Solution

$$\tilde{\dot{\mathbf{m}}}_A'' = \rho_A(\underline{\mathbf{V}}_A - \underline{\mathbf{V}}) = \dot{\mathbf{m}}_A'' - y_A \dot{\mathbf{m}}''$$

Thus, substituting into Eq. E.6,

$$\begin{aligned} \dot{\mathbf{m}}_A'' &= -\rho D_{AB} \nabla y_A + y_A(\dot{\mathbf{m}}_A'' + \dot{\mathbf{m}}_B'') \\ &= \text{(flux from diffusion)} + \text{(flux from bulk flow)} \end{aligned}$$

The reader will note the similarity between Fick's Law and Fourier's law of heat conduction,

$$\mathbf{q}'' = -k \nabla T \tag{E.7}$$

where $\mathbf{q}''$ is the heat flux vector. For constant ρ the right side of Eq. E.6 becomes $-D_{AB} \nabla \rho_A$, and for constant ρc_p the right side of Eq. E.7 can be written as

$-\alpha\ \nabla(\rho c_p T)$, where $\alpha = k/\rho c_p$. The ratio of the two diffusivities, α/D_{AB}, is the Lewis number. The ratio ν/D_{AB} is the Schmidt number. The transport properties α, ν, and D_{AB} all have the dimensions of length squared per unit time.

DIFFUSIVITY VALUES

The mass diffusivity for a binary system, D_{AB}, is a function of temperature and pressure for a given species pair. For nonpolar ideal gases one can predict D_{AB} with about 5% accuracy from kinetic theory [see Bird et al., pp. 508–513]. The computation is very simple if the appropriate molecular parameters are available in tabulated form. The theory shows that

$$D_{AB} = \frac{aT^n}{p} \tag{E.8}$$

where a is a constant and n is about 2 at low temperatures and about 1.65 at high temperatures. A simple rigid-sphere molecular model gives $n = \frac{1}{2}$. Note that $p = \rho RT$, so that ρD_{AB} in Eq. E.6 depends only on T for a given species pair:

$$\rho D_{AB} = \frac{aT^{n-1}}{R}$$

The effect of mixture strength on D_{AB} is very small. For real gases a generalized chart using corresponding states provides a rough guide to the correction needed. For combustion systems where pressure is high, such as diesel engines, the temperature is also high. For example, at 16:1 compression, nitrogen will be at $p/p_{\text{critical}} \approx 1.3$ and $T/T_{\text{critical}} \approx 6.3$. At these conditions the real gas corrections for diffusivity are negligible. Note, however, that for hydrocarbon vapor, T_{critical} is much higher, so that for the 16:1 compression of heptane, $p/p_{\text{critical}} \approx 1.6$ and $T/T_{\text{critical}} \approx 1.5$. Thus, we may expect that real gas corrections of perhaps 20% may be required. This is still a small correction given the ability to model mixing in real systems.

It is best to use experimental values of D_{AB} when available. Table E.1 gives some example D_{AB} values.

TABLE E.1
Diffusivities D_{12} (cm²/s) of some dilute gas pairs at 1 atm pressure

Gas pair	293 K	673 K
Ar-N_2	0.190	0.815
CH_4-N_2	0.220	0.8890
CO-N_2	0.231	0.878
CO_2-N_2	0.160	0.733
H_2-He	1.490	6.242
H_2-N_2	0.772	3.196
H_2-O_2	0.756	3.299
H_2O-N_2	0.242	
CO_2-SF_6	0.099	

Source: Reprinted with permission from Lide, *Handbook of Chemistry and Physics*, 74th ed., pp. 6-203–6-204, © 1993, CRC Press, Boca Raton, FL.

COMPUTATIONS FOR SYSTEMS WITH DIFFUSION

For very simple systems with no fluid motion other than caused by diffusion, the equations for mass diffusion are similar to those of heat conduction. However, in most practical systems bulk mass turbulent transport complicates the model. For flows along solid surfaces, boundary layer equations which include mass diffusion may be developed and solved analytically or numerically. As in the case of heat convection without mass transfer, the use of empirical film coefficients offers a simpler engineering approach. This is the approach used in this text for modeling droplet vaporization and solid fuel combustion.

For most practical combustion applications the models must include the complete equations of change in time and three dimensions. The equations of mass conservation for each species include both a source term due to diffusion and a generation term due to chemical reaction. The energy equation includes chemical energy terms and energy flux due to diffusion. This latter flux is caused by the fact that each diffusing species carries its own enthalpy in (and out) of the control volume. The presentation of these partial differential equations of change [see Bird et al.] is beyond the scope of this text, thus only some simple diffusion examples will be given to illustrate the theory.

EXAMPLES OF STEADY-STATE ONE-DIMENSIONAL DIFFUSION

EXAMPLE E.2. Consider a tube with insulated sides with the open top end exposed to a cross flow of a mixture of vapor A and gas B at temperature T_0 and pressure p_0. The open bottom end extends into a very large pan of liquid A. The liquid level is held fixed at $x = 0$. The solubility of gas B into liquid A is negligible for low pressures, and therefore the gas B is assumed to be stationary. The liquid A has a vapor pressure p_{A0} at temperature T_0. Assume that an electric heat control system keeps the liquid temperature at the interface constant at T_0. Calculate the vaporization rate of the liquid and the amount of energy that must be supplied by the heater.

Solution. Because $\dot{m}_B''$ is taken as zero here, the mass flux of vapor A in the tube is given by

$$\dot{m}_A'' = -\rho D_{AB}\frac{dy_A}{dx} + \dot{m}'' y_A \tag{E.9}$$

At steady state, mass conservation shows that $\dot{m}_A''$ is constant. Now convert the mass flux to a molar flux in terms of mole fraction x_A:

$$\dot{N}_A'' = \frac{\dot{m}_A''}{M_A} \tag{E.10a}$$

$$y_A = \frac{x_A M_A}{x_A M_A + x_B M_B} = \frac{x_A M_A}{M} \tag{E.10b}$$

$$dy_A = \frac{M_A M_B\, dx_A}{(x_A M_A + x_B M_B)^2} = \frac{M_A M_B\, dx_A}{M^2} \tag{E.10c}$$

$$\rho = \frac{p}{\hat{R}T/M} \tag{E.10d}$$

Thus

$$\dot{N}''_A = -\frac{pM}{\hat{R}T} D_{AB} \frac{M_B}{M^2} \frac{dx_A}{dx} \frac{1}{1 - \dfrac{x_A M_A}{M}} \tag{E.11a}$$

or

$$\dot{N}''_A = -\left(\frac{p}{\hat{R}T}\right) D_{AB} \frac{dx_A/dx}{1 - x_A} \tag{E.11b}$$

and Eq. E.9 becomes

$$\dot{N}''_A \, dx = -\frac{aT^{n-1}}{\hat{R}} \frac{dx_A}{1 - x_A} \tag{E.12}$$

where

$$D_{AB} = \frac{aT^n}{p} \tag{E.13}$$

For constant $T = T_0$ integrate recognizing that $\dot{N}''_A$ is constant for steady state:

$$\dot{N}''_A = \frac{(aT_0^{n-1}/\hat{R})}{L} \ln\left[\frac{1 - x_{AL}}{1 - x_{A0}}\right] \tag{E.14}$$

where $x_{A0} = p_{A0}/p_0$, the given mole fraction at $x = L$.

The solution is

$$\dot{m}''_A = \left(\frac{aT_0^{n-1}}{\hat{R}L}\right) \ln\left[\frac{1 - x_{AL}}{1 - x_{A0}}\right] \tag{E.15}$$

The energy supplied by the heater is $h_{f_g}(T_0)\dot{m}''_A$.

EXAMPLE E.3. Although somewhat contrived, let us consider the equimolal flow of isothermal gases A and B in a tube of length L connecting two infinite sources of gas A and gas B respectively. In this case $\dot{N}''_A = -\dot{N}''_B$, so that Eq. E.9 becomes

$$\dot{N}''_A = -\rho \frac{D_{AB}}{M} \frac{dx_A}{dx} \tag{E.16}$$

where

$$-\frac{L}{2} \le x \le +\frac{L}{2}$$

and the boundary conditions are $x_A(+L/2) = 0$ and $x_A(-L/2) = 1$. The solution is

$$\dot{N}''_A = -\left(\frac{aT_0^{n-1}}{\hat{R}}\right) \frac{x_A(L/2) - x_A(-L/2)}{L} \tag{E.17}$$

The flux of A has a negative sign because it flows in the negative x direction. The flux $\dot{N}''_B$ is given by

$$\dot{N}''_B = -\dot{N}''_A$$

MULTICOMPONENT DIFFUSION

Practical combustion problems involve multicomponent mixtures of fuel, products, air, and many intermediate species which include radicals. The mass flux of species relative to the mass average flux for ideal gases is

$$\tilde{\dot{m}}_i'' = \frac{\rho}{M^2} \sum_{j=1}^{n} M_i M_j \mathbf{D}_{ij} \nabla x_i \tag{E.18}$$

where

$$M = \sum_{i=1}^{n} x_i M_i$$

and the $\mathbf{D}_{ij}$ are the multicomponent mixture diffusivities for components i and j.

In binary mixtures ($n = 2$), $\mathbf{D}_{AB} = D_{AB}$ and $D_{AB} = D_{BA}$. However, $\mathbf{D}_{ij} \neq \mathbf{D}_{ji}$, and unlike their binary counterparts, they are composition-dependent. Equation E.18 is difficult to use, but fortunately Curtiss and Hirschfelder [1949] have shown that

$$\nabla x_i = \sum_{j=1}^{n} \frac{x_i \dot{\mathbf{N}}_j'' - x_j \dot{\mathbf{N}}_i''}{\rho D_{ij}/M} \tag{E.19}$$

where the D_{ij} are for a binary mixture of i and j.

In many combustion problems the most important diffusing species such as radicals are in only trace amounts while N_2 is in abundance. For this case one may approximate $\mathbf{D}_{ij} = D_{in}$ where n is nitrogen. Typically, the low–molecular weight species, such as H_2 in N_2, have a much higher diffusivity than for the higher–molecular weight species such as CO_2 in N_2. This causes the smaller molecular weight species to tend to diffuse away from the other species preferentially.

PROBLEM

E.1. Consider a porous rigid sphere of radius R through which gas A flows out radially into an infinite stagnant atmosphere of gas B. The mole fraction at R is $x_{A0} < 1$ and $x_A \to 0$ at large r. Neglect buoyancy effects.

(*a*) Show that $\dot{N}_A'' r^2$ is a constant for steady state.

(*b*) Using Example E.1 as a guide, show that, by integration,

$$\dot{N}_A''(r = R) = \left(\frac{aT_0^{n-1}}{\hat{R}}\right)\left(\frac{r/R}{r - R}\right) \ln\left[\frac{X_B(r)}{X_B(R)}\right]$$

where

$$D_{AB} = \frac{aT_0^n}{p_0}$$

and that for $r \to \infty$

$$\dot{N}_A''(r = R) = -\left(\frac{aT_0^{n-1}}{\hat{R}}\right)\left(\frac{1}{R}\right) \ln(1 - X_{A0})$$

and

$$4\pi R^2 \dot{N}_A''(R) = \dot{N}_A \propto R$$

REFERENCES

Bird, R. B., Stewart, W. E., and Lightfoot, E. N., *Transport Phenomena,* Wiley, New York, 1960.

Curtiss, C. F., and Hirschfelder, J. O., *J. Chem. Phys.,* vol. 17, pp. 550–555, 1949.

Hirschfelder, J. O., Curtiss, C. F., and Bird, R. B., *Molecular Theory of Gases and Liquids,* Wiley, New York, 1954.

Kee, R. J., Waranatz, J., and Miller, J. A., "A FORTRAN Computer Code Package for the Evaluations of Gas Phase Viscosities, Conductivities, and Diffusion Coefficients," SANDIA report SAND 83-8209 UC-32, March 1983.

Lide, D. R., *Handbook of Chemistry and Physics,* 74th ed., CRC Press, Boca Raton, FL, 1993, pp. 6-203–6-204.

Index

A

B

C

E

F

G

H

I

J

K

L

M

N

O

T

U

V

W

Y

Z